AA001058

2007 IFIP International Conference on Very Large Scale Integration

Atlanta, GA
15-17 October 2007

IEEE Catalog Number: CFP07LSI-PRT
ISBN 10:   1-4244-1709-0
ISBN 13:   978-1-4244-1709-4

Copyright © 2007 by The Institute of Electrical and Electronics Engineers, Inc.
All Rights Reserved

Copyright and Reprint Permissions: Abstracting is permitted with credit to the source. Libraries are permitted to photocopy beyond the limit of U.S. copyright law for private use of patrons those articles in this volume that carry a code at the bottom of the first page, provided the per-copy fee indicated in the code is paid through Copyright Clearance Center, 222 Rosewood Drive, Danvers, MA 01923.

For other copying, reprint or republications permission, write to IEEE Copyrights Manager, IEEE Operations Center, 445 Hoes Lane, Piscataway, New Jersey USA 08854. All rights reserved.

IEEE Catalog Number: CFP07LSI-PRT

ISBN 10: 1-4244-1709-0

ISBN 13: 978-1-4244-1709-4

LOC: 2007906648

Additional Copies of This Publication Are Available from:

IEEE Service Center
445 Hoes Lane
Piscataway, NJ 08854

Phone:	(800) 678-IEEE
	(732) 981-1393
Fax:	(732) 981-9667
E-mail:	customer-service@ieee.org

Message from the General and Program Chairs

On behalf of the Organizing and Program Committees, it is our great pleasure to welcome you to the 15th Annual IFIP International Conference on Very Large Scale Integration, VLSI-SoC'07, in the city of Atlanta, Georgia, USA. The conference is held in Technology Square Research Building of Georgia Institute of Technology. The Technology Square Research Building is home to more than 500 world class researchers and houses the Georgia Electronic Design Center, the Graphics, Visualization and Usability Center, and the Center for Research on Embedded Systems and Technology.

VLSI-SoC 2007 is the 15th in a series of international conferences, sponsored by the International Federation for Information Processing (IFIP, Technical Council 10, Working Group 10.5(VLSI)), that explores the state-of-the-art and the new developments in the fields of VLSI/ULSI, microelectronics design and test, and integrated mixed-technology devices. Previous conferences have taken place in Edinburgh, Trondheim, Tokyo, Vancouver, Munich, Grenoble, Gramado, Lisbon, Montpellier, Darmstadt, Perth, and Nice. This is the first time the conference is held in USA. Conferences are held annually since 2005.

The Technical Program Committee has the challenging task of selecting the best contributions from 109 submissions to the Conference Call. The Conference program includes 3 keynote speeches and 2 special sessions with 6 invited presentations, 46 scientific papers, 13 posters, 11 PhD Forum presentations, and one industrial demonstration. The Conference Proceedings that you have in your hands includes all scientific papers and invited presentations as regular papers, and the summary of the keynote speeches. In addition, a special PhD forum has been organized during the conference. This Ph.D. Forum includes 11 posters that have been submitted in a separate call, independently of the Conference Call. The papers describing these works have been included in the conference CD.

Many dedicated volunteers have helped in making possible this very rich program, including reviewers and supporting staff and students from Georgia Tech. We warmly thank them for their help. We also gratefully acknowledge and thank all authors who submitted papers and all participants who attended the conference. Particular thanks go to IFIP Working Group 10.5 for technical sponsorship and even providing travel grants for about 11 Ph.D. students from all over the world. Finally, we are grateful to our three sponsors, Mentor Graphics, Intel and Synplicity, for their support.

It is our most sincere wish that you find VLSI-SoC'07 a very high quality conference and we wish you will enjoy your stay in this highly attractive region.

Vincent Mooney
General Chair

Paul Hasler and Yung-Hsiang Lu
Program Co-Chairs

Committees

General Chair
Vincent Mooney
Georgia Institute of Technology, United States

Program Co-Chairs
Paul Hasler
Georgia Institute of Technology, United States
Yung-Hsiang Lu
Purdue University, United States

Steering Committee
Manfred Glesner
Technische Universitat Darmstadt, Germany
Ricardo Reis
Universidade Federal do Rio Grande do Sul, Brazil
Michel Robert
University of Montpellier, France
Luis Miguel Silveira
Instituto de Engenharia de Sistemas e Computadores Investigação e Desenvolvimento em Lisboa, Portugal

Publicity Chair
Matthew R. Guthaus
University of California-Santa Cruz, United States

Publications Co-Chairs
Sung Kyu Lim
Georgia Institute of Technology, United States
Ricardo Reis
Universidade Federal do Rio Grande do Sul, Brazil

Ph.D. Forum Chair
Wei-Chung "Wayne" Cheng
National Chiao Tung University, Taiwan

Exhibit Chair
Qing Wu
Binghamton University, United States

Special Session Chair
Hsien-Hsin "Sean" Lee
Georgia Institute of Technology, United States

Web Chair
Karthik Kumar
Purdue University, United States

Local Arrangements Chair
Lei Zhao
Georgia Institute of Technology, United States

Technical Program Committee

David Atienza Alonso
Ecole Polytechnique Federale de Lausanne, Switzerland

Juergen Becker
Universitaet Karlsruhe, Germany

Naehyuck Chang
Seoul National University, South Korea

Wei-Chung Cheng
National Chiao Tung University, Taiwan

Matthew Guthaus
University of California-Santa Cruz, United States

Byunghoo Jung
Purdue University, United States

Sung Kyu Lim
Georgia Tech, United States

Hsien-Hsin Sean Lee
Georgia Tech, United States

Tiziana Margaria-Steffen
University Potsdam, Germany

Subhasish Mitra
Stanford University, United States

Dimitrios Peroulis
Purdue University, United States

Ricardo Reis
Universidade Federal do Rio Grande do Sul, Brazil

Eric Rotenberg
North Carolina State University, United States

Donatella Sciuto
Politecnico di Milano, Italy

Dimitrios Soudris
Democritus University of Thrace, Greece

Flavio R. Wagner
Universidade Federal do Rio Grande do Sul, Brazil

Qing Wu
Binghamton University, United States

Additional Reviewers

Eduardo Briao
Michael Brown
Suparna Das
Vinodh Gopal
Himanshu Jain
Marcelo de Oliveira Johann
Pervez Khaled
José Carlos Sant'Anna Palma

Lazaros Papadopoulos
Huan Ren
Georgios Ch. Sirakoulis
Pablo G. Del Valle
Gilson Wirth
Dong Hyuk Woo

Conference Overview

How do you stay current in a fast-moving, ever-changing field? Chips are using newer technologies and integrating more components on a regular basis. Today's innovative technology will be obsolete tomorrow.

Staying one step ahead of innovation requires knowing about latest research and how it affects trends. Chip research responds to consumer demands and offers innovative technologies that help drive that demand.

Join us for **VLSI-SoC 2007 in Atlanta**, as we explore state-of-the-art technology and developments in Very Large Scale Integrated Systems and Systems on Chips.

Exchange ideas with leading experts and learn from both industrial and academic research results in the fields of integrated systems, microelectronic design and test, and novel applications.

See you **Monday–Wednesday, Oct. 15–17, at Georgia Institute of Technology's Technology Square Research Building** in Midtown Atlanta!

Why Attend?

- Make research contacts for future projects
- Discover the latest in chip research and development
- Float ideas to receive industry feedback
- Pose problems for the academic community
- Get technical solutions to today's challenges

Who Should Attend

- Logic designers
- Circuit designers
- Analog designers
- Mixed-Signal engineers
- DSP engineers
- Physical designers
- CAD developers
- Technical managers
- Academic Researchers

Conference Venue

Technology Square Research Building at Georgia Tech
85 Fifth St. N.W.
Atlanta, GA 30308
website: www.tsrb.gatech.edu

The Technology Square Research Building is the South's premier address for Next Generation Communications. The Technology Square Research Building is home to more than 500 world class researchers and houses the Georgia Electronic Design Center, the Graphics, Visualization and Usability Center, and the Center for Research on Embedded Systems and Technology.

Conference Hotel
Regency Suites Midtown Atlanta
975 W. Peachtree St. at 10th Street
Atlanta, GA 30309
1-800-642-3629
1-404-876-5003
website: www.regencysuites.com

Located just five blocks from the Technology Square Research Building, the Regency Suites is a beautifully renovated hotel in the heart of Midtown Atlanta. The hotel features a complimentary expanded continental breakfast each morning as well as a complimentary dinner if booked direct.

Located next to the *Midtown MARTA station*, the hotel is two blocks from Interstates 75/85. Secured underground parking is available for an additional charge.

Hotel rooms feature complimentary high speed Internet access, granite counter tops, and marble flooring and tub surrounds as well as a guest laundry and expanded fitness facility.

Keynote Speakers

Keynote 1

System-Level Implications of Nanometer Design: Commercial EDA Opportunity or Nano-Niche?

Serge Leef
General Manager, System-Level Engineering Division, Mentor Graphics

It is widely believed nanometer designs will become increasingly common. In addition to challenging physical design and verification tools, this trend will force engineers to face and address serious problems in system-level design.

While the emerging 100+ million gate chips will be manufacturable, they pose huge challenges in architectural specification, refinement and optimization; IP selection, evaluation and integration; system-wide analysis of performance and power; full system validation and implementation. How many design teams will really push the complexity to its limits?

Examine system-level design challenges as they apply to IP-dominant, bus-centric systems. Then, explore the economic forces that will either make these problems common to most designs or relegate them to a small number of teams focusing on ultra-high volume applications.

Biography

Serge Leef, General Manager of the System-Level Engineering Division at Mentor Graphics, leads two business units focused on markets where system-level design plays a pivotal role.

One unit houses product lines that focus on next generation EDA technologies for processor-based design, verification and test. Working with leading semiconductor IP companies, these products serve customers in wired and wireless communications, networking, switching, peripherals, consumer, and multimedia markets.

The second business unit concentrates on applying advanced system-level design automation techniques to the challenges associated with functional design of automotive electronics. In addition, Leef manages an "Emerging Products" program that identifies and develops promising, category-defining technologies.

Prior to joining Mentor Graphics in 1990, Leef was responsible for design automation at Silicon Graphics, where his team created revolutionary high speed simulation tools to enable design of high-speed 3D graphics chips that defined state-of-the-art in visualization, imaging, gaming and special effects for a decade.

Prior to 1987, Leef managed a CAE/CAD organization at Microchip Inc. From 1982 to 1987, he worked at Intel Corp. developing functional and physical design and verification tools for major 8- and 16-bit microcontroller and microprocessor programs. Leef holds a B.S.E.E. and M.S.C.S. from Arizona State University.

Keynote 2

Nanoelectronics: Retrospect and Prospect

James D. Meindl

Joseph M. Pettit, Chair, Professor of Microelectronics, Georgia Institute of Technology

The most important economic development of the 20th century was the information revolution. The principal driver of the information revolution is the silicon microchip for two salient reasons. From 1960 to 2006, the productivity of microchip technology increased by more than one billion times and simultaneously, performance increased by more that 100,000 times. Silicon microchip technology is approaching both physical and economic limits that are expected to severely curtail its rate of advance. Intensive exploratory research is aimed at discovering the next technology to propel the information revolution for the first half of the 21st century and beyond.

Biography

James Meindl, Director of the Joseph M. Pettit Microelectronics Research Center and the Joseph M. Pettit Chair Professor of Microelectronics, is the founding director of the new Nanotechnology Research Center. His research focuses on physical limits on gigascale integration and nanotechnology. A Life Fellow of IEEE and the American Association for the Advancement of Science, Meindl is a member of the American Academy of Arts and Sciences and the National Academy of Engineering.

He received the 2006 IEEE Medal of Honor, the IEEE's highest honor. In September 2004, he received the 2004 SRC Aristotle Award, recognizing outstanding teaching in its broadest sense. In 2003, Meindl was awarded first place on the IEEE International Solid State Circuits Conference 50-Year Anniversary Author Honor Roll.

He received the 2001 Class of 1934 Distinguished Professor Award from Georgia Institute of Technology, the 2000 IEEE Third Millenium Medal, the 1999 SIA University Research Award, and the 1997 Hamerschlag Distinguished Alumnus Award from Carnegie Mellon University.

Meindl received his bachelor's and master's degrees, as well as his Ph.D., in electrical engineering from Carnegie Institute of Technology (Carnegie Mellon University).

Keynote 3

The Third Wave of the Digital Revolution

Gene A. Frantz
Principal, Fellow for DSP Systems, Texas Instruments

We are in the middle of a digital revolution, resulting from multiple levels of innovation such as the transistor, integrated circuit, microprocessor, and DSP. Communications have been revolutionized by digital signal processors. Entertainment is being revolutionized. The question is simple: Is there a third wave of this revolution? Are we at the end of the digital revolution? Examine how we got here, then explore what might be the third wave of the digital revolution.

Biography

Gene Frantz, Principal Fellow at Texas Instruments, is responsible for finding new opportunities and creating businesses that utilize the company's digital signal processing technology. He has been with Texas Instruments for more than 30 years, most of it in digital signal processing.

A recognized leader in DSP technology, Frantz is a Fellow of the Institution of Electric and Electronics Engineers. He holds 40 patents in memories, speech, consumer products and digital signal processing. Frantz has written more than 50 papers and articles.

Programme-At-A-Glance

Monday, 15 October 2007

Time		
08:00–08:15	Registration/Breakfast	
08:15–08:30	WELCOME, Vincent Mooney, *General Chair* and Paul Hasler, *Program Co-Chair*	
08:30–09:30	Keynote by Serge Leef (Chair: Vincent Mooney)	
09:30–10:00	Morning Break	
10:00–11:30	**Session 1** **Analog Circuit Design** (Chair: Kenneth W. Hsu, *Rochester Institute of Technology*)	**Session 2** **CAD Tools** (Chair: Wayne Wolf, *Georgia Tech*)
11:30–13:00	Lunch	
13:00–14:30	**Session 3** **Modeling and Simulation** (Chair: Dimitrios Soudris, *Democritus University of Thrace*)	**Session 4** **Reconfigurable Systems** (Chair: Flavio Wagner, *UFRGS*)
14:30–15:30	Evening Break/Poster	
15:30–17:00	**Session 5** **Communication** (Chair: Subhasish Mitra, *Stanford*)	**Session 6** **High-Level Synthesis** (Chair: Amara Amara, *Institut Superieur D'Electronique de Paris*)

Tuesday, 16 October 2007

Time		
08:00–08:30	Registration/Breakfast	
08:30–09:30	Keynote by James Meindl (Chair: Vincent Mooney)	
09:30–10:00	Morning Break	
10:00–11:30	**Session 7** **New Devices** (Chair: Sung Kyu Lim, *Georgia Tech*)	**Session 8 (Special Session 1)** **Reconfigurable and Hybrid System** (Chair: Hsien-Hsin Sean Lee, *Georgia Tech*)
11:30–13:00	Lunch	
13:00–14:30	**Session 9** **Architecture and Compiler** (Chair: Qinru Qiu, *Binghamton University*)	**Session 10 (Special Session 2)** **Architecture Design Principles** (Chair: David Atienza Alonso, *Universidad Compultense de Madrid*)
14:30–15:30	Evening Break/Poster	
15:30–17:00	**PANEL**	
17:00–18:00	*Break (Buses leave at 17:30 hrs)*	
18:00–22:00	**Conference Party** @ *King Center*	

Wednesday, 17 October 2007

Time		
08:00–08:30	Registration/Breakfast	
08:30–09:30	Keynote by Gene Frantz (Chair: Paul Hasler)	
09:30–10:00	Morning Break	
10:00–11:30	**Session 11** **Physical Design and Test** (Chair: Frank K. Gurkaynak, *Ecole Polytechnique Federale de Lausanne*)	**Session 12** **Signal and Image Processing** (Chair: Wei-Chung Cheng, *National Chiao-Tung University*)
11:30–13:00	Lunch	
13:00–14:30	**Session 13** **Verification and Validation** (Chair: Byunghoo Jung, *Purdue University*)	**Session 14** **Estimation and Evaluation** (Chair: Yunsi Fei, *University of Connecticut*)
14:30–15:30	Evening Break/Ph.D. Forum	
15:30–17:00	**Session 15** **New Devices for Mixed Signal** (Chair: Salvador Mir, *TIMA*)	**Session 16** **Low Power Design** (Chair: Jaehwan (John) Lee, *Indiana University-Purdue University Indianapolis*)

Technical Program

Monday, 15 October 2007

Registration/Breakfast

Time 08:00–08:15 hrs

Time 08:15–08:30 hrs
Welcome
Vincent Mooney, General Chair and Paul Hasler, Program Co-Chair

Time 08:30–09:30 hrs
Keynote: Serge Leef
Chair: Vincent Mooney

Break	09:30–10:00 hrs

Session 1: Analog Circuit Design
Date/Time Monday, 15 October 2007 / 10:00 – 11:30 hrs
Chair(s) Kenneth W. Hsu *(Rochester Institute of Technology)*

1.1 Power Invariant Secure IC Design Methodology Using Reduced Complementary Dynamic and Differential Logic
Vijay Sundaresan, Srividhya Rammohan and Ranga Vemuri

1.2 Neuromorphic Building Blocks for Adaptable Cortical Feature Maps
C. M. Markan and Priti Gupta

1.3 An Analog Programmable Multi-Dimensional Radial Basis Function Based Classifier
Sheng-Yu Peng, Paul E. Hasler and David Anderson

Session 2: CAD Tools
Date/Time Monday, 15 October 2007 / 10:00 – 11:30 hrs
Chair(s) Wayne Wolf *(Georgia Tech)*

B2.1 ReCPU: A Parallel and Pipelined Architecture for Regular Expression Matching
Marco Paolieri, Ivano Bonesana and Marco D. Santambrogio

B2.2 Use of Gray Decoding for Implementation of Symmetric Functions
Osnat Keren, Ilya Levin and Radomir S. Stankovic

xviii   *Technical Program*

2.3   Parametric Structure-Preserving Model Order Reduction
Jorge Fernandez Villena, Wil H. A. Schilders and L. Miguel Silveira

> **Lunch**   **11:30–13:00 hrs**

Session
Date/Time
Chair(s)

3: Modeling and Simulation
Monday, 15 October 2007 / 13:00 – 14:30 hrs
Dimitrios Soudris *(Democritus University of Thrace)*

3.1   Hierarchical Statistical Analysis of Performance Variation for Continuous-Time
Delta-Sigma Modulators
Hua Tang

3.2   First Order Quasi-Static SoI MOSFET Channel Capacitance Model
Sameer Sharma and L. G. Johnson

3.3   Regression Based Circuit Matrix Models for Accurate Performance Estimation of
Analog Circuits
Almitra Pradhan and Ranga Vemuri

Session
Date/Time
Chair(s)

4: Reconfigurable Systems
Monday, 15 October 2007 / 13:00 – 14:30 hrs
Flavio Wagner *(UFRGS)*

B4.1   A Software-Supported Methodology for Designing High-Performance 3D FPGA
Architectures
Kostas Siozios, Kostas Sotiriadis, Vassilis F. Pavlidis and Dimitrios Soudris

B4.2   Estimating Design Time for System Circuits
*Cyrus Bazeghi, Francisco J. Mesa-Martínez, Brian Greskamp, Josep Torrellas
and Jose Renau*

B4.3   Transparent Acceleration of Data Dependent Instructions for General Purpose
Processors
Antonio Carlos Schneider Beck and Luigi Carro

> **Break/Poster**   **14:30–15:30 hrs**

Session
Date/Time
Chair(s)

5: Communication
Monday, 15 October 2007 / 15:30 – 17:00 hrs
Subhasish Mitra *(Stanford)*

5.1   VLSI Models of Network-on-Chip Interconnect
Dimitrios N. Serpanos and Wayne Wolf

5.2	Statistical Analysis of Systematic and Random Variability of Flip-Flop Race Immunity in 130 nm and 90 nm CMOS Technologies

5.2 Statistical Analysis of Systematic and Random Variability of Flip-Flop Race Immunity in 130 nm and 90 nm CMOS Technologies
Gustavo Neuberger, Fernanda Kastensmidt, Ricardo Reis, Gilson Wirth, Ralf Brederlow and Christian Pacha

5.3 AC-Coupling Strategy for High-Speed Transceivers of 10 Gbps and Beyond
Yikui (Jen) Dong, Steve Howard, Freeman Zhong, Scott Lowrie, Ken Paradis, Jan Kolnik and Jeff Burleson

Session	6: High-Level Synthesis
Date/Time	Monday, 15 October 2007 / 15:30 – 17:00 hrs
Chair(s)	Amara Amara *(Institut Superieur D'Electronique de Paris)*

6.1 SWORD: A SAT Like Prover Using Word Level Information
Robert Wille, Görschwin Fey, Daniel Große, Stephan Eggersglüß and Rolf Drechsler

6.2 Obtaining Delay Distribution of Dynamic Logic Circuits by Error Propagation at the Electrical Level
Lucas Brusamarello, Roberto da Silva, Gilson I. Wirth and Ricardo A. L. Reis

6.3 Minimizing Wire Delays by Net-Topology Aware Binding During Floorplan-Driven High Level Synthesis
Vyas Krishnan and Srinivas Katkoori

Tuesday, 16 October 2007

	Registration/Breakfast
Time	08:00–08:30 hrs
Time	08:30–09:30 hrs
	Keynote: James Meindl
	Chair: Vincent Mooney

Break	**09:30–10:00 hrs**

Session	7: New Devices
Date/Time	Tuesday, 16 October 2007 / 10:00 – 11:30 hrs
Chair(s)	Sung Kyu Lim *(Georgia Tech)*

7.1 A High-Driving Class-AB Buffer Amplifier with a New Pseudo Source Follower
Chih-Wen Lu, Yen-Chih Shen and Meng-Lieh Sheu

7.2	A New Analytical Approach of the Impact of Jitter on Continuous Time Delta Sigma Converters *Julien Goulier, Eric Andre and Marc Renaudin*
7.3	Transistor Level Automatic Layout Generator for Non-Complementary CMOS Cells *Adriel Ziesemer, Cristiano Lazzari and Ricardo Reis*

Session 8 (Special Session 1): Reconfigurable and Hybrid System
Date/Time Tuesday, 16 October 2007 / 10:00 – 11:30 hrs
Chair(s) Hsien-Hsin Sean Lee *(Georgia Tech)*

8.1	Computing and Design for Software and Silicon Manufacturing *Davide Pandini, Giuseppe Desoli and Alessandro Cremonesi*
8.2	An Adaptive Genetic Algorithm for Dynamically Reconfigurable Modules Allocation *Vincenzo Rana, Chiara Sandionigi, Marco Santambrogio and Donatella Sciuto*
8.3	New Tool Support and Architectures in Adaptive Reconfigurable Computing *Jürgen Becker, Adam Donlin and Michael Hübner*

> **Lunch** **11:30–13:00 hrs**

Session 9: Architecture and Compiler
Date/Time Tuesday, 16 October 2007 / 13:00 – 14:30 hrs
Chair(s) Qinru Qiu *(Binghamton University)*

9.1	Rate-Based Scheduling Policy for QoS Flows in Networks on Chip *Aline Mello, Ney Calazans and Fernando Moraes*
9.2	Parallelized Radix-2 Scalable Montgomery Multiplier *Nan Jiang and David Harris*
9.3	An Efficient Heterogeneous Reconfigurable Functional Unit for an Adaptive Dynamic Extensible Processor *Arash Mehdizadeh, Behnam Ghavami, Morteza Saheb Zamani, Hossein Pedram and Farhad Mehdipour*

Session 10 (Special Session 2): Architecture Design Principles
Date/Time Tuesday, 16 October 2007 / 13:00 – 14:30 hrs
Chair(s) David Atienza Alonso *(Universidad Complutense de Madrid)*

10.1	Simulation of Hybrid Computer Architectures: Simulators, Methodologies and Recommendations *Pranav Vaidya and Jaehwan John Lee*
10.2	New Parallel Programming Techniques for Hardware Design *Satnam Singh*

10.3   Efficient DSP Algorithm Development for FPGA and ASIC Technologies
Shiv Balakrishnan and Chris Eddington

> **Break/Poster**   **14:30–15:30 hrs**

Time   15:30–17:00 hrs
Panel

> **Break (Buses leave 17:30)**   **17:00–18:00 hrs**

> **Conference Party @ King Center**   **18:00–22:00 hrs**

Wednesday, 17 October 2007

Registration/Breakfast
Time   08:00–08:30 hrs

Time   08:30–09:30 hrs
Keynote: Gene Frantz
Chair: Paul Hasler

> **Break**   **09:30–10:00 hrs**

Session   11: Physical Design and Test
Date/Time   Wednesday, 17 October 2007 / 10:00 – 11:30 hrs
Chair(s)   Frank K. Gurkaynak *(Ecole Polytechnique Federale de Lausanne)*

11.1   Incremental Placement for Structured ASICs Using the Transportation Problem
Andrew C. Ling, Deshanand P. Singh and Stephen D. Brown

11.2   Test Data Compression and TAM Design
Julien Dalmasso, Marie-Lise Flottes and Bruno Rouzeyre

11.3   Dynamic Gates with Hysteresis and Configurable Noise Tolerance
Krishna Santhanam and Kenneth S. Stevens

Session	12: Signal and Image Processing
Date/Time	Wednesday, 17 October 2007 / 10:00 – 11:30 hrs
Chair(s)	Wei-Chung Cheng *(National Chiao-Tung University)*

12.1   A Low-Power Deblocking Filter Architecture for H.264 Advanced Video Coding
Jaemoon Kim, Sangkown Na and Chong-Min Kyung

12.2   The Hazard-Free Superscalar Pipeline Fast Fourier Transform Algorithm and Architecture
Bassam Jamil Mohd, Adnan Aziz and Earl E. Swartzlander Jr.

12.3   An Efficient H.264 Intra Frame Coder System Design
Ilker Hamzaoglu, Ozgur Tasdizen and Esra Sahin

> **Lunch**   **11:30–13:00 hrs**

Session	13: Verification and Validation
Date/Time	Wednesday, 17 October 2007 / 13:00 – 14:00 hrs
Chair(s)	Byunghoo Jung *(Purdue University)*

13.1   Qualification of Behavioral Level Design Validation for AMS & RF SoCs
Yves Joannon, Vincent Beroulle, Chantal Robach, Smail Tedjini and Jean-Louis Carbonero

13.2   Evaluating Memory Sharing Data Size and TCP Connections in the Performance of a Reconfigurable Hardware-Based Architecture for TCP/IP Stack
Jean Carlo Hamerski, Everton Reckziegel and Fernanda Lima Kastensmidt

13.3   Impact of Hardware Emulation on the Verification Quality Improvement
Youssef Serrestou, Vincent Beroulle and Chantal Robach

Session	14: Estimation and Evaluation
Date/Time	Wednesday, 17 October 2007 / 13:00 – 14:00 hrs
Chair(s)	Yunsi Fei *(University of Connecticut)*

14.1   Fast Estimation of Software Energy Consumption Using IPI(Inter-Prefetch Interval) Energy Model
Jungsoo Kim, Kyungsu Kang, Heejun Shim, Woong Hwangbo and Chong-Min Kyung

14.2   Power Optimization for Conditional Task Graphs in DVS Enabled Multiprocessor Systems
Parth Malani, Prakash Mukre and Qinru Qiu

14.3   A Minimum-Latency Block-Serial Architecture of a Decoder for IEEE 802.11n LDPC Codes
Massimo Rovini, Giuseppe Gentile, Francesco Rossi and Luca Fanucci

> **Break/Ph.D. Forum**   **14:30–15:30 hrs**

Technical Program   xxiii

Session **Date/Time** **Chair(s)**	15: New Devices for Mixed Signal Wednesday, 17 October 2007 / 15:30 – 17:00 hrs Salvador Mir *(TIMA)*

15.1   Full Custom Design of a Three-Stage Amplifier with 5500 MHz.pF/mW Performance in 0.18 μm CMOS
Run Chen, Liyuan Liu, Dongmei Li and Zhihua Wang

15.2   A 128 dB Dynamic Range 1 kHz Bandwidth Stereo ADC with 114 dB THD
YuQing Yang, Terry Sculley and Jacob Abraham

Session **Date/Time** **Chair(s)**	16: Low Power Design Wednesday, 17 October 2007 / 15:30 – 17:00 hrs Jaehwan (John) Lee *(Indiana University-Purdue University Indianapolis)*

16.1   A Bit-Sliced, Scalable and Unified Montgomery Multiplier Architecture for RSA and ECC
M. Sudhakar, R. V. Kamala and M. B. Srinivas

16.2   Low Power On-Chip Thermal Sensors Based on Wiresan
Basab Datta and Wayne P. Burleson

16.3   A Low-Power CAM Using a 12-Transistor Design Cell
Saleh Abdel-Hafeez, Shadi M. Harb and William R. Eisenstadt

Session **Date/Time**	Poster 15 – 16 October 2007 / 14:30 – 15:30 hrs

Improvement of Dual Rail Logic as a Countermeasure Against DPA
A. Razafindraibe, M. Robert and P. Maurine

A VHDL Based Approach for Fast and Accurate Energy Consumption Estimations
César A. M. Marcon, Sérgio Johann Filho and Fabiano P. Hessel

Circuit Prospects of DGFET: Variable Gain Differential Amplifier and a Schmitt Trigger with Adjustable Hysteresis
Srimoyee Sen, Urmimala Roy, Chaitanya Kshirsagar, Navakanta Bhat and Chandan Kumar Sarkar

High Speed SOC Design for Blowfish Cryptographic Algorithm
Brian Cody, Justin Madigan, Spencer MacDonald and Kenneth W. Hsu

Implementing Cellular Automata Modeled Applications on Network-on-Chip Platforms
N. Zompakis, L. Papadopoulos, G. Sirakoulis and D. Soudris

Optimum IR Drop Models for Estimation of Metal Resource Requirements for Power Distribution Network
Rishi Bhooshan and Bindu P. Rao

Impact of Task Migration in NoC-Based MPSoCs for Soft Real-Time Applications
Eduardo Wenzel Brião, Daniel Barcelos, Fabio Wronski and Flávio Rech Wanger

A Flexible Design Flow for a Low Power RFID Tag
José C. S. Palma, César Marcon, Fabiano Hessel, Eduardo Bezerra, Guilherme Rohde, Luciano Azevedo, Carlos Reif and Carolina Metzler

Co-Synthesis of Custom On-Chip Bus and Memory for MPSoC Architectures
Sujan Pandey, Christian Genz and Rolf Drechsler

An HDTV H.264 Deblocking Filter in FPGA with RGB Video Output
Vagner S. Rosa, Altamiro A. Susin and Sergio Bampi

Efficient Timing Closure with a Transistor Level Design Flow
Cristiano Lazzari, Cristiano Santos, Adriel Ziesemer, Lorena Anghel and Ricardo Reis

Hybrid Multiplierless FIR Filter Architecture Based on NEDA
J. Luis Tecpanecatl-Xihuitl, Ruth M. Aguilar-Ponce and Magdy Bayoumi

A Genetic Algorithm Based Heuristic Technique for Power Constrained Test Scheduling in Core-Based SOCs
Chandan Giri, Soumojit Sarkar and Santanu Chattopadhyay

-END-

Contents

Message from the General and Program Chairs	iii
Committees	iv
Technical Program Committee	v
Additional Reviwers	vi
Conference Overview	vii
Conference Venue	viii
Keynote Speakers	ix
Programme-At-A-Glance	xiii
Technical Program	xvii

Session 1: Analog Circuit Design

Power Invariant Secure IC Design Methodology Using Reduced Complementary Dynamic and Differential Logic   1
Vijay Sundaresan, Srividhya Rammohan and Ranga Vemuri

Neuromorphic Building Blocks for Adaptable Cortical Feature Maps   7
C. M. Markan and Priti Gupta

An Analog Programmable Multi-Dimensional Radial Basis Function Based Classifier   13
Sheng-Yu Peng, Paul E. Hasler and David Anderson

Session 2: CAD Tools

ReCPU: A Parallel and Pipelined Architecture for Regular Expression Matching   19
Marco Paolieri, Ivano Bonesana and Marco D. Santambrogio

Use of Gray Decoding for Implementation of Symmetric Functions   25
Osnat Keren, Ilya Levin and Radomir S. Stankovic

Parametric Structure-Preserving Model Order Reduction   31
Jorge Fernandez Villena, Wil H. A. Schilders and L. Miguel Silveira

Session 3: Modeling and Simulation

Hierarchical Statistical Analysis of Performance Variation for Continuous-Time Delta-Sigma Modulators   37
Hua Tang

First Order Quasi-Static SoI MOSFET Channel Capacitance Model 42
Sameer Sharma and L. G. Johnson

Regression Based Circuit Matrix Models for Accurate Performance Estimation of Analog Circuits 48
Almitra Pradhan and Ranga Vemuri

Session 4: Reconfigurable Systems

A Software-Supported Methodology for Designing High-Performance 3D FPGA Architectures 54
Kostas Siozios, Kostas Sotiriadis, Vassilis F. Pavlidis and Dimitrios Soudris

Estimating Design Time for System Circuits 60
Cyrus Bazeghi, Francisco J. Mesa-Martínez, Brian Greskamp, Josep Torrellas and Jose Renau

Transparent Acceleration of Data Dependent Instructions for General Purpose Processors 66
Antonio Carlos Schneider Beck and Luigi Carro

Session 5: Communication

VLSI Models of Network-on-Chip Interconnect 72
Dimitrios N. Serpanos and Wayne Wolf

Statistical Analysis of Systematic and Random Variability of Flip-Flop Race Immunity in 130 nm and 90 nm CMOS Technologies 78
Gustavo Neuberger, Fernanda Kastensmidt, Ricardo Reis, Gilson Wirth, Ralf Brederlow and Christian Pacha

AC-Coupling Strategy for High-Speed Transceivers of 10 Gbps and Beyond 84
Yikui (Jen) Dong, Steve Howard, Freeman Zhong, Scott Lowrie, Ken Paradis, Jan Kolnik and Jeff Burleson

Session 6: High-Level Synthesis

SWORD: A SAT Like Prover Using Word Level Information 88
Robert Wille, Görschwin Fey, Daniel Große, Stephan Eggersglüß and Rolf Drechsler

Obtaining Delay Distribution of Dynamic Logic Circuits by Error Propagation at the Electrical Level 94
Lucas Brusamarello, Roberto da Silva, Gilson I. Wirth and Ricardo A. L. Reis

Minimizing Wire Delays by Net-Topology Aware Binding During Floorplan-Driven High Level Synthesis 99
Vyas Krishnan and Srinivas Katkoori

Session 7: New Devices

A High-Driving Class-AB Buffer Amplifier with a New Pseudo Source Follower 105
Chih-Wen Lu, Yen-Chih Shen and Meng-Lieh Sheu

A New Analytical Approach of the Impact of Jitter on Continuous Time Delta Sigma Converters 110
Julien Goulier, Eric Andre and Marc Renaudin

Transistor Level Automatic Layout Generator for Non-Complementary CMOS Cells 116
Adriel Ziesemer, Cristiano Lazzari and Ricardo Reis

Contents xxvii

Session 8 (Special Session 1): Reconfigurable and Hybrid System

Computing and Design for Software and Silicon Manufacturing 122
Davide Pandini, Giuseppe Desoli and Alessandro Cremonesi

An Adaptive Genetic Algorithm for Dynamically Reconfigurable Modules Allocation 128
Vincenzo Rana, Chiara Sandionigi, Marco Santambrogio and Donatella Sciuto

New Tool Support and Architectures in Adaptive Reconfigurable Computing 134
Jürgen Becker, Adam Donlin and Michael Hübner

Session 9: Architecture and Compiler

Rate-Based Scheduling Policy for QoS Flows in Networks on Chip 140
Aline Mello, Ney Calazans and Fernando Moraes

Parallelized Radix-2 Scalable Montgomery Multiplier 146
Nan Jiang and David Harris

An Efficient Heterogeneous Reconfigurable Functional Unit for an Adaptive Dynamic Extensible 151
Processor
*Arash Mehdizadeh, Behnam Ghavami, Morteza Saheb Zamani, Hossein Pedram and
Farhad Mehdipour*

Session 10 (Special Session 2): Architecture Design Principles

Simulation of Hybrid Computer Architectures: Simulators, Methodologies and Recommendations 157
Pranav Vaidya and Jaehwan John Lee

New Parallel Programming Techniques for Hardware Design 163
Satnam Singh

Efficient DSP Algorithm Development for FPGA and ASIC Technologies 168
Shiv Balakrishnan and Chris Eddington

Session 11: Physical Design and Test

Incremental Placement for Structured ASICs Using the Transportation Problem 172
Andrew C. Ling, Deshanand P. Singh and Stephen D. Brown

Test Data Compression and TAM Design 178
Julien Dalmasso, Marie-Lise Flottes and Bruno Rouzeyre

Dynamic Gates with Hysteresis and Configurable Noise Tolerance 184
Krishna Santhanam and Kenneth S. Stevens

Session 12: Signal and Image Processing

A Low-Power Deblocking Filter Architecture for H.264 Advanced Video Coding 190
Jaemoon Kim, Sangkwon Na and Chong-Min Kyung

The Hazard-Free Superscalar Pipeline Fast Fourier Transform Algorithm and Architecture 194
Bassam Jamil Mohd, Adnan Aziz and Earl E. Swartzlander Jr.

An Efficient H.264 Intra Frame Coder System Design   200
Ilker Hamzaoglu, Ozgur Tasdizen and Esra Sahin

Session 13: Verification and Validation

Qualification of Behavioral Level Design Validation for AMS & RF SoCs   206
Yves Joannon, Vincent Beroulle, Chantal Robach, Smail Tedjini and Jean-Louis Carbonero

Evaluating Memory Sharing Data Size and TCP Connections in the Performance of a Reconfigurable   212
Hardware-Based Architecture for TCP/IP Stack
Jean Carlo Hamerski, Everton Reckziegel and Fernanda Lima Kastensmidt

Impact of Hardware Emulation on the Verification Quality Improvement   218
Youssef Serrestou, Vincent Beroulle and Chantal Robach

Session 14: Estimation and Evaluation

Fast Estimation of Software Energy Consumption Using IPI(Inter-Prefetch Interval) Energy Model   224
Jungsoo Kim, Kyungsu Kang, Heejun Shim, Woong Hwangbo and Chong-Min Kyung

Power Optimization for Conditional Task Graphs in DVS Enabled Multiprocessor Systems   230
Parth Malani, Prakash Mukre and Qinru Qiu

A Minimum-Latency Block-Serial Architecture of a Decoder for IEEE 802.11n LDPC Codes   236
Massimo Rovini, Giuseppe Gentile, Francesco Rossi and Luca Fanucci

Session 15: New Devices for Mixed Signal

Full Custom Design of a Three-Stage Amplifier with 5500 MHz.pF/mW Performance in 0.18 μm   242
CMOS
Run Chen, Liyuan Liu, Dongmei Li and Zhihua Wang

A 128 dB Dynamic Range 1 kHz Bandwidth Stereo ADC with 114 dB THD   248
YuQing Yang, Terry Sculley and Jacob Abraham

Session 16: Low Power Design

A Bit-Sliced, Scalable and Unified Montgomery Multiplier Architecture for RSA and ECC   252
M. Sudhakar, R. V. Kamala and M. B. Srinivas

Low Power On-Chip Thermal Sensors Based on Wiresan   258
Basab Datta and Wayne P. Burleson

A Low-Power CAM Using a 12-Transistor Design Cell   264
Saleh Abdel-Hafeez, Shadi M. Harb and William R. Eisenstadt

Session: Poster

Improvement of Dual Rail Logic as a Countermeasure Against DPA   270
A. Razafindraibe, M. Robert and P. Maurine

Contents

A VHDL Based Approach for Fast and Accurate Energy Consumption Estimations *César A. M. Marcon, Sérgio Johann Filho and Fabiano P. Hessel*	276
Circuit Prospects of DGFET: Variable Gain Differential Amplifier and a Schmitt Trigger with Adjustable Hysteresis *Srimoyee Sen, Urmimala Roy, Chaitanya Kshirsagar, Navakanta Bhat and Chandan Kumar Sarkar*	280
High Speed SOC Design for Blowfish Cryptographic Algorithm *Brian Cody, Justin Madigan, Spencer MacDonald and Kenneth W. Hsu*	284
Implementing Cellular Automata Modeled Applications on Network-on-Chip Platforms *N. Zompakis, L. Papadopoulos, G. Sirakoulis and D. Soudris*	288
Optimum IR Drop Models for Estimation of Metal Resource Requirements for Power Distribution Network *Rishi Bhooshan and Bindu P. Rao*	292
Impact of Task Migration in NoC-Based MPSoCs for Soft Real-Time Applications *Eduardo Wenzel Brião, Daniel Barcelos, Fabio Wronski and Flávio Rech Wanger*	296
A Flexible Design Flow for a Low Power RFID Tag *José C. S. Palma, César Marcon, Fabiano Hessel, Eduardo Bezerra, Guilherme Rohde, Luciano Azevedo, Carlos Reif and Carolina Metzler*	300
Co-Synthesis of Custom On-Chip Bus and Memory for MPSoC Architectures *Sujan Pandey, Christian Genz and Rolf Drechsler*	304
An HDTV H.264 Deblocking Filter in FPGA with RGB Video Output *Vagner S. Rosa, Altamiro A. Susin and Sergio Bampi*	308
Efficient Timing Closure with a Transistor Level Design Flow *Cristiano Lazzari, Cristiano Santos, Adriel Ziesemer, Lorena Anghel and Ricardo Reis*	312
Hybrid Multiplierless FIR Filter Architecture Based on NEDA *J. Luis Tecpanecatl-Xihuitl, Ruth M. Aguilar-Ponce and Magdy Bayoumi*	316
A Genetic Algorithm Based Heuristic Technique for Power Constrained Test Scheduling in Core-Based SOCs *Chandan Giri, Soumojit Sarkar and Santanu Chattopadhyay*	320
Author Index	325

Power Invariant Secure IC Design Methodology using Reduced Complementary Dynamic and Differential Logic

Vijay Sundaresan, Srividhya Rammohan and Ranga Vemuri

Department of ECE, University of Cincinnati

Cincinnati, OH 45219, USA

{sundarvy, rammohs, ranga}@ece.uc.edu

Abstract—**Security of cryptographic devices (secure ICs) like smart cards has come under threat from powerful side channel attacks like Differential Power Analysis (DPA). DPA uses power consumption information leaked from the secure IC in conjunction with statistical correlation techniques to retrieve the secret key stored in the secure IC. The most effective countermeasure to resist DPA attacks is to make the power consumption of the secure IC *invariant*, hence uncorrelated to the input data (secret key). In hardware implementations, this can be achieved by designing the secure IC using Dynamic and Differential Logic (DDL) style. In this paper, we present a novel methodology to design DPA-resistant power invariant secure ICs using Reduced Complementary Dynamic and Differential Logic (RCDDL). The proposed methodology involves strategies to design: 1) RCDDL gates, and 2) secure circuits using RCDDL gates. Experiments show significant improvements in security strength, average power consumption and area, when compared with a similar secure DDL and non-secure static-CMOS logic design styles.**

I. INTRODUCTION

In recent years, there has been a drastic increase in the use of cryptographic devices (secure ICs) like smart cards. Secure ICs are used to ensure security of increasing amounts of personal, and often confidential, data communicated. Secure ICs are hardware implementations of cryptographic algorithms like DES [1] and AES [2], which operate using a secret key. Recently, security of cryptographic devices has come under the threat from powerful side channel attacks, namely Differential Power Analysis (DPA) [3] [4]. This is because, secure ICs are typically implemented in static-CMOS (sCMOS) logic style. In the sCMOS implementation of the secure IC, as each CMOS transistor switches from one state to another, dynamic power is consumed (supply current is drawn). This power consumption occurs only during a $0 \rightarrow 1$ output transition. This property makes the power consumption of the secure IC correlated to the input data (secret key). Fig. 1 illustrates this property with an sCMOS XOR gate.

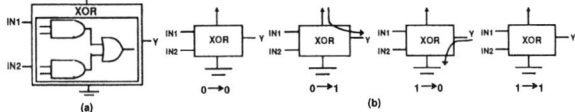

Fig. 1. sCMOS XOR gate (a) Logic Structure (b) Output Transitions

In DPA attacks, the attacker, exploits this basic property of sCMOS implementations, and by recording the power consumption and statistically extracting the correlation, retrieves the secret key stored in the secure IC. Many functionally secure cryptographic algorithms can be broken using DPA when implemented as secure ICs.

Countermeasures to thwart DPA attacks can be classified as [5]: *Algorithmic-level countermeasures* and *Circuit-level countermeasures*. Algorithmic countermeasures [6] [7] attempt to conceal power variations at the architectural or algorithmic-level. They are not really effective against DPA or its variants, as the power variations originate at the circuit level [5]. Circuit-level countermeasures attempt to make the power variations in the secure IC *independent* of the input data by using circuit or logic design techniques. Numerous

circuit-level countermeasures attempt to mask power variations by introducing random power variations during circuit operation [5] [8] [9]. When analyzing these techniques, masking merely lowers power information leaked, but does not eliminate it [4].

The most effective countermeasure to resist DPA attacks is to make the power consumption of the secure IC *invariant*, hence uncorrelated to the input data (secret key). Sense Amplifier Based Logic (SABL) [10], Simple Dynamic and Differential Logic (SDDL) [11], and Wave Dynamic and Differential Logic (WDDL) [11] propose *power invariant* circuit design techniques to counter DPA. Of these, we compare RCDDL results with WDDL style (as well as sCMOS logic style) as it is has been more widely analyzed circuit-level countermeasure. All power invariant circuit-level countermeasures design secure ICs in CMOS logic styles that *consume constant power for every input data cycle (cycle)*. To achieve constant power consumption such CMOS logic styles must charge/discharge a constant value of total capacitance for every cycle. This requirement is typically satisfied using a Dynamic and Differential Logic (DDL) style to design the secure IC. Fig. 2 illustrates DDL style, with an XOR logic function.

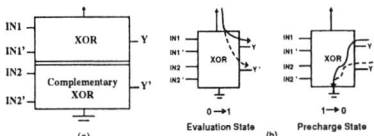

Fig. 2. DDL XOR gate (a) Structure (b) Output Transitions

In the DDL gate, refer Fig. 2 (a), for every input data either the uncomplemented output (Y) *or* the complemented output (Y') transitions. Further, dynamic logic introduces a *precharge* and *evaluation* state, thereby reducing the number of unique transitions possible in the gate to $1 \rightarrow 0$ (precharge state) and $0 \rightarrow 1$ (evaluation state), refer Fig. 2(b). Together, DDL style ensures that there is *exactly one* differential (uncomplemented or complemented) output transition during every cycle. Also, in order to achieve constant power consumption, fixed (constant) amount of charge must be charged/discharged in every cycle. This means that the total capacitance of the differential (uncomplemented and complemented) outputs must match. *All the existing DDL design techniques complement EVERY gate in the uncomplemented logic to generate the differential output and to match the capacitances. This is inefficient.*

In this paper, we propose a novel design methodology for DPA-resistant Secure ICs using Reduced Complementary Dynamic and Differential Logic (RCDDL) [12]. In RCDDL, the complemented output in the differential logic is generated by *reusing* part of the uncomplemented logic, while ensuring that a constant capacitance is charged/discharged in every cycle. Also, to ensure dynamic operation, wave (of '0's) propagation [11] is used to precharge all the RCDDL gates in the circuit. The proposed systematic design method involves strategies to design: 1) various types of RCDDL gates, and 2) circuits

978-1-4244-1709-4/07/$25.00 © 2007 IEEE

using the set of RCDDL gates. Further, we show that RCDDL gates could be used in conjunction with WDDL gates [11] to achieve improved logic functionality. Thus, when compared with circuits designed using only WDDL style gates and only sCMOS gates, we obtain significant improvements in security strength and average power consumption. It is important to note that although all power invariant countermeasures claim constant power consumption, minor power variations do persist due to second-order effects. An additional goal of any power invariant design technique must be to keep the residual (minor) power variations as minimal as possible.

This paper is organized as follows. Section II presents the RCDDL gate structure, operation and sizing requirements. Section III describes the types of RCDDL gates and the methods to design them. Section IV analyzes the RCDDL circuit design issues. Section V details the experimental results. Section VI concludes the paper.

II. REDUCED COMPLEMENTARY DYNAMIC DIFFERENTIAL LOGIC (RCDDL) STYLE [12]

In RCDDL style, dynamic and differential logic is generated by reusing part of the uncomplemented logic. This leads to significant improvements in security strength and average power consumption.

A. RCDDL - Gate Structure

Any logic function could be designed using the proposed RCDDL design method. Each RCDDL gate is divided into four segments, as shown in Fig. 3 (a).

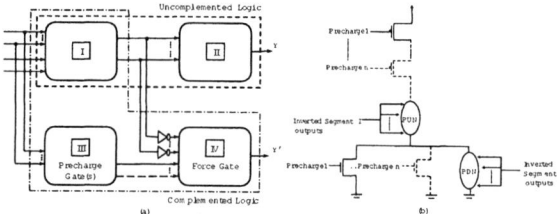

Fig. 3.   (a) RCDDL Gate Structure (b) Force Gate Structure

1) RCDDL Gate - Uncomplemented Logic: Segments I and II constitute the uncomplemented logic. They are designed as follows:

1) Segments I and II are designed from the Sum Of Products (SOP) expression of the logic function to be implemented. The SOP expression could be a two-level or multi-level expression.
2) The SOP expression implementing the logic function must ensure that the output Y is '0' during precharge state (to enable precharge wave propagation) i.e. when all differential inputs (e.g. A and A') are '0's, output Y must be '0'.
3) Segment I contains gates constituting all logic stages, excluding the last logic stage, of the logic function. Segment II contains the gate constituting the last stage of the logic function.

Further, Segments I and II can be designed using either negative gates (NAND/NOR) or positive gates (AND/OR). This makes the gate design conducive to logic synthesis and optimization strategies.

2) RCDDL Gate - Complementary Logic: Segments I, III and IV constitute the complementary logic. They are designed as follows:

1) The outputs from Segment I, are complemented (using inverters) and provided as inputs to Segment IV (force gate).
2) Segment III generates the precharge signal(s). Each precharge signal is generated using a NOR gate, which takes one set of differential inputs to the RCDDL gate as its inputs.

3) Segment IV contains the force gate, shown in Fig. 3 (b). PUN/PDN in the force gate corresponds to the complement of the function of the gate in Segment II. The complemented outputs from Segment I are provided as inputs to PUN/PDN. The precharge signals, from Segment III, are connected to PMOS transistors in series with PUN, and NMOS transistors in parallel with PDN.

The transistors in the four segments are then sized based on the sizing strategy presented in section II-C.

B. RCDDL - operation

RCDDL gates are dynamic and differential logic gates. The gates are designed in sCMOS logic.

Precharge state (wave propagation) is achieved by propagating a precharge wave (of '0's) through all the differential inputs of the circuit [11]. When all differential inputs of each RCDDL gate are '0's, output Y becomes '0'. The precharge signal(s) become '1'. Thus, forcing the complemented output Y' to become '0'. Hence, the name, "force" gate. Therefore, when all the differential inputs are '0's, all differential outputs of each RCDDL gate become '0's, so that all subsequent RCDDL gates in the circuit can be precharged. The precharge '0's pass through the circuit like a wave, forcing all the outputs in the circuit to '0's. Refer [11] for additional information on precharge wave generation.

Evaluation state is achieved when complemented differential inputs are provided to the circuit. When complemented inputs are provided to each RCDDL gate, output Y evaluates to the appropriate logic value. The precharge signal(s) become '0' and the complemented output Y' evaluates to the complement of output Y. This ensures normal mode of operation to produce complemented differential outputs in the circuit.

C. RCDDL - Sizing

After analyzing the RCDDL gate operation, assuming there are no timing requirements on inputs to various Segments, we can make the following observations on instantaneous current drawn.

During evaluate state, exactly half the PMOS transistors in Segment I transition to draw supply current. This is because exactly one of each of the differential inputs transitions, irrespective of the input data. Also, either Segment II (output Y) *or* Segment IV (output Y') will transition. Segment III does not contribute to current draw as all precharge gate(s) will discharge (turned OFF).

During precharge state, exactly half the PMOS transistors in Segment I that were turned OFF during evaluation state will transition, irrespective of the input. Thus, leading to a constant current draw from Segment I. Segment II (output Y) and Segment IV (output Y') does not contribute to current draw as exactly one of them will discharge (both outputs Y and Y' are '0's). Segment III (all precharge signal(s)) will always transition, leading to a constant current draw.

1) Force gate sizing strategy: The structure of the force gate imposes two restrictions on the arrival times of signals:

(i) *During precharge state, the precharge signal(s) must arrive earlier than the signals from Segment I.*
This is ensured, inherently, due to the presence of inverters connected to the inputs from Segment I, and can further be ensured by increasing the widths of NMOS transistor(s) connected to the precharge signal(s).
(ii) *During evaluation state, the precharge signal(s) must arrive later than the signals from Segment I.*

This can be ensured by decreasing the widths (or increasing the lengths) of NMOS transistor(s) connected to the precharge signal(s), and increasing the widths of PMOS transistors connected to the inputs from Segment I.

Now, based on the sizing changes made to the force gate, the PMOS transistors in the remaining segments are sized to ensure that the current drawn is within 10% of each other in every cycle.

III. TYPES OF RCDDL GATES

RCDDL gates can be classified based on the nature of the Sum-Of-Products (SOP) expression as: *Two-level gates* and *Multi-level gates*.

A. Two-level RCDDL Gates

Two-level RCDDL gates are designed from the two-level SOP expression of the logic function to be implemented. These expressions have all minterms in the expression. Segment I contains gates implementing all the PRODUCT terms, and Segment II contains the gate implementing the SUM term in the SOP expression. Each RCDDL gate can further be classified based on the nature of inputs in the SOP expression as: *Symmetric gates* and *Asymmetric gates*.

1) Two-level Symmetric RCDDL Gates: are gates that have all differential inputs (both uncomplemented and complemented inputs) available in the two-level SOP expression of the logic function. For a symmetric gate, a single precharge signal (Segment III) is generated from *any one* of set of differential inputs to the gate.

RCDDL XOR Gate: The XOR gate ($Y = A'.B + A.B'$) structure is shown in Fig. 4 (a). Note that we have used negative (NAND) gates to design the uncomplemented logic, in contrast with the design restrictions existing in WDDL style [11]. Fig. 4 (b) shows that instantaneous supply current for sCMOS, WDDL and RCDDL implementations of XOR. Notice that in sCMOS some inputs do not lead to current draw, whereas RCDDL exhibits power invariance by consuming *constant* current during every cycle.

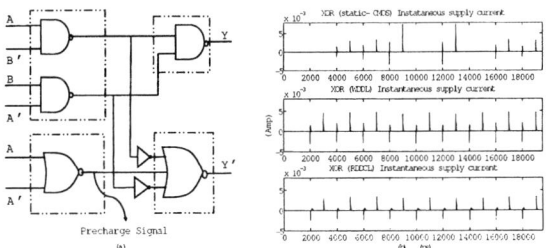

Fig. 4. (a) RCDDL XOR Gate Structure (b) I(vtstp) waveform for sCMOS, WDDL and RCDDL XOR Gates

RCDDL XOR Gate Operation: Precharge state: All inputs (A, A', B and B') are set to '0'. In Segment II, output Y generates '0'. The precharge signal from Segment III becomes 1. In Segment IV, the force gate forces the output Y' to '0'. Hence, all differential outputs are '0's when all differential inputs are '0's. *Evaluation state:* Normal complemented differential inputs are provided to the gate. Segments I and II operate normally, producing output Y. In Segment III, the precharge signal becomes '0'. Thus, the force gate (Segment IV) works normally, producing the complemented output Y'.

2) Two-level Asymmetric RCDDL Gates: are gates that do not have differential inputs (both uncomplemented and complemented inputs) in the SOP expression of the logic function.

In the case of asymmetric gates, the *asymmetric differential inputs*, i.e. input whose complementary or uncomplementary input is not part

of the SOP expression, are accounted for by introducing redundant gates. There are two methods to achieve this requirement. The first method is the *redundant precharging method*. In this method, a precharge signal is generated from *each* asymmetric differential input. Although Segment IV, producing the output Y', will function with only one precharge signal, redundant precharge signals are included to account for all asymmetric differential inputs. The second method is the *input-converter method*. In this method, the asymmetric input is passed through an input-converter logic, shown in Fig. 5. The input-converter logic takes in the differential input (e.g. A and A', Fig. 5) and generates the asymmetric input (A) that is required by the logic function (SOP expression).

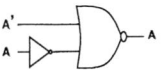

Fig. 5. Input-Converter Logic

Although, both methods work well, in our experiments, we found the redundant precharging method to have lesser instantaneous current consumption. This is because, the additional gates are generated in Segment III, and are used only during precharge state. Whereas, in the input-converter method, the additional gates are generated in Segment I and contribute to current draw during both precharge and evaluate states adding to the total current drawn in both states.

RCDDL Multiplexer (MUX) Gate: The RCDDL MUX gate ($Y = S'.A + S.B$) structure is shown in Fig. 6 (a). In the MUX gate, there are two asymmetric differential inputs A' and B' missing from the SOP expression. These are connected in the RCDDL gate, by generating a precharge signal from each asymmetric differential input. Although the force gate will function without the redundant precharge signals, they are included only to account for all differential inputs. Fig. 6 (b) shows that instantaneous supply current for sCMOS, WDDL and RCDDL implementations of MUX.

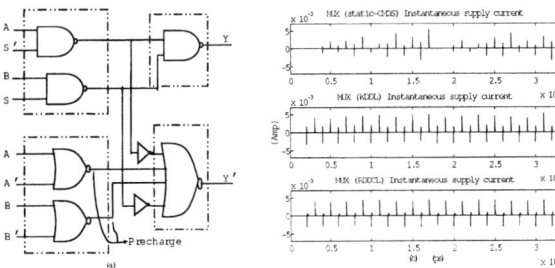

Fig. 6. (a) RCDDL MUX Gate Structure (b) I(vtstp) waveforms comparison between static-CMOS, WDDL and RCDDL MUX gates

RCDDL MUX Gate - Operation: Precharge state: All differential inputs (A, A', B, B', S and S') are set to '0'. In Segment II, output Y generates '0'. In Segment III, all precharge signals generated using asymmetric differential inputs (A and B) become '1'. In Segment IV, the force gate forces the output Y' to '0'. Hence, all differential outputs are '0's when all differential inputs are '0's. *Evaluation state:* Normal differential inputs are provided to the gate. Segments I and II operate normally, producing output Y. In Segment III, the precharge signal becomes '0'. Thus, in Segment IV, the force gate works normally, producing the complemented output Y'.

B. Multi-level RCDDL Gates

Multi-level RCDDL gates are designed from the optimized multi-level SOP expression of the logic function to be implemented. These expressions do not contain all minterms in the SOP expression. Each multi-level RCDDL gate can further be classified based on the nature of inputs in the SOP expression of the logic function as: *Symmetric gates* and *Asymmetric gates*. However, since most multi-level expressions are asymmetric, we analyze multi-level asymmetric RCDDL gates, with an example.

1) RCDDL Rop gate: The RCDDL Rop (Random operation) gate $(Y = A.C + B.(D + E))$ structure is shown in Fig. 7 (a). Segment I contains gate(s) constituting all logic stages, excluding the last logic stage, of the logic function. Segment II contains the gate constituting the last stage of the logic function. Fig. 7 (b) shows that instantaneous supply current for sCMOS, WDDL and RCDDL implementations of Rop.

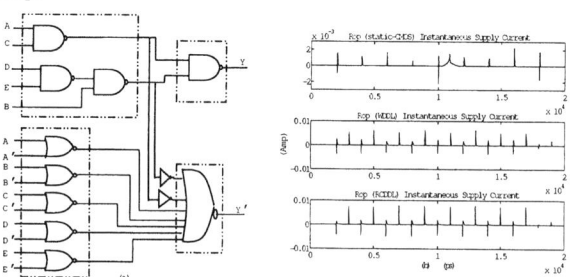

Fig. 7. (a) RCDDL Rop gate (b) I(vtstp) waveforms comparison between static-CMOS, WDDL and RCDDL Rop gates

RCDDL Rop Gate - Operation: Precharge state: All differential inputs (A, A', B, B', C, C', D, D', E and E') are set to '0'. In Segment II, output Y generates '0'. In Segment III, all precharge signals generated using asymmetric differential inputs (A, B, C, D, and E) become '1'. In Segment IV, the force gate forces the output Y' to '0'. Hence, all differential outputs are '0's when all differential inputs are '0's. *Evaluation state:* Normal differential inputs are provided to the gate. Segments I and II operate normally, producing output Y. In Segment III, all precharge signals become '0'. Thus, the force gate works normally, producing the complemented output Y'.

Delay, area and power analysis: Table in Fig. 8 shows the Max. Instantaneous Current Variation, Average Power and Area improvements (with delay penalty) of RCDDL XOR, RCDDL MUX and RCDDL Rop gates over sCMOS and WDDL implementations.

Gates	SCMOS				WDDL				RCDDL			
	Max. I_{inst} Variance (Amp)	Avg. Power (mW)	Area (λ^2)	Prop. Delay (ns)	Max. I_{inst} Variance (Amp)	Avg. Power (mW)	Area (λ^2)	Prop. Delay (ns)	Max. I_{inst} Variance (Amp)	Avg. Power (mW)	Area (λ^2)	Prop. Delay (ns)
XOR	9.52xe-6	0.1327	7020	0.2768	0.234xe-6	0.4039	22248	0.5070	0.0808xe-6	0.2944	19548	0.8325
MUX	2.05xe-6	0.0087	6264	0.1609	0.2042xe-6	0.4033	22356	0.5554	0.0169xe-6	0.3981	22140	0.7612
Rop	1.114xe-6	0.199	17604	0.838	0.338xe-6	0.487	35103	2.742	0.195xe-6	0.505	37044	3.441

Fig. 8. Max. Inst. Curr. Variation, Avg. Power, Area, Prop. delay of sCMOS, WDDL and RCDDL implementations of XOR, MUX and Rop functions

IV. RCDDL LOGIC & CHARACTERISTICS

A. Using RCDDL gates with WDDL gates - Analysis

It is important to note that improvements in RCDDL implementations are due to *reuse* of part of the RCDDL gate. Thus, building one-level gates (AND/OR) using RCDDL would prove to be inefficient. However, cryptographic algorithms (secure ICs) predominantly use operations such as permutation and substitution

[10] [13] that are mostly designed using XORs and MUXes. Hence, when designing secure ICs, the proposed RCDDL style would be an efficient alternative to existing logic styles.

Albeit, one-level logic gates (AND/OR) are required in many real-life circuits. To enable use of one-level logic gates, we designed these gates in WDDL style. Further, we performed experiments on *using RCDDL gates in conjunction with WDDL gates*. Our experiments varied from building a simple one-bit full adder using both RCDDL and WDDL gates, to building large circuits in various configurations. In this subsection, we introduce this notion with a simple one-bit full adder logic.

1) Full Adder: The one-bit Full Adder logic expression is $S = A \oplus B \oplus C_i; C_{i+1} = A.B + B.C_i + C_i.A$. The full adder logic structure is shown in Fig. 9 (a).

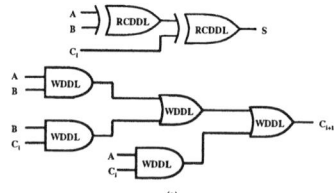

Logic Style	SCMOS				WDDL				RCDDL			
	Max. Var.	Avg. Power (mW)	Area (λ^2)	Prop. Delay (ns)	Max. Var.	Avg. Power (mW)	Area (λ^2)	Prop. Delay (ns)	Max. Var.	Avg. Power (mW)	Area (λ^2)	Prop. Delay (ns)
FA	136xe-7	0.288	14148	0.413	10.97xe-7	1.570	79056	0.949	7.6xe-7	1.065	55728	2.503

(b)

Fig. 9. RCDDL-WDDL one-bit full adder (a) Logic structure (b) Max. variation, average power and area Results

From the results shown in Fig. 9 (b), we can see that using RCDDL gates in conjunction with WDDL still provides improvements in max. instantaneous current variation, average power and area, when compared with one-bit full adder designed using *only* WDDL gates. The section V discusses experimental results for large circuits designed using both RCDDL and WDDL gates.

B. RCDDL Logic Design

Any circuit can be designed using RCDDL cells. To ensure *ideal* operation, the force gate timing requirement (subsection II-C) is satisfied by sizing the transistors appropriately. At circuit-level, this condition is satisfied by using the later arriving differential input signal(s) to generate the precharge signal(s). However, there is a drawback to using this approach. When several RCDDL cells are connected in series, there is a differential delay introduced between the inputs to the force gate, generated during evaluation state. For certain inputs, the inverted inputs from Segment I arrive earlier than precharge signals from Segment III. Although this condition poses no functionality issues, if the differential delay between the inputs becomes greater than the propagation delay of the gate, this causes a glitch in output Y'. This problem can be overcome by using glitch-reduction techniques like buffering, logic restructuring and pipelining, to equalize input arrival times, when the differential delay between input arrival times is greater than the propagation delay of the gate. Keeping differential delay between inputs signals lower than the propagation delay of the gate ensures that there will be no glitch produced from the design.

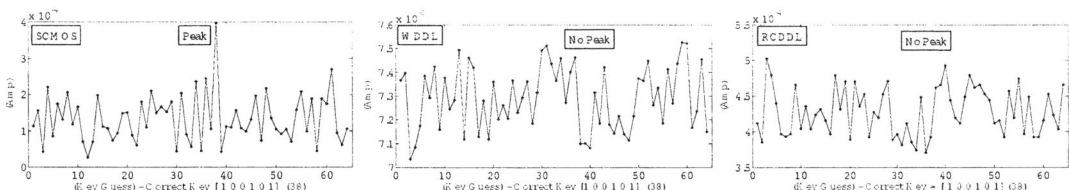

Fig. 10. DES traces from sCMOS, WDDL and RCDDL implementations respectively

V. EXPERIMENTAL RESULTS

A. Evaluation Metric

The *security strength* of any secure IC design can be quantified using the instantaneous power (supply current) variation ($Var(I_{inst})$) of the implementation [14]. We compared results for RCDDL circuits with the circuits implemented using WDDL and sCMOS design styles. We conducted 4 different experiments to measure the efficiency of our proposed RCDDL style. All experiments were performed on a Sun Blade 100 workstation with 512MB RAM. All gates were implemented in 0.25u technology, with 5V supply voltage. All gates were designed and extracted using Magic layout editor, and simulated using Synopsys Hspice transistor-level simulator. *All circuit experiments included layout parasitics.* Instantaneous power (supply current) measurements from Hspice were taken every 10ps, equaling a sampling rate of 100GHz. DPA attacks and analysis were performed using MATLAB/C++. The RCDDL library built consisted of XOR and MUX cells; combined with AND/OR gate implemented using WDDL logic; two sCMOS buffers with propagation delays equal to 0.3ns and 0.5ns, and a current-matched sCMOS D-FlipFlops.

B. Case Study I - DPA on Data Encryption Standard (DES) [1]

Our first experiment was to test the proposed logic style on a DES implementation. While DPA can be applied to many (types) cryptographic algorithms [1] [2], in this paper we evaluate the proposed RCDDL style using DES algorithm. DES is one of the most widely used and studied cryptographic algorithm. Our DES test circuit [14] consisted of a subset of a single round of DES encryption, shown in Fig. 11. This circuit has been shown to be a sufficient subset of the DES cryptographic algorithm [14]. We performed simulations by encrypting 2048 random input vectors, with a fixed secret key.

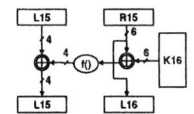

Fig. 11. DES Test Circuit [14]

Test Circuits	WDDL		RCDDL	
	Max. Curr. Var.	Avg Pwr (mW)	Max. Curr. Var.	Avg Pwr (mW)
DES	0.849xe⁻⁴	4.525	0.532xe⁻⁴	2.991
XOR-BT	1.630xe⁻⁴	10.520	0.849xe⁻⁴	8.056

Fig. 12. Experimental Results I & II - DES and XOR-BT.

We implemented the DES test circuit in all three logic styles. Fig. 10 shows the *DPA key guess vs peak differential instantaneous supply current (I_{inst})* trace. The secret key stored in DES (sCMOS) was retrieved using 495 texts, as indicated by the peak in Fig. 6 (sCMOS). In DES (WDDL) and DES (RCDDL), we were not able to recover the secret key even when using the exhaustive set of 2048 texts. No noticeable peaks are present in Fig. 10 (WDDL)

and (RCDDL). From the table in Fig. 12, DES (RCDDL) exhibits lower max. $Var(I_{inst})$ (better security strength) and average power consumption when compared with DES (WDDL).

C. Case Study II - Wiring Analysis

Our second experiment was performed to analyze the impact of wire delays on circuits designed using RCDDL style. We performed this experiment on a completely-balanced random-tree of logic depth 3. A random tree is a circuit where each node in the tree could be a gate of any cell-type from the library. The capacitance between two nodes (gates) in the tree was varied randomly, simulating varying wire lengths between the gates. The impact of increasing wire lengths on max. instantaneous current variation is shown in Fig. 13. As expected, after some minor variations, with increasing capacitance, max. instantaneous current variation increases. This is because increasing wire lengths introduce differential delays between inputs to gates that lead to introduction of glitches that consume current (power). However, this can be accounted for with buffering, as detailed in the next experiment.

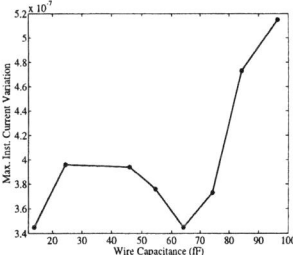

Fig. 13. Experimental Results III - Capacitance vs Max. Variation.

D. Case Study III - Buffering Analysis

Our third experiment was performed to analyze the RCDDL style by building a *realistic* circuit with buffering requirements due to unequal logic depths and wiring. Our test circuit consisted of a balanced XOR-tree (XOR-BT) [11] containing logic depths from 2 to 9. XOR-BT (RCDDL) was buffered to ensure that the differential delays between inputs for all gates were less than propagation delay ($\pm t_p$) of the gate. The results are presented in Fig. 12. We can see that the XOR-BT (RCDDL) has improved I_{inst} variation and average power consumption over WDDL implementations.

E. Case Study IV - Scalability Analysis

Our final experiment was performed to analyze the scalability of our proposed RCDDL style. We built 12 synthetic circuits with increasing number of gates and in different circuit configurations. The circuit configurations included completely balanced XOR-trees (XOR-1, XOR-2, XOR-3 and XOR-4), level-balanced random-trees (LB-1, LB-2, LB3 and LB-4), which contain one cell-type in each level of the tree, and completely-balanced random trees (RC-1, RC-2, RC-3 and RC-4). Experiments were conducted by simulating the circuits for 800 random input vectors, supplied at the rate of 100

MHz. Figures 14 and 15 show the maximum $Var(I_{inst})$, average power consumption, area from all circuit implementations.

Test Circuits	SCMOS			WDDL			RCDDL		
	Max. Var.	Avg. Pwr. (mW)	Area (λ^2) (Est.)	Max. Var.	Avg. Pwr. (mW)	Area (λ^2) (Est.)	Max. Var.	Avg. Pwr. (mW)	Area (λ^2) (Est.)
XOR-1	15.8xe⁻⁶	0.59	21060	0.603xe⁻⁶	1.33	66744	0.268xe⁻⁶	0.93	58644
XOR-2	51.22xe⁻⁶	3.61	105300	1.68xe⁻⁶	7.48	333720	1.94xe⁻⁶	5.56	293220
XOR-3	100xe⁻⁶	4.69	217620	3.32xe⁻⁶	10.33	689688	1.73xe⁻⁶	7.33	605988
XOR-4	178.9xe⁻⁶	9.70	442260	7.33xe⁻⁶	21.02	1401624	4.16xe⁻⁶	15.63	1231524
LB-1	2.36xe⁻⁶	0.16	21816	0.388xe⁻⁶	0.90	48276	0.142xe⁻⁶	1.12	48060
LB-2	122xe⁻⁶	5.17	185760	3.72xe⁻⁶	12.22	588060	4.13xe⁻⁶	10.51	537948
LB-3	340xe⁻⁶	6.98	284256	9.1xe⁻⁶	18.21	935928	4.92xe⁻⁶	14.4	857304
LB-4	676xe⁻⁶	5.85	383184	6.68xe⁻⁶	16.02	1179360	5.93xe⁻⁶	15.94	1140048
RC-1	10.9xe⁻⁶	0.31	18468	0.445xe⁻⁶	1.01	53244	0.428xe⁻⁶	0.97	50328
RC-2	42.2xe⁻⁶	2.61	108540	2.98xe⁻⁶	7.95	395172	1.27xe⁻⁶	7.72	374760
RC-3	56.5xe⁻⁶	3.03	234900	2.75xe⁻⁶	9.66	683532	2.16xe⁻⁶	9.16	645624
RC-4	133.82xe⁻⁶	7.05	526284	6.43xe⁻⁶	21.76	1585440	4.05xe⁻⁶	20.72	1500660

Fig. 14. Experimental Results III - Synthetic Circuits.

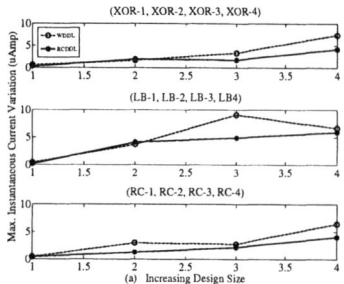

Fig. 15. Synthetic Benchmarks Results for RCDDL and WDDL - Max. Inst. Current Variation vs Increasing Design Size

For the 12 synthetic circuits used in this experiment, when compared with WDDL implementations, on average, we can see that using RCDDL style, reduced $Var(I_{inst})$ by 31.66%, reduced avg. power consumption by 14% and reduced area by 7.75%. On comparing with sCMOS implementations, $Var(I_{inst})$ reduced by 55.7X (average), thus drastically improving the security strength of the RCDDL implementations.

1) Power/Area/Delay characteristics: When comparing power consumption of circuits designed using WDDL and RCDDL gates, we can see that in WDDL gates, the number of logic gates required is at least twice the number of logic gates required in sCMOS gates. So, at gate-level, the average current drawn (average power) increases by at least 2X. Also, we observe that *the earliest arriving input signals control the current draw in each WDDL gate*. So, for a WDDL implementation of a circuit with n logic levels, all logic levels will draw current at the earliest input transition (same time). Therefore, the instantaneous supply current drawn (peak power) increases by at least nX when compared with a sCMOS implementation of the circuit. However, in RCDDL gates, the number of logic gates used is *less than* twice the number of logic gates in sCMOS gates. So, the average power consumed is *less than* WDDL gates. Also, *the current draw is controlled by the latest arriving signal(s)*, due to the force gate sizing requirements. Therefore, for an RCDDL implementation of the circuit with n logic levels, the supply current drawn *does not* occur at the earliest input transition. Hence, maximum supply current drawn (peak power) is *less than* nX, and hence, *less than* WDDL gates. Therefore, when compared with WDDL style, both average and peak power consumed in RCDDL decreases with increasing number of logic levels in the design.

When comparing area requirements, in RCDDL gates, due to *reuse* of a part of the logic, the number of gates required is less than 2X when compared with sCMOS implementations, therefore requiring lesser (or at least equal) area compared to WDDL gates.

The propagation delay of RCDDL implementations is more than that of WDDL implementations (on average less than 1.5X). This can be attributed to transistor sizing. The increase in delay is only due to increased rise time. The fall time of WDDL is more than that of RCDDL. However, when considering the cost of inadequate security strength in secure ICs, and also, since the size of secure IC circuits are usually small, the delay penalty incurred does not drastically affect the operation of the embedded applications using the secure IC.

Therefore, from all experimental results, we can summarize that RCDDL style circuit implementations consistently out-perform WDDL implementations by providing better security strength with improvements in average power consumption and area.

VI. CONCLUSIONS

In this work, we presented a novel design methodology using Reduced Complementary Dynamic and Differential Logic (RCDDL) [12] to design secure, DPA-resistant cryptographic devices (secure ICs) like smart cards. Experimental results on all 15 circuits (used in all experiments), including DES test circuit, show that circuits implemented using the proposed design methodology have improved security strength (reduced max. instantaneous current variation), when compared with WDDL implementations (42.29% average reduction) and sCMOS (55.7X average reduction) implementations. Also, RCDDL implementations show improvements in average power consumption and area, when compared with WDDL implementations.

REFERENCES

[1] "NIST - FIPS 46-3: Data Encryption Standard," National Institute of Standards and Technology (NIST), Tech. Rep., October 1999.
[2] "NIST - FIPS 197: Advanced Encryption Standard," National Institute of Standards and Technology (NIST), Tech. Rep., November 2001.
[3] P. Kocher, J. Jaffe, and B. Jun, "Differential Power Analysis," *Lecture Notes in Computer Science*, vol. 1666, pp. 388–397, 1999.
[4] T. S. Messerges, E. A. Dabbish, and R. H. Sloan, "Investigations of Power Analysis Attacks on Smartcards," in *In proc. of SMARTCARD-99*, May 1999, pp. 151–162.
[5] D. Suzuki et al, "Random Switching Logic: A New Countermeasure against DPA and Second-Order DPA at the Logic Level," *IEICE Transactions on Fundamentals of Electronics, Communications and Computer Sciences*, vol. E90-A, no. 1, pp. 160–168, 2007.
[6] J. S. Coron, ""Resistance against Differential Power Analysis for elliptic curves cryptosystems"," in *CHES '99*. LNCS 1717, 1999, pp. 292–302.
[7] T. Messerges and et al, ""Power Analysis Attacks of Modular Exponentiation in Smartcards"," in *CHES '99*. LNCS 1717, 1999, pp. 144–157.
[8] E. Trichina, "Combinational Logic Design for AES Subbyte Transformation on Masked Data," Cryptology eprint archive: Report 2003/236, IACR, nov 2003.
[9] S. Yang et al, "Power Attack Resistant Cryptosystem Design: A Dynamic Voltage and Frequency Switching Approach," in *Proc. of DATE 2005*.
[10] K. Tiri, M. Akmal, and I. Verbauwhede, "A Dynamic and Differential CMOS Logic with Signal Independent Power Consumption to Withstand Differential Power Analysis on Smart Cards," in *Proc. of European Solid-State Circuits Conference (ESSCIRC 2002)*, 2002, pp. 403–406.
[11] K. Tiri et al, "A Logic Level Design Methodology for a Secure DPA Resistant ASIC or FPGA Implementation," in *Proc. of DATE 2004*.
[12] S. Rammohan, "Reduced Complementary Dynamic and Differential Logic: a Circuit Design Methodology for DPA-resistant Cryptographic ICs," Master's thesis, University of Cincinnati, May 2007.
[13] A. J. Menezes, P. C. van Oorschot and S. A. Vanstone, *Handbook of Applied Cryptography*. CRC, October 1996, no. 978-0849385230.
[14] K. Tiri and I. Verbauwhede, "Securing Encryption Algorithms against DPA at Logic Level: Next Generation Smart Card Technology," *Lecture Notes in Computer Science*, vol. 2779, pp. 125–136, 2003.

Neuromorphic Building Blocks for Adaptable Cortical Feature Maps

C. M. Markan, Priti Gupta

Department of Physics & Computer Science
Dayalbagh Educational Institute
(*Deemed University*)
Dayalbagh, Agra - 282005, India
Email: markan_cm@hotmail.com, gupta.priti.84@gmail.com
Phone: 91-562-2121545, FAX: 91-562-2121226

Keywords: Floating Gate pFET, competitive learning, WTA, Feature maps, Ocular dominance, Orientation selectivity.

Abstract: 'Time-staggered Winner-Take-All' is a novel CMOS analog circuit that computes 'sum of weighted inputs" implemented as floating gate pFET 'synapse'[11]. Feedback circuit of the cell exploits adaptation dynamics of floating gate FETs refining its weights in response to stimulation by patterned inputs distributed over time. This paper discusses the application of 'ts-WTA' cell as a core learning circuit in designing adaptive neuromorphic feature selective cells for a variety of visual cortical features such as ocular dominance, orientation selectivity etc. An array of these ts-WTA cells when embedded on an RC network exhibits reaction-diffusion type clustering based on feature selective response. The cell's adaptive behavior resembles Stent's physiological variant of competitive Hebb learning [21] and hence has potential to act as a building block in design of adaptable feature maps in different cortices.

I. INTRODUCTION

Feature maps are critical interconnectivity patterns between hierarchically organized cortical layers by means of which different sensory features are extracted from a sensory image and mapped over the cortical sheet. Higher cortical layers successively extract more complex features from less complex ones represented by lower layers. In fact it has been shown that different sensory cortices are also an outcome of a mechanism by which a generic cortical lobe adapts to the nature of stimulus it receives so as to extract sensory features embedded in it. [1]. Thus feature extraction and hence formations of feature maps are fundamental underlying principles of parallel and distributed organization of information in the cortex. Any effort towards artificial or neuromorphic realization of cortical structure will have to comprehend these basic principles before any attempt is made to derive full benefit of cognitive algorithms active in the brain [15].

Neuromorphic cortical maps find useful applications in robotics designing adaptive generic machines that conserve resources by acquiring greater sensory acuity to more abundant features at the cost of others depending on the environment they are nurtured in [2,15]. Another emerging area of interest is that of neural prostheses, wherein implants such as retinal or cochlear are used to artificially stimulate sensory nerves. Such implants are effective only when cortical feature maps are intact [5]. Animals born with defunct sensory transducers find their representative cortical area encroached upon by competing active senses. In such animals or in those with damaged or diseased cortex, sensory implants are ineffective as the cortical apparatus (maps) to interpret

inputs from them is not in place. The only option in such cases is to revive cortical feature maps either biologically or through prosthetic neuromorphic realization [5].

Efforts towards artificial realization of feature maps [2-4] can be broadly classified in two categories: (a) 'Ice-cube' models [2,3] emulate developed or adult cortex wherein feed-forward feature selective connections of feature detectors and lateral circuitry interconnecting them is assumed to be pre-wired. The feed-forward inputs provide crude feature selectivity, while the recurrent lateral interaction enhances the tuning of response and hence provides greater acuity to feature detection. (b) The 'plastic' models emphasize on adaptable development of feature maps wherein unspecific feed-forward connections refine to form feature selective maps in response to neural activity. Such models realize competitive synaptic plasticity either as abstract algorithm [7] not suitable to model developing cortical feature maps [8], or simulate a biologically realistic algorithm on a computer interfaced to a neuromorphic vision sensors[4].

Thus this paper aims to build a biologically feasible framework for adaptive neuromorphic cortical feature maps. In section 2 we review our basic building block (ts-WTA) for competitive synaptic plasticity [10]. In section 3 we apply *ts*-WTA to extract visual features e.g. ocular dominance and investigate its ability to cluster on the basis of their feature preference i.e. form feature maps. In section 4 we show how an array of clustering *ts*-WTAs can be used to design an orientation selective cell. All circuits in this paper were simulated on TSpice11 with BSIM3 level 49 spice models for 0.35um CMOS Process.

978-1-4244-1709-4/07/$25.00 © 2007 IEEE

II. 'TIME STAGGERED WTA' CIRCUIT

In building adaptive neuromorphic structures, the biggest bottleneck has always been implementing adaptable connection strengths (weights) in hardware. While capacitor storage is volatile, the non-volatile digital storage is bugged with severe overheads of analog/digital domain crossing. However, improved quantum mechanical charge transfer processes across the thin oxide now offer a floating gate pFET whose threshold voltage and hence conductivity adapts much like Hebbian learning in a 'synapse' [6]. In fact, a competitive learning rule similar to Kohonen's unsupervised learning [7] and an adaptive WTA circuit with a refractory time period have also been developed using floating gate pFETs [13].

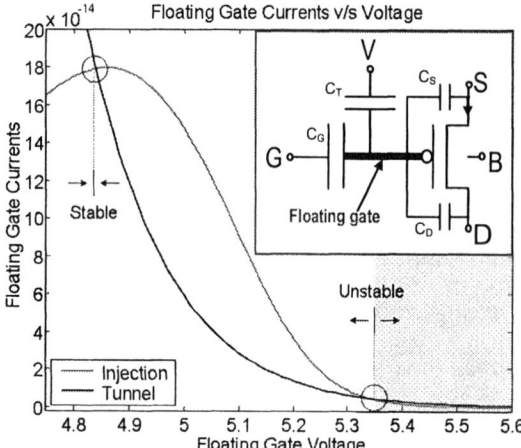

Figure-1: (Inset) Floating gate pFET ($V_B=V_{DD}$). Shows variation of Injection and Tunnel current as a function of V_{fg} for fixed tunnel voltage V_T and drain voltage V_d. The gradient on V_{fg} passes through two equilibrium (*crossover*) points, at higher values of V_{fg} the equilibrium point is unstable and one at lower values is stable.

Floating gate 'synapses': (*see inset figure 1*) The trapped charge on the floating gate of a double gate pFET can be altered using two antagonistic quantum mechanical transfer processes [4]:

Tunnelling: Charge is added to the floating gate by removing electrons from it by means of Fowler-Nordheim tunnelling across the oxide capacitor. The tunnel current is expressed in terms of terminal voltages across oxide capacitor i.e. tunnel voltage V_T, and floating gate voltage V_{fg} as [9], with $V_f=368.04$, $I_{to}=9.35 \times 10^8$ as fit parameters

$$I_{tunnel} = F_T(V_T, V_{fg}) = I_{to} \exp\left(-\frac{V_f}{V_T - V_{fg}}\right) \quad ..(1)$$

Injection: Charge is removed from the floating gate by adding electrons to it by current hot-electron injection (IHEI) from drain end of channel to the floating gate across thin gate oxide. Injection current is expressed as a semi-empirical equation in terms of pFET terminal voltages i.e. source (V_s), drain (V_d), floating gate voltage (V_{fg}) and source current Is as [9], with fit parameters η $=1.30 \times 10^{-5}$, $\beta =155.75$, $\delta =0.702$, for $Is = 10$nA and $\lambda=1$.

$$I_{injection} = F_I(V_d, Is, V_{fg}) = \eta.Is.\exp\left(-\frac{\beta}{(V_{fg} - V_d + \delta)^2} + \lambda.V_{sd}\right) \quad ..(2)$$

Adaptation in floating gate 'synapses': If the overall affect due to various terminals capacitatively connected to the floating gate is kept small enough as compared to the two tunnelling current [7,11] then the variation of charge on floating gate is expressed as

$$C_F \frac{\partial V_{fg}}{\partial t} = I_{tunnel} - I_{injection} = F_T(V_T, V_{fg}) - F_I(V_d, Is, V_{fg}) \quad ..(3)$$

With pFET in saturation its source current can be largely expressed as a function of V_{fg}, so that above equation is rewritten with magnitude of Is absorbed in F_I

$$C_F \frac{\partial V_{fg}}{\partial t} = I_{tunnel} - I_{injection} = F_T(V_T, V_{fg}) - F_I(V_d, V_{fg}).x \quad ..(4)$$

where a binary variable x is 1 when Is current flows and 0 otherwise. The variation of I_{tunnel} and $I_{injection}$ as a function V_{fg} is plotted in figure 1. Observe two equilibrium points characterize the floating gate dynamics of a pFET 'synapse' viz., (i) stable and (ii) unstable. Thus the floating gate pFET acts as a 'synapse' i.e. it has an adaptable conductivity or weight which grows on stimulation (*injection*) and saturates to an upper limit (*stable point*). In absence of stimulation the self decay term (*tunnelling*) prunes its weights till it finally goes dead from where it cannot recover (*unstable point*). Any variation in V_T and V_d merely relocates these equilibrium points.

'Time staggered Winner Take all': We now connect two such weighted inputs or 'synapses' to build a competitive learning cell as shown in figure 2(a) such that the voltage at common source terminal (V_s) is expressed as a weighted sum of the two branch currents Is_{i1}, Is_{i2}.

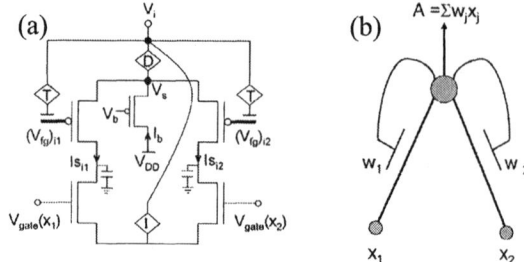

Figure 2: (a) Actual circuit of learning cell, (*Right*) its abstract model. In (a) $(V_{fg})_{i1}$, $(V_{fg})_{i2}$, and in (b) W_1, W_2 shows the floating gate based weighted connection. I, T and D are dependent voltage sources for injection, tunnel and activation nodes, respectively. x_1, x_2 are inputs and node voltage V_i is activation of the cell which is equivalent to A in (b). In (a) capacitor at floating gate terminal $C_F = 5$pF, at drain terminal is 500pF, V_{DD}=6V.

Then V_s can be written as

$$V_s = \left[V_{DD} - R_b. \sum_j Is_{ij}.x_j \right] \quad ..(5)$$

where R_b represents saturation resistance of a pFET with fixed bias voltage V_b (≈ 5.3V). Here input

$x_j=1 \Rightarrow j^{th}$ branch nFET is ON, i.e. $V_{gate}(x_j) = V_{DD}$
$x_j=0 \Rightarrow j^{th}$ branch nFET is OFF, i.e.$V_{gate}(x_j) = -1$V.

With floating gate pFET in saturation, source current largely depends on its gate-to-source voltage. Now with inputs applied one at a time, source voltage (V_s) itself depends on floating gate voltage of stimulated branch and so it won't be improper to express source current as a function of floating gate voltage i.e.

$Is_{ij} = f(V_{fg.ij}).x_j$. Rewriting (5), after substitution and suppressing constants, we have $V_s = \left[\sum_j f(V_{fg})_{ij}.x_j \right]$

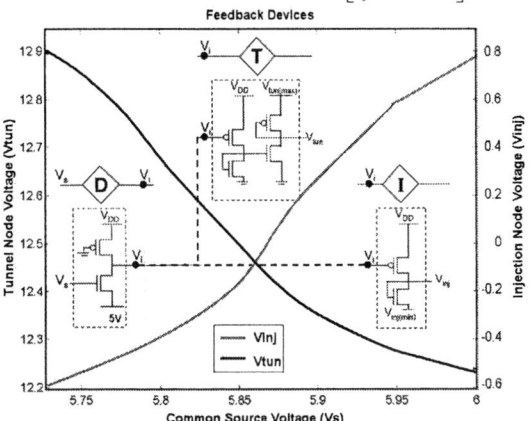

Figure 3: (*Inset*) Shows Feedback devices T, I. Both the devices act in conjunction with buffer device 'D' and their equivalent circuits are shown as inset. The graph shows variation of common tunnel node voltage V_{tun} and common Injection node V_{inj} w.r.t. common source voltage V_s. Here V_i is the activation node voltage at which influence from neighbouring cells will enter. While 'D' is inverting buffer, 'T' is dependent voltage device which translates V_i to higher tunnel voltage. 'I' is also an inverting device which translates V_i to appropriate injection voltage range. The overall effect of two inversions is non inverting relation between V_s and V_{inj}.

Feedback: Let us express tunnel feedback $V_T = T(V_i)$ and Injection feedback $V_d = I(V_i)$ as monotonically varying functions of activation node voltage V_i which itself is linked to common source voltage V_s through a buffer device D as $V_i = D(V_s)$ see figure 2. *In our design of feature selective cell we are guided by the basic principle: to accomplish WTA a self-excitation (injection) must accompany global nonlinear inhibition (tunnelling)* [10].

Rewriting adaptation equation (4) after substitution and suppressing the constant terms, we have:

$$\frac{d(V_{fg})_{ij}}{dt} = F_T \left[T \left(\sum_j f(V_{fg})_{ij} x_j \right), (V_{fg})_{ij} \right] - F_I \left[I \left(\sum_j f(V_{fg})_{ij} x_j \right), (V_{fg})_{ij} \right] x_j \quad .. (6)$$

With two inputs stimulated one at a time, there are two input patterns $\mathbf{x_1}$: $x_1=1$ & $x_2=0$ or $\mathbf{x_2}$: $x_1=0$ & $x_2=1$. If in every epoch all patterns occur once in a random order (*random-inside-epoch*) then at the end of an epoch the adaptation rate of floating gate voltage is given as

$$\frac{d(V_{fg})_{i1}}{dt} = \frac{d(V_{fg})_{i1}}{dt}\bigg|_{x_1} + \frac{d(V_{fg})_{i1}}{dt}\bigg|_{x_2} ; \text{ and } \frac{d(V_{fg})_{i2}}{dt} = \frac{d(V_{fg})_{i2}}{dt}\bigg|_{x_1} + \frac{d(V_{fg})_{i2}}{dt}\bigg|_{x_2} ;$$

In order to achieve winner-take-all it would be necessary that adaptation rates of the two floating gates in the above equation are of opposite sign. This is achieved by ensuring that overall gain of our feedback devices is greater than unity [11]. Initially if $(V_{fg})_{i1} > (V_{fg})_{i2}$ and the two arms are stimulated with equal probability i.e. as random-inside-epoch then this circuit will iteratively amplify the difference $\Delta(V_{fg})$ so that $(V_{fg})_{i1}$ rises to eventually switch off its pFET while $(V_{fg})_{i2}$ falls to reach the stable point so that its pFET stays on. Adaptability in

floating gates with time is shown in the figure 4.With I_{tunnel}, $I_{injection} \sim 10^{-13}A$ and C_{fg} in pF, V_{fg} varies slowly.

Though our circuit, which we call as *time-staggered WTA* (ts-WTA), is topologically similar to Lazzaro's WTA(*l*-WTA) circuit [12][13], yet it has some subtle and fundamental differences. *l*-WTA achieves positive feedback through common source terminal keeping gate voltages constant. In *ts*-WTA, positive feedback is achieved through special devices (I, T) that act on floating gates to achieve WTA. In *l*-WTA feedback through common source terminal requires all inputs be stimulated simultaneously and the competition is instantaneous, whereas *ts*-WTA requires stimulation of one input at a time and competition is staggered over several epochs. For clustering in self-organizing feature maps (SOFM), cells are required to communicate their feature selectivity through their output node. This cannot happen if all inputs are stimulated at the same time as it occurs for *l*-WTA. Therefore for map formation it is necessary that inputs be applied one at a time over all the participating cells. This clearly distinguishes the importance of '*ts*-WTA' w.r.t. *l*-WTA as a feature selective cell (see also sec V).

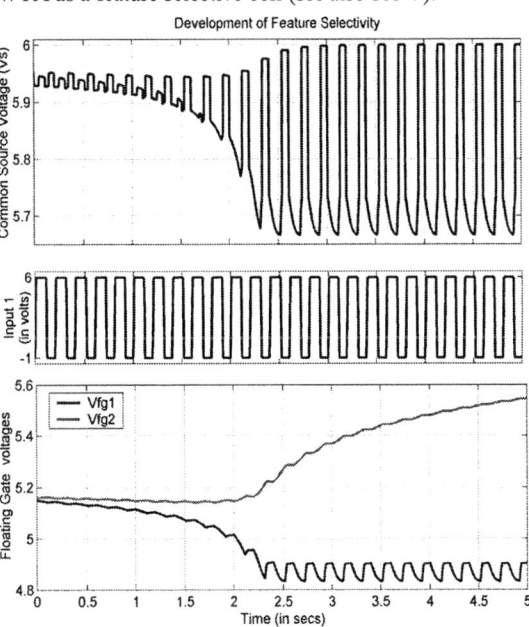

Figure 4: *(Top)* Development of feature selectivity observed as response of ts-WTA cell (V_s) to input patterns. (*Middle*) Input x_1, x_2 is invert of it. (*Bottom*) Evolution of V_{fg}s in ts-WTA circuit. Here we have used alternate stimulation instead of random-inside-epoch.

Another important property of adaptive feature selective cell is that it should (i) be adaptive during critical period, and (ii) exhibit normal behaviour *after* learning ceases. (i) *Learning rate parameter*: Since adaptability in our learning cell primarily depends on tunnel and injection currents, we gradually reduce feedback voltages i.e. decrease $V_{tunnel(max)}$ and increase $V_{injection(min)}$ to taper off learning rate to emulate the critical period effect see Table-1. (ii) *Operation after learning has ceased*: It may be noted that even after learning ceases, due to 1.6V

2007 IFIP International Conference on Very Large Scale Integration (VLSI-SoC 2007)

transition in injection voltage, the learning cell behaves normally. This is because with floating gate pFET in saturation small changes in drain voltage do not affect Is substantially and hence output V_s is generally unaffected.

VARIATION IN LEARNING RATE PARAMETER

Learning parameter $\gamma = \Delta V_{tun(max)} = \Delta V_{inj(min)}$	Unlearning Time (in secs)
0.0V	5
0.4V	20
0.8V	150
1.2V	1700
1.6V	>> 1700

Table-1: Shows the variation in learning parameter γ measured as change in feedback device source voltages.

III. VISUAL FEATURE EXTRACTION

We now illustrate applicability of ts-WTA to one of the least complex of the visual features i.e. ocular dominance.

Ocular Dominance: Cells of primary visual cortex of mammals have been observed to be dominated by inputs from one eye or the other, termed as ocular dominance. This domination is not genetically predetermined at birth rather develops in response to stimulus received from the two eyes. In animals whose one eye has been sutured or closed immediately after birth, cortical cells normally belonging to the closed eye are taken over or dominated by inputs from the active eye. This behavior is explained by the fact that at birth cortical cell receive nearly equal inputs from both eyes, slowly competition (similar to ts-WTA) between inputs from the two eyes leads to ocular domination. Thus ts-WTA cell can be directly used as ocular dominance feature selective cell provided its inputs represent inputs from the two eyes[21]. At the end of learning, ts-WTA cell responds only to the input that wins the competition i.e. it is either left eye dominated or right eye dominated.

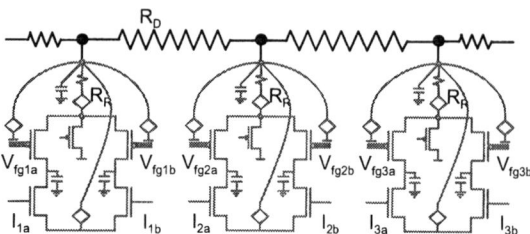

Figure 5: Shows diffusive interaction between V_i's of learning cells implemented in actual circuit by means of RC network. Note, as compared to figure 1, R_R is introduced between V_i and device D and a capacitor C is connected at V_i. R_D =10k, R_R =1k, C=10pF.

Map formation through diffusive interaction: Feature maps require that cells must cluster into groups with similar feature preferences such that there is both variety and periodicity in spatial arrangement of these cells across the cortical sheet. The basic mechanism that leads to clustering relies on the fact that neighboring cells have overlapping receptive fields and hence receive similar inputs. If somehow these neighboring cells can also be made to have similar responses then because of Hebbian learning individual cell's receptive field will also develop

similarly. This similarity will extend over a portion of cortex where a majority of cells have similar initial feature biases. With randomly chosen initial feature biases, clusters of feature selective cells will emerge distributed over the cortical sheet similar to those observed in cortical feature selectivity maps.

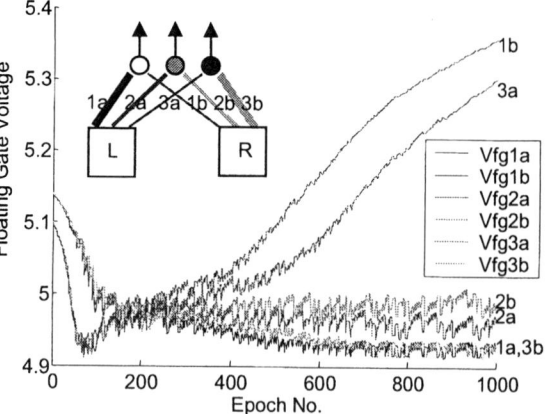

Figure 6: Shows development of V_{FG} for three cells of figure 5. (1) left dominated cell (*white*), (2) binocular cell (*gray*), (3) right dominated cell (*black*). Lower V_{FG} implies stronger connection strength. A dominated cell develops to have a large difference between two $V_{FG}s$, but a binocular cell has nearly equal $V_{FG}s$ which are relatively weaker as compared to a dominated cell.

Thus an important property that a learning cell must posses to achieve clustering is the ability to develop its feature selectivity under influence of its neighbours. More explicitly, it should exhibit three types of adaptive behaviour i.e. (i) if cell's bias is same as its neighbours, it must strengthen it, (ii) if cell's bias is opposite to its neighbours, it must reverse it, and (iii) if cell has equal number of neighbours with biases in favour and against, it must become unbiased i.e. respond equally to both inputs.

Though theoretical models employ these principles in different ways, yet not all are easy to realize in hardware. A class of models that are relatively easier to implement are those based on Turing's reaction-diffusion framework. Biologically such an interaction occurs in the form of chemicals leaking out of an active cell which enter neighbouring cells to lower their threshold thus encouraging them to also become active so that neighbouring cells, with similar inputs and outputs, develop similar feature selectivity. We model the diffusive interaction between feature selective cells by means of a RC network. Cells are connected to RC-grid through a diffusion device 'D', see figure 2,5. This diffusion device serves (i) to isolate neighbouring cells from directly altering response (V_s) of a cell, (ii) to drive the tunnel device that operates at higher voltage than voltages available at activation node.

We test this hypothesis for two cases (i) a row of 10 learning cells that are diffusively coupled on a 1-D RC network, see figure 5, and (ii) a 2-D RC grid of 100 x 100 cells, see figure 7. Here R_R and R_D refer to resistance values that balance the effect of Reaction term (activation of a cell to given input) and Diffusion term (effect of activation of neighbouring cell) respectively. To avoid any boundary effects we take periodic boundary conditions. Figure 7 shows simulation for 2-D for a period of 1000 epochs with each epoch of 60 ms.

Initially all cells were given faint random biases as either left or right dominated cells. During development all the cells are stimulated with identical patterns {(0,1), (1,0)} chosen in random order in every epoch. Under this *random inside epoch* stimulation cells perform 'time staggered WTA' i.e. strengthen their biases. Since these cells develop under influence of diffusive interaction they begin to clusters into left or right dominated cells. Cells which lie at the boundary of opposite biases turn binocular i.e. respond equally to both left and right input. This is reflected in the figure 7 as grey cells. The graph (figure 6) shows three cells chosen such that cells at the two ends represent clusters of opposite biases and the cell in the centre favours of one of them. After learning, cells at the two corners develop according to their own biases while the cell in the centre becomes unbiased with equal weights. The pattern of ocular dominance produced with this array of RC-coupled learning cells is shown in figure 7. While there are well distributed clusters of black (left dominated) and white (right dominated), observe the smooth change from one cluster to other occurs due to development of grey (binocular) cells. The creation of clusters and existence of binocular cells are the major strengths of this diffusive-Hebbian development.

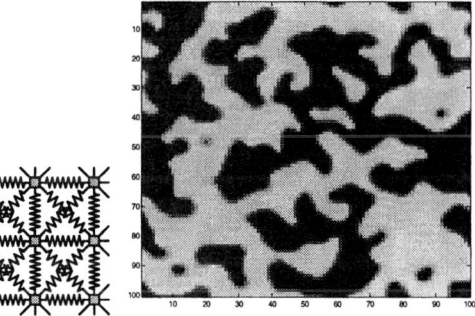

Figure 7: *Right*: Simulation results for 100x100 ts-WTA cells showing cell biases obtained as difference of Vfg$_{ia}$ and Vfg$_{ib}$. White (Black) represent cells dominated by left (right) eye, grey represents equal domination or binocular cell. *Left*: a 3x3 circuit.

IV. ORIENTATION SELECTIVITY

A relatively more complex visual feature to which cortical cells are known to respond is orientation selectivity. Orientation selective cells respond favorably to an oriented light bar shown within their receptive field. This preferential response, according to Hubel & Wiesel, depends upon the degree of alignment between the stimulus in the form of an oriented bar and the connection strengths from on- and off-center cells within the receptive field of the cortical cell. Though several different mechanisms have been proposed for orientation selective response of cortical cells, our design bears close similarity to the reaction-diffusion inspired competition and cooperation feed forward model proposed by [14].

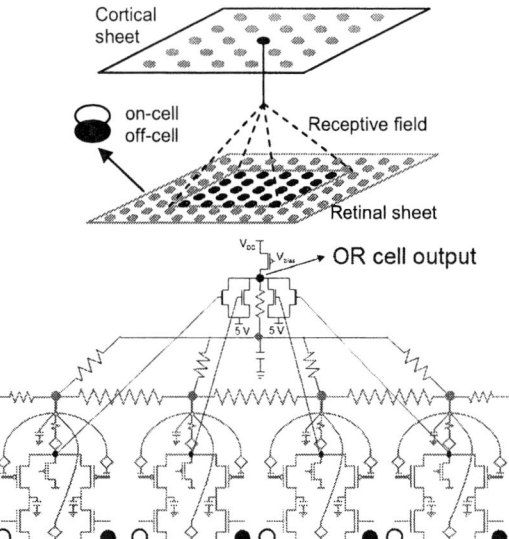

Figure 8: (*above*)A feed-forward description of organization of receptive field (5x5) for orientation selectivity. (*below*) Detailed circuit diagram (of only a portion of RF 4x1) with ts-WTA cells with on- (○) and off-center inputs (●).

An abstract sketch of feed forward model of an orientation (OR) selective cortical cell which connects to a 9x9 receptive field (RF) is shown in Figure 8. The figure also shows detailed circuit of OR-cell with only a portion (4x1) of the RF shown. Every feedforward connection is a ts-WTA cell with inputs from on-center (○) and off-center (●) cells of LGN/retina. As every connection must connect to only one of the two inputs, hence ts-WTA was a natural choice. Output of OR cell is linked to feedback of ts-WTA through a set of resistances to balance self activation v/s neighboring influence.

Results: The input patterns in the form of 4-pairs of on-center/off-center oriented bars were used to stimulate the receptive field of an OR-cell (see figure 9). Starting from a faint orientation bias the RF develops to strengthen its bias on being stimulated with the above patterns chosen in a random-inside-epoch manner.

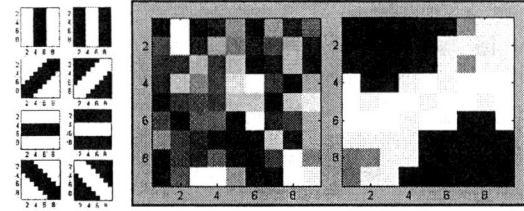

Figure 9: *Left*: Patterns used to stimulate OR selective cell. *Middle*: Initial RF (9x9) of OR selective cell. *Right*: Developed RF. Each box represents a ts-WTA cell with black => larger connection to off-center cell (●) as compared to on-center (○) cell, white => vice versa. The cell has developed orientation selective RF same as input pattern (2nd from right-top of *Left* figure). Initial ΔV_{fg} = ±1.2e-4 to ± 4.9e-3V, Final ΔV_{fg} = ±0.1874 to ±0.6441V

The detailed temporal development of V_{FG}s of all individual branches of the RF (see figure 10) are similar to figure 4, 6. Overall organization of RF of developed orientation selective cell is shown in figure 9 (right). The dark (white) spots imply connections where the off (on)-center inputs has won the ts-WTA competition. Gray cells imply connections of nearly equal strength to both on and off-center inputs.

We summarize the adaptive behavior of OR-cell as: (i) when we start with random initial bias, RF develops to be selective to one-of-eight patterns (*see figure 9*). (ii) Starting with faint orientation bias, if we stimulate the cortical cell with patterns other than its orientation bias than the cell develops selectivity to an orientation that is closest to its bias (*result not shown here*). Overall the behavior of the OR cell is once again like ts-WTA in the sense that there is time staggered competition between oriented patterns and the winner is determined by the initial bias and nature of stimulation

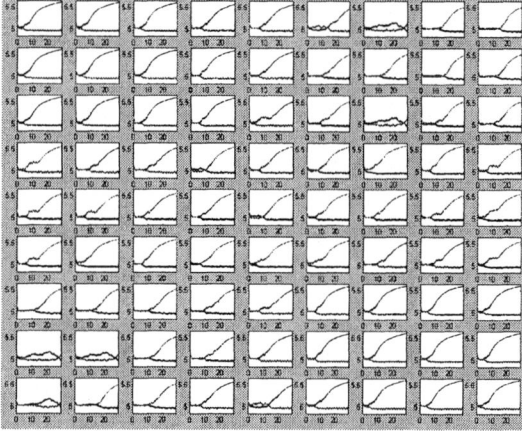

Figure 10: Temporal development V_{FG}s of OR-cell RF (*figure 9*). Blue represents V_{FG} for on- (○), and green as V_{FG} for off-centre (●) inputs. X-axis is epochs, and y-axis V_{fg}.

V. DISCUSSION

Neuromorphic design of ts-WTA highlights role of competition between asynchronous inputs such as left/right eye in ocular dominance or ON/OFF cells in orientation selectivity and is easily extendable to lagged/non-lagged for direction selectivity. Such competition brings about activity based refinement of afferent connections critical for development of feature selectivity [17-19] and is a widespread developmental process [16, 20]. In fact, synapse elimination due to non-coincident activity has been proposed as a physiological mechanism for Herb's postulate of learning [21]. This competition in presence of cooperation between synchronous patterned inputs leads to map formation.

VI. CONCLUSION

We present a basic building block for visual feature extraction and map formation in the form of analog CMOS cortical cell ts-WTA that computes 'sum of weighted inputs" implemented as floating gate pFET 'synapses'. Under appropriate feedback regime the cell refines its feature selectivity in response to stimulus patterns as a *time staggered*-WTA. An array of such cells

when diffusively coupled through an RC-network exhibits clustered development of feature selective cells similar to cortical ocular dominance maps. We illustrate the application of clustered development of *ts*-WTA cells with on/off-centre inputs, in design of an adaptive orientation selective cell.

Acknowledgements: This work was funded by research grant to CMM, (III.6 (74)/99-ST(PRU)) under SERC Robotics & Manufacturing PAC, Department of Science and Technology, Govt. of India.

References

1. Horng, S.H. and M.Sur., "Visual activity and cortical rewiring: activity-dependent plasticity of cortical networks.", *Progress in Brain Research* 157: 3-11, 2006.
2. Choi, TYW., Merolla, PA., Arthur, JV., Boahen, KW., Shi, BE. "Neuromorphic Implementation of Orientation Hypercolumns", *IEEE Tran. on Circuit & Systems*-I: Vol. 52(6) :1049-1060, 2005.
3. E. Chicca, Lichsteiner, P., Delbruck, T., Indiveri, G., Douglas, R., A "Multichip Pulse-Based Neuromorphic Infrastructure and Its Application to a Model of Orientation Selectivity", *IEEE Tran. on Circuits & Systems-I*, Vol. 54(5): 981-993, 2007.
4. Elliot, T., and Kramer, J., "Developing Topography and ocular dominance using two aVLSI vision sensors and a neurotrophic model of plasticity", *Neural Computation* Vol.14 :2353-2370, 2002
5. Merabet, LB., Rizzo, JF., Amedi, A., Somers, D., Leone, AP., "What blindness can tell us about seeing again: merging neuroplasticity and neuroprostheses", *Nature Reviews Neurosci.*, Vol 6 :71-77, 2005.
6. C. Diorio, P. Hasler, B. A. Minch, and C. A. Mead, "A Single-Transistor Silicon Synapse", *IEEE Trans. on Electron Devices*, vol. 43(11) : 1972-1980, 1996.
7. Hsu, D., Figueroa, M., Diorio, C., "Competitive learning with floating-gate circuits". *IEEE Trans. on Neural Networks*, vol. 13(3) : 732-744, 2002.
8. Kohonen, T., "Self-organizing neural projections", *Neural Networks*, Vol.19 :723–733, special issue, 2006.
9. Rahimi, K., Diorio, C., et.al., "A simulation model for floating gate MOS synapse transistors". *Proc. of ISCAS*, 2002.
10. Grossberg, S. "Adaptive pattern classification and universal recoding: I. Parallel development and coding of neural feature detectors". *Biological cybernetics* vol. 23: 121-134, 1976.
11. Bansal, M., & Markan, CM. "Floating gate 'Time staggered WTA' for feature selectivity", *Proc. of Workshop on Self-organizing Maps*, WSOM'03: 277-283, Kitakyushu, Japan, 2003.
12. J. Lazzaro, S. Ryckbusch, M.A.Mahowald, & C.A.Mead, "Winner-take-all Networks of O(N) complexity", *NIPS* 1: 703-711, Morgan Kaufman Publishers, San Mateo, CA, 1989.
13. Kruger, W., Hasler, P., Minch, B., Koch, C., "An adaptive WTA using floating gate technology". *Advances in NIPS* 9: 713-719, 1997.
14. Bhaumik, B., & Mathur, M. "A cooperation and competition based simple cell receptive field model and study of feed-forward linear and nonlinear contributions to orientation selectivity", *J. of Computational Neuroscience*, V14 :211-227, 2003.
15. Elliot, T., and Shadbolt, NR., "Developing a robot visual system using a biologically inspired model of neuronal development", *Robotics and Automation System*, Vol 45(2) : 111-130, 2003
16. Buffeli, M., et.al., "Perinatal switch from synchronous to asynchronous activity of motoneurons: link with synapse elimination", *Proc. Natl. Acad. Sci*, Vol.99(20): 13200-13205, 2002.
17. Zhang, L., Bao, S., & Merzenich, M., "Disruption of primary auditory cortex by synchronous auditory inputs during critical period", *Proc. Natl. Acad. Sci*, Vol.99(4) : 2309-2314, 2002.
18. Brickley SG., et.al., "Synchronising retinal activity in both eyes disrupts binocular map development in the optic tectum", *J. of Neurosci.*, 18(4): 1491-1504, 1998
19. Weliky, M., & Katz, L., "Disruption of orientation tuning visual cortex by artificially correlated neuronal activity", *Nature* 386: 680-685, 1997.
20. Linden, JF., & Schreiner, CE., "Columnar transformations in auditory cortex? A comparison to visual and somatosensory cortices", *Cerebral Cortex*, 13:83-89, 2003.
21. Stent, G. S., "A physiological mechanism for Hebb's postulate of learning", *Proc. Natl. Acad. Sci.*, 70(4): 997-1001, 1973.

An Analog Programmable Multi-Dimensional Radial Basis Function Based Classifier

Sheng-Yu Peng[*], Paul E. Hasler[†], and David Anderson[‡]

School of Electrical and Computer Engineering
Georgia Institute of Technology, Atlanta, Georgia 30332–0250
Email: [*]sypeng@ece.gatech.edu; [†]phasler@ece.gatech.edu; [‡]dva@ece.gatech.edu

Abstract— A compact analog programmable multi-dimensional radial basis function (RBF) based classifier is demonstrated. The probability distribution of each feature in the templates modeled by a Gaussian function is approximately realized by the transfer characteristics of a floating-gate bump circuit. The maximum likelihood, the mean, and the variance can be independently programmed. By cascading these floating-gate bump circuits, the transfer characteristics approximate a multivariate Gaussian function with a diagonal covariance matrix. An array of these circuits constitute a compact multi-dimensional RBF-based classifier. When followed by a winner-take-all circuit, the RBF-based classifier forms an analog vector quantizer. We use receiver operating characteristic curves and equal error rate to evaluate the performance of our analog classifiers. We show that the analog classifier performance is comparable to that of digital counterparts. The proposed approach can be at least two orders of magnitude more power efficient than the digital microprocessors at the same task.

A multivariate Gaussian response function is a fundamental building block in many classification applications, such as Gaussian mixture models (GMM), radial basis function (RBF) based classifiers, and vector quantizers. This paper discusses the development of an analog Gaussian response function having a diagonal covariance matrix and demonstrates its application to vector quantization.

Fig. 1 illustrates one possible application of this work as part of an analog speech recognizer [1] that includes a band-pass-filter bank based analog Cepstrum generator, an analog RBF-based classifier, and a continuous-time hidden Markov model (HMM) block built from programmable analog waveguide stages. By performing analog signal processing in the front end, not only the computational load of the subsequent digital processor can be reduced, but also the required specifications for the analog-to-digital converters can be loosened. As a result, the entire system can be more efficient.

When followed by a winner-take-all (WTA) stage, a RBF-based classifier forms a multi-dimensional analog vector quantizer. A vector quantizer compares distances or similarities between an input vector and the stored templates. It classifies the input data to the most representative template. Vector quantization is a typical pattern recognition and data compression technique. Crucial issues of the vector quantizer implementation concern the storage efficiency and the computational cost of searching the best-matching template. Analog storage and signal processing gives rise to better area and power efficiency. In previous analog designs [2], [3], to reduce the

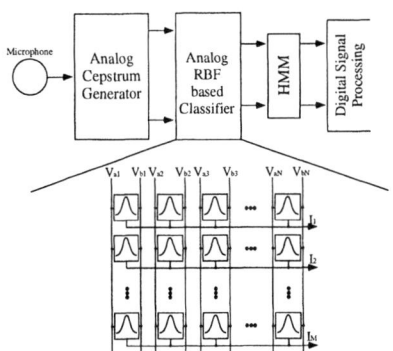

Fig. 1. Block diagram of a potential analog front-end system for speech recognition. Analog vector quantizer is viewed as a crucial component in this technology.

computational effort, only the distances from the input vector to the means of the templates are compared. To have better approximation of the distribution, many variations of analog RBF circuits with tunable width are designed but they are usually complicated and need extra circuits to store or to periodically refresh template data.

In this paper, we demonstrate a novel compact programmable analog RBF-based classifier that is composed of an array of two-input floating-gate bump circuits. The mean and the variance of each feature component distribution are stored in a floating-gate circuit. These two statistical moments can be programmed accurately and independently; therefore, the stored template information can be closer to the real distributions. An array of these floating-gate bump cells can also implement Gaussian mixture models (GMM). With a following winner-take-all circuit, the resultant analog vector quantizer can be applied to non-uniform, as well as uniform, variance scenarios. The whole classification system is compact and can be easily scaled up. A hardware implementation of a 7×2 analog vector quantizer was fabricated in $0.5\,\mu$m CMOS process and tested. We also fabricated a 16×16 highly compact low-power version of an analog vector quantizer occupying less than $1.5 \times 1.5 mm^2$ in $0.5\,\mu$m CMOS process.

I. PROGRAMMABLE FLOATING-GATE BUMP CIRCUIT

In our classifier, the Gaussian response function is approximated by the bell-shaped transfer characteristics of a proposed

13

978-1-4244-1709-4/07/$25.00 © 2007 IEEE

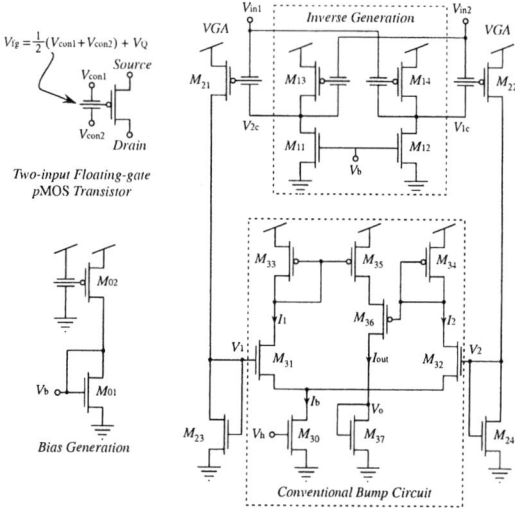

$$V_{\text{fg}} = \frac{1}{2}(V_{\text{con1}} + V_{\text{con2}}) + V_Q$$

Two-input Floating-gate pMOS Transistor

Bias Generation

Inverse Generation

Conventional Bump Circuit

Fig. 2. All floating-gate transistors in the schematics have two inputs with equal weights so the floating-gate voltage can be expressed as: $V_{\text{fg}} = \frac{1}{2}(V_{\text{con1}} + V_{\text{con2}}) + V_Q$, where V_Q is the charge-related voltage. The bump circuit is composed of an inverse generation block, fully differential amplifier, and a conventional bump circuit. The width and the center of the bump are set respectively by the common-mode and differential charges on the floating gates of M_{21} and M_{22}. The height is controlled by the tail current, I_b.

floating-gate bump circuit. The height, the width, and the center of the transfer curve represent the maximum likelihood, the variance, and the mean of a distribution respectively. The ability to program these three parameters individually empowers the classifiers to fit into different scenarios with the full use of statistic information up to the second moment. In addition, adjusting these parameters is equal to pre-scaling input signals in the analog fashion so that the circuit outputs can fall into the effective input range of the following stage. For example, in the analog vector quantizer implementation, despite the different distributions in different applications, the required precision of the following WTA circuit can remain relaxed if the input signals can be scaled properly.

The schematics of the proposed floating-gate bump circuit and its bias generation block are shown in Fig. 2. All floating-gate transistors have two input capacitances and all input capacitances are of the same size. The proposed bump circuit is composed of three parts: an inverse generation block, a conventional bump circuit [4], and in between a fully differential variable gain amplifier (VGA). The inverse generation block provides the complementary input voltages to the VGA so that the floating-gate common-mode voltage of M_{21} and M_{22} as well as the outputs of the VGA are independent of the input signal common-mode level. The width of the bell-shaped transfer curve can be adjusted by changing the VGA gain.

The inverse generation block has two floating-gate summing amplifiers. If the charges on M_{13} and M_{14} are matched and the transistors are in saturation region, we can have

$$V_{\text{in1}} + V_{\text{1c}} = V_{\text{in2}} + V_{\text{2c}} = V_{\text{const}}, \tag{1}$$

where V_{const} only depends on the bias voltage, V_b, and the charges on M_{13} and M_{14}. If the charge on M_{02} also matches that on M_{13} and M_{14}, the generated voltage, V_b, provides the summing amplifiers an operating range that is one V_{DSsat} away from the supply rails. The floating-gate voltages on M_{21} and M_{22} can be expressed as

$$V_{\text{fg},21} = \frac{1}{2}(V_{\text{in1}} + V_{\text{const}} - V_{\text{in2}}) + \frac{Q_{21}}{C_T}$$
$$= \frac{1}{2}\Delta V_{\text{in}} + V_{Q,\text{cm}} + \frac{1}{2}V_{Q,\text{dm}} \tag{2}$$

$$V_{\text{fg},22} = \frac{1}{2}(V_{\text{in2}} + V_{\text{const}} - V_{\text{in1}}) + \frac{Q_{22}}{C_T}$$
$$= -\frac{1}{2}\Delta V_{\text{in}} + V_{Q,\text{cm}} - \frac{1}{2}V_{Q,\text{dm}}, \tag{3}$$

where $\Delta V_{\text{in}} = V_{\text{in1}} - V_{\text{in2}}$, Q_{21} and Q_{22} are the amounts of charge on M_{21} and M_{22} respectively, C_T is the total capacitance seen from a floating gate, and

$$V_{Q,\text{cm}} = \frac{1}{2}\left(\frac{Q_{21} + Q_{22}}{C_T} + V_{\text{const}}\right)$$
$$V_{Q,\text{dm}} = \frac{Q_{21} - Q_{22}}{C_T}.$$

From (2) and (3), these two floating-gate voltages do not depend on the input signal common-mode level.

The nMOS transistors in the VGA are assumed in the transition between the above-threshold and the subthreshold regions. The pMOS transistors are assumed in the above-threshold region. Because the transfer characteristics of the two branches are symmetric, we can use the half circuit technique to analyze the VGA gain. By equating the currents flowing through the pMOS and nMOS transistors, we can have

$$I_{0,\text{p}}\left(\frac{W_\text{p}}{L_\text{p}}\right)\frac{1}{4U_T^2}[\kappa_\text{p}(V_{\text{DD}} - V_{\text{fg},21} - V_{\text{T0,p}})]^2$$
$$= I_{0,\text{n}}\left(\frac{W_\text{n}}{L_\text{n}}\right)\ln^2\left(1 + e^{\frac{\kappa_\text{n}}{2U_T}(V_1 - V_{\text{T0,n}})}\right) \tag{4}$$

where the subscripts of "p" and "n" refer to pMOS and nMOS transistors respectively, I_0 is the subthreshold pre-exponential current factor, κ is the subthreshold slope factor, V_{T0} is the threshold voltage, and U_T is the thermal voltage. At the peak of the bell-shaped transfer curve, $V_{Q,\text{dm}} = 0$ and

$$V_{\text{fg},21} = \frac{1}{2}\Delta V_{\text{in}} + V_{Q,\text{cm}}$$
$$V_1 = V_{\text{out,cm}} + \frac{1}{2}\Delta V_{\text{out}},$$

where $V_{\text{out,cm}} = (V_1 + V_2)/2$, $\Delta V_{\text{out}} = V_1 - V_2$. We can obtain the gain of the VGA by differentiating (4) with respect to $V_{\text{fg},21}$ and have

$$\frac{\Delta V_{\text{out}}}{\Delta V_{\text{in}}} = \frac{dV_1}{dV_{\text{fg},21}} = -\gamma\left(1 + e^{-\frac{\kappa_\text{n}}{2U_T}(V_1 - V_{\text{T0,n}})}\right)$$
$$= \frac{-\gamma}{1 - e^{-\frac{\gamma\kappa_\text{p}}{2U_T}(V_{\text{DD}} - V_{\text{fg},21} - V_{\text{T0,p}})}}$$
$$\approx -\gamma\left(1 + e^{-\frac{\gamma\kappa_\text{p}}{2U_T}(V_{\text{DD}} - V_{Q,\text{cm}} - V_{\text{T0,p}})}\right), \tag{5}$$

14   *2007 IFIP International Conference on Very Large Scale Integration (VLSI-SoC 2007)*

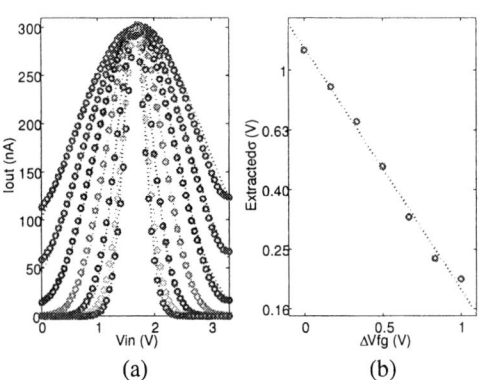

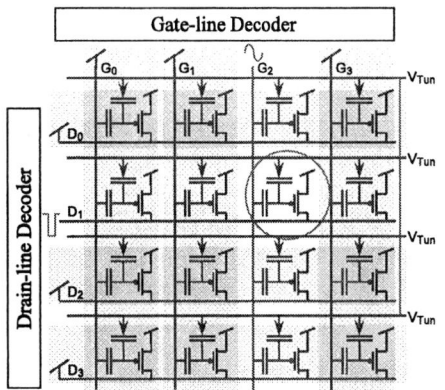

(a) (b)

Fig. 3. (a) Comparison of the measured 1D bumps (circles) and the corresponding Gaussian fits (dashed lines). One of the bump input voltages is fixed at $V_{DD}/2$ where V_{DD} is set to be 3.3V through the measurement. (b) Width and common-mode charge relation in semi-logarithmic scale. The width is characterized by the extracted σ. The shift of the programmed floating gate voltage, ΔV_{fg}, represents the common-mode charge level. The dashed line is the exponential curve fit.

Fig. 5. Programming an array of floating-gate transistors. Drain-lines and gate-lines are shared in rows and in columns respectively. By applying V_{DD} to unselected drain-lines and gate-lines, floating-gate transistors can be programmed individually. Decoders for programming are at the peripheries of the array.

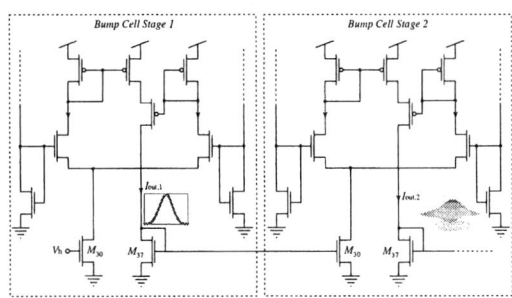

Fig. 4. By connecting the diode-connected output transistor to the tail transistor of the next stage bump cell, the resulting output current can approximate a multivariate Gaussian function with a diagonal covariance matrix.

a multivariate Gaussian function with a diagonal covariance matrix. Although the feature dimension can be increased by cascading more floating-gate bump cells, the bandwidth of the classifier decreases. The mismatches between the floating-gate bump circuits can be trimmed out by using floating-gate programming techniques. The output currents of an array of these floating-gate bump circuits can easily be summed up to implement GMMs.

II. PROGRAMMABLE ANALOG VECTOR QUANTIZER

Floating-gate transistors are not only used as computational devices but also as analog memories in this work. How to accurately programming an array of floating-gate transistors is, therefore, a critical technique in our classification system development. Fowler-Nordheim tunneling and channel hot electron (CHE) injection mechanisms are used to program charges on floating gates. The techniques of programming an array of floating-gate transistors have been detailed in many previous works [5], [6]. Fowler-Nordheim tunneling, which removes electrons from the floating gates through tunneling junctions, is used as a global erase. Accurate programming is performed by using CHE injection. An empirical model proposed in [6] is used to perform characterization and algorithmic programming. The precision of the programmed current level can be as accurate as 99.5%. As presented in [7], the retention time for the charges on floating gates can last over 10 years at room temperature.

To program an array of floating-gate bump cells, floating-gate transistors are arranged as in Fig. 5 in programming mode. Because there are two conditions required for CHE injection to put electrons on a device: a channel current and a high channel-to-drain field, we can deactivate the columns (or rows) of transistors by applying V_{DD} to the corresponding gate-lines (or drain-lines) so that there are no currents through (or no field across) the devices for injection. In this manner, each floating-gate transistor can be programmed individually.

where $\gamma = \frac{\kappa_p}{\kappa_n}\sqrt{\frac{I_{0,p}W_pL_n}{I_{0,n}L_pW_n}}$. Therefore, the gain increases approximately exponentially with the common-mode charge and, accordingly, we can expect the exponential relation between the extracted standard deviation of the transfer curve and the common-mode charge.

From (5), by adjusting the common-mode charge Q_{cm}, the gain of the VGA increases exponentially and hence the width of bump, which models the standard deviation of a distribution, decreases exponentially. Programming the common-mode charge to different levels generates bumps with different widths. We compare the resulting bump curves with the corresponding Gaussian fits in Fig. 3(a). In the semi-logarithmic plot of Fig. 3(b), the width depends on the common-mode charge exponentially as predicted by (5). The minimum width of the bump is set by the maximum gain of the VGA.

A diode-connected transistor, M_{37}, in the bump circuit converts the output current into a voltage. By feeding this voltage to the tail transistor, M_{30}, in the next stage bump circuit as shown in Fig. 4, the final output current approximates

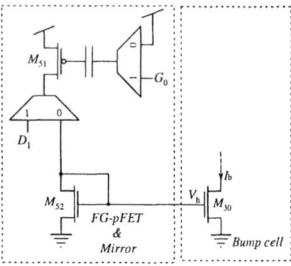

Fig. 6. Schematic of the *"FG-pFET & Mirror"* block. The charge on the *pMOS* transistor can be programmed to set the height of the bell-shaped transfer curve.

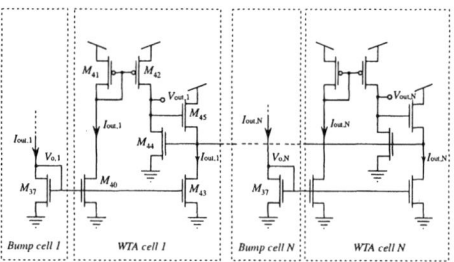

Fig. 7. Schematic of a current mode winner-take-all circuit. Only the output voltage of the winning cell will be high to indicate the best-matching template.

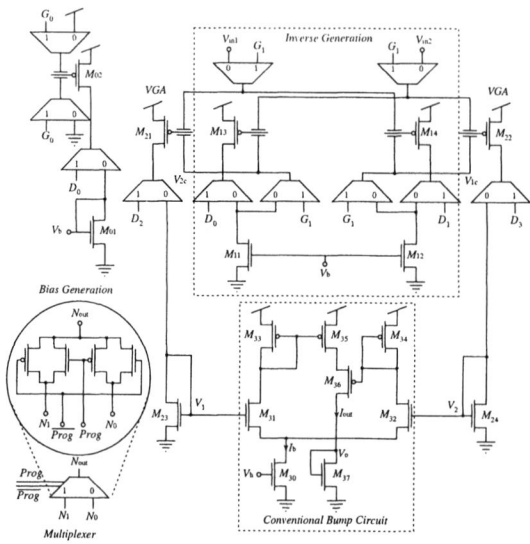

Fig. 8. Complete schematics of the floating-gate bump circuit. Multiplexers for programming are inserted into the original circuits. The "1" on the multiplexer indicates the connection in the programming mode and the "0" indicates the connection in the operating mode. The tunneling junction capacitors are not shown for simplicity. Most of the multiplexers are in bias generation and inverse generation blocks. Only two multiplexers are added in the bump cell that includes the VGA and the conventional bump circuit.

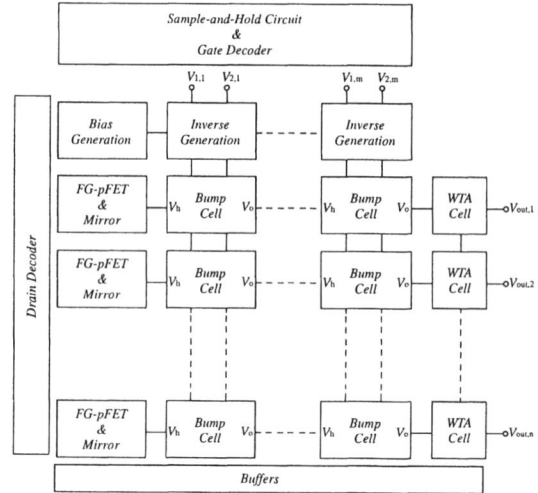

Fig. 9. Architecture of an analog vector quantizer. The core is the bump cell array followed by a WTA circuit. The main complexity from programming are at the peripheries and the system can be scaled up easily.

A *"FG-pFET & Mirror"* block shown in Fig. 6 is added in front of the first bump cell to program its tail current, which sets the height of the "bump." For the analog vector quantizer implementation, the final output currents of the RBF-based classifier are duplicated and are fed into a simple current mode winner-take-all circuit, for which the schematic is shown in Fig. 7. The output voltage of the winning cell only will be high to indicate the best-matching template.

To have the access to all drain and gate terminals of floating-gate transistors in the programming mode, multiplexers are inserted into the circuits as shown in Fig. 8. Most of the multiplexers are in the inverse generation and bias generation blocks. Because only one bias generation block is needed for the whole system, when the system is scaled up, the complexity of bias generation block does not cost. In the analog RBF-based classifier and vector quantizer, the same input voltage vector is compared with all stored templates. Therefore, the inverse generation can be shared by the same column of bump cells, each of which only includes a VGA and a conventional bump circuit. The number of inverse generation blocks is equal to the dimension of the feature space. Together with the gate-line and drain-line decoders, most of the programming overhead circuitries are at the peripheries of the floating-gate bump cell array; therefore the system can be easily scaled up and maintain high compactness. The compactness and the ease of scaling up are important issues in the implementation of an analog speech recognizer that requires more than a thousand of bump cells. The final architecture of our analog vector quantizer is shown in Fig. 9.

We use two examples to demonstrate the reconfigurability of our classifiers. Four templates are used as shown in Fig. 10. The floating-gate transistors of other unused templates are tunneled off. Four bell-shaped output currents emulate the bivariate Gaussian likelihood functions of four templates. The thick solid lines at the bottom, indicate the boundaries determined by the WTA outputs.

16   *2007 IFIP International Conference on Very Large Scale Integration (VLSI-SoC 2007)*

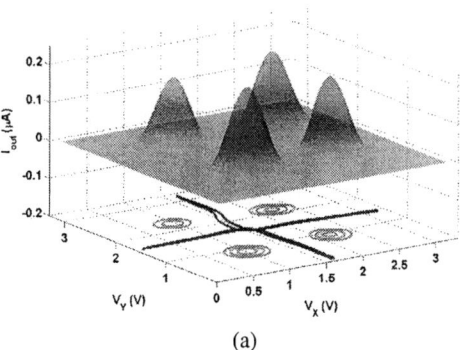

(a)

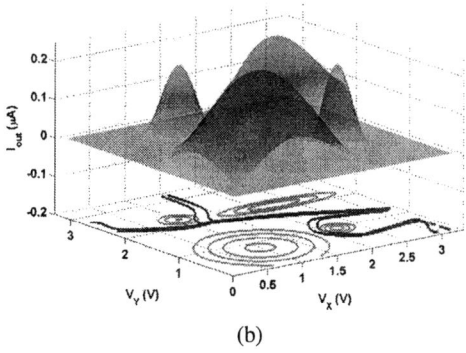

(b)

Fig. 10. Configurable classification results. The measured bump output currents (circle contours) and the WTA voltages (thick solid lines at the bottom) of four templates are superposed in a single plot. V_X and V_Y are the V_{in1} in the first stage and the second stage floating-gate bump circuits respectively. Both of their V_{in2} terminals are fixed at $V_{DD}/2$. (a) Four templates are programmed to have the same variance and evenly spaced means. (b) Four templates are programmed to have different variances with evenly spaced means.

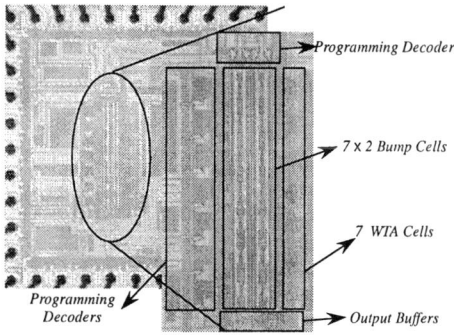

Fig. 11. Die photo of the analog VQ chip.

III. PERFORMANCE OF ANALOG VECTOR QUANTIZER

We have fabricated an analog VQ in $0.5\,\mu$m CMOS process and the die micrograph is shown in Fig. 11. We also fabricated a 16×16 highly compact low-power version of an analog vector quantizer in $0.5\,\mu$m CMOS process occupying less than $1.5 \times 1.5 mm^2$. Some important parameters and measured re-

Table of Parameters	
Size of VQ	7(templates)×2(components)
Area/Bump Cell	$42 \times 82\,\mu m^2$
Area/WTA Cell	$20 \times 35\,\mu m^2$
Power Supply Rail	$V_{DD} = 3.3V$
Power Consumption/Bump Cell	$90\mu W \sim 160\mu W$
Response Time	$20\mu \sim 40\mu$sec
Floating-gate Programming Accuracy	99.5%
Retention Time	10 years @ 25°C

sults are listed in the table. To measure the power consumption, we program several numbers of bumps with identical width and deactivate other bumps by tunneling their floating-gate transistors off. The power consumption is averaged over the entire 2-D input space. The slope of the curve in Fig. 12(a) indicates the average power consumption per bump cell with a specific value of width. The relation between the power consumption and the extracted standard deviation is shown in Fig. 12(b). We can also program the charge on M_{13} and M_{14} to optimize the speed and power trade-off.

We use receiver operating characteristic (ROC) curves and equal error rate (EER) to evaluate the classifier performance. Two separate 2D bumps are programmed to have the same variance with a fixed separation. The corresponding Gaussian fits are used as the actual probability density functions (pdf) of two classes. Comparing these two pdf's using different thresholds renders a ROC curve of these two Gaussian distributed classes. We use it as the evaluation reference. With the knowledge of the class distributions, comparing the output currents using different thresholds generates a ROC curve for the 2D bumps. Comparing each of the two WTA output voltages with different thresholds generates two ROC curves that characterize the classification results of the vector quantizer. The EER, which is the intersection of the ROC curve and the $-45°$ line as shown in Fig. 13(a), is the usual operating point of classifiers. In Fig. 13(b), both the ROC areas and the EER are plotted to investigate the effect of the bump width on the performance. At the EER point, the performance of our RBF classifier, which uses floating-gate bump circuits to ap-

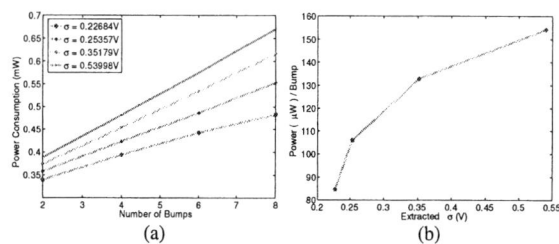

(a)   (b)

Fig. 12. (a) Power consumption of the analog vector quantizer with different number of bumps to be activated with fixed width. The slope of the curves indicate the power consumption per bump cell. (b) The relation between the power consumption per bump and the extracted variance of the bump. The wider the bump, the higher the power consumption.

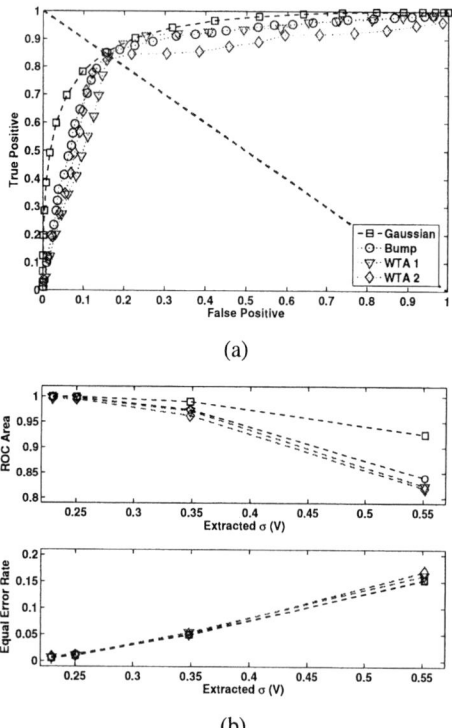

(a)

(b)

Fig. 13. ROC and EER performance of the classifiers. The effects of different bump widths on the receiver operating characteristic (ROC) area and the equal error rate (EER) performance. The separation of the means of two classes is 1.2V. The intercept plot shows the ROC curves of the Gaussian fits (squares), output currents of the 2D bumps (circles) and WTA output voltages (triangles and diamonds) with the extracted $\sigma = 0.55V$. The Gaussian fits are used as the actual pdf's of the two classes and the corresponding ROC curve is used as a reference. The intersection of the ROC curve and the $-45°$ line is the EER point, which is the usual operating point. The results show that the analog VQ is comparable to an ideal maximum-likelihood (ML) classifier.

proximate Gaussian likelihood functions, is undistinguishable from that of an ideal RBF-based classifier. Despite the finite gain of the WTA circuit, the performance of the analog vector quantizer is still comparable to an ideal maximum likelihood (ML) classifier. By optimizing the precision and speed of the WTA circuit, the performance can be improved but it is beyond the scope of this paper.

IV. POWER EFFICIENCY COMPARISON

To compare the performance of our analog system with DSP hardware, we estimate the metric of millions of multiply accumulates per second per milli-watt (MMAC/s/mW). When the system is scaled up, the bump cell efficiency dominates the performance. Therefore, we consider the performance of a single bump cell only. Each Gaussian function is estimated as 10 MACs and can be evaluated by a bump cell in less than $10\mu sec$ (which is still an overestimate) consuming about $120\mu W$. This is equivalent to 8.3 MMAC/s/mW. The performance of commercial low-power DSP microprocessors ranges

from 1 MMAC/s/mW to 10 MMAC/s/mW and a special designed high performance DSP microprocessor in [8] is below 50 MMAC/s/mW. If this comparison is expanded to include the WTA function, the efficiency of our analog system will improve even more relative to the digital system.

Although our power efficiency is comparable to the digital system, our classifier consumes much more power compared to other analog vector-matrix-multiplication system [9] of which efficiency is about 37 MMAC/s/μW. The reason is that the transistors M_{23} and M_{24} are way above threshold region. By making M_{21} and M_{22} long and raising the source voltage of M_{23} and M_{24}, from simulation, we can easily reduce the power consumption by at least two orders of magnitude. If the WTA circuit is also optimized, it is anticipated that future ICs can be three orders of magnitude more efficient than DSP microprocessors at the same task.

V. CONCLUSION

In this paper, we demonstrate a new programmable floating-gate bump circuit, of which the height, the center and the width of the bell-shaped transfer characteristics can be programmed individually. A multivariate radial basis function with a diagonal matrix can be realized by cascading these bump cells. Based on the new bump circuit, we build a novel compact RBF-based soft classifier and, by adding a simple current mode winner-take-all circuit, we implement an analog vector quantizer. The performance and the efficiency of the classifiers are comparable to the digital system. With slight modifications, the overall efficiency is anticipated to be improved by at least two to three orders of magnitude better than DSP microprocessors.

REFERENCES

[1] P. Hasler, P. D. Smith, D. Graham, R. Ellis, and D. V. Anderson,"Analog Floating-Gate, On-Chip Auditory Sensing System Interfaces," in *IEEE J. Sensors*, vol. 5, no. 5, pp.1027-1034, Oct. 2005.

[2] G. Cauwenberghs and V. Pedron i,"A low-power CMOS analog vector quantizer," in *IEEE J. Solid-State Circuits*, vol. 32, no. 8, pp.1278-1283, Aug. 1997.

[3] T. Yamasaki and T. Shibata,"Analog soft-pattern-matching classifier using floating-gate MOS technology," in *IEEE Trans. Neural Networks*, vol. 14, no. 5, pp.1257-1265, Sep. 2003.

[4] T. Delbruck, "Bump circuits for computing similarity and dissimilarity of analog voltage," in *Proc. Int. Neural Network Society*, Seattle, WA, 1991.

[5] M. Kucic, A. Low, P. Hasler, and J. Neff, "A programmable continuous-time floating-gate Fourier processor," in *IEEE Trans. Circuit and system II*, pp. 90-99, Jan. 2001.

[6] A. Bandyopadhyay, G.J. Serrano, and P. Hasler, "Adaptive Algorithm Using Hot-Electron Injection for Programming Analog Computational Memory Elements Within 0.2% of Accuracy Over 3.5 Decades," in *IEEE J. Solid-State Circuits*, vol. 41, no. 9, pp.2107-2114, Sept. 2006.

[7] V. Srinivasan, G. J. Serrano, J. Gray, and P. Hasler, "A precision cmos amplifier using floating-gates for offset cancellation," in Proc. CICC05, Sept. 2005, pp. 734737.

[8] J. Glossner, K. Chirca, M. Schulte, H. Wang, N. Nasimzada, D. Har, S. Wang; A. J. Hoane, G. Nacer, M. Moudgill, M., S. Vassiliadis, "Sandblaster low power DSP," in *IEEE Prec. Custom Integrated Circuits Conference*, pp.575-581, oct. 2004.

[9] R. Chawla, A. Bandyopadhyay, V. Srinivasan, and P. Hasler, "A 531nW/MHz, 128x32 current-mode programmable analog vector-matrix multiplier with over two decades of linearity," in *IEEE Prec. Custom Integrated Circuits Conference*, pp.651-654, oct. 2004.

ReCPU: a Parallel and Pipelined Architecture for Regular Expression Matching

Marco Paolieri, Ivano Bonesana
ALaRI, Faculty of Informatics
University of Lugano, Lugano, Switzerland
{paolierm, bonesani}@alari.ch

Marco D. Santambrogio
Dipartimento di Elettronica e Informazione
Politecnico di Milano, Milano, Italy
marco.santambrogio@polimi.it

ABSTRACT

Text pattern matching is one of the main and most computation intensive parts of systems such as Network Intrusion Detection Systems and DNA Sequencing Matching. Software solutions to this are available but often they do not satisfy the requirements in terms of performance. This paper presents a new hardware approach for regular expression matching: ReCPU. The proposed solution is a parallel and pipelined architecture able to deal with the common regular expression semantics. This implementation based on several parallel units achieves a throughput of more than one character per clock cycle (maximum performance of state of the art solutions) requiring just $O(n)$ memory locations (where n is the length of the regular expression). Performance has been evaluated synthesizing the VHDL description. Area and time constraints have been analyzed. Experimental results are obtained simulating the architecture.

1. INTRODUCTION

Searching for a set of strings that match a given pattern is a well known computation-intensive task, exploited in several different application fields. Nowadays there is an increasing need of high performance computing - as in the case of biological sciences. Matching a DNA pattern among millions of sequences is a very common and computationally expensive task in the Human Genome Project. In Network Intrusion Detection Systems - where regular expressions are used to identify network attack patterns - software solutions are not acceptable because they would slow down the entire system. Software solutions cannot always meet the requirements in terms of speed, therefore a different approach is required.

To move towards a full hardware implementation - overcoming the performance achievable with software - it is reasonable for these application domains.
Several research groups have been studying hardware architectures for regular expressions matching: mostly based on Non-deterministic Finite Automaton (NFA) as described in [1] and [2].

In [1] an FPGA implementation is proposed. It requires $O(n^2)$ memory space and processes a text character in $O(1)$ time (one clock cycle). The architecture is based on hardware implementation of Non-deterministic Finite Automaton (NFA); additional time and space are necessary to build the NFA structure starting from the given regular expression. The time required for a matching operation is not constant, it can be linear in best cases and exponential in

worst ones. We do not face with these limitations because we are able to store regular expressions using $O(n)$ memory locations. Furthermore, we do not require any additional time to start to process the regular expressions (from now on RE). In [2] an architecture that allows extracting and sharing common sub-regular expressions, in order to reduce the area of the circuit, is presented. It is necessary to re-generate the HDL description to change the regular expression. It is clear that this approach generates an implementation dependent on the pattern. In [3] a software that translates a RE into a circuit description has been developed. A Non-deterministic Finite Automaton has been used to dynamically create efficient circuits for pattern matching (that have been specified with a standard rule language).

The work proposed in [4] focuses on REs pattern matching engines implemented with reconfigurable hardware. A Non-deterministic Finite Automaton based implementation is used, and a tool for automatic generation of the VHDL description has been developed. All these approaches - [2], [3], [4] - require a re-generation of the HDL description whenever a new regular expression needs to be processed. In our solution we just require to update the instruction memory with the new RE. In [5] a parallel FPGA implementation is described: multiple comparators allow to increase the throughput for parallel matching of multiple patterns.

In [6] a DNA sequence matching processor using FPGA and Java interface is presented. Parallel comparators are used for the pattern matching. They do not implement the regular expression semantics (i.e. complex operators) but just simple text search based on exact string matching.

At the best of our knowledge this paper presents a different approach to the pattern matching problem: REs are considered the programming language for a dedicated CPU. We do not build either Deterministic or Non-deterministic Finite Automaton of the RE, hence we do not require additional setup time as in [1]. ReCPU - the proposed architecture - is a processor able to fetch an RE from the instruction memory and perform the matching with the text stored in the data memory. The architecture is optimized to execute computations in a parallel and pipelined way. This approach involves several advantages: on average it compares more than one character per clock cycle as well as it requires less memory occupation: for a given RE of size n the memory required is just $O(n)$. In our solution it is easily possible to change the pattern at run-time just updating the content of the instruction memory without modifying the underlying hardware. Considering the CPU-like approach a small *compiler* is necessary to obtain the machine code from the given

978-1-4244-1709-4/07/$25.00 © 2007 IEEE

RE (i.e. specified with a high-level description).

This paper is organized as follows: in Section 2 a brief overview of Regular Expressions focusing on the semantics - that have been implemented in hardware - is discussed. The idea of considering regular expressions as a programming language is fully described in Section 3. Section 4 provides a top-down description of the hardware architecture. *Data Path* is fully covered in 4.1 and *Control Path* in 4.2. Results of the synthesis process in terms of critical path and area are discussed in Section 5: a comparison of the performance with other solutions is proposed. Conclusions and future works are addressed in Section 6.

2. REGULAR EXPRESSION OVERVIEW

A *regular expression* [7] (RE), often called a *pattern*, is an expression that matches a set of strings. REs are used to perform searches over text data, and are commonly present in programming languages and text editors for text manipulation. In an RE single characters are considered regular expressions that match themselves and additionally several operators are defined. Let us consider two REs: a and b, the operators that have been implemented in our architecture follow:

- $a \cdot b$: it matches all the strings that match a and b;

- $a|b$: matches all strings that match either a or b ;

- $a*$: matches all strings composed by zero or more occurrences of a;

- $a+$: matches all strings composed by one or more occurrences of a;

- (a): parentheses are used to define the scope and precedence of the operators (e.g. to match zero or more occurrences of a and b, it is necessary to define the following RE: $(ab)*$).

3. PROPOSED APPROACH

The innovative idea we developed is to use REs as instructions for the ReCPU processor. ReCPU executes a program stored in the instruction memory, that is composed by a set of instructions, each of those part of the original RE. The approach followed is the same used in general purpose processors: a program is coded in a high-level language (e.g. C) and compiled (e.g. using gcc) into low level language (i.e. machine code). In our case an RE is defined using the common high-level language - as described in [7] or in [8] - it is compiled[1] into machine code and executed by ReCPU processor. Given a RE, a *compiler* is used to generate the instructions sequence that are executed by the core of the architecture on the text stored in the data memory. The compiler program does not need to perform many optimizations due to the simplicity of the RE programming language. The binary code produced by the compiler and composed of *opcode* and *operands* instructs the *execution unit* using the information about the number of parallel units of the architecture. Our compiler uses the idea behind the VLIW

---

[1]We developed a compiler that translates the high level description of the RE to the ReCPU machine code. As explained, the compiler takes advantage of some well known VLIW techniques.

architecture design style: the parallel units are exposed to the back-end (i.e. the compiler issues as many character comparisons as the number of parallel comparators present in the architecture).

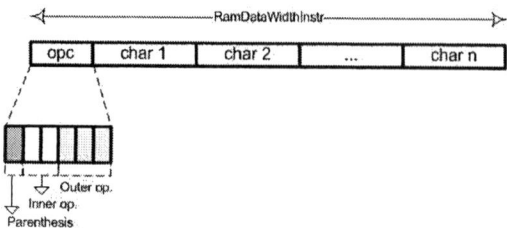

Figure 1: Instruction Structure.

Let us analyze some examples to clarify the mapping of complex RE operators into the programming language: operators like $*$ and $+$ correspond to *loop* instructions. Such operators find more occurrences of the same pattern (i.e. looping on the same RE instruction). This technique guarantees the possibility to handle complex REs looping on more than one instruction. The loop terminates whenever the pattern matching fails. In case of $+$ at least one valid iteration of the loop is required to validate the RE, while for $*$ there is no limitation on the minimum number of iterations.

Another feature of complex REs that can be perfectly managed considering the RE as a programming language is the use of nested parentheses (e.g. $(((ab) * (c|d))|(abc)))$. We mapped this to the *function call paradigm* of common programming languages, so that we can handle it as done in most processors. We consider an open parenthesis in an RE as a *call* instruction and respectively a closed parenthesis as a *return* instruction. Whenever an open parenthesis is encountered, the current context (i.e. all the internal registers of the *Control Path*) is pushed into an entry of a stack data structure and the execution continues normally. Whenever a close parenthesis is found, a pop operation is performed on the stack and the overall validity of the RE is checked by combining the outer operator of the the current instruction, the current context and the previously saved context popped from the stack. This way, our architecture can tackle very complex nested REs using a well and widely known function paradigm approach.

ReCPU binary instructions generated by the compiler are stored into the instruction memory. An instruction is composed by opcode and operands (i.e. the characters present in the pattern) as shown in Figure 1. The opcode is composed by three different parts: the MSB represents the use of a parenthesis (i.e. a function call), the next 2-bits represent the internal operands (i.e. *and* or *or* used to match the characters present in the current instruction) and the last bits select the external operand used to describe loops and close parenthesis (i.e. a *return* after a function call). A RE is completely matched whenever a NOP instruction is fetched from the instruction memory. A complete list of opcodes is shown in Table 1[2].

---

[2]Please notice that *don't care* values are expressed as "-".

Table 1: Bitwise representation of the opcodes.

opcode			RE
0	00	000	nop
1	--	---	(
0	01	---	and
0	10	---	or
0	--	001	)*
0	--	010	)+
0	--	011	)|
0	--	1--	)

4. ARCHITECTURE DESCRIPTION

The ReCPU has been designed applying some well known computer architectural paradigms to provide a high throughput by limiting the number of stall conditions (that are the largest waste of computation time during the execution). This section overviews the structure of ReCPU - shown in the block diagram of Figure 2 - focusing on the microarchitectural implementation. A more detailed description of the two main blocks, the *Data Path* and the *Control Path*, is provided.

ReCPU has a Harvard based architecture that uses two separate memory banks: one storing the text and the other one the instructions (i.e. the RE). Both RAMs must be dual port to allow parallel accesses. In the *Data Path* subsection the use of parallel buffers is described.

The main idea proposed by this paper is the execution of more than one character comparison per clock cycle. To achieve this goal several parallel comparators - grouped in units called *Clusters* - are placed in the *Data Path*. Each comparator compares an input text character with a different one from the pattern. The number of elements of the cluster is indicated as *ClusterWidth* and it represents the number of characters that can be compared every clock cycle whenever a sub-RE is matching. This figure is influencing the throughput whenever a part of the pattern starts matching the input text. The architecture is composed by several *Clusters* - the total number is indicated as *NCluster* - used to compare a sub-RE starting by shifted position of the input text. This influences the throughput whenever the pattern is not matching.

4.1 Data Path

In order to process more than one character per clock cycle, we applied some architectural techniques to increase the parallelism of ReCPU: pipelining, data and instructions prefetching, and use of multiple memory ports.

The pipeline is composed by two stages: *Fetch/Decode* and *Execute*. The *Control Path*, as explained in the next section, spends one cycle to precharge the pipeline and then it starts exploiting the prefetching mechanism. In each stage we introduced duplicated buffers to avoid stalls. This approach was advantageous because the blocks we replicated and the corresponding control logic are not so complex, leading to an acceptable increase in terms of area, while no overhead in terms of time constraints is present since they work in parallel. Hence, we have a reduction of the execution latency with a consequent performance improvement.

Due to the regular instruction control flow a good prediction technique with duplicated instruction fetching structures is able to avoid stalls. Indeed, considering the *Fetch/Decode* stages, the two instruction buffers load two sequential instructions: when an RE starts matching, one buffer is used to prefetch the next instruction and the other is used as *backup* of the first one. In case that the matching process fails (i.e. prefetching is useless) the first instruction (i.e. the backup one) can be used without stalling the pipeline. Similarly, the parallel data buffers reduce the latency of the access to the data memory.

According to this design methodology in the *Fetch/Decode* stage the decoder and the pipeline registers are duplicated. By means of a multiplexer, just one set of pipeline register values are forwarded to the *Execution* stage. As shown in the diagram, the multiplexer is controlled by the *Control Path*. The decode process extracts from the instruction the reference string (i.e. the characters of the pattern that must be compared with the text), its length - indicated as valid_ref and necessary because the number of characters composing the sub-RE can be lower than the width of the cluster - and the operators used.

The second stage of the pipeline is the *Execute*. It is a fully combinatorial circuit. The reference coming from the previous stage is compared with the data read from the RAM and previously stored in one of the two parallel buffers. Like in *Fetch/Decode* stage this technique (see Figure 2) reduces the latency of the access to the memory avoiding the need of a stall if a jump in the data memory is required[3].

We implemented the comparison using a configurable number of arrays of comparators. This is shown in details in Figure 3. Each cluster is shifted of one character from the previous in order to cover a wider set of data in a single clock cycle. The results of each comparator cluster are collected and evaluated by the block called *Engine*. It produces a match/not match signal going into the *Control Path*.

Our approach is based on a fully-configurable VHDL implementation. It is possible to modify some architectural parameters such as: number and dimensions of the parallel comparator units (*ClusterWidth* and *NCluster*), width of buffer registers and memory addresses. This way it is possible to define the best architecture according to the user requirements, finding a good trade-off between timing, area constraints and desired performance.

4.2 Control Path

We define an RE as a sequence of instructions that actually represent a set of conditions to be satisfied. If all the instructions of an RE are matched, then the RE itself is matched. The ReCPU *Data Path* fetches the instruction, decodes it and verifies whether it matches the current part of the text or not. But it cannot identify the result of the whole RE. Moreover, the *Data Path* does not have the possibility to request data or instructions from the external memories.

To manage the execution of the RE we designed a *Control Path* block containing some specific hardware structures

---

[3] A jump in the data memory is required whenever one or more instructions are matching the text and then the matching fails (because the current instruction is not satisfied). In this case a jump in the data memory address restarts the search from the address where the first match occurred.

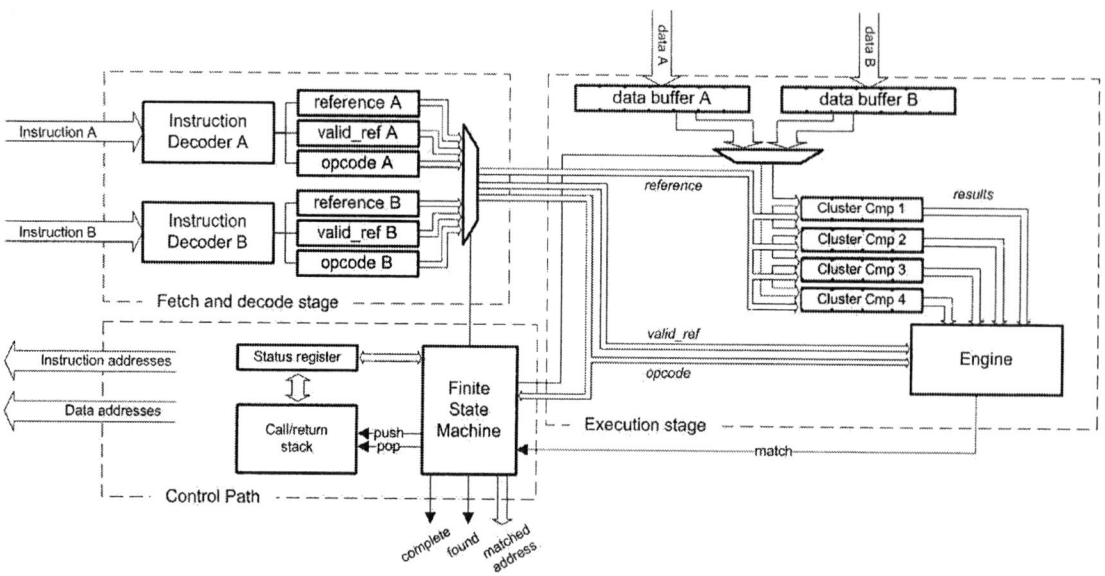

Figure 2: Block diagram of ReCPU with 4 Clusters, each of those has a ClusterWidth of 4. The main blocks are: Control Path and Data Path (composed by a Pipeline of Fetch/Decode and Execution stages).

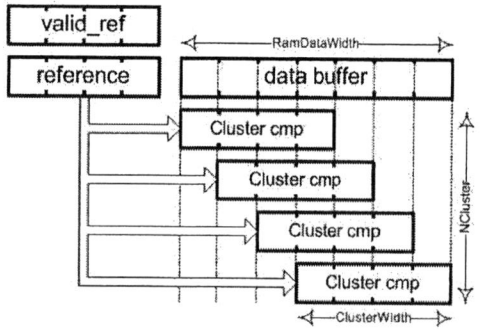

Figure 3: Detail of comparator clusters.

Figure 4: Finite state machine of the *Control Path*.

that we are going to describe in the current section. The core of the *Control Path* is the *Finite State Machine* (FSM) shown in Figure 4. The execution of an RE requires two input addresses: the RE start address and the text start address. The FSM is designed in such a way that after the preload of the pipeline, two execution cases can occur. When the first instruction of an RE does not match the text, the FSM loops in the EX_NM state, as soon as a match is detected the FSM goes into the EX_M state.

While not matching the text, the same instruction address is fetched and the data address advances performing the comparison by means of the clusters inside of the *Data Path*. If no match is detected the data memory address is incremented by the number of clusters. This way several characters are compared every single clock cycle leading to a throughput i.e. clearly more than one character/cc.

When an RE starts matching, the FSM goes into *EX_M* state and the ReCPU switches to the matching mode by using a single cluster comparator to perform the pattern matching task on the data memory. As for the previous case more than one character per clock cycle is checked by the different comparators of a cluster. When the FSM is in this state and one of the instructions composing the RE fails the whole process has to be restarted from the point where RE started to match.

In both cases (matching or not matching), whenever a NOP instruction is detected the RE is considered complete, so the FSM goes into the NOP state and the result is computed. The ReCPU returns a signal that indicates the match of the RE and the address of the memory location containing the first character of the matched string.

A particular case is represented by loops (i.e. + or * operators). We treat these operators with a call and return paradigm. When an open parenthesis is detected a call is performed: the *Control Path* saves the content of the status register (i.e. the actual matching value, the current program counter for instruction memory and the current internal operator) in a stack. The RE is then computed normally until a return operand is detected. A return is basically a closed parenthesis followed by +, * or |. It restores the old context and updates the value of the global matching. If a not matching condition is verified while the FSM is processing a call, the stack is erased and the whole RE is considered not matching. The process is restarted as in the simple not matching case.

Problems of overflow in the number of elements stored in the stack are avoided by the compiler. It knows the stack-size and computing the maximum level of nested parentheses is able to determine whether the architecture can execute the RE or not.

5. EXPERIMENTAL RESULTS

ReCPU has been synthesized using Synopsys Design Compiler[4] on the STMicroelectronics HCMOS8 ASIC technology library featuring $0.18\mu m$ silicon process. Validation of the proposed architecture has been exploited setting *NCluster* and *ClusterWidth* equal to 4. The synthesis results are presented in Table 2:

Table 2: Synthesis results for the ReCPU architecture with NCluster and ClusterWidth set to 4.

Critical Path (ns)	Area (μm^2)	Max Clock Frequency (MHz)
3.14	51082	318.47

Papers described in Section 1 show a maximum clock frequency between 100MHz and 300MHz. The results of the table show how our solution is competitive with the others having the advantage of processing in average more than one character per clock cycle (i.e. the case for all the other solutions like [1] and [2]).

We analyze different scenarios to figure out the performance of our implementation: whenever the input text is not matching the current instruction and the opcode represents a $\cdot$ operator, the maximum level of parallelism is exploited and the performance in terms of time required to process a character are up to:

$$T_{cnm} = \frac{T_{cp}}{NCluster + ClusterWidth - 1} \quad (1)$$

where the T_{cnm}, expressed in *ns/char*, depends on the number of clusters, the width of the clusters and the critical path delay T_{cp}. If the input text is not matching the current instruction and the opcode is a | then the performance are given by the following formula:

$$T_{onm} = \frac{T_{cp}}{NCluster} \quad (2)$$

[4]www.synopsys.com

If the input text is matching the current instruction then the performance depends on the width of one cluster (all the other clusters are not used):

$$T_m = \frac{T_{cp}}{ClusterWidth} \quad (3)$$

For each different scenarios, using the time per character computed using the formulas (1), (2) and (3) it possible to compute the corresponding bit-rate evaluating the achievable performance. The bit-rate B_x represents the number of bits[5] processed in one second and can be computed as follows:

$$B_x = \frac{1}{T_x} \cdot 8 \cdot 10^9 \quad (4)$$

where T_x represents one of the quantities resulting from (1), (2) and (3).

The numerical results for the implementation we have synthesized are shown in Table 3:

Table 3: Time necessary to process one character and corresponding bit-rate for the synthesized architecture.

T_{cnm} ns/char	T_{onm} ns/char	T_m ns/char	B_{cnm} GBit/s	B_{onm} GBit/s	B_m GBit/s
0.44	0.78	0.78	18.18	10.19	10.19

The results summarized in Table 3 represent the throughput achievable in different scenarios. Whenever there is a function call (i.e. nested parentheses) one additional clock cycle of latency is required. The throughput of the proposed architecture really depends on the RE as well as on the input text so it is not possible to compute a fixed throughput but just provide the performance achievable in different cases.

In our experiments we compared ReCPU with the popular software *grep*[6] using three different text files of 65K characters each. For those files we chose a different content trying to stress the behavior of ReCPU. We ran *grep* on a Linux Fedora Core 4.0 PC with Intel Pentium 4@2.80GHz, 512MB RAM measuring the execution time with Linux *time* command and taking as result the *real* value. The results are presented in Table 4.

Table 4: Performance comparison between grep software and ReCPU on a text file of 65K characters.

Pattern	*grep*	ReCPU	Speedup
$E|F|G|HAA$	19.1 *ms*	32.7 μs	584.8
$ABCD$	14.01 *ms*	32.8 μs	426.65
$(ABCD)+$	26.2 *ms*	393.1 μs	66.74

We notice that if loop operators are not present our solution performs equal either with more than one instruction and OR operators or with a single AND instruction (see the

[5]It is computed considering that 1 char = 8 bits.
[6]www.gnu.org/software/grep

first two entries of the table). In these cases the speedup is more than 400 times, achieving extremely good results with respect to software solution. In case of loop operators it is possible to notice a slow-down in the performance but still achieving a speedup of more than 60.

To prove performance improvements of our approach with respect to the other published solutions, we compare the bit-rates as described in the Table 5. It was not possible to compare the bit-rate for [6], [2] because this quantity was not published in the papers.

Table 5: Bit-Rate comparison between literature solutions and ReCPU.

Solution published in	bit-rate GBit/s	ReCPU GBit/s	Speedup factor (x)
[4]	$(2.0, 2.9)$	$(10.19, 18.18)$	$(5.09, 6.26)$
[3]	$(1.53, 2.0)$	$(10.19, 18.18)$	$(6.66, 9.09)$
[5]	$(5.47, 8.06)$	$(10.19, 18.18)$	$(1.82, 2.25)$
[1]	$(0.45, 0.71)$	$(10.19, 18.18)$	$(22, 25)$

In Table 5 the bit-rate range for different solutions is shown. We compared it with the one of ReCPU computing a speedup factor that underlines the speedup of our approach. It is shown that the performance achievable with our solution is n times faster than the other published research works. Our solution guarantees several advantages apart from the bit-rate improvement: $O(n)$ memory locations are necessary to store the RE and it is possible to modify the pattern at run-time just updating the program memory. It is interesting to notice - analyzing the results in the table - that in the worst case we are performing pattern matching almost two times faster.

6. CONCLUSIONS AND FUTURE WORKS

Nowadays the need of high performance computing is growing up. An example of this is represented by biological sciences (e.g. Humane Genome Project) where DNA sequence matching is one of the main applications. To increase the performance it is better to get advantage of hardware solutions for the pattern matching problem. In this paper we presented a novel architecture for hardware implementation of regular expression matching.

Our contribution regards a completely different approach of dealing with the regular expressions. REs are considered as a programming language for a parallel and pipelined architecture. This guarantees the possibility of changing the RE at run-time just modifying the content of the instruction memory and it involves a high improvement in terms of performance.

Some features, like the multiple characters checking, instructions prefetching and parallelism exposure to the compiler level are inspired to the VLIW design style.

The current state of the art guarantees a fixed performance of one character per clock cycle. Our goal was to figure out a way of extract some parallelism to achieve in average much better performance. We proposed a solution that has a bit-rate of at least 10.19 GBit/s with a peak of 18.18 GBit/s.

Future works are focused on the definition of a reconfigurable version of the proposed architecture based on FPGA-devices. This way, we could exploit the possibility to dynamically reconfigure the architecture at run-time. The study of possible optimizations of the *Data Path* to reduce the critical path and increase the maximum possible clock frequency is an alternative. We would also like to explore the possibility of adding some optimization in the compiler side.

7. REFERENCES

[1] R. Sidhu and V. Prasanna, "Fast regular expression matching using FPGAs," in *IEEE Symposium on Field-Programmable Custom Computing Machines (FCCM01)*, April 2001.

[2] C.-H. Lin, C.-T. Huang, C.-P. Jiang, and S.-C. Chang, "Optimization of regular expression pattern matching circuits on FPGA," in *DATE '06: Proceedings of the conference on Design, automation and test in Europe.* 3001 Leuven, Belgium, Belgium: European Design and Automation Association, 2006, pp. 12–17.

[3] Y. H. Cho and W. H. Mangione-Smith, "A pattern matching coprocessor for network security," in *DAC '05: Proceedings of the 42nd annual conference on Design automation.* New York, NY, USA: ACM Press, 2005, pp. 234–239.

[4] J. C. Bispo, I. Sourdis, J. M. Cardoso, and S. Vassiliadis, "Regular expression matching for reconfigurable packet inspection," in *IEEE International Conference on Field Programmable Technology (FPT)*, December 2006, pp. 119–126.

[5] I. Sourdis and D. Pnevmatikatos, "Fast, large-scale string match for a 10gbps fpga-based network intrusion," in *International Conference on Field Programmable Logic and Applications*, Lisbon, Portugal, September 2003.

[6] B. O. Brown, M.-L. Yin, and Y. Cheng, "DNA sequence matching processor using FPGA and JAVA interface," in *Annual International Conference of the IEEE EMBS*, September 2004.

[7] J. Friedl, *Mastering Regular Expressions*, 3rd ed. O'Reilly Media, August 2006.

[8] *Grep Manual*, GNU, USA, Jan 2002.

Use of Gray Decoding for Implementation of Symmetric Functions

Osnat Keren, Ilya Levin, and Radomir S. Stankovic

Abstract— This paper discusses reduction of the number of product terms in representation of totally symmetric Boolean functions by Sum of Products (SOP) and Fixed Polarity Reed-Muller (FPRM) expansions. The suggested method reduces the number of product terms, correspondingly, the implementation cost of symmetric functions based on these expressions by exploiting Gray decoding of input variables. Although this decoding is a particular example of all possible linear transformation of Boolean variables, it is efficient in the case of symmetric functions since it provides a significant simplification of SOPs and FPRMs. Mathematical analysis as well as experimental results demonstrate the efficiency of the proposed method.

Index Terms— Symmetric function, Gray code, linear transformation, autocorrelation.

I. INTRODUCTION

Linearization of switching functions based on linear transformation of variables is a classical method of optimization in circuit synthesis originating already in 1958 [22]. It has been recently efficiently exploited by several authors and discussed for different aspects due to its :

1) *Effectiveness.* When properly performed, the method provides considerable savings in complexity of the representation of functions with respect to different optimization criteria.

2) *Simplicity of the implementation.* The overhead comprises EXOR circuits required to perform the selected linear combination of variables. The overhead is usually quite negligible compared to the overall complexity of the implementation [12].

The linearization can be performed over different data structures used to represent functions. For example, it has been performed over Sum-of-Product (SOP) expressions [10], [13], [15], [28], AND-EXOR expressions [5], word-level expressions [27] as well as decision diagrams [7], [14], [18].

In spectral techniques, this method is studied as a mean to reduce the number of non-zero coefficients in spectral expressions for discrete functions [8], [12]. In [12], [20], and [21] the extensions to multiple-valued logic functions are discussed.

The complexity of determining an optimal non-singular binary matrix that defines the optimal linear transformation of variables is NP-complete. For this reason different strategies have been suggested in exploiting this method.

In searching for exact optimum, some restrictions should to be made on the number of variables in functions processed. For example, it has been reported in [7] that the complete

This work was supported by BSF under grant 2002259

search over all possible linear transformations is feasible for functions up to seven variables within reasonable space and time resources.

Another approach is to restrict considerations to particular classes of functions. For instance, in [10], [27] a method has been used for specific circuits, such as n-bit adders and an optimal linear transform has been found.

Alternatively, nearly optimal solutions can be provided by deterministic algorithms if analysis of additional information about the functions can be provided. In this direction, research has been reported by analyzing besides the functions their Walsh coefficients and autocorrelation coefficients, see, for instance, [8], [12] and [14] and references therein.

In this paper, we discuss a compromising approach. We show that when the class of functions is the totally symmetric Boolean functions, then an efficient linear transformation of variables can be determined analytically, it reduces to Gray decoding of input variables.

A justification to consider symmetric functions can be found in the following considerations. Symmetric Boolean functions represent an important fraction of Boolean functions. There are 2^{n+1} binary-valued symmetric functions out of 2^{2^n} functions. There are efficient circuit-based methods and complete BDD-based methods for identifying symmetries of completely and incompletely specified functions [11], [17], [19], [23], [29], [32].

In last several years, symmetric functions have been studied from different aspects. Optimal Fixed Polarity Reed-Muller (FPRM) expansions for totally symmetric functions are discussed in [4], [31] and references therein. A lower bound on the number of gates in conjunctive (disjunctive) normal form representation of symmetric Boolean functions is given in [30] and a method for generating a minimal SOP cover is presented in [3]. A multilevel synthesis of symmetric functions which exploits the disjoint decomposability and weight dependency of the functions is presented in [16] and a mapping of symmetric and partially symmetric functions to the CA-type FGPAs was suggested in [2]. A new expansion of symmetric functions and their application to non-disjoint functional decompositions for LUT-type FPGAs is presented in [25]. In this paper we show that the Gray decoding of the input variables almost always reduces the complexity in terms of the above three measures: the number of gates in two-level realization, the number of FPRM terms and the number of FPGA LUTs.

The paper is organized as follows. Section II gives basic definitions of symmetric Boolean functions and Gray codes. Section III presents the implementation of a symmetric function as a superposition of a Gray decoder and a non-linear function.

25

978-1-4244-1709-4/07/\$25.00 © 2007 IEEE

ection IV presents an illustrative example discussing in detail application of the proposed method. In Section V we discuss features of the proposed method and prove that the solutions produced can never increase complexity of representation of SOPs compared to the given initial representations. Section VI contains experimental results and Section VII concludes the paper.

II. PRELIMINARIES

A. Totally symmetric functions

Let $f(x) = f(x_{n-1}, \ldots x_0)$ a Boolean function of $n \geq 2$ inputs and a single output. The function f is *symmetric* in x_i and x_j iff

$$f(x_{n-1} \ldots x_i \ldots x_j \ldots x_0) = f(x_{n-1} \ldots x_j \ldots x_i \ldots x_0). \tag{1}$$

The function f is *totally* symmetric iff it is symmetric in all pairs of its variables.

A function $f(x) = S_i(x)$ is called an *elementary* symmetric function with working parameter i iff

$$S_i(x) = \begin{cases} 1 & ||x|| = i \\ 0 & otherwise \end{cases}$$

where $||x||$ is the Hamming weight of x. There are $n + 1$ elementary symmetric functions satisfying

$$\sum_x S_i(x) S_j(x) = \begin{cases} \binom{n}{i} & i = j \\ 0 & otherwise \end{cases}.$$

Any symmetric function can be represented as a linear combination of elementary symmetric functions, i.e. $f(x) = \oplus_{i=0}^{n} a_i S_i(x)$ where $a_i \in \{0, 1\}$. Hence, there are 2^{n+1} symmetric functions out of 2^{2^n} functions.

Example 1: Consider an elementary 5-inputs symmetric function $f(x) = S_3(x)$. The K-map of the function is given in Table I. The minimal SOP representation of the function consists of 10 minterms of 5 literals.

A Fixed Polarity Reed-Muller (FPRM) expansion is an EXOR of product terms, where no two products consists of the same variables and each variable appears in complemented or un-complemented form, but not in both [24]. In matrix notation [1], the FPRM expansion of a function $f(x_{n-1}, \ldots x_0)$ with a given polarity vector $h = (h_{n-1}, \ldots h_1, h_0)$, is defined as

$$f(x_{n-1}, \ldots x_0) = \left(\otimes_{i=0}^{n-1} [1, x_{n-1-i}^{h_{n-1-i}}] \right) \left(\otimes_{i=0}^{n-1} R^{h_{n-1-i}}(1) \right) F$$

where $\otimes$ is a Kronecker product,

$$x_i^{h_i} = \begin{cases} x_i & if \ h_i = 0 \\ x_i' & otherwise \end{cases}$$

and

$$R^{h_i}(1) = \begin{cases} \begin{pmatrix} 1 & 0 \\ 1 & 1 \end{pmatrix} & if \ h_i = 0 \\ \\ \begin{pmatrix} 0 & 1 \\ 1 & 1 \end{pmatrix} & otherwise \end{cases}$$

and F is the truth vector. The number of product terms in the FPRM depends on the polarity vector.

Example 2: The FPRM expansion of the *3-out-of-5* function in Example 1 with a positive polarity ($h = 0$) comprises 10 terms,

$$\begin{aligned} f &= x_4 x_3 x_2 \oplus x_4 x_3 x_1 \oplus x_4 x_2 x_1 \oplus x_3 x_2 x_1 \oplus x_4 x_3 x_0 \\ &\oplus x_4 x_2 x_0 \oplus x_3 x_2 x_0 \oplus x_4 x_1 x_0 \oplus x_3 x_1 x_0 \oplus x_2 x_1 x_0. \end{aligned}$$

The positive polarity produces the minimal number of terms, all the other 31 polarity vectors produces FPRM expansions of at least 16 product terms.

B. Gray code

The reflected binary code, also known as Gray code after Frank Gray [6], is used for listing n-bit binary numbers so that successive numbers differ in exactly one bit position. The definition of the Gray encoding and decoding is the following: Elements of a binary vector of length n, $z = (z_{n-1}, \ldots z_0)$ and the vector $x = (x_{n-1}, \ldots x_0)$ derived by Gray encoding are related as

$$x_i = \begin{cases} z_i & i = n - 1 \\ z_i \oplus z_{i+1} & otherwise \end{cases}$$

and

$$z_i = \begin{cases} x_i & i = n - 1 \\ x_i \oplus z_{i+1} & otherwise \end{cases}.$$

This relation can be written using matrix notation as $x = G_E z$ and $z = G_D x$ where $G_E = (\tau_{n-1}, \ldots, \tau_1, \tau_0)$ is a nonsingular matrix of the form

$$G_E = \begin{pmatrix} 1 & & 0 & \ldots & 0 & 0 \\ 1 & 1 & 0 & \ldots & 0 & 0 \\ 0 & 1 & 1 & \ldots & 0 & 0 \\ \vdots & \vdots & \vdots & \ldots & \vdots & \vdots \\ 0 & 0 & 0 & \ldots & 1 & 0 \\ 0 & 0 & 0 & \ldots & 1 & 1 \end{pmatrix}. \tag{2}$$

and $G_D = G_E^{-1}$. The matrices G_E and G_D are called the Gray encoding and the Gray decoding matrices, respectively. The implementation of the Gray encoder (decoder) requires $n - 1$ two-input EXOR gates.

Example 3: Let $n = 4$ and $z = (1, 1, 0, 1)$ then

$$\begin{aligned} x_3 &= z_3 = 1 \\ x_2 &= z_3 \oplus z_2 = 0 \\ x_1 &= z_2 \oplus z_1 = 1 \\ x_0 &= z_1 \oplus z_0 = 1 \end{aligned}$$

or

$$x = (\tau_3, \tau_2, \tau_1, \tau_0) z = \begin{pmatrix} 1 & 0 & 0 & 0 \\ 1 & 1 & 0 & 0 \\ 0 & 1 & 1 & 0 \\ 0 & 0 & 1 & 1 \end{pmatrix} \begin{pmatrix} 1 \\ 1 \\ 0 \\ 1 \end{pmatrix} = \begin{pmatrix} 1 \\ 0 \\ 1 \\ 1 \end{pmatrix}.$$

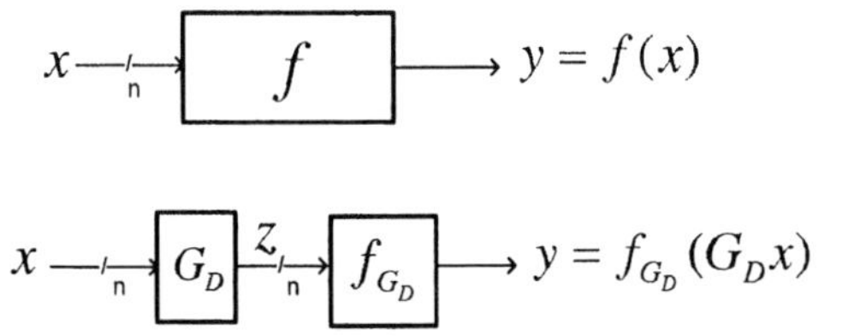

Fig. 1. Implementation of a Boolean function with a Gray decoding of the input variables

III. IMPLEMENTATION OF SYMMETRIC FUNCTIONS BY GRAY DECODED INPUTS

In this paper we introduce an implementation of a symmetric function as a superposition of two functions: a Gray decoder defined by the matrix G_D, and the corresponding function f_{G_D} whereas $f(x) = f_{G_D}(G_D x)$ (see Figure 1).

The main idea behind this approach is the following: A Boolean function maps elements of the vector space $\{0,1\}^n$ to $\{0,1\}$. The vector space $\{0,1\}^n$ is spanned by n base vectors, usually the binary vectors $\{\delta_i\}_{i=0}^{n-1}$ corresponding to the integer value 2^i are used. The set of δ_i's is called the initial basis. This basis is used in definition of SOP expressions.

Any set of n linearly independent vectors forms a basis, and in particular, the columns $\{\tau_i\}_{i=0}^{n-1}$ of the matrix G_E.

Since $Ix = G_E z$, the vector x can be interpreted as the coefficient vector that defines an element of $\{0,1\}^n$ using the initial basis, and z can be interpreted as the coefficient vector representing an element with the set of τ's. Thus, the matrices G_E and G_D define a linear transformation between the coefficient vectors.

Example 4: In Example 3, the element $(1,0,1,1) \in \{0,1\}^4$ can be represented as a linear combination of the initial base vectors $\delta_3 = (1,0,0,0), \delta_2 = (0,1,0,0), \delta_1 = (0,0,1,0)$ and $\delta_0 = (0,0,0,1)$, or as a linear combination of the columns of G_E. Namely,

$$(1,0,1,1) = 1 \cdot \delta_3 + 0 \cdot \delta_2 + 1 \cdot \delta_1 + 1 \cdot \delta_0 = 1 \cdot \tau_3 + 1 \cdot \tau_2 + 0 \cdot \tau_1 + 1 \cdot \tau_0,$$

thus, $x = (1011)$ and $z = (1101)$.

In theoretical considerations, complexity of circuit realization of a Boolean function is usually estimated without referring to a specific implementation technology. It is, therefore, often expressed in the number of two-input gates (AND/OR) that are required for the realization of the function considered. Formally, this criterion can be written in terms of a cost function [12], [26]

$$\mu(f) = |\{x | x, \tau \in \{0,1\}^n, f(x) = f(x + \tau), ||\tau|| = 1\}|$$

where $+$ stands for a bitwise EXOR of two binary vectors and $||\tau||$ is the Hamming weight of a binary vector τ. The autocorrelation function of f, is defined as $R(\tau) = \sum_{x \in \{0,1\}^n} f(x) f(x \oplus \tau)$. For a given function f, the value of μ can be related to the values of the autocorrelation function of f, at points corresponding to the base vectors,

$$\mu(f) = \sum_{i=0}^{n-1} R(\delta_i).$$

TABLE I
K-MAP OF A *3-out-of-5* FUNCTION

$x_4 x_3 x_2$ $x_1 x_0$	000	001	011	010	110	111	101	100
00						1		
01			1		1		1	
11		1		1				1
10			1		1		1	

TABLE II
K-MAP OF GRAY CODED *3-out-of-5* FUNCTION

$z_4 z_3 z_2$ $z_1 z_0$	000	001	011	010	110	111	101	100
00								
01		1	1	1	1	1		1
11			1		1		1	1
10								

In the case of initial basis, these are points 2^i, and linear transformation of variables performs the shift of these values.

There is a variety of minimization procedures that construct a linear transformation deterministically, see, for instance [14], [15] and [28] and references therein. It should be noticed that implementation of such procedures may be a space and time demanding task, and therefore, it is useful to take into considerations specific features of functions to be realized. In particular, we point out that for totally symmetric Boolean functions the linear transformation of variables derived from the Gray code almost always reduce the implementation cost. The same transformation often reduces the number of terms in Fixed polarity Reed-Muller expressions.

IV. MOTIVATION EXAMPLE

Consider the *3-out-of-5* function in Example 1. Let G_E and the G_D be the 5×5 Gray encoding and decoding matrices. The columns of G_E are binary vectors of length 5 corresponding to the integer values $1, 3, 6, 12$ and 24. Let $z = G_D x$ be the Gray decoded inputs. Table II shows the K-map of f_{G_D}. The minimal SOP representation of f_{G_D} consists of 5 products,

$$f_{G_D}(z_4, z_3, z_2, z_1, z_0) = z_3 z_2' z_0 + z_3 z_1' z_0 + z_4 z_2' z_0 + z_4' z_2 z_1' z_0 + z_4 z_3' z_1 z_0.$$

The FPRM expansion of f_{G_D} with a polarity vector $h = (11000)$ is

$$f_{G_D}(z_4, z_3, z_2, z_1, z_0) = z_0 \oplus z_2 z_1 z_0 \oplus z_3' z_2 z_0 \oplus z_4' z_3' z_0.$$

The values of the autocorrelation function of the original *3-out-of-5* function are shown in Figure 2 (top figure). The values of $R(\tau)$ at positions $\tau = 1, 2, 4, 8$ and 16 corresponding to the initial base vectors are all zero , thus, the minimal SOP comprises 10 minterms. The autocorrelation values at positions $\tau = 1, 3, 6, 12$ corresponding to the new base vectors $(\tau_0, \tau_1, \tau_2$ and $\tau_3)$ are equal to 6.

Applying the Gray decoding on the inputs is equivalent to permuting the autocorrelation values so that high autocorrelation values are now placed at positions 2^i. The autocorrelation function of f_{G_D} is shown at the bottom of Figure 2. The sum of the autocorrelation values of f_{G_D} at positions 2^i, $i =$

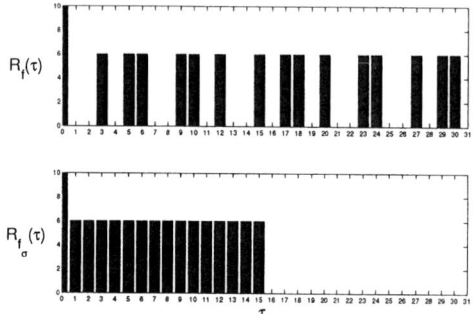

Fig. 2. Autocorrelation function values of the original *3-out-of-5* symmetric function f (top) and the values of the autocorrelation function corresponding to f_{G_D} with the Gray decoded inputs (bottom).

$0, \ldots, 4$ is $4 \cdot 6 + 0$, therefore, the number of pairs in the first merging step of the Quine-McClusky minimization algorithm is now 12 which leads to a minimal SOP representation.

V. ANALYSIS

Let $f(x) = f(x_{n-1}, \ldots x_0) \sum_{i=0}^{n} a_i S_i(x)$, $a_i \in \{0, 1\}$, a totally symmetric Boolean function of n variables and a single output. The autocorrelation function of $S_i(x)$ is [12]

$$
\begin{aligned}
R_{S_i}(\tau) &= \sum_{x \in \{0,1\}^n} S_i(x) S_i(x \oplus \tau) \\
&= \begin{cases} \binom{n - ||\tau||}{i - ||\tau||/2} \binom{||\tau||}{||\tau||/2} & ||\tau|| \text{ is even} \\ 0 & otherwise \end{cases}
\end{aligned}
$$

where $\binom{a}{b} = 0$ for $b < 0$.

The cross correlation between $S_i(x)$ and $S_j(x)$ is

$$
\begin{aligned}
R_{S_i,S_j}(\tau) &= \sum_{x \in \{0,1\}^n} S_i(x) S_j(x \oplus \tau) \\
&= \begin{cases} \binom{n - ||\tau||}{i - w} \binom{||\tau||}{w} & i - j + ||\tau|| \text{ is even} \\ 0 & otherwise \end{cases}
\end{aligned}
$$

where $w = (i - j + ||\tau||)/2$.

The autocorrelation function of f is

$$
\begin{aligned}
R_f(\tau) &= \sum_{x \in \{0,1\}^n} f(x) f(x \oplus \tau) \\
&= \sum_{i=0}^{n} a_i R_{S_i}(\tau) + \sum_{\substack{i,j = 0 \\ i \neq j}}^{n} a_i a_j R_{S_i,S_j}(\tau). \quad (3)
\end{aligned}
$$

Therefore, the autocorrelation values in positions corresponding the the initial set of base vectors $\{\delta_i\}_{i=0}^{n-1}$ is

$$
\begin{aligned}
R_f(\delta_i) &= 2 \sum_{k=1}^{n-1} a_k a_{k+1} R_{S_k, S_{k+1}}(\tau) \\
&= 2 \sum_{k=1}^{n-1} a_k a_{k+1} \binom{n-1}{k}. \quad (4)
\end{aligned}
$$

On the other hand, the autocorrelation values at positions corresponding to the base vectors $\tau_i = \delta_i + \delta_{i+1}$, $i =$

$0, \ldots n - 2$, defined by the columns of the Gray encoding matrix G_E, are

$$
R_f(\tau_i) = 2 \sum_{k=0}^{n} a_k \binom{n-2}{k-1} + 2 \sum_{k=1}^{n-2} a_k a_{k+2} \binom{n-2}{k} \quad (5)
$$

The following Theorem states that the realization cost of f_{G_D} with the Gray decoded inputs is less or equal to the realization cost of f for any totally symmetric function.

Theorem 1: Let $f(x) = \sum_{i=1}^{n} a_i S_i(x)$, $a_i \in \{0, 1\}$ a totally symmetric function, and let f_{G_D} the corresponding function with the Gray decoded inputs, i.e. $f(x) = f_{G_D}(G_D x)$. Then,

$$
\mu_f \leq \mu_{f_{G_D}}.
$$

Proof: The proof is based the fact that $R_{f_{G_D}}(\delta_i) = R_f(G_D^{-1} \delta_i) = R_f(\tau_i)$. Let $\Delta_i = R_f(\tau_i) - R_f(\delta_i)$, clearly, $\Delta_{n-1} = 0$ and for $0 \leq i < n - 1$, $\Delta_i = 2 \sum_{k=0}^{n} d_k$ where

$$
d_k = a_k \left(\binom{n-2}{k-1} - a_{k+1} \binom{n-1}{k} + a_{k+2} \binom{n-2}{k} \right). \quad (6)
$$

We now show that $\Delta_i \geq 0$ for all i. From 6, if $a_k = 0$ than $d_k = 0$, otherwise, there are four possible cases:
1) If $a_{k+1} = a_{k+2} = 0$ than $d_k > 0$.
2) If $a_{k+1} = a_{k+2} = 1$ than $d_k = 0$ since

$$
\binom{a}{b} = \binom{a-1}{b} + \binom{a-1}{b-1}.
$$

3) If $a_{k+1} = 0$ and $a_{k+2} = 1$ than $d_k > 0$.
4) If $a_{k+1} = 1$ and $a_{k+2} = 0$ than we may consider the sum $d_k + d_{k+1}$ and get

$$
\binom{n-2}{k-1} - \binom{n-1}{k} + \binom{n-2}{k} + a_{k+2} \binom{n-2}{k} \geq 0 \quad (7)
$$

Therefore, $R_{f_{G_D}}(\delta_i) = R_f(\tau_i) \geq R_f(\delta_i)$. From [12], the cost function μ_f of a function $f : \{0,1\}^n \to \{0,1\}$ equals to $\mu_f = 2^n - 2 R_f(0) + 2 \sum_{i=0}^{n-1} R_f(\delta_i)$, and thus $\mu_{f_{G_D}} \geq \mu_f$. ∎

VI. EXPERIMENTAL RESULTS

In this section, we compare the implementation cost of the original and Gray-coded functions in terms of:
a) The number of Look-Up-Tables (*LUT*s) required to implement the function by *SPARTAN3 xcs200ft256* as computed by LeonardoSpectrum.
b) The number of literals (L) in its minimal SOP representation as produced by ESPRESSO.
c) The number of nonzero terms in the optimal Fixed-Polarity Reed-Muller (*FPRM*) expansion.

Tables III and IV show the number LUTs for several totally symmetric functions of 8 and 12 input variables, the number of literals in the minimal SOP expression and the number of non-zero FPRM terms as computed with and without the Gray decoding. The improvement in those parameters is given in percentage. The symmetric functions $f = \sum_i a_i S_i(x)$ are specified by a set I, $I = \{i | a_i \neq 0\}$, of working parameters, I is written in the left column of Tables III and IV.

TABLE III
TOTALLY SYMMETRIC FUNCTIONS OF 8 INPUTS

I	LUT orig	LUT Gray	%	L orig	L Gray	%
3	12	7	41.7	448	92	79.5
4	13	9	30.8	560	106	81.1
3, 4	18	13	27.8	490	185	62.2
3, 5	15	8	46.7	896	45	95.0
3, 4, 5	18	15	16.7	336	123	63.4
2, 3, 5, 7	19	10	47.4	904	74	91.8
0, 2, 3, 5, 8	18	11	38.9	856	109	87.3

I	FPRM orig	FPRM Gray	%
3	64	24	62.5
4	107	15	86.0
3, 4	96	31	67.7
3, 5	104	17	83.6
3, 4, 5	162	49	69.7
2, 3, 5, 7	36	40	-11.1
0, 2, 3, 5, 8	107	25	76.6

TABLE IV
TOTALLY SYMMETRIC FUNCTIONS OF 12 INPUTS

I	LUT orig	LUT Gray	%	L orig	L Gray	%
3	65	26	60.0	2640	470	82.2
4	32	41	-28.1	5940	800	86.5
3, 4	143	71	50.3	5445	1225	77.5
3, 5	37	57	-54.1	12144	584	95.2
3, 4, 5	204	118	40.2	7920	1170	85.2
0, 2, 3, 5, 8	217	101	53.5	17876	1582	91.1

I	FPRM orig	FPRM Gray	%
3	232	200	13.8
4	794	166	79.1
3, 4	562	306	45.6
3, 5	1024	136	86.7
3, 4, 5	1354	356	73.7
0, 2, 3, 5, 8	738	328	55.6

Table V shows how the Gray decoding reduces the implementation cost of several totally symmetric LGSynth93 benchmark functions. Given a polarity vector, the number of non-zero FPRM terms of a k-output function is defined as the size of the union of the non-zero terms in the FPRM expansion of each one of the k single-output functions. For example, the original benchmark function *rd84* has four outputs, the number of non-zero FPRM terms of each is $28, 8, 1$ and 70 and the size of the union of these terms is 107. The number of non-zero terms of the corresponding Gray coded single-output functions is $14, 4, 1$ and 38 and the size of their union is 39.

TABLE V
BENCHMARK FUNCTIONS

	in	out	LUT orig	LUT Gray	L orig	L Gray
rd53	5	3	6	4	140	35
rd73	7	3	24	8	756	141
rd84	8	4	51	13	1774	329
9sym	9	1	36	36	504	135

	in	out	FPRM orig	FPRM Gray
rd53	5	3	20	12
rd73	7	3	63	24
rd84	8	4	107	39
9sym	9	1	173	33

VII. CONCLUSIONS

The problem of linearization of logic functions may be considered as a determining a linear transform for variables in a given function, which produces a representation of the function appropriate for particular applications. However, it is not always necessary to determine the best possible linear transformation for a class of functions. For many practical applications it is sufficient to find a suitable transform producing acceptable solutions.

In this paper we consider the class of symmetric functions and point out a suitable linear transformation of variables resulting in considerably reduced number of product terms in AND-OR and Reed-Muller expressions.

We propose a method to represent a symmetric logic function as a superposition of a linear portion that realize the Gray decoding of input vectors and a non-linear portion. Being a particular case of the linear transformation, the described Gray decoding transform enables to achieve very compact implementations of the initial symmetric function.

We have shown that the use of the Gray transform improves the complexity of the initial function implementation in terms of a specific cost function. Experimental results show that for majority of benchmarks the proposed method improves also a LUT based implementation of the function. The suggested approach can be extended to partially symmetric functions by the Gray decoding of each symmetry class separately.

REFERENCES

[1] J.T. Astola and R.S. Stankovic, *Fundamentals of Switching Theory and Logic Design: A Hands on Approach,* Springer-Verlag New York, 2006.

[2] M. Chrzanowska-Jeske and Z. Wang, " Mapping of symmetric and partially-symmetric functions to theCA-type FPGAs" *proc. of the 38th Midwest Symposium on Circuits and Systems*, vol. 1, pp. 290-293, Aug 1995.

[3] D. L. Dietmeyer, "Generating minimal covers of symmetric functions," *IEEE Transactions on Computer-Aided Design of Integrated Circuits and Systems*, Vol. 12, No. 5, pp. 710-713, May 1993.

[4] R. Drechsler, B. Becker, "Sympathy: fast exact minimization of fixed polarity Reed-Muller expressions for symmetric functions," *Proc. of the European Design and Test Conference*, pp. 91 - 97, March 1995.

[5] R. Drechsler and B. Becker, "EXOR transforms of inputs to design efficient two-level AND-EXOR adders," *IEE Electronic Letters*, vol. 36, no. 3, pp. 201-202, Feb. 2000.

[6] F. Gray, " Pulse code communication," March 17, 1953 (filed Nov. 1947). U.S. Patent 2,632,058.

[7] W. Günther, R. Drechsler, "BDD minimization by linear transforms", *Advanced Computer Systems*, pp. 525-532, 1998.

[8] S.L. Hurst,D.M. Miller, J.C. Muzio, *Spectral Techniques in Digital Logic*, Academic Press, Bristol, 1985.

[9] J. Jain, D. Moundanos, J. Bitner, J.A. Abraham, D.S. Fussell and D.E. Ross, "Efficient variable ordering and partial representation algorithm," *Proc. of the 8th International Conference on VLSI Design*, pp. 81-86, Jan. 1995.

[10] J. Jakob, P.S. Sivakumar, V.D. Agarwal, "Adder and comparator synthesis with exclusive-OR transform of inputs", *Proc. 1st Int. Conf. VLSI Design*, pp. 514-515, 1997.

[11] S. Kannurao and B. J. Falkowski, "Identification of complement single variable symmetry in Boolean functions through Walsh transform," *Proceedings of the International Symposium on Circuits and Systems (ISCAS)*, 2002.

[12] M.G. Karpovsky, *Finite Orthogonal Series in the Design of Digital Devices*, John Wiley, 1976.

[13] M.G. Karpovsky, E.S. Moskalev, "Utilization of autocorrelation characteristics for the realization of systems of logical functions," *Avtomatika i Telemekhanika*, No. 2, 1970, 83-90, English translation *Automatic and Remote Control*, Vol. 31, pp. 342-350, 1970.

[14] M.G. Karpovsky, R.S. Stankovic and J.T. Astola, "Reduction of sizes of decision diagrams by autocorrelation functions," *IEEE Trans. on Computers*, vol. 52, no. 5, pp. 592-606, May 2003.

[15] O. Keren, I. Levin and R.S. Stankovic, "Linearization of Functions Represented as a Set of Disjoint Cubes at the Autocorrelation Domain," *Proc. of the 7th International Workshop on Boolean Problems*, pp. 137-144, Sept. 2006.

[16] B. G. Kim, D.L. Dietmeyer, "Multilevel logic synthesis of symmetric switching functions," *IEEE Trans. on Computer-Aided Design of Integrated Circuits and Systems*, Vol.10, No. 4O, pp. 436-446, Apr. 1991.

[17] E.J., Jr. McCluskey, "Detection of group invariance or total symmetry of a Boolean function," *Bell Systems Tech. Journal*, vol. 35, no. 6, pp. 1445-1453, 1956.

[18] Ch. Meinel, F. Somenzi, T. Tehobald, "Linear sifting of decision diagrams and its application in synthesis," *IEEE Trans. CAD*, Vol. 19, No. 5, 2000, 521-533.

[19] D. Moller, J. Mohnke, and M. Weber, "Detection of symmetry of Boolean functions represented by ROBDDs," *Proc. of the International Conference on Computer-Aided Design (ICCAD)*, 1993, pp. 680684.

[20] C. Moraga, "Introducing disjoint spectral translation in spectral multiple-valued logic design", *IEE Electronics Letters*, 1978, Vol. 14, No. 8, pp. 248-243, 1978.

[21] C. Moraga, "On some applications of the Chrestenson functions in logic design and data processing", *Mathematic and Computers in Simulation*, Vol. 27, pp. 431-439, 1985.

[22] E.I. Nechiporuk, "On the synthesis of networks using linear transformations of variables", *Dokl. AN SSSR*. Vol. 123, No. 4, pp. 610-612, Dec. 1958.

[23] S. Panda, F. Somenzi, and B. Plessier, "Symmetry detection and dynamic variable ordering of decision diagrams," *Proc. of the International Conference on Computer-Aided Design (ICCAD)*, 1994.

[24] T. Sasao, *Switching Theory for Logic Synthesis*, Kluwer Academic Publishers, Feb. 1999

[25] T. Sasao, "A new expansion of symmetric functions and their application to non-disjoint functional decompositions for LUT type FPGAs", *IEEE Int. Workshop on Logic Synthesis, IWLS-2000*, May 2000.

[26] C. E. Shannon, "The Synthesis of Two-Terminal Switching Circuits," *Bell System Technical Journal*, Vol. 28, pp. 59-98, Jan. 1949.

[27] R.S. Stankovic, J.T. Astola, "Some remarks on linear transform of variables in adders," *Proc. 5th Int. Workshop on Applications of Reed-Muller Expression in Circuit design*, Starkville, Mississippi, USA, Aug. 10-11, pp. 294-302, 2001.

[28] D. Varma and E.A. Trachtenberg, "Design automation tools for efficient implementation of logic functions by decomposition," *IEEE Transactions on Computer-Aided Design of Integrated Circuits and Systems*, vol. 8, no. 8, pp. 901-916, Aug. 1989.

[29] K.H. Wang and J.H. Chen, "Symmetry Detection for Incompletely Specified Functions," *Proc. of the 41st Conference on Design Automation Conference, (DAC'04)*, pp. 434-437, 2004.

[30] G. Wolfovitz "The complexity of depth-3 circuits computing symmetric Boolean functions," *Information Processing Letters*, vol. 100, No. 2, pp. 41 - 46, Oct. 2006.

[31] S.N. Yanushkevich, J.T. Butler, G.W. Dueck, V.P. Shmerko, "Experiments on FPRM expressions for partially symmetric logic functions", *Proc.30th Int. Symp. on Multiple-Valued Logic*, Portland, Oregon USA, 141-146, May 2000.

[32] J. S. Zhang, A. Mishchenko, R. Brayton, M. Chrzanowska-Jeske, "Symmetry detection for large Boolean functions using circuit representation, simulation, and satisfiability", *Proceedings of the 43rd annual conference on Design automation* , pp. 510 - 515, 2006.

Parametric Structure-Preserving Model Order Reduction

Jorge Fernandez Villena [*]   Wil H. A. Schilders [†‡]   L. Miguel Silveira [*]

[*] INESC-ID/IST - Tech. U. Lisbon	[†] NXP Semiconductors Research	[‡] Dept. of Mathematics and Computer Science
Rua Alves Redol 9	High Tech Campus 37, 56656 AE	Technical University of Eindhoven
1000-029 Lisbon, Portugal	Eindhoven, The Netherlands	5600 MB Eindhoven, The Netherlands

jorge.fernandez@inesc-id.pt   w.h.a.schilders@tue.nl   lms@inesc-id.pt

Abstract—**Analysis and verification environments for next-generation nano-scale RFIC designs must be able to cope with increasing design complexity and to account for new effects, such as process variations and Electromagnetic (EM) couplings. Designed-in passives, substrate, interconnect and devices can no longer be treated in isolation as the interactions between them are becoming more relevant in the behavior of the complete system. At the same time variations in process parameters lead to small changes in the device characteristics that may directly affect system performance. These two effects, however, can not be treated separately as the process variations that modify the physical parameters of the devices also affect those same EM couplings. Accurately capturing the effects of process variations as well as the relevant EM coupling effects requires detailed models that become very expensive to simulate. Reduction techniques able to handle parametric descriptions of linear systems are necessary in order to obtain better simulation performance. In this work Model Order Reduction techniques able to handle parametric system descriptions are presented. Such techniques are based on Structure-Preserving formulations that are able to exploit the hierarchical system representation of designed-in blocks, substrate and interconnect, in order to obtain more efficient simulation models.**

I. INTRODUCTION

New coupling and loss mechanisms, including EM field coupling and substrate noise as well as process-induced variability, are becoming too strong and too relevant to be neglected, whereas more traditional coupling and loss mechanisms are more difficult to describe given the wide frequency range involved and the greater variety of structures to be modeled. The performance of each device in the circuit is strongly affected by the environment surrounding it. In other words, the response of each circuit part depends not only on its own physical and electrical characteristics, but to a great extent also on its positioning in the IC, i.e. on the devices to which it is directly connected to or coupled with. The high level of integration available in current RFIC designs leads to proximity effects between the devices, which induce EM interactions, that can lead to different behaviors of the affected parts. In any manufacturing process there is always a certain degree of uncertainty involved given our limited control over the environment. For the most part this uncertainty was previously ignored when analyzing or simulating complete systems, or assumed to be accounted for in the individual device models. However, as we step towards the nano-scale and higher frequency eras, such environmental, geometrical and electromagnetic fluctuations become more significant. Nowadays, parameter variability can no longer be disregarded, and its effect must be accounted for in early design stages so that unwanted consequences can be minimized. This leads to parametric descriptions of systems, including the effects of manufacturing variability, which further increases the complexity of such models. Reducing this complexity is paramount for efficient simulation and verification. However, the resulting reduced models must retain the ability to capture the effects of small fluctuations, in order to accurately predict behavior and optimize designs. This is the realm of *Parametric Model Order Reduction* (pMOR). Furthermore, these parametric fluctuations of the physical characteristics of the devices can affect not only the performance of such device, but the coupling between devices. For this reason the parametric models of the individual blocks of a system can no longer be simulated in isolation but must be treated as one entity and verified together. Such reduction must take advantage of the hierarchical description of those systems namely to account for designed-in elements as well as interconnect effects. To this end, structure-preserving techniques must be used which not only retain structural properties of the individual systems but also its connections and couplings.

The goal of this paper is therefore to discuss and present techniques for model order reduction of interconnect, substrate or designed-in passives, taking into account their dependence on relevant process or fabrication parameters and their coupling and connections. The paper is structured as follows: in Section II an overview of several existing pMOR techniques will be discussed. In Section III an introduction to two-level hierarchy MOR will be done, and an extension to improve the reduction will be presented. In Section IV the proposed methodology for combining the parametric techniques with the hierarchical reduction will be proposed. To illustrate the procedure, its pros and cons, in Section V some reduction results will be presented for several real-life structures. Finally conclusions will be drawn in Section VI.

II. PARAMETRIC MODEL ORDER REDUCTION

Actual fabrication of physical devices is susceptible to the variation of technological and geometrical parameters due to deliberate adjustment of the process or from random deviations inherent to the manufacturing procedures. This variability leads to a dependence of the extracted circuit elements on several parameters, of electrical or geometrical origin. This dependence results in a parametric state-space system representation, which in descriptor form can be written as

$$C(\lambda)\dot{x}(t,\lambda)(\lambda) + G(\lambda)x(t,\lambda) = Bu(t)$$
$$y(t,\lambda) = Lx(t,\lambda) \tag{1}$$

where $C, G \in \mathbb{R}^{n \times n}$ are respectively the capacitance and conductance matrices, $B \in \mathbb{R}^{n \times m}$ is the matrix that relates the input vector $u \in \mathbb{R}^m$ to the inner states $x \in \mathbb{R}^n$ and $L \in \mathbb{R}^{n \times p}$ is the matrix that links those inner states to the outputs $y \in \mathbb{R}^p$. The elements of the matrices C and G, as well as the states of the system x, depend on a set of P parameters $\lambda = [\lambda_1, \lambda_2, \ldots, \lambda_P]$ which model the effects of the mentioned uncertainty. Usually the system is formulated so that the matrices related to the inputs and outputs (B and L) do not depend on the parameters. This time-domain descriptor yields a parametric dependent frequency response modeled via the transfer function

$$H(s,\lambda) = L(sC(\lambda) + G(\lambda))^{-1}B \tag{2}$$

for which we seek to generate a reduced order approximation, able to accurately capture the input-output behavior of the system for any point in the multidimensional frequency-parameter space.

$$\hat{H}(s,\lambda) = \hat{L}(s\hat{C}(\lambda) + \hat{G}(\lambda))^{-1}\hat{B} \tag{3}$$

In general, one attempts to generate a *Reduced Order Model* (ROM) whose structure is as much similar to the original as possible, i.e. exhibiting a similar parametric dependence and retaining as much of the original structure as possible. Many techniques have been proposed to tackle this problem and in the following we review some of the most commonly used.

A. Multi-Dimensional Moment Matching

These techniques appear as extensions to nominal moment-matching techniques [1], [2], [3]. Moment matching algorithms have gained a well deserved reputation in nominal MOR due to their simplicity and efficiency. The extensions of these techniques to the parametric case are usually based in the implicit or explicit matching of the moments of the parametric transfer function (2). This type of algorithms assumes small fluctuations of the parameters, so that an affine model based on the Taylor Series expansion can be used for approximating the behavior of the conductance and capacitance, $G(\lambda)$ and $C(\lambda)$, expressed as a function of the parameters. The Taylor series can be extended up to the desired (or required) order, including cross derivatives, for the sake of accuracy. Some schemes, denoted as Multi-Parameter Moment Matching use this idea to match, via different approaches, the multi-parameter moments of the parametric transfer function (2) (for details see [4], [5],

[6]). However these methods usually suffer of oversize when the number of moments to match is high.

A slightly different approach, that provides more compact ROMs, is presented in [7], which relies on the computation of several subspaces, built separately for each dimension, i.e. the frequency s and the parameter set λ. Given a parametric system (1), the first step of the algorithm is to obtain the k_s block moments of the transfer function with respect to the frequency when the parameters take their nominal value (for example, via [1]). This block moments will be denoted as Q_s. The next step is to obtain the subspace which matches the k_{λ_i} block moments of x with respect to each of the parameter λ_i, and will be denoted by Q_{λ_i}. Once all the subspaces have been computed, an orthonormal basis can be obtained so that its columns spans the union of all previously computed subspaces. Applying the resulting matrix in a projection scheme ensures that the parametric ROM matches k_s moments of the original system with respect to the frequency, and k_{λ_i} moments with respect to the parameter λ_i.

B. Variational PMTBR

A novel approach was recently proposed that extends the PMTBR [8] algorithm to include variability [9]. This approach is based on the statistical interpretation of the algorithm (see [8] for details) and enhances its applicability. In this interpretation, the approximated Gramian is seen as a covariance matrix for a Gaussian variable, $x(0)$, obtained by exciting the underlying system description with white noise. Rewriting the Gramian as

$$X_{\lambda} = \int_{S_{\lambda}} \int_{-\infty}^{\infty} (j\omega C_{\lambda} + G_{\lambda})^{-1}BB^T(j\omega C_{\lambda} + G_{\lambda})^{-H}p(\lambda)dwd\lambda \tag{4}$$

where $p(\lambda)$ is the *Probability Density Function* (PDF) of λ in the parameter space, S_{λ}. Just as in the original PMTBR algorithm, a quadrature rule can be applied in the parameter plus frequency space to approximate the Gramian via numerical computation (see [9] for details). The accuracy of the resulting ROM does not depend on the accuracy of the approximation of the integral, but on the projection subspace. After the quadrature is performed in the overall variational subspace, the deterministic procedure is followed and the most relevant vectors are selected via *Singular Value Decomposition* (SVD) in order to build a projection matrix meant to be used as a congruence transformation on the parametric system matrices (1). As in the deterministic case, an error analysis and control can be included, via the eigenvalues of the SVD, but in this variational case, only an expected error bound can be given The complexity and computational cost is generally the same as that of the deterministic PMTBR plus the previous quadrature operations, and, it has been shown that the size of the reduced model is less sensitive to the number of parameters in the description, or how this parameter dependence is modeled.

III. Block Hierarchical Model Order Reduction

A. Structure Preservation

As pointed out, individual blocks inside an RFIC can no longer be treated in isolation, and for this reason the complete system must be treated as an entity. Considering the linear component blocks including designed-in passives, interconnect, etc, the system description has an interesting structure, where the diagonal blocks correspond to the individual block matrices, whereas the off-diagonal blocks correspond to the static interconnections (in the G matrix) and dynamic couplings (C matrix). Standard model order reductions techniques can be applied to this joint, global system and while the resulting reduced model will usually be able to accurately capture the input-output behavior of the complete set of blocks, this approach leads to full reduced matrices. Furthermore, the original two-level hierarchy with interconnections and couplings can no longer be recovered.

An alternative approach is to perform the reduction of the individual models in a hierarchical fashion, i.e to reduce each model independently without taking into account the rest of the models or the environment. Hence every model is reduced separately and thus the hierarchy and structure of the global system is maintained. However, to apply MOR to each model means to capture its individual behavior, not the global one. This can be inefficient as too much effort may be spent capturing some local behavior that is not relevant for the global response (maybe filtered by another model). Furthermore certain aspects of the global response might be missed as it is not clear at the component level how relevant they are. To avoid these problems, one can reduce each component block separately but oriented to capture the global input-output response. This approach will provide us with more control in the reduction stage while also preserving the structure of the interconnections. The transfer function to match is the global one, so the most relevant behavior for the complete RF system is captured. What is more, only the global inputs and outputs of the complete RF block are relevant, so the inefficiencies caused by the large number of ports of the individual component blocks is avoided.

Some recent methods have advocated this approach. In [10] a control theoretic viewpoint of reduction of interconnected systems was presented, but it has the disadvantage that it is unable to treat capacitive couplings. The *Block Structure Preserving* (BSP) technique was first presented in [11] and later generalized in [12].

$$G = \begin{bmatrix} G_{11} & \dots & G_{1N_b} \\ \vdots & \ddots & \vdots \\ G_{N_b1} & \dots & G_{N_bN_b} \end{bmatrix} \quad C = \begin{bmatrix} C_{11} & \dots & C_{1N_b} \\ \vdots & \ddots & \vdots \\ C_{N_b1} & \dots & C_{N_bN_b} \end{bmatrix}$$

$$B = \begin{bmatrix} B_1^T & \dots & B_{N_b}^T \end{bmatrix}^T \quad L = \begin{bmatrix} L_1 & \dots & L_{N_b} \end{bmatrix} \quad (5)$$

The main idea was to retain the system block structure, i.e. the two-level hierarchy, after reduction via projection, allowing for a more efficient reduction and the maintenance of certain system properties, such as the degree of sparsity,

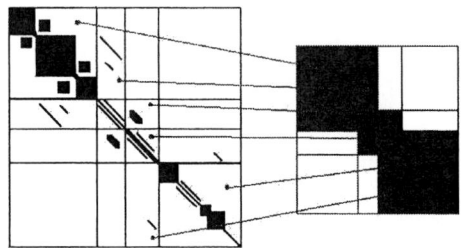

Fig. 1. Illustration of block hierarchy in the system matrix and effect of reduction using BSP.

and the block hierarchical structure. The procedure relies on expanding the projector of the global system (obtained via any classical MOR projection technique) into a block diagonal matrix, with block sizes equal to the sizes of its N_b individual component blocks (5). A basis that spans a suitable subspace for reduction via projection is then computed (for example a Krylov subspace). The projector built from that basis can be split and restructured into a block diagonal one so that the 2-level structure is preserved under congruence transformation.

$$\begin{bmatrix} V_1 \\ \vdots \\ V_{N_b} \end{bmatrix} \equiv colsp \left[Kr \left\{ A, R, q \right\} \right] \rightarrow \begin{bmatrix} V_1 & & \\ & \ddots & \\ & & V_{N_b} \end{bmatrix} = \check{V} \quad (6)$$

where $Kr\{A, R, q\}$ is the q column Krylov subspace of the complete system ($A = G^{-1}C$ and $R = G^{-1}B$). The block-wise congruence transformation is (see Figure 1)

$$\begin{align} \hat{G}_{ij} = V_i^T G_{ij} V_j & \qquad \hat{B}_i = V_i^T B_i \\ \hat{C}_{ij} = V_i^T C_{ij} V_j & \qquad \hat{L}_j = L_j V_j \end{align} \quad (7)$$

It should be noticed that the above projection matrix $\check{V}$ has N_b (number of blocks) times more columns than the original projector . This leads to an N_b times larger reduced system. On the other hand, this technique maintains the block structure of the original system and gives us some flexibility when choosing the size of the reduced model depending on the block layout and relevance. The reduced system will be able to match up to N_b times q block moments of the original complete transfer function (see [12] for details) under the best conditions (i.e. with very weak entries in the off-diagonal blocks). Under the worst conditions, only q block moments are matched, i.e. the same number than in the *flat* reduction. This technique is applicable to the global system, composed of the individual blocks and their connections (including both resistive as well as capacitive or inductive couplings between the blocks). The BSP technique therefore preserves the block structure of the system. However, the inner structure of the blocks themselves is lost since the procedure turns any non-empty block in the original system into a full block, but it is still possible to identify the blocks and relate them to the original device or interaction block. Nevertheless, if any block is empty in the global system matrix, it remains empty after reduction, increasing the sparsity.

B. PMTBR in Block Structure MOR

Any projection-based MOR procedure can be extended in the BSP manner to maintain the hierarchical structure of a system. In the case of the PMTBR algorithm, additional characteristics of the procedure can be further taken advantageous of in the current framework. The PMTBR algorithm links the rational projection methods with the *Truncated Balanced Realizations* (TBR) framework [13]. The procedure is based on the estimation, via a quadrature rule, of the frequency-based integral expression for the controllability Gramian, (4),

$$\bar{X} = \sum_k z_k z_k^H = ZZ^H \qquad (8)$$

where $Z = [z_1 \ z_2 \ \dots]$ and $z_i = (jw_i C + G)^{-1} B$. In [8] it was shown that if the quadrature scheme (8) is accurate enough, then the estimated Gramian $\bar{X}$ converges to the original one X, which implies that the dominant eigenspace of $\bar{X}$ converges to the dominant eigenspace of X. If the system has some internal structure, then the matrix Z that is computed from the vector samples of the global system can be split into blocks. The estimated Gramian can be written block-wise as

$$\begin{bmatrix} Z_1 \\ \vdots \\ Z_{N_b} \end{bmatrix} \rightarrow ZZ^H = \begin{bmatrix} Z_1 Z_1^H & \dots & Z_1 Z_{N_b}^H \\ \vdots & \ddots & \vdots \\ Z_{N_b} Z_1^H & \dots & Z_{N_b} Z_{N_b}^H \end{bmatrix} = \bar{X} \qquad (9)$$

But if we expand the matrix Z into diagonal blocks

$$\breve{Z} = \begin{bmatrix} Z_1 & & \\ & \ddots & \\ & & Z_{N_b} \end{bmatrix} \rightarrow \breve{Z}\breve{Z}^H = \begin{bmatrix} Z_1 Z_1^H & & \\ & \ddots & \\ & & Z_{N_b} Z_{N_b}^H \end{bmatrix} = \breve{X}. \qquad (10)$$

From (9) it can be seen that $Z_i Z_i^H = \bar{X}_{ii}$, i.e. the matrix $\breve{X} = \breve{Z}\breve{Z}^H$ is a block diagonal matrix whose entries are the block diagonal entries of the matrix $\bar{X}$. Under a good quadrature scheme, the matrix $\bar{X}$ converges to the original X, and therefore $\breve{X}$ will converge to the block diagonals of X. This means that the dominant eigenspace of $\breve{X}$ converges to the dominant eigenspace of the block diagonals of X. We can then apply an SVD to each block of the Z matrix

$$Z_i = V_i S_i U_i \ \rightarrow \ \breve{X}_{ii} = \bar{X}_{ii} = V_i S_i^2 V_i^T \qquad (11)$$

where S_i is real diagonal, and V_i and U_i are unitary matrices. The dominant eigenvectors of V_i corresponding to the dominant eigenvalues of S_i can be used as a projection matrix in a congruence transformation over the system matrices for model order reduction. The elements of S_i can also be used for a priori error estimation in a way similar to how Hankel Singular Values are used in TBR procedures. Using these block projectors V_i, a structure preserving projector for the global system can be built (6) which will capture the most relevant behavior of each block (revealed by the SVD) with respect to the global response (recall that Z is composed of sample vectors of the complete system). This approach provides us with more flexibility when reducing a complete system composed of several blocks and the interactions between them, as it allows to control the reduced size of each device via an error estimation on the global response.

IV. PARAMETRIC BLOCK STRUCTURE MOR

From the two-level hierarchical description of a system it is possible to have some extra block information that allows us to perform a more efficient MOR. But the behavior of the individual blocks that compose the system is subject to the effect of process variations, both geometrical and electrical. Such variations, as previously pointed out, also affect the interactions and couplings between these blocks. Any system-wide EM simulations must address these effects. Therefore, the variability study must be done over the complete system, and after model generation, a two-level parametric system will be obtained, with the block matrices in the block diagonals and the interactions between them in the off-diagonals. All these blocks will be functions of the relevant process and geometrical parameter. For instance, for conductivity,

$$G = \begin{bmatrix} G_{11}(\lambda_{\{11\}}) & \dots & G_{1N_b}(\lambda_{\{1N_b\}}) \\ \vdots & \ddots & \vdots \\ G_{N_b 1}(\lambda_{\{N_b 1\}}) & \dots & G_{N_b N_b}(\lambda_{\{N_b N_b\}}) \end{bmatrix} \qquad (12)$$

where $\lambda_{\{ij\}}$ represents the set of parameters affecting block G_{ij}. From (12) is clear that we have a parametric system depending on $\lambda = [\lambda_{\{11\}} \dots \lambda_{\{N_b N_b\}}]$. Therefore we can apply parametric MOR reduction. Note that any parameter affecting several blocks (diagonal blocks and their interactions) is treated as a single parameter (this reduces the number of parameters). However, in order to maintain the system structure, BSP techniques can be applied. This is possible as long as the selected pMOR technique is based in a projection scheme, which is the case for most of the existing procedures. The extension is very simple: obtain a suitable basis for projection from the ***complete system***, and then split and expand it into a block structure preserving projector. If the basis spans the most relevant behavior of the parametric system, then the expanded BSP projector will capture those as well. All the advantages and disadvantages mentioned in Section III hold here. But there is an extra and important advantage in the parametric case: **the BSP technique maintains the block parametric dependance**, i.e. if a block C_{ij} depends on a set of parameters $\lambda_{\{ij\}}$, then the reduced block $\hat{C}_{ij} = V_i^T C_{ij} V_j$ will depend on the same parameter set and no other.

On the other hand, as previously discussed some pMOR algorithms yield a very large ROM, and therefore their combination with BSP techniques will lead to an extremely large ROM. However, it was shown in Section II-B that the ROM size of the Variatinal PMTBR method is less sensitivity to the number of parameters. Furthermore, this method has a direct relation with PMTBR: the only difference is in the sampling scheme for obtaining the matrix whose columns spans the desired subspace, the rest of the procedure being exactly the same. Therefore, the results obtained in Section III-B are applicable to the variational case. The advantage of the control and error estimation still remains, although in this case only an expected error bound can be given. Such control is very useful when the models of a complete entity have very different sizes: if the same ROM size is applied to every

Algorithm I: Block Structure Preserving VPMTBR

Starting from a Block Structured System C, G, B, L with N_b blocks:

1: Select a quadrature rule of K points in the space $[s, \quad \lambda]$
2: For each point compute: $\quad z_i = (s_i C(\lambda_i) + G(\lambda_i))^{-1} B$
3: Form the matrix columns $Z = [z_1 \ldots z_k]$
4: Split it into N_b blocks,

$$Z = \begin{bmatrix} Z_1 \\ \vdots \\ Z_{N_b} \end{bmatrix}$$

5: For each block Z_j obtain the SVD: $\quad Z_j = V_j S_j U_j$
6: For each matrix V_j drop the columns whose singular values falls bellow the desired global tolerance
7: Build a Block Structure Preserving Projector from the remaining columns

$$\breve{V} = \begin{bmatrix} V_1 & & \\ & \ddots & \\ & & V_{N_b} \end{bmatrix}$$

8: Apply $\breve{V}$ in a congruence transformation on the Block Structured System C, G, B, L

block, the reduction may grow unnecessarily large. In contrast, the complexity of the proposed methodology is exactly the same as that for the non-structure-preserving techniques. The only difference is that the SVD (or orthonormalization in the moment matching approaches) must be done block-wise in order to avoid numerical errors. This can turn into an advantage, because for some blocks the number of vectors needed is lower, so less computational effort is required.

V. RESULTS

To illustrate the proposed procedure we present results from two examples to which several pMOR techniques were applied. These included [9] denoted as VPMTBR, [7] denoted as PPTDM, and two Block Structure preserving methods: *Algorithm I*, denoted as BS VPMTBR, and block struture based on [7], denoted as BS PPTDM. The non-reduced model response will be denoted as Original or Perturbed, depending on whether a parameter variation has been applied.

A. Example 1 - Spiral

The first example system is composed of three blocks: a *Multiple Input Multiple Output* (MIMO) RC ladder of size 101, with 2 ports, a MIMO Spiral Inductor of size 4961, with 2 ports, and another RC ladder of size 101 and 2 ports. The three systems are connected in series as shown in Figure 2, so the global input is the input of the first RC and the output is the output of the second RC. The Spiral has each of its ports conected to each ladder. The system depends on five parameters, affecting different blocks. Figure 3 shows the frequency response of the self-admittance Y_{11} of the nominal system, the pertubed response of the non-reduced system, and the responses of the PMTBR-based models (the PPTDM and BS PPTDM models do not produce competitive results sizewise,

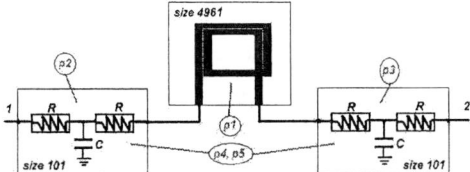

Fig. 2. Interconnection scheme for Example 1, with original sizes and parameter indication.

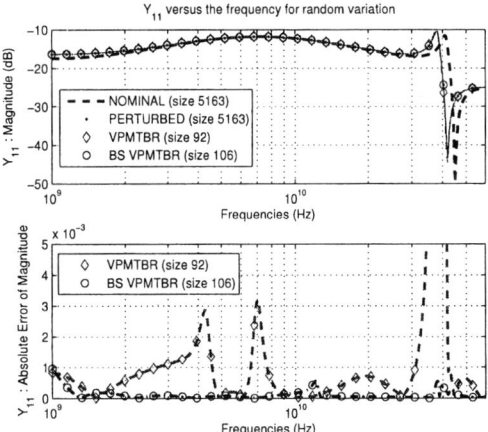

Fig. 3. (Up) Magnitude in dB of Y_{11} versus the frequency of Example 1 for the nominal, the pertubed and the parametric ROMs for a random parameter variation. (Down) Error of the Magnitude of Y_{11} for the ROMs w.r.t. the perturbed response.

TABLE I
CHARACTERISTICS OF THE PMOR METHODS APPLIED

MOR Method	Example 1		Example 2	
	Size	NNZ (G C)	Size	NNZ (G C)
NONE	5163	22545 6631	1600	4768 12588
VPMTBR	92	8464 8464	66	4356 4356
PPTDM	150	22500 22500	544	295936 295936
BS VPMTBR	106	11108 8228	96	722 5438
BS PPTDM	352	103502 42856	160	1600 17200

as seen from Table I, and therefore were omitted). Table I shows the main characteristics of the obtained ROMs. The moment matching techniques are less efficient, as the Spiral requires a high-order model. The PMTBR-based techniques obtain a better compression overall: BS VPMTBR yields a sligthy bigger ROM, but it maintains the block structure of the original system, and is able to control the size of each reduced block depending on its relevance on the global response.

B. Example 2 - Coupled Buses

This example, depicted in Figure 4, is composed of 16 blocks: 2 buses of 8 parallel lines each (each line modeled as an RC ladder of 100 segments), are on different metal layers, and cross at a square angle. The inputs and outputs are taken at the edges of each line of the first bus, so the system will have 16 ports. In this case there is no interconnection, just coupling

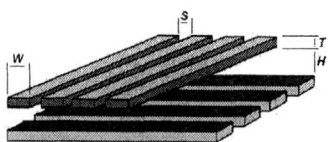

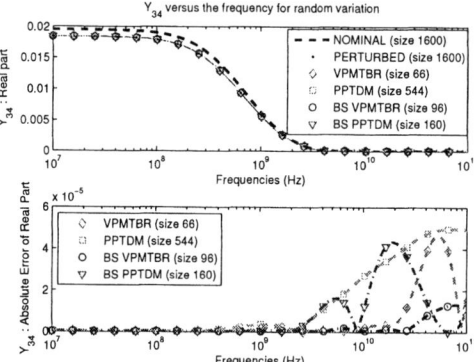

Fig. 4. Bus topology for Example 2.

Fig. 5. (Up) Y_{34} versus the frequency for Example 2 for the nominal, pertubed and parametric ROMs with random parameter variation set. (Down) Absolute Error of the ROMs w.r.t. the perturbed response.

effects. Each line is assumed coupled to the previous and the next line of their bus, and to every line of the other bus in the crossing area. Each line has its width (W) as a parameter, which implies 16 independent parameters. The width variation affects the line model, as well as the in-bus coupling (width variation also affects the interline spacing), and the inter-bus coupling (the crossing area varies). Figure 5 shows the frequency response of the nominal system, the pertubed response of the non-reduced system, and the responses of the ROMs for VPMTBR, PPTDM, BS VPMTBR and BS PPTDM. Again, the main characteristics of the resulting ROMs are shown in Table I. The PPTDM based algorithms result in very large ROMs even for small number of moments to match (2 w.r.t. the frequency and 2 w.r.t. each parameter). For these reasons each block moment from PPTDM was truncated to 10 vectors to keep the size manageable (otherwise no reduction would be possible). While this seems to produce acceptable results, there is little control over the result. On the other hand, the PMTBR based techniques leads to more compressed ROMs, as the SVD reveals the most relevant vectors. In the case of the BS VPMTBR, the control of each block allows different reduction sizes for each bus: since the ports of the 2nd bus are not taken into account, less effort is needed to capture its behavior. In fact, the models for the 1st bus are of sizes 8 to 10, while models for the 2nd bus are all size 3. The ability to control reduction locally is clearly an advantage of the method.

VI. CONCLUSION

In this paper we have presented a block structure-preserving parametric model order reduction technique, as an extension of existing pMOR techniques in order to improve the reduction

when a two-level hierarchical structure is available in the system description. This type of structure is common in coupled or interconnected systems, and can lead to simulation advantages. The methodology presented here is general as it can be used with any projection pMOR technique to mantain the two-level hierarchy and the block-parameter dependance. The presented extension of the PMTBR-based procedures into the Block Structure Preserving framework, allows more control on the reduction, provided by the inclusion of estimated error bounds on the single blocks oriented to the global response.

ACKNOWLEDGMENT

This work has been performed under the EU STREP project CHAMELEON-RF (contract no. 027378) and EC SERA project COMSON (contract no. 019417) under a Marie Curie Fellowship. The authors gratefully acknowledge the support from the EU, and would like to thank Wim Schoenmaker (MAGWEL), Nick van der Meijs (TU Delft), Daniel Ioan, and Gabriela Ciuprina (PU Bucharest) for many helpful discussions and for providing some of the simulation examples.

REFERENCES

[1] A. Odabasioglu, M. Celik, and L. T. Pileggi, "PRIMA: passive reduced-order interconnect macromodeling algorithm," *IEEE Trans. Computer-Aided Design*, vol. 17, no. 8, pp. 645–654, August 1998.

[2] P. Feldmann and R. W. Freund, "Efficient linear circuit analysis by Padé approximation via the Lanczos process," *IEEE Transactions on Computer-Aided Design of Integrated Circuits and Systems*, vol. 14, no. 5, pp. 639–649, May 1995.

[3] I. M. Elfadel and D. L. Ling, "A block rational arnoldi algorithm for multipoint passive model-order reduction of multiport rlc networks," in *International Conference on Computer Aided-Design*, San Jose, California, November 1997, pp. 66–71.

[4] L. Daniel, O. C. Siong, S. C. Low, K. H. Lee, and J. K. White, "A multiparameter moment-matching model-reduction approach for generating geometrically parametrized interconnect performance models," *IEEE Trans. Computer-Aided Design*, vol. 23, pp. 678–693, May 2004.

[5] P. Li, F. Liu, X. Li, L. Pileggi, and S. Nassif, "Modeling interconnect variability using efficient parametric model order reduction," in *Proc. Design, Automation and Test in Europe Conference and Exhibition*, February 2005.

[6] X. Li, P. Li, and L. Pileggi, "Parameterized interconnect order reduction with Explicit-and-Implicit multi-Parameter moment matching for Inter/Intra-Die variations," in *International Conference on Computer Aided-Design*, San Jose, CA, November 2005, pp. 806–812.

[7] P. Gunupudi, R. Khazaka, M. Nakhla, T. Smy, and D. Celo, "Passive parameterized time-domain macromodels for high-speed transmission-line networks," *IEEE Trans. On Microwave Theory and Techniques*, vol. 51, no. 12, pp. 2347–2354, December 2003.

[8] J. R. Phillips and L. M. Silveira, "Poor Man's TBR: A simple model reduction scheme," *IEEE Trans. Computer-Aided Design*, vol. 24, no. 1, pp. 43–55, Jan. 2005.

[9] J. Phillips, "Variational interconnect analysis via PMTBR," in *International Conference on Computer Aided-Design*, San Jose, CA, USA, November 2004, pp. 872–879.

[10] A. Vandendorpe and P. V. Dooren, "Model reduction of interconnected systems," in *Proc. of 16th International Symposium on Mathematical Theory of Networks and Systems (MTNS 2004)*, Leuven, Belgium, July 2004, pp. THP3–4.

[11] R. W. Freund, "Sprim: Structure-preserving reduced-order interconnect macro-modeling," in *International Conference on Computer Aided-Design*, San Jose, CA. U.S.A, November 2004, pp. 80–87.

[12] H. Yu, L. He, and S. X. D. Tan, "Block structure preserving model order reduction," in *BMAS - IEEE Behavioral Modeling and Simulation Wokshop*, September 2005, pp. 1–6.

[13] B. Moore, "Principal Component Analysis in Linear Systems: Controllability, Observability, and Model Reduction," *IEEE Transactions on Automatic Control*, vol. AC-26, no. 1, pp. 17–32, February 1981.

Hierarchical Statistical Analysis of Performance Variation for Continuous-time Delta-Sigma Modulators

Hua Tang

Electrical and Computer Engineering Department
University of Minnesota Duluth, Duluth MN 55812
Email: htang@d.umn.edu

Abstract—**Statistical analysis has become increasingly important with increasing process parameter variations in manufacturing. Monte Carlo method has been most popular for statistical analysis, but it is not efficient for complex circuits/systems due to overwhelming computational time. In this paper, we present a general hierarchical method for efficient statistical analysis of performance parameter variations for complex circuits/systems and conduct a case study on a 4th order continuous-time Delta Sigma modulator. At circuit-level, we use response surface modeling method to extract quadratic models of circuit-level performance parameters in terms of process parameter variations. Then, at system-level, we use behavioral models to extract statistical distribution of the overall system performance parameter. The method can achieve a good tradeoff between computational efficiency and accuracy.**

Index Terms—**statistical analysis, process variation, Delta-Sigma modulator, response surface modeling, behavioral modeling**

I. INTRODUCTION AND PREVIOUS WORK

Semiconductor technology has migrated to nanometer design, which has result in increasing process parameter variations since precise process parameter control is not possible [1] [2]. Due to variations of these process parameters introduced in the manufacturing, the performance parameters of circuits/systems also vary. Designers are interested in finding the exact (or well approximated) statistical distribution of the performance parameters to get an idea of the yield and to fully validate the design, thus comes the need of statistical analysis.

One popular approach to statistical analysis has been the Monte Carlo method [3], which iteratively simulates the circuits/systems to collect a large enough number of samples of the performance parameters by random sampling the process parameters. Whereas Monte Carlo method is affordable with simple circuits, such as simple analog amplifiers, it becomes infeasible for more complicated circuits [2]. For example, for some complex analog circuits such as the Delta-Sigma modulator, each simulation for one performance parameter sample takes even days [8]. As a result, directly applying the Monte Carlo method for statistical analysis of performance parameters in such cases, though most accurate, is not realistic from computational time point of view.

[5] performs statistical analysis of filter circuits by considering only the variation of resistors and capacitors while treating OpAmp as ideal blocks. Similar work has been also reported on statistical analysis of Delta-Sigma modulators. [12] [13] [14] simulate ideal behavioral models of the modulators using the Monte Carlo method, in which all signal path coefficients are considered to have independent Gaussian or uniform distribution. While the above methods are generally computationally efficient and can be useful in a top-down design flow for topology selection [14], they may not be acceptable in realistic statistical analysis for final system validation and characterization. One way to perform realistic statistical analysis for complex circuits/systems is to do it hierarchically. Hierarchical statistical analysis has been used in a few examples [4] [6] [7]. [4] considers multiple levels of design and at each design level uses design equations between performance parameters and design parameters and then propagates the variance of process parameters (at the lowest design level) to the performance parameter for the overall system. [6] uses a hierarchical method for statistical analysis of lock time variation of (PLL) Phase Locked Loop. But these methods may work well only in the case of linear equations or models between performance parameter and design parameters.

Recently, a hierarchical method for statistical analysis of discrete time Delta-Sigma modulators is reported in [7]. First, the key building block of the modulator, the integrator, is statistically characterized. In this case, the state output of each integrator is the performance parameter and response surface modeling method is applied to generate a quadratic model for the state output in terms of process parameters. But note that, the above procedure has to be repeated for $M \times N$ times, where M represents the number of input state values between the signal range In the second step, a lookup-table-based simulation approach is used to compute the SNR (Signal-to-Noise Ratio) of the modulator. By varying the process parameters, the table is checked depending on the input state and feedback state value. If at the sampling time instant, the input state and feedback state value is not found in the table, then an linear interpolation scheme is used to generate the output corresponding to that case. At this step, table-based simulation is very efficient and acceptable accuracy between the method and realistic circuit simulation is reported for a simple second order discrete time Delta-Sigma modulator [7].

While the above approach works for discrete-time Delta-Sigma modulators due to its sampled data operation (only the final value at the end of the clock cycle is needed), it

978-1-4244-1709-4/07/$25.00 © 2007 IEEE

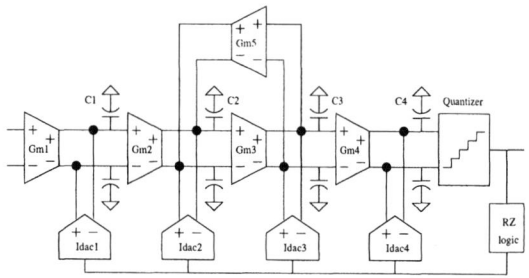

Fig. 1. **Circuit implementation of the $\Delta\Sigma$ modulator**

may not work for continuous-time Delta-Sigma modulators. Thus, the method may not be generalized to other complex circuits/systems [7]. In this paper, we present a general and hierarchical method for statistical analysis of performance parameters for complex circuits/systems. At circuit-level, we extract quadratic models for performance parameters in terms of process parameters using the response surface modeling method. At system-level, a behavioral model, which has its design parameters mapped from circuit-level performance parameters, is used for efficient simulation of the overall system. Then Monte Carlo simulations are performed for statistical analysis of the performance parameter of the overall system. Also, it is not limited to linear models and it randomly samples the process parameters instead circuit-level performance parameters, therefore avoids the computation of mean, variance and co-variance of circuit-level performance parameters [4] [6].

The rest of the paper is organized as follows. Section 2 illustrates the hierarchical statistical analysis method in the context of a fourth order continuous-time Delta-Sigma modulator with experimental results. Finally, conclusion is given in Section 3.

II. HIERARCHICAL STATISTICAL ANALYSIS

In this Section, we illustrate the hierarchical method for statistical analysis of performance parameters in the context of a 4th order continuous-time Delta-Sigma modulator. Our goal is to find the statistical distribution of SNR for the modulator with full transistor-level design. The targeted is shown in Figure 1 using OTA-C based integrator. The modulator was designed respecting the specification for WCDMA standard in wireless communication, that is at least 70dB of SNR in a signal bandwidth of 1.92MHz [9]. The OSR (Oversampling Ratio) was selected to be 40 considering some safety margin in the case of circuit-level non-idealities (clock period T_s is thus $6.51ns$). The NTF (Noise Transfer Function) was designed from [8]. We designed the overall modulator circuit in Cadence Analog Design Environment [15] with IBM CMRF7SF $0.18\mu m$ technology [16]. One full transistor-level simulation for 9,192 sampled outputs took 20 hours (first 1,000 transient samples not used to compute SNR). It gave a nominal SNR of 72.4dB for the modulator. Note that for simplicity, we designed the modulator with single-bit quantizer.

Simulating such a modulator in the flat transistor-level is very time-consuming, therefore there have been extensive

efforts to construct behavioral models for fast simulation. Recently, extensive work has been attempted on system-level behavioral modeling and simulation of both DT (Discrete Time) and CT (Continuous Time) $\Delta\Sigma$ modulators considering circuit-level non-idealities. A design tool implemented in MATLAB/SIMULINK for DT $\Delta\Sigma$ modulator is reported in [10]. Amaya *et al* proposes to use MATLAB/SIMULINK to simulate both DT and CT $\Delta\Sigma$ modulators [11]. These behavioral models are very efficient in terms of computational time and are reasonably accurate. While they are used to determine the building block design specifications in a top-down hierarchical design flow [20], we use it for statistical analysis of SNR variation in a bottom-up hierarchical method. For example, the behavioral model for the fourth order continuous-time Delta-Sigma modulator in Figure 1 is shown in Figure 2 (implemented in MATLAB/SIMULINK [21]). The considered circuit-level non-idealities are clock jitter at the quantizer, OTA non-linearity, OTA noise, DAC (Digital-to-Analog) feedback variation and loop delay. For detailed modeling techniques, we refer the reader to [10] [11].

A. Overview of the hierarchical method for statistical analysis

The design parameters of the behavioral model in Figure 2 are signal path coefficients and non-ideality specifications. They can all be accurately extracted from circuit-level simulations. This is straightforward for non-ideality specifications in the model. For signal path coefficients, the following equation holds for modulator design comparing Figure 2 to Figure 1, $\frac{Gm_1}{sC_1} = \frac{b_1}{sT_s}$, where Gm_1 is the transconductance of the first OTA, T_s the clock period and s the Laplace variable. Then we obtain $b_1 = \frac{Gm_1 T_s}{C_1}$. So, for each case of process parameter variation, we compute Gm_1 and C_1 and then map them to b_1. Similarly for t_{12}, t_{23}, t_{32} and t_{34}. Note that t_{45} does not affect the modulator at all and is not implemented in real circuit design. For coefficient a_1, a_2, a_3 and a_4, Gm_1 in the above equation is replaced by $Idac_i$ $(i = 1, ..., 4)$, where $Idac_i$ denotes the DAC feedback current at the corresponding stages.

Now, we can plan the statistical analysis of SNR variation for the overall modulator as follows. At circuit-level, there are a total of 11 building blocks, 5 OTA circuits, 4 feedback DAC circuits, 1 latched comparator, and 1 RZ (Return-to-Zero) logic circuit (refer to Figure 1). We need to statistically characterize the performance parameters that appear in the behavioral model it terms of process parameters. We apply response surface modeling method so that each performance parameter of a circuit is characterized by a quadratic equation in terms of process parameters. Then, we plug these equations in the system-level behavioral model. Subsequently, we randomly vary the process parameters, generate the performance parameters using the quadratic equations, map them to the design parameters in the behavioral model and then run the model for SNR evaluation. At system-level, since behavioral model simulation is very fast, it is affordable to use Monte Carlo analysis to evaluate the statistical distribution of SNR of the modulator. We discuss quadratic model extraction of each circuit-level performance parameter in the next subsection.

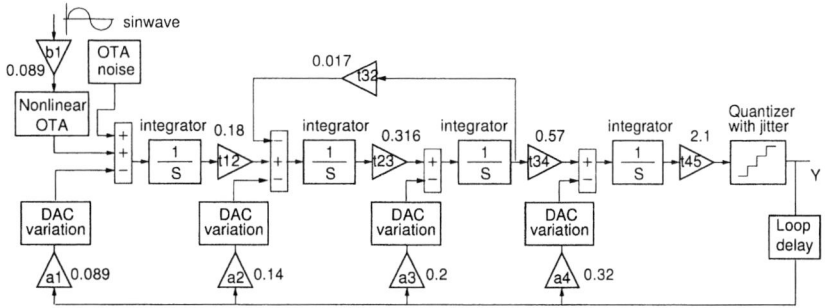

Fig. 2. **4th order CT $\Delta\Sigma$ modulator with major circuit-level non-idealities**

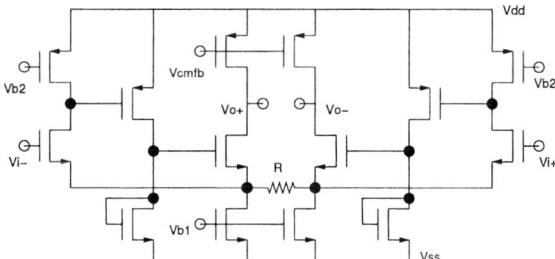

Fig. 3. **Transistor-level schematic of the OTA circuit**

B. Circuit-level performance parameter extraction

For the first building block OTA in the modulator, its circuit schematic is shown in Figure 3. Since the linearity of the OTA is very important, the OTA uses a resistor for source degeneration [17]. Thus, we would expect both the subset of dominant process parameters related to the transistor (NMOS and PMOS) and the resistor should be considered. The dominant process parameters for transistors include the following seven parameters, the width variation (ΔW), length variation (ΔL), gate oxide thickness (T_{ox}), NMOS threshold voltage (V_{thp}), PMOS threshold voltage (V_{thn}), NMOS base mobility (μ_n) and PMOS base mobility (μ_p). Since the global variation by the above seven process parameters dominates the local variation (or the mismatch) between transistors, local variation was neglected as in [7] [19]. Suppose that we use poly resistor, the main dominant process parameter is d_r, the resistance per square area. Note that all the process parameters considered are assumed to be independent [16]. With the defined process parameters, we can then extract quadratic models of performance parameters in terms of process parameters [2]. We apply the DoE and response surface modeling method for the purpose [18]. CCD (Central Composite Design) is used to collect the data [18]. The CCD used here requires a resolution V 2^{8-2} fractional factorial DoE, which includes 64 runs at different combinations of the process parameters, and 16 axial runs and a few center runs (we used 4 center runs in the paper). So the total number of simulation runs amounts to 84. The '± 1' level in the fractional factorial DoE was defined as $\pm\sigma$ of the process parameters in order to reduce bias error [18]. Also, for the 16 axial runs, the process parameters took levels

of '± 2.828' ($\pm 2.828\sigma$) to give approximate rotatability for the CCD (rotatability gives equal variance for design points on the same sphere) [18]. Note that the performance parameters we consider for the OTA are THD (Total Harmonic Distortion) in order to characterize the OTA non-linearity, transconductance and RMS (Root Mean Square) noise that need to be mapped to the behavioral model in Figure 2.

The overall simulation time was about 40 minutes to collect the data for the OTA in Cadence. Among the three performance parameters, the THD seemed to be most sensitive to process variations. At nominal, the THD was 75dB, and it was in the range of (54,78)dB among the data. When linear regression was applied to generate quadratic models for THD, it was found that the accuracy between the simulated data and predicted data was not satisfying. To improve the accuracy, a *log* transformation was applied to it and the resulting quadratic model was much better. The coefficient of determination R^2 [18] was improved from 0.79 to 0.93, and the correlation of the simulated data and predicted data was improved to 0.96 compared to 0.78 before. Figure 4 shows a plot of simulated data versus predicted data. Similarly, the transconductance and RMS noise were extracted and the linear regression worked quite well in these two performance parameters without the need of transformation. The correlation for simulated data and predicted data was almost 1. Thus, the response surface modeling together with the DoE was efficient and accurate for modeling purposes.

The above extraction process discussed for OTA needs to be conducted on other circuit blocks. The difference is mainly on the subset of dominant process parameters to be considered for the specific performance parameter. For the other four OTA circuits, the process parameters are exactly the same, since the same circuit topology with source degeneration shown in Figure 3 is re-used. For the latched comparator, RZ logic and DAC feedback circuit, we considered them together to extract the loop delay and DAC current as a function of process parameters. There are four DAC feedback circuits, therefore each DAC feedback circuit is combined with latched comparator and RZ logic to extract the DAC current of the corresponding stage. In this case, we used the same set of dominant process parameters as the OTA except that the process parameter for the resistor was thrown away since it would not have any effect here. A resolution VII of (2^{7-1})

2007 IFIP International Conference on Very Large Scale Integration (VLSI-SoC 2007) 39

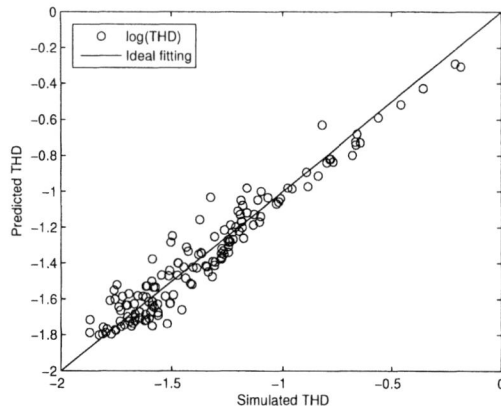

Fig. 4. **Comparison of simulated data and predicated data**

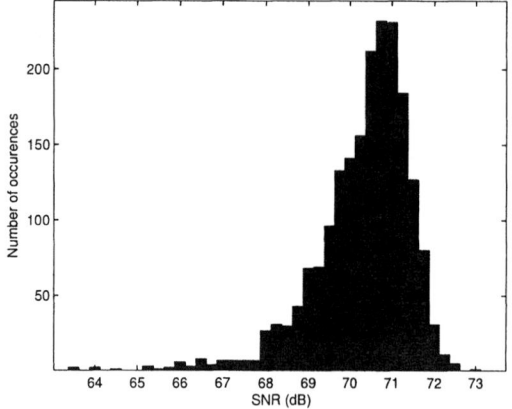

Fig. 5. **Statistical distribution of SNR for the overall modulator**

runs was used plus another 14 axial runs and 4 center runs to model quadratic equations [18]. The overall simulation time for all four DAC circuits were 100 minutes.

In this Section, we have extracted quadratic models of performance parameters in terms of process parameters for all building circuit blocks (more accurately, for all non-ideality specifications and design parameters in the behavioral model in Figure 2). The total simulation time to collect the data points for all the building blocks was about 5 hours.

C. Monte Carlo simulation of the behavioral model

In this Section, we use Monte Carlo method to evaluate the statistical distribution of SNR for the overall modulator. Since we had obtained quadratic models for all circuit-level performance parameters in last Section, they could now be properly mapped to design parameters in the behavioral model, the signal path coefficients and the non-idealities shown in Figure 2. By randomly varying the process parameters, we compute the performance parameters using quadratic models, map them to design parameters in the behavioral model and

run the model to evaluate SNR in a Monte Carlo method. Note that here we directly vary the process parameters. This is quite convenient since the process parameter are all assumed to be independent following Gaussian distribution (normalized to zero mean and unity variance).

Note that the Monte Carlo method is feasible here because the behavioral model simulation is very fast. Each simulation is in the order of seconds. Using the above presented approach, the SNR distribution is shown in Figure 5 with 2,000 samples, which were obtained in 6 hours in MATLAB/SIMULINK. Note that there are 30 negative SNR samples that are not shown in the Figure, which represents instable modulators. From the Figure, it can be seen that the modulator could potentially lose more than 2 bits of resolution (6dB) under process variation. we compute the performance yield by counting the percentage of instances that still satisfy the required performance specification (SNR larger than 70dB). In this case, the yield is 67%. We compared the simulation results to those obtained in full transistor-level simulation. We randomly select five samples of process parameters, run full transistor-level simulation to obtain five SNR samples and then compare them to those obtained from the hierarchical method. The simulation results from both methods matched up reasonably well, with maximum error of less than 3.0dB.

Finally, we compare the proposed hierarchical method for statistical analysis to other methods in literature. Compared to the method in [7], which is limited to sampled data systems, our method is general to many types of systems since the proposed method is to characterize the circuit-level performance parameters and these performance parameters become design parameters in a higher level behavioral model. Also, the computational efficiency is comparable for both methods. The behavioral models used in the proposed method is the same as those used in the top-down design flow for Delta-Sigma modulators [11]. On the other hand, it is possible that the method in [7] is more accurate for specifically sampled data systems. Compared to other hierarchical methods [4] [6], the proposed method is free of mean, variance and correlation computation and is not limited to linear models.

III. CONCLUSION

This paper presents a general and hierarchical method for statistical analysis of performance parameters for complex mixed-signal circuits/systems. We first model the circuit-level performance parameters in terms of the dominant set of process parameters with quadratic equations using DoE and the response surface modeling method. Then at system-level, we plug the quadratic equations in the behavioral model so that performance parameters are mapped to design parameters and simulation is performed on the behavioral model to evaluate the overall system performance parameter. Due to efficiency of behavioral simulation, simple and accurate Monte Carlo method is applied to evaluate a large enough number of samples of system performance parameter. We illustrate the method with an example of a fourth order continuous-time Delta-Sigma modulator. Good accuracy is observed when comparing the results of the proposed method to those obtained in real transistor-level simulations.

REFERENCES

[1] S. Nassif, "Modeling and Analysis of Manufacturing Variations", *Proc. of IEEE Custom Integrated Circuits Conf.*, 2001, pp. 223-228.

[2] X. Li, J. Le, P. Gopalakrishnan, L. Pileggi, "Asymptotic probability extraction for non-Normal distributions of circuit performance", *IEEE Trans. on Computer Aided Design of Integrated Circuits and Systems*, Vol. 1, No, 1, Jan 2007, pp. 21-42.

[3] R. Y. Rubinstein, "simulation and the Monte Carlo Method", *Wiley-Interscience*, 1991.

[4] F. Liu, J. Flomenberg, D. Yasaratne, S. Ozev, "Hierarchical Variance Analysis for Analog Circuits based on Graph Modeling and Correlation Loop Tracing", *Proc. of Design, Auto. and Test Europe Conf.*, 2005.

[5] A. Graupner, W. Schwarz, R. Schuffny, "Statistical Analysis of Analog Structures Through Variance Calculation", *IEEE Trans. Circuits and Systems I*, Vol 49, No. 8, Aug 2002, pp. 1071-1078.

[6] E. Felt, S. Zanella, C. Guardiani, A. Sangiovanni-Vincentelli, "Hierarchical Statistical Characterization of Mixed-Signal Circuits Using Behavior Modeling", *Proc. of Inter. Conf. Computer Aided Design*, 1996.

[7] G. Yu and P. Li, "Lookup table based simulation and statistical modeling of Sigma-Delta ADCs", *Proc. of IEEE/ACM Design Automation Conference*, pp.1035-1040, Jul 2006.

[8] S. Norsworthy, R. Schreier, G. Temes, "Delta-Sigma Data Converters: Theory, Design, and Simulation", *IEEE Press*, 1996.

[9] Robert H.M. Van Veldhoven, "A Triple-Mode Continuous-Time Delta-Sigma Modulator with Switched-Capacitor Feedback DAC for a GSM-EDGE/CDMA2000/UMTS Receiver", *IEEE Journal of Solid-State Circuits*, Vol. 38, No. 12, Dec 2003, pp. 2069-2076.

[10] P. Malcovati, S. Brigati, F. Francesconi, F. Maloberti, P. Cusinato, A. Baschirotto, "Modeling Sigma-Delta Modulator Non-idealities in Simulink", *IEEE Trans. Circuit and Systems I*, Vol. 50, No. 3, Mar 2003, pp. 352-364.

[11] J. Ruiz-Amaya, J.M. de la Roas, F. Medeiro, F.V. Fernandez, R. del Rio, D. Perez-Verdu, A. Rodriguez-Vazquez, "High-level synthesis of switched-capacitor, switched-current and continuous-time Sigma-Delta modulators using SIMULINK-based time-domain behavioral models", *IEEE Trans. on Circuits and System I*, Vol. 52, No. 9, Sep 2005, pp. 1795-1810.

[12] O. Bajdechi, G. Gielen, J. Huijsing, "Systematic Design Exploration of Delta-Sigma ADCs", *IEEE Trans. Circuits and Systems I*, Vol. 51, No. 1, Jan 2004, pp. 86-95.

[13] H. Wang, T. Kuo, "An automatic coefficient design methodology for high-order bandpass sigma-delta modulator with single-stage structure", *IEEE Trans. on Circuits and Systems II*, Vol. 53, No. 7, Jul 2006, pp. 580-584.

[14] H. Tang, A. Doboli, "High-level Topology Synthesis for Delta-Sigma Modulators Optimized for Complexity, Sensitivity and Power Consumption", *IEEE Trans. on CAD of Integrated Circuit and System*, Vol. 25, No. 3, pp. 597-607, Mar 2006.

[15] "Analog Design Environment User Guide", *Cadence Inc.*, 2005.

[16] "IBM CMRF7SF Model User Guide", *IBM Corporation*, Nov 2004.

[17] A. Leuciuc, Y. Zhang, "A Highly Linear Low Voltage MOS Transconductor", *Proc. Inter. Symposium on Circuits and Systems*, 2002.

[18] R. H. Meyers, D. C. Montgomery, "Response Surface Methodology: Process and Product Optimization using Designed Experiments", *2nd Edition, Wiley*, 2002.

[19] B. P. Harish, N. Bhat, M. B. Patil, "On a Generalized Framework for Modeling the Effects of Process Variations on Circuit Delay Performance Using Response Surface Methodology", *IEEE Trans. on Computer Aided Design of Integrated Circuits and Systems*, Vol. 26, No. 3, Mar 2007, pp. 606-614.

[20] G. Gielen, R. Rutenbar, "Computer Aided Design of Analog and Mixed-signal Integrated Circuits", *Proc. of IEEE*, Vol. 88, No. 12, Dec 2000, pp. 1825-1852.

[21] "Using MATLAB Version 7", *The MathWorks Inc*, 2002.

FIRST ORDER QUASI-STATIC SOI MOSFET CHANNEL CAPACITANCE MODEL

Sameer Sharma (*sameer.sharma@okstate.edu*) and L. G. Johnson (*lgjohn@okstate.edu*)

Oklahoma State University, Stillwater, OK 74075

Abstract - Conventional MOS models for circuit simulation assume that the channel capacitances do not contribute to net power dissipation. Numerical integration of channel currents and instantaneous terminal voltages however shows the existence of higher order dissipating terms. To overcome these limitations, we present a self-consistent, first order quasi-static charge model that is able to predict dissipative (transport) and conserved (charging) current components. Charge conservation is insured by using the current continuity equation. An analytical expression for energy stored in the channel is derived by separating out current components that contribute to net power dissipation. The power dissipation estimation is made computationally efficient by leaving out energy conserving terms.

I. INTRODUCTION

The accurate modeling of the MOSFET channel capacitance has been an ongoing effort for many decades. First, Meyer's [1] reciprocal capacitive model, then Ward's [2] charge-based non-reciprocal capacitance model have been used. Many papers have been written on the comparison of these models. Some [3-5] claim that Meyer's model fails due to charge non-conservation which justifies the usage of charge-base models, while others claim [6] that the charge non-conservation is mainly due to the incorrect mathematical modeling of non-linear capacitance by the simulation software. Ward's model artificially partitions the channel charge into source and drain components. As pointed out by Fossum [7], it is not clear whether we have explored all other possibilities; we may be able to achieve a better result with a different channel partition or may be with no partition at all. Many ideas have also been suggested for estimation of energy and power taking into consideration the input slew dependency [8], propagation delay [9], short circuit power [10] and power supply current measurements [11-13].

Many models have also been put forward to analyze the charging and the trans-capacitive current components. One of such models by Lim-Fossum [14] has a first order transient transport current and suggests the difference between non-reciprocal capacitive elements to be responsible for this current. We show that this model is correct for transistor current computation; however it is inconsistent and has some drawbacks when used to predict power consumption. These drawbacks are:

- Lim-Fossum's equations use Ward's channel charge partition model.
- The MOS capacitors dissipate power and the trans-capacitive term used in the charge model includes both dissipating as well as conserved components which are not separated.

The charge partition model puts a constraint only on charge conservation. Even though the model predicts the channel charge correctly to first order, the device power is only predicted to zero order. The model may not include complete first order trans-capacitive currents due to the redistribution of the charges in the channel. This could cause the actual output waveform and delay to deviate from the simulation results [15]. In reality, the MOS channel is not purely an energy storage device [16]. Thus, it is not appropriate to leave out higher order dissipating terms due to charge redistribution as they become significant at higher frequencies.

Though many papers [3-5] and books [18-20] have been written on the transient transport current, no one has found a solution to separate the transport and charging current components. This makes our model and the closed form expressions for the dissipative and conserved currents significant. We have developed a self consistent, first order, quasi-static charge distribution model. It analyzes the first order power dissipation and computes the energy function for the conserved component of the charge storage. The existence of the energy function makes it possible to exclude energy conserving terms that do not contribute to power dissipation, making the total power estimation computationally efficient.

The rest of the paper is organized as follows. In the second section, we have used a one dimensional MOSFET transistor model with the current continuity equation to compute the channel currents and the channel charges as well as the currents at the source and the drain ends. In the third section, we have computed the static and the first order power by integrating the power density over the entire channel. This leads to the derivation of closed-form analytical expressions for the conserved and dissipative current components from the first order drain and source currents in the forth section. Using the conserved components, we have proved the existence of an energy function. Finally, we have developed an equivalent circuit by following the method used by Lim-Fossum and verified the results for current and charge. Even though they used a charge partition instead of solving exactly as we have, both models predict the same source and drain currents, and hence the same terminal capacitances. However, we are able to separate out these capacitances into conserved and dissipative components.

II. CHARGE DISTRIBUTION CALCULATION

In order to obtain an analytical solution, the current flow is considered in one dimension parallel to the surface of the device. It is assumed that the region under the channel is completely depleted of mobile charges. This fully depleted assumption for SOI MOSFET's helps us to make use of a linear relationship between the body and the surface potential to compute the stored energy function without partitioning the channel charge. The linear relation also provides a simplified charge model and terminal currents. It should be noted that solving these equations involves complicated algebraic calculations that are practically impossible without modern mathematics tools like "Mathematica" [21].

Fig 1: SOI MOSFET Structure

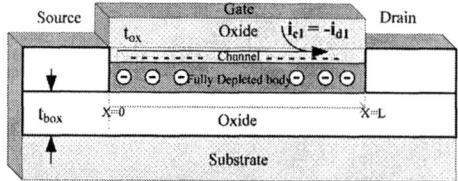

Fig. 1 shows nMOS SOI transistor. The charge per unit length (q_c) at a position x along the channel is given by

$$q_c(x) = -c_{ox}(v_{gb} - v_{fb} - v_{cb}(x) - \phi + q_b(x)/c_{ox}) \qquad (1)$$

Similarly, the body charge (back gate) per unit length (q_b) at x can be written as

$$q_b(x) = -c_{ox}(k_1 + \alpha v_{cb}(x)) \tag{2}$$

where v_{fb}, v_{gb} and v_{cb} are the flat band, gate and channel voltages with respect to the body. k_1 and α are body effect coefficients. $c_{ox} = W(c_{ox}/A)$ is the oxide capacitance per unit length, where W is the channel width. Charge conservation is insured by defining the gate charge per unit length q_g as

$$q_g = -(q_b + q_c) \tag{3}$$

It will be convenient to define the channel charge per unit length at the source $(x = 0) \, q_s$ and the drain $(x = L) \, q_d$ and their time derivatives as

$$q_s = -c_{ox} v_{gst} \tag{4}$$

where $v_{gst} = v_{gb} - v_t - v_{sb}$

$$\frac{d}{dt} q_s = -c_{ox} \frac{d}{dt} v_{gst} \tag{5}$$

In equation (4), v_t is the threshold voltage. The body effect requires including the dependence of the threshold voltage on source terminal voltage and the substrate charge parameter [22].

$$v_t(v_{sb}) = v_{t0} + \alpha v_{sb} \tag{6}$$

where $v_{t0} = v_{fb} + k_1 + \phi$ is the threshold voltage at zero v_{sb}, and ϕ is the fermi potential. At the drain end,

$$q_d = -c_{ox} v_{gdt} \tag{7}$$

where $v_{gdt} = v_{gb} - v_t - v_{sb} - (1 + \alpha)(v_{db} - v_{sb})$

$$\frac{d}{dt} q_d = -c_{ox} \frac{d}{dt} v_{gdt} \tag{8}$$

It is assumed that velocity saturation effects can be neglected. Assuming strong inversion, diffusion current in the channel is small. Drift current at a distance x along the channel can be written as

$$i_c(x,t) = q_c(x,t) \mu \frac{d}{dx} v_{cb}(x) \tag{9}$$

where μ is the charge carrier mobility in the channel. Charge conservation is assured using the one dimensional continuity equation

$$\frac{d}{dx} i_c(x,t) = -\frac{d}{dt} q_c(x,t) \tag{10}$$

Using (9) in (10) gives

$$\frac{d}{dx} [q_c(x,t) \mu \frac{d}{dx} v_{cb}(x)] = -\frac{d}{dt} q_c(x,t) \tag{11}$$

Taking charge per unit length as a linear function of potential along the channel as in equation (1) and (2) gives

$$\frac{d}{dx} q_c(x,t) = (1 + \alpha) c_{ox} \frac{d}{dx} v_{cb}(x) \tag{12}$$

Substituting $\frac{d}{dx} v_{cb}(x)$ in (11) and rearranging terms gives

$$\frac{d}{dx} [q_c(x,t) \frac{d}{dx} q_c(x,t)] = -\frac{(1 + \alpha)}{\mu} c_{ox} \frac{d}{dt} q_c(x,t) \tag{13}$$

In the quasi-static approximation, equation (13) can be solved iteratively to compute the current and the charge in the channel by expanding $q_c = q_{c0} + q_{c1} + ...$ where q_{c0} is a function of terminal voltages only and q_{c1} is a linear function of first order time derivatives of terminal voltages. In terms of the steady state (zero

order) charge per unit length at any position x along the channel, equation (13) reduces to

$$\frac{d}{dx} (q_{c0} \frac{d}{dx} q_{c0}) = 0 \tag{14}$$

Performing integration from the source $(x = 0)$ to the drain $(x = L)$, the zero order charge along the channel becomes

$$q_{c0} = -\sqrt{(q_s^2(1 - x/L) + q_d^2 x/L)} \tag{15}$$

and the steady state drift current component of equation (9) simplifies to

$$I_{c0} = \frac{\mu}{(1 + \alpha) c_{ox}} q_{c0} \frac{d}{dx} q_{c0} \tag{16}$$

Equation (16) gives the usual equation for steady current neglecting velocity saturation, which is shown in Table I. The first order current and charge can be found by keeping terms of first order in time derivatives in equation (13)

$$\frac{d}{dx} (q_{c0} \frac{d}{dx} q_{c1} + q_{c1} \frac{d}{dx} q_{c0}) = -\frac{(1 + \alpha)}{\mu} c_{ox} \frac{d}{dt} q_{c0} \tag{17}$$

This equation can be solved for first order channel charge per unit length as

$$q_{c1} = -\frac{(1 + \alpha)}{\mu} c_{ox} \frac{1}{q_{c0}} \frac{d}{dt} \int (\int q_{c0}[x] dx) dx) \tag{18}$$

Values of q_{c0} and q_{c1} are substituted in equation (11) to compute the first order current along the channel

$$i_{c1} = \frac{\mu}{(1 + \alpha) c_{ox}} (q_{c0} \frac{d}{dx} q_{c1} + q_{c1} \frac{d}{dx} q_{c0}) \tag{19}$$

Finally, equation (19) can be solved to compute the first order channel current at the source $i_{s1} = i_{c1}(x = 0)$ and the drain $i_{d1} = -i_{c1}(x = L)$ ends in all regions of operation. We have assumed pinch-off saturation which occurs when $q_d = 0$ for $v_{ds} \geq \frac{(v_{gs} - v_t)}{(1 + \alpha)}$. In cut-off, it is assumed that both $q_d = 0$ and $q_s = 0$. Table I summarizes the charge and current in all regions of operations.

III. CALCULATION OF MOSFET POWER

The static current is usually used to determine power dissipation for MOS transistors. Charge redistribution in the channel causes additional power dissipation. In the quasi-static model, charge redistribution is assumed to happen instantaneously with no propagation delays. However, the channel charge density still changes as an indirect function of time through the dependence on time varying terminal voltages. This allows the use of the quasi-static model to predict the charge redistribution and the associated power dissipation.

The conventional charge model is based on the presumption that the MOSFET capacitors do not contribute any net power dissipation in the channel. But, the channel capacitances are not energy conserving [16]. They do have some higher order power dissipative terms due to the charge redistribution in the channel. These dissipative terms become significant at higher frequencies, which make it essential to include their effects for accurate power dissipation prediction.

The instantaneous total power going into the transistor channel P_c can be estimated by integrating the power per unit length,

$$P_c = \int_0^L \frac{d}{dx}(i_c(x)v_{cb}(x))dx$$
$$= \int_0^L v_{cb}(x)(\frac{d}{dx}i_c(x))\,dx + \int_0^L i_c(x)(\frac{d}{dx}v_{cb}(x))\,dx \quad (20)$$

where the first integral represents change in stored energy and second term represents power dissipation. Keeping non-zero terms to first order in time derivatives, equation (20) can be expanded as:

$$P_c = P_{c0} + P_{c1,diss} + P_{c1,cons} \quad (21)$$

where

$$P_{c0} = \int_0^L I_{c0}(\frac{d}{dx}v_{cb0}(x))\,dx = I_{c0}(v_{db} - v_{sb}) \quad (22)$$

$$P_{c1,diss} = \int_0^L i_{c1}(\frac{d}{dx}v_{cb0}(x))\,dx \quad (23)$$

$$P_{c1,cons} = \int_0^L v_{cb0}(\frac{d}{dx}i_{c1})\,dx \quad (24)$$

The total instantaneous power P into the transistor is the sum of channel power P_c and gate power $P_{g1,cons}$.

$$P = P_c + P_{g1,cons} \quad (25)$$

Where the gate power is

$$P_{g1,cons} = i_{g1}v_{gb} \quad (26)$$

Equation (22) represents the zero order power dissipation. Equation (23) represents the first order power dissipation due to the trans-capacitive transient current components and equation (24) represents the first order conserved power in the channel. Since the gate power estimated using equation (26) is assumed to be purely reactive and leakage free, it becomes necessary to add its contribution together with the conserved components from the channel to obtain a closed form solution for the stored energy function. Table II summarizes the power components. We have used $v_{gbt0} = v_{gb} - v_{t0}$.

IV. FIRST ORDER CURRENT, CAPACITANCE AND ENERGY FUNCTION CALCULATIONS

As seen from Table I, first order current is a function of terminal voltages and their time derivatives, and as mentioned above, the coefficient of dv/dt instead of representing purely storage capacitance, is also responsible for some of the power dissipation in the channel. This suggests that the first order drain and the source currents consist of two separate components, one that contributes to power dissipation in the channel, and another that is responsible for the energy storage. Taking this approach, i_{d1} and i_{s1} can be expanded as

$$i_{d1} = i_{d1,cons} + i_{d1,diss} \quad (27)$$

$$i_{s1} = i_{s1,cons} + i_{s1,diss} \quad (28)$$

The dissipative current component is computed by dividing the power dissipated in the channel (23) by the drain to source potential

$$i_{d1,diss} = \frac{P_{c1,diss}}{v_{ds}} = i_{tt,diss} = -i_{s1,diss} \quad (29)$$

$i_{tt,diss}$ in equation (29) is the trans-capacitive transport current that is responsible for the extra power dissipation in the channel, and is defined as positive going into the drain. The conserved drift component can now be computed by subtracting the dissipated component from the total first order current.

$$i_{d1,cons} = i_{d1} - i_{tt,diss} \quad (30)$$

$$i_{s1,cons} = i_{s1} + i_{tt,diss} \quad (31)$$

Separation of currents into conserved and dissipative terms helps to compute energy conserving capacitances. This is one of the most important findings, as all other capacitive models can not find stored energy because both the conserved and the dissipative terms are mixed together. However, in our model, the capacitances are estimated simply from the conserved components that were calculated using equations (30) and (31).

$$i_{i1,cons} = C_{cii}\partial_t v_{ib} - \sum_{j \neq i,b}(C_{cij}\partial_t v_{jb}) \; ; \; i,j = g,d,s,b. \quad (32)$$

where C_{cii}, C_{cij} are the conserved components of the capacitor. In equation (32) and all the subsequent equations, the subscript notation 'c' or 'd' stands for conserved or dissipative components. In terms of conserved current, total conserved power can be written as

$$P_{cons} = \sum_{i \neq b} v_{ib} i_{i1,cons} \quad (33)$$

In terms of capacitors, using equation (32) in equation (33) gives

$$P_{cons} = \sum_{i \neq b} v_{ib}(C_{cii}\partial_t v_{ib} - \sum_{j \neq i,b} C_{cij}\partial_t v_{jb}) \quad (34)$$

Conserved power can also be defined as the rate of change of energy

$$P_{cons} = \frac{dE}{dt} = \sum_{j \neq b} \frac{\partial E}{\partial v_{jb}} \frac{d}{dt} v_{jb} \quad (35)$$

Comparing equations (34) and (35)

$$\frac{\partial E}{\partial v_{jb}}(v_{gb}, v_{sb}, v_{db}) = \sum_{i \neq b}(C_{cii} v_{ib} - \sum_{j \neq i,b} C_{cij} v_{jb}) \quad (36)$$

For the energy function to exist as an analytic function of the terminal voltages and for equation (36) to have a closed form solution, the second order partial derivatives must be equal [16].

$$\frac{\partial}{\partial v_{ib}}(\frac{\partial E}{\partial v_{jb}}) = \frac{\partial}{\partial v_{jb}}(\frac{\partial E}{\partial v_{ib}}) \; ; \; i,j = g,d,s,b.$$

Table III, IV and V summarize the storage and dissipative current components, conserved and dissipative capacitances and the energy function.

V. EQUIVALENT CIRCUIT

In this section, we develop an equivalent circuit by following the method used by Lim-Fossum [14]. Table IV showed that the capacitances were not reciprocal, which makes the capacitance representation using two terminal reciprocal capacitances impossible if these capacitances are made to represent the total first order drain current. However, equation (30) can be rewritten with reciprocal capacitors as

$$i_{d1,cons} = C_{gd}\partial_t v_{dg} + C_{bd}\partial_t v_{db} + i_{tt,cons} \quad (37)$$

where

$$i_{tt,cons} = (C_{gd} - C_{cdg})\partial_t v_{gb} + (C_{csd} - C_{cds})\partial_t v_{sb} + C_{csd}\partial_t v_{ds} \quad (38)$$

The dissipative component of current from equation (29) can also be written in terms of dissipative capacitances as

$$i_{d1,diss} = C_{ddd} \frac{\partial v_{db}}{\partial t} - C_{ddg} \frac{\partial v_{gb}}{\partial t} - C_{dds} \frac{\partial v_{sb}}{\partial t}$$
$$= i_{tt,diss} = -i_{s1,diss} \tag{39}$$

Fig. 2: Equivalent Circuit of a four terminal SOI MOSFET

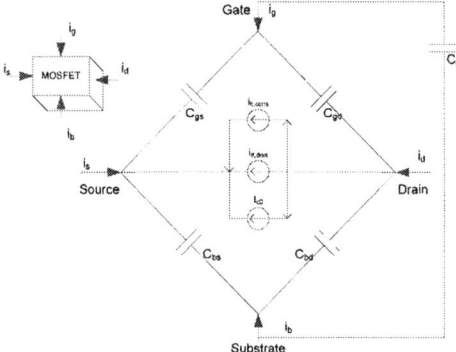

Fig. 2 shows an equivalent circuit of a four terminal MOSFET. The model is equivalent to Lim-Fossum [14], but we have broken the trans-capacitive transport current, i_{tt} into conserved and dissipative components. There are three current components flowing from the drain to the source terminal. The current component responsible for the first order power dissipation in the channel is represented by $i_{tt,diss}$ the conserved current component is represented by $i_{tt,cons}$. I_{c0} represents the steady state zero order current. The two terminal reciprocal capacitances C_{gd}, C_{gs}, C_{bd}, C_{bs} and C_{gb} represents the conserved gate to drain, gate to source, substrate to source, substrate to drain and gate to substrate capacitances respectively. The two terminal capacitances do not conserve energy by themselves; the conserved component of i_{tt} must be included. C_{ddd}, C_{ddg}, C_{dds} in equation (39) represents the dissipative drain to drain, drain to gate and drain to source capacitances respectively.

VI. COMPARISION AND DISCUSSION

Fig 3: Total, Conserved and Dissipative Capacitances vs v_{ds}

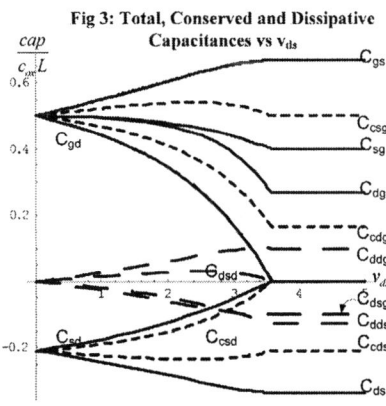

Our model verifies that Ward's method of channel charge partitioning works correctly when the charge has a linear dependence on position (x). Our model also verifies Lim-Fossum's [14] equations

for a fully depleted SOI MOSFET that uses Ward's partition scheme. It predicts the same source and drain currents, and hence the same terminal capacitances (C_{ij}). However, we are able to partition these C_{ij}'s into conserved (C_{cij}) and dissipated (C_{dij}) capacitive components, as shown in Fig. 3.

The partitioning approach to capacitances offers several advantages over conventional trans-capacitances.

- The energy stored in the conserved capacitances can be predicted.
- They can be made to agree with Meyer's capacitances [1] if the body effect and body bias are ignored.

Our other significant contribution has been in the power estimation. Our models have improved the device power measurements by implementing two important concepts:
a) First order terms have to be included for power dissipation estimation as they become significant at higher frequencies.
b) Stored components can be ignored for computationally efficient power dissipation estimation.

The average device power, $\overline{P}$, is then possible by taking dissipative current times voltage and integrating them over time. A simple simulation can be used to show the importance of first order power.

Fig. 4: v_{gb} and v_{db} waveforms

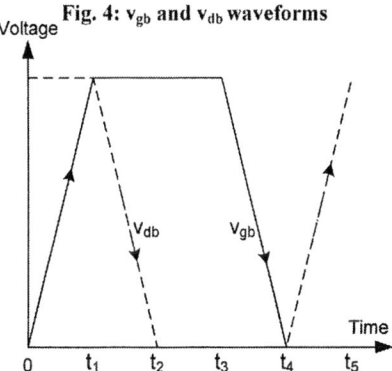

Fig. 4 shows the idealized voltage waveforms for the drain and the gate terminals used for simulation of turning a transistor on then off. The average first order dissipative power from the first transition ($v_{ds}=v_{db}$) when v_{gb} goes from low at t_0 to high at t_1 is computed by

$$\overline{P}(t_0 \to t_1) = \frac{1}{(t_1-t_0)} \int_{t_0}^{t_1} (i_{d1,diss} v_{db} + i_{s1,diss} v_{sb}) dt \tag{40}$$

If we assume the source and the substrate are at the same potential ($v_{sb}=0$), equation (40) can be rewritten as

$$\overline{P}(t_0 \to t_1) = \frac{1}{(t_1-t_0)} \int_{t_0}^{t_1} (i_{d1,diss} v_{db}) dt \tag{41}$$

In the second power dissipating transition, when the gate terminal is high, the drain swings from high at t_1 to low at t_2. The dissipative power equation (40) reduces to

$$\overline{P}(t_1 \to t_2) = \frac{1}{(t_2-t_2)} \int_{t_1}^{t_2} i_{d1,diss} v_{db} dt \tag{42}$$

During the interval t_2 to t_4, there is no power dissipation in the channel ($v_{ds}=0$). The final power transition occurs when the drain waveform swings from low at t_4 to high at t_5. As the gate voltage has already reached a steady low value, the power equation becomes

$$\overline{P}(t_4 \to t_5) = \frac{1}{(t_5 - t_4)} \int_{t_4}^{t_5} i_{d1,diss} v_{db}\, dt \qquad (43)$$

The total dissipative power for a complete cycle is computed taking the sum of all these powers as

$$\overline{P} = \overline{P}(t_0 \to t_1) + \overline{P}(t_1 \to t_2) + \overline{P}(t_4 \to t_5) \qquad (44)$$

For a complete cycle, energy is conserved. This allows us to leave out the conserved component from the power equation for computationally efficient power dissipation predictions [16]. Nonetheless, the total dissipative power includes the first order terms as predicted by equation (44). These first order dissipative components become significant at higher frequencies and modify the total power dissipated in the channel as shown in Fig 5. The total power is no longer constant, and at high frequencies becomes dependent on the switching frequencies.

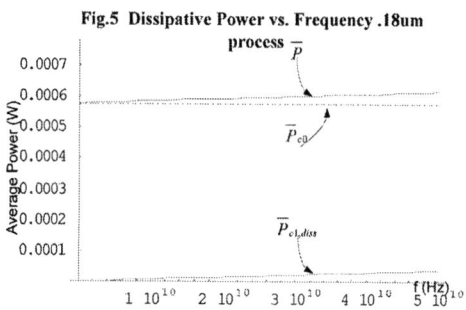

Fig.5 Dissipative Power vs. Frequency .18um process

The result also shows that we need to be extra careful while doing the power measurements. It is not appropriate to look only at the channel dissipation; the first order power dissipation does have contributions from the gate. If the power dissipation is estimated by just considering the total channel power, there would be an extra negative component from the conserved energy. In that case, the channel could act as an energy generator. In reality, that is not the case. Power is pumped from the gate to the channel and when the contribution from the gate is added, the conserved terms cancel out (Fig. 6).

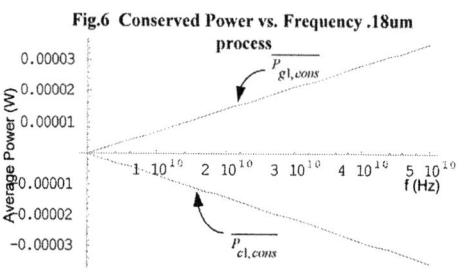

Fig.6 Conserved Power vs. Frequency .18um process

VII. CONCLUSIONS

The development of a self consistent, first order, quasi-static charge distribution model for a fully depleted SOI MOSFET has been described. The Lim-Fossum current and charge model has been verified as correct to first order even though the Ward partition of source and drain charge was used. The transient current is separated into conserved and dissipative components. Significance of higher order power dissipation at higher frequencies is discussed. The existence of energy function also is validated to make the power dissipation estimation computationally efficient.

REFERENCES

[1] J.E. Meyer, "MOS models and circuit simulation," RCA Rev., vol. 32, pp. 42-63, 1971.
[2] D.E.Ward, "Charge-Based Modelling of Capacitance in MOS Transistors", Stanford Electronics Lab., Stanford University, Tech. Rep. F201-11, June 1981.
[3] A.D. Snider, "Charge conservation and the trans-capacitance element: an exposition," IEEE Transaction on Education. Vol.38, no.4, November 1995.
[4] K.A. Sakallah, Yao-Tsung Yen, and S.S.Greenberg, "A first-order charge conserving MOS capacitance model," IEEE Transactions on Computer-Aided Design, vol. 9, pp. 99-108, January 1990.
[5] S. S. Chung, "A charge-based capacitance model of short-channel MOSFET's," IEEE Transactions on Computer-Aided Design, vol. 8, no. 1, January 1989.
[6] M. A. Cirit, "The Meyer model revisited: why is charge not conserved?" IEEE Transactions on Computer- Aided Design, vol. CAD-8, pp. 1033-1037, October 1989.
[7] J.G. Fossum, H.Jeong, and S. Veeraraghavan, "Significance of the channel-charge partition in the transient MOSFET model," IEEE Transactions on Electron Devices, vol. Ed-33, no.10, October 1986.
[8] S. Turgis, D. Auvergne, "A Novel Macromodel for Power Estimation in CMOS Structures", IEEE Transactions on Computer-Aided Design., Vol. 17, No. 11, November 1998.
[9] L. Bidounis, S. Nikolaidis, "Propagation Delay and Short-circuit Power Dissipation Modeling of the CMOS Inverter", IEEE Transactions on Circuits and Systems, Vol. 45, No. 3, March 1998.
[10] K. Nose, T. Sakurai, "Analysis and Future Trend of Short-Circuit Power", IEEE Transactions on Computer Aided Design., Vol 19, NO 9, September, 2000.
[11] K.A. Jenkins, R. L. Franch, "Measurement of VLSI Power Supply Current by Electron-Beam Probing", IEEE Journal of Solid-State Circuit, Vol 27, No 6, pp.948-950, June 1992.
[12] P. Gupta, B. Kahng, "Quantifying Error in Dynamic Power Estimation of CMOS Circuits", Proceedings of Quality Electronic Design, pp. 273-278, March 2003.
[13] F. N. Najm, "A survey of Power Estimation Techniques in VLSI circuits", IEEE Trans. VLSI syst, Vol. 2, pp 664-649, Dec 1994.
[14] H. Lim and J. Fossum, "A charge-based large-signal model for thin-film SOI MOSFET`s", IEEE Journal of Solid-State Circuits, Vol. Sc-20, no.1, February 1985.
[15] W. Lie, M. Chang, "Transistor Transient Studies Including Trans-capacitive Current and Distributive Gate Resistance for Inverter Circuits", IEEE Transactions on Circuits and Systems, Vol. 45, No. 4, April 1998.
[16] S. Sharma, "Quasi-Static Energy and Charge Conserving MOSFET Channel Capacitance Model," PhD. Dissertation, Oklahoma State University, May 2007.
[17] BSIMSOI3.1 MOSFET MODEL, Users' Manual
[18] Yannis Tsividis, "Operation and the Modeling of The MOS Transistors", Oxford University Press, June 2003.
[19] William Liu, "MOSFET models for SPICE simulation, including BSIM3v4 and BSIM4", John Wiley and Sons, Inc., 2001.
[20] Yuhua Cheng and Chenming Hu, "MOSFET modeling and BSIM3 user's guide", Kluwer Academic Publishers, 1999.
[21] Wolfram Reseach, Inc., Mathematica, Version 5.2, Champaign, IL (2005).
[22] H. Lim and J. Fossum, "Threshold Voltage of Thin-Film Silicon-on-Insulator (SOI) MOSFET's", IEEE Transactions on Electron Devices, Vol. Ed-30, No. 10, October 1983.

TABLE I

ZERO AND FIRST ORDER CHARGES AND CURRENTS FOR NMOSFET

	Linear	Saturation	Cut-Off
Conditions	$q_c < 0$ $v_{gdt} > 0$ $v_{gst} > 0$	$q_c < 0$ $v_{gdt} = 0$ $v_{gst} > 0$	$q_c = 0$ $v_{gdt} = 0$ $v_{gst} = 0$
q_s	$-c_{ox}v_{gst}$	$-c_{ox}v_{gst}$	0
q_d	$-c_{ox}v_{gdt}$	0	0
I_{c0}	$\dfrac{\mu c_{ox}}{2L(1+\alpha)}(v_{gst}^2 - v_{gdt}^2)$	$\dfrac{\mu c_{ox}}{2L(1+\alpha)}v_{gst}^2$	0
i_{s1}	$\dfrac{2c_{ox}L}{15(v_{gdt}+v_{gst})^3}[2v_{gdt}(\tfrac{d}{dt}v_{gdt})(v_{gdt}^2+3v_{gdt}v_{gst}+v_{gst}^2)$ $+v_{gst}(\tfrac{d}{dt}v_{gst})(8v_{gdt}^2+9v_{gdt}v_{gst}+3v_{gst}^2)]$	$\dfrac{2}{5}c_{ox}L\dfrac{d}{dt}v_{gst}$	0
i_{d1}	$\dfrac{2c_{ox}L}{15(v_{gst}+v_{gdt})^3}[v_{gdt}(\tfrac{d}{dt}v_{gdt})(3v_{gdt}^2+9v_{gdt}v_{gst}+8v_{gst}^2)$ $+2v_{gst}(\tfrac{d}{dt}v_{gst})(v_{gdt}^2+3v_{gdt}v_{gst}+v_{gst}^2)]$	$\dfrac{4}{15}c_{ox}L\dfrac{d}{dt}v_{gst}$	0

TABLE II

POWER EQUATIONS

	Linear	Saturation	Cut-Off
P_{c0}	$\dfrac{\mu c_{ox}}{2L(1+\alpha)}v_{ds}(v_{gst}^2 - v_{gdt}^2)$	$\dfrac{\mu c_{ox}v_{ds}v_{gst}^2}{2L(1+\alpha)}$	0
$P_{c1,diss}$	$c_{ox}L\dfrac{v_{ds}(v_{ge}-v_{ce})}{30(v_{ge}+v_{gst})^3}[3v_{gdt}^2\tfrac{d}{dt}v_{gdt}+3v_{gst}^2\tfrac{d}{dt}v_{gst}$ $+7v_{gdt}(\tfrac{d}{dt}v_{gdt}+\tfrac{d}{dt}v_{gst})v_{gst}]$	$\dfrac{c_{ox}L}{10}v_{ds}$ $\cdot(\tfrac{d}{dt}v_{gst})$	0
$P_{c1,cons}$	$\dfrac{-c_{ox}L}{6(1+\alpha)}\{-3(v_{gst}\tfrac{d}{dt}v_{gst}+v_{gdt}\tfrac{d}{dt}v_{gdt})+$ $\dfrac{4(v_{gdt}\tfrac{d}{dt}v_{gdt}(v_{gdt}+2v_{gst})+v_{gst}\tfrac{d}{dt}v_{gst}(2v_{gdt}+v_{gst})v_{gbt0})}{(v_{gdt}+v_{gst})^2}\}$	$-\dfrac{c_{ox}L}{6(1+\alpha)}\tfrac{d}{dt}v_{gst}\cdot$ $(4v_{gbt0}-3v_{gst})$	0

TABLE III

STORED AND DISSIPATED CURRENT COMPONENTS

	Linear	Saturation	Cut-Off
$i_{d,diss} =$ $-i_{s1,diss}$	$\dfrac{c_{ox}L(v_{gst}-v_{gdt})}{30(v_{gdt}+v_{gst})^3}[3(v_{gdt}^2\tfrac{d}{dt}v_{gdt}+v_{gst}^2\tfrac{d}{dt}v_{gst})$ $+7v_{gdt}v_{gst}(\tfrac{d}{dt}v_{gdt}+\tfrac{d}{dt}v_{gst})]$	$\dfrac{c_{ox}L}{10}\dfrac{d}{dt}v_{gst}$	0
$i_{d1,cons}$	$\dfrac{c_{ox}L}{6(v_{gdt}+v_{gst})^2}[v_{gdt}\tfrac{d}{dt}v_{gdt}(3v_{gdt}+5v_{gst})$ $+v_{gst}\tfrac{d}{dt}v_{gst}(3v_{gdt}+v_{gst})]$	$\dfrac{c_{ox}L}{6}\dfrac{d}{dt}v_{gst}$	0
$i_{s1,cons}$	$\dfrac{c_{ox}L}{6(v_{gdt}+v_{gst})^2}[v_{gdt}\tfrac{d}{dt}v_{gdt}(v_{gdt}+3v_{gst})+$ $v_{gst}\tfrac{d}{dt}v_{gst}(5v_{gdt}+3v_{gst})]$	$\dfrac{c_{ox}L}{2}\dfrac{d}{dt}v_{gst}$	0

TABLE IV

DISSIPATIVE CAPACITANCES

	Linear	Saturation	Cut-Off
C_{ddg}	$\dfrac{c_{ox}L(v_{gst}-v_{gdt})(3v_{gdt}^2+14v_{gdt}v_{gst}+3v_{gst}^2)}{30(v_{gdt}+v_{gst})^3}$	$\dfrac{1}{10}c_{ox}L$	0
C_{dds}	$-\dfrac{(1+\alpha)c_{ox}L(v_{gst}-v_{gdt})v_{gst}(7v_{gdt}+3v_{gst})}{30(v_{gdt}+v_{gst})^3}$	$-\dfrac{1+\alpha}{10}c_{ox}L$	0
C_{ddb}	$\dfrac{c_{ox}\alpha L(v_{gst}-v_{gdt})(3v_{gdt}^2+14v_{gdt}v_{gst}+3v_{gst}^2)}{30(v_{gdt}+v_{gst})^3}$	$-\dfrac{\alpha}{10}c_{ox}L$	0
C_{dsg}	$-C_{ddg}$	$-C_{ddg}$	0
C_{dsd}	$\dfrac{(1+\alpha)c_{ox}L(v_{gdt}-v_{gst})v_{gdt}(3v_{gdt}+7v_{gst})}{30(v_{gdt}+v_{gst})^3}$	0	0
C_{dsb}	$-C_{ddb}$	$-C_{ddb}$	0

TABLE V

CONSERVED CAPACITANCES

	Linear	Saturation	Cut-Off
C_{gb}	$\dfrac{\alpha}{3(1+\alpha)}c_{ox}L\dfrac{(v_{gdt}-v_{gst})^2}{(v_{gdt}+v_{gst})^2}$	$\dfrac{\alpha}{3(1+\alpha)}c_{ox}L$	$\dfrac{\alpha c_{ox}L}{(1+\alpha)}$
C_{gd}	$\dfrac{2}{3}c_{ox}Lv_{gdt}\dfrac{(v_{gdt}+2v_{gst})}{(v_{gdt}+v_{gst})^2}$	0	0
C_{gs}	$\dfrac{2}{3}c_{ox}Lv_{gst}\dfrac{(2v_{gdt}+v_{gst})}{(v_{gdt}+v_{gst})^2}$	$\dfrac{2}{3}c_{ox}L$	0
C_{csg}	$\dfrac{c_{ox}L}{6}\dfrac{v_{gdt}^2+8v_{gdt}v_{gst}+3v_{gst}^2}{(v_{gdt}+v_{gst})^2}$	$\dfrac{1}{2}c_{ox}L$	0
C_{csb}	αC_{csg}	αC_{csg}	0
C_{csd}	$-\dfrac{(1+\alpha)}{6}c_{ox}L\dfrac{v_{gdt}(v_{gdt}+3v_{gst})}{(v_{gdt}+v_{gst})^2}$	0	0
C_{cdg}	$\dfrac{c_{ox}L}{6}\dfrac{3v_{gdt}^2+8v_{gdt}v_{gst}+v_{gst}^2}{(v_{gdt}+v_{gst})^2}$	$\dfrac{1}{6}c_{ox}L$	0
C_{cdb}	αC_{cdg}	αC_{cdg}	0
C_{cds}	$-\dfrac{(1+\alpha)}{6}c_{ox}L\dfrac{v_{gst}(3v_{gdt}+v_{gst})}{(v_{gdt}+v_{gst})^2}$	$-\dfrac{1+\alpha}{6}c_{ox}L$	0

TABLE VI

ENERGY FUNCTION

	Linear	Saturation	Cut-Off
Q_g	$c_{ox}L(v_{gb}-v_{fb}-\phi-v_{sb}-\dfrac{v_{ds}-v_{sb}}{2})$ $+\dfrac{(1+\alpha)(v_{gb}-v_{sb})^2}{12(v_{gst}-(1+\alpha)(v_{ds}-v_{sb})/2)}$	$c_{ox}L(v_{gb}-v_{fb}-$ $\phi-v_{sb}-\dfrac{v_{gst}}{3(1+\alpha)})$	$c_{ox}L(v_{gb}$ $-v_{fb}-\phi$ $-\dfrac{v_{gbt0}}{(1+\alpha)})$
E_f	$\dfrac{1}{4}c_{ox}L\{\alpha(v_{db}^2+v_{sb}^2)$ $+(v_{db}-v_{gbt0})^2+(v_{gbt0}-v_{sb})^2\}$ $+Q_g v_{t0}$	$\dfrac{1}{4}c_{ox}L\{\alpha(\dfrac{v_{gbt0}^2}{1+\alpha}+v_{sb}^2)$ $+(v_{gbt0}-v_{sb})^2\}$ $+Q_g v_{t0}$	$\dfrac{c_{ox}\alpha L}{2}\cdot$ $\dfrac{v_{gbt0}^2}{1+\alpha}$ $+Q_g v_{t0}$

Regression based Circuit Matrix Models for Accurate Performance Estimation of Analog Circuits

Almitra Pradhan and Ranga Vemuri
Department of ECE, University of Cincinnati, Cincinnati, OH 45221, USA
Email: {pradhaa, ranga}@ececs.uc.edu

Abstract— Automated analog circuit synthesis techniques depend on fast and reliable estimation of circuit performance. This paper presents a highly accurate method of estimating performances by constructing models of the circuit matrix instead of the traditionally used performance models. Device matching in analog circuits is utilized to identify identical elements in the circuit matrix and reduce the number of elements to be modeled. Experiments conducted on three operational amplifier topologies demonstrate the effectiveness of the method in achieving correct performance prediction. Results show that the performances can be predicted within a mean error of 0.1% compared to a SPICE simulation.

I. INTRODUCTION

Circuit sizing is the process of determining device dimensions and biasing of a given topology to achieve the desired performance goals. In automated circuit sizing, an optimization algorithm proposes device sizes and bias from a search range and the evaluator checks if the performance goal is met. The evaluator has to be both fast and accurate. Spice simulation, symbolic analysis and macromodeling have been used by researchers for performance evaluation. Spice simulation is slow but speedup can be achieved by parallelism [1]. Symbolic analysis is fast but suffers from term explosion for larger circuits. With macromodels, the relation between the design variables and circuit performance is captured by a black box abstraction. These evaluate much faster than direct simulation and can achieve speedy synthesis. However, the performance parameters are extremely difficult to model. Macromodels can suffer from inaccuracies and research efforts are directed at using complex modeling strategies to achieve good accuracy. In [2], Gielen *et al.* review a number of performance macromodeling approaches.

To obtain performance parameters of an analog circuit at a given point in the search space, the system matrix of the circuit is generated. This matrix, also called the circuit matrix, is derived based on the Modified Nodal Analysis (MNA) formulation. The circuit matrix can be represented as follows:

$$(G + sC)x = B; \; y = L^T x$$

where, G: conductance submatrix, C: susceptance submatrix, B: input vector, L: output vector, x: unknown state vector, y: output

This work was supported in part by National Science Foundation under award number CCF-0429717.

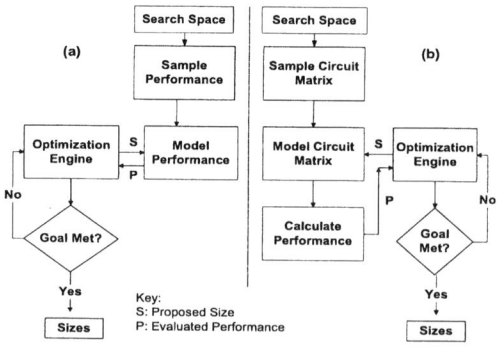

Fig. 1. (a)Performance Modeling Approach (b)Matrix Modeling Approach

Pre-defined MNA stamps for all circuit elements allow circuit matrix generation to be quite straightforward. The MNA stamp of a mosfet is written in terms of its small signal values such as transconductance (gm), output conductance (gds), capacitance (cgs, cgd, cgb) etc., whereas for other circuit elements stamps are in terms of the component values. The small signal values of mosfets are obtained by linearizing the circuit around the operating point. The circuit matrix is solved to obtain the frequency response of the circuit. Performance parameters such as the low frequency gain, Unity gain frequency (UGF), Gain Margin (GM), Phase Margin (PM) are calculated from the frequency response. In simulation based synthesis, the spice engine generates and solves the circuit matrix. Macromodeling approaches use fast evaluating models and eliminate the use of spice. As shown in fig. 1 macromodeling is possible at *two* places in the synthesis flow:

1) Modeling the performance parameters
2) Modeling the circuit matrix

Most of the existing macromodeling techniques use the first approach i.e. they model the performance parameters directly. Such methods greatly concentrate on the performance estimation speed, but suffer a tradeoff with accuracy. This paper presents an *alternative method* of estimating performance characteristics of linear analog circuits by constructing a model of the circuit matrix. The advantage, as will be seen, is that the matrix can be very accurately modeled even with simpler modeling approaches such as multivariate polynomial regression. Since it is possible to accurately estimate performance

48

978-1-4244-1709-4/07/$25.00 © 2007 IEEE

values, true design convergence is obtained by this method.

Performance is not directly modeled but it is calculated from the matrix model. Although this requires some extra computation time, the speed loss is not significant and is offset by the gain in accuracy and advantage of true convergence. The matrix model generation time is dependent on the circuit size. We have significantly reduced the number of models to be built by utilizing device matching properties of analog circuits. When matrix models are used in optimization based synthesis, partial model evaluation is done to speed up the matrix computation in successive iterations.

This paper is organized as follows. Section II compares modeling of the circuit matrix versus performance macromodeling. Section III describes the procedure used for efficient generation of the matrix models. Section IV presents the performance estimation results. Section V discusses speeding up synthesis by partial evaluation of matrix models. Section VI concludes the paper.

II. MOTIVATION

Performance estimation of analog circuits can use either system level models or performance level models. It is known that the relation between performance parameters such as UGF, PM and device sizes is extremely nonlinear [3], [4]. Sophisticated modeling approaches such as posynomials, neural networks are needed for modeling these severely nonlinear responses. However, these approaches too give significant errors [5]. We have observed that system matrix elements have lesser nonlinearity and can be accurately modeled.

Consider the operational Transconductance Amplifier (OTA) in fig. 5(a) as an example. We generated plots of performance parameters against device sizes and matrix elements against device sizes. Figures 2, 3 are representative plots of performance (PM) and matrix element (gds_M4). We can intuitively state from the figures that the matrix element is less nonlinear. The qualitative observation that matrix elements have less nonlinearity is now backed with two quantitative measures:

1) entropy of response curves
2) variance of local differentials

Entropy measures the complexity of a response curve [6], higher the entropy more complex the response. Entropy is calculated by definition from [7]. Variance of local first order differentials measures smoothness of a response, with lesser variance indicating greater smoothness. A response that has

TABLE I
ENTROPY AND LOCAL DIFFERENTIAL VARIATION OF OTA

Response Variable	Variance of Local Differential	Entropy
Matrix Element (Worst Case)	0.0316	0.8565
Gain	0.0369	0.8072
UGF	0.0422	0.7076
Gain Margin	0.2318	1.5674
Phase Margin	0.4248	3.8017
Results are on a dataset of 2000 points		

low entropy and is smooth is less complex to model. Worst case entropy and local variance values among all matrix elements is shown in Table I. The table also shows the entropy and local variance for performance parameters. Phase and Gain Margins have entropy and local variance an order greater than the matrix elements. From these qualitative and quantitative measures we can infer that matrix elements are less nonlinear and can be modeled with greater accuracy than their performance counterparts.

III. MODELING METHODOLOGY

The matrix elements show a linear or curvilinear variation with respect to design variables. We model the response matrix by polynomial regression. The input variables of the model, usually the transistor widths, are normalized on a [0,1] range using eq.(1), since for polynomial regression it is important that higher order terms do not have high collinearity with lower order terms [8].

$$x_{transformed} = \frac{x - x_{min}}{x_{max} - x_{min}} \quad (1)$$

The response is modeled using a least squares (LS) polynomial fit given by the following equation:

$$Y(x_1...x_n) = \beta_0 + \beta_1 x_1 + .. + \beta_n x_n + \beta_{11} x_1^2 + \beta_{12} x_1 x_2 + .. \quad (2)$$

(where β_is are coefficients of the polynomial fit.)

It is observed that the capacitance sub-matrix terms are highly collinear with respect to the design variables, and lower order polynomials are sufficient for modeling them. The conductance sub-matrix containing terms such as gm, gds etc are more nonlinear and are modeled by higher order polynomials. Once the response model within acceptable error

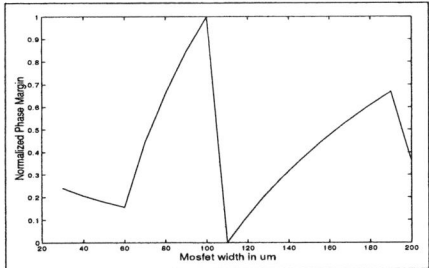

Fig. 2. Phase Margin vs. Device Width of OTA

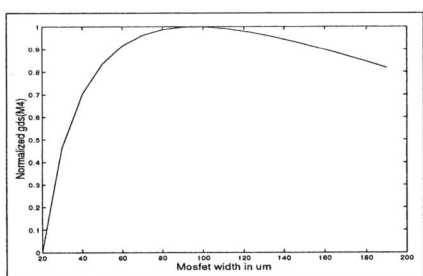

Fig. 3. Matrix Element vs. Device Width of OTA

limits is obtained by a LS fit, the regression coefficients are saved. The response at any unknown design point within the model bounds can now be predicted by simply plugging the input variable values in the model given by eq.(2). This makes response prediction extremely fast.

A. Circuit Matrix Generation

The first step in matrix macromodeling is generation of the circuit matrix. Subsequently values for the matrix are obtained in the design space and the matrix is modeled. Since we want to model the circuit matrix in terms of its elements, we would like to reduce the number of matrix elements to be modeled to as few as possible. To enable this reduction, we take advantage of:

- *matched* element identification
- *reverse* element identification

In the OTA circuit fig 5(a), we can see that the transistor pairs $M0 - M1, M2 - M3, M4 - M5$ and $M6 - M7$ are matched. Using the half circuit concept [9] we know that the small signal values of the matched pairs will be equal. Thus, if the matrix elements are linear combinations of small signal values of matched elements, even these matrix elements will be identical. As a simple example, in the OTA the pairs $M0 - M1$ and $M2 - M3$ are matched and $gm0 = gm1$ and $gm2 = gm3$. If the circuit matrix has two elements, one being $gm0 + gm2$ and the other being $gm1 + gm3$, we know that these two elements will always have the same value. Thus a single model will be sufficient for both these matrix elements. With the MNA formulation we have seen that such identical elements occur at many places in the circuit matrix.

It is also observed that in the MNA matrix, some elements appear only with a reversal of polarity. For example, one matrix element is $gm4$ and the other is $-gm4$. It is possible to use a single model for elements that occur with opposite signs. Thus, we observed two properties of the circuit matrix elements which will help us reduce the number of elements to be modeled.

When the circuit matrix is generated through its MNA formulation, the matrix coefficients are first generated in a symbolic form to identify identical and reverse polarity elements. For our benchmarks, the number of non-zero coefficients in the original matrix versus the number of coefficients that need modeling after reduction is depicted in Table II. The achievable reduction depends on the topology and the number of matched elements.

TABLE II

REDUCTION OF MATRIX ELEMENTS

Benchmark	Original Matrix Elements	Elements after Reduction	Percentage Reduction
TSO	43	24	44
OTA	39	14	64
Differential Amplifier	145	61	58

TABLE III

MODELING ACCURACY FOR OTA

Matrix Element	Polynomial Model Order	Max Error (%)	Mean Error (%)	Std Dev (%)
C_{11}	2	0.0667	0.0129	0.0094
C_{13}	2	0.0439	0.0107	0.0085
C_{16}	3	0.0467	0.0122	0.0089
C_{33}	3	0.0908	0.0153	0.0155
C_{35}	1	0.0409	0.0102	0.0085
C_{55}	1	0.0430	0.0104	0.0081
C_{66}	3	0.0488	0.0115	0.0087
G_{11}	2	0.0641	0.0179	0.0119
G_{13}	6	0.2016	0.0207	0.0266
G_{15}	6	0.1879	0.0198	0.0252
G_{51}	7	0.3574	0.0602	0.0508
G_{55}	6	0.1888	0.0196	0.0254
G_{61}	4	0.1423	0.0202	0.0192
G_{66}	4	0.1256	0.0268	0.0206

B. Data Generation and Modeling

As with any modeling approach, we first need to generate raw data on which the model will be built. The data is obtained by performing a spice operating point analysis at a number of design points and storing values of circuit matrix elements. We have used random numbers drawn on a uniform distribution of the device ranges to sample the entire design space. About 2000 random data points are sampled for circuits with smaller design space such as the two stage amplifier, OTA and about 4000 points for circuits such as the differential amplifier with a larger design space . We have used high order polynomial response surface models for the circuit matrix as these give adequate accuracy. For polynomial models it is important to choose the order appropriately since choosing a lower order than necessary will give an erroneous model, whereas choosing a higher order will cause overfitting. In our benchmark circuits we find that polynomials with order 8 and beyond tend to overfit. We predefine the maximum order as 7 for our models. The model error is calculated using eq.(3). We define an error of 0.5% as the allowable model error. Starting with a linear model, if the model error is less than the allowable error, that order is chosen, else we fit a polynomial

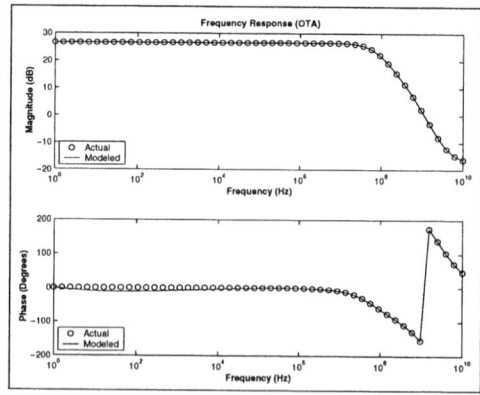

Fig. 4. Actual vs. Modeled Frequency Response of OTA

with one higher order. This is done till the maximum order of 7 is reached. In some cases, increasing the order, gives very little return in terms of error reduction (the adjusted R^2 regression criterion), in which we use a lower order model to avoid complexity. Algorithm 1 shows the entire modeling procedure. Table III shows the modeling accuracy for each matrix element of the OTA matrix. The frequency response of the OTA with the original system matrix versus the modeled matrix at a random design point is shown in fig. 4. It is seen that the two frequency responses match extremely well.

$$ModelError = \left| \frac{ActualValue - PredictedValue}{ActualValue} \right| * 100\%$$

(3)

After the model has been generated using sample data, the next step is model validation. Validation is necessary to ensure that the regression model obtained holds good for the entire design space and not just the sample data used to build the model. Validation of the model involves assessing the effectiveness of the model against an independent set of data and is essential if confidence in the model is to be expected [10]. For the purpose of validation we generate an independent set of random data points, 1000 data points for smaller circuits and 2000 points for larger circuits. The validated matrix model is used for estimating the performance of the analog circuit.

Algorithm 1 Generate Matrix_Model

Input: circuit.spice
Output: regression coefficients for all matrix elements
Generate_System_Matrix();
Identify_Unique_Elements();
Generate_Data();
∀ Unique Elements do:
order = 1;
done = *false*;
while (!done) **do**
  polyfit(response, variables, order);
  Error(order) = Calc_Model_Error(order);
  if (error(order) <= max allowed err) **then**
    Save_Reg_Coeffs(element);
    done = true; break;
  end if
  if ((error(order)-error(order-1)) <= 1%) **then**
    Save_Reg_Coeffs(element);
    done = true; break;
  end if
  if (order <= max allowed ord) **then**
    Increment(order,1);
  else
    Save_Reg_Coeffs(element);
    done = true;
  end if
end while

IV. EXPERIMENTS AND RESULTS

We have used three benchmark circuits: the two stage amplifier (TSO), the operational transconductance amplifier (OTA) and the high gain differential amplifier (DA) shown in fig. 5 for testing the accuracy of our models. The TSO [4] is a 8 transistor circuit with five design variables, the OTA is a 9 transistor circuit with four variables and the differential

TABLE IV
DESIGN VARIABLE RANGES FOR BENCHMARKS

Benchmark	Mosfet Count	Number of Variables	Ranges
TSO	8	5	M1-M5,M7:20-80um, Cc:2-10pF
OTA	9	4	M2-M5:20-200um M0,M1,M6,M7:20-35um
DA	33	5	M1-M10, 4*M25, 4*M26, 2*M23, 2*M27, 2*M28, 2*M32: 40-200um, Cc = 10-50pF

amplifier [11] is a 33 transistor circuit (biasing circuit is not shown in figure) with five variables. Design space reduction was done as explained in [4] to obtain the design variables. The design variables and their ranges used for the experiments (Table IV) are selected similar to earlier published performance macromodeling work of [4] to enable a comparison of results for the two methods. The design variable ranges are such that all design points lie in a valid pocket i.e. all transistors are in saturation in the given range.

Operating point analysis is done using Synopsys®Hspice and values for the elements of the matrix are obtained. For generating and evaluating the polynomial regression models the Matlab®Statistics Toolbox running on a 1.7GHz Pentium®M with 512 MB RAM is used.

Table VI shows the time required to build the models and the time to estimate performance values for a given size. Table V shows the maximum matrix modeling errors for the benchmarks. It is seen that the elements are modeled very accurately with the maximum error about 0.5-4%. Figure 8 is a plot of the element with worst case error of 4.37%. Figures 6, 7 compare the ac frequency response obtained from actual circuit matrix and the modeled circuit matrix for the TSO and Differential amplifier for a random circuit size. The modeled response matches the actual response extremely well. We compare our model building and estimation time with a performance macromodeling approach [12] that uses support vector machines. As performance is directly modeled in the second case the estimation time is lower, but the maximum error is 10.1%.

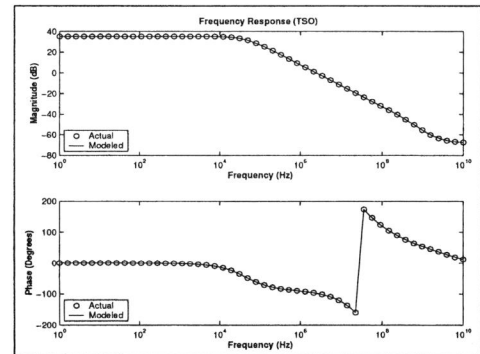

Fig. 6. Actual vs. Modeled Frequency Response of TSO

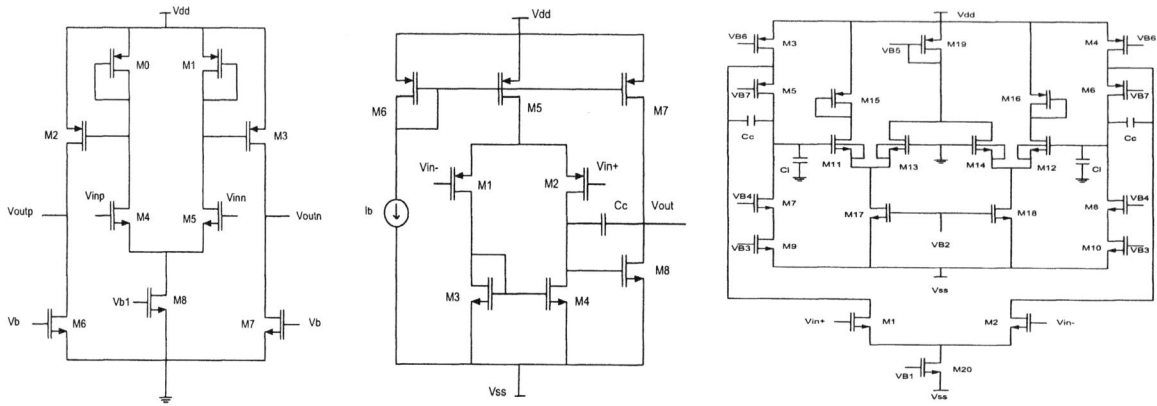

Fig. 5. Schematic of (a) Operational Transconductance Amplifier (OTA); (b) Two-Stage Op-Amp (TSO); (c) Differential Amplifier (DA)

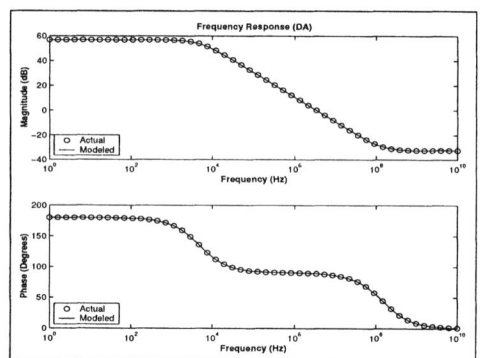

Fig. 7. Actual vs. Modeled Frequency Response of Differential Amplifier

TABLE V
WORST CASE VALIDATION ERROR

Benchmark	Validation Dataset Size	Worst case Error (%)
TSO	1000	1.8
OTA	1000	0.51
DA	2000	4.37

The performance parameters are calculated from the generated matrix models and results are compared with a spice simulation. Table VII shows the maximum, mean and standard deviation of the performance estimation error for all benchmarks. The maximum error with matrix models is about 3% and the highest mean error is about 0.1%. To enable a comparison with performance macromodeling, polynomial regression models were built on the performance parameters directly. Table VIII comprises the results of directly modeling the performance. As would be expected, the errors are higher. The TSO and Differential Amplifier circuits are identical to the work of [4] which uses neural networks for performance estimation. The maximum performance estimation error in [4] is 45% and highest mean error is about 5%.

As the circuit matrix is modeled, the time for an operating point analysis is saved which can be upto 70% of the total

TABLE VI
MODELING AND ESTIMATION TIME

Benchmark	Modeling Time	Performance (All) Estimation Time
Matrix Modeling Approach		
TSO	$3.7min$	$0.033sec$
OTA	$16sec$	$0.021sec$
DA	$31.7min$	$0.104sec$
Competing Approach [12]		
TSO	$131.15min$	$0.01sec$
OTA	$50.085min$	$0.001sec$

analysis time [13]. Although the performance is not available directly and needs an extra step for its computation, the performance calculation time is much smaller than a spice evaluation. The added advantage with our method is that since the entire ac behavior is modeled, any related performance can be evaluated. Thus, if a performance parameter is required, it simply needs to be evaluated from the matrix model and a new model need not be generated for that parameter.

V. PARTIAL MODEL EVALUATION

This section describes speeding up model evaluation during circuit sizing by partial model computation. An optimization algorithm such as Simulated Annealing (SA) used in sizing works by perturbing the current solution to propose a new solution. With incremental perturbation a single parameter of the current solution is varied in every iteration. An important

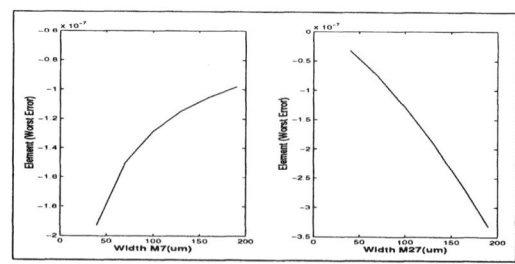

Fig. 8. Variation of Matrix Element having Maximum Error

TABLE VII

PERFORMANCE ESTIMATION ACCURACY WITH PROPOSED APPROACH

Benchmark	Max Error (%)	Mean Error (%)	Std Dev
Two Stage Op-Amp			
Gain	0.3231	0.0301	0.0370
UGF	0.6234	0.0544	0.0623
GM	1.2944	0.0900	0.1220
PM	0.7848	0.0521	0.0833
CMRR	0.7372	0.0668	0.0818
Operational Transconductance Amplifier			
Gain	0.1555	0.0134	0.0168
UGF	0.3111	0.0232	0.0320
GM	0.3199	0.0363	0.0367
PM	1.2864	0.0597	0.1068
Differential Amplifier			
Gain	3.2670	0.1214	0.1490
UGF	2.2863	0.1473	0.1820
PM	0.7970	0.0648	0.0666

TABLE VIII

ESTIMATION ACCURACY BY DIRECT PERFORMANCE MODELING

Benchmark	Polynomial Order	Max Error (%)	Mean Error (%)	Std Dev
Two Stage Op-Amp				
Gain	4	0.2523	0.0285	0.0272
UGF	7	14.06	1.4307	1.2449
GM	5	3.4894	0.6363	0.4933
PM	6	2.6748	0.3855	0.3463
CMRR	5	0.4704	0.0512	0.0555
Operational Transconductance Amplifier				
Gain	5	0.7582	0.0784	0.0878
UGF	6	8.4583	1.3731	1.0731
GM	7	7.1955	0.6016	0.7836
PM	7	3.3e3	186.31	411.50
Differential Amplifier				
Gain	4	6.2527	1.2365	1.1651
UGF	2	35.8732	9.9997	9.3995
PM	4	0.5234	0.0490	0.0567

TABLE IX

DIFFERENTIAL AMPLIFIER SYNTHESIS WITH PARTIAL MODEL
EVALUATION

Performance Specification	Estimated	Actual
Gain $\geq$ 78 dB	78.86	79.04
UGF $\geq$ 25 MHz	25.38	24.95
Phase Margin $\geq$ 88 Deg	88.43	88.72

SA iteration, a total of 14,364 solutions are evaluated in 30 minutes, whereas with partial matrix computation in each iteration a total of 15,219 solutions are evaluated. Thus, by partial model evaluation 855 more candidate solutions are evaluated in the same time. Circuit sizing results for the Differential Amplifier by Simulated Annealing with partial model evaluation are given in Table IX.

VI. CONCLUSION

Two methods for performance estimation of analog circuits, performance modeling and circuit modeling, are compared. It is demonstrated that the circuit matrix can be accurately modeled using polynomial regression. The number of coefficients that need to be modeled are significantly reduced by taking advantage of transistor matching. The accuracy of the proposed method is validated through experiments on three operational amplifier benchmarks. Partial model evaluation is used to speed up performance computation during synthesis.

REFERENCES

[1] M. Krasnicki, R. Phelps, R. A. Rutenbar, and L. R. Carley, "MAEL-STROM: Efficient simulation-based synthesis for custom analog cells," in *Proc. - 36th Design Automation Conference*, 1999, pp. 945–950.

[2] G. Gielen, T. McConaghy, and T. Eeckelaert, "Performance space modeling for hierarchical synthesis of analog integrated circuits," in *Proc. - 42nd Design Automation Conference*, 2005, pp. 881–886.

[3] P. Mandal and V. Visvanathan, "Macromodeling of the A.C. characteristics of CMOS op-amps," in *Proc. International Conference on Computer-Aided Design.*, vol. 7, no. 11, Nov. 1993, pp. 334–340.

[4] G. Wolfe and R. Vemuri, "Extraction and use of neural network models in automated synthesis of operational amplifiers," *Computer-Aided Design of Integrated Circuits and Systems, IEEE Transactions on*, vol. 22, no. 2, pp. 198–212, Feb. 2003.

[5] T. McConaghy and G. Gielen, "Analysis of simulation-driven numerical performance modeling techniques for application to analog circuit optimization," in *Proc. - ISCAS*, no. 6, May 2005, pp. 1298–1301.

[6] France, Michel Mendes and Henaut, Alain, "Art, therefore entropy," *Leonardo*, vol. 27, no. 3, pp. 219–221, 1994.

[7] A. Denis and F.Cremoux, "Using the entropy of curves to segment a time or spatial series," in *Mathematical Geology*, vol. 33, no. 8, Nov. 2002, pp. 899–914.

[8] M. Kutner, C. Nachtsheim, W. Wasserman, and J. Neter, *Applied Linear Regression Models.* McGraw-Hill/Irwin, 2003.

[9] B. Razavi, *Design of Analog CMOS Integrated Circuits.* McGraw-Hill, Inc., 2000.

[10] J. Rawlings, S. Panstula, and D. Dickey, *Applied Regression Analysis : A Research Tool.* Springer, 2001.

[11] J. M. Cohn, *Analog Device-Level Layout Automation.* Norwell, MA, USA: Kluwer Academic Publishers, 2000.

[12] M. Ding and R.Vemuri, "A combined feasibility and performance macromodel for analog circuits," in *Proc. 42nd Design Automation Conference*, 2005, pp. 63–68.

[13] M. Ranjan, W.Verhaegen, A.Agarwal, H.Sampath, R.Vemuri, and G.Gielen, "Fast, layout-inclusive analog circuit synthesis using precompiled parasitic-aware symbolic performance models," in *Proc. DATE*, 2004, p. 10604.

observation is that a design parameter affects only some elements of the circuit matrix. Thus during an SA move only the affected matrix elements need to be re-evaluated.

Partial model evaluation during synthesis is explained using the Differential Amplifier as an example. The Differential Amplifier has 61 matrix elements and 5 design variables. The design variables are four mosfet widths $(w_1 - w_4)$ and capacitance C_c. The correlation coefficient between matrix elements and design variables is calculated. If the p value of the correlation is less than 0.1, the correlation is considered significant. Based on the p values it is seen that w_1 affects 8 matrix elements, w_2 affects 30, w_3 and w_4 affect 30 and 36 elements respectively whereas 11 elements are dependent on C_c. Thus, with partial matrix evaluation a maximum of 36 elements are evaluated when w_4 changes and only 8 elements need to be evaluated if w_1 changes.

As the partial model evaluation is used in the context of SA, we compute the number of solutions the SA engine can evaluate in a given time when partial evaluation is used and when full evaluation is used. The SA engine is allowed to run for 30 minutes. With full matrix computation at each

A Software-Supported Methodology for Designing High-Performance 3D FPGA Architectures

Kostas Siozios, Kostas Sotiriadis, Vassilis F. Pavlidis[†] and Dimitrios Soudris

Department of Electrical and Computer Engineering
Democritus University of Thrace, 67100, Xanthi, Greece
{ksiop, kostsot, dsoudris}@ee.duth.gr

[†] Dept. of Electrical and Computer Engineering
University of Rochester, USA
email: pavlidis@ece.rochester.edu

Abstract — **A software-supported systematic methodology for exploring and evaluating alternative 3D reconfigurable FPGA architectures is introduced. Two new software tools were developed: *(i)* a placement and routing tool for 3D FPGAs (3DPRO) and *(ii)* a power/energy consumption estimation tool for such architectures (3DPower). Both of them are part of the new Design Framework, named 3D-MEANDER. We mainly focus our exploration on parameters that dominate the maximum operation frequency of the 3D FPGAs (*i.e.* vertical interconnections, number of layers, etc.). We evaluate the efficiency of the proposed methodology by making an exhaustive exploration for device delay, power consumption and utilized number of vertical connections for alternative 3D interconnection schemes. Experimental results demonstrate the effectiveness of our methodology, considering the 20 largest MCNC benchmarks. We achieve an average decrease in the delay, the wire length, and the energy consumption of 27%, 26%, and 34%, respectively, as compared to traditional 2D FPGAs, considering 3D architectures with 50% and 70% of fabricated vias. Also, we proved that actually-utilized via links are practically independent from the number of fabricated vias of a 3D FPGA architecture.**

I. INTRODUCTION

In the real estate market, an often-stated truism is that as land becomes more expensive, there is a tendency to build upward, rather than outward. This idea has some resonance in the domain of silicon ICs, where the sizes of the die are limited by yield and performance constraints, among other things. Three-dimensional integration can mitigate many of these limitations. For example, a considerable reduction in the number and length of the global interconnects can be achieved. This decrease results, in turn, in performance enhancements and decreased power consumption for 3D ICs as compared to 2D circuits. Existing 3D fabrication technologies are surveyed in [1]. This work focuses on available interconnection architectures among the layers of 3D ICs and emphasizes the open issues for current and upcoming 3D technologies.

Recently many research groups from academia [2, 3, 4, 5, 6, 7, 8], industry [9], and research institutes [1] have spent significant effort on designing and manufacturing applications in 3D technologies. Several companies [9] develop 3D ICs for commercial purposes by wafer stacking, where the distance between the layers is determined by the wafer thickness. Note that the existing industrial research primarily concerns the manufacturing and fabrication processes rather than the development of CAD tools to support the design of emerging 3D technologies.

A fabrication technology for 3D ICs is presented in [5]. In this work a number of wafers are integrated with a high density inter-wafer interconnects. A 3D integration process flow is also presented.

This paper is part of the 03ED593 research project, implemented within the framework of the "Reinforcement Program of Human Research Manpower" (PENED) and co-financed by National and Community Funds (75% from E.U.-European Social Fund and 25% from the Greek Ministry of Development-General Secretariat of Research and Technology).

In this approach, an application can be implemented either by minimizing the total wire length or by minimizing the number of the interlayer interconnects. Other design objectives, such as power consumption or delay, however, are ignored. Furthermore, the exploration or modification of any architecture characteristic is not possible with the described approach.

In [4], the potential of a 3D FPGA technology, based on 3D routing switches, is evaluated using analytic models. The interlayer interconnections are formed by high aspect ratio vias and Cu-Cu bumps. A tool, called PR3D, is also developed for designing a 3D FPGA architecture exclusively consisting of 3D switch boxes.

Although 3D integration promises considerable benefits, several challenges need to be satisfied. Among others, design space exploration is essential to build high-performance and/or low energy systems and architectures that exploit all of the advantages offered by 3D integration. In addition, CAD tools that facilitate the design of 3D circuits are required. Up to date there are only a few academic CAD tools [4, 6] for mapping applications on 3D FPGA technologies, while there is no complete CAD flow in order to promote the commercialization of this potent design paradigm. Furthermore, there is no commercial CAD tool for realizing applications on 3D FPGAs, similar to the standalone tools and/or design flows provided by Cadence, Mentor Graphics, and Xilinx for 2D technologies. Consequently, there is an absolute necessity to develop algorithms and software tools to exploit the advantages of the third dimension, and to solve time consuming and complex tasks, such as floorplanning, placement, and routing (P&R) for 3D reconfigurable architectures.

In [6], a P&R approach for island style 3D FPGA architectures is described. A partitioning-based placement and simulated annealing-based refinement tools are used, which target on the reduction of the interconnection length. The authors report gains in wire lengths compared to 2D architectures, without considering, however, the wire power consumption and delay. Hence, these tools cannot be used for exploring alternative 3D architectures.

In [4], a similar P&R approach for 3D FPGAs is described. The reconfigurable architecture consists of multiple stacked functional layers, while the communication among layers is realized by using 3D Switch Boxes (SBs). A tool, named TPR, for placement and routing in 3D FPGA architectures based on the VPR [10] tool was developed. Although TPR is one of the first attempts in academia to develop tools for 3D FPGA, it suffers from many limitations. The target architecture utilized in this tool initially assumes an unlimited number of vertical interconnections, while the TPR aims at minimizing the number of these interconnect. Such a scenario, however, is not realistic, since the total number and the spatial distribution of the vertical interconnects (via) between layers are important problems that need to be addressed. The authors in [4] reported gains of about 25% in wire length and 35% in delay for an FPGA architecture with 10 layers. Such a number of layers are not feasible, however, for the existing 3D integration technologies [1].

978-1-4244-1709-4/07/$25.00 © 2007 IEEE

Also, these tools cannot estimate/calculate other important design parameters, such as, the energy/power consumption.

In this paper a design methodology for exploring several parameters of 3D-FPGA reconfigurable architectures is introduced. This methodology is supported by two novel software tools, namely 3DPRO (3D Placement and Routing Optimizer) and 3DPower, which belong to the 3D MEANDER Framework. In order to evaluate the proposed methodology, we quantify a number of cost factors, such as delay (or performance), energy consumption, and total wire length over a plethora of 3D FPGA architectures. We perform architecture exploration for different number and various locations of the vertical interconnects (i.e., vias) that connect circuits located on different layers. To best of our knowledge, the software-supported architecture methodology for exploring/evaluating 3D FPGAs with different number of vias is presented for the first time in the literature. In order to evaluate the effectiveness of our methodology, the 20 largest MCNC benchmarks are used. We proved that we can design 3D architectures with smaller fabrication costs. Furthermore, we provide a qualitative comparison between the proposed 3DPRO tool and the existing TPR.

The rest of the paper is organized as follows. In Section 2, we describe the modeling approach of the 3D FPGA architecture, while the proposed tool framework for exploring and realizing applications on 3D FPGAs is presented in Section 3. The evaluation results that demonstrate the efficiency of the proposed methodology under the energy-delay product criterion are presented in Section 4. Finally, the main points of the work are summarized in Section 5.

II. MODELING OF THE 3D FPGA ARCHITECTURE

In order to realize the vertical connections among different layers, we have to extend some conventional 2D Switch Box (SBs) to employ connections to the other layers of the 3-D FPGA. Although the SBs utilized are based on the pattern found in Xilinx XC-4000 FPGA architecture, the results are applicable for any other SB pattern found in bibliography. Different SB topologies utilize a different number of pass transistors leading to different interconnection delay and power consumption values. For example, in a 2D SB an incoming routing track can be connected to three other wires ($F_s = 3$). Similarly, for a 3D SB, the incoming routing track is possible to be connected to five other tracks ($F_s = 5$). In the first case, the SB is formed by 6 transistors, while in the 3D approach 10 transistors are required. As we target to reconfigurable architectures, the power consumption (especially the static power component) is one of the upmost parameters for reduction and, therefore, the selection of the appropriate connectivity across the 3D device layers is essential to achieve a high performance and low power implementation of 3D FPGAs. Furthermore, a high density of vertical vias can occupy large portion of a layer, where active circuits and interconnects must be excluded. This situation limits the available Si-area and wiring resources in each layer counteracting the benefits of 3D integration. In addition, the effect of the distribution and length of these vias on the performance and power consumption of 3D FPGAs has not been explored.

The proposed 3D FPGAs can be constructed by placing a number of identical 2D FPGAs on individual layers, providing communication by interlayer vias among vertically adjacent switch-blocks. Hence, the switch-block configuration is extended to the third dimension, while the structure of the individual logic blocks remains unchanged.

Based on the required connectivity for the successful implementation of an application onto FPGAs, the nets can be routed by using various channel segments to enhance both the delay/power efficiency and resource utilization. For all of the simulation/evaluation experiments presented in this work, we use a multi-segment routing architecture similar to the one that appears in the Xilinx Virtex architecture for horizontal tracks (composed from routing segments of lengths $L1$, $L2$, $L6$, and long lines, while the distribution of the

segments in each channel is 8%, 20%, 60%, and 12% respectively). For the vertical wires we use segment tracks of length $L1$.

In order to model the vertical wires we assume that these vias are electrically equivalent to horizontal routing tracks with the same length. This means that the vertical tracks of our architecture has the same delay and power consumption as the horizontal segments with length $L1$. This assumption is based on the fabrication process [5], where the fabrication of interlayer vias with length 5 μm -10 μm is feasible. For these segments, the delay of the wires dominates the delay of the switches (similarly to 2D architectures).

III. MEANDER FRAMEWORK FOR 3D FPGAS

New software tools are developed to support the proposed methodology for exploring 3D FPGAs architectures. The main advantage of these tools is that they can be easily integrated onto the existing MEANDER design flow [12, 13, 14] (Figure 1). The 3D flow adopts some existing CAD tools from the 2D toolset, which do not need to be aware of the third-dimensional FPGA topology. In other words, we reuse in the 3D design flow all those tools, which are independent of the technology platform. Only the tools which are related with P&R and power calculation tasks should be replaced by the new tools, because these tools consider the particular traits of the 3D FPGAs, i.e., technology platform dependent. More specifically we replaced the current version of P&R tool of the 2D flow, i.e., EX-VPR [13], with the proposed P&R tool, named 3DPRO. We also replaced the existing PowerModel tool, with the 3DPower for modeling and calculating power consumption in 3D architectures. To best of our knowledge, this toolset is the first complete framework in academia for mapping applications onto 3D reconfigurable devices starting from a hardware description language (HDL) description of the application and ending up to configuration file generation. The synthesis step involves the functional translation of a circuit to appropriate gate level format (.BLIF) for the remaining tools of the flow. Next, we map the gate level description to the logic of the target 3D FPGA architecture. To this step, the used tools are FPGA technology independent and can be used in both 2D and 3D design flows. Next, the new platform-dependent tools 3DPRO and 3DPower used in the remaining stages of the design flow are described.

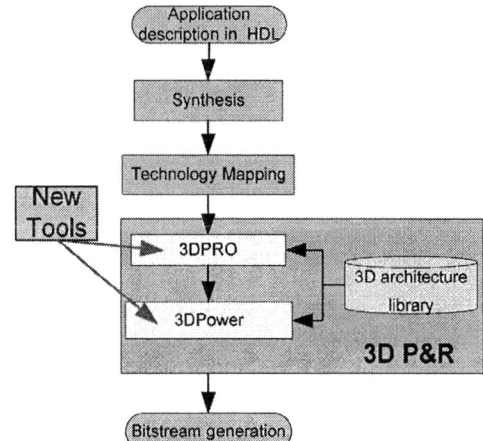

Figure 1. The 3D MEANDER Framework

A. Description of the new tools

The philosophy of the 3DPRO tool follows that of EX-VPR [13] and TPR [4], while the corresponding pseudo-code is shown in Figure 2. The tool is divided into three distinct stages. Initially, the application is partitioned into different layers having as criterion the interlayer communication. In the second stage, these partitions are assigned to different layers. As we intend to explore different communication scenarios of via placement (distribution) in each layer of the 3D FPGA, we utilize as much as possible the available vertical connections among different layers. This constraint is implemented inside the partitioning algorithm, which is an extended version of hMetis [2].

The placement step assigns each logic block to a specific (x, y) position on each layer. The placement procedure takes place for every layer individually, and then, the simulated annealing placement (similar to the EX-VPR tool) is applied to improve the routability and wire length of the whole 3D design. The cost function for each network i used for this simulated annealing placement is given by:

$$Cost_{3D}(i) = q(i) \times \left[\frac{bb_x(i)}{C_{ac,x}(i)^a} + \frac{bb_y(i)}{C_{ac,y}(i)^a} + \frac{bb_z(i)}{C_{ac,z}(i)^a} \right]$$

For each net, i, $bb_x(i)$, $bb_y(i)$, $bb_z(i)$ denote the spans of its bounding box on three directions (x,y,z), while $q(i)$ is a correction factor to 3D bounding box computation. The $C_{ac,x}(i)$, $C_{ac,y}(i)$ and $C_{ac,z}(i)$ are the average channel capacities (in routing tracks) in the x, y and z directions, respectively, over the bounding box of network i. The exponent, a, allows the relative cost of using narrow and wide channels to be adjusted.

Both the partitioning and placement algorithms build structures by taking advantage of any locality and regularity in the computational tasks in order to minimize resource usage (energy, area, routing time) and maximize performance (maximize throughput, minimize latency). The routing algorithm connects different logic elements through the available horizontal wire segments and vertical vias. The penalty of using each routing wire (horizontal or vertical) is parameterized through the architecture description file. In this work, we investigate the vias distribution across the system, which means that the objective is to maximize the usage of the available via for each scenario. Based on this objective, we employ an architecture file where all of the routing segments (horizontal and vertical tracks) have the same penalty value for usage in the hardware resource graph. Both the placement and routing procedure are based on [4, 10] and they are significantly extended to handle domain-specific 3D FPGAs architectures.

The second developed tool, named 3DPower, concerns the modeling and calculation of energy/power consumption of an application mapped onto a 3D FPGA device. This tool adopts some principles from the tool proposed by [11] for conventional (2D) FPGAs. More specifically, the 3DPower is aware about the number of layers and the number and location of the designer-specified vertical interconnects among layers.

```
read technology mapped netlist;
read architecture file;
for i = 1 to Total Networks
 {
  {
   partition application on N layers; {
    calculate required vertical connections;
    if vertical connections > critical try again;
    if vertical connections <= critical accept partitioning;}
 }
for layer = 1 to N {
 assign partition to each layer;
 place layer;
 if succeed placement = FALSE {
  reorder placement over layers;
  try again; }
 }
write placement file;
```

```
for layer = 1 to N {
 route application with h/v (horizontal/vertical) tracks;
 if succeed routing {
  H/V = current horizontal/vertical widths;
  reduce horizontal/vertical tracks;
  try again; }
 if failed routing {
  if h/v < H/V then {
   increase horizontal/vertical tracks;
   try again; }
  else {
   restore H/V values;
   route application with H/V values; }
  }
 }
write routing file;
calculate delay and area;
for i = 0 to Total Networks {
 for each LUT that form Network i {
  calculate static probability;
  calculate transition density; }
 calculate activity of net i;
 net power = 0;
 for each segment used to route this net {
  calculate capacitance of segment;
  net power = net power + switching power for this network;
 total power = total power + net power;
 write power file;
```

Figure 2. Pseudo code for P&R on 3D FPGAs

IV. EXPLORATION AND COMPARISON RESULTS

Since the P&R steps are critical for the efficient realization of applications onto reconfigurable architectures, we perform a qualitative comparison between 3DPRO and the TPR (which is the only known tool for P&R on 3D FPGAs) tool. The comparison results are listed in Table 1. We can explore a diversity of parameters such as delay, energy/power consumption, leakage power, area, channel width of horizontal and vertical tracks, and location of vias. Furthermore, the exploration of vertical connections is significant due to the impact of these via on the silicon area and the interconnect resources. Consequently, the 3DPRO provides larger flexibility to the designer to perform architecture exploration for many design parameters.

The effectiveness of the proposed methodology is exhibited by exploring several 3D architectures for various parameters. We performed exploration with the following assumptions: (i) Total number of layers is equal to four, (ii) Percentage of vertical interconnects per layer ranges from 0% (i.e., conventional 2D FPGA) to 100%, (iii) The location (x, y, z) of each vertical connection per layer remains invariant, (iv) A via connection between adjacent layers (with length L_v) is electrically equivalent to $L1$ wires formed on the 2D FPGA plane, (v) The via width is $W=4$ in any layer, (vi) The hardware resources of each layer are identical (i.e., identical number of Basic Logic Elements (BLEs)) among different layers and (vii) The applications are implemented onto the smallest number of BLEs per FPGA layer that can be mapped.

Table 1. Qualitative comparison between TPR and our proposed solution

Feature	TPR [4]	3DPRO
Architecture exploration	Yes	Yes
Measure Delay	Yes	Yes
Measure Wire length	Yes	Yes
Measure Power	No	Yes
Supported switch boxes	Subset Wilton Universal	Designer specified
Heterogeneous interconnect (simultaneously 2D/3D SBs)	No	Yes
Vias exploration	No	Yes
Temperature-aware P&R	No	Yes
Part of complete framework	No	Yes
Full custom 3D interconnections	No	Yes

Assuming a layer of size $X \times Y$ and $K\%$ is the available 3D SBs per layer, the pattern of placement of a 3D SB is derived as follows: Assigning first a 3D SB to a location (x, y) of a certain layer, then the neighboring 3D SBs are assigned to the locations $(x+r+1, y, z)$, $(x-r-1, y, z)$, $(x, y+r+1, z)$ or $(x, y-r-1, z)$, where r is the smallest integer that is derived by:

$$r = \left\lfloor \left(\frac{1}{K\%} - 1\right) \right\rfloor \qquad (1)$$

In addition, the parameter r indicates the number of 2D SBs between two neighboring 3D SBs. This assumption assists the router of 3DPRO to efficiently employ the vertical connections. For example, a layer from a 3D FPGA architecture with $K = 25\%$, *i.e.*, $r = 3$ in Figure 3 is depicted.

Since we map applications onto a multi-layer FPGA architecture, the resulting implementation with respect to delay, power, and area is tightly firmed to the partitioning process. Based on this, we can claim that if another partitioning algorithm is employed, the evaluation results may be different, but proposing a new partitioning algorithm is beyond the scope of this paper. Alternatively, the architecture exploration procedure should be performed for any possible setup of 3D architectures.

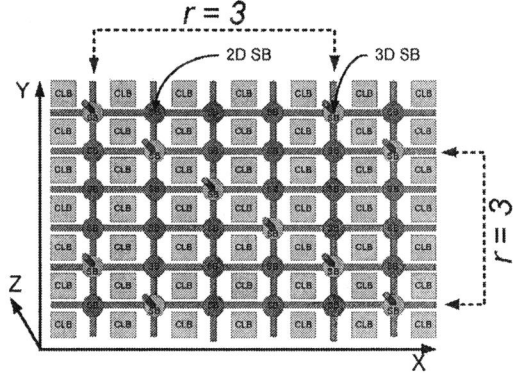

Figure 3. A layer from a 3D FPGA architecture with r = 3

In order to evaluate our methodology we performed an exhaustive exploration with the MCNC benchmarks. The results are summarized in upcoming Figures. All the benchmarks are mapped to the smallest square FPGA (or layer) that fits, in terms of logic elements and narrowest routing channel. The horizontal axis corresponds to the percentage of via connections in each layer of the 3D FPGA (which is identical to the percentage of 3D SBs of an FPGA layer), while the vertical axis shows the normalized value of each design parameter (*i.e.*, delay, power, Energy×Delay Product,

etc.) in relation to single layer FPGA. These points correspond to Pareto points showing all of the possible solutions. For each figure we normalized the results with the corresponding design parameter of a conventional (*i.e.*, 2D) FPGA. The Pareto points for 100% vias per layer correspond to the TPR solutions.

Since we propose a high performance 3D architecture, the selection criterion for designing FPGAs is to maximize the operation frequency. Figure 4 gives the average variation of this parameter (in normalized manner) over the MCNC benchmarks for different number of layers and vias percentage. The solution with 5 layers (bold line) outperforms almost all the other implementations, achieving to increase the device operation frequency up to 30% compared to 2D architectures. Moreover, this solution seems to be realistic, as with current 3D integration technologies [1] we can build FPGAs composed up to 5 layers. Next, we will evaluate architectures with three maximal values 30%, 50% and 100% 3D of SBs in terms of the operation frequency curve, considering the five layers FPGA.

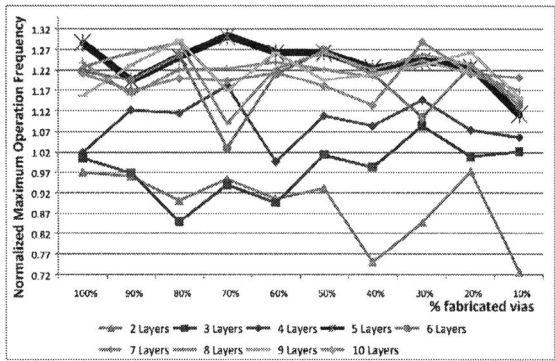

Figure 4. Average maximum operation frequency over the MCNC benchmarks for different number of layers and vias percentage

One of the goals of stacking layers on 3D architectures is to reduce the total wirelength, which has multiple impacts on all the design parameters that affect the performance. Among others, shorter wires leads to smaller resistance/capacitance, and hence to reduced delay, power/energy consumption and silicon area requirements. Figure 5 shows the average wirelength for different vias percentage and number of layers over the MCNC benchmarks. Our proposed solution with 5 layers (bold line) has reasonable gains over the 2D solution (up to 40% wirelength reduction). Also, for FGPAs composed with more than 5 layers, there are no significant gains in wirelength reduction.

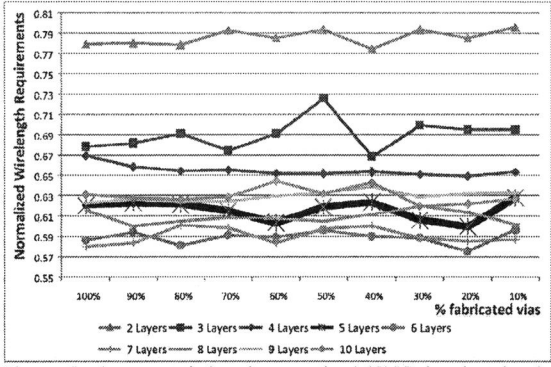

Figure 5. Average wirelength over the MCNC benchmarks for different number of layers and vias percentage

Due to the fact that reconfigurable devices exhibit high energy consumption, we explore the energy requirements over the MCNC benchmarks for different number of layers and vias percentage. The results are summarized on Figure 6. Based on this graph, the solution with 5 layers (bold line) reduces the energy consumption up to 37%, compared to 2D FPGAs. Moreover, from this graph we can see that the energy savings for devices composed by more than 5 layers are negligible, as these devices exhibits similar wirelength requirements (Figure 5).

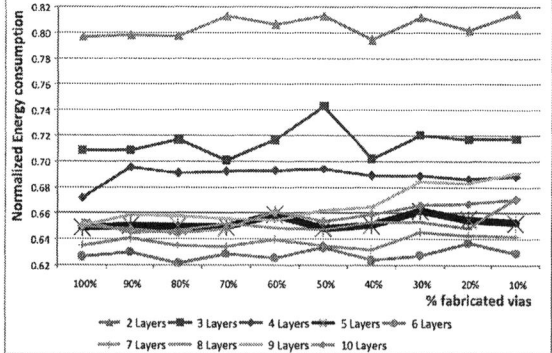

Figure 6. Average energy requirements over the MCNC benchmarks for different number of layers and vias percentage

Several points can be made from the previous graphs. As we increase the number of layers, the applications are realized more efficiently (smaller delay for critical nets and energy consumption) in 3D FPGAs compared to 2D reconfigurable architectures. For the kind of applications that examined here, these gains seem to be negligible for 3D FPGAs composed by more than 5 layers. This depends mainly on the chosen partitioning and placement algorithms. Secondly, we can claim that the developed P&R tools provide promising results for 3D architectures, where only a percentage of SBs forms 3D via connections. More specifically, we can conclude that as we vary the number of fabricated vias on each layer, significant reduction on design parameters may be achieved, leading to more efficient 3D architectures.

Choosing the 3D architecture with the three local maxima for operation frequency (50%, 70% and 100%) values from Figure 4, we performed detail exploration in terms of the delay, the wirelength and the energy requirement for the 20 largest MCNC benchmarks, as shown in Table 2. We performed a comparison between 2D (conventional) FPGA and 3D FPGA architecture consisting of 5 layers with 50%, 70% and 100% of the SBs of each layer to form 3D connections. Considering number of vias 50% the average reduction in the delay, the wirelength, and the energy consumption is 25%, 25%, and 28%, respectively. Similarly, the corresponding values for 70% vias, are 27%, 26%, and 34%, respectively. Indeed, the wirelength reduction (*i.e.*, capacitance reduction) due to 3D integration results in remarkable improvements in delay and energy consumption.

Furthermore, in Table 2 the columns with 100% vias give the calculated values of delay, wire length, and energy consumption, which correspond to the 3D architectures of [4]. It can be seen that these average values is similar to the ones of the explored 3D FPGA architecture results (*i.e.*, 50% and 70% vias). Specifically, a decrease up to 80% in the utilized vertical interconnects is observed. The last point is very important because we achieved the same improvements employing less hardware resources *i.e.*, vias. To best of our knowledge, it is the first time in the literature where the efficiency of a 3D FPGA architecture remains unchanged with less hardware resources, *i.e.*, fewer vias.

From 3D fabrication/manufacturing point of view, the smaller number of vias means: *(i)* smaller fabrication costs and *(ii)* larger useful silicon area in each layer (a via contact occupies much more

silicon area than a simple metal contact). The increased number of vias means more silicon and eventually greater cost.

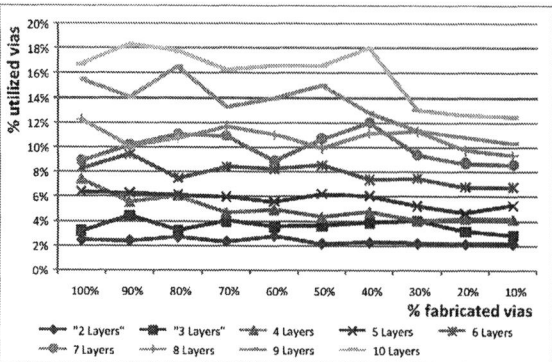

Figure 7. Vertical interconnects utilization

In Figure 7, it can be seen the utilization degree of the fabricated vias. We can easily infer that the number of actually-utilized vertical interconnects deviates a small fraction around the average utilization degree, for given number of layers. Considering the number of layers (from 2 to 10), the corresponding average values range from 2.31% (2 layers) up to 15.85% (10 layers), while the largest deviation from the average values range between 0.44% (2 layers) and 3.17% (9 layers), respectively. More specifically, given a certain number of layers, the via utilization degree remains almost invariant. *i.e.*, it is relatively independent from the percentage of vias per layer. In addition to that, this utilization degree does not seem to be relevant to total number of active layers. In Figure 7, it can be observed that the percentages of the utilized via links of the 5-layers architectures with fabricated 50% and 70% are 6.21% and 5.92%, respectively, which means both architectures utilized almost the same number of vias.

More details about the vias utilization on the 20 biggest MCNC benchmarks can be found on Table 3. As we can conclude, the percentage of utilized via for 5 layer FPGA architectures, is almost invariable of the percentage of fabricated vertical links (vias) between layers. More specifically, the average utilization ratio of vias for FPGAs composed by 50%, 70% and 100% 3D SBs (vias) are 15.36%, 16.57% and 18.03%, respectively. Consequently, we proved that the design of efficient 3D FPGA architectures with smaller number of vias than 100%, is feasible with reduced fabrication costs. In contrary, all the existing designs supports fully-populated vias 3D FPGA designs only.

V. CONCLUSIONS

A systematic methodology for exploring alternative 3D reconfigurable FPGA architectures is presented. This methodology is software supported by two new tools, namely 3DPRO and 3DPower, which belong to the first complete 3D FPGA Design Framework (3D MEANDER) in academia. Comparison results indicate average improvements about 20% in the delay, 23% in the wirelength, and 34% in the energy consumption for the proposed 3D FPGAs as compared to existing 2D FPGA architectures.

VI. ACKNOWLEDGMENTS

The authors would like to acknowledge the support from Prof. Kia Bazargan and Hushrav Mogal (University of Minnesota, USA) for the fruitful discussions about specific parts of TPR tool.

Table 2. Comparison results about MCNC benchmarks: Implementation in 2D and 3D FPGA architecture (with 50%, 70% and 100% via links, 5 layers and maxima operation frequency)

Benchmark	Delay ($\times 10^{-8}$ sec)				Wire Length ($\times 10^3$)				Energy ($\times 10^{-9}$ Joule)			
	2D	3D (50%)	3D (70%)	3D (100%)	2D	3D (50%)	3D (70%)	3D (100%)	2D	3D (50%)	3D (70%)	3D (100%)
alu4	14.8	9.72	10.0	10.1	41.40	36.63	38.96	38.36	8.56	6.14	6.10	6.09
apex2	17.1	13.0	12.0	14.0	78.36	63.02	62.84	64.08	12.9	7.68	7.68	9.16
apex4	14.7	10.2	10.3	11.3	44.06	36.06	34.98	34.53	5.50	3.68	3.46	3.42
bigkey	10.8	6.46	7.39	7.00	59.03	49.44	49.88	49.34	13.6	9.88	9.92	9.95
clma	63.2	32.7	34.5	38.3	379.42	315.52	292.86	293.53	72.6	52.2	48.5	48.5
des	14.7	8.63	9.11	9.37	94.07	55573	58.57	56.05	22.6	13.1	13.3	13.0
diffeq	18.7	17.2	16.5	16.8	51.98	34.29	34.29	34.09	24.3	11.8	8.30	8.41
dsip	8.19	5.51	6.93	7.77	53.70	37.89	38.26	37.60	13.3	7.09	7.12	6.98
elliptic	31.7	29.1	25.7	24.7	147.66	86.96	86.37	86.41	20.1	18.5	16.3	16.7
ex1010	25.3	24.0	21.0	19.8	181.30	149.25	146.67	142.23	18.5	11.7	11.4	10.9
ex5p	10.5	10.6	8.81	10.4	42.53	33.13	32.15	32.14	5.45	3.79	3.70	3.70
frisc	44.3	32.9	32.0	32.6	174.10	105.16	105.70	109.54	35.6	30.7	25.8	26.8
misex3	10.9	11.1	9.22	9.58	48.83	38.60	39.75	40.56	8.37	5.89	5.77	5.65
pdc	27.4	21.8	20.1	24.3	257.77	195.58	192.61	203.38	25.7	17.2	17.4	18.2
s298	26.0	22.9	20.6	24.0	62.12	57.11	58.51	58.17	14.8	10.0	9.92	9.86
s38417	48.5	26.7	32.2	25.4	297.09	238.98	233.26	235.52	53.2	34.4	33.4	34.2
s38584	29.3	20.1	19.2	19.1	310.79	200.84	200.84	200.72	43.4	35.2	30.9	31.2
seq	15.6	11.4	13.5	12.0	64.36	54.82	51.35	53.32	9.84	8.07	6.85	6.94
spla	21.4	23.0	19.4	17.4	169.22	134.85	135.50	131.76	15.9	14.2	11.8	11.1
tseng	19.0	18.4	16.4	17.2	34.17	24.86	24.92	24.88	27.9	24.3	22.7	24.5
Average:	23.6	17.8	17.2	17.6	130.00	97.41	95.90	96.34	22.6	16.3	15.0	15.3
Ratio:	1.00	0.75	0.73	0.74	1.00	0.75	0.74	0.74	1.00	0.72	0.66	0.68

Table 3. Comparison results about MCNC benchmarks: Via utilization in 3D FPGA architecture (with 50%, 70% and 100% via links, 5 layers and maxima operation frequency).

Benchmark	50% 3D SBs			70% 3D SBs			100% 3D SBs		
	Total vias	Utilized	(%)	Total vias	Utilized	(%)	Total vias	Utilized	(%)
alu4	1742	1476	23.06	4066	1637	25.58	5808	1704	26.63
apex2	2250	2241	28.94	5250	2322	29.98	7500	2836	36.62
apex4	1300	1558	30.05	3032	1701	32.81	4332	1669	32.20
bigkey	2074	264	2.86	4838	406	4.41	6912	424	4.60
clma	14746	2883	5.18	34406	3235	5.81	49152	3502	6.29
des	2250	566	6.14	5250	842	9.14	7500	679	7.37
diffeq	2624	567	5.67	6124	510	5.10	8748	640	6.40
dsip	1742	334	4.73	4066	307	4.35	5808	376	5.33
elliptic	5760	1341	6.47	13440	1965	9.48	19200	1373	6.62
ex1010	5476	3317	17.93	12776	3109	16.81	18252	3503	18.94
ex5p	1300	1498	32.40	3032	1648	35.64	4332	1741	37.65
frisc	6350	1404	6.41	14818	1406	6.42	21168	1722	7.86
misex3	1588	1610	22.82	3704	1725	24.45	5292	1983	28.10
pdc	5760	6228	30.03	13440	6158	29.70	19200	8345	40.24
s38417	13396	1413	2.63	31256	1787	3.32	44652	1183	2.20
s38584	13396	717	1.48	31256	771	1.59	44652	878	1.81
seq	2074	2178	25.73	4838	2294	27.10	6912	2428	28.69
spla	4162	4704	32.67	9710	5098	35.40	13872	5392	37.44
tseng	2074	469	6.65	4838	549	7.78	6912	536	7.60
Average via utilization (%)	15.36			16.57			18.03		

VII. REFERENCES

[1] Eric Beyne, "The Rise of the 3rd Dimension for System Integration", 8th Electronics Packaging Technology Conf., 2006

[2] G. Karypis, et. al., "Multi-level Hypergraph Partitioning: Applications in VLSI Design", Proc. ACM/IEEE DAC, pp. 526-529, 1997.

[3] Alexander, et. al., "Placement and Routing for Three-Dimensional FPGAs", 4th Canadian Workshop on Field-Programmable Devices, pp. 11-18, 1996

[4] Cristinel Ababei, et. al., "Placement and Routing in 3D Integrated Circuits", IEEE Design and Test, Vol. 22, No. 6, pp. 520-531, Nov-Dec 2005.

[5] R. Reif, et. al., "Fabrication Technologies for Three-Dimensional Integrated Circuits", Proc. Int. Symp. on Quality Electronic Design (ISQED), 2002.

[6] Shamik Das, et. al., "Technology, Performance, and Computer Aided Design of Three Dimensional Integrated Circuits", Proc. of the 2004 Int. Symp. on Physical Design, pp. 108-115, 2004.

[7] Arifur Rahman, et. al., "Wiring Requirement and Three-Dimensional Integration Technology for Field Programmable Gate Arrays", IEEE Trans. on VLSI Systems, Vol. 11, No. 1, pp. 44-54, February 2003.

[8] V. Pavlidis and E. Friedman, "Interconnect Delay Minimization through Interlayer Via Placement in 3-D ICs", Proc. of ACM Great Lakes Symp. on VLSI, pp. 20-25, 2005.

[9] 3D IC Industry Summary, available at "http://www. tezzaron.com/technology/3D%20IC%20Summary.htm".

[10] V. Betz, J. Rose and A. Marquardt, "Architecture and CAD for Deep-Submicron FPGAs", Kluwer Academic Publishers, 1999.

[11] Kara K.W. Poon, et. al., "A Flexible Power Model for FPGA's", 12th Int. Conf. on Field Program. Logic and Appl. (FPL), 2002.

[12] http://vlsi.ee.duth.gr/amdrel

[13] K. Siozios, et.al., "An Integrated Framework for Architecture Level Exploration of Reconfigurable Platform", Proc. of 15th FPL, pp. 658-661, 2005.

[14] K. Siozios, et. al., "A Novel FPGA Architecture and an Integrated Framework of CAD Tools for Implementing Applications", IEICE Trans. on Information and Systems, Vol. E88-D, No. 7, pp. 1369-1380, July 2005.

Estimating Design Time for System Circuits

Cyrus Bazeghi   Francisco J. Mesa-Martínez   Brian Greskamp[†]   Josep Torrellas[†]   Jose Renau

Dept. of Computer Engineering, University of California Santa Cruz

http://masc.soe.ucsc.edu

[†]Dept. of Computer Science, University of Illinois at Urbana-Champaign

http://iacoma.cs.uiuc.edu

ABSTRACT

System design complexity is growing rapidly. As a result, current development costs are constantly increasing. It is becoming increasingly difficult to estimate how much time it will take to design and verify these designs, which are getting denser and increasingly more complex. To compound this problem, circuit design cost estimation still does not have a quantitative approach. Although designing a system is very resource consuming, there is little work invested in measuring, understanding, and estimating the effort required.

To address part of the current shortcomings, this paper introduces μPCBComplexity, a methodology to measure and estimate PCB (printed circuit board) design effort. PCBs are the central component of many systems and require large amounts of resources to properly design and verify. μPCBComplexity consists of two main parts; a procedure to account for the contributions of the different elements in the design, and a non-linear statistical regression of experimental measures in order to determine a good design effort metric. We use μPCBComplexity to evaluate a series of design effort estimators for twelve PCB designs. By using the proposed μPCBComplexity metric, designers can estimate PCB design effort.

1 Introduction

Printed circuit board (PCB) design effort keeps growing due to such constraints as rising clock frequencies, thermal issues, reduced area, increasing number of layers, mixed signal devices, and the ever increasing component count and density. All of these factors combined have led to a steady rate of increase in development costs for current systems. As we design ever larger, denser and more complex systems, it is becoming increasingly difficult to estimate how much time would be required to design and verify them. To compound this problem, PCB design effort estimation still does not have a quantitative approach. We present in this paper a first step toward creating a design effort metric that is highly correlated with design effort for PCB layout. We follow a similar approach taken in [1] as the principles that are applicable to microprocessors are also applicable to PCBs. In this paper, design effort corresponds to the number of engineering-hours required for implementation (layout) of a PCB design.

This paper analyzes and proposes various statistics to estimate the layout effort required to develop PCBs. We investigate and quantify statistics such as area, component count, pin count and device types and sizes for many PCBs. We analyze several of these statistics, and propose a metric, obtained after applying non-linear regression over the different statistics, which we call μPCBComplexity. In addition, we provide insights on the correlation between several statistics and the design effort for many systems with known layout times.

Different designs have different constraints, leading to specific challenges; typical design constraints being area, frequency, and manufacturing cost. For example, having area being a primary design constraint, may lead to a requirement for additional layers, more expensive package types, and more complex placement and routing. A design constrained by cost, on the other hand, may require a balance between number of layers, area, drill density, types of packages and possibly the number of different drill sizes. Having clear constraints is necessary in estimating layout effort as it can drastically affect complexity.

We define design effort to be the required time in man-months to produce the layout for a given system. Design effort is equivalent to layout time when the project has a single developer, which is frequent even for complex PCBs. Nevertheless, for a given effort requirement, it is possible to reduce the design time by increasing the number of workers. Nevertheless, increasing the number of workers decreases the productivity per worker. The relationship between these two elements has been widely studied in software metrics and business models. Since the conversion between design effort and design time can be approximated, the remainder of this paper focuses only on design effort.

The rest of the paper is organized as follows. Section 2 covers other work in this area; Section 3 describes the statistical techniques that allow us to calibrate and evaluate the μPCBComplexity regression model; Section 4 describes the setup for our evaluation; Section 5 evaluates several statistics for the boards in our analysis; and Section 6 presents conclusions and future work.

This work was supported in part by the National Science Foundation under grants 0546819 and 720913; Special Research Grant from the University of California, Santa Cruz; and gifts from SUN.

2 Related Work

The capability to rapidly develop complex PCBs is a tremendous competitive advantage, since high development productivity is essential for the success of any design team. Although some companies have used statistical methods to estimate PCB design time, those methods are considered trade secrets [9]. Other companies do not release details because they provide competitive advantage over other companies. As a result, we are unaware of any published work on the topic of predicting the engineering hours required for a PCB design.

[1] focuses on microprocessor design effort. While the work described in this paper focuses on PCB design metrics, [1] uses a similar regression model, but both papers analyze different sets of statistics and targets.

There exists some published work that aids in the layout effort. A useful model for wiring density is called "Rent's rule", after an IBM engineer who popularized it [7]. This model attempts to calculate the required trace spacing on a board using the dimensions, number of routing layer, and the number of connections (assuming they are distributed according to Rent's rule).

Another paper that looks at productivity is [6] which identifies the need for standards or infrastructures for measuring and recording the semiconductor design process. They propose improving design technology, time-to-market, and quality-of-result by addressing the Design Productivity Gap and the Design "Technology" Productivity Gap. However, this previous work focused mostly on the problems associated with the infrastructure and design tools related to the physical implementation of semiconductor designs, while the focus of this paper is layout effort associated with PCB designs.

In [8] a factor similar to the productivity factor is described. They use the "process productivity parameter" to tune the estimating process for software projects. They contend that if you know the size, time, and the process productivity parameter you can use it to make estimates for a new project. So long as the environment, tools, methods, practices, and skills of the people have not changed dramatically from one project to the next.

Much research has been done in Design for Manufacturing (DFM) and Design for Production (DFP) which attempts to improve the production and manufacturing times of PCB assemblies. This paper seeks to develop a metric that can aid in predicting the layout effort, based on analysis of characteristics of PCBs at a low-level so as to better plan for future generations of systems. In [2] the issue of embedded passive components is discussed as a necessity to the smaller electronic devices requiring ever smaller PCBs. They note that board area is becoming so critical that to keep pace with the size constraints new techniques are required. Our goal would be to eventually develop a set of metrics and a model that estimates design effort by also taking into account manufacturing times.

3 Approach

The goal of this paper is to develop not only a quantitative approach but also produce a model that quickly estimates design effort based on several easily gathered statistics. This is important because being able to predict design effort is advantageous in helping to reduce design costs. To build the model, we analyze many commercial computer/electronic devices and gather data from their PCBs. The layout times for these PCBs were well documented which was a requirement for this analysis. Table 1 lists the critical components of PCB designs as determined by [2]. These parameters contribute to the complexity of a design, and hence the time required to do layout.

1.	Board dimensions (length and breadth)
2.	Total wiring requirements
3.	Number of layers
4.	Number of embedded resistors (if used)
5.	Number of embedded capacitors (if used)
6.	Set of active component types and their number
7.	Thickness of the board
8.	Number of discrete resistors
9.	Number of discrete capacitors

Table 1: Critical design parameters for a PCB.

Some design parameters listed in Table 1 are dependent on other factors. For example, the size of the board is defined by the number of embedded and discrete passive components and total wiring requirements. However, the total wiring requirements are governed by the number of embedded and discrete passive components in the PCB. And further more, the total number of layers in the PCB depends on the size of the board, the number of embedded and discrete resistors and bypass capacitors [2].

These critical design parameters are focused towards manufacturability, not design effort estimation. We used them as a starting point in determining what parameters or metrics to analyze and include for correlation with design effort. None of the boards in our study have embedded passive components; instead we focus on the total number of all components (passive and discrete) and the pin count for them. These are easily obtainable values.

Since the routing data is not easily obtainable, the number of pins for all the components in the design is taken into account instead. While this is not an ideal metric since not all pins are used or have very short traces (VDD or GND), it is readily obtainable and does not hamper the focus of this paper, namely effort prediction starting from higher level design descriptions, such as a bill of materials (BOM) or schematics.

In order to find a metric highly correlated with design effort, several statistics were gathered from the existing designs. For each isolated board with a known design effort, we look at several statistics and apply non-linear regression to find a highly correlated metric.

We present our design effort model as the aggregate of a set of statistics (S_i). Each of which has a specific constant (w_i), associated with it, which assigns a weight to the importance of every statistic used as input in the model. The aggregate of the statistics is inversely proportional to the productivity of a specific design team which is represented by a constant (ρ). The model is presented in Equation 1. In order to find suitable values for each of the data weights (w_i) we perform mixed non-linear regressions on this equation. The design team productivity factor (ρ) is constant per design group, and it needs to be adjusted on a per company or design team basis. If the ρ is unknown, then the absolute design effort is invalid and only the breakdown inside the project is correct. Obtaining the value of ρ is simple; all that is needed is to have the design effort for a single project. Alternatively, it is possible to develop a productivity benchmark suite that calibrates ρ for a given company.

$$\text{Design Effort} = \frac{1}{\rho} \times \sum_{k=1}^{n} (w_k \times S_k) \qquad (1)$$

In order to determine the weights that give a generalized solution to Equation 1, [1] proposes to use a mixed non-linear regression model. If there are no productivity adjustments, it is possible to use a simpler non-linear regression model. While the sum of a large number of random variables is distributed normally, the product of a number of random variables is distributed *lognormally* — a distribution where the logarithm of the variable is normally distributed [3]. Therefore, since the random variables have a log normal distribution an even simpler linear regression model can not be used.

To evaluate the accuracy of the model (Section 5), we use σ as a measure of error associated with the fit. Consequently, it is important to understand what different values of σ tell us about the quality of the estimate. For a given σ, we can find a *confidence interval* for the estimated effort. The $x\%$ confidence interval for a metric is defined to be the range of efforts ($Estimate_{low}, Estimate_{high}$) such that $P(Estimate_{low} <$ metric prediction $< Estimate_{high}) = x/100$. For example, the 90% confidence interval gives us two values a and b such that there is a 90% chance that the actual effort is between metric prediction $\times a$ and metric prediction $\times b$.

3.1 Productivity Adjustments

In software development projects, it is well known that different development teams have different productivities. For example, it has been shown that the productivity difference between teams can be up to an order of magnitude [4]. We believe that a similar effect occurs between PCB design teams. The productivity differences may be due to multiple factors, including the average experience of the designers in the team and the tools used. In our model, ρ captures this effect.

The boards under study in this analysis either all came from one manufacturer, or we only had one board from the manufacturer, so the use of a productivity factor was not necessary.

3.2 Team Size Dynamics

Although some board designs require long periods of time, it is very rare to find multiple developers doing different sections of the same board. The PCB layout effort by nature is a linear task done by one engineer at a time. To reduce the design time, we have found two approaches: multi-timezone working environments, and "surgical" teams.

A multi-timezone team has different designers working in multiple time zones, this is, once a designer stops working a new designer can continue and pick up where the previous designer left. A "surgical team" [5] follows an alternative design organization, with the surgeon, or chief designer, at the helm and a supporting staff that has their tasks allocated by the chief of staff. In the PCB case, we may have other designers doing such tasks as making footprint images for components, which can be a tedious effort.

3.3 R-Language

This section provides the R-language [10] code to fit the non-linear mixed-effects model and the non-linear regression model. The mixed-effects model is needed when productivity adjustments (ρ) are required, a simpler model is used when no productivity adjustments are required.

Recall that our model has a multiplicative lognormal error and also a lognormal distribution for the random effect ρ. Simply taking the logarithm of both sides of the equation gives us the requisite additive normal error and normal random effect as follows. Hence we have the need for a non-linear model.

```
# mixed-effects non-linear model
nlme(model=log(Effort) ~
   (log_rho) + log(w1*stat1 + w2*stat2)
  ,random = log_rho ~ 1 | team
  ,fixed  = list(w1 ~ 1, w2 ~ 1)
  ,start  = c(0.1, 0.1)
  ,data=(traw)
  ,method="ML")

# non-linear model
nls(log(Effort) ~
   log(w1*stat1 + w2*stat2)
  ,start=list(w1=0.1,w2=0.1)
  ,data=traw)
```

The R-language is also used to compute the confidence intervals. To obtain a 90% confidence interval for a given σ (s) generated, the following R-language code $c(exp(s * qnorm(0.05)), exp(s * qnorm(0.95)))$ is used.

4 Evaluation Setup

We gathered data from a number of PCB designs for the analysis done in this paper. Table 3 shows the types of statistics gathered for each of the boards analyzed. When calculating the area consumed for each component we did not consider the cases where routing, or in the more rare case placement, could be done underneath a component. Several board designers pointed out that the component and pin density of the board was one of the crucial factors to estimating design effort. To

Board	Description	Engineering Notes
B1	Signal Conditioning	Many thru-hole components. Analog board with many important signal paths
B2	AE RMS	Many thru-hole components. Analog board with many important signal paths
B3	PMD Motor Controller	Many high density components
B4	Motor Driver	New footprints
B5	Enviro Controller	Forgot reasons why it took so long
B6	Current Source	Many components on a small board. Mechanical constraints
B7	Arbitrary Waveform Generator/Amplifier	Placement constraints due to noise reduction
B8	ACDC Monitor	Cost major factor. Time consuming to keep to a 2 layer board
B9	Tank Monitor	Cost major factor. Time consuming to keep to a 2 layer board
B10	Air spring remote	Very small. RF constraints
B11	Air Spring Controller	2 Isolated grounds with placement constraints
B12	Network Appliance	Electrical/mechanical design challenges and thermal concerns

Table 2: Description of boards analyzed.

capture component and pin density, we define them with equation 2 and equation 3 respectively.

$$\text{Component Density} = \frac{\text{\# Components}}{\text{PCB Area} \times \text{\# Sides w/ components}} \quad (2)$$

$$\text{Pin Density} = \frac{\text{\# Pins}}{(\text{PCB Area})} \quad (3)$$

Table 2 gives a description of the boards along with the engineering notes that we were able to gather from the designers. Boards B7-B12 used SPECCTRA for OrCAD which is a common auto-router used in industry. No data was available on the use of an automatic router for the other boards but it can be safely assumed that some auto-route tool was used.

Board Statistic	Description
PCB Size (mm^2)	Physical size of the PCB
# of Sides w/ Comp	Either 1 or 2 sides has components
# of Routing Layers	Layers used for routing traces
# of Layers	The total number of layers in the PCB
Components	
# Passive	Passive components (resistors...)
# Digital	Digital integrated circuits (IC)
# Analog	Analog ICs or devices (opamps...)
Total #	Total count of all components on PCB
Total Area (mm^2)	Total area of all components on PCB
Density	Ratio of component area to area
Pins	
# Passive	Pins for all passive components
# Digital	Pins for all digital components
# Analog	Pins for all analog components
Total	Pins for all devices on PCB
Density	Ratio of number of pins to area

Table 3: Description of the statistics gathered from the PCBs.

In discussions with the designer of boards B8 and B9 the size of the LCD in the system dictated the size of the PCB and the housing that contained it. The LCD was counted as a component in our analysis and took one complete side of both of these boards, forcing the placement and routing of all other components to one side. Cost was the main consideration for both these boards also and this forced the designer to route everything using only 2 layers.

Among boards B7 through B11 the smallest board, B10,

was judged to be the most difficult to layout, whereas boards B7 and B11 were the easiest. This was attributed to the areas available to do the placement and routing. B7 and B11 were two of the largest boards reviewed and they were not area constrained, this gives much latitude to the designer for placement and makes the autorouter produce better results. With a more constrained area more human intervention is required during the routing phase which was the case for B10.

Board B12 had the longest system development time which extended the layout time due to the many system changes. At times in a PCB design there exist situations in which the PCB designer can not make forward progress due to electrical and mechanical design choices and issues, hence idle times. For this particular project it actually took approximately 5 months to resolve all issues (which were not related to PCB design effort, for example, cosmetic, placement of I/O). The actual layout time is estimated to be about 10 weeks but would be shorter if starting from a complete (and unchanging) specification.

For the placement stage we only had to consider the number of sides of the board on which components were mounted. Most of the boards in this study had the components all on one side, though a few had bypass capacitors mounted on one side, which accounted for a negligible amount of space. Again, thru-hole devices would affect the available placement area as it did the available routing area as space would be lost on both sides of the board, unlike with surface mounted components. This was not a factor in this study since most boards only used one side for placement. Boards B8 and B9 had components on both sides but one side was populated by only one component, the LCD. Board B10, the only other board with components on both sides, did not have any thru-hole devices present.

5 Evaluation

Our evaluation analyzes 12 different printed circuit boards from two seperate companies. Table 4 shows the main results and characteristics for each of these. The first column corresponds to each of the statistics or metrics presented in Table 3 (Section 4). Columns B1 to B12 correspond to each of the boards (Table 2). The last column corresponds to the σ between the row and design effort. Since the boards either were designed by the same team, or we only had one board from a particular company, we do not evaluate the productivity factor (ρ). This simplifies the analysis, and we can use non-linear regression instead of the mixed-effects non-linear regression model. With σ we can compute the confidence interval. For the lognormal distribution used, the mapping between σ and the 90% confidence interval is shown in Figure 1. We will use this chart to compare the accuracy of different estimators.

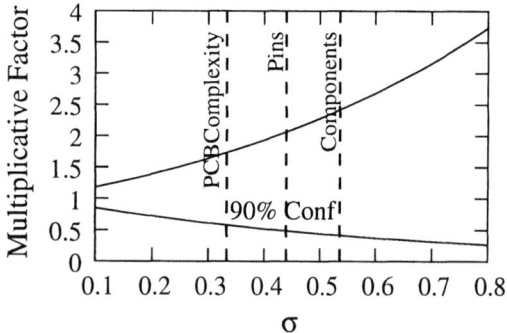

Figure 1: Mapping between the standard deviation of the error (σ) and the 90% confidence interval for the lognormal error distribution used.

The design effort values were obtained by interviewing the original designers. Obviously, there is perfect correlation with itself so $\sigma = 0$. A zero σ results in a perfect $(1, 1)$ confidence interval. We now proceed to analyze easily available statistics like number of components and pin count. These two sets of statistics are easily available before the PCB design starts. They are part of the PCB specification.

From the boards analyzed, we observe that it is best to use the total number of components to estimate design effort ($\sigma = 0.53$). Although traces for analog components and digital components are more difficult than traces for passive components, the low amount of digital and/or analog components on several of the boards make it difficult to use them as a method to estimate effort. Figure 1 shows the confidence interval for a $\sigma = 0.53$ as the intersection between the components line and the confidence interval line $(0.41, 2.39)$. This means that using the number of components on the specification, we have a 90% confidence that the design effort would be between 0.41 and 2.39 times the prediction.

Statistics about the pins are as easily available as components even before the design starts. The number of pins is a better predictor ($\sigma = 0.45$) than the number of components. The resulting 90% confidence interval for the number of pins is $(0.47, 2.09)$. This means that just by using the pins, we have a 90% confidence that the prediction is around half or double the expected design effort. Not shown in the table is the result of combining the number of pins and the components to predict design effort. The results did not improve because there is a high correlation between pins and components.

Area is not such an effective metric. Even assuming a per-

fect knowledge if the final dimension of the board, we can just estimate design effort with a $(0.21, 4.61)$ confidence interval. Table 4 also shows other statistics such as number of sides used, routing layers, and number of layers. Those statistics are not so useful by themselves because they are highly quantized, and this makes them difficult to use to predict effort.

The proposed $\mu PCBComplexity$ metrics are now evaluated. To obtain $\mu PCBComplexity$ shown in Table 4, we analyzed multiple combinations of parameters and followed suggestions from experienced board designers. The best results were achieved when using the following equation:

$$Effort \propto \#\ Passive\ Comp. + Comp.\ Density + Pin\ Density \quad (4)$$

Section 4 explains how to compute component density and pin density. To obtain the factors on equation 4, we perform non-linear regression as explained in Section 3. Although neither pin nor component density can achieve better predictions than the number of pins, when integrated together in the $\mu PCBComplexity$ metric we achieve a $0.33\ \sigma$. As Figure 1 shows, this represents a $(0.58, 1.72)$ confidence interval. This roughly means that by using the proposed $\mu PCBComplexity$ metrics, with a 90% confidence designers can predict design effort with less than 40% error.

Figure 2 shows a scatter-gather plot between design effort and our $\mu PCBComplexity$ metric. Each point corresponds to a different board. The plot does not include the B12 board to zoom on the area where most of the boards are located. This plot is an intuitive way to see that there is a high correlation between design effort and the metric proposed.

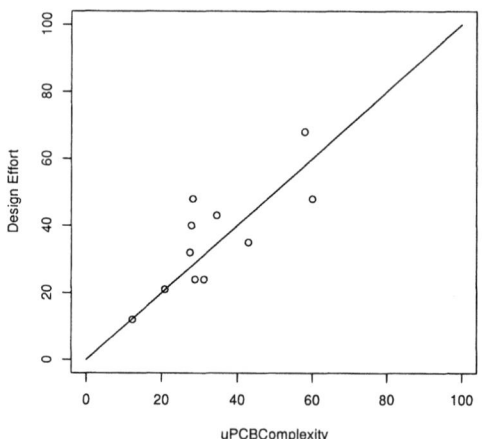

Figure 2: Scatter-gather plot of design effort vs. PCB metric

$\mu PCBComplexity$ works well because PCB design complex-

	B1	B2	B3	B4	B5	B6	B7	B8	B9	B10	B11	B12	σ
Design Effort (hours)	68	35	43	21	48	48	24	40	32	24	12	400	–
Components													
# Passive	213	165	101	80	108	222	116	86	83	19	47	2643	0.56
# Digital	15	0	17	0	8	2	0	11	8	4	4	94	1.79
# Analog	35	24	8	10	24	53	28	4	16	1	11	91	1.18
Total #	263	189	126	90	140	277	144	101	107	24	62	2828	0.53
Total Area (mm2)	6214	9053	6964	2719	9144	6579	8104	12193	12296	777	5430	38611	0.75
Pins													
Passive	563	429	365	182	414	578	414	194	188	39	109	5843	0.62
Digital	154	0	518	0	107	32	0	175	173	88	32	6889	1.88
Analog	360	208	216	98	72	448	150	25	53	14	65	924	1.10
Total	1077	637	1099	280	593	1058	564	394	414	141	206	13647	0.45
PCB Size (mm2)	22194	22194	22194	16256	38710	20430	22190	10943	10943	1277	25473	72600	0.93
# of Sides w/ Comp	1	1	1	1	1	1	1	2	2	2	1	2	0.81
# of Routing Layers	2	2	3	2	2	2	3	2	2	4	2	6	0.66
# of Layers	4	4	6	4	4	4	4	2	2	4	2	8	0.67
Component Density (x1000)	70	50	33	33	21	80	38	27	29	55	14	115	0.60
Pin Density	54	32	55	19	17	57	28	40	42	122	9	207	0.64
$\mu PCBComplexity$ (hours)	58	43	35	21	28	60	31	28	28	29	12	682	0.33

Table 4: Statistics, design effort, and correlation results of study boards.

ity increases as the component and pin density increases. Designers can increase the number of layers on the PCB to decrease the pin density or increase the area to reduce both densities. The problem is that both approaches require more costly boards. As a result, designers trade off between time to market and density.

6 Conclusions & Future Work

The goal of this paper is to introduce an initial exploration of the correlation between some easily obtained metrics from the design of a PCB, and the design effort required during the layout stage of development. To do so we extend a previously proposed complexity model [1] to the PCB domain. Furhter, many simplifications have been made; we do not account for traces of differing sizes, we ignore hole sizes or density, the frequency of the boards are also not considered, nor the extra considerations required for analog noise filtering. Also, additional PCBs from more companies with teams of differing sizes may be needed in order to develop a more general model for a general prediction of design effort.

Many factors and constraints affect the design effort required for a board to be successfully placed and routed. Some difficulty metric would be helpful but guidelines need to be established as "difficulty" is a fairly subjective term. Being able to analyze different options for a board would be useful, such as being able to change the size of the board to see what effect it would have on the estimated design effort. This can be expanded to also include the number of layers since this would ease routing congestion.

The evaluation shows that a simple statistics like PCB area size and number of components yield some correlation with design effort. With a 90% confidence, pins has a (0.47, 2.09) confidence interval. This means that roughly by looking at the number of pins, the typical design time error is half/double with a 90% confidence. Much better results can be achieved

with the proposed $\mu PCBComplexity$ metric. In that case the confidence interval for a 90% confidence is (0.58, 1.72). This roughly means that less than 40% estimation error is done with a 90% confidence.

Despite the good initial results, we believe that much work needs to be done in gathering relevant design metrics in order to evaluate (with associated known design times) and to refine relevant metrics and models for the design effort of modern PCBs. A major goal to our work is to define a set of equations, that given some easily obtainable design parameters, can generate an accurate estimators for design time.

REFERENCES

[1] C. Bazeghi, F. Mesa-Martinez, and J. Renau. μComplexity: Estimating Processor Design Effort. In *International Symposium on Microarchitecture*, Nov 2005.

[2] M. Chincholkar and J. Herrmann. Modeling the impact of embedding passives on manufacturing system performance. September 2002.

[3] E.L. Crow and K. Shimizu. *Lognormal Distributions: Theory and Application*. Dekker, 1988.

[4] T. DeMarco and T. Lister. *Peopleware Productive Projects and Teams*. Dorset House Publishing, 1999.

[5] JR. Frederick P. Brooks. *The Mythical Man-Month*. Addison-Wesley, 1995.

[6] A. B. Kahng. Design technology productivity in the dsm era (invited talk). In *Conference on Asia South Pacific Design Automation*, pages 443–448. ACM Press, 2001.

[7] B.S. Landman and R.L. Russo. On a pin versus block relationship for partitions of logic graphs. *IEEE Transactions on Comput.*, (12):1469–1479, Dec 1971.

[8] L. H. Putnam and W. Myers. *Five Core Metrics: The Intelligence Behind Successful Software Management*. Dorset House Publishing, May 2003.

[9] Numetrics Management Systems. Design Complexity and Productivity. Technical report, Numetrics Management Systems, Inc., 2004. http://www.numetrics.com.

[10] The R Development Core Team. *The R Reference Manual - Base Package*. Network Theory Limited, 2005.

Transparent Acceleration of Data Dependent Instructions for General Purpose Processors

Antonio Carlos Schneider Beck   Luigi Carro

Instituto de Informática – Universidade Federal do Rio Grande do Sul, Porto Alegre, Brazil

{caco, carro}@inf.ufrgs.br

ABSTRACT

Although transistor scaling keeps following Moore`s law, and more area is available for designers, the clock frequency and ILP rate do not present the same level of growth anymore. This way, new architectural alternatives are necessary. Reconfigurable fabric appears to be one emerging possibility: besides exploiting the parallelism among instructions, it can also accelerate sequences of data dependent ones. However, coarse grain reconfiguration wide spread usage is still withhold by the need of special tools and compilers, which clearly do not sustain the reuse of legacy code without any modification. Based on all these facts, this work proposes a new Binary Translation algorithm, implemented in hardware and working in parallel to the processor, responsible for transforming sequences of instructions at run-time to be executed on a dynamic coarse-grain reconfigurable array, tightly coupled to a traditional RISC machine. Therefore, we can take advantage of using pure combinational logic to optimize even control-flow oriented code in a totally transparent process, without any modification in the source or binary codes. Using the Simplescalar Toolset together with the embedded benchmark suite MIBench, we show performance improvements and area evaluation when comparing against traditional superscalar architectures.

1. INTRODUCTION

The possibility of increasing the number of transistors inside an integrated circuit year by year, following Moore´s Law, has been pushing performance at the same level of growth. However, high performance architectures as the diffused superscalar machines are now challenging well known limits of the ILP [1]: considering the Intel's family of processors, the IPC rate has not increased since the Pentium Pro [2]. This way, recent speed-ups in performance occurred mainly thanks to boosts in clock frequency through the employment of deeper pipelines. Even this approach, though, is reaching a limit. For example, the clock frequency of Intel's Pentium 4 processor only increased from 3.06 to 3.8 GHz between 2002 and 2006 [3].

Because of these reasons, companies are migrating to chip multiprocessors to take advantage of the extra area available, even though there is still a huge potential to speed up a single thread software. Hence, new architectural alternatives that can take advantage of the integration possibilities and that can address the performance issues stated before become necessary.

Reconfigurable fabric appears to be a serious candidate to be one of these solutions. By translating a sequence of operations into a combinational circuit performing the same computation, one could gain performance and reduce energy consumption at the price of extra area [4][5]. Furthermore, at the same time that reconfigurable computing can explore the ILP of the applications, it also speeds up sequence of data dependent instructions, which is its main advantage when comparing to traditional architectures.

Another characteristic of reconfigurable architectures is their regularity: it is common sense that as the more the technology shrinks to 65 nanometers and below, the harder it will be to print the geometries employed today, directly affecting the yield [6]. Furthermore, because circuit customization is a very expensive process, regular circuits customized in the field are also considered as the new low cost solution.

However, even with all these positive aspects cited before, reconfigurable architectures are still not largely used. The major problem precluding their usage is the necessity of special tools and compilers, modifying in somehow the source or binary code. As the old X86 ISA has been showing, keeping legacy binary code reuse and traditional programming paradigms are key factors to reduce the design cycle, allowing one to deploy the product as soon as possible on the market.

Based on all these facts, our work proposes the use of a technique called Dynamic Instruction Merging, which is a new binary translation approach implemented in hardware, used to detect and transform sequences of instructions at run time to be executed in a reconfigurable array, in a totally transparent process: there is no necessity of changing the code before its execution at all.

The employed array is coarse-grained and tightly coupled to the processor, composed of simple functional units and multiplexers. Therefore, it is not limited to the complexity of fine-grain configurations, making possible its implementation in any future technology, not just in FPGAs. Consequently, we can take all the advantages of the reconfigurable systems cited before, maintaining independence of technology and binary code reuse.

In this work we show some results concerning the potential of using such technique, demonstrating the binary translation algorithm, the structure of the reconfigurable hardware and how they interact with each other. Besides presenting the performance improvements and area overhead, we also compare our technique against a superscalar processor based on MIPS R10000.

This paper is organized as follows. Section 2 shows a review of the existing reconfigurable processors, some other approaches regarding dynamic translation of instructions and what is our contribution considering the whole context. Section 3 demonstrates the system, looking at the structure of the reconfigurable array and the algorithm itself. Section 4 presents the simulation environment and results. Finally, the last section draws conclusions and introduces future work.

2. RELATED WORK

2.1 Reconfigurable Architectures

The well known ASIP circuits have specialized hardware that accelerates the execution of the applications they were designed for. A system with reconfigurable capabilities would have almost the same benefit without having to commit the hardware into silicon. A reconfigurable processor can be adapted after design, in

978-1-4244-1709-4/07/$25.00 © 2007 IEEE

the same way programmable processors can adapt to application changes. That is why reconfigurable systems have already shown to be very effective, implementing some parts of the software in a hardware reconfigurable logic, as shown in Figure 1. Huge software speedups [4] as well as a reduction in system energy have been achieved [5].

Reconfigurable systems can be classified in different ways and aspects, considering coupling, granularity and instructions type [7]. A large range of systems with reconfigurable logic has already been proposed. For instance, processors like Chimaera [8], have a tightly coupled reconfigurable array in the processor core. The array is, in fact, an additional functional unit in the processor pipeline, sharing the same resources of the other units.

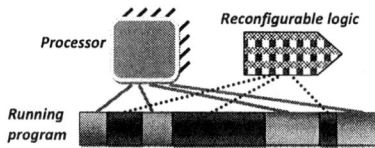

Figure 1. An example of a reconfigurable system

Reconfigurable fabric has also been applied in other levels of the architecture, imposing radical changes to the programming paradigm, involving the development of new compilers and tools. Putting this concept to the edge, an example of total dataflow architecture is the Wavescalar processor [11].

2.2 Binary Translation

The concept of binary translation (BT) [12] is very ample and can be applied in various levels. BT is based on a system, which can be implemented in hardware or software, responsible for monitoring the running program. After the analysis, some transformation is done in the code, with the purpose of adapt an existing binary to be executed in a specific ISA, to provide means to enhance the performance or even both.

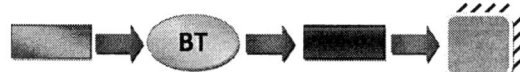

Figure 2. The Binary Translation (BT) process

Existing optimizations include dynamic recompilation and caching of previous binary translation results. For instance, the Daisy architecture is based on a VLIW processor that uses binary translation at runtime to better exploit the ILP of the application [13]. One of the advantages of using this technique is that this process is transparent, since there is no need for any modifications in the binary code. Consequently, it requires no extra designer effort and causes no disruption to the standard tool flow used during the software development.

2.3 Reuse of Instructions

The principle of Trace reuse [14] relies on the idea that sequence of instructions with the same operands will be repeated a large number of times during the program execution. Hence, instead of executing this sequence using the ordinary functional units, its result is fetched from a special memory. For this, there are the input and output contexts, which are saved in the Reuse Trace Memory (RTM). A context is composed by the program counter, registers and memory addresses. The input context is the

processor state considering the first instruction of a given sequence. The output context is, in turn, the set of results of all the instructions which belong to that sequence. Each time that an instruction with the same input context previously found is fetched again, the processor state is updated with the output context, avoiding the execution of all instructions that compose that trace (Figure 3).

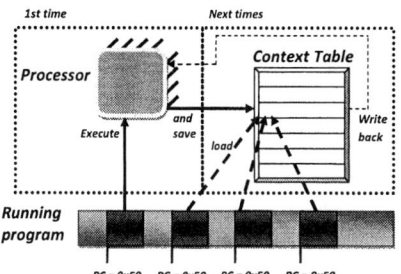

Figure 3. The Trace Reuse Technique

However, the context and trace sizes usually become huge, limiting the field of action of such approach, and increasing the complexity of the reuse detection algorithm. Good results are achieved just when using very optimistic assumptions, such as one cycle per trace reuse and the use of huge Reuse Trace Memories, not feasible even in future technologies because of power issues. The memory size grows too fast mainly because identical sequences of instructions, but with different contexts, must occupy different slots in this special memory.

2.4 Dynamic Detection and Reconfiguration

Trying to unify some of these ideas, Stitt et al. [15] presented the first studies about the benefits and feasibility of dynamic partitioning using reconfigurable logic, producing good results for a number of popular embedded system benchmarks. The structure of this approach, called warp processing, is a SOC. It is composed by a microprocessor to execute the software, another microprocessor where the CAD algorithm runs, a dedicated memory and an FPGA. Firstly, the microprocessor executes the binary, and a profiler monitors the instructions in order to detect critical regions. After that, the CAD software decompiles it to a control data flow graph, synthesizes it and maps the circuit onto a simplified FPGA structure.

However, although the CAD system is very simplified comparing to conventional ones, it remains complex: it does decompilation, CFG analysis, place and route etc, and, according to the work, 3.6 MB of memory are necessary for its execution, which is still huge for nowadays on-die memories. Another issue is the use of the FPGA itself: besides area consuming, it is also power inefficient because of the excessive switches and the considerable amount of static power. As a consequence, this technique is just limited to critical parts of the software working well just in very particular programs, such as filters.

In [16] it is also presented a very similar reconfigurable structure used in this work: a coarse-grain array, composed by very simple functional units, tightly coupled to an ARM processor. This array is called CCA. However, in the same way of the technique above, it relies on complex graph analysis, which is performed statically with compiler help. Moreover, it does not support memory operations or shifts, and has a very small number of input and outputs allowed, limiting its field of application.

2.5 Our Approach

Our work is based on a special hardware (Dynamic Instruction Merging Machine), designed in order to detect and transform of instructions to be executed in the reconfigurable hardware. This is done concurrently while the main processor fetches valid instructions. When this unit realizes that there is a certain number of instructions that are worth being executed in the array, a binary translation is applied to this sequence. This translation transforms the original sequence of instructions to a configuration of the array, which performs exactly the same function. After that, this configuration is saved in a special cache, indexed by the PC register.

The next time the saved sequence is found, the dependence analysis is no longer necessary: the processor just needs to load the configuration from the special cache and the operands from the register bank, setting the reconfigurable hardware as active functional unit. Then, the array executes the configuration with that context and writes back the results, instead of executing everything in the normal flow of the processor. Finally, the PC is updated, in order to continue the normal operation.

Depending on the size of the special cache used to keep these configurations, the increase in performance can be extended to the whole software, not being limited to loop centered applications. By transforming any sequence of opcodes into a single combinational instruction in the array one can achieve great gains, since less access to program memory and less iterations on the datapath are required.

In a certain way, the approach saves the dependence information of the sequences of instructions, avoiding performing the same job for the same sequence of instructions as superscalar processors do. It is interesting to point out that almost half of the number of pipeline stages of the Pentium 4 processor is related to dependence analysis [3]; and half of the power consumed by the core of the Alplha 21264 processor is also related to extraction of dependence information among instructions [17]. Moreover, both the DIM machine as the reconfigurable array work parallel to the processor, bringing no delay overhead or increasing the critical path of the pipeline structure.

Comparing to the techniques cited before, our approach also takes advantage of a reconfigurable system, but a coarse grain one, so it can be implemented in any technology, not just FPGAs. Together with that, we use binary translation to avoid the need for code recompilation or the utilization of extra tools, making the optimization process totally transparent to the programmer. The algorithm for the detection and transformation of binary code is very simple, in the sense that it takes advantage of the hierarchal structure of the reconfigurable array. Hence, the use of complex on-chip CAD software or graph analyzers is not necessary, saving another processor in the system just to perform this task.

Moreover, the proposed technique relies on the same basic idea of trace reuse, where sequences of instructions are repeated. However, it presents the advantage that just one entry in the special memory is needed for the same sequence of instructions, even when they have different contexts. This takes the pressure off from the cache system, making possible its implementation with a small memory footprint, with realistic assumptions concerning execution and accesses times, even for present day

technologies. Figure 4 summarizes the technique and its similarities with the previous ones.

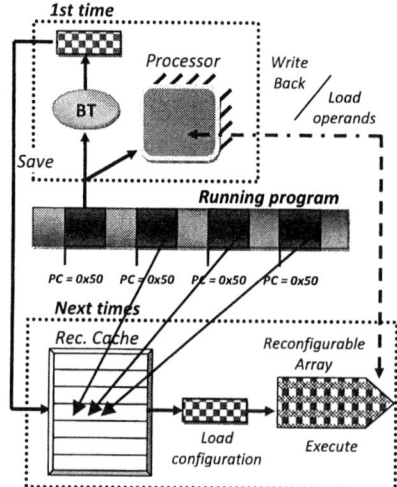

Figure 4. The proposed approach

3. THE RECONFIGURABLE SYSTEM

3.1 Architecture of the Array

The reconfigurable unit is a dynamic coarse-grain array tightly coupled to the processor, working as another functional unit in the execution stage, using the same approach of Chimaera [8]. This way, no external accesses to the array are necessary (which in turn could increase the delay and power consumption). Furthermore, this makes the control logic simpler, diminishing the overhead required in the communication between the reconfigurable array and the rest of the system. The array is two dimensional, composed by lines and columns, where each collum is represented by a set of ordinary functional units (ALU, shifter, multiplier, etc). Each instruction is allocated in a row. If two instructions do not have data dependence, they can be executed in parallel, in the same line.

The columns are divided in groups, where each group takes a determined number of cycles to be executed, depending on the delay of each functional unit. The delay can vary depending on the technology and the way the functional unit was implemented. The detection algorithm can be adapted to different delays. For instance, depending on the critical path of the processor, more sequential ALUs can be put together to be executed at the same cycle.

An overview of the general structure of the array is shown in Figure 5. Basically, there is a set of buses that receive the values from the registers. These buses will be connected to each functional unit, and a multiplexer is responsible for choosing which value will be used (Figure 5a). As can be observed, there are two multiplexers that will make the choice of which operand will be issued to the functional unit. We call them as input multiplexers.

After that, there is a multiplexer for each bus line that will choose what result will continue through that line. These are the output multiplexers (Figure 5b). As some of the values of the input context or previous results generated by previous operations can be used by other functional units after it was already used, the first input of each output multiplexer is the previous result of that bus.

Note that in the example used in Figure 5, the first group supports up to two loads to be executed in parallel, while in the second group three simple logic/arithmetic operations are allowed. The reconfigurable array can not afford any kind of floating point operation.

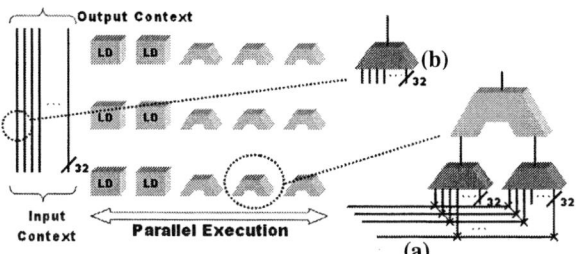

Figure 5. The structure of the Reconfigurable Array

3.2 Reconfiguration and Execution

As the detection for the address that will be used in the reconfiguration is done in the first stage of the pipeline, and the reconfigurable array is in the fifth stage, there are 4 cycles available between the detection and the use of the array. As one cycle is necessary to find the cache line that has the array configuration, three cycles are available for the reconfiguration, which involves the load of the values of all registers that will be used by that configuration, the load of immediate values, the configuration for the multiplexers and functional units and so on.

During the execution of the operations in the array, one issue is the load instructions. They stay in a different group in the array as shown in figure 5, and the number of columns of this group depends on the number of read ports available in the memory (which means the number of loads that can occur simultaneously). Operations that depend on the result of a load have already been allocated in the array during the detection phase, considering a cache hit as the total load delay. If a miss occurs, the whole array stops until it is resolved.

Finally, the results that need to be written back either in the memory or in the local registers are allocated in a buffer. The values will be allowed to be written back just when they are not used anymore for that configuration of the array. For instance, if there are two writes in the same register in a determined configuration, just the last one will be performed, since the first one was already consumed inside the array by other instructions.

3.3 The Binary Translation Algorithm

Some tables are necessary in order to perform the routing of the operands inside the reconfigurable array as well as the configuration of the functional units. Other intermediate tables are also needed, however, they are used just during the detection phase. These tables are:

Dependence table: Saves the information about data dependence of the instructions in a small bitmap, informing what registers will be written for the instructions that belong to each line. Since there it is no necessity of keeping this information for each separated instruction, one can reduce the hardware necessary to check true data dependencies (RAW – read after write).

Resource Table: Stores what function each functional unit must perform.

Read Table: Informs what operand from the input context must be read. This table has two inputs, since there are two source operands for each functional unit. It is important to point out that the input context is basically an indirect table. In other words, not necessarily the first slot of the bus needs to store the value of the register R1.

Write table: In this table one can find what value each context slot will receive. Note that this table is different when comparing to the read one. In the previous table the multiplexers were responsible for choosing what values from the context slots would be issued to each functional unit. This table informs what values from the whole set of the functional units that compose each line will continue in each slot of the context bus.

Context table: It has two lines, the first one representing the input context, and will be used in the reconfiguration phase, and the second one called current table, that will be used during the detection phase. Its final state represents what values will be written when the execution of the array finishes.

Summarizing the algorithm, for each incoming instruction, the first task is the verification of RAW (read after write) dependences. The source operands are compared to a bitmap of target registers of each line. If the current line and all above do not have that target register equal to one of the source operands of the current instruction, this instruction can be allocated in that line, in a row as left as possible, depending on the group, as explained before.

When this instruction is allocated in that line, the bitmap of target registers is updated. This way, for each instruction just one bitmap per line is necessary to be analyzed. Indirectly, such technique increases the size of the window of instructions, which is one of major limiting factors of ILP, exactly due to the number of comparators that is necessary [19]. For each line there is also the information about what registers can be written back or saved to the memory. This way, it is possible to write results back that will not be used anymore in the array in parallel to the execution of other operations.

The algorithm supports functional units with different delays and functions, and the use of immediate values in the input context; handles with false data dependencies among instructions; and performs speculative execution. For the speculative execution, each operand that will be written back has a flag indicating its depth concerning speculation. When the branch is taken, it triggers the writes of these correspondent operands.

The speculative policy is one of the simplest ones, based on bimodal branch predictor. For each level of the tree of basic blocks, the counter must achieve the maximum or minimum value (indicating the way of the branch). When the counter equals to this value, the instructions corresponding to this basic block are added to that configuration of the array. The configuration is always indexed by the first PC of the whole tree. If miss speculation occurs a determined number of times, achieving the opposite value of the respective counter, that entire configuration is flushed out and another one begins, starting everything again.

4. RESULTS

4.1 Performance

The Simplescalar toolset was employed for our experiments. We used the PISA instruction set, which is based on the MIPS IV ISA. Although the out-of-order simulator has some differences when comparing to the MIPS R10000 processor, we configured it to behave as close as possible to this processor. The configuration is summarized in Table 1a.

In Table 1b, we show three different configurations for the array that we used in the experiments. The last configuration was used in order to try to figure out what is the real potential of our technique. For each array configuration we also vary the size of the reconfiguration cache: 2 to 512 slots. Moreover, for each one of these configurations we evaluate the impact of doing speculation, up to three basic blocks ahead. Furthermore, we increased the cache memory in order to achieve almost no cache misses, so we can evaluate our results without the influence of it.

Table 1: Configurations

Out of Order
Fetch, decode and commit = up to 4 instructions
Register Update Unit = 16 Entries
Load/Store Queue = 16 entries
Functional Units = 2 Integer ALU, 1 multiplier, 2 memory ports
Branch Predictor = Bimodal/512 entries

(a)

	Reconfigurable Array		
	C #1	C #2	C #3
#Lines	27	54	99
#Columns	11	16	30
#ALU / line	8	8	11
#Multipliers /	1	2	3
#Ld/st / line	2	6	8

(b)

Table 2a shows the IPC of the out-of-order processor cited before. This table can be used to compare the IPC of this processor against the IPC of the instructions that are executed inside the array, in different configurations. Figure 3 shows this analysis. For each configuration, we vary the speculation: no speculation, 1 and 2 basic blocks ahead. We also change the number of slots available in the reconfigurable cache (4, 16, 64, 128 and 512). We are using a subset of the MIBENCH set [10].

As it is shown in Figure 6, we can achieve a higher IPC when executing instructions in the reconfigurable array in comparison to the out-of-order superscalar processor in almost all variations. However, the overall optimization when using our technique depends on how many instructions are executed in the reconfigurable logic instead of using the normal flow of the processor. Table 3 (at the end of the article) shows the overall speedup obtained when coupling the reconfigurable array to the out-of-order processor against the out-of-order without it.

Table 2: IPC in the Out-of-Order and BPI rate

Algorithm	IPC - Out-of-Order		Branch per Instr.
Basicmath	1.43		5.8751
CRC	2.13		7.9954
dijkstra	1.76		5.6011
Jpeg decode	1.86		6.2554
patricia	1.40		4.4255
qsort	1.79		4.6243
sha	1.94		7.9381
stringsearch	1.60		4.8709
Susan Smoothing	1.64		15.8098
Susan Corners	1.83		13.4952
tiff2bw	1.90		22.5567
tiff2rgba	1.92		13.4952
tiffdither	1.56		18.9188
tiffmedian	1.91		30.686

(a)   *(b)*

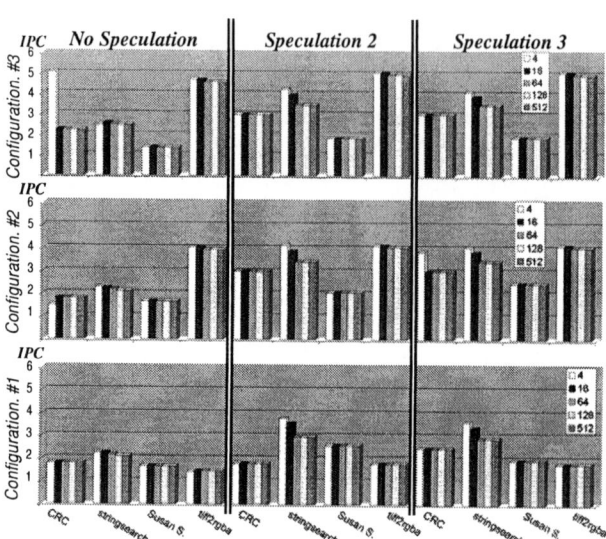

Figure 6. IPC rate in the reconfigurable array considering different configurations and cache sizes.

The four benchmarks were chosen because they represent a very control-oriented algorithm, a dataflow one and a midterm between both, plus the CRC, which is the biggest benchmark in the set. In Table 2b the benchmarks are classified according to the average number of branches per instructions. It is important to notice that reconfigurable systems in general can just show improvements when the programs are very dataflow oriented. The proposed technique, on the other hand, can optimize control and data oriented programs, as it ca be observed by the results.

4.2 Area Evaluation

In order to give an idea of the area overhead, we implemented the hardware detection and the reconfigurable array in VHDL. The tool used was the Mentor Leonardo Spectrum [9], with the library TSMC 0.18u. As we do not have available any implementation of a superscalar processor in any Hardware Description Language, we took the data about its number of transistors from [18] and other measurements from [19]. Although this comparison will not give us exactly values, it will present realistic measurements about the implementation of our approach.

Table 4a shows how many functional units and multiplexers would be necessary to implement the configuration #1 of table 1, and what are the number of gates they take. In this same table one can also observe the number of gates taken by the Dynamic Instruction Merging hardware. In table 4b it is shown the number of bits necessary to keep one configuration in the reconfigurable cache. Note that, although 256 bits are necessary for the Write Bitmap Table, they are not counted in the final total. This table is temporary and is used just during detection. This way, there is no need to save its values in the special cache. Finally, in table 4c, the number of Bytes needed for different cache sizes is presented, depending on how much configurations they can store.

Table 4: Area evaluation for the reconfigurable array

Unit	#	Gates
ALU	216	337,824
LD/ST	36	5,904
Multiplier	6	20,067
Input Mux	510	327,420
Output Mux	216	66,096
DIM Hardware		1,024
Total		735,223

(a)

Table	#bits
Write Bitmap Table	256
Resource Table	903
Reads Table	1,896
Writes Table	648
Context Start	40
Context Current	40
Immediate Table	128
Total	3655

(b)

#Slots	#Bytes
2	7,566
4	14,620
8	30,143
16	58,480
32	118,856
64	233,920
128	468,488
256	935,680

(c)

Finally, Figure 7a presents the layout of the Superscalar MIPS processor. According to [18], the total number of transistors of core in the MIPS R10000 is 2.4 million. As presented in table 4a, the array together with the hardware detection occupies 735,223 gates. We are considering that one gate (result given by the synthesis tool) is equivalent to 4 transistors, which would be the amount necessary to implement a NAND or NOR gates. This way, the reconfigurable array and DIM hardware would take 2,940,892 transistors. The area overhead is represented in Figure 6b. In this figure is also presented the area overhead concerning the reconfigurable cache, in number of different configurations supported.

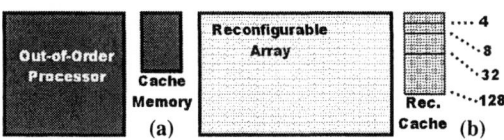

Figure 7. Area overhead presented by the reconfigurable array and its special cache

5. CONCLUSIONS AND FUTURE WORK

Although there are some improvements concerning the algorithm and the structure of the reconfigurable array the need to be done, this work demonstrated that it is possible to keep advantage of a reconfigurable architecture to speed up the system, in a totally transparent process and with a feasible area overhead. Using speculation in the array, we have obtained a mean speedup of up to 30% in the IPC using configuration 3, when comparing against a MIPS R10000 based superscalar processor. Now, we are working on finding the best shape for the reconfigurable array.

Another future work will be the measurement of the energy consumption of the system. Similar techniques applied to an embedded processor have already shown that such structures bring a huge energy saving [20] since, besides the fact that this technique trades sequential logic for combinational one to execute instructions, less accesses to the instruction memory are required, as well as less dependence analysis between instructions are necessary.

6. REFERENCES

[1] David W. Wall, "Limits of instruction-level parallelism", In Proceedings of the fourth international conference on Architectural support for programming languages and operating systems, p.176-188, April 08-11, 1991

[2] Sima, D, "Decisive aspects in the evolution of microprocessors". In Proceedings of the IEEE, vol. 92, pp. 1896-1926, 2004

[3] Intel Pentium 4 Homepage – http://www.intel.com/products/processor/pentium4/index.htm

[4] Venkataramani, G., Najjar, W., Kurdahi, F., Bagherzadeh, N., Bohm W., "A Compiler Framework for Mapping Applications to a Coarse-grained Reconfigurable Computer Architecture. Conf. on Compiler". In Architecture and Synthesis for Embedded Systems (CASES), 2001.

[5] Stitt, G., Vahid F., "The Energy Advantages of Microprocessor Platforms with On-Chip Configurable Logic". In IEEE Design and Test of Computers, 2002

[6] Z. Or-Bach, "Panel: (when) will FPGAs kill ASICs?". In 38th Design Automation Conference, 2001.

[7] K. Compton and S. Hauck, "Reconfigurable computing: A survey of systems and software". In ACM Computing Surveys, vol. 34, no. 2, pp. 171-210, June 2002.

[8] Hauck, S., Fry, T., Hosler, M., Kao, J., "The Chimaera reconfigurable functional unit". In Proc. IEEE Symp. FPGAs for Custom Computing Machines, Napa Valley, CA, pp. 87–96. 1997.

[9] Leonardo Spectrum, available at homepage: http://www.mentor.com

[10] Guthaus, M.R., Ringenberg, J.S., Ernst, D., Austin, T.M., Mudge T., Brown, R.B., "MiBench: A Free, Commercially Representative Embedded Benchmark Suite". In 4th Workshop on Workload Characterization, Austin, TX, Dec. 2001

[11] Swanson, S., Michelson, K., Schwerin, A., Oskin. M., "WaveScalar. MICRO-36", Dec. 2003

[12] Gschwind, M., Altman, E., Sathaye, P., Ledak, Appenzeller, D., "Dynamic and Transparent Binary Translation". In IEEE Computer, vol. 3 n. 33, pp. 54-59, 2000

[13] K. Ebcioglu, E. A., "DAISY: Dynamic compilation for 100% architectural compatibility". IBM T. J. Watson Research Center - Technical Report, Yorktown Heights, NY, 1996.

[14] González, A., Tubella, J., Molina, C., "Trace-Level Reuse". In Int'l Conf. on Parallel Processing, Sep. 1999.

[15] Stitt, G., Lysecky, R., Vahid, F., "Dynamic Hardware/Software Partitioning: A First Approach". In Design Automation Conference, 2003.

[16] Clark, N., Tang, W. Mahlke, S., "Automatically Generating Custom Instruction Set Extensions". In Workshop on Application Specific Processors (WASP). Turkey, 2002.

[17] Wilcox K., Manne, S., "Alpha processors: A history of power issues and a look to the future". In CoolChips Tutorial An Industrial Perspective on Low Power Processor Design in conjunction with Micro-33, 1999.

[18] Yeager, K.C., "The Mips R10000 Superscalar Microprocessor,"; IEEE Micro, pp. 28-40, Apr. 1996.

[19] Burns, J.; Gaudiot, J.-L., "SMT layout overhead and scalability". In Parallel and Distributed Systems, IEEE Transactions on On page(s): 142-155, Volume: 13, Issue: 2, Feb 2002

[20] Beck, A. C. S., Carro, L., "Dynamic Reconfiguration with Binary Translation: Breaking the ILP barrier with Software Compatibility". In: Design Automation Conference (DAC), 2005

Table 3: Speedups using the reconfigurable array coupled to the out-of-order processor

Algorithm	#Cycles in the Out-Of-Order	% of Speed Up - Out-of-Order coupled to array with configuration 1									% of Speed Up - Out-of-Order coupled to array with configuration 3								
		No Speculation			Speculation 2			Speculation 3			No Speculation			Speculation 2			Speculation 3		
		4	64	256	4	64	256	4	64	256	4	64	256	4	64	256	4	64	256
Basicmath	111169924	5.03	13.75	17.85	3.52	14.49	21.79	3.40	15.22	23.31	5.76	19.27	26.40	4.63	19.83	30.33	4.86	20.52	32.14
CRC	399531928	-16.01	-16.03	-16.03	-5.20	-5.21	-5.21	9.03	9.03	9.03	3.97	3.97	3.97	8.12	8.14	8.14	20.75	20.77	20.77
dijkstra	31094638	-22.29	-24.31	-24.33	1.30	1.25	1.25	8.45	8.46	8.46	-21.96	-20.08	-20.04	1.00	4.34	4.36	4.13	7.65	7.67
Jpeg decode	3942226	-9.15	-9.72	-9.77	4.63	3.24	3.29	7.11	7.45	7.61	9.76	11.92	12.05	16.55	18.94	19.06	16.77	19.51	19.68
patricia	95927575	4.41	13.30	13.72	3.99	14.42	21.52	3.26	14.22	21.96	5.06	17.97	18.89	5.25	18.80	29.07	4.57	18.58	29.80
qsort	23435690	-8.76	-11.69	-11.69	4.18	4.18	4.18	0.37	-30.41	-30.21	24.29	38.95	38.95	16.79	43.74	43.74	16.44	40.72	40.72
sha	6800950	11.56	13.07	13.07	27.22	33.45	33.45	26.30	31.29	31.29	22.57	25.48	25.48	39.91	48.66	48.66	41.27	50.28	50.28
stringsearch	115917	16.32	20.16	21.23	28.95	35.20	35.24	28.50	35.39	35.38	21.02	27.05	30.57	31.25	41.02	41.17	31.04	42.61	42.63
Susan Smoothing	15628090	-0.94	-3.22	-3.22	0.31	-0.99	-1.00	2.13	1.59	1.59	25.35	35.66	35.69	26.87	37.95	37.96	23.73	32.05	32.04
Susan Corners	533870	2.16	1.79	1.79	4.40	4.29	4.28	1.13	4.29	4.28	32.69	41.44	41.44	37.53	41.44	41.45	33.89	37.13	37.12
tiff2bw	27391803	-4.24	-4.38	-4.42	0.88	0.82	0.82	-0.20	-0.20	-0.20	-5.65	-5.42	-5.39	19.08	19.60	19.60	24.41	25.22	25.22
tiff2rgba	23796384	-10.94	-11.39	-11.40	-1.53	-1.75	-1.75	-1.19	-1.39	-1.40	57.19	57.83	57.83	58.29	59.69	59.69	47.30	48.87	48.87
tiffdither	188757828	1.48	8.88	8.92	6.65	9.34	9.41	4.47	-21.46	-23.52	4.33	18.15	18.30	10.73	19.33	19.57	7.95	14.31	14.60
tiffmedian	93254386	3.95	3.74	3.73	12.91	12.82	12.82	7.42	7.38	7.38	14.13	14.11	14.13	27.23	27.43	27.43	27.36	27.72	27.72

VLSI Models of Network-on-Chip Interconnect

Dimitrios N. Serpanos[*]

Dept. of Electrical & Computer Engineering
University of Patras, Greece
serpanos@ece.upatras.gr
[*] Also, with the Industrial Systems Institute
(ISI), Patras, Greece

Wayne Wolf

School of Electrical & Computer
Engineering
Georgia Institute of Technology
USA
wayne.wolf@ece.gatech.edu

Abstract

We use VLSI circuit models to analyze the relative delay of interconnect subsystems for networks-on-chips (NoCs). Most work in NoCs has selected a network topology based on higher-level performance models, such as packet delay. Our model parameterizes the interconnect subsystem size by N, the number of IP cores (processors, memories, etc.) to be connected. This paper analyzes busses, crossbars, and some multi-stage networks. We compare the delay required transfer a specific amount of information (bits) between two cores. Considering the data transfer parallelism in crossbars, we make 2 different comparisons: (i) transfer between 2 devices, and (ii) parallel transfers between all devices.

1. Introduction

Networks-on-chips (NoCs) are important subsystems in systems-on-chips and chip multiprocessors. Although many NoCs have been designed, the network topology was in most cases decided fairly early in the design process based largely on architectural considerations; several studies [6,5,9], for example, design NoCs based primarily on traffic considerations with technology parameters used primarily for tuning. However, circuit characteristics play an even bigger role in the design of on-chip networks than they do in the networks that connect processors on boards or racks. We know of no VLSI analysis that compares the characteristics of different network topologies.

In this paper, we develop models for the delay of networks-on-chips. These models take into account circuit characteristics such as interconnect and device capacitance. The models are parameterized by N, the number of cores to be connected by the network. N is related to the length and area characteristics that are so important to the circuit behavior of interconnect. We can build some common model elements, but we will develop separate models for each network topology. Because those models are based upon a common set of design and technology parameters, we can compare the characteristics of different topologies as a function of the size of the multiprocessor.

This paper develops models for several topologies: bus, crossbar, and some multi-stage networks. We know of no other work with a similar comparison of VLSI interconnection networks that relates technology and architectural parameters.

The next section introduces some basic notation for our models. We then develop a model for a single-cycle transaction followed by a model for multi-cycle transactions. We follow that by an analysis of multi-stage networks.

2. Problem Definition & Notation

We assume that N **cores** (IP modules) are connected through the interconnection network. The N cores are capable of both I/O operations, thus the number of inputs (n_i) to the interconnect is the same as the number of outputs (n_o) and $n_i = n_o = N$.

In the bus configuration, control of the bus is determined by a bus arbiter, while, in the case of the crossbar switch, the decision of the transfers to be made is performed by a scheduler.

Our comparison must take into account physical characteristics of the layout. In the

978-1-4244-1709-4/07/$25.00 © 2007 IEEE

case of the bus interconnect, we consider a layout in which the N devices are connected on the bus with a spacing that is normalized to distance 1; thus, the length of the bus (L_b) is N. The bus is composed of control (CL), address (AD) and data (WB) lines. The cycle of the bus is denoted, in general, as T_B.

In the case of the crossbar switch, we consider a layout in which the N cores are connected in an NxN configuration, where the inputs to the switch are the interconnected device (core) outputs and the outputs of the switch are the interconnected core inputs; this is the only configuration that allows the any-to-any connectivity through the switch as is possible with the bus. Each connection through the switch is composed of data lines only (WS lines in width). The switch is assumed to be synchronous operating with a clock whose cycle is denoted as T_S.

3. Analytical Comparison

In this section we compare interconnects for single cycle transmissions (data transfers).

3.1. Bus performance

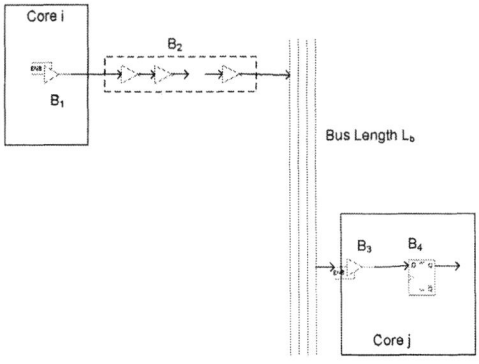

Figure 1: Bus model

Figure 1 shows the bus model used for the analysis. Bus-attached cores have input and output connections to the bus, which has a normalized length L_b= N. In our model we consider a long bus with short core-to-bus connections, rather than a short bus with long core-to-bus connections. Our decision is based on the fact that the typical practice in NoC system design follows our model, in contrast to the use of multiple long core-to-

bus connections that lead to an area-inefficient design that would simply replace tri-state I/O buffers in the core-to-bus connections with multiplexers/de-multiplexers.

The main bus wires have length L_b. Each input has its own tri-state buffer/driver (B_1) and a sequence of cascaded buffers to drive the long bus wire (B_2); each one of the outputs has its own input tri-state buffer (B_3), with capacitive load C_L, and is connected to a register (B_4) that stores the incoming data. The total bus delay is the sum of these three components:

$$\delta_b = \delta_{in} + \delta_w + \delta_{out} \qquad \text{(EQ-1)}$$

where δ_{in} is composed of the delays to drive the tri-state buffer B_1 and the cascaded buffers B_2, δ_w is the wire delay on the bus and δ_{out} is the delay of the tri-state buffer B_3 driving the register B_4.

Considering the long bus wires and the large capacitive load driven by a transmitter on the bus, we calculate the delays as follows. As shown in Figure 1, we consider the path from one transmitter on the bus. The buffers B_1, B_2 and B_3 need to be sized so that buffers B_1 and B_2 constitute a cascade, i.e. we consider B_1 and B_2 as a cascade altogether (i.e., we view B_1 as part of B_2). The cascade drives a large load composed of N buffers B_3, which is $N*C_L$. Thus, disregarding the delay of the long wire, the delay $(\delta_{in}+\delta_{out})-\delta_{B4}$, where δ_{B4} is the delay of the last buffer B_4, is computed using the formula for exponentially tapered buffers [1] to:

$$\delta_{in} + \delta_{out} - \delta_{B4} = k\left(N\frac{C_L}{C_g}\right)^{1/N} t_{min} \text{ (EQ-2)}$$

In this delay, k is the number of stages in the cascaded buffers (B1 and B2), t_{min} is the time to drive a minimum size load, which has a capacitance C_g. Thus, in EQ-2, C_g and t_{min} are constants that depend on the technology.

The wire delay δ_w is a function of the length L_b (which is normalized to N) and is given by the Elmore delay:

$$\delta_w = \frac{1}{2}RCN(N-1) \qquad \text{(EQ-3)}$$

Finally, the delay δ_{B4} is determined using a simple τ model by the load capacitance of buffer B_4 and the register (buffer B_4) delay:

$$\delta_{B4} = 0.69(R_n + R_l)C_L \qquad \text{(EQ-4)}$$

where R_n is the effective resistance of a transistor for the technology used and R_l is the resistance of the gate directly connected to buffer B_4 , i.e. the resistance of the first gate of the "core" after the register B_4. Since all interconnection configurations (bus and crossbar) have a delay δ_{B4}, which is the same in value, for simplicity we will disregard this delay in the remaining calculations as well as in the crossbar case.

Thus, the above give a total bus delay as:

$$\delta_b = \delta_{in} + \delta_{out} - \delta_{B4}$$

$$= k\left(N\frac{C_L}{C_g}\right)^{1/k} t_{min} + \frac{1}{2}RCN(N-1)$$

$$= k_1 C_L^{1/k} N^{1/k} + k_2 N + k_3 N^2 \qquad \text{(EQ-5)}$$

Thus, delay δ_b , i.e. the bus cycle T_B, is $O(N^2)$, where N is the number of interconnected systems (cores).

3.2. Crossbar performance

For the crossbar, we will consider two different approaches: (a) an inverter-based and (b) a multiplexer-based design. The first approach leads to an easier and effective design for small switches, while the multiplexer approach provides several advantages over alternatives, including a more compact design that leads to shorter wire lengths [3].

Buffer-based design

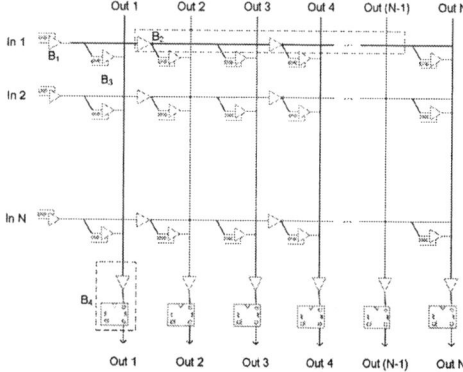

Figure 2: Inverter-based switch

Figure 2 shows the crossbar model. Importantly, in the case of crossbar, we

assume that the transmissions are uni-directional, i.e., since we have N cores and the switch is NxN, each core is connected to both an input and an output and data transmission occurs only in one direction (from inputs to outputs). This is a significant difference from the bus model where wires are used for bidirectional transmission, because unidirectional transmission allows us to insert buffers (repeaters) in the connections and thus, obtain lower transmission times. As Figure 2 shows, a transmitting core is connected to the switch with a buffer (B_1) that drives N tri-state buffers B_3 (one per output, but only one will be Enabled by the switch scheduler) and each tri-state buffer B_3 will drive a register B_4 (similarly to the bus), which will latch the transmitted data at the receiver's side. The transmission between buffers B_1 and B_3 is performed through a line with repeating buffers, as explained above, which is modeled as an RC transmission line with inserted buffers; the collection of repeaters is mentioned in the Figure as one buffer B_2.

The transmission delay (δ_c) through the crossbar is the sum:

$$\delta_c = \delta_{ic} + \delta_x + \delta_{oc} \qquad \text{(EQ-6)}$$

where δ_{ic} is the crossbar input delay, i.e. the delay of the input tri-state buffer (B_1), δ_x is the transmission delay through the internal crossbar (RC line with inserted buffers B_2), and δ_{oc} is the output delay through buffer B_3 and to latch B_4. Similarly to the case of the bus above, we will not include the delay of latch B_4 in our calculations.

In this case, the following hold:
1. the length of the wires of the crossbar is N in both axes, x and y, considering that we have N inputs and N outputs to the switch;
2. the connections through the crossbar are uni-directional.

So, in this case, we design the switch by inserting buffers in the lines (in the x and y directions) so that the overall transmission delay is reduced and the buffer that switches the signal from the x direction to the y direction in an established path constitutes a buffer of the sequence in the overall path. Thus, in order to calculate the delay of the switch we need to calculate the delay of the longest path, i.e. a path length $L_c = 2 * N$ (normalized, as in the case of the bus). Considering that the longest path passes through (2N-1) cross-points, the total capacitance of the longest path is at most $(2N-1)*C_L$.

Based on the above, the total delay of the longest path can be approximated with Bakoglu's formula [9]:

$$\delta_c = 2.5\sqrt{R_0 C_0 R_{int} C_{int}}$$
$$= 2.5\sqrt{N}\sqrt{R_0 C_0 R_{int} C_L} \qquad \text{(EQ-7)}$$

Thus, the delay of the buffer-based switch is $O(N^{1/2})$, resulting in a switch cycle $T_S = O(N^{1/2})$.

Multiplexer-based design

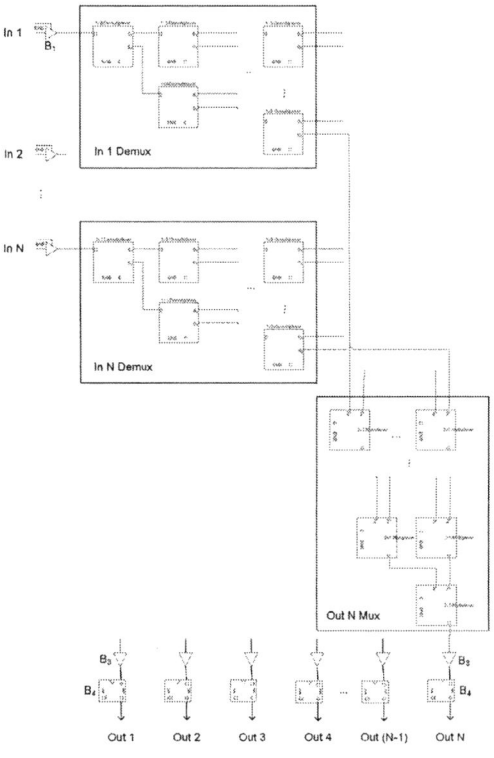

Figure 3: Multiplexer-based switch

For a multiplexer-based switch, the layout is shown in Figure 3. As the figure shows, a transmitting core is connected to the switch with a buffer (B_1) that drives an 1-to-N de-multiplexer (or a tree of 1-to-2 de-multiplexers, as shown in Figure 3), which in turn is connected to N N-to-1 multiplexers (possibly implemented as a tree of 2-to-1 multiplexers, as shown in Figure 3). At each output there exists a buffer B_3 that drives a register B_4 (similarly to the bus), which will latch the transmitted data at the receiver's side.

According to Dutta et al. [3], the crossbar switch's vertical control lines (the control lines of the multiplexers and de-multiplexers) grow as a function of N logN, where N is the size of the switch (N inputs and N outputs). Thus, assuming normalized lengths, similarly to the case of the bus, this leads to a data path in the crossbar (input-to-output) to have length L_s= N*logN. This longest path, which is the worst case scenario for a bit transfer through the switch, is the one that specifies the clock cycle of the switch and thus, we will use it for our analysis. Importantly, the path contains a sequence of multiplexers (or a single mux) and a sequence of de-multiplexers (or a single de-mux); Figure 3 shows the case of the sequence of (de)multiplexers, which is 2logN in a path (logN in the tree of de-multiplexers at the input and logN in the tree of multiplexers at the output).

The transmission delay (δ_c) through the crossbar is the sum:

$$\delta_c = \delta_{ic} + \delta_x + \delta_{oc} \qquad \text{(EQ-8)}$$

where δ_{ic} is the crossbar input delay, i.e. the delay of the input buffer (B_1), δ_x is the transmission delay through the internal crossbar (RC line with inserted buffers B_2), and δ_{oc} is the output delay through buffer B_3 and to latch B_4.

Based on EQ-3, we can calculate delays δ_{ic} as:

$$\delta_{ic} = 0.69(R_n + R_L)C_L \qquad \text{(EQ-9)}$$

while $\delta_{oc} = \delta_{out}$, similarly to the case of the bus.

The capacitance seen from the input of a transmission gate (n-type + p-type switch) is [7]:

$$C_{tg} = C_L + \left(\frac{C_{ox}}{2} + \frac{C_{ox}}{2}\right) \qquad \text{(EQ-10)}$$

The delay through the mux element is the sum of the transmission gate and inverter delays:

$$\delta_c = 0.7(C_L + C_{ox}) + 0.69(R_n + R_L)C_L \qquad \text{(EQ-11)}$$

We calculate δ_x as a delay through a de-multiplexer tree and a multiplexer tree:

$$\delta_x = \delta_{mux}\log N + \delta_{mux}\log(N-1) \qquad \text{(EQ-12)}$$

The input and output buffer delays are negligible compared to the de-multiplexer and multiplexer delays. So, we can approximate the total crossbar delay as:

$$\delta_c \approx \delta_x = 2\delta_{mux}\log(N-1) \qquad \text{(EQ-13)}$$

Thus, the delay of a multiplexer-based crossbar design is logarithmic, leading to a crossbar cycle $T_S = O(\log N)$.

4. Interconnection networks

The bus and the crossbar switch constitute the two ends of the spectrum of interconnections among N systems. A wide range of multi-stage interconnection networks have been proposed for interconnection of N systems. Multi-stage networks implement multiple parallel data paths, achieving lower average data transfer delay than busses; however, their parallel data paths are fewer than the N parallel paths of a crossbar switch, leading to lower implementation complexity, since their complexity grows as a smaller function of the $O(N^2)$ of a crossbar switch. The complexity of typical multistage interconnections networks, such as the Omega, Butterfly, etc., grows as $O(N\log N)$. Thus, among other applications, they constitute attractive interconnects for NoC interconnections.

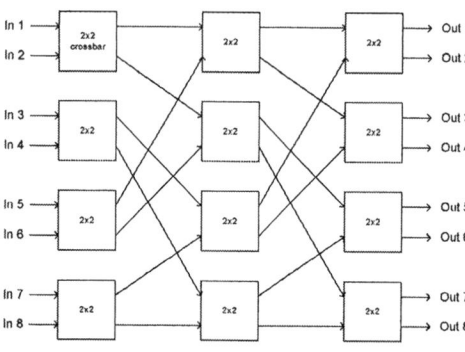

Figure 4: Omega interconnection network

A typical multistage interconnection network, such as the Omega network shown in Figure 4, is characterized by the use of $\frac{N}{2} \times \log N$ crossbar switches (2x2 switches) organized in logN stages, each

with (N/2) crossbars; Figure 4 shows an 8x8 Omega network with 3 stages of 4 2x2 switches each. Thus, a path through the interconnection network includes lg N 2x2 crossbar switches. Importantly, the wires that are included in the interconnection network are unidirectional and connect switches between successive stages of the interconnection network. Thus, the wire length grows proportionally to the size of a stage (N/2 switches) and their layout leads to a maximum size proportional to N/4, i.e. maximum length wires connect switches that are located half a stage away. So, for the maximum length wires, we can consider that they constitute transmission lines that are designed as RC lines with buffer insertion.

Considering the above, we can calculate the maximum delay over a path through the interconnection network as:

$$\delta_i = \log N \times \delta_w \delta_c \qquad \text{(EQ-14)}$$

where δ_w is the delay of a maximum length wire and δ_c is the delay of a 2x2 crossbar switch. Assuming a buffer-based switch, we can use the formulas from Subsection 2.2 to calculate both δ_w and δ_c. The delay δ_w can be calculated as the delay of a buffered RC line (since, for large N, the length of the wires is proportional to N/4 and the wires are unidirectional):

$$\delta_w = 2.5\sqrt{R_0 C_0 R_{int} C_{int}} \qquad \text{(EQ-15)}$$

where R_{int} and C_{int} are the total resistance and capacitance of the wire. Also, δ_c is

$$\delta_c = 2.5\sqrt{R_0 C_0 R_{int-w} C_L} \qquad \text{(EQ-16)}$$

where C_L is the load of the outgoing wire from the switch. This load is a constant, considering the the outgoing switch is a buffered RC line and thus, C_L is, basically, the load of the first buffer of the wire.

So, finally, the delay δ_i is:

$$\delta_i = \log N \times \delta_w \delta_c$$
$$= 2.5^2 \log N \times R_0 C_0 \sqrt{R_{int} R_{int-w} C_{int} C_L} \qquad \text{(EQ-17)}$$

leading to a path delay that is $O(\log N)$.

The above analysis can be easily extended to interconnection networks that are not multistage networks, such as meshes, trees, etc. Importantly, in most of these networks the wire lengths are constant and not a function of the size of interconnected systems. In such networks, the maximum delay through the network can be easily calculated by identifying the longest path (critical path) of the interconnection and taking into account the delays of the wires (constant) and the delays of the used switches.

Clearly, the above analysis applies to both buffered and non-buffered interconnection networks; the differences in the calculations originate only from the different resistance and capacitive loads in each case.

5. Conclusions

This work provides a modeling approach for the VLSI characteristics of networks-on-chips as well as specific results for busses, crossbars, and some multi-stage networks. Our modeling methodology allows network-on-chip designers to relate technology parameters directly to the characteristics of network components and topologies.

An important aspect of this work is to demonstrate that the use of crossbar switches and interconnection networks leads to improved performance not only due to the exploitation of parallelism, but because of higher clock rates as well. The significant difference between clock cycle lengths in busses, $T_B = O(N^2)$, and crossbar switches, $T_S = O(N^{1/2})$ for buffer-based and $T_S = O(logN)$ for multiplexer-based switches, indicates that parallelism in data transfers is not only desirable for multiple parallel transfers, but necessary for higher clock rates that benefit the interconnected cores as well.

Our future work will extend our analysis of multi-stage networks as well as power consumption. We also hope to relate application-level parameters to VLSI design characteristics.

Acknowledgments

This work was supported in part by the National Science Foundation under grant CNS-0509463.

References

[1] Wolf W., *Modern VLSI Design – System-on-Chip Design* (3rd Ed.), Prentice Hall, 2002.

[2] Dutta S. and Wolf W., *Asymptotic Limits of Video Signal Processing Architectures*, IEEE Transactions on Circuits and Systems for Video Technology, 5(6), Dec. 1995, pp. 545-561.

[3] Dutta S., O'Connor K.J. and Wolfe A., *High-Performance Crossbar Interconnect for a VLIW Video Signal Processor*, in Proceedings, Ninth Annual ASIC Conference and Exhibit, IEEE, 1996, pp. 45-49.

[4] Giovanni De Micheli and Luca Benini, eds., *Networks on Chips : Technology and Tools*, Morgan Kaufman, 2006.

[5] Axel Jantsch and Hannu Tenhunen, eds., *Networks on Chip*, Kluwer Academic Publishers, 2003.

[6] Neil H. E. Weste and David Harris, *CMOS VLSI Design: A Circuits and Systems Perspective*, 3rd Edition, Addison Wesley, 2005.

[7] R. Jacob Baker, *CMOS: Circuit Design, Layout, and Simulation*, Wiley-Interscience, 2005.

[8] H. B. Bakoglu, *Circuits, Interconnections, and Packaging for VLSI*, Addison-Wesley, 1990.

[9] K. Goossens, J. Dielissen, O. P. Gangwal, S. G. Pestana, A. Radulescu, and E. Rijpkema, *A design flow for application-specific networks on chip with guaranteed performance to accelreate SoC design and verification*, in Proceedings of the Conference on Design Automation and Test in Europe, vol. 2, IEEE Computer Society Press, 2005, pp. 1182-1187.

Statistical Analysis of Systematic and Random Variability of Flip-Flop Race Immunity in 130nm and 90nm CMOS Technologies

Gustavo Neuberger, Fernanda Kastensmidt, Ricardo Reis
Universidade Federal do Rio Grande do Sul (UFRGS)
Instituto de Informática
Programa de Pós-Graduação em Microeletrônica (PGMicro)
Porto Alegre, RS, Brazil
{neuberg, fglima, reis}@inf.ufrgs.br

Gilson Wirth
Universidade Federal do Rio Grande do Sul (UFRGS)
Departamento de Engenharia Elétrica
Porto Alegre, RS, Brazil
wirth@inf.ufrgs.br

Ralf Brederlow +, Christian Pacha
Infineon Technologies
Munich, Germany
Christian.Pacha@infineon.com
+) since October 2006 with Texas Instruments, Freising, Germany

Abstract

Statistical process variations are a critical issue for circuit design strategies to ensure high yield in sub-100nm technologies. In this work we investigate the variability of flip-flop race immunity in 130nm and 90nm low power CMOS technologies. An on-chip measurement technique with resolution of ~1ps is used to characterize hold time violations of flip-flops in short logic paths, which are generated by clock-edge uncertainties in synchronous designs. Statistical die-to-die variations of hold time violations are measured various register-to-register configurations and show overall 3σ die-to-die standard deviations of 12-16%. Mathematical methods to separate the measured variability between systematic and random variability are discussed, and the results presented. They show that while systematic variability is the major issue in 130nm, it is significantly decreased in 90nm technology due to better process control. Another important point is that the race immunity decreases about 30% in 90nm, showing that smaller clock skews can lead to violations in 90nm.

1. Introduction

Modern synchronous digital designs necessarily include a large amount of flip-flops (FF) in pipeline stages to improve data throughput. FF timing is determined by the CLK-Q propagation time, setup time and hold time. Complying with the specified setup and hold times is a pre-requisite for a stable sampling of the data signal around the clock edge. Due to the increasing relevance of process, voltage and temperature variations for robust circuit operation in modern CMOS technologies on the one hand and the frequent use of FFs in microprocessor, DSP cores and dedicated hardware on the other hand, a precise statistical characterization of FF is mandatory. This has motivated investigations of variability of the FF propagation time using Monte Carlo simulation [1]. Statistical variations of setup and FF propagation times in critical paths are essential for maximum chip performance. In contrast to this, a violation of the hold time in short FF-logic-FF paths lead to complete chip failure. In this case races in short pipeline stages are generated by a combination of clock skew and jitter between sending and receiving FFs, and process variations within the circuits. The internal race immunity is a figure of merit to characterize the robustness of a FF against race conditions and is defined as the difference between clock-to-Q delay and hold time. Hence, the race immunity strongly depends on the specific FF type [2].

Especially scan chains for DFT schemes [3] are sensitive circuit structures since no logic is placed between the FFs. Several techniques for diagnosis of hold time failures in scan chains [3-6] as well as in generic short logic paths [7] are proposed. These techniques are applied for buffer insertion, i.e. hold time fixing, to increase the delay of these paths during chip design [8]. However, depending on the design and FF properties, without detailed analysis of the critical clock skew and process variability, the extra delay introduced during hold-time fixing can be over or under estimated. In this work, we therefore present a statistical analysis of the race immunity in several test paths, due to process variability in 130nm and 90nm CMOS technologies. The experimental data is obtained using a precise on-wafer measurement technique with ~1ps resolution. This measurement technique has been presented in [9] for a 130nm CMOS technology and is here transferred to 90nm CMOS to facilitate a comparison between both technologies.

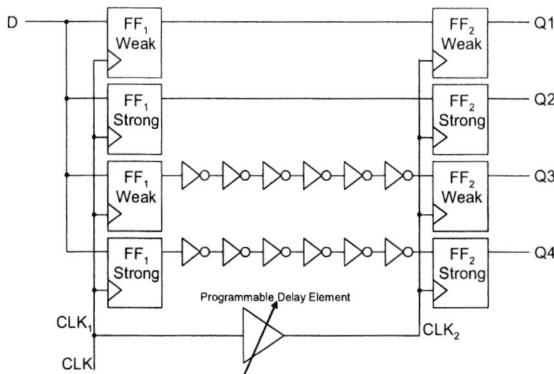

Figure 1. Different test circuits with sensitivity to race conditions.

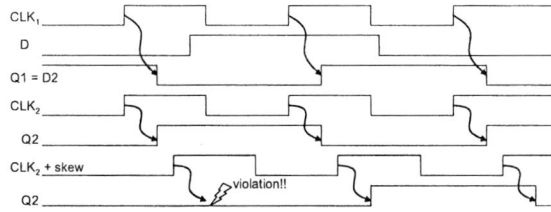

Figure 2. Timing diagram showing hold time violation.

The paper is organized as follows: section II describes the test circuits, timing issues, and measurement scheme for a precise FF characterization. Section III discusses different mathematical methods to separate experimental data between systematic and residual random variability. Section IV contains experimental results measured on test circuits fabricated in 130nm and 90nm low power CMOS technology. Finally, the paper is concluded in section V.

2. Test Circuit and Measurement

To evaluate the impact of statistical variations on hold time violations four different logic paths are considered, the same used in [9]. The two basic configurations are two simple pipeline stages with two master-slave edge-triggered FFs without logic between them, representing one stage of a scan chain. Further pipelines including six small inverters between the FFs, represent short logic paths. The FFs used in this work are conventional rising edge-triggered master-slave FFs composed of CMOS transmission gates in the forward propagation path and C2MOS latches in the feedback loops [10] with typical library extensions such as input and output node isolations and local clock buffers.

For each configuration a version with the weakest FF of the standard cell library, i.e. smallest transistor sizes and hence largest sensitivity to process variations, and a version with 8x increased driving strength is used. Comparing the results of both it is possible to analyze the impact of different transistor dimensions on the variability. The inverters used in both versions are of the minimum size, since these configurations represent typical non-critical paths where large driving capability is not required.

To emulate clock uncertainties, the sending and receiving FFs are controlled by different clock signals. The clock signal CLK2 of the receiving FFs is generated by a programmable delay line as shown in fig. 1. If this artificial clock skew is large enough, i.e. CLK2 arrives after CLK1 and exceeds the internal race immunity t_{CLK-Q}-t_{HOLD} of the FF, a race is produced and detected if the output of both FFs are of same value at same time (Q1(t)=Q2(t)). The violation can be detected by initializing the FFs with opposite values, and applying a pulse in the data input, as shown in fig. 2. As long as Q1(t)≠Q2(t) pipeline operation is correct. Equation (1) describes the timing conditions in the case of a violation. Especially, fast FFs with large hold times are sensitive to hold time violations. Δt_{var} includes variations from different sources.

$$t_{CLK-Q} - t_{hold} - t_{CLKskew} - \Delta t_{var} < 0 \quad (1)$$

Hence, the occurrence of a hold time violation depends on the FF race immunity, the process variations as well as on the maximum clock uncertainty.

If the clock uncertainty is very well controlled and race immunity is large enough, process variability plays a minor role. However, this is not the case in the majority of semi-custom designs that have to meet a short time-to-market.

The measured critical clock skew (race immunity) is obtained used the technique described in [9]. It generates an artificial skew that is programmable over a wide range of 80 steps corresponding to a resolution of ~1ps. The complete scheme is able to precisely measure the race immunity in the 4 test circuits. Moreover, a ring oscillator is included to compare the variabilities of the FF race immunity and ring oscillator frequency.

3. Separation of Systematic and Random Residual Variations

With the discussed measurement technique, it is possible to measure the overall variability on the wafer. However, for a deeper analysis, it is necessary to make mathematical transformations in the obtained data. Several methods to make the separation between the different components of the variability are present in the literature [11, 12]. In this work, we will focus in how to separate the data between systematic (over the wafer) variability and residual (within-die, local, or residuals due to imperfection in the measurement) variability.

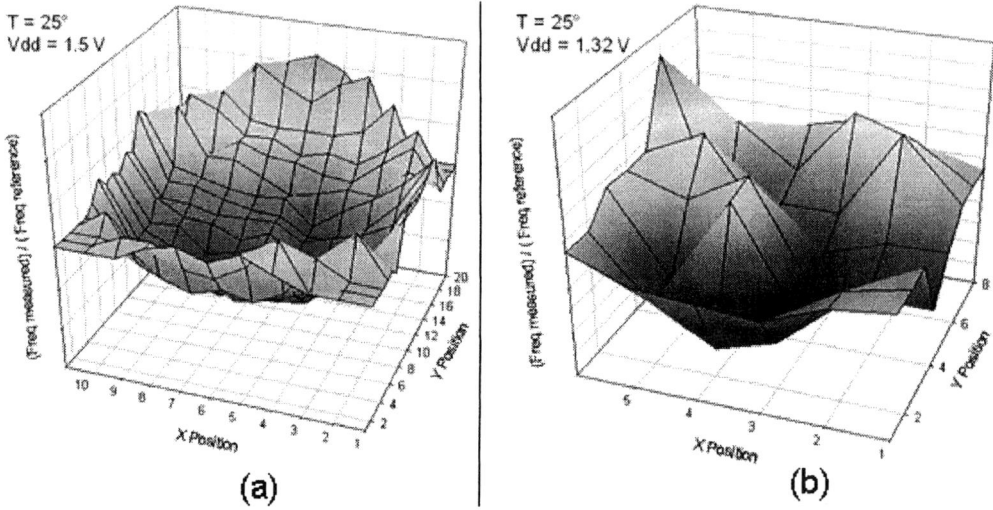

Figure 3. Normalized frequency variability of the RO (a) over the 130nm wafer, (b) over the 90nm wafer.

A simple but widely used method is the moving average. In this method, the measured value in each die is substituted by the average of the value in the die itself with the values of the neighbor dies. If the number of dies is large, the average window can be expanded. We will analyze the results using a 3x3 window (the die with its direct adjacent neighbors) and a 5x5 window (with neighbors up to 2 dies of distance). The drawback of this method is some deterioration in the borders, since we do not have all neighbors available for the average.

Another common method is curve fitting. In this method, we take the measured data and apply a linear regression to find the curve that it approximates better. The curve can be a paraboloid, a plane, a Gaussian, and many others, depending on specific issues of the fabrication process. This is a more complex method, and requires a mathematic intensive computation.

4. Experimental Results

The circuits are fabricated in 130nm and 90nm low power CMOS technologies using regular-VT core devices. For the 130nm CMOS technology, 182 chips are measured on one wafer, while only 36 dies are available in 90nm CMOS due to a larger reticle size. Nominal supply voltages are Vdd = 1.5V for 130nm CMOS, and Vdd = 1.32V for in 90nm CMOS, respectively. The temperature was 25°C in both cases.

First, the variability of the ring oscillator frequency over the wafers is analyzed, with different results (fig. 3). The 130nm wafer shows a typical global wafer variation with slower dies in the center of the wafer, while in 90nm the distribution seems to be more random, with smaller systematic variability, probably due to the larger reticle size and better controlled manufacturing process. The frequencies are normalized to omit confidential technology data. The faster circuits achieve resolutions less than 1ps, while none of the chips had a resolution of more than 1.2ps. It is important to note that the 90nm wafer was a test and not a production wafer, and the systematic variability was further reduced before the technology entered in production, even though the test wafer presented an improvement in systematic variability, if compared to 130nm.

Fig. 4 shows the die-to-die distribution of the critical clock skew for 0-1 transitions in all 4 test circuits in 130nm wafers. The expected Gaussian curve for normal distributions is observed. The 90nm wafer shows similar Gaussian curve. Based on this data and repeating the measurement procedure for 1-0 transitions, the mean critical clock skew and the standard deviation are extracted. The first 4 columns of table 1 summarize the results for 130nm. The results are normalized again.

The 3σ deviation of the delay can be up to 15% of the nominal value in 130nm. The critical skews are in the range of the clock skew that can be expected in circuits using the same technology, showing that these statistical effects have to be considered during hold-time fixing at the end of the layout generation. It is important to note that using larger FFs, the absolute variation of the critical skew decreases, but the relative value remains similar, since these circuits are faster. This indicates that larger FFs have an increased probability of violation, since the clock skew needed to provoke the failure is smaller.

The test circuits with extra inverters have an expected larger absolute variability, but relatively it is smaller, showing that the FFs are more sensitive to process variations than the inverters, or a large number of inverters average the variability.

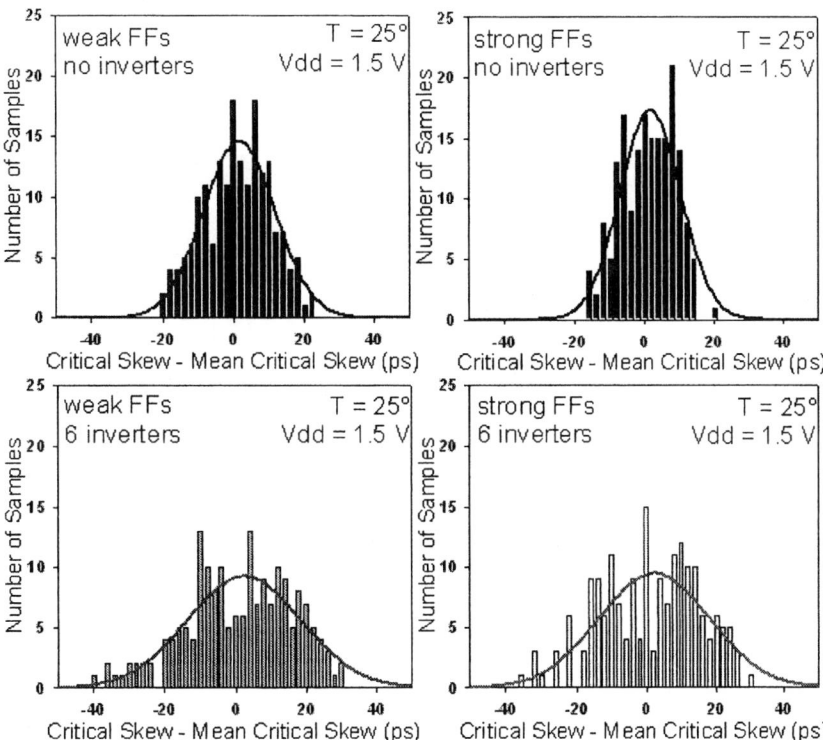

Figure 4. Measured distribution of the critical clock skews for rising transitions in 130nm wafer. The mean critical skew is set to 0ps.

Figure 5 shows a graphical comparison of the results of race immunity found for both technologies. It is possible to see that the race immunity decreases about 30% from 130nm to 90nm. This is an expected value, since it is the speed-up from one technology to another. However, it is much more difficult to scale the clock skew in the same percentage in the scaling. It shows that the problem of hold time violations becomes more critical, and the clock skew and variability must be better controlled in newer technologies.

The next step in the analysis was to apply the separation methods described in the previous section in the RO frequency variability. The three methods were compared: moving average with a 3x3 window, moving average with a 5x5 window, and curve fitting. In 130nm, the curve obtained was a paraboloid, what could be observed already in the original data. However, in the 90nm wafer, the original data was very random and difficult to see any systematic dependence, but the mathematical methods showed a slightly inclined plane, with ring oscillator frequency increasing slightly from one side of the wafer to the other.

Regarding the numerical results, the standard deviation calculated with the 3x3 moving average method was very close to the one found with the curve fitting method. However, the 5x5 moving average

method presented results more than 20% different from the other, always decreasing systematic variability while increasing the random residuals, showing that a 5x5 window may be too large for the available data, masking part of the systematic variability, and especially leading to a deformation at the corners.

Table 1. Total, systematic and residual variability in the critical clock skew using the 3x3 moving average method in 130nm wafer.

Circuit	Type	μ	3σ	Syst.	Rand.
weak FFs,	Rising	100.00	14.9%	12.1%	7.5%
no inverters	Falling	109.95	13.3%	10.9%	6.7%
strong FFs,	Rising	88.87	13.6%	10.4%	7.6%
no inverters	Falling	95.00	12.9%	10.1%	7.1%
weak FFs,	Rising	181.70	13.1%	10.7%	6.4%
6 inverters	Falling	192.71	12.0%	10.0%	5.7%
strong FFs,	Rising	170.54	13.1%	10.9%	5.8%
6 inverters	Falling	177.52	12.5%	10.4%	5.6%

Based on these results, we decided to continue the analysis using only the 3x3 moving average method, due to its simplicity and very close results compared to curve fitting. The final step was to apply the method in the data obtained for the critical clock skew distribution in all circuit configurations. The last 3 columns of table I show the results of the total measured variability, and the

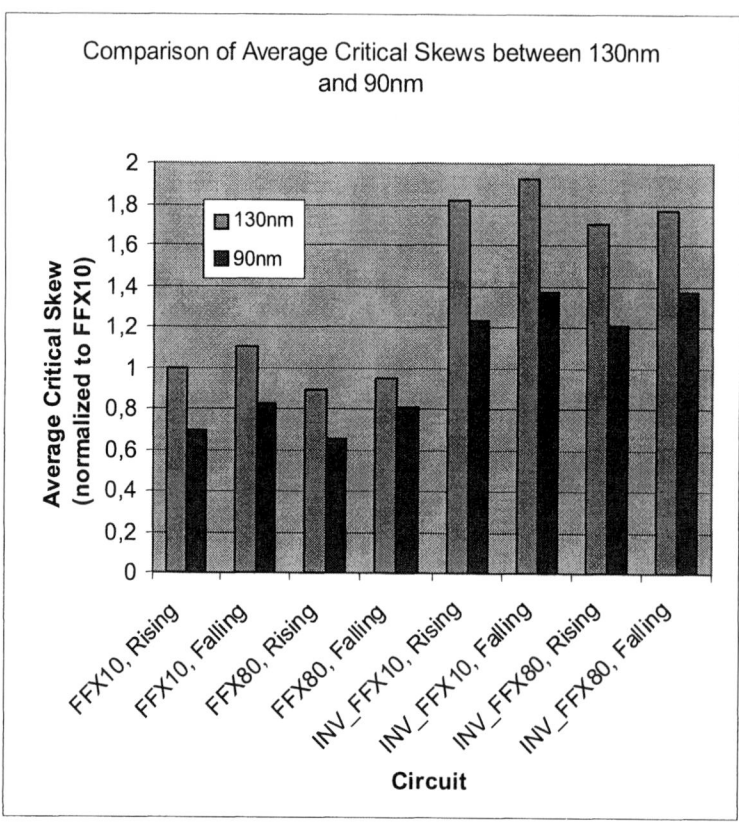

Figure 5. Comparison of race immunity absolute value.

systematic and residual variability calculated with the method, in the 130nm wafer.

The results show that the systematic variability is dominant in 130nm technology, while it is decreasing due to a better process control in 90nm.

5. Conclusion

This work presents an experimental analysis of the variability of hold time violations of edge-triggered master-slave FFs due to process variations in 130nm and 90nm low power CMOS technologies. For accurate on-wafer characterization, a test circuit and a measurement technique with ~1ps resolution are used. The presented methodology provides detailed information about the circuit robustness of FFs under realistic operating conditions. This precise FF characterization then enables designers to perform hold-time fixing for short paths considering statistical variations of FFs as well as delay increasing inverters during buffer insertion. Moreover, during standard cell library development, the methodology is beneficial to optimize the FF portfolio, i.e. to balance race immunity and clock-to-Q propagation delay for various cell driving strengths and different FF topologies.

Statistical variation of race immunity in edge-triggered master-slave flip flops is investigated experimentally.

Mathematical methods to isolate systematic and random residual variations from the experimental data are discussed and compared. Results show that from 130nm to 90nm, the systematic variability is decreasing.

The results also show that the absolute race immunity reduces by about 30% from 130nm to 90nm CMOS technology due to speed improvement, leading to a faster CLK-Q delay. This indicates that hold time violations are a harder problem in newer technologies if the clock skew is not expected to scale in the same way.

Future work includes the investigation of the impact of different temperature and supply voltage on variability, and the analysis of the normality of the measured distributions.

References

[1] H. Q. Dao, K. Nowka, and V. G. Oklobdzija, "Analysis of clocked timing elements for dynamic voltage scaling effects over process parameter variation", Proc. Intl. Symposium on Low Power Electronics and Design (ISLPED), 2001, pp. 56-59.

[2] D. Marković, B. Nikolić, and R. W. Brodersen, "Analysis and design of low-energy flip-flops", Proc. Intl. Symposium on Low Power Electronics and Design (ISLPED), 2001, pp. 52-55.

[3] Y. Huang, W. T. Cheng, S. M. Reddy, C. J. Hsieh, and Y. T. Hung, "Statistical diagnosis for intermittent

scan chain hold-time fault", Proc. Intl. Test Conf, 2003, pp. 319-328.

[4] S. Edirisooriya, and G. Edirisooriya, "Diagnosis of scan failures", Proc. VLSI. Test Symposium, 1995, pp. 250-255.

[5] R. Guo, and S. Venkataraman, "A technique for fault diagnosis of defects in scan chains", Proc. Intl. Test Conf, 2001, pp. 268-277.

[6] J. C.-M. Li, "Diagnosis of multiple hold-time and setup-time faults in scan chains", IEEE Transactions on Computers, Nov. 2005, Vol 54 Issue 11, pp. 1467-1472.

[7] Z. Wang, M. Marek-Sadowska, K.-H. Tsai, and J. Rajski, "Diagnosis of hold time defects", Proc. IEEE Intl. Conf. on Computer Design (ICCD), 2004, pp. 192-199.

[8] N. V. Shenoy, R. K. Brayton, and A. L. Sangiovanni-Vincentelli, "Minimum padding to satisfy short path constraints", Proc. IEEE/ACM Intl. Conf. on Computer Aided Design (ICCAD), 1993, pp. 156-161.

[9] G. Neuberger, F. Kastensmidt, R.Reis, G. Wirth, R. Brederlow, C. Pacha, "Statistical characterization of hold time violations in 130nm CMOS technology", IEEE European Solid-State Circuits Conference (ESSCIRC), 2006.

[10] G. Gerosa et al., "A 2.2W, 80 MHz superscalar RISC Microprocessor", IEEE J. Solid-State Circuits, 1994, pp. 1440-1454.

[11] B. E. Stine, D. S. Boning, and J. E. Chung, "Analysis and decomposition of spatial variation in integrated circuit processes and devices", IEEE Transactions on Semiconductor Manufacturing, Feb. 1997, Vol 10 Issue 1, pp. 24-41.

[12] D. S. Boning, and J. E. Chung, "Statistical metrology: understanding spatial variation in semiconductor manufacturing", Proc. Microelectronic Manufacturing Yield, Reliability and Failure Analysis II: SPIE 1996 Symposium on Microelectronic Manufacturing, 1996, pp. 16-26.

AC-Coupling Strategy for High-Speed Transceivers of 10Gbps and Beyond

Yikui (Jen) Dong, Steve Howard, Freeman Zhong, Scott Lowrie, Ken Paradis, Jan Kolnik, Jeff Burleson

LSI Logic Corp.
Milpitas, CA 95035 USA

Abstract—AC coupling in a transmission link is preferred and often required for the functioning of high speed transceivers. But at data rate of 10Gbps and beyond, both the external AC coupling and the conventional on-chip AC coupling approaches bring in heavy burden that pushes to the fundamental limits and are difficult to afford. This paper examines the AC-coupling methods for multi-Gb/s transceivers, and points out the impairments in the existing implementations. A hybrid structure offering both the signal-bump and the AC-capacitor functions under the stringent return-loss requirements of a 10Gb/s+ I/O is proposed and implemented in 65nm standard CMOS. A sizeable 5.1pF AC capacitor is measured with ultra low parasitic expense ratio of less than 120fF.

I. INTRODUCTION

In high-speed transceiver design, AC coupling in the channel between the transmitter and the receiver connection is preferred, and often, required for the functioning of the link. In the DC coupled links, the signal is sensitive to duty cycle distortion due to the ineluctable common-mode voltage mismatch between the transmitter and the receiver, and is problematic for the receiver to recover at very high-speed. And, the AC coupling isolates the transmitter and receiver common-mode voltages for independent optimization, which provides maximum freedom to commercialize the transceiver.

Although highly preferred and often essential, AC coupling is well-known to be difficult to implement for 10Gbps+ high speed links, especially for long-reach high-loss channels with reflection. The traditional way of implementing AC coupling by adding discrete components on the board between the two communicating chips fails. The on-chip AC coupling solution also faces great challenge. In Section II, an overview of the available AC coupling schemes is given. The basic configuration, the fundamental issues, and the design criteria are discussed. In Section III, we examine the existing on-chip AC capacitor implementation and point out its impairments that preclude the application in ultra high speed transmissions. In Section IV, we propose a novel hybrid structure that provides both the signal bump and the AC cap functions. The new structure is customized for high speed application and can simultaneously offer several other key features. The experimental results and concluding remarks are given in Section V and VI respectively.

II. CHALLENGES IN AC COUPLING SOLUTION

A. External AC Coupling Approach

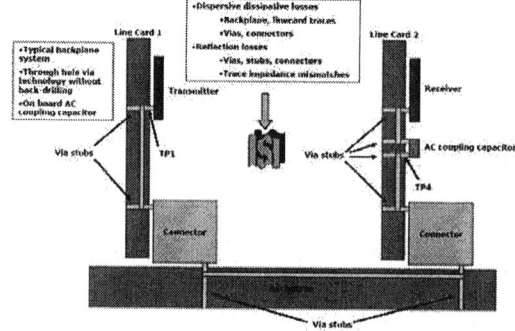

Figure 1. Typical backplane system with external AC coupling (Graph reproduced from reference [1]).

A typical external AC coupled backplane transmission system is shown in Figure 1. The AC cap is placed on the line card at the receiver side. The primary losses can be partitioned into two categories, the dispersive dissipative loss and the reflection loss. Both of them are affected by the number and the severity of the via-stubs. The amount of signal loss and reflection associated with the via stub impedance mismatch are strong functions of frequency. At the transmission rate of 10Gbps and beyond, the additional signal loss resulted from the two via stubs mounting the capacitor are significant, and could easily exceed 3dB at baud rate and the severity depends on the quality of the line card design and manufacturing. In long-reach high-loss serial-link, the high-frequency signals often barely reach the receiver sensitivity threshold. The extra loss would cut down the already marginal signal strength and brings it even closer to the fundamental limit set up by the circuit offsets, cross talk, electrical noise floor and other non-idealities. As a result, the data recovery is often no longer reliable. In addition, the via-stubs also bring in extra high-frequency reflection that makes the channel response bumpy. It stresses, and often, breaks the equalization scheme. In a word, the external AC coupling method works well for low to mid-rate data transmission, but with the increase of the data rate, the damages made to signal integrity rise up quickly. After the 10Gbps node, especially for the long-reach legacy channels, the external AC-coupling solution becomes a heavy burden to afford.

B. On-Chip AC Coupling Approach

On the other hand, the alternative approach -- on-chip AC coupling -- is also next to impossible to implement. The major difficulty lies in the struggle of obtaining an AC coupling capacitor of adequate size that is necessary to pass through enough information without contributing substantial amount of parasitic capacitance killing the return-loss performance at the high data rate. A detail examination would reveal that at 10Gbps transmission rate and beyond, the traditional way of implementing on-chip AC coupling scheme stops to work and new methods that can restraint the amount of parasitic capacitance is required.

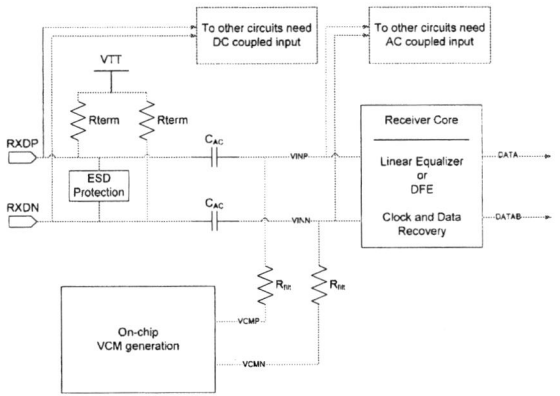

Figure 2. On-chip AC coupling scheme.

Figure 2 shows the simplified schematic of an on-chip AC-coupled receiver. In multi-Gbps high-speed transceiver design, the impedance matching and signal reflections at the receiver and transmitter inputs/outputs are of vital importance. The reflection is closely related to the total parasitic capacitance seen at the bump. In the case of the on-chip AC coupled receiver, this includes the parasitic contributed by the bump, the ESD protection circuitry, the termination resistors, the AC coupling cap, the input transistors and the routing lines. The two most significant parasitic sources are usually the bump and the on-chip AC coupling capacitor. With existing implementations, their combined contribution could easily exceed 400fF even with a small AC capacitor size of 1pF.

Figure 3 shows the return-loss performance as a function of parasitic capacitance. The total parasitic budget drops quickly with the increase of transmission rate. As shown in the plot, at the data rate of 3.2Gbps, even with 1.18pF parasitic, the receiver still meets the -8dB reflection bench mark. At 12.8Gbps, even with the favorite exact resistive matching, the total parasitic budget is merely 280fF. With this diminished allowance, not much circuit or routing can be attached to the I/O bumps without destroying the link performance. The bump, the ESD protection, the termination resistors, and the

receiver input transistors are all essential elements that cannot be omitted. It is already next to unmanageable to fit all of them within the 280fF budget at 12.8Gbps transmission. The AC coupling capacitor must be made 'free' with negligible parasitic capacitance before being able to fit into the system.

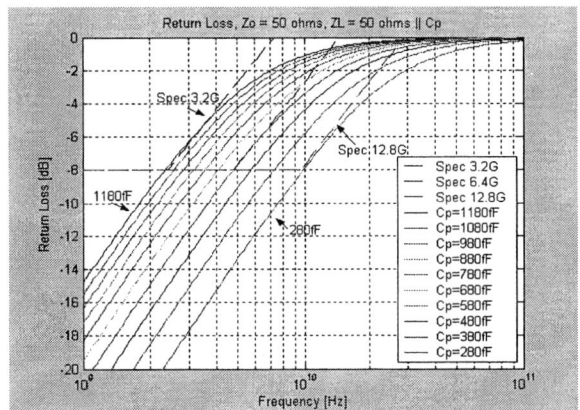

Figure 3. Return-loss as a function of parasitic capacitance.

III. IMPAIRMENTS IN EXISTING ON-CHIP CAPACITOR IMPLEMENTATION

A. Previous on-chip capacitor

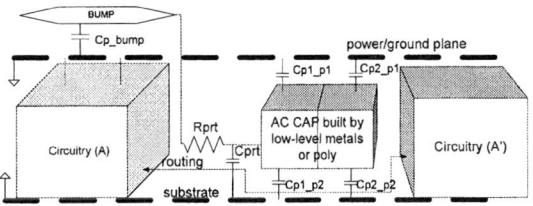

Figure 4. Cross section view with the conventional on-chip AC capacitor.

With the previous implementation, the bump and the AC coupling capacitor are two separate entities, and horizontally, can be moved freely relative to each other. The AC coupling capacitor is usually implemented using lower layer metals or poly. The layout cross section view is shown in Figure 4.

B. Impairments for high-speed or high-density application

The existing on-chip cap implementation is cumbersome and is fundamentally impaired for the challenge. 1st, the capacitor is usually implemented with lower layer metals or poly. It requires footprint in the floor-plan that expels other circuits. A typical coupling capacitor of a few pF could easily translates to significant silicon cost of thousands of um^2. In many cases it actually dominates the silicon budget for a receiver. The area penalty becomes difficult to accommodate, especially for high channel count designs; 2nd, the bump and the AC capacitor each contribute its own set of parasitic capacitance. As the largest two consumers of the total parasitic budget, the penalty is substantial and directly relates to return-loss failure at high data rate; 3rd problematic and expensive

2007 IFIP International Conference on Very Large Scale Integration (VLSI-SoC 2007)

sensitive high-speed signal routing is required from the bump to the AC capacitor. Often, it can be long distance of 100um or more depending on the placement priorities of various circuit blocks. The routing presents large distributed R and C loading to the high speed signal. Differential signal routings are normally required. It is especially difficult to simultaneously achieve multiple goals of keeping the transmission bandwidth over long distance and matching the P and N routes in the presence of different neighboring circuits; 4th, the capacitor made of lower level metal creates blockage for crossover routing. Because of its bulky nature, this is especially problematic and creates routing channel congestion.

IV. PROPOSED HYBRID BUMP CAP

A. The bump-cap

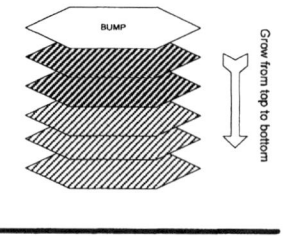

Figure 5. Hybrid bump / AC capacitor.

We propose a hybrid integrated circuit structure that serves both the bump/pad and AC-coupling capacitor functions for significantly reduced combined parasitic capacitance. It is particularly suitable in the environment of large scale integrated circuit implementation in modern deep submicron CMOS technology where multiple layer metal options are usually available and required. Shown in Figure 5, the structure consists of a solid piece of top layer metal, and a few adjacent lower layer metals formed under it. The top metal follows the attributes as a common bump, but serves as both the bump and one of the capacitor plates. The bump-cap is one unified entity and grows vertically downwards lower layers of metal with the same confining contour. The exact number of metal layers needed is decided by the minimum capacitance required by the application. Fence, or other structures which mingle the existence of both plates of the cap at each and every layer of metals below the top, are used to increase the unit capacitance. Different fence spacing may be used for thick and thin metal layers. The fence structure can be made of comb or any other shapes. It can also be made of one unit or a collection of multiple units. Each of the two capacitor plates at every metal layer is connected to itself at other layers through metal-to-metal contacts. And as a result, both plates of the cap are accessible at the bottom of the hybrid structure.

B. Applications

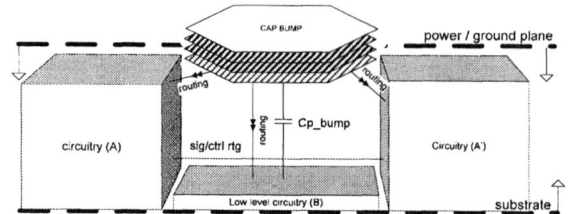

Figure 6. Layout cross section views with the proposed hybrid bump-cap.

Figure 6 shows the proposed hybrid bump-cap is 'free' without silicon footprint requirement. Since the bump-cap is made of high level metals of M3 or above, other active or passive circuits can be placed under it, and no silicon budget has to be allocated to the bump-cap. In addition, the parasitic capacitance contribution is significantly smaller. Compared with its counterpart shown in Figure 4, the hybrid bump-cap eliminates the large pcaps Cp1_bump, Cp1_p1, Cp2_p1, and the routing parasitics Rprt and Cprt. The pcap Cp1_p2 and Cp2_p2 are also reduced significantly. Furthermore, the signal routings are much easier. Both plates of the bump-cap can be accessed from the bottom. The horizontal size of the bump-cap is large, and since it is made of many units of mini capacitors that are bundled together, the signals at the both ends of the capacitor are delivered to a wide projected area and readily accessible by the circuits from below or close by. Instead of paying a price for the signal distribution, the complex routing connecting the mini-caps is actually good and contributes to the capacitance needed for AC-coupling purpose. With the capacitor moved away from lower layer metals, the difficulty resulted from the metal-path blockage from the conventional implementation is eliminated, and the crossover signal routing, shown in Figure 6 from circuitry (A) to (A'), would be direct and much easier. Furthermore, in the case of ultra high-speed applications where aggressively low parasitic capacitance is sought, the circuits under the bump can be removed to further reduce the parasitic capacitance.

C. Key features

The key attributes of the proposed structure including: 1) Hybrid structure, achieve both the bump and AC coupling capacitor function simultaneously. 2) Substantially reduced combined parasitic capacitance, usually more than half. 3) Eliminated the AC coupling cap footprint requirement. 4) Eliminate the problematic high speed signal routing from the bump to the cap. 5) Moved the AC capacitor away from lower layer metals, and cleared out the crossover routing congestion. 6) Provide other critical circuit shorter access to the signals at the both sides of the AC cap.

V. EXPERIMENTAL RESULTS

The hybrid bump-cap was implemented in TSMC 65nm CMOS. A picture of the implementation is shown in Figure 7. The circuit started with the bump pad, and an array of M6M7 fence-caps are built under and integrated to it. The QuickCap extraction of the entire structure reveals that an AC coupling capacitance of 5.2pF is obtained, with total parasitic capacitance at each side to the substrate of merely 40fF. A silicon die micrograph is shown in Figure 8.

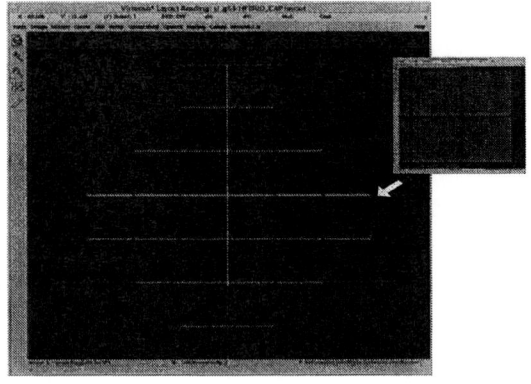

Figure 7. The hybrid bump-cap inplemented in 65nm standard CMOS.

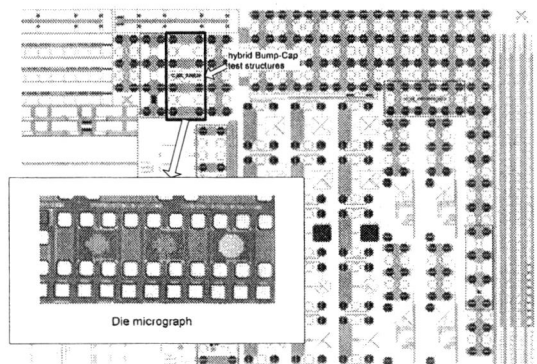

Figure 8. Die micrograph.

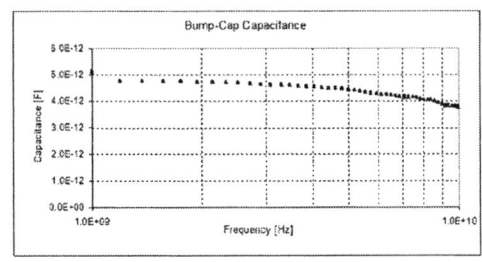

Figure 9. Measured hybrid bump-cap capacitance as function of frequency.

Electrical properties of the hybrid bump-cap were investigated by analyzing the S-parameter data collected by measuring the structure in 2-port Ground-Signal-Ground configuration. To properly account for the parasitics associated with probe-pads and leads, Open and Short de-embedding structures were also included on the test-chip. OPEN/SHORT de-embedding sequence was adopted to adjust the measured S-parameter data for proper extraction of the equivalent parameters. Figure 9 shows the extracted

capacitance as a function of frequency from 1 to 10GHz. It is evident that the capacitance calculated based on RF measurements correlates well with the values predicted by extractions. Figure 10 shows the Smith-chart comparison of the S11 parameters measured from the silicon from 1 to 20GHz with simulated values from a simple RC Π model of the bump-cap. As illustrated in the figure, this circuit consists of a 5.1pF coupling capacitor whose value comes from the RF measurements, and was consistent with the QuickCap extractions; and the parasitic capacitance and resistance, representing coupling between each port of the device and the substrate. Values of these parasitic components were obtained through curve-fitting, and the capacitance value is in reasonable agreement with the QuickCap extractions. This simple model provides fairly satisfactory correlation to the S11-measurements of the bum-cap structure in the frequency range of interest.

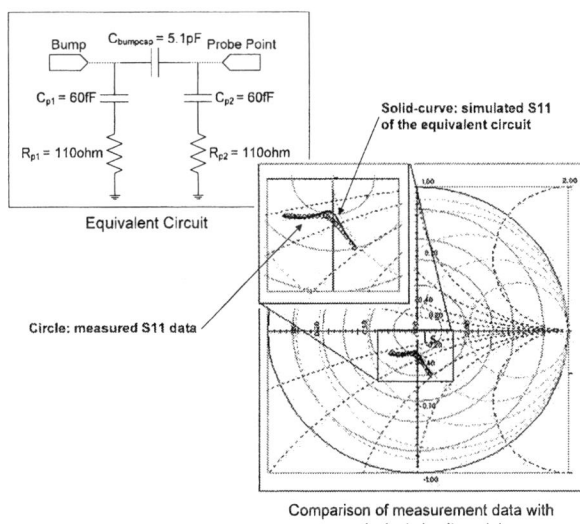

Figure 10. The reflection coefficient S11 of the bump-cap in Smith-chart.

VI. CONCLUSIONS

In this paper, the ac coupling methods for high speed transceivers are examined. The impairments of the existing methods are pointed out, and a novel hybrid structure that offers both the bump and the AC cap functions is proposed. The new integrated circuit structure offers many key features and made it possible, for the first time, to implement a sizable cap on-chip for 10Gbps I/Os without major setback of return loss performance. The hybrid structure was implemented in 65nm standard CMOS, and the measurement results confirmed the theoretical prediction.

REFERENCES

[1] W. Peters et al, "ATCA Channel Data for Backplane Ethernet Task Force", IEEE *P802.3ap Task Force Meeting Presentation*, Sept. 2003.

SWORD: A SAT like Prover Using Word Level Information

Robert Wille   Görschwin Fey   Daniel Große   Stephan Eggersglüß   Rolf Drechsler

Institute of Computer Science, University of Bremen, 28359 Bremen, Germany

{rwille,fey,grosse,segg,drechsle}@informatik.uni-bremen.de

Abstract—Solvers for Boolean *Satisfiability* (SAT) are state-of-the-art to solve verification problems. But when arithmetic operations are considered, the verification performance degrades with increasing data-path width. Therefore, several approaches that handle a higher level of abstraction have been studied in the past. But the resulting solvers are still not robust enough to handle problems that mix word level structures with bit level descriptions.

In this paper, we present the satisfiability solver SWORD – a SAT like solver that facilitates *word* level information. SWORD represents the problem in terms of modules that define operations over bit vectors. Thus, word level information and structural knowledge become available in the search process. The experimental results show that on our benchmarks SWORD is more robust than Boolean SAT, K*BMDs or SMT.

I. INTRODUCTION

The number of elements integrated within digital circuits grows exponentially and this trend is going to continue for at least another 10 years. Already today millions of gates are integrated in a single circuit. Throughout the design flow for such complex systems, techniques to represent and manipulate the function are needed. In particular, to formally verify the correctness of a circuit with respect to all design states and input sequences, techniques for symbolic function manipulation are applied.

Current state-of-the-art tools for formal verification use Boolean techniques like *Binary Decision Diagrams* (BDDs) [1], *AND-Inverter-Graphs* [2] and provers for *Boolean Satisfiability* (SAT) [3], [4]. No word level information such as knowledge about arithmetic operations or structural knowledge is directly used for function manipulation. As a result, the performance of verification tools degrades with increasing data-path width.

For this reason, approaches to exploit such high level information have been proposed in the past [5], [6], [7]. But pure word level approaches suffer from complexity problems when irregularities in the word level structure occur, e.g. bit slicing [8]. The recent concept of *Satisfiability Modulo Theories* (SMT) [9], [10], [11], [12] is more powerful since multiple provers are combined, but still structural information is not available. Related work is discussed in more detail in Section II and empirically compared in Section V.

In this paper, we propose SWORD – a *SAT*-like prover that uses *word* level information and also resembles the structure of the original problem. Internally, the problem is represented as a composition of modules; each module is defined over bit vectors and enforces the constraints for a word level operation

on the corresponding Boolean variables. The main advantages of this approach are the following:

- *Compact problem representation:* The composition of word level modules is a much more compact representation than the transformation to Boolean constraints.
- *Knowledge about structure and semantics:* This knowledge is determined by the position of a module within the problem instance and the type of a module. Such information helps to predict the impact of a decision or of learned information during the search process more accurately.
- *Efficient reasoning:* Different types of modules require different reasoning procedures and decision heuristics to allow for an efficient search procedure. These procedures are designed for each type of module individually in the proposed framework.

Thus, SWORD combines the advantages of a Boolean proof procedure with the power of word level knowledge. The proposed solver is empirically compared to K*BMDs [6] as a word level decision diagram, the Boolean SAT solver MiniSat [4] and the SMT solver Yices [12].

II. RELATED WORK

Several approaches to incorporate word level information in the proof process have been proposed so far. BDDs have been generalized to the word level quite early [5] resulting in K*BMDs [6] as a very general form. These diagrams can represent word level multiplications very efficiently, but whenever bit nibbling occurs – as is common practice in circuit descriptions – the performance degrades. In fact, *BMDs may be exponentially large for certain functions [8].

A different approach is the transformation of the problem into *Integer Linear Programming* (ILP) constraints [7]. But the same limitations to pure word level descriptions have been observed. A pure ILP-based approach is often too slow for real world applications.

Combining Boolean provers and word level provers seems to be more promising. The framework proposed in [13] is based on an ATPG engine that is enhanced by arithmetic word level primitives. An arithmetic constraint solver is applied to validate bit level assignments on the circuit. But the powerful learning concepts known from Boolean SAT are not incorporated.

Due to the tremendous improvements in the performance of provers for Boolean SAT in the recent past [14], [15], [16], several researchers investigated the combination of SAT with other proof techniques, i.e. *Satisfiability Modulo Theories* (SMT) [9], [10],

[11], [12]. An SMT solver integrates a Boolean SAT solver with another solver (or multiple solvers) for specialized theories. Usually, the SAT solver works on an abstract representation of the problem and steers the overall search process. Each satisfiable assignment for the Boolean SAT problem has to be validated on the concrete problem using the theory solver. The solver proposed in [17] can be seen as a specialized SMT solver for bit vector logic. Tightly coupling the different solvers, especially to enforce learning due to conflicts resulting from partial assignments and to efficiently carry out implications, is a challenge in this area. Usually, validating a given SAT assignment by using the theory solver is very time consuming. Therefore the overall performance is limited by the performance of the theory solver. In our framework no theory solvers are needed. Moreover, structural information about the original problem is available.

A very general theoretical framework for hierarchical SAT solving was presented in [18]. There, the problem is also decomposed into modules, where each module may have different implication procedures. But no experimental evidence was given and no hints for an implementation were provided.

Nonetheless our solver works similar to such a hierarchical solver. Besides specialized implication procedures also dedicated decision heuristics are applied for different types of modules.

III. BOOLEAN SAT SOLVING

Our algorithm inherits the basic structure of a classical algorithm to solve a problem instance of Boolean *Satisfiability* (SAT) [14]. Therefore we briefly review the techniques applied in Boolean SAT solvers.

A. Basic Algorithm

The SAT instance is represented as a Boolean formula in *Conjunctive Normal Form* (CNF), which is given as a set of clauses; each clause is a set of literals and each literal is a propositional variable or its negation.

The basic search procedure to find a satisfying assignment is shown in Figure 1 and has the structure of the DPLL algorithm [3]. Instead of simply traversing the complete space of assignments, intelligent decision heuristics, conflict based learning and sophisticated engineering of the implication algorithm lead to an effective search procedure. The description follows the implementation of the procedure in modern SAT solvers. While there are free variables left (a), a decision is made (c) to assign a value to one of these variables. Then, implications are determined due to the last assignment by *Boolean Constraint Propagation* (BCP) (d). This may cause a conflict (e) that is analyzed. If the conflict can be resolved by undoing assignments from previous decisions, backtracking is done (f). Otherwise the instance is unsatisfiable (g). If no further decision can be done, i.e. a value is assigned to all variables and this assignment did not cause a conflict, the CNF is satisfied (b). In the following the *decision level* d denotes the number of

variables assigned by decisions in the current partial assignment, i.e. neglecting variable assignments due to implications.

B. Limits of Boolean SAT

Due to the translation of the problem into CNF, the power of BCP as an implication engine and the efficiency of learning are limited. In the verification domain, the original problem is usually given at the word level. Operations are defined over bit vectors. Each Boolean variable that is visible in a bit vector at this level is called *module variable* in the following. The translation of word level operations over bit vectors of *module variables* into CNF involves the creation of a large number of *auxiliary variables* [19]. The dependencies between these variables are modeled by constraints in terms of clauses.

Example 1. *Consider an $n \times n$-multiplier. On the word level, $4n$ module variables are needed for the bit vectors of the operands and the result.*

On the other hand, the multiplier can be represented by n^2 AND gates [20], i.e. the number of auxiliary variables is in $\theta(n^2)$. A single gate can be modeled by three clauses for each element. Therefore the multiplier can be represented by a CNF with $\theta(n^2)$ clauses[1].

Simplified, all these auxiliary variables have to be considered during BCP; but implications on auxiliary variables do not yield a reduction of the search space for the original problem. Moreover, conflict clauses may be derived, that are defined over auxiliary variables only – again without pruning the search space of the original problem. In principle, this problem can be prevented by introducing additional clauses, that describe the implications on module variables directly, but then the translation becomes inefficient due to a large number of clauses.

IV. USING WORD LEVEL INFORMATION

In this section we describe the architecture of SWORD and how word level information can be used during the solve process. Therefore, we first explain the representation of the problem and present the overall algorithm. Afterwards the utilization of word level information in decision making, the implication engine and conflict analysis are explained in more detail.

A. Representation

SWORD represents the problem in terms of so called *modules*. Each module defines an operation over bit vectors of *module variables*. Each module variable is a Boolean variable.

Example 2. *Figure 2 shows an equivalence checking problem in terms of a miter circuit. A multiplier is compared to a realization that sums up the partial products.*

---

[1]More efficient translations may be available, but the problem instance still grows.

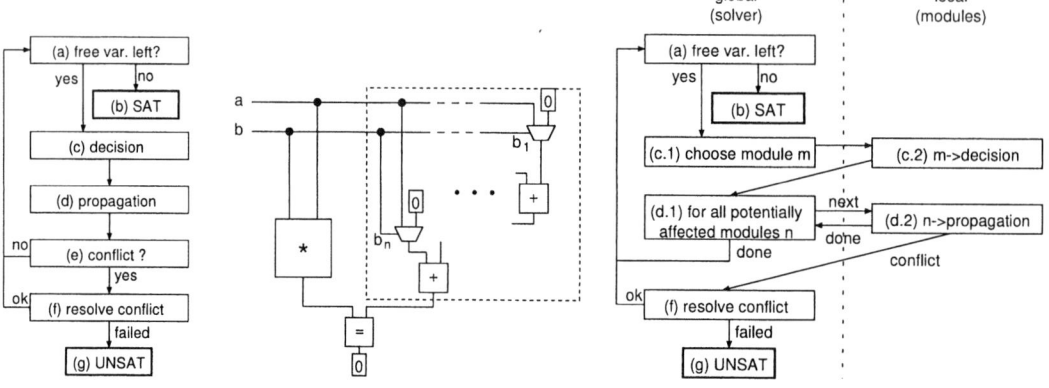

Fig. 1. DPLL   Fig. 2. Miter for a multiplier   Fig. 3. Algorithm

SWORD represents this problem by using one module representing a multiplier, $n-1$ modules representing an adder, n modules representing a multiplexor and one module representing a comparator. No auxiliary variables are needed.

B. Overall Algorithm

The overall algorithm of SWORD is shown in Figure 3. This algorithm is similar to the DPLL procedure as applied in standard SAT solvers: While free variables remain (a) a decision is made (c), implications resulting from this decision are carried out (d), and if a conflict occurs, it is analyzed (f). The important difference is that SWORD has two operation levels: the *global* algorithm controls the overall search process and calls the *local* procedures of modules for decision and implication. Thus, decision making and implication engine can be adjusted for each type of module.

In more detail, the solver first chooses a particular module based on a *global decision heuristic* (c.1). Then, this module chooses a value for one of its variables according to a *local decision heuristic* (c.2). Afterwards, the solver calls the *local implication procedures* (d.2) of all modules that are potentially affected (d.1) by the previous decision or implication. Here a *variable watching scheme* similar to the one presented in [15] is used, which can efficiently determine these modules. The chosen modules imply further assignments and detect conflicts.

C. Decision Strategies

1) Global Decision: The global decision procedure chooses a module, that assigns a value to one of its connected module variables. So the global decision procedure has to decide, which module will make the best decision, i.e. which decision of a module leads to as many implications as possible. Therefore a (global) heuristic is employed to decide which modules are "more important" than others. To determine the importance of a particular module, semantic information such as the type or structural information such as the position within the overall problem are available.

Example 3. *Again, consider the miter circuit shown in Figure 2. In this example the primary inputs and the outputs of the multiplier module are considered more important than, for example, the select input of one of the multiplexors. Therefore, the global decision heuristic selects the multiplier module first.*

To realize this efficiently, the global decision heuristic currently uses a static priority based on the type of the module. Here, more complex modules (e.g. multipliers) are considered as being more important and, therefore, are selected for a decision with a higher priority than less complex modules. The complexity is measured in the number of two-input gates needed to describe a module. Furthermore the priority of a particular module can be increased/decreased when it is located near to the primary inputs/outputs or the objective. By this, each global decision can be done very efficiently, because no complex data manipulation is necessary.

2) Local Decision: The local decision procedure of a module assigns a value to one of its module variables. The impact of a particular decision depends on the type of a module. Therefore different strategies are applied for different types of modules. For example, a module representing a multiplier uses a different heuristic than a module representing an AND gate. In the following an adder exemplifies the local decision procedures of SWORD. This type of module is simple enough to be explained within the page limitation, but provides some interesting insight.

An n-bit adder $ADD : \mathbb{B}^n \times \mathbb{B}^n \to \mathbb{B}^{n+1}$ is considered, which is represented by a module in SWORD. The module variables connected to this module are given by $a_{n-1}, \ldots, a_0$ and $b_{n-1}, \ldots, b_0$ that represent the inputs of the adder and $o_n, \ldots, o_0$ that represent the outputs.

For an adder, assigning some variables a_i, b_i or o_i (with $n > i \geq 0$) while variables a_j, b_j or o_j (with $i > j \geq 0$) are still unassigned, often does not allow to imply values for the outputs. In contrast, when all of the least significant bits of both operands are given, the

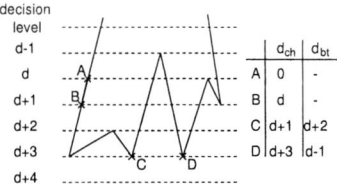

Fig. 4.   Search tree and decision levels

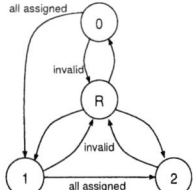

Fig. 5.   FSM for an adder

corresponding bits of the outputs can be determined. Therefore the variable representing the least significant unassigned bit is assigned first.

From an implication point of view, the local decision procedure is realized as a *Finite State Machine* (FSM). This allows to carry out decisions efficiently. The FSM has $n + 1$ states and is in state i ($n > i \geq 0$) when all variables with lower significance than i are assigned, i.e. a_j, b_j and o_j ($i > j \geq 0$) are assigned. Thus, if the FSM is in state i, only the variables a_i, b_i and o_i are considered. If all of these variables are assigned, the FSM proceeds to state $i + 1$. Otherwise at least two of these variables are unassigned (because an implication is carried out when only one variable is unassigned, as explained in Section IV-D.2).

An additional state R is needed to recalculate the state when it was invalidated: Due to backtracking the state of the local FSM of a module may be invalidated because currently assigned variables may become unassigned. This is recognized by tracking the decision level. The decision level of the last state transition, i.e. since the last change of a state, is stored in d_{ch} and the lowest decision level that has been reached after a backtrack intermediately is stored in d_{bt}. The state of the FSM may only be invalidated when $d_{bt} < d_{ch}$.

Example 4. *Figure 4 illustrates this mechanism. The search tree is indicated by the plain line and the decision levels that are reached are also shown. A transition of the FSM is indicated by a cross. The table shows the values of d_{ch} and d_{bt} before the transition is done. The first transition occurs at A and d_{ch} is changed from 0 to d; d_{bt} is uninitialized. At B the decision level has increased; the state is still valid; d_{ch} is updated to $d + 1$. Due to a backtrack d_{bt} is set to $d+2$. Thus, at C the state from decision level $d+1$ is still valid. In contrast, when transition D is done, the state is potentially invalid and has to be recalculated.*

The resulting FSM for a 3-bit adder is shown in Figure 5; only state transitions are indicated, internal variables are not shown.

D. Implication Engine

The implication engine is also divided into a global part and local procedures that are dedicated to the type of a module.

1) Detection of Affected Modules: Globally those modules that may be affected by a previous decision or implication have to be identified. This is done by a variable watching scheme. Currently, a conservative

approach is applied: the local propagation procedure of each module that contains a variable that has been assigned is called. Such a static scheme is efficient, because module variables usually only connect to a few modules – often only two modules.

2) Local Implication: The local implication procedures only consider the connected module variables for the propagation of values. For efficiency these procedures do not determine all implications that are possible, but only those that can be derived efficiently. Again, the local implication procedure of an adder exemplifies the local implication procedures.

The implication procedure works similar to the decision procedure: If, for example, the input bit a_i and the output bit o_i and all less significant input bits (a_j and b_j with $i > j \geq 0$) are assigned, the third variable (b_i in the example) can be implied. This implication procedure does not guarantee to detect implications on higher significant bits and is therefore not too powerful. But in most cases implications on these bits are improbable.

The implication procedure relies on the same FSM that is used for decisions. Additionally, the carry bits $c_{n-1}, \ldots, c_0$ are internally updated at each state transition. In state i ($n > i \geq 1$) carry bit c_{i-1} is also given. Therefore an implication can be carried out efficiently based on the current state i, the value of the carry bit c_{i-1} and the values of the module variables a_i, b_i, o_i.

Note, due to the implication procedure a conflicting assignment may not be detected directly. But when the FSM reaches state n, i.e. all module variables are assigned, the consistency of the assignment will be validated. However, due to the order of decisions conflicts are usually detected early. The mechanisms for conflict analysis are explained in detail in the next section.

E. Conflict Analysis

In SWORD conflict analysis and learning are quite similar to the classical approach of a SAT solver. Upon detection of a conflict, the module returns the conflicting variables to the global solve process. Then, conflict analysis is carried out. Currently we adapted the implementation of MiniSat [4]. Because SWORD does not work in terms of clauses, a separate *implication graph* is stored globally. Each module updates this graph when an implication is carried out. The learned information is stored in terms of clauses as in standard SAT solvers. Therefore an additional clause module exists, which handles all clauses generated by conflict

2007 IFIP International Conference on Very Large Scale Integration (VLSI-SoC 2007)

analysis (and applies the known state-of-the-art SAT techniques).

The conflict graph keeps track of the reasons for a particular assignment. Thus, the identification of a reason is crucial in this context. The smaller the reason, the smaller the conflict clauses and the more effectively the search space is pruned. Again, an adder is facilitated to give an idea of how the implication graph is created.

Example 5. *Assume, o_i is implied based on the internal value of c_{i-1} and the module variables a_i and b_i. Furthermore, due to previous assignments $a_{i-1} = 0$ and $b_{i-1} = 0$, the reasons for these assignments are already stored in the implication graph. In this case input bits with lower significance than $i - 1$ do not influence the value of o_i, because no carry bit is propagated beyond $i - 1$. Thus, the four variables a_i, b_i, a_{i-1} and b_{i-1} are identified as the reason for the implication on o_i. The four edges (a_i, o_i), (b_i, o_i), (a_{i-1}, o_i) and (b_{i-1}, o_i) are added to the implication graph. Note, that the reasons for $a_{i-1} = 0$ and $b_{i-1} = 0$ are already stored in the graph.*

Like in standard SAT solvers, only conflict clauses up to a certain length are learned. The ratio behind this heuristic is that short clauses prune a large part of the search space while longer clauses are less valuable.

Semantical knowledge is also exploited in this process. For example, a conflict clause is not learned if it contains variables that are associated to a complex module like a multiplier – in this case only backtracking is carried out. This heuristic is motivated by the observation that usually a large number of clauses is learned that describe the behavior of a multiplier which causes memory overhead but does not speed up the search.

V. EXPERIMENTAL RESULTS

This section provides experimental results for SWORD in comparison to the Boolean SAT solver MiniSat [4], K*BMDs [6] using the package of [21] as a representative of pure word level approaches, and the SMT solver Yices [11], [12].

All experiments were carried out on an AMD Athlon64 3500+ (Linux, 2.2 GHz, 1 GB). We considered different benchmark problems. In the following, the name indicates the type of the problem. The prefix $ec\_$ indicates equivalence checking of a multiplier ($mul\_$) on the word level with another multiplier that is given as word level module ($mul\_$), as sum of partial products ($pp\_$), or as gate level description ($gt\_$), respectively. Thereby, a miter circuit is used. In some cases the least significant bit was ignored in the miter (indicated by $li\_$) and in other cases a fault was injected at the gate level to create a satisfiable instance (indicated by $ft\_$). The prefix pc_arith indicates a property checking problem that contains arithmetic modules. Finally, a number indicates the bit width of the data path.

Table I provides run times for K*BMDs, SWORD and Yices, while Table II shows results in comparison

TABLE I

COMPARISON TO WORD LEVEL SOLVER

circuit	K*BMD	SWORD	SMT
ec_mul_mul_7	<0.01s	0.35s	<0.01s
ec_mul_mul_8	<0.01s	1.67s	<0.01s
ec_mul_mul_9	<0.01s	8.02s	<0.01s
ec_mul_mul_10	<0.01s	37.09s	<0.01s
ec_mul_pp_7	0.01s	0.62s	15.83s
ec_mul_pp_8	0.01s	3.10s	105.56s
ec_mul_pp_9	0.01s	15.54s	>500s
ec_mul_pp_10	0.01s	59.85s	>500s
ec_mul_gt_7	3.48s	0.91s	10.93s
ec_mul_gt_8	13.60s	4.69s	82.40s
ec_mul_gt_9	53.45s	23.20s	>500s
ec_mul_gt_10	202.31s	113.48s	>500s
ec_mul_mul_li_7	>500s	0.34s	0.29s
ec_mul_mul_li_8	>500s	1.66s	1.96s
ec_mul_mul_li_9	>500s	7.95s	58.15s
ec_mul_mul_li_10	>500s	37.01s	>500s
pc_arith_a_6	0.5s	0.36s	<0.01s
pc_arith_a_7	2.1s	1.72s	<0.01s
pc_arith_a_8	8.7s	8.21s	<0.01s
pc_arith_a_9	35.8	37.83s	<0.01s
pc_arith_b_10	1.69s	1.42s	0.07s
pc_arith_b_11	3.18s	4.68s	0.15s
pc_arith_b_12	6.36s	12.24s	0.34s
pc_arith_b_13	12.82	30.91s	0.96s

to MiniSat. An x in column sat indicates whether the problem instance is satisfiable. For each benchmark the number of variables to represent the problem, the number of clauses for MiniSat and the number of modules for SWORD are given in columns var, cls and mod, respectively. The memory requirements and the CPU time in seconds are provided in columns mem and $time$. Finally, the improvement in run time of SWORD over MiniSat is shown in column imp.

As expected K*BMDs performs very well on pure word level problems and outperform SWORD in this case (e.g. benchmark set ec_mul_mul). But when the description is provided at the bit level the performance degrades significantly (ec_mul_gt). Furthermore, bit level operations cannot be handled efficiently ($ec\_mul\_mul\_li$). Yices also handles the pure word level problems extremely efficient. But again, when word level and lower level descriptions are mixed, the performance degrades. On these benchmarks SWORD is more robust. In comparison to MiniSat SWORD requires less memory and is significantly faster (except benchmark set pc_arith_b). In the best case up to three orders of magnitude can be achieved.

VI. CONCLUSIONS

We presented the satisfiability solver SWORD that uses a SAT like algorithm and exploits word level information in the search process. SWORD works on a representation of the problem in terms of modules. This yields a powerful framework for decision making, implications and conflict analysis. Experimental results show on our benchmarks, that SWORD is more robust than other approaches that were considered here.

A task for future work is developing techniques for automating the creation of modules for SWORD. Furthermore, the application to other problem domains

TABLE II

COMPARISON TO BIT LEVEL SOLVER

circuit	sat	MiniSat				SWORD				imp
		var	cls	mem	time	var	mod	mem	time	
ec_mul_mul_7		519	1766	3.98MB	2.02s	43	3	2.73MB	0.35s	5.77
ec_mul_mul_8		687	2348	4.50MB	10.79s	49	3	2.73MB	1.67s	6.46
ec_mul_mul_9		879	3014	5.65MB	54.96s	55	3	2.73MB	8.02s	6.85
ec_mul_mul_10		1095	3764	8.45MB	461.44s	61	3	2.73MB	37.09s	12.44
ec_mul_pp_7		1012	3381	4.24MB	3.98s	228	17	2.73MB	0.62s	6.41
ec_mul_pp_8		1331	4460	5.00MB	25.76s	292	19	2.73MB	3.10s	8.30
ec_mul_pp_9		1694	5689	6.93MB	189.24s	364	21	2.73MB	15.54s	12.17
ec_mul_pp_10		2101	7068	>10.16MB	>500s	444	23	2.86MB	59.85s	>8.35
ec_mul_gt_7		519	1766	3.98MB	2.02s	274	246	2.73MB	0.91s	2.21
ec_mul_gt_8		687	2348	4.50MB	10.79s	360	328	2.86MB	4.69s	2.30
ec_mul_gt_9		879	3014	5.65MB	54.96s	458	422	2.86MB	23.20s	2.36
ec_mul_gt_10		1095	3764	8.45MB	461.44s	568	528	2.86MB	113.84s	4.05
ec_mul_mul_li_7		518	1761	3.99MB	2.03s	43	3	2.73MB	0.34s	5.97
ec_mul_mul_li_8		686	2342	4.36MB	7.95s	49	3	2.73MB	1.66s	4.78
ec_mul_mul_li_9		878	3009	5.90MB	88.88s	55	3	2.73MB	7.95s	11.17
ec_mul_mul_li_10		1094	3759	8.11MB	409.51s	61	3	2.73MB	37.01s	11.06
ec_mul_gt_ft_18	x	3687	12788	17.16MB	70.58s	1880	1808	3.12MB	<0.01s	>7058.00
ec_mul_gt_ft_19	x	4119	14294	16.84MB	54.88s	2098	2022	3.29MB	0.01s	5488.00
ec_mul_gt_ft_21	x	4575	15884	20.10MB	73.91s	2328	2248	3.30MB	<0.01s	>7391.00
ec_mul_gt_ft_22	x	5055	17558	24.91MB	111.03s	2570	2486	3.43MB	0.03s	3701.00
pc_arith_a_6		572	1980	4.11MB	3.78s	55	10	2.73MB	0.36s	10.50
pc_arith_a_7		740	2562	5.00MB	28.52s	61	10	2.73MB	1.72s	16.58
pc_arith_a_8		932	3228	6.93MB	196.98s	67	10	2.73MB	8.21s	23.99
pc_arith_a_9		1148	3978	>10.16MB	>500s	73	10	2.73MB	37.83s	>13.21
pc_arith_b_10		250	852	3.60MB	0.01s	77	17	3.89MB	1.42s	<0.1
pc_arith_b_11		268	911	3.61MB	0.01s	82	17	4.68MB	4.68s	<0.1
pc_arith_b_12		286	970	3.59MB	0.01s	87	17	6.70MB	12.24s	<0.1
pc_arith_b_13		304	1029	3.59MB	0.01s	92	17	7.70MB	30.91s	<0.1

than verification is an important topic. As one example logic synthesis for reversible circuits with SWORD was introduced in [22].

ACKNOWLEDGMENTS

We wish to thank João Marques-Silva and Paulo Jorge Matos for many helpful discussions in the area of SMT.

REFERENCES

[1] R. Bryant, "Graph-based algorithms for Boolean function manipulation," *IEEE Trans. on Comp.*, vol. 35, no. 8, pp. 677–691, 1986.

[2] A. Kuehlmann, V. Paruthi, F. Krohm, and M. Ganai, "Robust Boolean reasoning for equivalence checking and functional property verification," *IEEE Trans. on CAD*, vol. 21, no. 12, pp. 1377–1394, 2002.

[3] M. Davis, G. Logeman, and D. Loveland, "A machine program for theorem proving," *Comm. of the ACM*, vol. 5, pp. 394–397, 1962.

[4] N. Eén and N. Sörensson, "An extensible SAT solver," in *SAT 2003*, ser. LNCS, vol. 2919, 2004, pp. 502–518.

[5] R. Bryant and Y.-A. Chen, "Verification of arithmetic functions with binary moment diagrams," in *Design Automation Conf.*, 1995, pp. 535–541.

[6] R. Drechsler, B. Becker, and S. Ruppertz, "K*BMDs: A new data structure for verification," in *European Design & Test Conf.*, 1996, pp. 2–8.

[7] R. Brinkmann and R. Drechsler, "RTL-datapath verification using integer linear programming," in *ASP Design Automation Conf.*, 2002, pp. 741–746.

[8] J. Thathachar, "On the limitations of ordered representations of functions," in *Computer Aided Verification*, ser. LNCS, vol. 1427. Springer Verlag, 1998, pp. 232–243.

[9] S. A. Seshia, S. K. Lahiri, and R. E. Bryant, "A hybrid SAT-based decision procedure for separation logic with uninterpreted functions," in *Design Automation Conf.*, 2003, pp. 425–430.

[10] H. Ganzinger, G. Hagen, R. Nieuwenhuis, A. Oliveras, and C. Tinelli, "DPLL(T): Fast decision procedures," in *Computer Aided Verification*, ser. LNCS, vol. 3114, 2004, pp. 175–188.

[11] B. Dutertre and L. Moura, "A Fast Linear-Arithmetic Solver for DPLL(T)," in *Computer Aided Verification*, ser. LNCS, vol. 4114, 2006, pp. 81–94.

[12] B. Dutertre and L.Moura, *The YICES SMT Solver*, 2006, available at http://yices.csl.sri.com/.

[13] C.-Y. Huang and K.-T. Cheng, "Using word-level ATPG and modular arithmetic constraint-solving techniques for assertion property checking," *IEEE Trans. on CAD*, vol. 20, no. 3, pp. 381–391, 2001.

[14] J. Marques-Silva and K. Sakallah, "GRASP: A search algorithm for propositional satisfiability," *IEEE Trans. on Comp.*, vol. 48, no. 5, pp. 506–521, 1999.

[15] M. Moskewicz, C. Madigan, Y. Zhao, L. Zhang, and S. Malik, "Chaff: Engineering an efficient SAT solver," in *Design Automation Conf.*, 2001, pp. 530–535.

[16] E. Goldberg and Y. Novikov, "BerkMin: a fast and robust SAT-solver," in *Design, Automation and Test in Europe*, 2002, pp. 142–149.

[17] G. Parthasarathy, M. Iyer, K.-T. Cheng, and L.-C. Wang, "An efficient finit-domain constraints solver for circuits," in *Design Automation Conf.*, 2004, pp. 212–217.

[18] Y. Novikov and R. Brinkmann, "Foundations of hierarchical sat-solving," in *Int'l Workshop on Boolean Problems*, 2004, pp. 103–141.

[19] G. Tseitin, "On the complexity of derivation in propositional calculus," in *Studies in Constructive Mathematics and Mathematical Logic, Part 2*, 1968, pp. 115–125, (Reprinted in: J. Siekmann, G. Wrightson (Ed.), Automation of Reasoning, Vol. 2, Springer, Berlin, 1983, pp. 466–483.).

[20] M. M. Mano and C. R. Kime, *Logic and Computer Design Fundamentals*, 3rd ed. Pearson Education, 2004.

[21] M. Herbstritt, *wld: A C++ library for decision diagrams*, Institute of Computer Science, Albert-Ludwigs-University, Freiburg im Breisgau, 2000, http://ira.informatik.uni-freiburg.de/software/wld.

[22] R. Wille and D. Große, "Fast Exact Toffoli Network Synthesis of Reversible Logic," in *Int'l Conf. on CAD*, 2007.

Obtaining delay distribution of dynamic logic circuits by error propagation at the electrical level

Lucas Brusamarello[1], Roberto da Silva[1], Gilson I. Wirth[2], Ricardo A. L. Reis[1]

{lucas, rdasilva, wirth, reis}@inf.ufrgs.br

[1]Instituto de Informática - UFRGS

[2]Departamento de Engenharia Elétrica - UFRGS

Abstract— In deep-sub-micron technologies, process variability challenges the design of high yield integrated circuits. While device critical dimensions and threshold voltage shrink, leakage currents drastically increase, threatening the feasibility of reliable dynamic logic gates. Electrical level statistical characterization of this kind of gates is essential for yield analysis.

This paper proposes a yield model for dynamic logic gates based on error propagation using numerical methods. We study yield of a dynamic-NOR using static keeper. The analytical formulations can be extended to a wide range of dynamic gates (for example pre-charge dynamic gates using dynamic keeper) because we use numerical approach for the calculation of derivatives required by error propagation. The proposed methodology presents errors less than 2% as compared to Monte Carlo simulation, while increasing computational efficiency up to 50×.

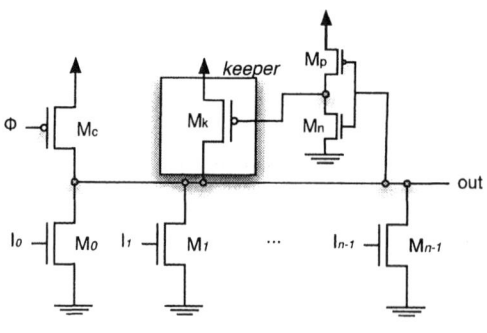

Fig. 1. Dynamic NOR with keeper

I. INTRODUCTION

Performance and reliability of deep submicron technologies are being increasingly affected by process variations and leakage current [1]. Prediction of the percentage of manufactured circuits that will achieve timing constraints becomes a major problem for the circuit designer. Therefore, the use of statistical methods in circuit design is of increasing relevance. When considering electric level simulations, the statistical characterization of circuits must be related to the microscopic features that cause device performance variability and affect circuit yield.

Electrical parameters variability may be decomposed into die-to-die variations (D2D) and within-die variations (WD) [2] [3]. Within-die variations may be due to many sources, for instance the discreteness of matter and energy (dopant atoms, photo resist molecules, and photons). A well known example of WD parameter is threshold voltage (Vt) variability due to the Random Dopant Fluctuations (RDF) [4].

Die-to-die variations may be originated from equipment asymmetries (like asymmetries in chamber gas flows, thermal gradients and so on) or imperfections in equipment operation and process flow. These asymmetries and imperfections affect the average value of the parameters from die to die, wafer to wafer, and lot to lot. In the case of a D2D parameter k, transistors close to each other are affected by the same constant fluctuation δk.

MOSFET sub-threshold leakage currents have been increasing exponentially across successive technology generations due to threshold voltage and channel length reduction [5].

Furthermore, with decreasing device dimensions and supply voltages, the charge amount at the circuit nodes reduces. These effects impact negatively the robustness and feasibility of wide (high fan-in) domino logic gates [6].

In order to increase the circuit noise margin (reducing sensitivity to leakage current, charge sharing effect and coupling noise), pre-charge dynamic gates can be designed using the traditional static keeper [7], as shown in figure 1. This circuit is composed by a dynamic output, a static inverter and a static keeper transistor. If the output is at VDD, the keeper provides a path from the power supply to the output preventing the output to be discharged by leakage currents. During a transition, the keeper and pull-down network transistors compete to determine the logical state of the dynamic node. The time delay of a transition is inverse to the keeper transistor sizing, while noise margin is proportional to it.

In [8] cell characterization using numerical error propagation is proposed. However, their cell modeling methodology does not include D2D variation, although their proposed SSTA algorithm considers spatial correlation at gate level. Furthermore, the quantitative contribution of each random parameter to the circuit performance variance is not analyzed. As shown further in our work, this analysis may help to improve yield. Finally, in that work only first order approximation for numerical derivatives is employed, and algorithm complexity and accuracy is not analyzed.

This work presents a probabilistic model to compute yield

978-1-4244-1709-4/07/$25.00 © 2007 IEEE

of a pre-charge dynamic gate. We study the variability in the delay of a pre-charge dynamic NOR with a static keeper. The methodology presented in this paper is intended to be generic enough to model response (delay, contention time, power, leakage, among others) variability of any kind of pre-charge circuits, including circuits employing dynamic keeper [9] [6]. This generality is true because we employ numerical techniques for computing circuit sensitivities. In [10] we propose yield analysis of SRAM memory using error propagation and numerical derivatives.

This paper is organized as follows. Section II extends error propagation and numerical approach to the problem of delay and contention time of dynamic gates considering or not the static keeper. Section III presents a probabilistic model to describe probability density function (PDF) and yield of a dynamic gate with n inputs. The results of the methodology as compared to Monte Carlo simulations are shown in section IV. Finally our conclusions are exposed in section V.

II. ERROR PROPAGATION AND MC SIMULATIONS

Statistical estimates of electric characteristics of digital and analog circuits are often obtained by Monte Carlo simulations [11] considering a large sample of simulations at electric level [12]. Since MC requires a huge number of runs to obtain reasonable approximation for variance and error estimates, error propagation (EP) emerges as a suitable way to compute variance of a electrical response of circuits with small number of random variables. The error propagation method works by computing the variance of the measure of interest as a function of its partial derivatives in relation to the dependent variables and their variances. Variances of the dependent variables are quantities given by the foundry.

In this work the partial derivatives are computed numerically, using electrical simulations [12]. The method is generic enough to work with an arbitrary circuit response (time delay, power, noise margin,...) and is independent of the electric parameters (Vt, L, W, Tox,...) as well as the transistors topology.

Subsection II-A shows numerical approximations to compute partial derivatives. Subsection II-B presents the formulas to compute variance of a dynamic logic circuit and subsection II-C shows a way to obtain the sensitivity of the variance in function of each electric parameter.

A. Numerical derivatives

This work employs linear approximation for computing sensitivities required by error propagation, using 1 and 2 points around the mean value of each parameter. The difference between these formulas is the accuracy in the numerical estimates.

Considering a function of m variables $f = f(x_1, x_2, ..., x_m)$, according to Fundamental Theorem of Calculus for $\varepsilon \ll 1$ follows

$$\frac{\partial f}{\partial x_i}(x_1, ..., x_m)\Big|_{x_i = \bar{x}_i} = \frac{f(x_1, .., \bar{x}_i + \varepsilon, ...x_m) - f(x_1, .., \bar{x}_i, ...x_m)}{\varepsilon} + O(\varepsilon) \tag{1}$$

In this case we need two simulations to calculate each numerical derivative: one to compute $f(x_1, .., \bar{x}_i + \varepsilon, ...x_m)$ and other to compute $f(x_1, .., \bar{x}_i, ...x_m)$. As $f(x_1, .., \bar{x}_i, ...x_m)$ is computed only once, for computation of all derivatives the total number of simulations is $m + 1$.

In order to obtain a more precise approximation, algebraic manipulations over Taylor expansion results in a formula with accuracy $O(\varepsilon^2)$:

$$\frac{\partial f}{\partial x_i}(x_1, ..., x_m)\Big|_{x_i = \bar{x}_i} = \frac{f(x_1, .., \bar{x}_i + \varepsilon, ...x_m) - f(x_1, .., \bar{x}_i - \varepsilon, ...x_m)}{2\varepsilon} + O(\varepsilon^2) \tag{2}$$

however in this case it is necessary to compute $2m$ runs. If we are working with a small m this higher order approximation still means a significant reduction in the number of electrical simulations, if compared to the usual sampling statistical processes.

Compared to traditional Monte Carlo, for which a reasonable number of runs must be performed to obtain a suitable estimate of variance, here the numerical estimates of derivatives can avoid thousand of runs of the electrical simulator by performing only 1 or 2 runs multiplied by the number of variables of the function. Moreover, derivatives can be analyzed and it emerges as an efficient methodology to improve critical parameters of the circuit, in order to optimize yield.

B. Delay variance of the dynamic logic circuit

The delay of a dynamic gate with keeper and n inputs (see Figure 1) can be written as a function of the parameters associated to the $(n + 4)$ transistors, which are labeled as Mc, Mn and Mk and $\{Mi\}_{i=0}^{n-1}$, Mp. Voltage thresholds are represented by Vt_{Mc}, Vt_{Mp}, Vt_{Mn}, Vt_{Mk} and $\{Vt_{Mi}\}_{i=0}^{n-1}$. Channel length variations of these transistors, i.e., L_{Mc}, L_{Mp}, L_{Mn}, L_{Mk} and $\{L_{Mi}\}_{i=0}^{n-1}$, are divided into two components: D2D and WD.

D2D components, represented by $L_{Mc}^{(d2d)}$, $L_{Mp}^{(d2d)}$, $L_{Mn}^{(d2d)}$, $L_{Mk}^{(d2d)}$ and $\{L_{Mi}^{(d2d)}\}_{i=0}^{n-1}$, are synchronized random Gaussian variables, such that

$$L_x^{(d2d)} = \alpha_x \bar{L} + \xi \cdot \sigma \tag{3}$$

where x denotes the indexes Mc or Mi with $i = 0, ..., n-1$. The α_x's and σ are constants while ξ is standard normal variable ($N(0,1)$). In other words these variables are the same random variable, although they can present different mean values. From that, defining $L^{(d2d)} = \bar{L} + \xi \cdot \sigma$ we have $L_x^{(d2d)} = (\alpha_x - 1)\bar{L} + L^{(d2d)} =$ constant $+L^{(d2d)}$.

The delay of the circuit is defined as the maximum time required to propagate a transition in the input to the output. In the beginning of the evaluation phase ($\phi = 1$) the dynamic output is VDD and every transition $0 \to 1$ at one or more inputs will cause a transition $1 \to 0$ at the output. In the dynamic gate all inputs are symmetric, so that for a n-input circuit the probability of the maximum delay to be given by the input i is $1/n$.

Assume, without loss of generality, a transition $000...0 \to 100...0$ at the input. In order to analyze the variability of a dynamic gate, first compute the average and standard deviation of the time delay of this transition, which is given by

$$t\left(L^{(d2d)}, L_{Mc}^{(wd)}, L_{Mp}^{(wd)}, L_{Mn}^{(wd)}, L_{Mk}^{(wd)}, \left\{L_{Mi}^{(wd)}\right\}_{i=0}^{n-1}, Vt_{Mc}, Vt_{Mp}, Vt_{Mn}, Vt_{Mk}, \left\{Vt_{Mi}\right\}_{i=0}^{n-1}\right)$$

and the uncertainty in t, using error propagation taking into account circuit symmetry, is given by

$$
\begin{aligned}
\sigma_t^2 &= \left(\frac{\partial t}{\partial L_{Mc}^{(d2d)}} + \frac{\partial t}{\partial L_{M0}^{(d2d)}} + (n-1)\frac{\partial t}{\partial L_{M1}^{(d2d)}}\right)^2 \sigma_{L^{(d2d)}}^2 \\
&+ \left(\frac{\partial t}{\partial L_{Mn}^{(d2d)}} + \frac{\partial t}{\partial L_{Mp}^{(d2d)}} + \frac{\partial t}{\partial L_{Mk}^{(d2d)}}\right)^2 \sigma_{L^{(d2d)}}^2 \\
&+ \left(\frac{\partial t}{\partial L_{Mc}^{(wd)}} + \frac{\partial t}{\partial L_{M0}^{(wd)}} + (n-1)\left(\frac{\partial t}{\partial L_{M1}^{(wd)}}\right)\right)^2 \sigma_{L^{(wd)}}^2 \\
&+ \left(\frac{\partial t}{\partial L_{Mn}^{(wd)}} + \frac{\partial t}{\partial L_{Mp}^{(wd)}} + \frac{\partial t}{\partial L_{Mk}^{(wd)}}\right)^2 \sigma_{L^{(wd)}}^2 \\
&+ \left(\frac{\partial t}{\partial Vt_{Mc}} + \frac{\partial t}{\partial Vt_{M0}} + (n-1)\left(\frac{\partial t}{\partial Vt_{M1}}\right)\right)^2 \sigma_{Vt}^2 \\
&+ \left(\frac{\partial t}{\partial Vt_{Mn}} + \frac{\partial t}{\partial Vt_{Mp}} + \frac{\partial t}{\partial L_{Mk}^{(wd)}}\right)^2 \sigma_{Vt}^2
\end{aligned}
\tag{4}
$$

The dynamic-logic NOR with a static keeper considering variations in Vt and L requires the computation of 18 partial derivatives, regardless the number of inputs. For instance, to obtain σ_t^2 for a circuit with a static keeper using 1 point around mean for numerical derivatives, 19 electrical simulations are required.

From $\bar{t}$ (delay calculated at the nominal values) and σ_t (computed using formulation 4) one can obtain the Gaussian Probability Density Function (PDF). Error propagation is independent of probability distribution, once the inputs parameter are only errors and function sensitivity for each parameter.

C. Contribution of each parameter to circuit variability

When dealing with the challenges imposed by design for manufacturability, it is essential to have a methodology capable of identifying what parameters contribute most to the circuit variability. Error propagation uncovers the quantitative contribution of each transistor to the variability in circuit performance.

Revisiting equation 4, one can obtain the sensitivity of the circuit variance to WD parameter Vt of transistor Mt, which is given by:

$$K(Vt_{Mt}) = \left(\frac{\partial t}{\partial Vt_{Mt}}\right)^2 \sigma_{Vt}^2. \tag{5}$$

In order to compute the sensitivity of variance to D2D parameters, a re-weighted function can be defined as in the expression for $L^{(d2d)}$:

$$r(L_{Mt}^{(d2d)}) = \left(\frac{|\partial t/\partial L_{Mt}^{(d2d)}|}{\sum_{j=1}^{k}|\partial t/\partial L_{Mj}^{(d2d)}|}\right) \tag{6}$$

where $\sum_{i=1}^{k} r(L_{Mt}^{(d2d)}) = 1$ for the k synchronized variables ($k = n + 4$). The contribution of channel length variations of the transistor Mt to the delay variance is given by:

$$K(L_{Mt}^{(d2d)}) = r(L_{Mt}^{(d2d)}) \times \left(\frac{\partial t}{\partial L^{(d2d)}}\right)^2 \sigma_{L^{(d2d)}}^2 \tag{7}$$

III. PROBABILISTIC/STATISTICAL ANALYSIS OF LOGIC GATES

Consider the schematic shown in figure 1, let t_i denote the time delay of a transition $I_i = 0 \to 1$ being i arbitrary, i.e. $0 \le i \le (n-1)$. The probability of the time delay of a transition in the input I_i to be less or equal to τ_{max} is the cumulative probability density function $f_i(t_i \le \tau_{max}) = \int_{-\infty}^{\tau_{max}} p_{\bar{t}_i, \sigma_i}(t_i) dt_i$, where $p_{\bar{t}_i, \sigma_i}(t)$ is a Gaussian PDF with average $\bar{t}_i$ and standard deviation σ_i. Supposing all inputs t_i are independent random variables then the probability of the dynamic logic gate time delay to be less than τ_{max} is

$$P(T < \tau_{max}) = f_0(t_0 \le \tau_{max})f_1(t_1 \le \tau_{max})...f_{n-1}(t_{n-1} \le \tau_{max}).$$

If all inputs are symmetric i.e., $\bar{t}_i = \bar{t}$ and $\sigma_i = \sigma$ for all $i = 0, ..., (n-1)$ then:

$$P(T < \tau_{max}) = f(T \le \tau_{max})^n = \left(\int_{-\infty}^{\tau_{max}} p_{\bar{t}, \sigma}(\tau)d\tau\right)^n \tag{8}$$

and the distribution for maximum delay (delay of the dynamic gate), in a first approximation and for $h \ll 1$, is given by

$$F(T) = \left(\int_{-\infty}^{\tau_{max}} p_{\bar{t}, \sigma}(T)dT\right)^n - \left(\int_{-\infty}^{\tau_{max}-h} p_{\bar{t}, \sigma}(T)dT\right)^n. \tag{9}$$

IV. RESULTS

Consider the pre-charge dynamic-NOR schematic shown in figure 1. Let W_{Mi} be the channel width of the pull-down transistors, W_{Mclk} the channel width of transistor M_{clk}, W_{Mp}, W_{Mn} and W_{Mk} the width of transistors M_p, M_n and M_k respectively. Consider $W_{Mi} = W_{Mn} = 1\mu m$ and $W_{Mclk} = W_{Mp} = 2.5\mu m$.

We suppose threshold voltage (Vt) and channel length (L) are random variables with Gaussian distribution. For each transistor i, L_i is decomposed into one spatially correlated component and one spatially uncorrelated component, i.e $L_i = L^{(d2d)} + L_i^{(wd)}$ so that $L^{(d2d)} = N(70nm, 5nm)$ and $L_i^{(wd)} = N(0, 5nm)$ — where $N(m, s)$ denotes a random variable with mean m and standard deviate s. Threshold voltages are random variables given by $Vt_{PMOS} = N(-0.22V, 13.3mV)$ and $Vt_{NMOS} = N(0.2V, 13.3mV)$. These values are in accordance to ITRS [13] and [14]. The transistor model employed is

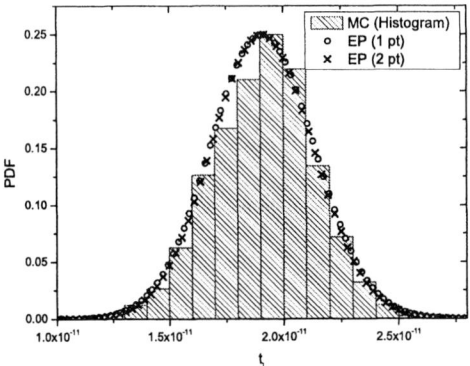

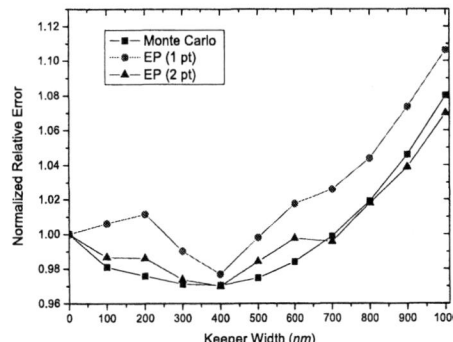

Fig. 2. PDF obtained by EP compared to histogram obtained by MC for time delay of a transition in 8-input dynamic-NOR with keeper ($W_k = 500nm$)

Fig. 3. Relative error for 8-input dynamic-NOR in function of keeper width, normalized by relative error of 8-input dynamic-NOR without keeper

Berkley BSIM3 Predictive Technology Model for the 70nm node (BPTM70) [15].

Equation 4 gives variance using error propagation applied to a dynamic-NOR with a static keeper. Again, this simulation consists in the study of delay variability for a transition $00...00 \rightarrow 10..00$, i.e. a transition $0 \rightarrow 1$ at one input (without loss of generality we consider transition at input I_0, because inputs are symmetric). The partial derivatives for 6 transistors must be computed. Since each transistor has 3 random parameters (Vt, $L^{(d2d)}$ and $L^{(wd)}$), 18 partial derivatives must be computed. Figure 2 shows EP using 1 or 2 points numerical derivatives as compared to MC using 10^3 samples for an 8-input dynamic-NOR with $W_{Mk} = 500nm$. Error propagation using 1 point around mean for derivative requires 19 Spice simulations while the approach using 2 points requires 36 Spice simulations.

Figure 3 shows the normalized relative error (σ/μ) of the delay of the 8-input dynamic-NOR as a function of keeper channel width. The curve indicates that there is an optimal size for keeper strength. Dynamic-NOR designed with $W_{Mk} = 400nm$ presents a 3% smaller relative error compared to a dynamic-NOR without keeper, while a design using $W_{Mk} = 1\mu m$ presents a 6% increase in delay variability. For all simulations, error propagation using 1 point around mean for numerical derivatives presents an error up to 2% compared to MC, while the approach using 2 points for derivatives presents an error smaller than 1%.

From the standard deviate obtained and average previously computed, probability of gate delay T to be less than a constant τ_{max} is can be computed using equation 8. Figure 4 shows the yield of an 8-input dynamic-NOR with static keeper ($W_{Mk} = 500nm$) in function of the time constraint τ_{max}.

Distribution for the gate delay is computed using equation 9, where σ_τ and $\overline{\tau}$ can be computed using the proposed methodology or MC. Figure 5 exposes the Gaussian PDF of the dynamic-NOR. In order to derive the PDF using EP, we

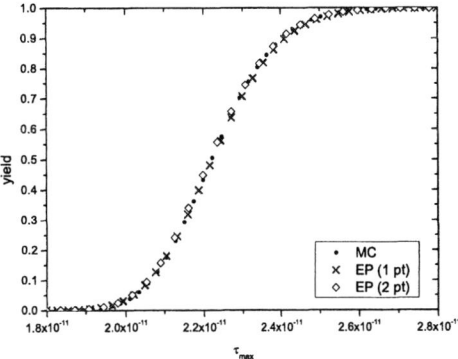

Fig. 4. $p(t_0 < \tau_{max}, t_1 < \tau_{max}...,t_{(7)} < \tau_{max})$ for a 8-input dynamic-NOR, computed by MC and EP using 1 or 2 points for derivatives

run 19 simulations in the case of derivatives using 1 point around mean ($50\times$ faster as compared to Monte Carlo) and 36 electrical simulations considering 2 points ($27\times$ faster as compared to Monte Carlo).

Figure 6 presents the contribution of each parameter to the delay variability, as discussed in in section II-C. The Contribution of $L^{(wd)}$ and $L^{(d2d)}$ of the transistor which has a transition at the input are orders of magnitude more significant than other parameters.

V. CONCLUSIONS

A novel methodology for variability analysis in dynamic logic is presented. The methodology shows results that are statistically equivalent to the usual sampling techniques, like Monte Carlo simulation, while increasing simulation speed by orders of magnitude. Our theoretical approach is generic and can be extended to other kinds of dynamic or static gates with

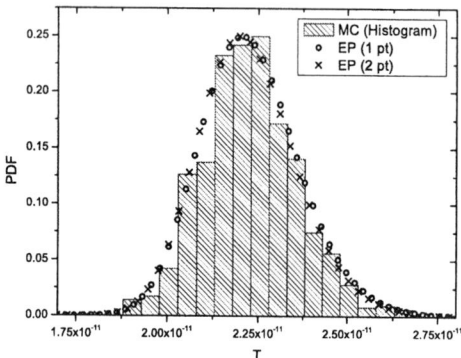

Fig. 5. 8-input dynamic-NOR with keeper ($W_k = 500nm$) delay PDF, computed by MC and EP using 1 or 2 points for derivatives

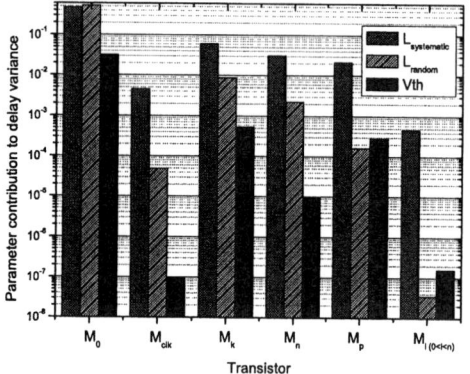

Fig. 6. Contribution of each parameter to delay variability of a dynamic-NOR

minor changes. The proposed methodology allows quantifying the contribution of each component to the variability in circuit behavior. The components that contribute more for circuit behavior variability may then be selected for optimization, aiming to decrease variability at minimal cost.

For the case study circuits and process technology here analyzed, it is clear that the contribution of channel length is orders of magnitude more relevant than threshold voltage. Moreover, in the case of a dynamic gate with static keeper, we verified an optimal strength for the keeper transistor. These results are relevant for a yield enhancement phase.

REFERENCES

[1] Y. Taur, D. Buchanan, W. Chen, D. Frank, K. Ismail, S.-H. Lo, G. Sai-Halasz, R. Viswanathan, H.-J. Wann, S. Wind, and H.-S. Wong, "Cmos scaling into the nanometer regime," in *Proceedings of the IEEE*, vol. 85, Apr 1997, pp. 486–504.

[2] P. S. Zuchowski, P. A. Habitz, J. D. Hayes, and J. H. Oppold, "Process and environmental variation impacts on asic timing," in *Computer Aided Design, 2004. ICCAD-2004. IEEE/ACM International Conference on*, 2004, pp. 336–342.

[3] M. Orshansky, L. Milor, P. Chen, K. Keutzer, and C. Hu, "Impact of spatial intrachip gate length variability on the performance of high-speed digital circuits," *Computer-Aided Design of Integrated Circuits and Systems, IEEE Transactions on*, vol. 21, no. 5, pp. 544–553, 2002.

[4] H. Mahmoodi, S. Mukhopadhyay, and K. Roy, "Estimation of delay variations due to random-dopant fluctuations in nanoscale cmos circuits," *Solid-State Circuits, IEEE Journal of*, vol. 40, no. 9, pp. 1787–1796, 2005.

[5] R. Rao, A. Srivastava, D. Blaauw, and D. Sylvester, "Statistical analysis of subthreshold leakage current for vlsi circuits," *IEEE Transactions on Very large Scale Integration (VLSI) Systems*, vol. 12, no. 2, pp. 131–139, February 2004. [Online]. Available: http://www.gigascale.org/pubs/527.html

[6] A. Alvandpour, R. K. Krishnamurthy, K. Soumyanath, and S. Y. Borkar, "A sub-130-nm conditional keeper technique," *Solid-State Circuits, IEEE Journal of*, vol. 37, no. 5, pp. 633–638, 2002.

[7] S.-J. Shieh, J.-S. Wang, and Y.-H. Yeh, "A contention-alleviated static keeper for high-performance domino logic circuits," *Electronics, Circuits and Systems, 2001. ICECS 2001. The 8th IEEE International Conference on*, vol. 2, pp. 707–710 vol.2, 2001.

[8] K. Kang, B. C. Paul, and K. Roy, "Statistical timing analysis using levelized covariance propagation," *Design, Automation and Test in Europe, 2005. Proceedings*, pp. 764–769 Vol. 2, 2005.

[9] V. Kursun and E. G. Friedman, "Domino logic with variable threshold voltage keeper," *Very Large Scale Integration (VLSI) Systems, IEEE Transactions on*, vol. 11, no. 6, pp. 1080–1093, 2003.

[10] L. Brusamarello, R. D. Silva, R. A. L. Reis, and G. I. Wirth, "Yield analysis by error propagation using numerical derivatives considering wd and d2d variations." in *ISVLSI*. IEEE Computer Society, 2007, pp. 86–91.

[11] J. G. Amar, "The monte carlo method in science and engineering," *Computing in Science and Engineering*, vol. 8, no. 2, pp. 9–19, 2006.

[12] S. Inc., *HSPICE Simulation and Analysis User Guide*, 2005.

[13] *The International Technology Roadmap for Semiconductors*, Semiconductor Industry Association, 2005 Edition.

[14] S. Nassif, "Design for variability in dsm technologies [deep submicron technologies]," in *Quality Electronic Design, 2000. ISQED 2000. Proceedings. IEEE 2000 First International Symposium on*, 2000, pp. 451–454.

[15] Y. Cao, T. Sato, D. Sylvester, M. Orchansky, and C. Hu, "New paradigm of predictive mosfet and interconnect modeling for early circuit design," in *Custom Integrated Circuit Conference*, June 2000, pp. 201–204.

Minimizing Wire Delays by Net-Topology Aware Binding during Floorplan-Driven High Level Synthesis

Vyas Krishnan and Srinivas Katkoori

Department of Computer Science & Engineering,
University of South Florida, Tampa, Florida, USA
{krishnan, katkoori}@cse.usf.edu

Abstract

With shrinking feature sizes in deep sub-micron technologies, interconnect delays play a dominant role in the cycle time of digital circuits. It is essential to consider the impact of physical design during high-level synthesis. No prior work exists in literature that accounts for the topology of nets resulting from binding decisions during high-level synthesis. This paper presents a novel floorplan-aware high-level synthesis technique that uses accurate net topologies and distributed wire-delay models to guide resource allocation and binding decisions during design-space exploration. The proposed approach tightly integrates a floorplanner with a high-level synthesis binding algorithm. The location of data path modules in the floorplan is used to determine the minimal length RSMT of every net, to which the delay model is applied to accurately estimate delays of multi-terminal nets. Our results show that, when compared to previous approaches, the synthesis technique proposed in this paper reduces wire delays by as much as 48.9% in 70nm technology, with an average improvement of 38.6%, and an overhead of only 3.6% in chip area.

1. Introduction

With deep sub-micron technologies, interconnect delays between modules is increasingly becoming a major part of the cycle time [1]. Interconnect delays strongly depend on the number and sizes of the modules used in a data path, and the relative locations of these modules in the floorplan. The area of the floorplan is determined by the number of modules present in the floorplan, and their positioning, which is partly determined by the binding decisions taken during high-level synthesis. Thus, it is important to consider binding information during floorplanning. Conversely, during binding, the effect of floorplanning should be considered. The floorplan information is necessary to accurately predict the structure interconnect of the interconnects in a data path.

Taking interconnect costs into account during high-level synthesis has attracted significant attention. In some of the early work on HLS [2]-[6], a simple estimate of the interconnect cost was determined by counting the number of wires and multiplexers required by a design. These estimates did not use any physical-level information and were used mainly to compare the wiring complexity of different data path design alternatives. However, when interconnect delays began to be comparable to (and even exceed) gate delays, these simple estimates were no longer adequate, since it is not possible to accurately predict the performance of a design without first knowing enough about its floorplan and the structure of its interconnect. In response to this, a number of researchers have considered the impact of physical details (such as floorplanning information) on high-level synthesis [7]-[12]. Others have used floorplan-level information for interconnect-driven HLS [13],[14]. Most of these approaches use a loosely-coupled floorplanner, where the floorplanning and binding decisions are made independent of each other. However, previous work has shown that tightly coupling high-level and physical synthesis, by simultaneously performing HLS and floorplanning, improves their combined performance [15]-[17]. The work proposed in this paper integrates a floorplanner with a high-level synthesis binding algorithm.

The main contribution of this work is an integrated floorplan-aware high-level synthesis technique that uses accurate net topologies and distributed wire-delay models to guide resource allocation and binding decisions.

The rest of this paper is organized as follows. In Section 2, we describe the nature of the problem addressed in this paper. In Section 3, we introduce the timing model used to estimate net delays. In Section 4, we describe our wire-topology aware binding algorithm. In Section 5, we present experimental results, and in Section 6 we conclude the paper.

2. Estimating Wire Delays during High-Level Synthesis

The primary motivation for using floorplan-level information during high-level synthesis is to more accurately predict the interconnect structure of a data path. Wire length estimates from derived floorplans are used to drive the HLS synthesis. The underlying premise is that minimizing wire length often leads to corresponding reductions in interconnect delays. To simplify layout-driven high-level synthesis algorithms, most of the previous work have used simple interconnect models to estimate interconnect delays such as linear wire-delay models applied to two-terminal nets [10][11][13].

Resource sharing is a commonly used technique to reduce hardware requirements in high-level synthesis resulting in multi-terminal nets. A common simplification made in previous work is to break multi-terminal nets into two-terminal nets (point-to-point nets) [9]-[13],[15],[16]. This simplification has three major drawbacks:

- it ignores the net topology of a multi-terminal net during delay estimation,
- it assumes that all source-to-sink delays on a multi-terminal net are separable and independent, and
- it ignores the downstream capacitances during the source-to-sink delay computation.

Another common simplification in most of previous work is to use simple linear delay models to estimate wire delays. This simplification results from the fact that interconnects are treated as simple two-terminal point-to-point nets.

978-1-4244-1709-4/07/$25.00 © 2007 IEEE

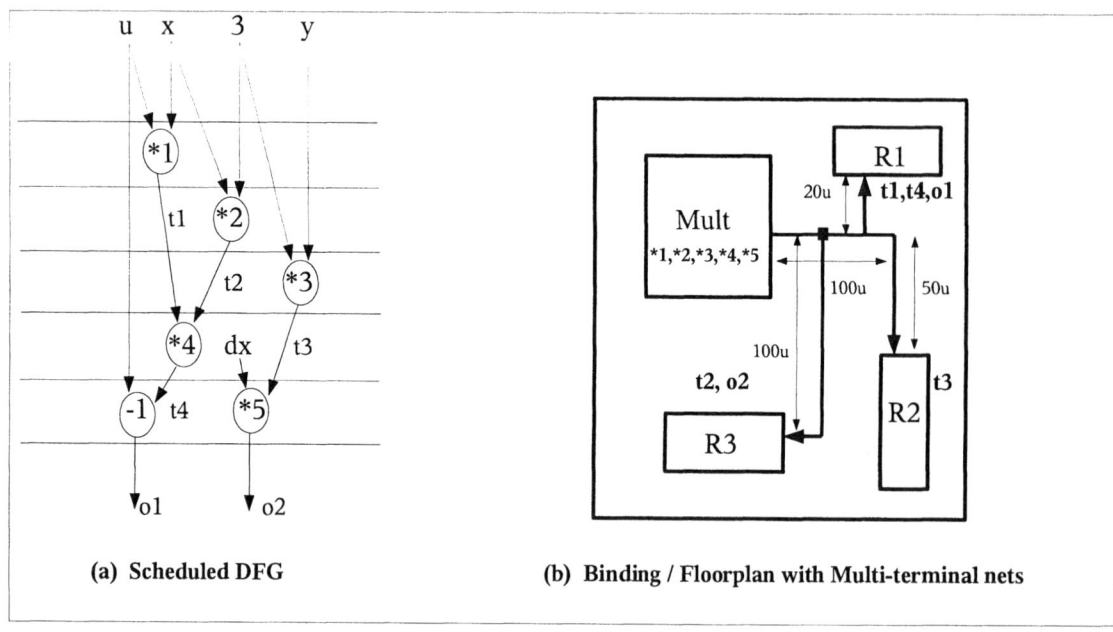

(a) Scheduled DFG	(b) Binding / Floorplan with Multi-terminal nets

Figure 1. Motivating example

An illustrative example:

Treating multi-terminal nets as equivalent to a set of two-terminal nets often leads to gross under-estimation of the net delays, especially when wire parasitics can no longer be ignored in DSM technology. This can be illustrated with a motivating example (Figure 1), showing the importance of accounting for the wire topology of a multi-terminal net during delay estimation. The data flow graph (DFG) [2] in Figure 1(a) is mapped to an RTL data path with one multiplier and three registers, and a corresponding floorplan in Figure 1(b).

Table I. Comparison of arrival-time computations with and without accounting for wire topology

Arrival times for module	Point-to-point delay model with lumped wire delays	Point-to-point delay model with distributed wire delays	Delay model using net topology with distributed wire delays
R1	941.4 ps	1247.9 ps	1783.9 ps
R2	1136.5 ps	1267.5 ps	1787.4 ps
R3	1078.2 ps	1233.7 ps	1778.9 ps

Table I shows the estimated net delays for the data path in Figure 1(b) using ITRS technology parameters from [18] for the 70nm technology. The table compares delays of datapath nets modeled as traditional two-point nets, with delays obtained by taking the topologies of the nets into account. The Elmore delay model with distributed wire parasitics is used to compute the net delays for both the techniques. Please note that most of the earlier work use a simpler linear delay model for the wires using lumped R and C values for the wire parasitics.

The table clearly demonstrates that

- to correctly estimate the signal arrival times at all the sink nodes of a multi-terminal net requires an estimation of the topology of the net, and the use of this information in an accurate delay estimation model,

- one cannot treat the different sink paths on a multi-terminal net as being "separable" and independent, and
- ignoring the presence of other downstream loads on a net while computing the arrival time on a sink pin can result in significant errors.

Buffers can be used on different sink paths on a net, thus allowing them to be treated as separable. However, buffers consume silicon real estate and dissipate power. A design flow that accounts for the wire parasitics of an entire multi-terminal net when computing the signal arrival times, could potentially reduce the need for extensively buffering signal nets to achieve timing closure.

3. Timing Model

To estimate net delays we first determine the topology of each net from its minimal wirelength RSMT, and then compute the wire RC delays using a distributed π-model for the wire segments in a net. The main advantage of this model is that it allows us to accurately estimate the delay between the source pin and each sink pin of a net. This is important since sink pin delays among pins on a net can differ significantly, especially for long nets, or for nets with large fanouts. The computation of individual delays results in much greater fidelity between estimated net delays and actual wire delays obtained after detailed routing, improving success in timing-closure.

During high-level synthesis, decisions made during binding determine the number of sink pins driven by a source pin. For example, a register shared by a large number of variables in the scheduled DFG may require the register to drive a large number of data path modules that consume this variable. Binding also determines the number and types of multiplexers needed in a data path. The lengths of the wire segments to different sink pins in multi-terminal nets are determined by the data path floorplan. To be effective, a timing-driven binding algorithm for high-level synthesis must consider the impact of binding, floorplanning, net fanouts, and net topologies, on the the estimated net delays.

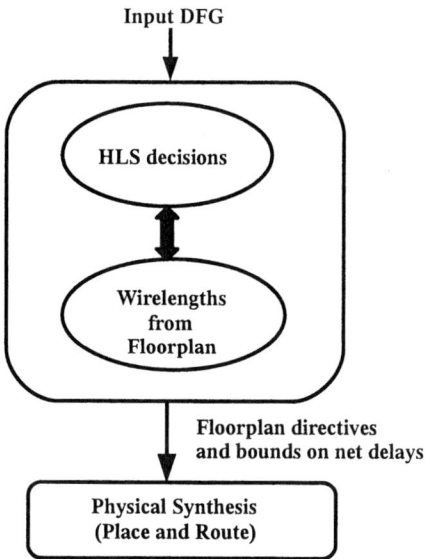

Input DFG

HLS decisions

Wirelengths from Floorplan

Floorplan directives and bounds on net delays

Physical Synthesis (Place and Route)

Figure 2. Traditional physically-driven High-level Synthesis

4. A Framework for Wire-topology aware Binding for High-level Synthesis

A traditional floorplan-driven high-level synthesis technique, as proposed in previous approaches, is shown in Figure 2. In these approaches, simple wire delay models (such as wire delays with lumped R and C values applied to 2-terminal nets) derived from floorplans are used to guide HLS-level decisions. However, as shown in the motivating example in Section 2, these models often under-estimate net delays, thus providing either incorrect or overly constrained time-bounds to the physical synthesis step that follows HLS. This often leads to either over-design, or long convergence times with multiple design re-spins, or in the worst case, a failure to achieve timing closure.

To provide better bounds for net delays during physical synthesis, it is necessary to determine the topology of multi-terminal nets present in a design, to more accurately estimate net delays that correlate better with actual delays present in the final layout. To the best of our knowledge none of the previous work found in the literature has addressed this during HLS.

This work proposes a net-topology aware binding algorithm for HLS. The algorithm used in this work is illustrated in Figure 3. A Simulated Annealing based iterative improvement is used to tightly couple the HLS binding and floorplanning phases of synthesis. This enables a simultaneous search of the design spaces of module bindings, register bindings, floorplans, and net topologies, for solutions with smaller net delays. A unique feature of our algorithm is the use of two interleaved chains of moves that alternate between a sequence of bindings and floorplan moves. This allows the algorithm to perform independent neighborhood search of the binding and floorplan search spaces.

The binding algorithm accepts a scheduled data flow graph and a resource allocation for the data path. A compatibility graph for each resource type is maintained to ensure that the algorithm always generates a legal resource binding. The algorithm returns the best RTL binding and floorplan found by the SA.

```
Inputs: (1) Module Compatibility Graph from DFG
        (2) Module Allocation

while ( curr_temp < Stop_Temperature ) do
{
    for every temperature do
    {
        for pre-set number of binding iterations do
        {
            Select Binding Move
            Execute move
            Compute COST
            if accept( COST, curr_temp)
                update current solution
            else undo Binding Move
        }
        for pre-set number of floorplan iterations do
        {
            Select Floorplanning Move
            Execute move
            Compute COST
            if accept( COST, curr_temp)
                update current solution
            else undo Floorplanning Move
        }
    }
}
return best solution
```

Figure 3. Proposed Net-topology aware timing-driven synthesis algorithm

In the SA, floorplans are represented using the Sequence Pair notation [19]. Three floorplans moves are used: (a) module rotate, (2) module move, and (3) module swap. Three SA moves are used to explore the space of resource bindings: (1) re-assign a module binding, (2) swap assignments of two compatible modules, and (3) swap the input variable assignments of commutative DFG operations.

The binding moves are guided by the resource compatibility graph maintained by the SA. During these binding moves, if the number of sources at the input of a data path resource (module or register) changes as a result of a change in the binding, multiplexers can vanish, appear, or change size. Changes in binding can significantly affect the netlist topology in a data path, and thus the resulting floorplan and wire length statistics. By tightly integrating the HLS step of resource binding with floorplanning, any changes in the netlist topology are immediately reflected in the actual floorplan. Due to the incremental floorplan update feature of our algorithm, the SA need not generate a new floorplan from scratch after each move. This makes our SA very efficient, since the new floorplan usually has only a small difference from the previous one.

In this work, we treat the layout-driven HLS as an optimization problem with the aim of minimizing wire delays. Figure 4 illustrates the steps performed during the SA cost function computation. In addition to the traditional objectives of chip area and wire length, the cost function also minimizes the net delays among RTL modules in the floorplan.

2007 IFIP International Conference on Very Large Scale Integration (VLSI-SoC 2007)

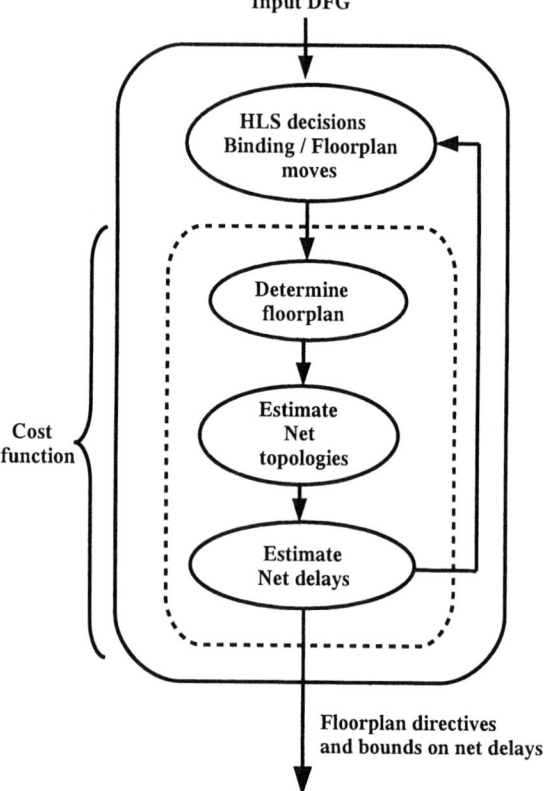

Input DFG

HLS decisions
Binding / Floorplan
moves

Determine
floorplan

Estimate
Net
topologies

Estimate
Net delays

Cost
function

**Floorplan directives
and bounds on net delays**

Figure 4. Cost function computation

The proposed algorithm is generic enough to allow any technique to determine net topologies and estimate net delays. For example, any of the accurate RSMT estimators from the literature could be used. Similarly, any accurate delay estimation engine (such as higher-moment models [20], *RLC*-models [21], statistical delay models [22]) could be used to determine net delays. The choice of the net topology and net delay estimators is a trade-off between computation efficiency and required accuracy.

In this work, FLUTE [23], a recently proposed RSMT estimation algorithm is used for determining net topologies. FLUTE determines the net topology (RSMT) that is optimal (in terms of wire length) for nets with net-degree up to 7, and to within 5% optimal for nets with higher net-degrees. Additionally, FLUTE is very efficient in terms of run-times when compared to other algorithms. The choice of FLUTE was mainly dictated by its runtime efficiency, since the net-topology estimation engine is used thousands of times within our SA iteration loop.

Once the topology is obtained for each net, the widely used Elmore-delay model based delay calculation engine is used to compute the signal arrival times for all the sink pins on a net. The Elmore-delay model was used primarily for its run-time computation efficiency, and its excellent fidelity with actual net delays [24]. The Elmore-delay model is widely used in VLSI physical synthesis literature as a delay estimator [25].

5. Experimental Results and Discussion

To validate the effectiveness of the proposed approach, an SA-based net-topology aware HLS Binding algorithm was developed

and tested on four standard data flow intensive DSP benchmarks drawn from the HLS literature, namely – the 16-point FIR filter, the 8-point IIR filter, the Elliptic Wave (EWF) filter, and the 1-point 8X8 DCT filter. Each of these benchmarks was specified as a scheduled data flow graph (DFG), capturing the behavioral description of the architecture to be synthesized. The algorithm was implemented in C and executed on a PC using a 1.86GHz Intel Core2 Duo CPU with 2GB of RAM, and running under Linux.

In this set of experiments, the objective was to minimize the clock period of a time-step of a DFG schedule. Two algorithms were compared:

(1) **Method-1**: an interconnect-centric floorplan-aware HLS synthesis that back-annotates the wire lengths derived from an integrated floorplanner to guide the high-level synthesis decisions (this is similar to that proposed in previous work), and

(2) **Method-2**: our proposed net-topology driven high-level synthesis flow that uses net topologies to estimate delays computed with an accurate distributed wire-delay model.

Table 2 compares the maximum wire delay among all data transfers in a scheduled data flow graph, for 70nm technology, using both methods. From the table, it can be observed that the proposed technique leads to significant improvements in the maximum wire delay, with corresponding reductions in the clock period. An average improvement in wire delays of 38.6% was achieved over a traditional floorplan-driven technique, with a maximum reduction of up to 48.9%.

Table 2. Comparison of net delays for 70nm technology

Benchmark	Max. Wire Delay Method-1 (traditional)	Max. Wire Delay Method-2 (proposed)	% improvement
IIR	335 ps	224 ps	34.1%
EWF	540 ps	322 ps	40.4%
FIR	504 ps	369 ps	31.1%
DCT	937 ps	569 ps	48.9%
		Avg:	38.6%

Table 3. Comparison of total wire length for 70nm technology

Benchmark	Total WL using Method-1 (traditional)	Total WL using Method-2 (proposed)	% difference
IIR	24892	23680	-4.9%
EWF	36471	32305	-11.4%
FIR	64213	64800	+9.1%
DCT	115982	122002	+5.1%
		Avg:	-0.53%

Table 4. Comparison of chip area for 70nm technology

Benchmark	Area using Method-1 (traditional)	Area using Method-2 (proposed)	% difference
IIR	106624	107030	+0.4%
EWF	71808	74400	+3.5%
FIR	180840	189680	+4.7%
DCT	170404	180560	+5.6%
		Avg:	+3.6%

Tables 3 and 4 respectively compare the total wire length and chip area of the same benchmarks using both techniques. From the tables it can be seen that the chip area and wire lengths are comparable when using both techniques. These results demonstrate that the improvements in wire delays were accomplished by our technique with little overhead in terms of increased chip area or wire length.

Table 5. Pertinent design data and execution time

Benchmark	Size |V| + |E|	Number of RT-level modules	Execution time
FIR (8 point)	28	20	7s
IIR	28	26	9s
EWF	77	29	22s
FIR (16 point)	99	64	1m:58s
DCT	120	74	3m:05s
FIR (32 point)	296	102	8m:43s

Table 5 lists the execution times for the proposed net-topology driven high-level synthesis methodology. The table compares the execution times for benchmarks of different sizes. The size of each benchmark is measured in terms of the number of operations and variables in a scheduled data flow graph. In column 2 of the table, $|V|$ represents the number of nodes, and $|E|$ represents the number of edges, of a benchmark data flow graph. In addition to the benchmark sizes, the table also lists the number of RT-level modules needed to implement the data path for each benchmark. The RT-level datapath modules include datapath computational units as well as multiplexers used to steer data between these units. The number of data path modules provides a measure of the size of the floorplanning problem instance addressed by the algorithm. The execution times in column 4 is listed in terms of minutes and seconds.

The advantages offered by using wire-topology to estimate net delays during HLS synthesis decisions is evident when we back-annotate the computed module-to-module delays on the floorplan, to the scheduled DFG. The module-to-module delays are used to determine the contribution of the wire delays during execution of each of the DFG operations on the scheduled DFG. Figures 5, 6, 7, and 8 illustrate this for the IIR, EWF, DCT, and FIR filter benchmarks respectively. In these figures, the x-axis represents the wire delay between modules for all signal paths in a datapath, and the y-axis represents the number of nets that correspond to these delays. These wire delays are computed using the ITRS projected values for copper interconnect used in intermediate level nets for the 70nm technology.

In our experiments, we assume that DFG operations can be completed in a single clock period (*i.e.*, single cycle operations). During each DFG operation, data transfers occur between several modules (input registers, multiplexers, data path units, and output registers). These wire delays are strongly dependent on the HLS binding and floorplan decisions made during HLS synthesis, and the resulting topologies of the nets created from these decisions.

These experiments demonstrate that incorporating accurate wire delay computations that take the wire topologies into account lead to significant improvements in wire delays and provide more realistic delay constraints to the subsequent physical synthesis steps. The delay distribution plots indicate that using the total wirelength metric and point-to-point net delays to guide HLS binding decisions does not necessarily lead to solutions with optimal clock cycle times. Use of the topologies of the nets, with accurate delay estimates often provide better feedback to the HLS steps during layout-driven synthesis.

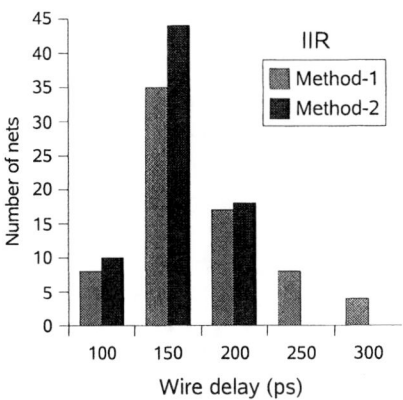

Figure 5. Comparison of net delay distributions for the IIR filter benchmark using 70nm technology

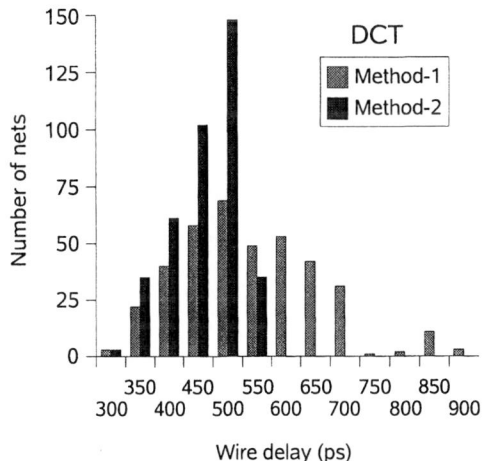

Figure 6. Comparison of net delay distributions for the DCT filter benchmark using 70nm technology

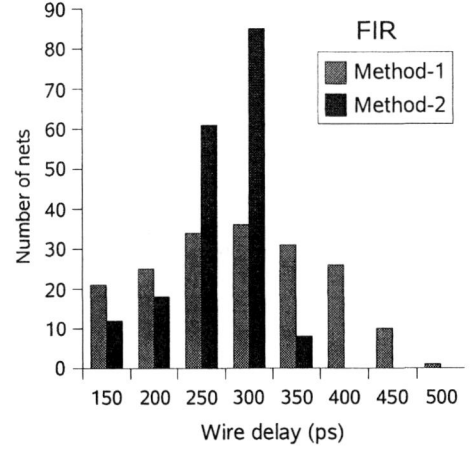

Figure 7. Comparison of net delay distributions for the FIR filter benchmark using 70nm technology

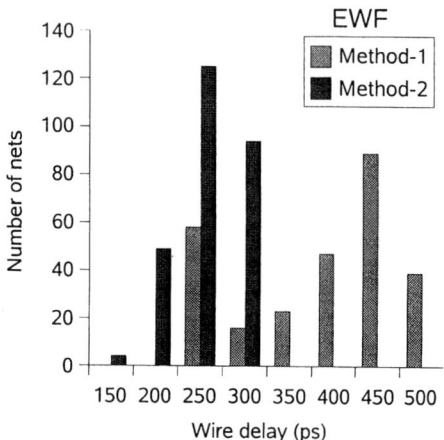

Figure 8. Comparison of net delay distributions for the EWF filter benchmark using 70nm technology

The smaller delay values obtained with the proposed net topology-aware synthesis could be explained by examining the way the HLS binding and the floorplanning steps use this information during synthesis. Using information on module locations, the proposed algorithm is able to determine the impact of a HLS binding decision on the resulting topology of the net driven by a datapath module, and the source-to-sink delays for all the pins on the net. This way, the algorithm is better able to recognize binding decisions that lead to nets with large fanouts, or nets with large downstream loads, which a traditional approach may miss altogether, since they only consider two-terminal point-to-point nets. From our experiments, it is evident that estimating net delays from net topologies results in smaller and more predictable net delays when compared with previous layout-aware high-level synthesis approaches that treat the source-to-sink delays in a multi-terminal net as separable and independent.

6. Conclusions

With increasing wire parasitics in deep sub-micron technologies, it is important to use more accurate delay estimation models during high-level synthesis. The signal paths between the source and multiple sink pins in a multi-terminal net are not separable and independent. This must be taken into account when estimating net delays during high-level synthesis. We have presented a new algorithm that computes the wire topology of multi-terminal nets from floorplans during high-level synthesis, and uses in an accurate Elmore-based model using distributed wire delays for signal nets. Experimental results show that using our accurate delay estimate is quite effective in reducing wire delays. We used this to minimize contribution of wire delays in the cycle time of data paths synthesized during high-level synthesis, with reductions in wire delays of 38.6% on average, when compared to a synthesis flow that minimizes floorplan wire lengths in an attempt to minimize cycle time. These improvements in cycle times were achieved with minimal overhead in terms of chip area or total wirelength.

References

[1] J. Cong, "An Interconnect-Centric Design Flow for Nanometer Technologies," in *Proc. IEEE*, pages 505-528, April 2001.

[2] P.G. Paulin and J.P. Knight, "Scheduling and binding algorithms for high-level synthesis," in *Proc. DAC 1989*.

[3] C.A. Papachristou and H.Konuk, "A linear program driven scheduling and allocation followed by an interconnect optimization algorithm," in *Proc. DAC 1990*.

[4] T.A. Ly, W.L. Elwood, and E.F. Girczyc, "A generalized interconnect model for data path synthesis," in *Proc. DAC 1990*.

[5] S. Tarafdar and M.Leeser, "The DT-model: High-level synthesis using data transfer," in *Proc. DAC 1998*.

[6] C. Jego, E. Casseau, and E. Martin, "Interconnect cost control during high-level synthesis," in *Proc. Int. conf. Design Circuits Integrated System*, Nov 2000.

[7] M.C. McFarland and T.J. Kowalski, "Incorporating bottom-up design into hardware synthesis," *IEEE Trans. CAD*, vol. 9, no. 9, pp. 938-950, Sept.1990.

[8] D.W. Knapp, "Fasolt: A program for feedback-driven data-path optimization," *IEEE Trans. CAD*, vol. 11, no. 6, pp. 677-695, June. 1992

[9] J.P. Weng and A.C. Parker, "3D Scheduling: High-level synthesis with floorplanning," in *Proc. DAC 1992*.

[10] Y.M. Fang and D.F. Wong, "Simultaneous functional-unit binding and floorplanning," in *Proc. ICCAD 1994*.

[11] M. Xu and F.J. Kurdahi, "Layout-driven RTL binding techniques for high-level synthesis using accurate estimators," in *ACM TODAES*, vol. 2, no. 4, pp. 312-343, Oct. 1997.

[12] W.E. Dougherty and D.E. Thomas, "Unifying behavioral synthesis and physical design," in *Proc. DAC 2000*.

[13] A. Stammerman, et. al., "Binding, allocation, and floorplanning in low power high-level synthesis," in *Proc. ICCAD 2003*.

[14] L. Zhong and N.K. Jha, "Interconnect-aware low power high-level synthesis," in *IEEE Trans. CAD*, vol. 24, no. 3, pp. 336-351, Mar. 2005.

[15] Z. Gu and R.P. Dick, "Unified Incremental Physical-Level and High-Level Synthesis," in *Proc. DAC 2005*.

[16] R. Kastner, et. al., "Layout Driven Data Communication Optimization for High Level Synthesis," in *Proc. DATE 2006*.

[17] V. Krishnan and S. Katkoori, "3D-Layout aware binding algorithm for high-level synthesis of three-dimensional integrated circuits," *ISQED 2007*.

[18] International Technology Roadmap for Semiconductors 2006, *www.itrs.net*

[19] H. Murata *et al.*, "VLSI Module Placement based on Rectangle-Packing by the Sequence-Pair," in *IEEE Trans. CAD*, vol. 15, no.12, pp. 1518-1524, December 1996.

[20] B. Tutuianu, F. Dartu, and L. Pileggi, "An Explicit RC-Circuit Delay Approximation Based on the First Three Moments of the Impulse Response," in *DAC 1996*, pp. 611-616.

[21] A.B. Kahng and S. Muddu, "An Analytical Delay Model for RLC Interconnects," in *IEEE Trans. CAD*, vol. 16, no.12, pp.1507- 1514, December 1997.

[22] K. Agarwal *et.al.*, "Statistical Interconnect Metrics for Physical-Design Optimization," in *IEEE Trans. CAD*, vol. 25, no.7, pp. 1273-1288, July 2006.

[23] C. Chu, "FLUTE: Fast Lookup Table Based Wirelength Estimation Technique," in *Proc. ICCAD 2004*.

[24] K.D. Boese et al., "Fidelity and near optimality of Elmore-based delay constructions," in *Proc. ICCAD 1993*, pages 81-84.

[25] J. Cong and D.Z Pan, "Interconnect Performance Estimation Models for Design Planning," in *IEEE Trans. CAD*, vol. 20, no. 6, pp. 739-752, June 2001.

A High-Driving Class-AB Buffer Amplifier with a New Pseudo Source Follower

Chih-Wen Lu, Yen-Chih Shen and Meng-Lieh Sheu

Abstract—A high-driving class-AB buffer amplifier, which consists of a high-gain input stage and a pseudo source follower, is proposed. The pseudo source follower consists of two same types of differential pairs rather than two complementary error amplifiers. The high-driving capability is mainly provided by the folded amplifiers. An experimental prototype buffer amplifier implemented in a 0.35-μm CMOS technology demonstrates that the circuit dissipates an average static power consumption of only 660 μW at a power supply of 3.3 V, and exhibits the slew rates of 2.7 V/μs and 3.8 V/μs for the rising and falling edges, respectively, under a 300Ω/150 pF load. The second and third harmonic distortions (HD2 and HD3) are -67 dB and -65 dB, respectively, at 20 KHz under the same load.

Index Terms—class-AB, buffer amplifier, pseudo source follower, error amplifier.

I. INTRODUCTION

THE class-AB buffer amplifier is widely used for driving heavy resistive or capacitive loads [1-5]. To achieve the extended voltage swing, the output transistors should be connected in a common-source configuration. The quiescent current of the output transistors should be small, while the dynamic current should be as large as possible. The gates of the two output transistors are normally driven by two in-phase ac signals separated by a dc voltage. For example, Hogervorst et al. [4-5] proposed a two-stage, compact, power-efficient 3 V CMOS operational amplifier, in which the output stage is biased by a floating class-AB control. Langen et al. [6] also proposed compact low-voltage power-efficient operational amplifier cells for VLSI, in which a class-AB control is used to bias the output transistors. The above amplifiers are compact and power-efficient. Another approach, which employs a pseudo source follower shown in Fig. 1 to realize class-AB CMOS buffer amplifiers, have been widely used for a large voltage swing output. It is composed of a pair of complementary common-source transistors with two feedback loops consisting of a pair of complementary error amplifiers [7-10]. This offers a wide output voltage swing and a large ratio between the maximum transient current (class B current) and the quiescent current. However, when the input voltage of the pseudo source follower is near to VDD/VSS, the gate-to-source voltage of the output transistors cannot reach a

The authors are with the Department of Electrical Engineering, National Chi Nan University, Taiwan, R.O.C. (e-mail: cwlu@ ncnu.edu.tw).

large value [5]. Fig. 2 shows the schematic of the conventional pseudo source follower. When the input voltage is near to VSS, the maximum gate-to-source voltage of M12 is only $V_{SG8} - V_{SD8}$ where V_{SG8} and V_{SD8} are the source-to-gate and source-to-drain voltages of M8, respectively. This limits the drive capability of the pseudo source follower. Also, the NMOS input error amplifier (M1~M5) conducts no current for a low level input. Then the node at the gate of M11 is high impedance and is easy to be disturbed. In this work, a new pseudo source follower is proposed to overcome this problem. The pseudo source follower employs two same types of error amplifiers rather than the complementary ones. Each error amplifier uses a folded amplifier to obtain a driving capability. The proposed buffer amplifier is designed to work at voice band frequencies for the telecommunication and audio applications.

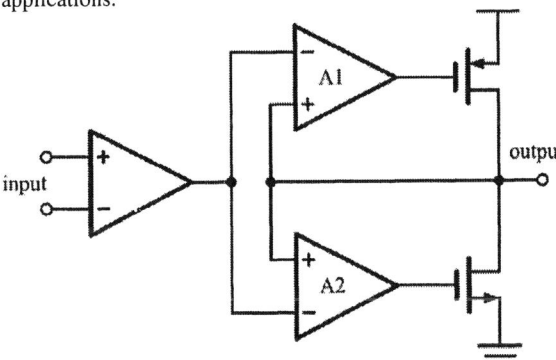

Fig. 1 The conventional architecture of class-AB buffer amplifier.

II. PROPOSED CLASS-AB BUFFER AMPLIFIER

The proposed class-AB buffer amplifier has the same architecture in Fig. 1 but its pseudo source follower employs two same types of differential pairs. Fig. 3 shows the first error amplifier, A1, which drives the output PMOS transistor, M23. This error amplifier, which is consisted of M1~M9, is a folded amplifier. M2 and M3, which are biased by M1, form the input differential pair. The active load, M4 and M5, is folded by the constant current sources, M8 and M9. M6-M7 are two common-gate amplifiers. When the voltage of the inverting input terminal, in1-, is increased, the current flowing in M3 is reduced but the current in M5 is increased. Then the gate voltage of the PMOS output transistor, M23, is pulled down to a lower level, so M23 starts to charge the output node.

105

978-1-4244-1709-4/07/$25.00 © 2007 IEEE

Since the gate voltage of M23 can be pulled down to minimum level of $V_{DS7(triode)} + V_{DS9(triode)}$, which is very close to VSS, M23 has a large driving capability.

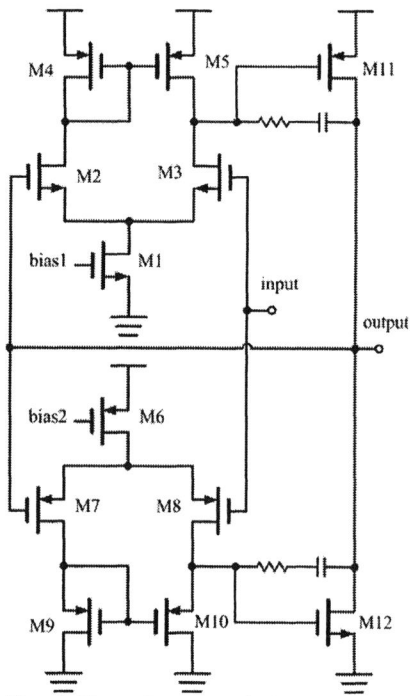

Fig. 2 The schematic of the conventional pseudo source follower.

The second error amplifier, A2, is shown in Fig. 4. The differential pair is biased by M10 and its currents are mirrored to the folded active load by the current mirrors of M13-M16. M21-M22 are the folded active load of the differential pair. When the voltage of the inverting input terminal is reduced, the currents in M12, M14, and M15 are increased. However, the current flowing in M21 is reduced. This increases the gate voltage of the output NMOS transistor, M24. M24 then discharges the output node. Since the gate voltage of M24 can be raised up to a maximum value of VDD $- (V_{SD18(triode)} + V_{SD20(triode)})$, which is very close to VDD, M24 has a large discharge capability. This means that the proposed pseudo source follower has a large driving capability.

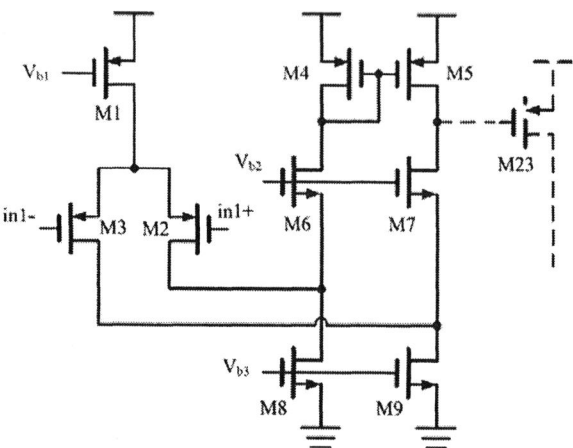

Fig. 3 The proposed first error amplifier, A1.

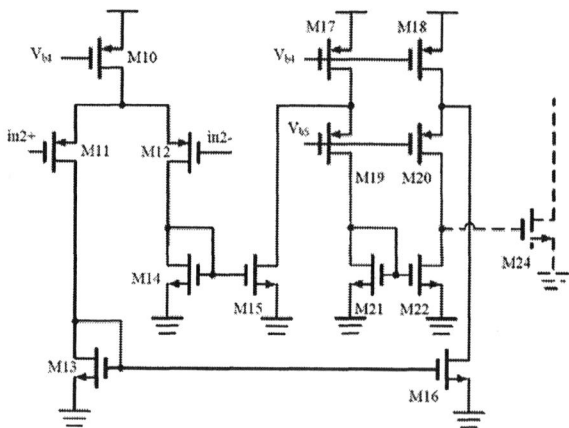

Fig. 4 The proposed second error amplifier, A2.

Fig. 5 shows the complete schematic of the proposed class-AB buffer amplifier where M25-M31 consists of a two-stage amplifier, M1-M9 and M10-M22 form the first and second error amplifier, respectively, and M23-M24 are two output transistors. The resistors of Rcs1-Rcs3 and capacitors of Ccs1-Ccs3 are used for the Miller compensation. Since all of the differential pairs are the same types, they have the same input common-mode range. When the input voltage is near to VSS, the gate voltage of M23 is still well controlled. This overcomes the above problem of the conventional circuit.

106   *2007 IFIP International Conference on Very Large Scale Integration (VLSI-SoC 2007)*

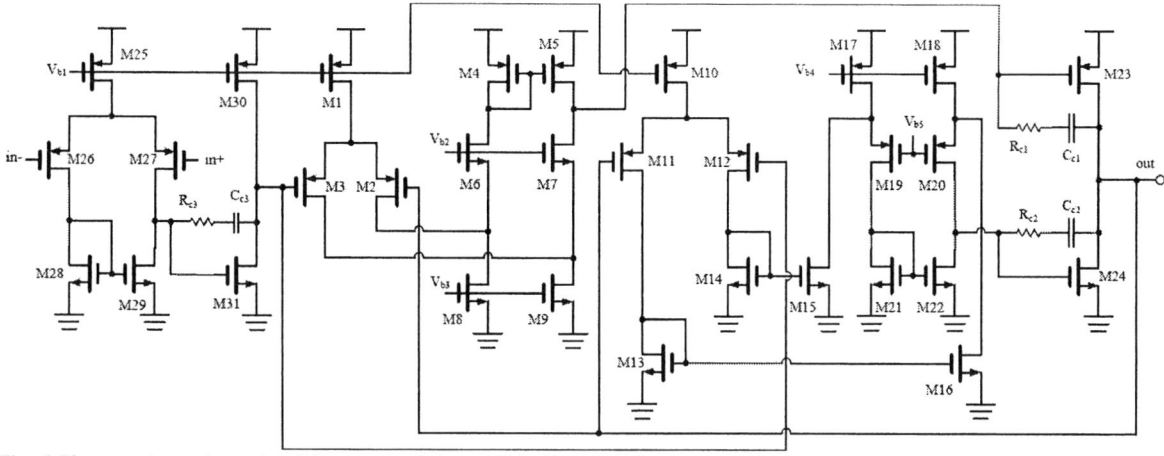

Fig. 5 The complete schematic of the proposed class-AB buffer amplifier.

III. EXPERIMENTAL RESULTS

The proposed output buffer amplifier was fabricated using a 0.35-μm CMOS technology. The die photograph is shown in Fig. 6. The active area of the buffer is only 242×47 μm^2. Fig. 7 shows the measured result of the output with the input of 2.4 V swing of a 20 KHz triangular wave of the unity-gain buffer amplifier loaded with 300 Ω resistor in parallel with a 150 pF capacitor. It can be seen that the output basically follows the input. The step responses of the unity-gain buffer amplifier with the same load with the voltage swings of 20 mV and 2.4 V are shown in Fig's. 8 and 9, in which Fig. 8 shows the small signal response with 20 mVpp and Fig. 9 shows the large signal response with 2.4 Vpp. The slew rates are 2.7 V/μs and 3.8 V/μs for the rising and falling edges, respectively. Fig. 10 shows the measured results of the output with the input of a large dynamic range (2.4 Vp-p) of a 20 KHz sinusoidal wave for the unity-gain buffer amplifier. Fig. 11 shows its magnitude spectrum of the proposed buffer amplifier with a 2.4 Vp-p output swing into a 300Ω/150pF load. The second and third harmonic distortions (HD2 and HD3) are -67 dB and -65 dB, respectively, at 20 KHz under the same load. All of the measured results are summarized in Table 1.

IV. CONCLUSION

A high-driving class-AB buffer amplifier, which employs two pseudo source followers, has been presented. An experimental prototype buffer amplifier implemented in a 0.35–μm CMOS technology demonstrates that the circuit dissipates an average static power consumption of only 660 μW at a power supply of 3.3 V, and exhibits the slew rates of 2.7 V/μs and 3.8 V/μs for the rising and falling edges, respectively, under a 300Ω/150 pF load. The second and third harmonic distortions (HD2 and HD3) are -67 dB and -65 dB, respectively, at 20 KHz under the same load.

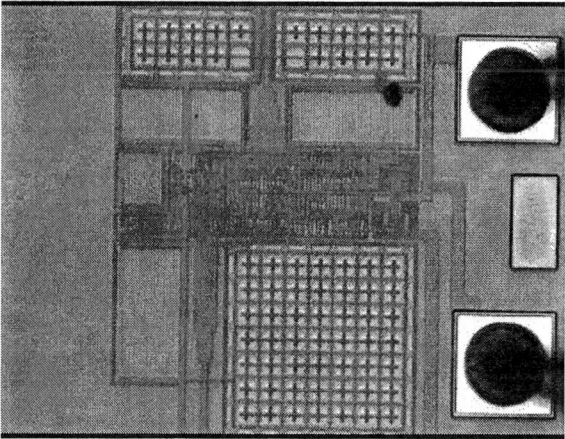

Fig. 6 The die photograph.

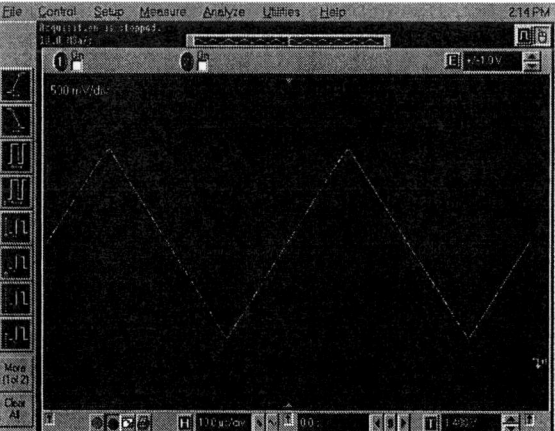

Fig. 7 The measured result of the output with the input of 2.4 V swing of a 20 KHz triangular wave of the unity-gain buffer amplifier loaded with 300 Ω resistor in parallel with a 150 pF

2007 IFIP International Conference on Very Large Scale Integration (VLSI-SoC 2007)

capacitor.

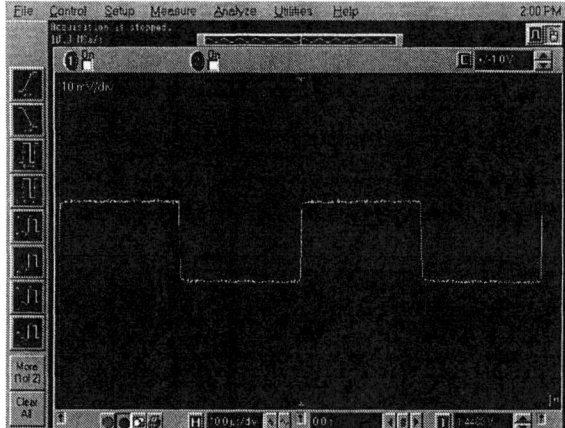

Fig. 8 The measured step response of the unity-gain buffer amplifier loaded with 300 Ω resistor in parallel with a 150 pF capacitor with the voltage swing of 20 mV.

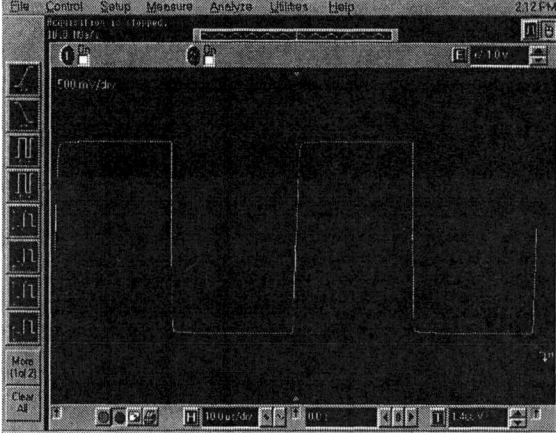

Fig. 9 The measured step response of the unity-gain buffer amplifier loaded with 300 Ω resistor in parallel with a 150 pF capacitor with the voltage swing of 2.4 V.

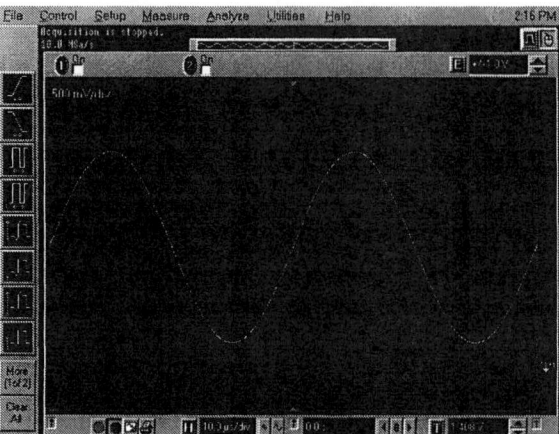

Fig. 10 The measured results of the output with the input of a large dynamic range (2.4 Vp-p) of a 20 KHz sinusoidal wave for the unity-gain buffer amplifier.

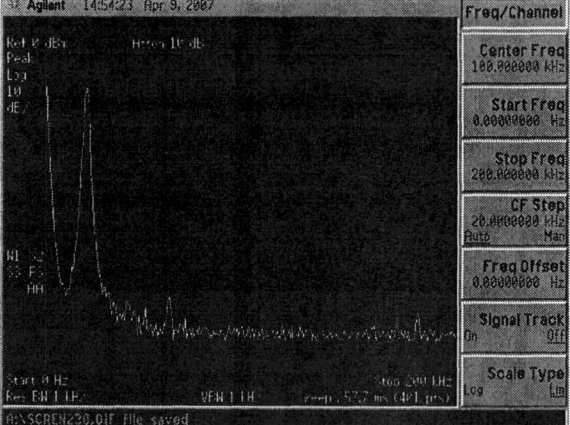

Fig. 11 The measured magnitude spectrum of the proposed buffer amplifier with a 2.4 Vp-p output swing into a 300Ω/150pF load.

ACKNOWLEDGMENT

The authors would like to thank the Chip Implementation Center of National Science Council for their support in chip fabrication.

REFERENCES

[1] Chih-Wen Lu and Chung Len Lee "A Low Power High Speed Class-AB Buffer Amplifier for Flat Panel Display Application", IEEE Transactions on VLSI Systems, Vol 10, No. 2, pp. 163-168, April 2002.

[2] Chih-Wen Lu "A Low Power High Speed Class-AB Buffer Amplifier for Flat Panel Display Driver Application", Digest of SID, pp. 281-283, 2002.

[3] Fan You, S.H.K. Embabi and Edgar Sanchez-Sinencio, "Low-Voltage Class AB Output Amplifiers with Quiescent Current Control," IEEE

Journal of Solid-State and Circuits. Vol.33, No.6, pp. 915-920, June 1998.

[4] Ron Hogervorst, John P. Tero, Ruud G. H. Eschauzier, and Johan H. Huijsing, "A Compact Power-Efficient 3V CMOS Rail-to-Rail Input/Output Operational Amplifier for VLSI Cell Libraries," IEEE ISSCC pp.244-245, 1994.

[5] Ron Hogervorst, John P. Tero, Ruud G. H. Eschauzier, and Johan H. Huijsing, "A Compact Power-Efficient 3V CMOS Rail-to-Rail Input/Output Operational Amplifier for VLSI Cell Libraries," IEEE Journal of Solid-State and Circuits. Vol. 29, No.12, pp. 1505-1513, December 1994.

[6] Klaas-Jan de Langen, Johan H. Huijsing, "Compact Low-Voltage Power-Efficient Operational Amplifiers Cells for VLSI," IEEE Journal of Solid-State and Circuits. Vol. 33, No.10, pp. 1482-1496, October 1998.

[7] A. Torralba, R.G. Carvajal, J. Martinez-Heredia and J. Ramirez-Angulo," Class AB output stage for low voltage CMOS op-amps with accurate quiescent current control," ELECTRONICS LETTERS Vol. 36, No. 21, 12th October 2000.

[8] Kevin E. Brehmer, and James B. Wieser, "Large Swing CMOS Power Amplifier" IEEE Journal of Solid-State and Circuits. Vol. SC-18, No.6, pp. 624-629, December 1983.

[9] John A. Fisher, "A High-Performance CMOS Power Amplifier" IEEE Journal of Solid-State and Circuits. Vol. SC-20, No.6, pp. 1200-1205, December 1985.

[10] Joongsik Kih, Byungsoo Chang, Deog-Kyoon Jeong, and Wonchan Kim, "Class-AB Large-Swing CMOS Buffer Amplifier with Controlled Bias Current," IEEE Journal of Solid-State and Circuits. Vol. 28, No.12, pp. 1350-1353, December 1993.

[11] Gaetano Palumbo, and Salvatore Pennisi, "High-Frequency Harmonic Distortion in Feedback Amplifiers: Analysis and Applications," IEEE Transactions on Circuits and Systems-E Fundamental Theory and Applications, Vol. 50, No. 3, pp. 328-340, March 2003.

[12] P. Gray and R. Meyer, Analysis and Design of Analog Integrated Circuits, 3rd ed. New York: Wiley, 1993.

[13] Frank N. L. Op't Eynde, Patrick F. M. Ampe, Lode Verdeyen, and Willy M. C. Sansen, "A CMOS Large-Swing Low-Distortion Three-Stage Class AB Power Amplifier," IEEE Journal of Solid-State and Circuits. Vol. 25, No.1, pp. 265-273, February 1990.

Table 1 Performance summary.

	This work	Brehmer [6]	Fisher [7]	Kih [8]
Technology	0.35-μm 1P4M CMOS	Proprietary P^2CMOS	5-μm CMOS	1.2-μm 2P2M CMOS
Die area	242 × 47 μm^2 (active)	1500 mils^2	1000 mils^2	103 mils^2
Power supply	3.3 V	± 5 V	± 5 V	5 V
Open loop gain	85 dB	83 dB	93 dB	85 dB
Slew rate	2.7 V/μs (rise) 3.8 V/μs (fall)	0.6 V/μs	1.5 V/μs	0.65 V/μs
Harmonic distortion	(freq. = 20 KHz, $V_{out,pp}$ = 2.4 V R_L = 300 Ω, C_L = 150 pF) HD2 = -67 dB HD3 = -65 dB	(freq. = 4 KHz, V_{IN} = 3.3 V_P R_L = 300 Ω, C_L = 1000 pF) -49.9 dB	(freq. = 3 KHz, V_{IN} = 3 V_P R_L = 200 Ω) HD2 = -73 dB HD3 = -78 dB	(freq. = 5 KHz, $V_{out,pp}$ = 3.5 V R_L = 300 Ω, C_L = 150 pF) THD = -63.6 dB
Input common mode range	2.4 V	NA	+3.3 ~ -5.5 V	4.25 V
Power dissipation	660 μW	5 mW	12.7 mW	4.7 mW

A new analytical approach of the impact of jitter on continuous time delta sigma converters

Julien Goulier, Eric Andre

STMicroelectronics
850 Rue Jean Monnet
38926 Crolles Cedex, France
julien.goulier@st.com

Marc Renaudin

TIMA Laboratory
46 Avenue Felix Viallet
38031 Grenoble Cedex, France

Abstract—The performances of continuous time delta sigma converters are severely affected by clock jitter and no generic technique to predict the corresponding degradations is nowadays available. This paper presents a new analytical approach to quantify the power spectral density of jitter errors. This generic computational method can be applied to all kind of delta sigma converters. Furthermore, clock imperfections are described by means of phase noise spectrum, consequently all possible type of jitters can be taken into account. This paper also describes the temporal non ideal clock models that have been created to simulate the impact of jitter on delta sigma converters and validate the theoretical results.

I. INTRODUCTION

The current attractiveness for continuous time delta sigma converter is largely due to the fact that it is possible to make them work at higher frequencies than their equivalent discrete time implementation. This specificity is widely used in order to increase the bandwidth or the resolution of the converters. This uninterrupted augmentation of sampling frequency induces an amplification of the ratio between jitter and clock period, making less and less negligible the influence of jitter on the converter performances.

Jitter impact on continuous time delta sigma converter is a tricky problem. The need of a better comprehension of the phenomena and an accurate estimation of the jitter degradations is nowadays still high. In the present paper, our new approach of the jitter problem will be described.

In section II, after a quick reminder of the jitter impact on discrete time delta sigma converters, we will focus on the specificity of continuous time implementation regarding clock jitter and explain our approach to analyze this problem. Hence, we will derive the complete set of equations describing the impact of jitter on a 2nd order modulator and discuss about the possibility to extend this result to more complex architectures. Finally in section IV, the equations accuracy will be verified via some numerical comparisons with simulations.

II. INFLUENCE OF JITTER ON DELTA SIGMA CONVERTERS

In a discrete time delta sigma (DT$\Delta\Sigma$), the input signal is sampled before being converted. So analyzing clock jitter on those converters is equivalent to the investigation of irregular sampling problem [1]. This assumption can be done as long as the imperfections of the clock do not perturb the transfer function of the converter loop. Under the assumption of a white phase noise for the clock signal, the maximum achievable signal to noise ratio (SNR) of a discrete time converter is given by:

$$SNR_{dB} = 10 \log \left(\frac{OSR}{\left(2\pi f_{max}\right)^2 \sigma^2} \right) \quad (1)$$

In this formula, σ is the standard deviation of the Gaussian distribution of the jitter at each clock edge; OSR is the oversampling ratio, and f_{max} is the maximal input frequency. Despite the important restrictions for the application of this equation, white phase noise and sinusoidal input, this formula is widely used for the design of DT$\Delta\Sigma$.

Unfortunately, in continuous time delta sigma converters (CT$\Delta\Sigma$) the jitter impact can not be reduced to the irregular sampling problem, and so (1) is inappropriate. The main reason why this equation is not valid any more is the fact that the sampling element in CT$\Delta\Sigma$ is not in front of the loop but inside it, see Figure 1. Moreover, in continuous time implementation the quantization noise introduced by the inner ADC is also responsible of losses linked to the clock imperfections.

Several articles have already been published on the specific topic of jitter in CT$\Delta\Sigma$ [2]-[4], giving us some interesting clues to understand the phenomena. In our approach of the jitter problem, we have decided not to make any initial assumption on the impact of this imperfection. So the first step of the study is to identify all possible errors introduced by jitter; only after this phase a mathematical estimation of the errors will be practicable.

978-1-4244-1709-4/07/$25.00 © 2007 IEEE

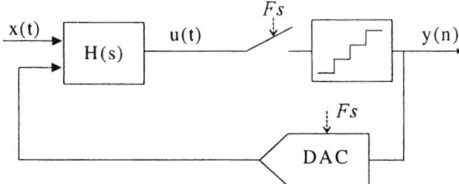

Figure 1. Typical block diagram of a $\Delta\Sigma$_CT

If we consider that clock jitter has an impact on every continuous time function or signal, two kinds of jitter errors can be identified in a CT$\Delta\Sigma$. The first error, called sampling error, relates to the continuous input signal x(t) whereas the second one is introduced by the continuous time loop filter H(s). This second type of error is called integration error.

1) Sampling error

The source of this error is the continuous time input signal x(t). Thus this error happens in both discrete and continuous $\Delta\Sigma$. However the quantity of noise introduced by sampling errors is quite different whether the implementation is continuous or discrete. In a CT$\Delta\Sigma$, the input signal is processed by the loop filter before being sampled.

2) Integration error

This kind of error is specific to continuous time delta sigma converters and is related to the couple DAC/loop filter. Indeed, the processing of the jittered DAC output by the loop filter is responsible for the introduction of errors.

It is obvious that every clock non ideality modifies the timing diagram provided by the DAC. Those slight timing variations, normally processed by the continuous time filter, introduce voltage errors on every stage of the loop filter. The errors introduced in the loop filter by the variation of the integration period are defined by the term "integration errors". The number of integration errors is equal to the modulator order since there is one voltage error at each integrator output.

In spite of the localization of integration errors inside the loop filter, the DAC implementation has a strong influence on those errors. Indeed the DAC is the triggering element of integration errors, so every modification of its implementation induces important changes in the resulting errors. It is for example well known that CT$\Delta\Sigma$ using switched capacitor DAC are less sensitive to jitter than those with non return-to-zero (NRZ) DAC.

To conclude this phase of identification of the jitter errors, the impact of clock imperfections can be summarized as the introduction of N+1 errors for an Nth order modulator: one sampling error and N integration errors.

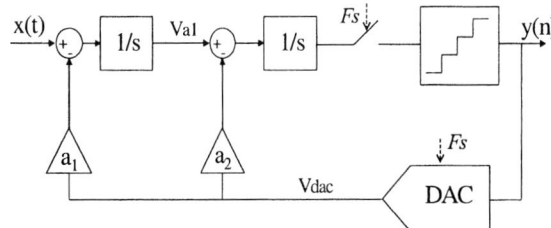

Figure 2. Second order $\Delta\Sigma$ with NRZ feedback

III. ANALYTICAL EVALUATION OF JITTER DEGRADATION

In the previous section, the errors introduced by jitter have been identified; we now need to quantify them in order to derive a mathematical expression of the performance degradations. First the complete set of equations for a 2nd order $\Delta\Sigma$ modulator with NRZ feedback will be established. Then we will show that it is possible to extend those formulas to other architectures.

A. Second order $\Delta\Sigma$ with NRZ feedback

The architecture of the considered converter and the localization of the jitter errors are given on Figure 2. For the following calculations the classical linear model of $\Delta\Sigma$ modulators will be used. This means that the non-linear quantizer is replaced by a white noise adder.

1) Estimation of integration errors

The input signal is continuous and directly applied to the loop filter; it is thus correctly processed by all the continuous time blocs preceding the sampler without introducing integration errors. Therefore to estimate the integration errors we simply assume that the input signal is null.

Consider Δt the jitter error during the Nth clock period, that is to say from the instant $t=nT_S$ to $t=(n+1)T_S+\Delta t$. Throughout the period, the voltage $Vdac$ is constant and sent back to the loop filter trough $a1$ and $a2$, which is the principle of NRZ feedback. The perturbation of the integration time due to the jitter Δt introduces two integration errors, $e1$ in the first integrator and $e2$ in the second stage of the modulator.

The error $e1$ is due to the fact that a_1*V_{dac} is integrated during $T_S+\Delta t$ instead of T_S. This error is generated within the first stage of the $\Delta\Sigma$ thus an equivalent voltage error $Ve1$ at the input of the first integrator can be computed.

$$Ve1 = a_1 V_{dac} * \frac{\Delta t}{T_S} \qquad (2)$$

In order to derive the power spectral density (PSD) of this error $S_{Ve1}(f)$, we can calculate the Fourier transform of its autocorrelation function. The autocorrelation function r_{Ve1} of the error $e1$ is given by:

$$r_{Ve1}(mT) = E\left[Ve1(T) \cdot Ve1(T+mT) \right]$$
$$= \left(\frac{a_1}{T_s}\right)^2 r_{Vdac}(mT) \cdot r_{\Delta t}(mT) \qquad (3)$$

where E denotes the expectation operator, r_{Vdac} and $r_{\Delta t}$ are respectively the autocorrelation functions of the feedback voltage and timing jitter. By applying the Fourier transform to (3), the error spectrum can be found:

$$S_{Ve1}(f) = \left(\frac{a_1}{T_s}\right)^2 S_{Vdac}(f) \otimes S_{\Delta t}(f) \qquad (4)$$

The symbol $\otimes$ represents the convolution operator.

If we multiply this spectrum by the signal transfer function STF of the modulator and replace the temporal jitter spectrum $S_{\Delta t}$ by the phase noise spectrum S_θ, the equation of the PSD of the error e1 at the output of the converter can be derived. Knowing that $S_{\Delta t}(f) = \left(\frac{T_s}{2\pi}\right)^2 S_\theta(f)$, the expression of S_{Ve1} at the output of the converter is:

$$S_{Ve1}(f) = \left(\frac{a_1}{2\pi}\right)^2 \left[S_{Vdac}(f) \otimes S_\theta(f)\right] * STF(f) \qquad (5)$$

Of course, the same calculation method can be applied to the error $e2$, introduced within the second stage of the modulator. The equation is just a little bit more complex because $e2$ has got two components, the first part of the error is due to the single integration of $a_2 * V_{dac}$, and the second one to the double integration of $a_1 * V_{dac}$.

$$e2 = a_2 V_{dac} * \Delta t \left(1 + \frac{a_1}{a_2} T_s + \frac{a_1}{a_2}\frac{\Delta t}{2}\right) + V_{a1} * \Delta t \qquad (6)$$

where V_{a1} is the output voltage of the first integrator, which is the integral of V_{dac}.

From this equation an equivalent second stage voltage error $Ve2$ can be derived. Furthermore, the quantities Ts and Δt are quite smaller than 1; consequently two terms of (6) can be neglected and $Ve2$ approximated to:

$$Ve2 \approx \left(a_2 V_{dac} + V_{a1}\right)\frac{\Delta t}{T_s} \qquad (7)$$

Finally, if the Fourier transform of the autocorrelation function of $Ve2$ is calculated and multiplied by the transfer function TF_{e2} between the input of the second stage and the output of the modulator, we can derive an expression

for the PSD of Ve2. Knowing that $S_{Va1}(f) = \dfrac{1}{(2\pi f)^2} S_{Vdac}(f)$, the PSD of $Ve2$ is given by:

$$S_{Ve2}(f) = \frac{1}{(2\pi)^2}\left[\left[\left(a_2^2 + \frac{1}{(2\pi f)^2}\right) S_{Vdac}(f)\right] \otimes S_\theta(f)\right] * TF_{e2}(f) \qquad (8)$$

In this chapter, the PSD expressions of the two integration errors have been calculated.

2) Estimation of the sampling error

In section II, we have stated that one part of the jitter error is linked to the discretization of the input signal by the CT$\Delta\Sigma$. Even though this jitter degradation is easily understandable, the input signal being sampled when it gets through the modulator, we lack a detailed explanation of the phenomenon allowing us to analytically define an exact formula of the sampling error PSD.

From extensive observations and simulations of jitter in CT$\Delta\Sigma$ it comes out that, in NRZ feedback $\Delta\Sigma$, the errors introduced by jitter in relation with the input signal are equal to the errors that would happen if the input signal was filtered by the STF of the modulator before being sampled. This behavioral analysis has no physical meaning since modeling a CT$\Delta\Sigma$ by a STF equivalent block followed by a sampler is irrelevant. However it allows us to quantify the sampling error and to give an easy and understandable equation.

The PSD of the errors introduced by an isolated sampler is given by [5]:

$$S_{error}(f) = \left[\left(\frac{f}{F_s}\right)^2 S_x(f)\right] \otimes S_\theta(f) \qquad (9)$$

If this equation is applied to our specific case, the following mathematical equation is obtained. This formula gives us the PSD of the errors introduced by clock jitter in relation with the input signal.

$$S_{Vin+jitter}(f) = \left[\left[\left(\frac{f}{F_s}\right)^2 S_{Vin}(f) \cdot STF(f)\right] \otimes S_\theta(f)\right] \qquad (10)$$

From the three PSD equations, (5) (8) and (10), two essential remarks can be made. First, the dependency of jitter degradations to quantization noise, which is a specificity of CT$\Delta\Sigma$, is confirmed by (5) and (8). The second remark relates to the importance of phase noise profile. All formulas present a convolution involving phase noise, so the knowledge of the clock imperfections is a prerequisite for a good estimation of jitter degradations.

With the estimated PSD of all the errors introduced by the jitter in the CTΔΣ, it is quite simple to find the SNR degradation. Indeed, we just have to integrate (5), (8) and (10) on the right range of frequencies. In section IV, we will express in figure some examples in order to attest of the formulas accuracy. First, the possible extension of those equations to generic converter architectures is discussed.

B. Possible extensions of the method and results

The above calculations have been conducted in the special case of a 2nd order converter to facilitate the comprehension of the phenomena; it is obviously possible to do exactly the same work with other architectures. However, in higher order modulators, order greater than 2, the errors introduced by the integration stages that are close to the quantizer have a small influence on performances because there are shaped by the loop. Thus the set of equations defined in the preceding section can be considered as a good approximation of the impact of jitter for every modulator with NRZ feedback DAC.

In the last decade, different methods have been proposed to reduce the jitter sensitivity of CTΔΣ. Switched-capacitor (SC) DAC [6] and FIRDAC [7] are two techniques which have proven their efficiency. If the computation principle previously described is applied to ΔΣ using those correction systems, the resultant benefit can be evaluated.

We have analyzed in details the case of switched capacitor DAC and computed the new set of equations providing the jitter errors PSD. This study has shown that the calculations are comparable to those detailed in paragraph III.A. We are not going to detail them in this paper but some numerical results for a 3rd order modulator with a SC DAC will be given in the next section.

IV. VALIDATION OF THE ANALYTICAL JITTER ERRORS ESTIMATION

In the previous sections, our approach to estimate the impact of clock jitter on the output signal of CTΔΣ has been explained. To prove the accuracy of the given formulas, they will now be compared with simulations.

A. Clock jitter modeling

In order to simulate the impact of jitter on CTΔΣ, temporal models of non-ideal clocks are needed. To realize clock signals presenting different phase noise profile, a voltage controlled oscillator (VCO) has been modeled. This frequency synthesis circuit has been chosen because it is simple enough to be accurately modeled and it allows us to generate a wide range of jittered clocks. This non ideal clock

model has been created with Matlab Simulink blocks and used to drive CTΔΣ modulators.

The phase noise profile of our VCO model is characterized by a -20dB/decade slope and a phase noise floor. The decreasing phase noise slope is a classical feature of an oscillator while the phase noise floor represents the bufferization of the clock signal. Thus, this model possesses two tuning parameters, the levels of the noise slope and noise floor, allowing us to generate different non ideal clocks. Moreover this VCO has been included in a phase locked loop (PLL) to create a more complex jittered clock.

The VCO phase noise profile can be easily translated to temporal imperfections using the classical relations between phase noise and temporal jitter [8]. In fact, the phase noise slope of the VCO corresponds to an accumulated Gaussian timing error while phase noise floor relates to an independent Gaussian temporal error. It is those two temporal imperfections that have been used to create the Matlab Simulink model of VCO. The model accuracy has been validated using phase noise profile comparisons. Figure 3 shows a validation example of the VCO model. The black curve is the theoretical phase noise level while the grey one is the phase noise profile extracted from the simulation of the Matlab VCO model.

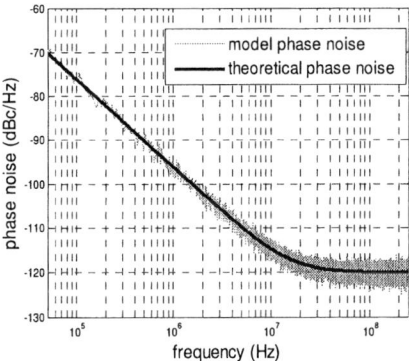

Figure 3. VCO phase noise model validation

B. Jitter equations comparisons with simulations

From the equations stated in section III, we know that jitter degradations are related to the architecture of the converter, the phase noise profile and the input signal PSD. To prove the precision of our jitter impact computation, formulas and simulations have been compared for different CTΔΣ architecture and several phase noise profile. The comparisons have focused on two criterions, the converter output PSD and the SNR value. To simulate the impact of jitter on the performances of CTΔΣ, the VCO model described in the preceding paragraph has been used to drive different converters, see Figure 4.

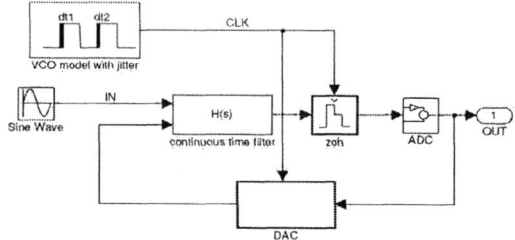

Figure 4. CTΔΣ simulation with non ideal clock

To explore the architecture dependency, three different converter architectures have been used: a 2nd order feedback modulator with NRZ DAC, a 4th order feedback with NRZ DAC and a 3rd order feedback modulator with Switched capacitor DAC. All modulators used a 4-bits internal quantizer. Moreover, two sinusoidal signals with the same amplitude but different frequencies, *Fin1=5MHz* and *Fin2=25MHz*, have been used to illustrate the relation between the jitter degradation and the input PSD.

Finally, to demonstrate how the clock phase noise profile modifies the errors introduced by jitter, two dissimilar clocks have been defined. The frequency of both clocks is 500MHz. The first clock has a flat phase noise profile at -120dBc/Hz, whereas the second clock is a type 1 PLL, with a 500KHz cut off frequency. The PLL phase noise is equal to -90dBc/Hz at 500 KHz and the phase noise floor is located at -120dBc/Hz. The phase noise profiles of those two clocks are represented on Figure 5.

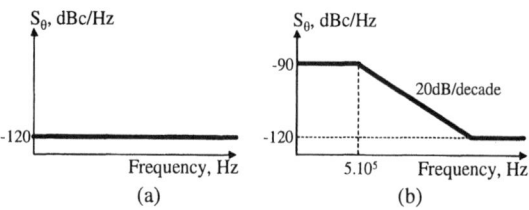

Figure 5. Clocks phase noise profiles, (a) white noise, (b) PLL

For all the possible combinations of architecture, input signal and clock, the converter output PSD and the SNR from 0 to 10 MHz have been derived from equations and simulations.

For each test case, the correct superposition of the simulated PSD with the calculated one demonstrates the reliability of our jitter impact estimation method. PSD comparison examples, with the two non ideal clocks, are shown in Figure 6 and 7. The out of band PSD is not shown on those Figures because it is dominated by quantification noise. The curves correspond to the output signals of the 2nd order feedback modulator with NRZ DAC and a sinusoidal

input signal at 5MHz. The PSD superpositions are evident, and they are confirmed by the calculation of SNR values. For the white phase noise clock comparison case, the SNR achieved by the simulated converter is equal to 64.82dB and the SNR given by the equation is 64.50dB. In the second test case, the SNR values are respectively 62.63dB and 62.59dB.

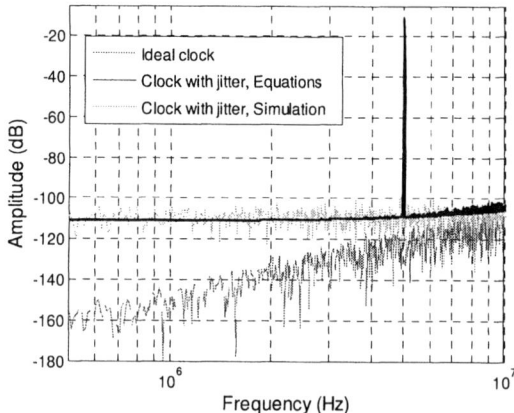

Figure 6. Output spectrum comparison of a 2nd order CTΔΣ controlled by the white phase noise clock

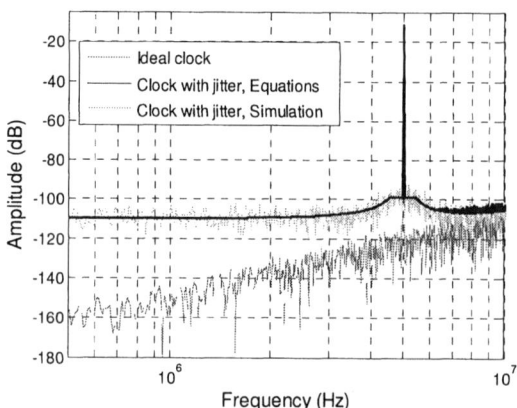

Figure 7. Output spectrum comparison of a 2nd order CTΔΣ controlled by the PLL clock

The same PSD and SNR comparisons have been done with the others converters and clocks and resulted in comparable conclusions on the accuracy of the jitter estimation method. The SNR values of the 12 test cases described above are summarized in Table I. The SNR from simulations are in regular characters, while those from formulas are in bold font. For information, the SNR value of the input signal sampled by non ideal clocks is also given in Table I. Those numbers correspond to the degradations introduced by a jittered clock if a DTΔΣ was used.

The SNR comparison, encapsulated in table I, illustrates the accuracy of the mathematical jitter error estimation method presented in this paper. The discrepancies between calculated and simulated SNR values are indeed really small, always less than 1 dB.

Moreover, the jitter degradation dependence to the three key parameters (modulator architecture, phase noise and input signal) is highlighted by both simulations and equations. The validity of our approach of the jitter problem and the accuracy of the equations are clearly demonstrated by the given results.

TABLE I.   SNR COMPARISONS

	Ideal Clock	Clock 1 : white noise		clock 2 : PLL	
		Fin1	Fin2	Fin1	Fin2
Sampled input signal	∞	87.06dB	73.17dB	66.81dB	72.21dB
		87.02dB	**73.04dB**	**67.10dB**	**72.99dB**
2nd order modulator with NRZ feedback	72.5dB	64.82dB	63.73dB	62.63dB	63.64dB
		64.50dB	**63.94dB**	**62.59dB**	**63.91dB**
4th order modulator with NRZ feedback	95dB	71.05dB	69.10dB	65.20dB	68.90dB
		70.65dB	**68.96dB**	**65.53dB**	**68.92dB**
3rd order modulator with SC return	86.6dB	80.80dB	83.58dB	66.17dB	83.25dB
		81.25dB	**83.96dB**	**66.81dB**	**83.91dB**

V.   CONCLUSION

In this paper, a new analytical approach to solve the problem of clock jitter in CTΔΣ is presented. By focusing on continuous time components and signals, two kinds of jitter errors have been identified and mathematical equations of those errors PSD have been derived. Finally, the accuracy of the jitter errors formulas has been proven with exhaustive comparisons with simulated converters controlled by non ideal clocks.

The provided results quite clearly confirm the relation between the jitter errors and the converter architecture. This strong relationship automatically draws aside the possibility to derive a single and simple jitter error equation as it is the case for discrete time converters. However, the presented work provides an efficient mathematical method to specify the clock phase noise profile needed to achieve the targeted performances of CTΔΣ converters.

REFERENCES

[1] B. E. Boser, B. A. Wooley, "The design of sigma-delta modulation analog-to-digital converters", IEEE Journal of Solid-State Circuits, vol . 23, December 1988.

[2] J. A. Cherry, W.M. Snelgrove, "Clock jitter and quantizer metastability in continuous-time delta-sigma modulators", IEEE Transaction on Circuits and Systems-II, vol. 46, June 1999.

[3] E. J. Van Der Zwan, E. C. Dijkmans, "A 0.2 mW CMOS ΣΔ modulator for speech coding with 80 dB dynamic range", IEEE Journal of Solid-State Circuits, vol 31, December 1996.

[4] M. Ortmanns, F. Gerfers, Y. Manoli, "Fundamental limits of jitter insensitivity in discrete and continuous-time sigma delta modulators", International Symposium on Circuits and Systems, ISCAS 2003.

[5] N. Da Dalt, M. Harteneck, C. Sandner, A. Wiesbauer "On the jitter requirements of the sampling clock for analog-to-digital converters", IEEE Trans. Circuits and Systems-I, vol. 49, September 2002.

[6] M. Ortmanns, F. Gerfers, Y. Manoli, "A continuous-time ΣΔ modulator with reduced sensitivity to clock jitter through SCR feedback", IEEE Trans. Circuits and Systems-I, vol 52, May 2005.

[7] O. Oliaei, "Continuous-time sigma-delta modulator incorporating semi-digital FIR filters", International Symposium on Circuits and Systems, ISCAS 2003.

[8] T. C. Weigandt, "Low phase noise, low timing jitter design techniques for delay cell based VCOs and frequency synthesizers", Ph.D. thesis, University of California, Berkeley, 1998, pp 17-30.

Transistor Level Automatic Layout Generator for non-Complementary CMOS Cells

Adriel Ziesemer, Cristiano Lazzari, Ricardo Reis

Instituto de Informática
Universidade Federal do Rio Grande do Sul (UFRGS)
Av. Bento Goncalves, 9500. Bloco IV. CP 91501-970. Porto Alegre/RS, Brasil
{amziesemerj,clazz,reis}@inf.ufrgs.br

Abstract—This paper presents a tool that makes it possible to generate full layouts of CMOS cells from its transistor level netlist in SPICE format. The tool generates the cells under a linear matrix (1D) similar layout style and is able to support unrestricted circuit structures, continuous transistor sizing and folding. It features a transistor placement algorithm for width reduction that aims the reduction of the number of diffusion gaps and the wirelength of the internal connections. The circuit nets are routed using a negotiation-based algorithm, and an Integer Linear Programming (ILP) solver is used to compaction. The experimental results show that our methodology produces layouts competitive to exact methods. The runtimes were kept low even for cells with more than 30 transistors.

I. INTRODUCTION

With the constant increase in complexity of VLSI circuits, the layout design is also becoming more complex, error-prone and time-consuming. Current circuits that require an efficient area and performance achievement, often employ a cell-based methodology that needs the generation of large cell libraries (with different logics and drive strengths) whose layouts must be optimized. To address this problem, cells synthesis tools can be used to quickly generate physical layouts for a given transistor-level netlist, accordingly to the design rules and some constrains specified by the designer.

Uehara [1] was the first to propose a method and a general layout style for width minimization of static dual series-parallel cells. Methods based on this style are also known as one-dimensional (1D) or linear, since transistors are placed exclusively in one direction along two P/N diffusion rows. Since then, several transistor placement algorithms for the 1D style were proposed. Maziasz and Hayes [2] have shown an exact algorithm computationally feasible for both width and height minimization of CMOS series-parallel cells.

There is a huge set of decisions to make in the physical design of a circuit, as position and layer of power lines, position and direction of transistors, position of contacts and

vias, routing tracks management [3], transistor sizing, ... This requires an excessive run time to the layout automation tool exercise all the possible solutions. The goal is to obtain a tool that can provide the best possible solution in a viable running time. Cells with unequal number of P and N transistors introduces additional degrees of freedom that makes exact algorithms practicable only for small circuits. So, to solve big cells, heuristics are used to reduce the complexity. Gupta and Hayes [4] proposed a method that uses a combination of heuristics and exact algorithms to generate cells with unrestricted circuit structure and support to transistor sizing via folding. It constructs clusters of strongly connected transistors as a first step in the physical synthesis. IIzuka [5], in a similar way, use hierarchy to reduce the complexity before applying satisfiability to place the transistors. Their heuristics performed very well and the results were similar to the found in the flatten approach for most of the cells.

However, exact methods [1][2][4][5][8][11] have failed in use other important quality metrics, like the length of the internal connections, during transistor placement. This can increase the routing complexity, leading to bigger capacitances and wider cells.

This paper presents a cell synthesis tool for automatic one-dimensional layout generation of CMOS cells from its transistor level netlist description, allowing different transistor sizes and no restrictions on the transistor network organization. The generator supports transistor folding and uses Threshold Accept [7] to determine a placement that maximizes the diffusion sharing and minimizes the interconnection length. The routing and compaction step is included in the tool so that it can produce a complete and usable cell layout as result. The support for different fabrication technologies is provided by the use of a custom rules file.

II. OVERVIEW

The tool receive as input a file containing a SPICE netlist of the cells (with their respective and individually sized

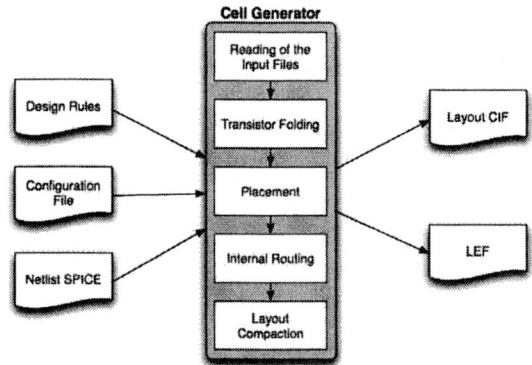

Figure 1. Cell synthesis flow

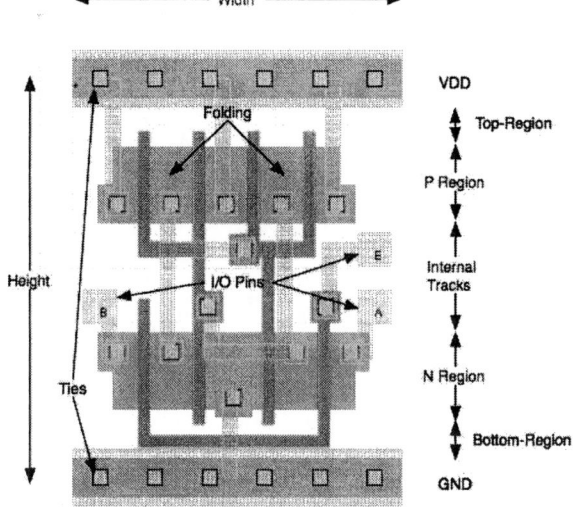

Figure 2. Layout style of our approach

transistors and interconnections), a configuration file (which defines the layout topology and control parameters to the generator), and a technology file which contains a description of the design rules.

The flow is illustrated in Figure 1 and it is discussed in the following sections. The design flow objective is, for a given transistor network, to place and route the transistors using the proposed layout style in such a way that the cell width and interconnections length are minimized. At the end, the circuit is compacted to produce a design error-free layout in CIF (Caltech Intermediate Format) and LEF (Library Exchange Format) formats.

III. LAYOUT STYLE

While the 1D style is well defined in [1], our layout style have some modifications to better support recent fabrication technologies and adequate itself to produce layouts in the standard cell format. An illustration of our layout style is shown in Figure 2. From this model, we also defined some assumptions listed in Table I.

As the height of the cell is fixed, the maximum size of the diffusion rows is given by the space available between the internal tracks and the ties/supply row. Polysillicon routing in the top/bottom regions is just allowed in the case of exist enough space inside these regions.

The routing tracks over the diffusions have the minimal metal pitch allowing contacts, so that no rule violation occurs and the total number of tracks at each diffusion row is maximized. The cell boundaries, as well as the input/output ports, are placed aligned to the routing grid to facilitate the routing step, which frequently uses a gridded router.

TABLE I.   ASSUMPTIONS ABOUT OUR LAYOUT STYLE

1.	Support to unrestricted transistor structure and individually sized transistors.
2.	Transistors placed into two parallel horizontal rows, one for the PMOS transistors and other for the NMOS transistors.
3.	Intra-cell routing made exclusively with: polysilicon, metal 1 and diffusion (for connection of adjacent transistors with the same gate/drain signal).
4.	Internal tracks with adjustable size between the P/N regions for polysilicon and metal 1 routing.
5.	Additional tracks over the P/N diffusion regions for metal routing over the active area.
6.	Supply rows in the top and bottom limits of the cell using metal 1 for VDD and GND connections.
7.	Ties placed under the supply rows in the cell boundary and aligned to the routing grid to avoid abutment problems with neighbor cells.
8.	Single metal-to-diffusion contact in the active areas.
9.	Input/output ports placed aligned to the routing grid.
10.	Jogs are allowed along the tracks.
11.	Cell height and size/position of the internal track provided by the configuration file.
12.	No re-ordering in the transistor structure is made.

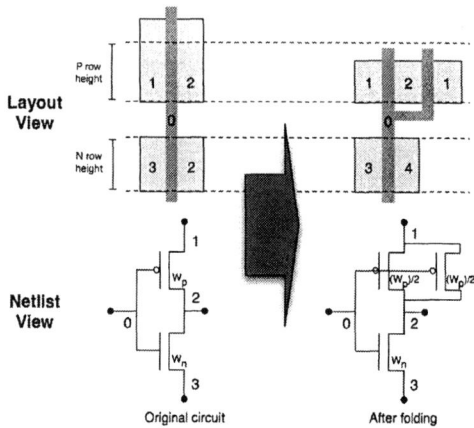

Figure 3. Transistor folding in a large transistor

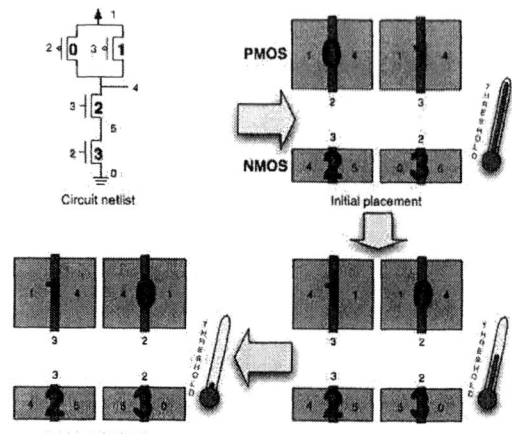

Figure 4. Execution of the transistor placement algorithm

IV. TRANSISTOR FOLDING

Transistor sizing is essential to produce high performance circuits. Many recently tools are able to perform individual transistor sizing to optimize timing and power consumption [14]. Layouts produced in the 1D layout style with different sized transistors tend to waste area since the height of each diffusion row is adjusted accordingly to its tallest transistor.

To solve this problem, one of the most used methods is the transistor folding as illustrated in the Figure 3. It consists of break bigger transistors into smaller ones connected in parallel to keep short the cell height, at the expense of a little increase in the cell width.

Accordingly to Gupta [8], the folding problem can be classified as static/dynamic placement with static/dynamic folding. In our approach we addressed the dynamic placement with static folding problem. Given the diffusion rows limits, we fold the transistors by directly modifying the cell netlist, creating new transistors in parallel, before the execution of the placement step. The advantage of this approach is that it gives more freedom to the placement algorithm so that it can achieve better results than if the folding is executed after the placement as in Lib [11].

V. PLACEMENT

At this design step, we suppose that all transistors met the maximal diffusion row height. So, no further modification on the transistor netlist structure is made.

The aim of the placement step is to find out a transistor ordering that minimizes important constraints like diffusion breaks and gates mismatches. Several exact algorithms are able to find out the optimal result [4] or near optimal [5] for this problem in a computationally feasible runtime. However, these methods don't consider the use of some quality metrics that can lead to a better quality solution - like

total wirelength and channel density – since it significantly rises the algorithm complexity. Cellerity [9] was the first tool to successfully use all these metrics in a Simulated Annealing transistor placement.

In our approach, we implemented transistor placement by using the Threshold Accept (TA) algorithm from [12]. We were able to find the optimal or very nearly optimal result to practically all our test cases in a feasible time. An illustration of a possible execution of the TA algorithm is shown in Figure 4. The details of our implementation are described bellow.

Given a folded transistor-level netlist, we create two different lists of P and N transistors and make P/N pairs accordingly to its absolute position in both lists as illustrated in Figure 5. Eventually unpaired transistors are associated with new blank elements, which are created in the same position in the opposite list.

The minimization algorithm is implemented by creating a perturbation and a cost procedure over this structure. The perturbation function performs transistor movements in both lists to explore the solution space while the cost function returns a score to each partial solution. Smaller scores lead to a better solution.

As first step in our approach, an initial random placement without regard to the optimal placement solution is performed. We attempted to use the Euler Path algorithm to produce the initial placement and run the TA algorithm with a smaller initial threshold but the results were not as good as using a random input. It seems that the algorithm tends to get stuck in a local minimum with ease.

Subsequently, a perturbation function incrementally modifies the initial placement by moving a set of contiguous transistors over the P/N lists at each iteration. The movements can be made one list at a time or in both

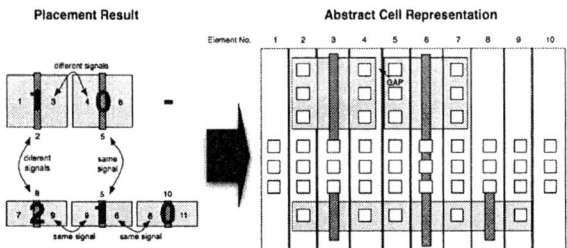

Figure 5. Illustration of the cell structure created before the routing

transistor lists. Prior to performing the move operation, a window with variable size and a move type are selected (adjacent transistors doesn't necessary have to share diffusion inside the window). Two different types of moves were implemented. The first one shifts the selected window inside the list. The other inverts the transistor ordering inside the window as well as its orientation. All parameters are randomly selected: window position/size, move type and list that will experience the perturbations.

The cost function determines the quality of the solutions along the iterations. A good move provided by the perturbation function gives a better score, while a bad move relaxes this constraint. To measure the cost, several quality factors were implemented.

The first one is the gate mismatch. It counts the number of transistors located in the same position in both lists with different gate signals. These transistors, when placed together, tend to minimize the routing complexity inside the cell and have better electrical characteristics than when placed far one from each other.

The second one is the total number of diffusion gaps in both lists. Gap is the space between two non-continuous diffusion regions. It inserts extra space in the cell width producing a shorter transistor density and for these reason its use is avoided.

The third measure is the number of elements needed to transpose the given placement to the abstract cell representation. It counts as element each gate, diffusion and spaces inserted in the cell according to the proposed transistor ordering. It can also be referred as a width measure of the cell since the width is directly related to this value. This way, diffusion gaps inserted in the same place in both diffusions doesn't count twice and have an overall better score than when placed separately.

The latest quality factor is the wirelength cost, which contributes to reduce the routing complexity and can improve the electrical characteristics of the cell. To calculate the cell wirelength, we measure the horizontal distance, in number of elements, from the beginning of each net until the end. The sum of the individual values obtained for each net of the cell, excluding VDD and GND, is the score.

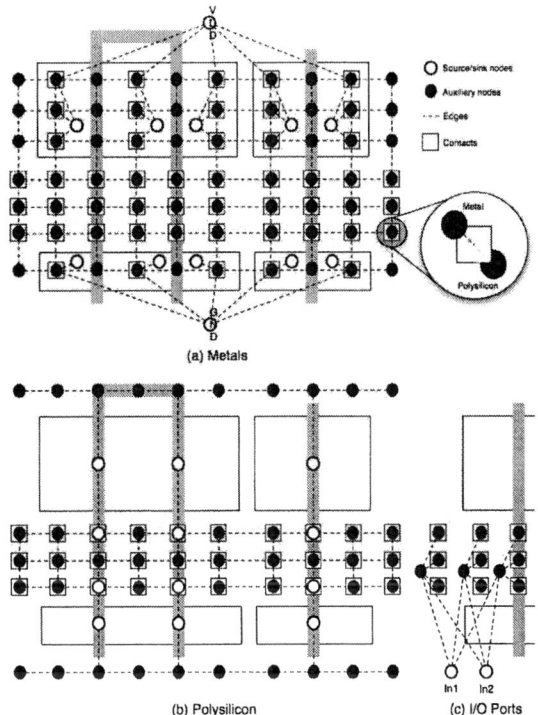

Figure 6. Routing graph

Some of these metrics contribute more than others to the overall cell quality. The weighting is necessary to balance the contribution of each of these factors in the final cost. The weights used in our cost function was experimentally determined and the score is given by the following equation

$$W = 4W_{gm} + 2W_g + 3W_c + W_{wl}$$

where W_{gm} is the number of gate mismatches, W_g is the number of diffusion gaps, W_c is the cell width (in number of elements) and W_{wl} is the total length of the connections.

VI. ROUTING AND PORTS PLACEMENT

Once the transistors have been placed, it is necessary to connect the transistors and the input/output ports accordingly to the netlist. Routing can have a large impact on the cell width and also, in its electrical characteristics. According to Guruswanny [9], improperly connections introduces crossover of wires, extra contacts and other issues that impact in the circuit performance, yield and area. The placement of the ports can also impact the cell width and its position must be optimized.

In our approach, both cell routing and ports placement are problems that are solved together and an explanation of the algorithm is made in the following paragraphs.

As happen with standard cells, we reserved two layers for intra-cell routing: polysilicon and metal 1. Some cell synthesis tools report the use of channel routing algorithms to perform interconnection along the P/N region [15]. We adopt a graph-based methodology because its versatility which allows routing in other cell regions like over the active areas.

Prior to the execution of the detailed routing algorithm, we create a structure with the abstract representation of the cell components according to the 1D placement result. A list of the vertical cell components (called of "elements") is created. Each element holds a P and N transistor part (source/drain, gate, or a blank indication - if there is no transistor associated), a diffusion gap indicative flag and a structure to store the nodes of the graph (in the points where can cross connections according to our layout style). At this point, the adjacent transistors that share some connection using diffusion are already represented. Two additional blank elements are inserted in the beginning and in the end of this list to allow routing in the transistor boundaries.

Next, a graph is created over this structure as illustrated in Figure 6. The algorithm used to solve the graph is a negotiation-based router similar to PathFinder [10]. This algorithm can route multi-net weighted graphs and solve the signals conflicts efficiently.

To achieve a better performance, we set different weights to the graph edges according to its layer and position. Connections in polysilicon mean a bigger cost than in metal and contacts have an even bigger cost. Connections in metal over the transistor gates have an increased cost since it frequently inserts additional space in the cell width and therefore must be avoided.

Input/output port connections require additional space to be placed and also there are fixed rules that must be followed. A chain of two or more serial transistors in the same diffusion row can be placed without diffusion contacts between the gates. This allows an area reduction if no port is placed in the closest track to the transistors. For this reason, we achieved better area results when increasing the routing cost of the graph edges that lead to these ports.

VII. COMPACTION

Compaction is used to produce the final layout, according to the result provided by the placement/routing steps and regarding the technology design rules. We used an ILP (Integer Linear Programming) solver [13] to perform compaction in the X direction. A set of equations representing the layout geometries is created and the solver is called to find a feasible solution that minimizes some constraints like the cell width. The values of the variables obtained from the solver represent the final position of the rectangles in each layer.

TABLE II. COMPARISON RESULTS OF THE CELL SYNTHESIS TOOL

	#Trans.	#Gaps		Width (µm)	Run Time (s)
		Our tool	[6]		
INV	2	0	0	2.8	0.08
BUFFER	4	0	0	5.6	0.32
MUX2:1 (PTL)	4	1	-	7	0.23
NAND2	4	0	0	4.2	0.28
MUXI2:1 (PTL)	6	1	-	5.6	0.66
NOR3	6	0	0	5.6	0.71
MUX4:2 (PTL)	8	1	-	11	13.01
NOR4	8	0	0	8.4	1.64
OAI211	8	0	0	8.4	2.80
XOR2	10	0	-	8.4	0.94
MUX2	12	0	-	11.2	2.96
OA33	14	0	-	12.6	14.96
XNR30	20	3	-	16.8	13.82
AOI444	24	1	-	21	71.44
OAI444	24	1	-	21	59.00
MUX41	26	2	-	21	55.32
FAD1	28	1	1	21	85.61
JK1	34	3	-	30.8	229.20

- not compared

To achieve a better performance, we introduced other minimization functions with different weights. In our tests, we put the cell width constraint with the biggest weight followed by (in descending order) the width of the P/N diffusion islands, size of the polysilicon doglegs and the size of the metal doglegs. The weights can be adjusted by modifying the configuration file of the tool.

Most standard-cell place and route tools use grid routers. A better efficiency is achieved when the I/O ports are placed on the routing grid and the cell boundaries on the placement grid. For this reason we managed the compactor to support this restrictions.

VIII. RESULTS

The tool was implemented in C++ on a PowerPC G5 workstation with 1Gb of RAM under MAC OSX operating system. Table II presents a comparison between our tool and the results obtained from Iisuca [6] - which uses satisfiability for transistor placement of dual circuits - for a number of typical cells in a 0.35µm technology.

Analyzing the results, we can verify that our method produced results with equal number of gaps in the test cases we were able to compare. Additionally, our methodology has the advantage of generate cells with non-complementary logic and use the total wirelength in the minimization function. According to the full adder picture from [5], our layout presented 15.4% wirelength improvement on this cell.

Our run time was quite fast for cells with a large number of transistors taking less than 4 minutes to generate a cell with 34 transistors.

Figure 6 shows some of the cells we have automatically generated by our tool.

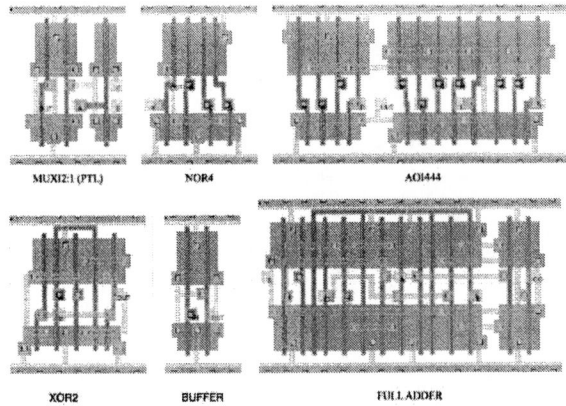

Figure 7. Examples of cell layouts automatically generated by our tool

IX. CONCLUSIONS

We presented a new layout generator tool of CMOS cells that is capable of support circuits with unrestricted transistor structure by using Threshold Accept to place the transistors in a one-dimensional layout style. A negotiation based router and ILP were also successfully applied to produce the full layout of the cells. Our results demonstrate that our methodology is very efficient and computationally feasible for a wide range of cells, including the pass transistor logic (PTL) family. The run times were keep small even for large cells without loss in quality when compared to an exact method.

REFERENCES

[1] T. Uehara and W. M. vanCleemput, "Optimal Layout of CMOS Functional Arrays", in IEEE Transactions on Computers, Vol. C-30, No. 5, pp.305-312, May 1981.

[2] R. L. Maziasz and J.P.Hayes, "Exact Width and Height Minimization of CMOS Cells", in Proc. ACM/IEEE 28th Design Automation Conference, pp.487-493, 1991.

[3] R. Reis, "Power and Timing Driven Physical Design Automation," PATMOS2003 – 13th International Workshop on Power and Timing

Modeling", Optimization and Simulation, Torino, September 10-12, 2003. LNCS Springer Verlag (keynote speaker).

[4] A. Gupta, S-C. The and J. P. Hayes, "XPRESS: A Cell Layout Generator with Integrated Transistor Folding", Proceedings of the 1996 European Design and Test Conference, Washington, DC, USA. Anais. IEEE Computer Society, 1996, pp.393.

[5] T. Iizuca, M. Ikeda and K. Asada, "Exact Minimum-Width Transistor Placement for Dual and Non-dual CMOS Cells", IEICE Transactions on Fundamentals of Electronics, Communications and Computer Sciences, pp.3485-3491, December, 2005

[6] T. Iizuca, M. Ikeda and K. Asada, "High speed layout synthesis for minimum-width CMOS logic cells via Boolean satisfiability", ASP-DAC '04: Proceedings of the 2004 conference on Asia South Pacific design automation, pp.149-54, 2004.

[7] G. Dueck et al., "Threshold accepting: A general purpose optimization algorithm appear superior to simulated annealing", Journal of Computational Physics, 1990.

[8] A. Gupta and J. P. Hayes, "Optimal 2-D cell layout with integrated transistor folding", in proceedings of the 1998 IEEE/ACM International Conference on Computer-Aided Design, New York, NY, USA., pp.128–135, 1998.

[9] M. Guruswamy et al., "CELLERITY: a fully automatic layout synthesis system for standard cell libraries", in Proc. ACM/IEEE 34th Design Automation Conference, California, United States, pp.327-332, 1997.

[10] L. Mcmurchie and C. Ebeling, "PathFinder: A Negotiation-Based Performance-Driven Router for FPGAs", Field-Programmable Gate Arrays, 1995. FPGA '95. Proceedings of the Third International ACM Symposium, pp. 111- 117, 1995.

[11] Y. C. Hsieh et al., "LiB: a cell layout generator", In: DAC '90: Proceedings of the 27th ACM/IEEE Conference on Design Automation, New York, NY, USA. Anais. ACM Press, 1990, pp.474–479.

[12] R. Hentschke, M. Johann and R. Reis, "New Place and Route Algorithms for Wire Length Improvement With Concern to Critical Paths", in 10th Annual ACM/SIGDA Ph.D. Forum at DAC.

[13] Lpsolve, "Mixed Integer Linear Programming (MILP) Solver", http://lpsolve.sourceforge.net/5.5/

[14] C. Santos et al., "Effects of Using a Pin-to-Pin Delay Model on a Library-Free Transistor/Gate Sizing Scheme", in Proc. of the IEEE International Midwest Symposium on Circuits and Systems (MSCAS2005), USA, pp. 315-318 , 2005.

[15] K. Tani et al., "Two-dimensional layout synthesis for large-scale CMOS circuits", Proceedings of ICCAD 91, pp.490-493, 1991.

Computing and Design for Software and Silicon Manufacturing

Davide Pandini[1], Giuseppe Desoli[2], and Alessandro Cremonesi[3]

[1] STMicroelectronics, Central CAD and Design Solutions, Agrate Brianza, 20041 Italy
[2] STMicroelectronics, Advanced System Technology, Cornaredo, 20010 Italy
[3] STMicroelectronics, Advanced System Technology, Agrate Brianza, 20041 Italy
davide.pandini@st.com, giuseppe.desoli@st.com, alessandro.cremonesi@st.com

Abstract — An increasing demand for higher performance, for lower power density, and for greatly expanded functionalities will determine radical changes in the future computing architectures. These widely acknowledged emerging trends are however insufficient to address all the challenges introduced by advanced silicon nanometer technologies. It is well known that manufacturability for high yield, along with design productivity and predictability and system reconfigurability for reduced NRE costs and faster time-to-market, are major problems in gigascale SoC design. Therefore, only focusing the design efforts on performance, power consumption, and throughput can hinder the potentials of the new computing architectures and limit the silicon yield. In this paper, we introduce an innovative *architecture-to-silicon* platform that by exploiting the concept of regularity at different levels of abstraction addresses the emerging challenges for the new computing architectures, and links system and architecture definition with silicon fabrication.

I. INTRODUCTION

The microprocessor that has sustained the Moore's Law and pushed the technology scaling into the nanometer regime, and more in general computing architectures based on the microprocessor, are perhaps undergoing the most significant transformations since their introduction. To achieve the predictions of Moore's Law, increased transistor density is of course important, but the next key challenge is to integrate the basic foundations such as process technology with architecture and software, to drastically reduce the development cycle and NRE costs, to meet tighter *time-to-market* windows, and to achieve *high yield* to compensate for the soaring economic investments necessary to develop the next nanometer technology nodes and build the manufacturing facilities. Such trend will dictate a deep rethinking of the computing platforms and design methodologies, to address the following challenges:

- *Scalability*: Next generation of computing systems need to be scalable, i.e., derived from basic building blocks, where components of different throughput and capacities can be statically but more so, dynamically configured;
- *Performance*: Future computing architectures will have workloads orders of magnitude greater than current systems. Additionally, they might be required to be capable of rapid context switching (*flexibility*), and to support multiple different workloads simultaneously (*parallelism*);
- *Memory latency*: Memory requirements must be further expanded to handle more than one application at a time,

with increased data traffics and gaps between memory and processor speed;
- *Low power and power management*: Power consumption will be even more critical because of the high-performance requirements, and increased leakage current at 65nm and below. Workloads should be scheduled to reduce peak power consumption;
- *Multitasking*: Intrinsically non-deterministic and unpredictable multiple applications must be supported simultaneously. Hence, dynamic resources management mechanisms will have to be an integral part of the software and hardware;
- *Fault tolerance*: The reliability of the components plus the ability of the total system to operate effectively with degraded elements are essential;
- *Real time*: For several applications guaranteed response within acceptable timing window is critical for correct system operations;
- *Predictability*: An early evaluation of process-related effects and variability at the architectural level will provide a direct path from architecture to silicon manufacturing, thus avoiding costly design re-spins;
- *Manufacturability*: A fast yield ramp with high-performance, cost-effective, and predictable manufacturing process is necessary to optimize the profitability of the overall design-to-silicon flow.

The new computing architectures addressing the challenges inherent in performance and technology scaling will be based on *divide-and-conquer*, breaking large applications into several smaller functions, and distributing them across small and low-power computing units. This approach is leading to the development of a multi-core on a single chip technology, to achieve higher performance at lower clock rates by means of parallelism built upon a multi-core architecture. Rather than relying on one large power-hungry microprocessor, distributed-core chips will activate only the cores necessary to carry out specific functions, while the idle functional units are switched off. Obviously, the power management strategy is a critical component in this approach. These architectures could ideally use specialized cores for various classes of computations (graphics, speech recognitions, communication protocol processing, etc.), and a dynamic reconfiguration of the cores, interconnects, and caches. Such chip-level multiprocessing architectures will be able to deliver massive performance while managing power and heat. Moreover, technology and frequency scaling are running into some fundamental physical barriers, such as the increase of leakage current, and the benefits of higher clock rates will be impaired by interconnect parasitic delay and memory latency. Memory access time does not scale with operating frequencies, and traditional architectures will be limited by the *Von Neumann bottleneck*.

However, in spite of all the technology-related effects, the design efforts must be focused at the architectural level to manage the complexity of modern multi-processor System-on-Chip (MPSoC) designs. Hence, an alternative design paradigm based on forms of regularity at different levels of abstraction, on structured on-chip communication, and with some application-specific customization, would replace the traditional standard cell-based ASIC design style for a wide range of applications. By exploiting regularity at the architectural, logical, and physical level, we will obtain an early interconnect and process-related effect predictability, thus drastically reducing the number of iterations necessary to achieve the design closure. In addition, more regularity in the physical implementation will introduce a *correct-by-construction* approach that greatly facilitates the verification phase. The building block and enabler of this *architecture-to-silicon* design platform will be a reduced library of configurable and regular logic components that are implemented by means of a small number of regular layout shapes. These highly predictable and manufacturable *regular fabrics* will allow focusing the design efforts at the system level, providing a direct path to physical implementation and silicon fabrication.

II. THE CHALLENGES AHEAD

Power dissipation and increasing variability are emerging as first-order concerns that must be effectively addressed in process development as well as in circuit and architecture design. Traditionally, MOSFET scaling efforts were focused on extending performance, both by improving device speed as well as integrating more devices and functionality on the same die, but in the last few years, chip power and power density have become a major challenge. From all the technology scaling projections, it is clear that leakage current will increase dramatically as more transistors will be integrated within the same die size, causing more heat and power dissipation. At 65nm several researchers predict that static and dynamic power dissipation will be comparable [7]. Indeed this is being confirmed by early physical implementations for high-performance devices built with high-speed libraries; hence, efficient on-chip power management is mandatory. Transistor variability of design-related parameters, resulting either from manufacturing fluctuations, or from the device intrinsic atomistic nature affecting, for example, channel doping, will increase as technology scales down. As CMOS approaches the 22nm technology, where there could be less than 50 Si atoms in the device channel, the stochastic threshold voltage variations could be around 100mV. Random dopant variation has no known remedy; it is simply inherent to the process and cannot be mitigated.

In addition, delay variation induced by spatial process parameter spreading and by temporal supply voltage and temperature fluctuations, will profoundly impair parametric yield, timing analysis precision, and design predictability. At the technology level, over the past few years, Resolution Enhancement Techniques (RET) and worst-case design have mitigated these effects. Although several process and lithography advances are necessary to continue shrinking device features sizes, and scaled devices will face several electrical and reliability problems, it is clear that the technology roadmap will progress as long as it is economically advantageous. However, in the future, cooperative circuit and technology co-design, and architectures developed concurrently with innovative devices will be necessary to continue scaling silicon technologies.

III. DESIGN FOR MANUFACTURING

Today it is widely accepted that manufacturability is one of the major challenges for gigascale SoC design in nanometer technologies at 65nm and below. Such trend has lead to growing interest in *Design for Manufacturing* (DFM). Starting from approximately $1 million in 1994, the average cost for one SoC design is estimated to be in the range of $20-50 million in 2010, when the 45nm technology node is forecasted to be in production [1]. The cost raise is partially due to the increasing difficulty of first-pass design success. Mask cost, driven by the complexities of sub-wavelength lithography, has been on a similar upward trend: a conservative estimate for a mask set cost shows an increase from about $100 thousand dollars for a $0.35\mu m$ design to almost $9 million for a 45nm design (source: EETimes, Synopsys, VLSI Research). It is important to notice that an important factor impacting mask cost that has escalated dramatically is the mask writing time, which is directly proportional to the shapes on the mask and has increased from a few days to over a month due to RET complexity. With such economic pressure, it is no surprise that concerns about device yield dominate within the semiconductor and EDA industry. A primary, well-established, and effectively implemented facet of DFM originated from physical design characterization and involves techniques to improve layout robustness against random as well as systematic process defects and variations. A second DFM facet of rapidly growing importance for 65nm and beyond focuses on layout optimization for increasingly challenging lithography resolution. One successfully demonstrated technique to this aspect of DFM employs a set of restrictive design rules to improve manufacturability and performance [2]. However, a much more promising approach is based on the *regular fabrics* [3], which not only enforces a restricted number of litho-friendly layout shapes to address the printability challenges introduced by sub-wavelength lithography and to reduce the mask-writing time, but also allows to limit the set of logic primitives necessary to map a circuit and to define a top-down predictable design methodology [4].

Traditionally, the concept of DFM has been bounded within the perimeter of physical design and post-layout optimization. However, the complexity of the new computing systems, along with the ultimate goal of every IDM to fabricate complex chips with high yield, dictates a *DFM-driven* approach at different levels of abstraction [5]. The discontinuity between software, architecture, and manufacturing primarily stems from the current technology platforms and design flows based on HDL (Hardware Description Language) synthesis and std. cell implementation. As a consequence, we believe that the link between software and architecture on one side, and silicon implementation on the other hand has been neither completely analyzed, nor properly addressed so far. In order to overcome this fundamental limitation, a new *architecture-to-silicon* platform centered on the concept of regularity needs to be defined and developed.

A. Regularity For Achievable Designs in Nanometer Technologies

One of the main sources of variability is the decreasing ability of lithography to control printed layout structures, and the related unpredictability in relevant design parameters like device performance, wire delay, and leakage is increasing. Hence, for design methodologies that depend on worst-case properties, the benefits of new technologies are small, if any at all, and the escalating yield erosion will not justify the massive investments necessary to build new manufacturing facilities.

A promising approach to develop a litho-friendly design methodology is based on the *regular fabrics*, which enforce regularity in a bottom-up fashion at the physical design stage (micro regularity), by significantly limiting the total number of layout patterns, and also impose a top-down regularity by reducing the number of logic components used to implement a given design (macro regularity) [3][4]. Moreover, this approach allows developing a design flow based on application-specific logic synthesis and structured physical design, which will facilitate the introduction of structured and configurable paradigms for on-chip communication. We believe that the introduction of regularity in the design-to-silicon flow will be mandatory at 45nm and below, to guarantee a high yield for increasingly complex SoC designs. Therefore, we propose to extrapolate the concept of regularity at higher levels of abstraction, during the micro- and macro-architectural definition. Regular fabrics ideally will allow the concurrent creation of platforms, structured and predictable design, components, and the associated synthesis/design automation flows, thus enabling:

- New forms of synthesis for generic, programmable, and reconfigurable components;
- Regular global interconnects to support wiring reuse and globally asynchronous communications;
- New manufacturing design rules that guarantee printability, yield, and control process variations;
- Platform-specific system-level performance prediction;
- Structured software, whose embedded regularity could be exploited for a more optimized mapping on the new computing platforms, including coping with local process variations dynamically and offering intrinsic (transient) fault tolerance.

IV. VARIATION-TOLERANT AND RELIABLE ARCHITECTURES

Device design, supply voltage reduction, and tolerance control have limited ability to solve the power and variability crisis. The rest must be accomplished by new circuit design techniques and computing architectures. Not only must circuits and archi-

tectures allow more power and variability management, but they must improve throughput without introducing more complexity. A crucial task to manage complexity is first-pass working silicon to avoid costly design re-spins. The predictability of the regular fabrics will enable a system-level sign-off, linking architectural definition with silicon implementation. Chip variability is of great importance, influencing yield and overall chip-level performance, and it can be global (die-to-die, wafer-to-wafer, etc.) or local (intra-die). Advanced process control can effectively address chip-to-chip, wafer-to-wafer, and lot-to-lot variability by minimizing global variations. Local within-chip variations resulting from processing pattern density effects require process design modifications. Across-wafer and across-chip systematic variations are also typically corrected by process modifications. In contrast, non-systematic within-chip variations cannot be addressed by process control, process design, or process modification; they require far more advanced solutions. In the future, self-correcting circuits will be necessary to address this type of variability. These circuits rely upon on-chip monitors to continuously test functionality and performance, self-correcting by turning on and off peripheral circuits. While circuits themselves will be capable of self correction, the entire micro and macro architecture should also be able to react to coarser level events, gathering necessary feedbacks from embedded monitoring of critical key parameters and working conditions, dynamically assisted by software-controlled global heuristics.

The three major challenges in dealing with random variability are characterization, reduction, and accommodation. The design of appropriate high-speed test structures on dedicated test chips is an important aspect of variability characterization. In addition, appropriate test structures embedded in the chip design itself can be used to directly evaluate variability within the product and even to provide inputs for decisions on dynamic change in the chip operation (i.e., self-adaptive circuits). Design can play an important role in variability reduction, and DFM can optimize layout robustness against both random and systematic process yield degradation. Enormous opportunities exist to consider variability through design practices. A design technique to deal with variability is a dynamic approach where appropriate test structures embedded in the chip are continuously monitored during the operation of the chip. Feedback is provided to the power supply voltage or back-bias to accommodate initial variations across the chip and to compensate for workload variations, while remaining within allowed power-performance limits. In summary, future transistor sizes will be below 20nm, and to compensate for the intrinsic variability introduced by the atomistic nature of these nano-devices, thus guaranteeing robust systems, innovative variation-aware techniques at the circuit and architectural level need to be explored.

V. HIGH-SPEED AND SCALABLE ON-CHIP COMMUNICATION

New on-chip interconnection architectures are required to address the inefficiencies introduced by the communication overhead between cores, caches, and other computational units in MPSoCs. Although several improvements will be achieved with new materials like copper, low-permittivity dielectrics, and eventually optical interconnects, we believe that for distributed systems, the challenges will not be entirely met at the technology level, and both innovative interconnect architectures overcoming the scalability limitations of bus-centric communication, and new paradigms removing the restrictions inherent with synchronous communication are necessary. A promising approach is based on *self-adaptive pleisio-chronous design*, where the frequency of different modules may vary, thus meeting performance constraints by coordinating the different modules with a system-level feedback loop. These techniques can be combined into a delay-insensitive architecture template, where the modules are microprocessors and memories, and the interconnections are derived from a pleisio-chronous Network-on-Chip (NoC). In this frame, NoC and Globally-Asynchronous Locally-Synchronous (GALS) can be on-chip communication architectures and paradigms capable of supporting several cores and caches, and to be self-adaptive to limit the effects of process and environmental variability.

A. Influence of Interconnect Scaling on Computing Architectures

While transistor speeds are scaling approximately linearly with feature size, wires are getting slower with each new technology. Even assuming higher aspect ratios, the absolute delay for a fixed-length wire in top-level metal layers with optimally placed repeaters is increasing with each technology generation, and due to faster clock frequencies, wire delays are growing at an even higher rate. If, on the other hand, the wire length itself is scaled with technology, its delay appears to become smaller, and approximately tracks the decreasing delay of transistor gates. Hence, the complexity of future chips could be reduced by partitioning a design into small domains, each of which has interconnections using short local wires internal to the domain, with long multi-cycle global wires exclusively for inter-domain (global) communication.

A similar conclusion about the need to partition logic on a chip is reached when considering the clock skew problem in high-speed circuits. As mentioned earlier, even at fixed lengths, the delay, and hence the clock skew, becomes worse with smaller feature size. Thus, a smaller fraction of each clock cycle remains available for useful computation. A potential solution to this problem is to have multiple clock domains, each with its own clock distribution network that is independent of the others. Communication between domains may be either synchronized at lower clock rates, or may be self-timed through some form of handshaking (GALS). Both solutions embody the modular design concepts needed to overcome these wiring and clock distribution limitations. A distributed shared-memory multiprocessing architecture inherently partitions the chip into areas that contain multiple processors and multiple memory banks. Each processor and each bank could form a natural domain both for local wiring and clock distribution. The overall chip performance will be determined by the amount of states and logic that can be reached in a sufficiently small number of clock cycles. *Increases in instruction-level parallelism will be limited by the amount of states reachable in a cycle, not by the number of transistors that can be manufactured on a chip.*

For conventional microarchitectures, when wire delays grow relative to gate delays, improvements in clock rate and instruction throughput (IPC) become directly antagonistic. Designers must select among deeper pipelines, smaller structures, or slower clocks, and none of these choices, nor their best combination, will result in scalable performance. The inherent trade-off between access time and capacity will force designers to limit or even decrease the size of the structures to meet clock rate expectations. It was projected that in a 35nm technology with a 10GHz clock, accessing even a 4KB level-one cache will require 3 clock cycles. Following these scaling projections, *memory latency* is a critical bottleneck for advanced high-performance multiprocessor computing architectures. The large memory-oriented elements, such as caches and register files, will be unable to continue increasing in size while remaining accessible within one clock cycle. For a given cache capacity, the transition to smaller feature sizes decreases the cache access time, but not as fast as projected by the faster clock rates. One alternative to slower clocks or smaller caches is to pipeline cache accesses and allow each access to complete in multiple cycles. Due to the nonlinear scaling of capacity with access time, adding a small number of cycles to the cache access time may substantially increase the available cache capacity. Therefore, the designer is faced with two interacting choices: how aggressively to push the clock rate by reducing the number of levels of logic per cycle, and how to scale the size and pipeline depth of different microarchitectures. Moreover, memory access time will remain a critical bottleneck for the overall system performance and throughput.

VI. PROCESSOR IN MEMORY

Memory latency (i.e., the Von Neumann bottleneck) is one of the most critical problems in multi-core computing architectures, since in order to keep many high-performing cores processing a large amount of data it is important having the on-chip memory close to the cores. *Processor-In-Memory* (PIM) [6] is a computing architecture proposed to reduce the memory latency. PIM combines logic and memory on the same chip with the possibility of incorporating multiple cores/memory blocks on a single die to allow a direct and fast access to memory row buffers, thus increasing the effective memory bandwidth, reducing overhead and latency, and improving power efficiency. All these issues must be addressed in the new distributed computing architectures, and the trade offs analyzed more in depth.

A. Analyzing Processor-In-Memory Computing Architecture

The Von Neumann bottleneck is originated by the isolation of the memory from the processor. The bandwidth internal to the memory and to the processor data path is extraordinary high, while the interconnections between memory and processor can be relatively slow. By integrating multiple processors and multiple independent memory blocks on the same die, PIM can establish a new relationship between both classes of devices. Each processor is associated with a memory block from which it acquires data and possibly instructions. Floating point operations can be shared between different processor-memory pairs, rather than being dedicated to a single processor-memory pair. It is important to notice that this simple architectural description is by no means constraining the concept of PIM, which is a general template for distributed computing architectures.

In PIM, processors are usually small and relatively simple execution units, possibly comprising bit-level ALUs and DSPs, relying on scalability for speed. Simple operations that may be the majority are performed without incurring in the power penalty usually present in larger cores. Power management is facilitated with PIM architectures, since PIM processors provide a degree of computational granularity that may be dynamically controlled to deliver the requested performance. Only the processors necessary to deal with the current workload need to be active, while all the other processors can be shutdown. Moreover with data partitioned for locality, such that a given data set is kept in specific memory array, these memories may be put in stand-by with only the occasional refresh required thus minimizing the power consumption of the memory.

A PIM chip is a multiprocessor system, and therefore requires to exchange messages, data, and signals between memory-processor pairs, where the internal local communication channels satisfy the high bandwidth and low latency requirements. PIM can implement different interconnect topologies and on-chip communication schemes like NoC other than busses, and can provide higher bandwidth with redundancy and fault tolerance. Processor-to-processor, memory-to-memory, and processor-to-memory communications can be supported within PIM between processor-memory pairs. Direct processor-to-memory communication can overcome the Von Neumann bottleneck of conventional systems. Reliability through active fault tolerance is enabled by PIM architectures. Ordinarily, with a hard on-chip failure the device is unusable. With PIM the fine-grain structure means that a hard fault may be isolated to a portion of the chip allowing the remaining processor-memory units to continue operations. Additionally, transient faults could also be dealt with by error detection schemes associated with memory operands and computational units results, coupled with self-adaptive logic as previously described.

Multithreading in a PIM framework can achieve very high memory bandwidth efficiency by overlapping the data row access for one thread from the memory array of a given processor, while executing the operations of another thread. Real-time operations can also be facilitated by PIM, since entire processors can be dedicated to time-critical functions. In conclusion, PIM is a computing architecture that allows embedding logic within memory to reduce the memory latency, and the PIM architecture can be considered as a suitable candidate to address the challenges introduced by the new multiple-core/distributed-memory computing systems.

B. Regular Fabrics: the Architecture-to-Silicon Technology Platform for PIM and Distributed Multi-Core Systems

On-chip memory is largely exploited in SoC design, where up to 80% of the overall circuit can be made of memories. The macro-regularity of the memory array allows pushing the layout shapes of the bit-cells below the design rules that must be applied to the standard cells. A tight integration of logic and memories within the same die is a complex task: the memory blocks tend to be pre-allocated at the chip boundary during floorplanning, while logic is accommodated towards the center

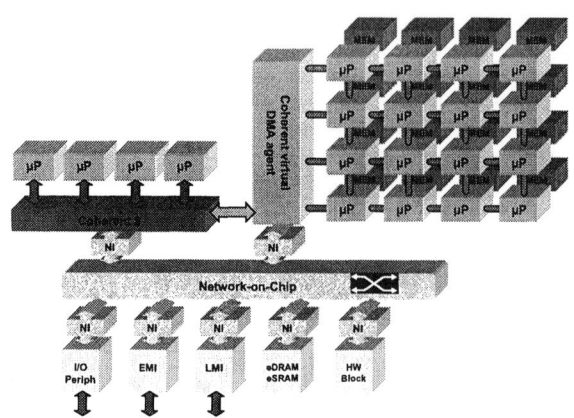

Figure 1. High-level view of ST xSTream computing fabric

of the die. This approach is far from optimal for latency, where memory should be as close as possible to the computing units. This problem will be aggravated in multiple core architectures, when several processing elements will have to be integrated on the same chip and will access the memory concurrently. With the current technology platforms and design flows based on standard cells, embedding logic within memory could be achieved with potentially low manufacturability and consequently yield loss. In contrast, the regular fabrics that can be implemented with the same pushed rules as bit-cells, allow a much tighter integration between logic and memory with high manufacturability; therefore, regularity will consent a better design and fabrication of the new multi-core/distributed-memory architectures.

VII. xSTREAM: A COMPUTING ARCHITECTURE FOR HIGH-PERFORMANCE EMBEDDED APPLICATIONS

The STMicroelectronics xSTream computing architecture is based on the convergence of communication and computation to address scalability and programmability of high-performance embedded functionalities, such as graphics, multimedia and radio subsystems. It is a coarse/middle grain parallel-distributed and shared-memory architecture combining a uniform layered communication network based on a NoC backbone infrastructure (STNoC), and several HW IPs supporting higher levels of the communication abstraction. Moreover, xSTream addresses some of the increasingly challenging design and silicon fabrication issues at the architecture, microarchitecture, communication, and design level, through the use of advanced techniques such as: a) voltage and frequency scaling (by means of a GALS model); b) local clock generators adapted to local silicon process variations (to increase yield and fault tolerance); c) skew insensitive design (with mesochronous and delay insensitive NoC links); d) regular fabrics for design predictability and silicon manufacturability. The xSTream architecture is illustrated in Figure 1, where the system is composed of the traditional "host processing" part on the left side, which is a Symmetric Multiprocessing (SMP) subsystem for future scalability, while the entity on the top-right of the picture is the "streaming engine". xSTream can address the needs of data-flow-dominated and highly computationally intensive semi-regular tasks, typical of embedded products. The streaming nature of the mapped kernels makes it possible to design a semi-regular fabric of programmable functional units interconnected via a relatively simple network of point-to-point channels. The processing elements of the streaming fabric are relatively simple programmable processors or functional units with a general purpose but simple basic ISA. They execute instructions fetched from local memories instead of caches. Local memory is also used for wide data accesses. The interconnect structure plays a critical role in connecting the computing units, providing a self-synchronizing support for software pipelines. This is achieved with a set of lightweight routers, very similar to the ones defined for NoC replacements of standard bus infrastructures, but with more freedom for simplification, due to the constrained nature of the communication patterns versus a generic NoC system backbone. The fabric is not limited to the exploitation of programmable cores. It is also possible to use hybrid approaches were some elements of the array can be fixed functions, implemented in a classic ASIC design style, if such functions are critical and fixed. The xSTream architecture template can accommodate different computing nodes, from programmable ones to hardwired functions. However, in order to exploit the benefits of regularity, the template includes a suitable programmable element, a highly parametric programmable engine called xPE (i.e., xSTream processing element). The xPE is optimized for stream-oriented, data-flow-dominated, and performance-intensive computation, with the target on embedded multimedia, telecom, and signal processing applications. The requirements considered for the definition of the xPE architecture and microarchitecture are:

- Low silicon area or more precisely, high computing density in terms of MIPS per physical gate;
- High computational power with outstanding power figures in term of MIPS/mW, MIPS/MHz, associated with relatively high (for the embedded world) operating frequencies;
- Support for high-level programming languages and compiler friendliness;
- Design time configurability with a wide range of tuning knobs enabling extensive trade-offs for area, power, and computing density for a single xSTream node to adapt the fabric granularity to the specific application domain.

The xPE microarchitecture is a highly streamlined and "minimalist" core architecture to tune system frequency and limit core size by shaving off most of the complexity required for general purpose microprocessors and media processors. Local memories and interfacing are optimized for managing and accessing data streams, as well as wide vector instructions operating on packed data words. xPE supports a fine-grained multithreading to exploit task-level parallelism, and to simplify the data-flow application mapping to achieve the more conventional latency hiding benefits of multithreading.

One of the xSTream architecture main goals is the validation of emerging and innovative design methodologies based on regularity at different levels of abstraction, and focused on design predictability, yield improvements, dynamic and static power management, and variation-tolerant design. This project is part of an internal multi-divisional R&D program at STMicroelectronics called *Computing and Design for Software and Silicon Manufacturing.*

VIII. REGULARITY EXTRACTION FROM SOFTWARE

In order to exploit the architectural regularity of next multicore/distributed-memory computing architectures, the system should be customized towards the special set of tasks that it has to perform. Hence, we believe that regularity should be extracted and preserved not only from *architecture-to-silicon*, but ideally also at the software level. Given the subset of functions that a system would execute frequently, it is possible to map some of these functions on specific functional units. If well chosen, they could enhance the performance of the whole system, providing the benefits of frequent hard-logic execution, with simultaneous flexibility and reconfigurability.

By extracting functional regularity from software applications, and consequently generating more complex instruction candidates for mapping on small processors (or other functional units such as DSPs), it will be possible to derive a candidate set of *templates* (or macro-functions) for the compiler, which are repeated occurrences of possibly interdependent simple sequences of operations. The templates will expand the compiler technology with some level of architecture awareness, potentially allowing to improve latency. Although regularity extraction from SW is still at an early stage, ideally it will be part of the future distributed computing architectures, and we expect more research work in this emerging field in the near future.

ACKNOWLEDGEMENTS

The authors would like to thank Prof. Larry Pileggi of Carnegie Mellon University for valuable technical discussions, and for sharing his pioneering work on regular fabrics.

REFERENCES

[1] http://public.itrs.net/

[2] L. Liebmann et al., "High-performance Circuit Design for the RET-enabled 65nm Technology Node," in *Proc. of SPIE*, vol. 5379, Feb. 2004, pp. 20-29.

[3] V. Kheterpal et al., "Design Methodology for IC Manufacturability Based on Regular Logic-Bricks," in *Proc. of Design Automation Conf.*, Jun. 2005, pp. 353-358.

[4] T. Jhaveri, L. T. Pileggi, V. Rovner, and A. J. Strojwas, "Maximization of Layout Printability/Manufacturability by Extreme Layout Regularity," in *Proc. of SPIE*, vol. 6156, Feb. 2006, pp. 615609-1-615609-15.

[5] L. Liebmann et al., "Integrating DfM Into a Cohesive Design-To-Silicon Solution," in *Proc. of SPIE*, vol. 5756, Feb. 2005, pp. 1-12.

[6] T. Sunaga, H. Miyatake, K. Kitamura, P. M. Kogge, and E. Retter, "A Parallel Processing Chip with Embedded DRAM Macros," *IEEE J. Solid-State Circuits*, vol. 31, no. 10, pp. 1556-1559, Oct. 1996.

[7] K. Roy, S. Mukhopadhyay, and H. Mahmoodi-Meimand, "Leakage Current Mechanisms and Leakage Reduction Techniques in Deep-Submicrometer CMOS Circuits," in *Proc. of IEEE*, vol. 91, no. 2, pp. 305-327, Feb. 2003.

An adaptive genetic algorithm for dynamically reconfigurable modules allocation

Vincenzo Rana, Chiara Sandionigi, Marco Santambrogio and Donatella Sciuto
Politecnico di Milano - Dipartimento di Elettronica e Informazione
Via Ponzio 34/5 - 20133 Milano, Italy
chiara.sandionigi@dresd.org,
{rana, santambr, sciuto}@elet.polimi.it

ABSTRACT

This paper aims at defining an adaptive genetic algorithm tailored for the allocation of dynamically reconfigurable modules. This algorithm can be tuned at run-time with a set of parameters to best characterize different architectural scenarios (i.e., single device or multi-FPGAs characterized by several kinds of communication infrastructures) and to adapt the performance of the algorithm itself to the scenario in which it has to operate.

The proposed approach has been validated with a large set of meaningful combinations of parameters (i.e. changing the mutation or the crossover probability), in order to demonstrate the possibility of performing either a fast or an accurate allocation phase.

1. INTRODUCTION

Nowadays, thanks to reconfigurable devices (such as FP-GAs), it is possible to dynamically tailor the hardware to a specific application, in order to dramatically improve the performance. One of the most suitable approaches in the development of reconfigurable systems is the module-based approach (see [1]), in which the original application is partitioned into several functions, each one of them implemented as a single module. These modules, thus, can be either dynamically loaded into the system or removed from the system, in order to change its overall functionality. The most recent Xilinx design flow, the Early Access Partial Reconfiguration (EAPR) flow, is based on the same approach, as described in [2].

One of the most interesting challenges in such a scenario is the allocation of requested modules in the free space of the reprogrammable device. The allocation phase has to take into account the fragmentation of the device in order to keep the maximum set of contiguous free slots, able to contain bigger modules. On the other hand, this phase has to be executed in a very short time, since it is not desirable to further increase the overhead due to the reconfiguration processes.

The approach presented in [5] trades the execution time for quality of placement, introducing a placement algorithm that is a hybrid solution of the best-fit and first-fit algorithm. Another feasible solution to this problem is represented by the adaptive genetic algorithm proposed in this paper. This algorithm can be tuned for different scenarios of dynamic reconfiguration. In fact, since it can be executed with a different combination of parameters, it can perform the al-

location task either in a very short time or in a very accurate way, as shown by the presented experimental results.

This paper deals with the application of an adaptive genetic algorithm to the allocation of dynamically reconfigurable modules, introducing a very flexible approach to perform the allocation phase. In particular, the next section presents the scenario in which the genetic algorithm can be applied. Section 3 introduces the genetic algorithm on which the adaptive genetic algorithm presented in this paper is based. Section 4 describes the details of the adaptive genetic algorithm and all the parameters that it is possible to tune in order to achieve different levels of performance. Section 5 presents the experimental results that prove the effectiveness of the proposed approach. Finally, conclusions are drawn in Section 6.

2. RECONFIGURABLE SCENARIO

One of the most general platforms on which a configurable or reconfigurable system can be developed is a multi-FPGA scenario where the reconfigurable resources are distributed on several interconnected FPGAs. In such a scenario it is common to have a master FPGA able to reconfigure, partially or totally, other slave FPGAs. These slave FPGAs can be divided into several slots that can be filled with IP-Cores (or modules) by the master FPGA.

Figure 1 presents a collection of different scenarios. In all these scenarios, each master FPGA is characterized by the presence of an embedded PowerPC processor, on which the Operating System runs, in addition to the static hardware components such as a memory controller, general purpose inputs/outputs, and a reconfiguration manager.

The slave FPGAs, instead, hold the reconfigurable resources used to dynamically load hardware modules into the system. These resources are used according to a 1D-placement with a granularity of four CLB (Configurable Logic Block) columns [6]. This means that dynamic modules always use the full height of the FPGA, while their width is a multiple of four CLB columns, even if this scenario can be easily extended to the 2D scenario realized using Xilinx Virtex-4 [3] and Virtex-5 FPGAs [4].

In the first scenario, called Scenario A in Figure 1, there is one FPGA that is used both as a master and as a slave FPGA. An example of such a scenario can be found in [8]. This FPGA is logically divided into two different parts:

- a **fixed part**, that is the part of the FPGA that contains the PowerPC processor and that acts as a single

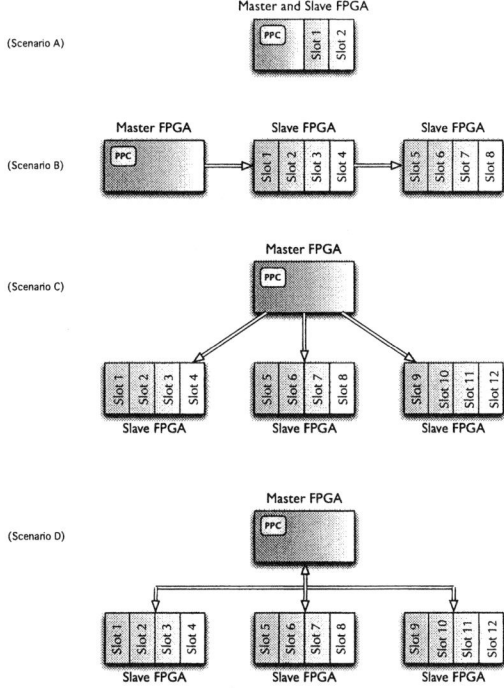

Figure 1: Multi-FPGA scenarios

master FPGA

- a **reconfigurable part**, that is handled as a single slave FPGA, even if the number of slots that it is possible to configure is smaller

On the other hand, in all the remaining scenarios each FPGA of the system acts either as a master or as a slave FPGA, without logical internal divisions.

The differences between these scenarios reside in the different ways in which the communication infrastructure is implemented. The second scenario (Scenario B) presents a chain communication in which the master FPGA can communicate with just one slave FPGA, and each slave FPGA can communicate just with the following one, for instance by using a communication module in the last slot.

Scenario C and Scenario D, instead, represent a point to point connection and a bus-based connection, respectively. In both these scenarios the master FPGA is able to communicate directly with each slave FPGA. [7] presents an architecture that can be represented using Scenario D.

Even if the presented scenarios differ in the logical partitioning of master and slave FPGAs sets and in their communication infrastructures, they can be reduced to the same class of platforms from the software point of view. For this reason they can be handled by the same software solution, as described in the following.

3. THE GENETIC ALGORITHM

A first version of the genetic algorithm that can be used for the allocation of dynamically reconfigurable modules has

been first proposed in [9]. This approach proposes the encoding of a single chromosome (that has to contain the information about the solution that it represents) as a pair of arrays, the *Slots* and the *Modules* arrays:

- The *Slots* array consists of a collection of genes, which contain the information on which module is configured on each slot of the reprogrammable device. In particular, each gene directly corresponds to a single slot of a slave FPGA. Since on a device of n slots it is possible to configure not more than n modules (this is possible only when each configured module requires just one slot), the alleles of these genes are represented by a number between *0* and *n-1*. The numbers contained in the *Slots* array correspond to the position of a gene in the second array.

- The *Modules* array consists of a set of genes that represent hardware IP-Cores. The following numbers represent the coding of the alleles for this kind of gene:

  - 0: this number means that the module is not configured on the reprogrammable device, since it has not been placed yet or it has already been deleted from the system;

  - 1: this number indicates that the module has been already configured on the FPGA and it is still running, so at this time it cannot be directly unloaded from the system;

  - -2: a module characterized by this number is a cached IP-Core. In other words it is a module that has already been placed on the reprogrammable device but it is not currently used, thus it is possible to unload it to overwrite its slots with the configuration of a more useful IP-Core.

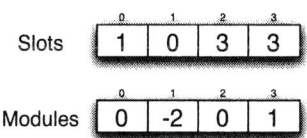

Figure 2: Genetic algorithm chromosome

The example shown in Figure 2 represents a status of the system in which the second module (module 1) is configured on the first slot of the FPGA (slot 0) and the fourth module (module 3) is placed on the third and on the fourth slot (slot 2 and slot 3), while the second slot (slot 1) is free (since the first module, module 0, is not configured).

The *Slots* array gives further information, indicating that the second module (module 1) is cached, while the fourth module (module 3) is still running. This means that the largest module that is possible to configure starting from this status is a module that requires two slots, since it can be configureb on the first two slots of the FPGA (slot 0 and slot 1), by unloading the second module (module 1) that is currently cached.

The proposed genetic algorithm is performed each time a set of new modules have to be configured on the reprogrammable devices of the system. If each module can be

2007 IFIP International Conference on Very Large Scale Integration (VLSI-SoC 2007)   129

placed in n positions, an exhaustive search with a set of m IP-Cores requires n^m evaluations of feasible solutions. With a genetic algorithm it is possible to considerably decrease the time required by the allocation process, since it works on a smaller set of solutions, trying to modify them to reach a good sub-optimum solution in a reasonable time.

In particular, the first step of the algorithm is the creation of an initial set of randomly generated chromosomes. Then, after the fitness evaluation, a subset of chromosomes is chosen to create a new population. These chromosomes are called parents of the offspring, that it formed through the crossover process.

The crossover task is performed by randomly choosing two parents. The new chromosome is generated by keeping the genes of the first part of the first parent, while the other genes are directly taken from the second parent. During this phase it is possible to introduce, with a random probability, a mutation. This is defined as a change in the partial solutions found by the parents. In other words it means that the location inherited by the parents can be randomly modified, to prevent that all solutions in the population fall into a local optimum.

4. ADAPTIVE GENETIC ALGORITHM

The genetic algorithm described in Section 3 has been extended with a set of configurable parameters that make the algorithm dynamically adaptive with respect to the platform scenario where it has to work. These parameters provide the possibility to choose either a fast or a very accurate allocation phase, depending on the timing performance and on the space constraints.

The parameters that can be tuned to tailor the solution onto a specific scenario are:

- **initial population size**, that is the initial size of the randomly generated population, as described in Section 4.1;

- **selection size**, the number of chromosomes that are chosen to create the new population, described in Section 4.2;

- **maximum number of rounds**, introduced in Section 4.3, that is the maximum number of generations that can be performed before stopping the execution of the algorithm;

- **minimum fitness**, described in Section 4.4, that is the fitness threshold;

- **crossover probability**, that is the probability of performing a crossover of two parents in order to generate a new offspring (otherwise the first parent is not modified), as presented in Section 4.5;

- **neutral mutation probability**, described in Section 4.6, that is the probability of performing a neutral mutation on the new chromosome;

- **positive mutation probability**, described in Section 4.7, that is the probability of performing a positive mutation on the new chromosome;

- **negative mutation probability**, described in Section 4.8, that is the probability of performing a negative mutation on the new chromosome.

Each parameter can be tuned in order to achieve the desired performance, both in terms of time and in terms of refused modules.

It is possible, in fact, that a particular scenario requires a fast allocation phase, for instance when the module that has to be deployed has to be available in a very short time. In this case it is possible to run the genetic algorithm with a set of parameters that provides a feasible position for the module in a fast way. The execution of the algorithm with this set of parameters affects the performance of the algorithm itself and increases the fragmentation of the reconfigurable device, but this negative effect can be kept under control by choosing the most suitable set of parameters, as shown in Section 5.

On the other hand, when a module is requested in advance with respect to its real utilization time (for instance when pre-fetching is performed), it is possible to execute the genetic algorithm with a set of parameters that allows the search for a solution that minimizes the fragmentation of the reprogrammable device. To achieve this result, it is necessary to know the right set of parameters that are able to reduce the average number of refused modules during the whole life of the system.

For these reasons, each parameter has been tested with a large set of significant values, as described in the following sections.

4.1 Initial population size

Each time a module is requested, the genetic algorithm has to create an initial population that consists of randomly generated individuals. Each one of these individual has to satisfy all the constraints, since it has to represent a feasible solution. The single chromosome within the population will change its characteristics, but the total number of chromosomes will not change, since the population size is fixed to the value of the size of the initial population. The initial population size, then, will affect the whole execution of the genetic algorithm, since it represents the size of the population on which each operation (such as crossover and mutations) will be performed. The genetic algorithm has been tested with three different values, that are 10, 50 and 100 chromosomes.

4.2 Selection size

When the fitness of each chromosome of the population is evaluated in order to choose the chromosomes that will act as parents (that are, in other words, the chromosomes with the maximum fitness value) during the generation of the new population, it is possible to select a set of these chromosomes that will be kept, without any changes, in the next generation. The selection size is hence the number of chromosomes that will be kept without any changes, while the difference between the initial population size and the selection size represents the number of chromosomes that have to be created during the offspring generation phase. The selection size depends on the initial population size: for this reason the values of the selection size has been chosen as 1/4, 1/2 and 3/4 of the initial population size, that represent three different situations, in which few, half or a lot is preserved from the previous generation.

4.3 Maximum number of rounds

The generation (that consists of the evaluation of the fitness, in the selection of the most suitable solutions and in the generation of the children) has to be performed either for the maximum number of rounds or until the minimum fitness is reached. In the first case, in which the minimum fitness is never reached, the value that represents the maximum number of rounds has to be chosen keeping into account that a big value requires a large execution time, while a small value can lead to a solution that is not optimal and that increases the fragmentation of the reconfigurable device. In particular, in our experiments, we used for this parameter the following values: 10, 20 and 50.

4.4 Minimum fitness

The minimum fitness represents the threshold that has to be exceeded in order to accept a chromosome as a final solution. This parameter is very important since it allows an early-stop of the algorithm when a good solution has been found. Obviously, with a small minimum fitness value, the final solution will be not optimized, while a big value of this parameter will probably bring the algorithm to execute for the maximum number of rounds, as described in Section 4.3. The minimum fitness is hence a measurement of the goodness of the desired soution. The goodness index will be explained more in details in Sextion 5. For our experiments we used three different values: 100, 1000 and 2000.

4.5 Crossover probability

The crossover task is performed by randomly choosing two parents within the set of the selected chromosomes, as introduced in Section 4.2. Each new chromosome is generated by keeping the genes of the first part of the first parent, while the other genes are directly taken from the second parent. When the crossover is not performed, the new chromosome is equal to one of the two parents, chosen randomly. In both cases, children always represent valid solution for the given problem. The crossover parameter is hence responsible for the generation of an offspring that mixes the good characteristics of the more suitable solutions of the previous generation, in the hope to form a better one. In our experiments we tested this probability with the following values: 25%, 50% and 75%.

4.6 Neutral mutation probability

Each time a new chromosome is generated it is possible to perform a neutral mutation by modifying the position of the requested module within the reconfigurable device (for this reason it has been called neutral mutation, since it preserves the status of the modules configured on the reconfigurable device). This mutation allows the generation of a new solution that was not present in the initial population, so it is an index of the difference between the solutions achieved by a population and the following one. The new location of the requested module has to be a feasible position, since each chromosome has always to represent a feasible solution. As with the other probabilities, we tested this parameter with the following values: 25%, 50% and 75%.

4.7 Positive mutation probability

With a positive mutation it is possible to free space on the reprogrammable device by deleting a module that was previously kept in cache. This mutation allows the increase of the number of positions where the requested module can be placed (as described in Section 5) without any penaliza-

tion. The slots occupied by the deleted module are marked in a special way, since they have to be recognized at the end of the algorithm, when slots that have been deleted but that are not used by the requested module can be simply reintroduced without introducing any overhead and increasing the goodness of the final solution. Also this probability has been tested with the following values: 25%, 50% and 75%.

4.8 Negative mutation probability

A negative mutation, in which a module that has been removed from the cache will be reintroduced in the cache, can be introduced to increase the goodness of the solution at run-time. This kind of modules, in fact, can be reintroduced in the cache in order to avoid the placement of the requested module, without any penalization, in a location that will lead to delete a cached module. In our experiments, we used the following values for this probability: 25%, 50% and 75%.

5. EXPERIMENTAL RESULTS

Each combination of the values of the parameters presented in Section 4 has been tested in order to achieve the performance characterization of all the possible sets of parameters.

The base scenario on which these tests have been performed consists of a reconfigurable device that has been divided in fifty reconfigurable slots. Furthermore, the size of the single module that can be deployed on the system ranges from one to three slots.

Table 1: Parameters values

Parameter	First value	Second value	Third value
Initial population size (IPS)	10	50	100
Selection size	$\frac{1}{4} * IPS$	$\frac{1}{2} * IPS$	$\frac{3}{4} * IPS$
Maximum number of rounds	10	20	50
Minimum fitness	100	1000	2000
Crossover probability	25	75	100
Neutral mutation probability	25	75	100
Positive mutation probability	25	75	100
Negative mutation probability	25	75	100

Table 1 presents all the possible values for each parameter. Since there are eight parameters and each parameter presents three different values, it is necessary to perform $3^8 = 6561$ experiments in order to evaluate all the possible combinations of parameters' values.

For each combination of parameters an experiment has been performed that consists of the following steps:

- fifty tests consisting of fifty module requests each have been performed. In particular, each test performs the following tasks:

- a random module request is given as input to the genetic algorithm;
- the result (success/fail) of this process and the time required for its execution are stored to calculate the fitness of the current solution;
- randomly a module is deleted from the reconfigurable device (in order to avoid the saturation of the device itself);

- at the end of each test the status of the reprogrammable device has been reset and the average results of the simulations (number of refused modules, cash index and timing performance) have been updated.

Figure 3 shows the average goodness index (that represent the fitness of a given solution) for each combination of parameters (the test flow previously described has been performed two times in order to avoid erroneous results). The goodness has been evaluated as follows:

$$Goodness = \frac{CI}{ET*RM}$$

where:

- CI is the Cash Index: this index is inversely proportional to the fragmentation of the reprogrammable device ($CI = \frac{1}{Fragmentation}$);

- ET is the Elapsed Time: it represents the time that is necessary to perform a whole experiment, that consists of 2500 module requests;

- RM is the number of Refused Modules. In other words, this index represents the number of modules that have not been placed during the execution of the algorithm.

Table 2 shows four combinations whose goodness index exceeds 100, that are also shown in Figure 3. The goodness index, as previously hinted, is directly proportional to the cash index and inversely proportional to both the number of refused modules and the elapsed time. It is also possible to tune this goodness function in order to give more importance to the first two components (for instance, by using this function for the goodness index, $Goodness = \frac{CI^2}{ET*RM^2}$, the result will be a solution optimized in terms of the number of refused modules) or to the last one (for instance, by using the following function, $Goodness = \frac{CI}{ET^2*RM}$, the result will be a solution optimized with respect to timing performance).

Table 3 presents two combinations of parameters that lead to a very small number of refused modules (both combinations have achieved less then 200 refused modules). In both these combinations the maximum number of rounds has been set to 50 and in the second one the initial population size has been set to 50 too.

Finally, Table 4 shows the top three combinations that are able to perform the allocation of a requested module in a very short time. By using these combinations, in fact, it is possible to accomplish a single module request in less than 0.2 milliseconds, since 2500 modules requests require less than 0.5 seconds. All the combinations presented in Table 4 are characterized by an initial population size of 10, by a selection size of 7, by a maximum number of 10 and by a minimum fitness of 100.

Table 2: Top four experimental results

Combination number	22	942	1554	2289
Initial population size	10	10	10	10
Selection size	2	7	5	5
Maximum number of rounds	10	20	50	10
Minimum fitness	100	100	100	1000
Crossover probability	25	50	25	25
Neutral mutation probability	75	75	50	75
Positive mutation probability	50	50	50	25
Negative mutation probability	25	75	75	75
Number of refused modules	250	230	175	262
Elapsed time (s)	0.5465	0.544	0.6885	0.5295
Cash index	15450	13540	12977	14399
Goodnes index	113	108	107	104

Table 3: Refused modules optimization

Combination number	1554	1891
Initial population size	10	50
Selection size	5	37
Maximum number of rounds	50	50
Minimum fitness	100	100
Crossover probability	25	50
Neutral mutation probability	50	25
Positive mutation probability	50	25
Negative mutation probability	75	25
Number of refused modules	175	180
Elapsed time (s)	0.6885	1.792
Cash index	12977	17381
Goodnes index	107	54

6. CONCLUSIONS

Figure 3 proves that the goodness index (Y-axis), eval-

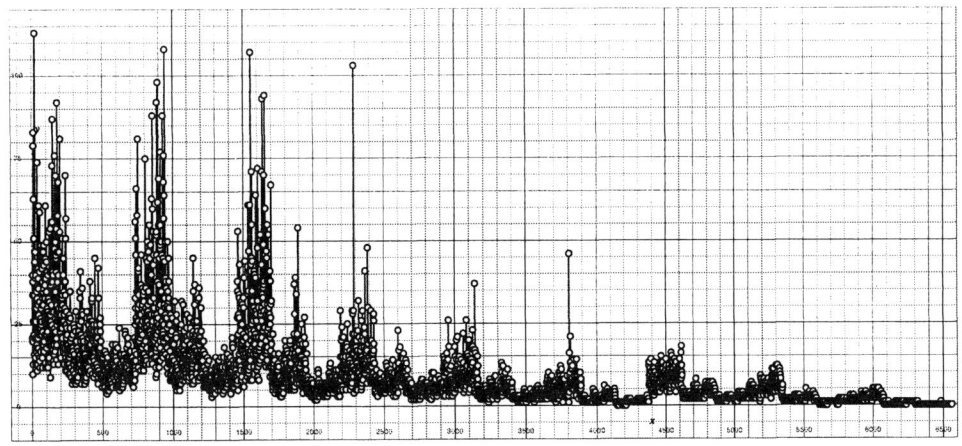

Figure 3: Goodness index for all the solutions

Table 4: Timing optimization

Combination number	166	169	179
Initial population size	10	10	10
Selection size	7	7	7
Maximum number of rounds	10	10	10
Minimum fitness	100	100	100
Crossover probability	25	25	25
Neutral mutation probability	25	25	50
Positive mutation probability	50	75	75
Negative mutation probability	25	25	50
Number of refused modules	522	546	335
Elapsed time (s)	0.473	0.476	0.477
Cash index	12579	11854	14770
Goodnes index	51	46	92

uated for all the possible combinations of the parameters (X-axis), is a cyclic function and that it is significantly affected by the changes in the parameters value.

Furthermore, results presented in Section 5 have shown how it is possible to perform an allocation of a requested module with a different combination of parameters in order to achieve different optimizations. It is possible either to minimize the number of refused modules or to reduce the time required for the computation. It is also possible, finally, to use a combination of parameters that optimizes the goodness index; this makes it possible to achieve an optimal

compromise between the three presented metrics.

The genetic algorithm presented in Section 3 and extended as described in Section 4 has been proved to be an effective solution for dynamically reconfigurable modules allocation.

7. REFERENCES

[1] Xilinx Inc., Two Flows of Partial Reconfiguration: Module Based or Dif- ference Based, Tech. Report XAPP290, Xilinx Inc., November 2003.

[2] Xilinx Inc., Early Access Partial Reconfiguration User Guide, Tech. Report UG208, Xilinx Inc., March 2006.

[3] Xilinx Inc., Virtex-4 User Guide, Tech. Report UG070, Xilinx Inc., April 2007.

[4] Xilinx Inc., Virtex-5 User Guide, Tech. Report UG190, Xilinx Inc., February 2007.

[5] K. Bazargan, R. Kastner, M. Sarrafzadeh *Fast template placement for reconfigurable computing systems*, IEEE design and test - Special issue on reconfigurable computing, pages 68-83, Volume 17, Issue 1, January 2000.

[6] H. Kalte, M. Porrmann, U. Ruckert *System-programmable-on-chip approach enabling online fine-grained 1D-placement*, IPDPS'04, Workshop 3, page 141.

[7] H. Kalte, M. Porrmann, U. Ruckert *A Prototyping Platform for Dynamically Reconfigurable System on Chip Designs*, Proceedings of the IEEE Workshop Heterogeneous reconfigurable Systems on Chip (SoC), 2002.

[8] Alberto Donato and Fabrizio Ferrandi and Marco D. Santambrogio and Donatella Sciuto *Coperating system support for dynamically reconfigurable SoC architectures.*, IEEE-SoCC, 2005.

[9] Vincenzo Rana, Chiara Sandionigi and Marco Domenico Santambrogio, *A genetic algorithm based solution for dynamically reconfigurable modules allocation*, Southern Conference on Programmable Logic, pages 183-186, 2007.

New Tool Support and Architectures in Adaptive Reconfigurable Computing

Jürgen Becker[1], Adam Donlin[2], Michael Hübner[1]

[1] Universität Karlsruhe (TH), Germany
[2] Xilinx Inc., San Jose, USA
{becker, huebner}@itiv.uni-karlsruhe.de
adam.donlin@xilinx.com

Abstract

Novel methods and reconfigurable architectures provide an increased design space by exploiting the dynamic and partial reconfiguration of hardware. The multi-adaptivity of this heterogeneous reconfigurable architectures reaches from adaptation to performance requirements over adaptation to power consumption in relation to an available amount of energy to adaptation to not predictable requirements from the user. Especially the unpredictable demands and requirements to a computing architecture require a high and filigree adaptivity in order to find an optimized point of operation while run-time. Additional to this issue the increased availability of electronic systems comes by introduction of novel methods for failure redundancy which can be seen as an application of this multi-adaptive system. In this contribution the ideas for a novel system approach will be presented in three parts. First the hardware and methods providing the multi-adaptivity will be presented. This is the basis for higher level design tools and opens a variety of parameters for adaptivity. The mechanisms of reconfigurability will be introduced in detail from basic knowledge to advanced mechanisms and methods. In addition the abstraction levels for manipulation the reconfigurable architecture and points to the tool support for novel reconfigurable FPGA architectures from Xilinx are sketched.

Keywords: Dynamic and partial reconfiguration, reconfigurable hardware, FPGA

1. Introduction

Run-time adaptive electronic systems enable the possibility to adjust the system's behavior and functionality according to non predictable events from their environment and inner status. Adaptivity "On-Demand" in relation to unpredictable requirements of the user and the system's environment, means the optimization of power consumption and performance while run-time by providing capacity for computation.

Traditional microprocessor based electronic systems can be adapted to the application by changing their program but have the lack that sequentially based processing of data decreases the performance and, additionally, the fixed hardware structure cannot be adapted to internal or external requirements. Reconfigurable hardware allows adapting the system architecture (the hardware) and the software at design-time but also while run-time. This new system approach, enabling the computation in time and space, provides and opens new degrees of freedom for hardware / software Co-Design which was originally developed for the design of Application Specific Circuits (ASIC). Those methodologies were used in former times to design static systems and can now be exploited for hardware / software partitioning while run-time. They now can be used to optimize the system status because the hardware provides a flexible physical adaptation of system resources. In this work a new system approach for dynamic and partial reconfigurable systems will be presented. Parametrizable functional patterns were placed On-Demand on a dynamic and partial reconfigurable architecture. The required partitioning methods as well as the communication primitives were analyzed and developed and integrated on real hardware. Further more the placement of 2 dimensional functional blocks and the wiring were presented and are the basis of new degrees of freedom for the flexibility while run-time in comparison to the former one dimensional method. This approach allows optimizing the exploitation of the reconfigurable chip area. The parallel data processing on the hardware provides novel methodologies for system integration of embedded electronic systems. The paper starts with the description of the hardware methods and the multi-adaptive architectures in section 2. Section 3 describes the novel tool-flow which is required for the proposed system design. Section 4 closes the paper with the conclusion and outlook.

2. Hardware and methods providing the multi-adaptivity architecture

The requested multi-adaptivity for actual and future electronic systems requires the filigree adaptation of hardware resources (logic cells and interconnection) in order to fulfill the requests from an unpredictable runtime scenario as described above. The benefit for the system is, that the system's status can be adjusted which leads to a reduction of power dissipation while keeping the requested performance for fulfilling real time requirements. The manipulation of the resources is enabled by an exchangeable configuration memory which controls the resources of the chip. The basic of this methodology will be described in the next section.

2.1. Mechanisms of reconfigurability

Figure 1 shows the very basic elements of an FPGA. The exemplarily structure of the FPGA consists of two layers. The configuration layer with its memory includes the information of the actual configuration of the second layer, the hardware layer. The memory is programmed after synthesis, place and route procedure. The configurable logic is connected to the configuration memory. As an easy example figure 1 shows a gate with three inputs and one output which is programmed as logic AND operation. Internal wires of the hardware layer, select one address of the configuration memory. Dependently to the content of the memory, a buffer is switched to 0 or 1. This element, responsible for logic operations is called: Configurable Logic Block (CLB). Additional to logic resources, the hardware layer provides routing resources for connecting the logic elements. With switch matrixes, horizontal and vertical signal lines can be connected. The switches (CMOS-Transistors) are controlled by the configuration memory.

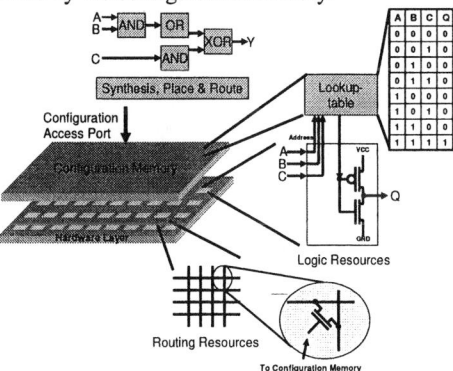

Figure 1. Basic elements of the internal FPGA structure

Of course this is a very easy example but it can be used to explain how dynamic and partial reconfiguration works. While run-time, different configurations were sent via the configuration access port to the configuration memory. It is clear what happens: Both logic elements and routing resources can be influenced and adapted to a new functionality and routing. It becomes clear that

changes influence the behavior of functions in the neighborhood if signal lines cross the area where partial reconfiguration occurs.

2.2. 1- and 2- dimensional reconfiguration

The possibility to perform dynamical and partial hardware reconfiguration of FPGAs increases their flexibility and ability of run-time adaptation. This feature is provided for example by Xilinx Virtex II FPGAs, which can be dynamically reprogrammed using the Internal Configuration Access port (ICAP, see [5]). By reconfiguring parts of the chip's architecture on-demand, the application bitstreams that are currently not needed can be stored on an off-chip memory, for example on a FLASH memory device, in order to reduce the necessary chip area. Since very often not all functionality of an application is required at the same time, this approach enables the realization of more functionality on one chip while keeping the chip size smaller as it would be necessary if all functionality is integrated on the device at the same time in parallel. An example of a system which uses this feature is presented in [4] and [10]. Due to this approach, it is possible to save costs and reduce power consumption since not actually used modules of a complete system do not allocate configuration memory and corresponding power consuming hardware [6]. Introducing the paradigms of [7] into a reconfigurable system and exploiting the adaptivity while run-time, opens a wide spectrum for power and performance aware designs. This approach in terms of "Systems-on-Chip" is the very promising way to overcome problems with traditional used technologies. Designing chips with the features of processors, reconfigurable architectures and even micromechanical parts can help to shrink the size of devices and optimize the performance and power consumption.

The 1- dimensional reconfigurable system approach leads to designs, where fixed rectangular shaped modules of hardware, were substituted. The size of these modules normally is related to the maximum required area of the biggest hardware module. This means, the biggest hardware module sets the size of the partial reconfigurable areas. Considering this, enables the allocation of every configurable area for each hardware module. Figure 2 shows the hardware reconfigurable system which is briefly described in [4]. The four reconfigurable modules have exactly the same size of resources on the FPGA (here Xilinx Virtex-II). Certainly this approach is adequate for a variety of applications (e.g. automotive, see [8]). The lack of this approach is that the resources of the FPGA cannot be utilized in an optimal manner since not all hardware modules exploit all logic which is within the related area. Therefore a new approach was developed which allows the placement of

2007 IFIP International Conference on Very Large Scale Integration (VLSI-SoC 2007)   135

rectangular hardware modules with a variable size. This approach motivates for the development of 2-dimensional reconfigurable systems.

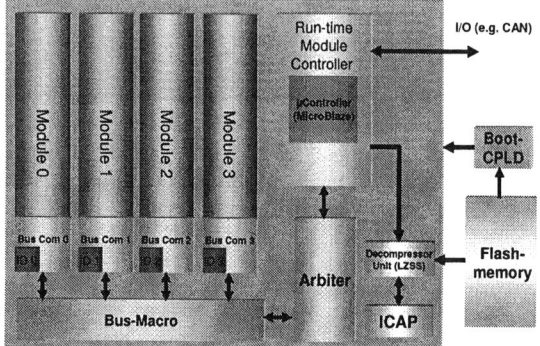

Figure 2.   1-dimensional Dynamic Reconfigurable System

The 2-dimensional reconfigurable hardware system architecture approach benefits from the improved exploitation of the resources of the FPGA. Figure 3 illustrates the method for this approach.

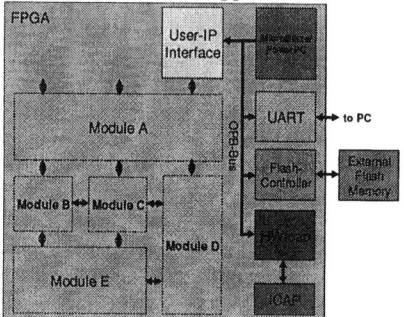

Figure 3.   2-dimensional reconfigurable hardware system architecture

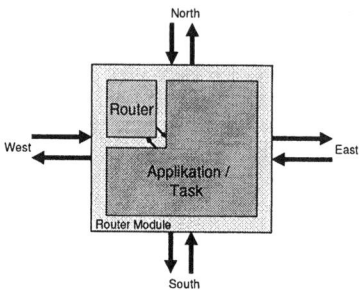

Figure 4.   Router and Application / Task Module

In opposite to the 1-dimensional reconfigurable system, modules with a non-fixed rectangular shape were placed while run-time to the reconfigurable area. Certainly, this approach requires a special design of the modules which

includes a router for communication between the modules itself, but also with the static system including the microcontroller (here the Xilinx MicroBlaze).Figure 4 shows the abstract view to one reconfigurable module with its 4 communication directions and one connection to the included functionality. The more detailed view to the functionality which is illustrated in figure 5 shows that the hardware structure is constructed efficiently in terms of utilized resources in comparison to the functionality within the module block. The overhead of the integrated router is in the design example described in [9] is 62 configurable logic blocks which is around 7% of the complete module size (880 configurable logic blocks).The approach of 2 dimensional placement on Xilinx FPGAs can be achieved by exploiting the read-, modify-, write-back method which can be processed as visualized in Figure 6. In the first phase, an area is selected for modification. The information content of this area represents the actual configuration of this place on the reconfigurable architecture. Via the ICAP interface the data were read from the configuration memory for further processing.

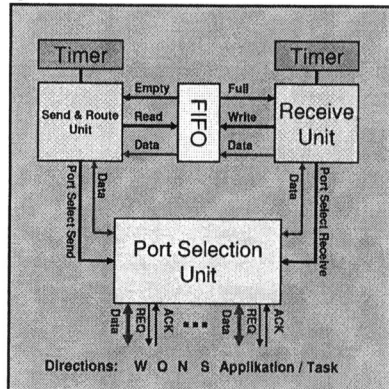

Figure 5.   Schematic view of the router

In the merging (or modification) phase, the pre-routed IP-core will be restored from an external memory in order to update the existing rectangular shaped partial reconfiguration data. After merging this two bit stream data, the write-back method is used to transfer the configuration back to the FPGA's configuration memory via the ICAP port within the device. In order to find the correct position for the merging process, the organization of the configuration memory has to be taken into account. Figure 7 shows one CLB columns which consists of 22 frames with the width of one bit. The length of one frame depends on the size of the selected FPGA. The different CLB blocks can be addressed by clustering the respective fractions all 22 frames. It has to be considered, that the frames content 2 I/O-Blocks which are aligned to each CLB column.

By using this method is has to be taken into account, that areas utilizing run-time changeable RAM-based resources like shift registers (e.g. SLR16 in Virtex) are affected if the content of this resources change while this process. In that case it has to be decided if parts of the design have to be stopped while read-, modify-, write-back method. This can be achieved by a simple clock-gating method which can also be processed with dynamic and partial reconfiguration.

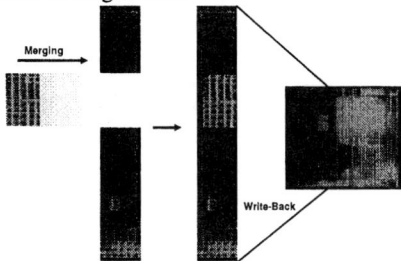

Figure 6. Merging (Modification) and Write-Back process

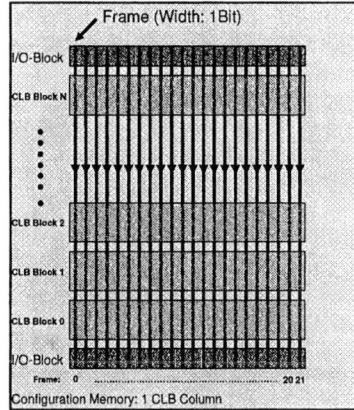

Figure 7. Organization of the configuration memory

The exploitation of the novel 2-dimensional approach for run-time adaptive systems is described in [11] where the filigree access to the resources of the FPGA is exploited to reduce power dissipation and adapt the system performance by manipulating the wiring of the integrated application. This feature of On-line Routing is currently the most filigree run-time manipulation method for such kind of systems and can now be exploited by coupling to higher level design methodologies (see [13] and [12]) which were now be realizable while run-time.

3. Design abstractions and tool support for reconfigurable systems

In this section we will look at the design considerations and tools that are relevant to designers building reconfigurable systems on Xilinx FPGAs. Abstract descriptions of systems are important to increasing the productivity of designers. In the best case, an abstraction allows the system designer to capture all of the relevant details of their design, reason about the qualities of that design and transform the design's abstract representation into a working implementation of the system.

3.1. Standard Design Abstractions

The artful step in creating a design abstraction is the selection of which details of the system can be omitted from the abstract specification. For example, most readers will be familiar with the register-transfer level (RTL) design abstraction employed in contemporary digital design. RTL design descriptions can be simulated for verification and synthesized for implementation. The synthesis process calculates the missing implementation details and optimizes the system to pre-determined performance constraints. For an abstraction to be useful to the designer, the amount of effort required to capture the system description at that level of abstraction must be balanced with the quality of the implementation that can be created from it. Until recently, there was balance between the effort of creating an RTL design description and the quality of results achieved from synthesizing it to an implementation. The same was not true when RTL was first introduced because higher quality designs could be assembled from design primitives in schematic tools. Today, system complexity threatens to overwhelm the value of RTL as a design abstraction. Attempts to create more abstract design descriptions to replace RTL have only had limited success because they have quality of results from the automatic synthesis tools for those abstractions have not been sufficient.

3.2. Reconfiguration Design Abstractions

Today, reconfigurable FPGA designs must leverage existing RTL design tools to create implementations. This means that the primary design abstraction for reconfiguration in FPGAs today is also RTL. Alternative design abstractions for reconfiguration have been proposed in the past. JBits [1] and JHDL [3], for example, were design languages built on top of Java and employing a structural hardware design abstraction. Structural design consumes more design effort for most designs and, because of that, the JBits and JHDL design environments were only used to describe the "system level" connectivity of IP blocks and the "design level" implementations of blocks with simple, regular microarchitectures.

The RTL abstraction in itself does not capture all of the design information that is relevant to a reconfigurable system design. Indeed, there are aspects of designing a reconfigurable system that violate the basic rules of RTL designs. Crucially, RTL asserts that only one design module may be instantiated per module interface in the design while reconfigurable systems require the ability to instantiate multiple design modules at the same signal interface. This is depicted in figure 8 where both "Module A" and "Module B" are instantiated at the "Data In" and "Data Out" interfaces. It could be argued that VHDL, by virtue of its ability to define entities (module interfaces) separate from its implementation and architectures allows multiple reconfigurable modules to be specified. However, VHDL allows multiple architectures for an entity to be declared but not instantiated.

Tools for regular RTL design also require signal paths in the implementation of a design module to originate at one or more output ports and terminate at one or more input ports. Reconfigurable designs require the ability to route incomplete signal paths in the implementation of the design and reconcile them later when the design is actually running on the FPGA. This is analogous to the ability of software programs to dynamically resolve the precise memory locations for symbols in software libraries at runtime. If software were restricted in the same way that RTL tools restrict reconfigurable designs, it would not be possible to compile a source code file without knowing the exact memory location of every function call and variable in all of the libraries that the source file depends on. Clearly, the existing design abstractions (and the tools that service them) have opportunity to evolve better support for reconfigurable systems.

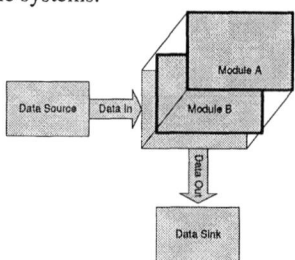

Figure 8. A simple reconfigurable system

3.3. Design Flows for Reconfiguration

In this section we provide a short review of the main design considerations in the current design flow for reconfigurable systems on FPGAs. A full discussion of this design flow can be found in [2]. To support reconfiguration, the RTL description of the system must be separated into one design for the "fixed" part of the system and separate designs for "reconfigurable" modules that will be configured into the fixed design during runtime. At least one region is reserved in the "fixed" system so that the reconfigurable modules can be loaded

into the live design safely. When the fixed design is implemented, the designer may want to pre-load it with one of the reconfigurable modules or leave the area "blank". As reconfigurable modules may be of different sizes, the area reserved in the fixed system must be no smaller than the dimensions of the largest reconfigurable module. This creates a cyclic dependency in the design flow between the implementation of the fixed system and the implementation of the reconfigurable modules: the physical floorplan of the reserved areas and the reconfigurable modules must match exactly if the design is to operate correctly at runtime. The reconfigurable design flow includes a planning/estimation phase where the size, floor-plan and performance requirements for the reconfigurable modules are iteratively resolved with respect to size and performance requirements of the fixed system. The Xilinx PlanAhead floor-planning tool offers support at this phase of the design flow by letting the designer graphically floor-plan the fixed system and constrain the area of the reconfigurable module. PlanAhead gives early estimates of the area and resource requirements for the reconfigurable modules, making it simpler to constrain and reconcile the regions that must be reserved in the fixed system.

Signals that cross the interface between a reconfigurable module and the fixed system must be managed carefully. The design flow must guarantee that the signal path in the fixed system will connect to the correct signal in the reconfigurable module, when that module is configured into the actual FPGA. In addition, the tools must also insure that signal paths within the reconfigurable module do not arbitrarily cross into the fixed region. Signals from the fixed region may pass into the reconfigurable module region provided none of the reconfigurable modules also attempt to use those signals. The design flow guarantees this by remembering which routing resources are used in the fixed region and preventing the reconfigurable module from using those signals when the module is going through the design flow.

To insure the interface signals are consistent between the reconfigurable module and the fixed region, a special design macro is instantiated at the boundary between the two. The reconfiguration design flow understands the special role these macro components play in the operation of the reconfigurable system and makes sure that the placement of the macros is exactly consistent between the fixed system and the corresponding reconfigurable modules.

Once the physical design flow has mapped the fixed system and the instances of the reconfigurable modules to physical resources in the target FPGA, it is possible to generate programming bitstreams for the target FPGA. As mentioned earlier, the flow may generate multiple versions of the fixed system with a reconfigurable module preloaded in the reconfigurable regions. The flow also generates "partial" bitstreams that only configure the parts of the FPGA where the reconfigurable module should be placed. One partial bitstream is generated for each of the module implementations and an optional "blank" bitstream that erases the contents of the reconfigurable region can be generated. As discussed in [2], blank bitstreams are useful for managing the power consumption of a reconfigurable system.

3.4. Future reconfiguration design flows

Future design flows for reconfigurable systems will extend the reconfiguration design abstraction further than the "reconfigurable module" systems characterized in **Fehler! Verweisquelle konnte nicht gefunden werden.**. These advanced reconfigurable design abstractions will allow modules to be relocated between reconfigurable regions automatically and allow regions themselves to have different interfaces over time. The preservation of state when a module is "deconfigured" from its target FPGA is of interest in certain classes of reconfigurable system. As more reconfigurable applications are designed on FPGAs, developers can anticipate a greater automation of the RTL design partitioning and transformation steps that, today, are performed manually.

4. Conclusions and future work

This paper describes the approach for multi-adaptive reconfigurable systems, which allows a higher degree of freedom for system adaptation. Performing dynamic reconfiguration offers the possibility of on-line routing and placement of communication primitives and functionality in order to optimize the power / performance trade-off in run-time adaptive systems. The proposed methods can directly be connected to higher-level design methodologies which are presented in [12] and [13]. The described methods are used in real applications for driver assistant systems in the automotive domain (see [14]).

Since dynamic and partial reconfiguration found its way from academic labs to industry research and development groups, investigation for exploitation of the proposed approaches in real-world applications is a challenging task. Novel trends show, that the exploitation of dynamic and partial reconfiguration in embedded systems provide a manifold flexibility in different kind of applications. Reliability, failure-redundancy and run-time adaptivity by usage of real-time hardware reconfiguration are important keywords for actual and future embedded systems.

5. References

[1] C. Patterson, S. Guccione, "*JBits Design Abstractions*", In FCCM 2001, pp 251-252.

[2] A. Donlin, "*Applications, design tools and low power issues in FPGA reconfiguration*", chapter 21 in "Designing Embedded Processors", edited by J. Henkel and S. Parameswaram, Springer, 2007, pp 513-541.

[3] P. Bellows, B. Hutchings, "JHDL –An HDL for Reconfigurable Systems", In FCCM 1998, pp. 175-184.

[4] M. Ullmann, M. Huebner, B. Grimm, J. Becker: "An FPGA Run-Time System for Dynamical On-Demand Reconfiguration", RAW04, Santa Fee

[5] B. Blodget, S. McMillan: "A lightweight approach for embedded reconfiguration of FPGAs", DATE´03, Munich Germany

[6] J. Becker, M. Hübner, M. Ullmann: "Power Estimation and Power Measurement of Xilinx Virtex FPGAs: Trade-offs and Limitations", SBCCI03, Sao Paulo, Sep. 03

[7] L. Benini, G. De Micheli: "Networks on Chip: A New Paradigm for Systems on Chip Design", Date 02, March 3~7, Paris France

[8] Becker, M. Hübner, G. Hettich, R. Constapel, J. Eisenmann, J. Luka: "Dynamic and Partial FPGA Exploitation", Proceedings of the IEEE Special Issue "Advanced Automobile Technologies" (February 2007, Volume 95, Number 2)

[9] M. Hübner, C. Schuck, M. Kühnle, J. Becker: "New 2-Dimensional Partial Dynamic Reconfiguration Techniques for Real-Time Adaptive Microelectronic Circuits", ISVLSI2006, Karlsruhe, Germany

[10] C. Bobda, M. Majer, A. Ahmadinia, T. Haller, A. Linarth, J. Teich, S. Fekete and J. van der Veen :"The Erlangen Slot Machine: A Highly Flexible FPGA-Based Reconfigurable Platform", Proceedings of the IEEE Symposium on Field-Programmable Custom Computing Machines (FCCM), Napa, USA, pp. 319-320, April 17-20, 2005

[11] J. Becker, M. Hübner, K. Paulsson: "Physical 2D Morphware and Power Reduction Methods for Everyone", Dagstuhl Seminar, April 2006, Schloss Dagstuhl

[12] Dittmann, Florian; Götz, Marcelo; Rettberg, Achim: "Model and Methodology for the Synthesis of Heterogeneous and Partially Reconfigurable Systems". In: Proceedings of the Reconfigurable Architecture Workshop, Long Beach, CA, USA 2007

[13] Dittmann, F.; Rammig, F. J.; Streubühr, M.; Haubelt, C.; Schallenberg, A.; Nebel, W.: "Exploration, Partitioning and Simulation of Reconfigurable Systems". it -Information Technology, 3(7), 1. Jan. 2007

[14] Claus, C.; Zeppenfeld, J.; Müller, F.; Stechele, W.: " Using Partial-Run-Time Reconfigurable Hardware to accelerate Video Processing in Driver Assistance Systems", Proceedings of DATE 2007, Nice, France, April 16-20, 2007

Rate-based Scheduling Policy for QoS Flows in Networks on Chip

Aline Mello, Ney Calazans, Fernando Moraes

Faculdade de Informática - Pontifícia Universidade Católica do Rio Grande do Sul, PUCRS - Porto Alegre, Brazil
{alinev, calazans, moraes}@inf.pucrs.br

ABSTRACT

Several propositions of NoC architectures claim providing quality of service (QoS) guarantees, which is essential for e.g. real time and multimedia applications. The state-of-art in NoC literature provides QoS at design time, using circuit switching and/or priority-based scheduling. Both methods optimize a given network template to achieve the QoS requirements after traffic generation and network simulation. However, modern SoCs may execute applications not devised at design time, and these may easily have its QoS requirements violated by a previously fixed NoC structure. This paper proposes a method to achieve QoS requirements in NoCs at execution time. The proposed rate-based scheduling policy is employed to determine the priority of each QoS flow being transmitted through the network. The basis of this scheduling method is the difference between the rate required by a given flow and the rate currently used by this flow. This difference corresponds to the flow *priority* used by the scheduler. Differently from traditional priority-based scheduling, the priority is dynamically adjusted. Preliminary results show the efficiency of the rate-based scheduling to meet QoS requirements, by comparing the proposed scheduling to priority-based scheduling.

1. INTRODUCTION

Network-on-Chip (NoC) is an emerging paradigm for communications within large VLSI systems implemented on a single silicon chip, known as System on Chip (SoC). In a NoC based SoC, modules such as processor cores, memories and other specialized IP blocks exchange data encoded in packets, using the NoC as a subsystem for data transport.

Distributed multimedia applications (e.g. 3G phones), need to communicate in real-time and are sensitive to the quality of services (QoS) they receive from the NoC. The term QoS refers here to the capacity to control the communication infra-structure to meet the application design requirements in what concerns the communication among modules of the SoC.

Usually, NoC literature employs two services class definitions: best effort (BES), and guaranteed (GS). BE services guarantee the transmission of all packets from a given source to a given target without any temporal bound guarantee. GS provide rigid bounds on one or on a subset of performance figures such as throughput (GT), latency, jitter and packet losses. This paper proposes to add a new service class, named Quality Services (QS). QS is defined as a service class where the network actively tries to reach application requirements without guaranteeing rigid bounds

for performance figures. Three reasons can be advanced to propose this new service class. First, the typical workload of SoC applications is tolerant to limited variation in performance figures, possibly not requiring GS. Second, although GS can locally provide the best possible level of services, in general QS is capable of achieving a better level of global performance. Third, QS allows a better use of resources, leading naturally to a better dimensioning of the NoC.

Most current NoC implementations only provide support to BE services [1], including commercial products such as Arteris [2]. BE services are inadequate to satisfy QoS requirements for applications/modules with tight performance requirements.

NoC implementations providing support to QoS ([2]-[12]) try to achieve performance requirements at *design time*. This requires application traffic modeling, system simulation and NoC optimization and/or sizing. The internal router architecture of such NoCs employs circuit switching and/or priority-based scheduling to attain performance requirements for a given application. Circuit switching allows implementing GS and priority-based scheduling is a technique to meet QS.

Several modern SoCs may execute applications not devised at design time, and these may easily have its QoS requirements violated by a previously fixed NoC structure. Also, designing a NoC to support any traffic scenario is often unfeasible in terms of power and area.

The objective of this paper is to propose and evaluate a method to achieve QoS requirements at *execution time* for NoCs using QS. The proposed method uses a rate-based scheduling policy, being a two-step process. First, a data flow requiring QoS (called a *QoS flow*) is admitted in the NoC if and only if the NoC can transmit the rate required by the specific flow end-to-end, in what is called admission control. Next, each router dynamically defines the priority of each QoS flow locally, as a function of the rate used by this flow.

This paper is organized as follows. Section 2 presents related work in NoCs that offer support to obtain QoS, discussing limitation of current methods. Section 3 details the proposed scheduling method for QoS flows. Section 4 evaluates the proposed method, comparing it to priority-based scheduling method. Finally, Section 5 presents conclusions and directions for future work.

2. RELATED WORK

Current NoC designs employ at least one of three methods to provide QoS: (*i*) dimensioning the network to provide enough bandwidth to satisfy all IP requirements in the system; (*ii*) providing support to circuit switching for all or for selected IPs; (*iii*) making available priority-based scheduling for packet transmission.

Harmanci et al. [3] present a quantitative comparison between circuit switching and priority-based scheduling, showing that the prioritization of flows on top of a connectionless communication network is able to guarantee end-to-end delays in a more stable form than circuit switching. However, the reference does not quantify results numerically. A possible explanation for this is the use of a TLM SystemC modeling, instead of clock cycle accurate models advanced here. Additionally, the structural limitations of circuit switching and priority-based scheduling are not depicted.

The first method to provide QoS mentioned above is advocated e.g. by the Xpipes NoC [4]. A designer sizes Xpipes according to application requirements, adjusting each channel bandwidth to fulfill the requirements. However, applying this method alone does not guarantee avoidance of local congestions (hot spots), even if bandwidth is largely increased. This fact, coupled to ever-increasing performance requirements [5], render the method inadequate to satisfy requirements for a wide range of distinct applications.

The second method, support to circuit switching[1], provides a connection-oriented distinction between flows. This method is used in Æthereal [6], aSOC [7], Octagon [8], Nostrum [9] and SoCBUS [10] NoCs. For example, the Nostrum NoC [9] employs virtual circuits (VC), with the routing of QoS flows decided at design time. The communications on the physical channels are globally scheduled in time slots (TDM). The VCs guarantee throughput and constant latency at execution time, even with variable traffic rates. Circuit switching NoCs create connections for each or to selected flows. The establishment of connections requires allocation of resources such as buffers and/or channel bandwidth. This scheme has the advantage of guaranteeing tight temporal bounds for individual flows. However, this method has two main disadvantages: (*i*) poor scalability [3]; (*ii*) inefficient bandwidth usage. Here, router area grows proportional to the number of supported connections, penalizing scalability. Resource allocation for a given flow is based in worst case scenarios. Consequently, network resources may be wasted, particularly for bursty flows.

QNoC [11], DiffServ-NoC [3] and RSoC [12] are examples of NoCs adopting the third method, packet switching with priorities. This connectionless technique groups traffic flows into different classes, with different service levels for each class. The method requires separate buffering to manipulate packets according to the services levels. To each service level corresponds a priority class. The network always serves first non-empty higher priority buffers. Packets stored in lower priority buffers are transmitted only when there are no higher priority packets waiting to be served. This scheme offers better adaptation to varying network traffic and a potentially better utilization of network resources. However, end-to-end latency and throughput cannot be guaranteed, except to the higher priority flows. Also, it is necessary to devise some form of starvation prevention for lower priority flows. When flows share resources, even higher priority flows can have an unpredictable behavior. Consequently, this method often provides a poorer QoS support than circuit switching.

Neither circuit switching nor priority methods guarantee QoS for *concurrent multiple flows*. When using the circuit switching method, the network may reject a number of flows, due to limited amount of simultaneously supported connections, even if network bandwidth is available. When multiple flows with the same priority compete for the same resources, priority-based networks have behavior similar to BE service networks [13]. As mentioned before, networks using any of the three above described methods employ techniques at design time to guarantee QoS, through traffic modeling, simulation-based network sizing (topology, buffer depth, flit width) and network synthesis. The drawbacks of sizing the network at design time are: (*i*) the complexity of traffic modeling and system simulation is very high, being thus error-prone; and (*ii*) the network designed in this way may not guarantee QoS for new applications. The first drawback may force the use of simplified application/environment models, which can in turn lead to incorrect dimensioning of the NoC parameters for synthesis. The second drawback may arise if new applications must run on the system after some initial implementation, as occurs with reconfigurable or programmable systems.

The main performance figures used in the above reviewed NoCs are end-the-end latency and throughput. Nonetheless, when QoS is considered, another concept can be of relevance, *jitter*. Jitter can be defined as the variation in latency, caused by network congestion, or route variations [14]. In connectionless networks, buffers introduce jitter. When packets are blocked, latency increases. Once the network can release packets from blocking, latency reduces, due to burst packet diffusion. Therefore, networks using only priorities cannot guarantee controlled jitter.

Some other works advocate different methods to enhance QoS. For example, Andreasson and Kumar proposed a *slack-time aware* routing [15][16], a source routing technique to improve overall network utilization by dynamically controlling the injection of BE packets in the network in specific paths, while guaranteed throughput (GT) packets are not employing those paths. This work is not directly related to QoS achievement.

---

[1] In this paper, the term *circuit switching* is used to refer to both, networks providing physical level structures to establish connection between source and destination, as well as to packet switched networks that employ higher level services (such as virtual circuits) to establish connections.

3. RATE-BASED SCHEDULING POLICY

The proposed scheduling policy assumes the following NoC features: wormhole packet switching, deterministic routing, and physical channels multiplexed in at least two virtual channels (VC). BE flows are transmitted using only one VC, while QoS flows may use any VC. This resource reservation for QoS flows is necessary to avoid that multiple BE flows block the possibility of using the channel for some QoS flow. The proposed policy is a two-step process: admission control followed by dynamic scheduling.

The admission step determines if the network may accept a new QoS flow without penalizing performance guarantees already assigned to other QoS flows. The admission step starts by sending a control packet from the source router to the target router, containing the rate required by the IP. The QoS flow is admitted into the network if and only if all routers in the path to the target can transmit at the required rate. When the control packet arrives at the target, an acknowledgment signal is back propagated to the source router. This process is similar to the connection establishment in circuit switching, but differently from circuit switching there is no static resource reservation.

When the QoS flow is admitted, a *virtual connection* is established between the source and target router, as in ATM [17] networks. This virtual connection corresponds to a line in the *flow* table (see Figure 1) of each router in the connection path. Each line of the flow table identifies the QoS flow using the following fields: source router, target router, required rate, and used rate. The *flow table* depth determines how many simultaneous QoS flows can be admitted by each router. The virtual connection is released by the source router with another control packet.

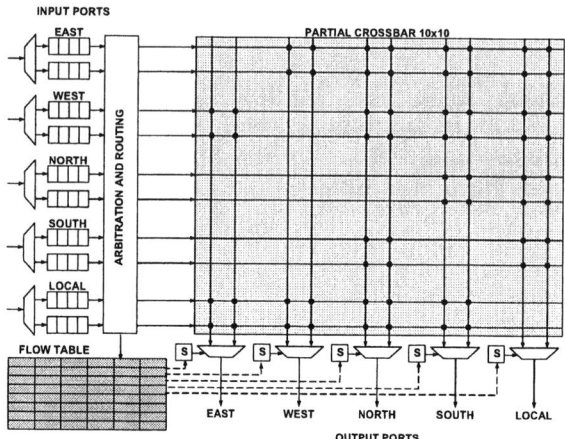

Figure 1. Router architecture with support for rate-based scheduling.

Once the virtual path is established, the source router may start sending QoS flow packets. When packets arrive at a router input port they are stored in input buffers, arbitrated (e.g. using round robin) and routed (e.g. XY deterministic algorithm) to an output port (Figure 1). Packets assigned to the same output port are served according to the proposed scheduling policy.

The implemented scheduling policy adopts a work-conserving mechanism, which is idle only when there is no packet awaiting to be served. BE flows are transmitted only when no QoS flows require the physical channel. When two or more Qos flows compete, the higher priority flow is scheduled first.

As illustrated in Figure 1, the flow table is read by the scheduler (blocks named S in Figure 1) to find the priority of each QoS flow assigned to a same output port. The QoS flow *priority* is the difference between the required rate and the rate currently used by the QoS flow. The flow priority is periodically updated according to Equation 1. A positive priority means that the flow used less rate than required in the considered sampling period. A negative priority means that the flow is violating its admitted rate in the considered sampling period.

$$priority_i = required\ rate - used\ rate_i \qquad (1)$$

The required rate is fixed during the admission control step. The used rate (UR) is *periodically* computed according to Equation 2:

$$UR_i = \begin{cases} CR_i, & if\ UR_{i-1}=0 \\ \dfrac{UR_{i-1}+CR_i}{2}, & if\ UR_{i-1} \neq 0 \end{cases} \qquad (2)$$

where:

- CR: is the *current rate* used during the current period;
- UR: is the average of the previous used rate and the current used rate.

Figure 2 illustrates packets of a given QoS flow being transmitted. Timestamps T0 to T4 designate when the rates are sampled, assuming in this example 10 time units in each interval. The table in the Figure corresponds to the behavior of one flow table from T0 to T4.

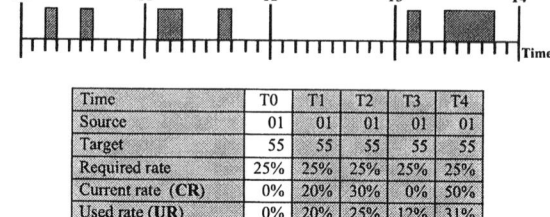

Time	T0	T1	T2	T3	T4
Source	01	01	01	01	01
Target	55	55	55	55	55
Required rate	25%	25%	25%	25%	25%
Current rate (CR)	0%	20%	30%	0%	50%
Used rate (UR)	0%	20%	25%	12%	31%
Actual rate	0%	20%	25%	16%	25%
Priority	25	5	0	13	-6

Figure 2. Transmission of packets for a given QoS flow.

In this example, the 4th line of the table contains the required rate (25%) for this flow. At timestamp T1 the current rate (5th line) is 20%, corresponding to the channel bandwidth used by the flow in the previous interval (T0-T1). According to the Equation 2 it is possible to obtain the used rate (6th line of the table). The 8th line of the table contains the flow priority, which is updated according to Equation 1.

The interval between timestamps is an important parameter of the proposed method. The 7th line contains the actual flow rate (shown here for comparison purposes, not present in the flow table). If the chosen interval is too short, the computed used rate may not correspond to the actual rate, compromising the scheduling method. If the interval is too long, the computed used rate will be accurate, but the flow priority will remain fixed for a long period, also compromising the method.

To minimize the error induced by the sampling period, the method in fact employs two sample intervals. In the previously presented example, consider a second current rate (CR2) and a sample interval 4 times larger than the original one. In this example, CR2 will be equal to 100% (summation of CR from T0 to T4) in T4. Dividing CR2 by 4, the corrected used rate is obtained (CUR, Equation 3). It can be observed that applying CUR to UR each n intervals (4 in this example), the error is minimized.

$$CUR = \frac{CR2}{n} = \frac{\sum_{i=0}^{n-1} CR_i}{n} \qquad (3)$$

Consequently, in Equation 1 UR_i receives CUR when i mod n is equal to zero, where n corresponds to the result of dividing the long sample interval value by the short sample interval value.

It is important to mention that if only the used rate were considered in the priority computation ($priority_i = 100 - UR_i$), the scheduling policy tends to balance physical channel use. This implies disregarding the fact that distinct QoS flows may require distinct rates. Consider two QoS flows with instantaneous UR of 20% and 30%, respectively. The first flow would be scheduled first. Assume that the required rates are 10% and 40% to each flow. The first flow scheduled first does not consider that it is using more bandwidth than the required. Using Equation 1 the correct schedule is obtained.

4. EXPERIMENTAL RESULTS

This Section compares the performance of the priority-based scheduling with the proposed rate-based scheduling, since both support QS. Traffic injection and results capture is modeled with SystemC, while the NoC is modeled through RTL VHDL [18]. The NoC parameters are: 8x8 mesh topology; XY routing; 16-bit flits; 2 virtual channels; 8-flit buffers associated to each virtual channel.

4.1 Experimental Setup

Table 1 presents the flows used in the experiments. Flow A is characterized as a CBR service with QoS requirements, as latency and jitter. Nodes generating flows A transmit 2000 packets. The results do not take into account the first 50 packets and the last 50 packets. They are discarded from results, since the traffic at the beginning and the end of the simulation does not correspond to regular load operation. Flow B is a BE flow modeled using a Pareto distribution. This flow is used to disturb QoS flows, being considered as noise traffic. For this reason, results for the B flow are not discussed.

Table 1. Flows Characterization.

Type	Service	QoS	Distribution	Number of Packets	Packet Size	Target
A	CBR	Yes	Uniform (20%/30%)	2000	50, 100, 200, 500	Fixed
B	BE	No	Pareto (20% on)	Random	20	Random

Figure 3 presents the spatial distribution of source and target nodes. In this scenario, two QoS flows (F1 and F2) originated at different nodes share part of the paths to targets. The remaining network nodes transmit B flows, disturbing the QoS flows.

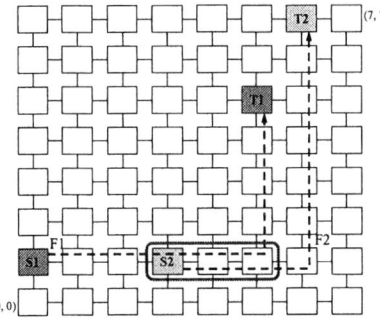

Figure 3. Spatial distribution of source and target nodes for flows with QoS requirements. Dotted lines indicate the path of each flow. Rounded rectangles highlight the area where flows compete for network resources.

Equation 4 gives the ideal latency to transfer a packet from a source to a target, in clock cycles.

$$ideal\ latency = 5N + P \qquad (4)$$

where:
- 5: is the router minimal latency (arbitration and routing);
- N: is the number of routers in the communication path (source and target included);
- P: is the packet size.

When the packet size is 50 flits, the ideal latency for the scenario presented in the Figure 3 is 100 (5x10+50) clock cycles for both scheduling.

4.2 Results

Table 2 presents the latency values, jitter and throughput when the packet size is 50 flits and the inserted rate is 20%. Both scheduling policies guarantee throughput close to the inserted rate (20%).

Analyzing the priority-based scheduling, F2 has average latency near to ideal, while F1 flow has higher latency (average latency 44% far from the ideal latency). F1 and F2 are CBR flows with the same priority, competing for the same resources. They insert packets in the network at fixed intervals. As the F2 source node is closer to the region disputed by the flows, it is always served first. This experiment demonstrates that priority-based scheduling is inefficient for QS when flows with the same priority compete for the same resources.

Table 2. Results for flows F1 and F2, 50 flits, 20% load.

Performance Figures		Priority-based		Rate-based	
		F1	F2	F1	F2
Latency	**Ideal (ck)**	**100,00**	**100,00**	**100,00**	**100,00**
	Minimum (ck)	141,00	100,00	119,00	119,00
	Average (ck)	**144,23**	**101,78**	**148,95**	**121,93**
	Maximal (ck)	154,00	133,00	174,00	133,00
Jitter (ck)		2,66	3,04	18,63	3,03
Average throughput (%)		19,21	19,21	19,35	19,20

In the rate-based scheduling, the priority is dynamically updated according to the used rate, not as a function of the arrival time of the packets in the router. Therefore, as both flows have the same required rate, the bandwidth is equally divided between the flows, reducing the difference between the F1 and F2 average latency from 42% (when priority-based scheduling is used) to 22%. The jitter is increased when compared to priority-based scheduling because F1 and F2 are not served always in the same order.

Table 3 displays the latency values, jitter and throughput when the packet size is 50 flits and the inserted rate is 30%. As presented in Table 2, both scheduling policies guarantee throughput close to the inserted rate (30%).

Table 3. Results for flows F1 and F2, 50 flits, 30% load.

Performance Figures		Priority-based		Rate-based	
		F1	F2	F1	F2
Latency	**Ideal (ck)**	**100,00**	**100,00**	**100,00**	**100,00**
	Minimum (ck)	141,00	100,00	119,00	119,00
	Average (ck)	**143,82**	**101,44**	**137,31**	**121,94**
	Maximal (ck)	156,00	112,00	184,00	131,00
Jitter (ck)		2,75	2,77	21,20	2,47
Average throughput (%)		28,80	28,81	29,60	28,80

It is possible to observe that average latency and jitter of QoS flows (F1 and F2) presented in Table 3 have similar behavior of Table 2. The reasons for the similar behavior are: (1) F1 and F2 are CBR flows with the same inserted rate (fixed intervals between packets); (2) the total used load by these flows is inferior to 100%, allowing them to be transmitted with the same delay; (3) F1 and F2 are priority flows that only compete between themselves for the same resources. In priority-based scheduling, the disturbing traffic does not interfere the QoS flows due to its lower priority. However, in rate-based scheduling, F1 average latency is slightly reduced when the injection rate has increased. The reason is the amount of conflicts between BE and QoS flows in the shared virtual channel[2], which changes with the injection rate.

---

[2] One virtual channel is reserved for QoS flows, while the second one is shared between QoS and BE flows. QoS flows may use the shared virtual channel when no BE packet is being transmitted.

Figure 4 presents the F1 and F2 average latency when the packet size is 50, 100, 200, and 500 flits. Figure 4(a) shows the behavior of the priority-based scheduling. In this Figure it is possible to observe the difference between the F1 and F2 average latency increases according to the packet size. As mentioned before, the F2 average latency is smaller because F2 source node is closer to the region disputed by the flows. Analyzing the rate-based scheduling presented in Figure 4(b), the F2 average latency is slightly higher when compared to the priority-based scheduling. However, the difference between the F1 and F2 average latency is significantly reduced.

This results shows the superiority of the rate-based scheduling over the priority-based scheduling, allowing to deliver QoS packets with similar latency, independently of the packet size.

(a)
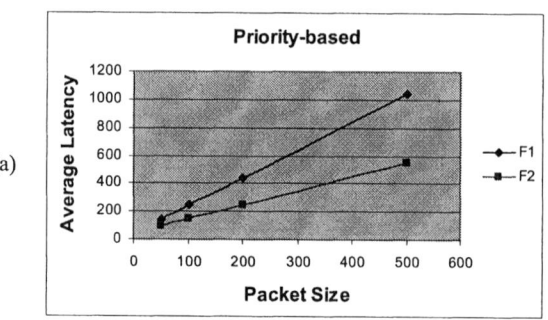

(b)
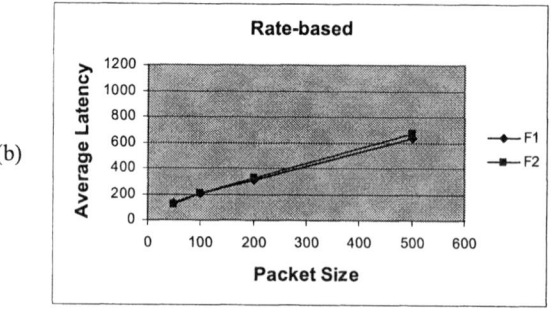

Figure 4. Average latency for F1 and F2 flows, Experiment I, CBR Traffic. (a) priority-based scheduling; (b) rate-based scheduling.

Sample periods are a critical factor in the rate-based method, since they define how frequently priorities are updated. Rate-based results, presented in Table 2 and Table 3, employed short and long sample periods equal to 1000 and 8000 clock cycles, respectively. Small differences (less than 1%) were observed when samples periods were reduced to 250/2000 clock cycles and packet size equals to 50 flits. However, reducing the samples periods to 250/2000 clock cycles and packet size equals to 200 flits increased the overall latency. These results show that a trade-off between packet size and samples period should be established, as discussed in Section 3.

5. CONCLUSIONS AND FUTURE WORK

As discussed, the state of the art in NoCs still does not provide efficient solutions to achieve QoS for applications when the network traffic is not known in advance. The main drawback of circuit switching and priority-based scheduling is the performance unpredictability when QoS flows compete for the same network resources. This paper presented a rate-based scheduling policy, which adjusts the flow priority w.r.t. the required flow rate and current rate used by the flow.

Good results were obtained with CBR flows. When QoS flows with the same priority compete for resources, priority-based scheduling favors the flow that reaches the shared resources first, penalizing the latency of the second arriving flow. Rate-based scheduling overcomes this problem, balancing flows according to their required rates.

As future work it is possible to enumerate: (*i*) evaluating the proposed method with more experiments; (*ii*) evaluating area overhead of the approach, which is expected to be small, because only a small table and few counters were added to the NoC router; (*iii*) implementing congestion control mechanisms; (*iv*) developing services in superior layers to the network layer, allowing to include the requirements specification and to verify if the network is supporting these requirements.

6. ACKNOWLEDGMENTS

This research was supported partially by CNPq (Brazilian Research Agency), project 300774/2006-0.

7. REFERENCES

[1] Di Micheli, G.; Benini, L. *"Networks on Chips: Technology and Tools"*. Morgan Kaufmann, 2006, 304 p.

[2] Arteris. *"Arteris Network on Chip Company"*. 2005. Available at http://www.arteris.net.

[3] Harmanci, M.D.; Escudero, N.P.; Leblebici, Y.; Ienne, P. *"Quantitative Modelling and Comparison of Communication Schemes to Guarantee Quality-of-Service in Networks-on-Chip"*. In: ISCAS, 2005, pp. 1782-1785.

[4] Bertozzi, D.; Benini, L. *"Xpipes: A Network-on-chip Architecture for Gigascale Systems-on-Chip"*. IEEE Circuits and Systems Magazine, v.4(2), 2004, pp. 18-31.

[5] Shin, J.; Lee, D.; Kuo, C.-C. *"Quality of Service for Internet Multimedia"*. Prentice Hall, 2003, 204 p.

[6] Goossens, K.; Dielissen, J.; Radulescu, A. *"Æthereal Network on Chip: Concepts, Architectures, and Implementations"*. IEEE Design and Test of Computers, v.22(5), 2005, pp. 414-421.

[7] Liang, J.; Swaminathan, S.; Tessier, R. *"aSOC: A Scalable, Single-Chip communications Architecture"*. In: IEEE International Conference on Parallel Architectures and Compilation Techniques, 2000, pp. 37-46.

[8] Karim, F.; Nguyen, A.; Dey S. *"An interconnect architecture for network systems on chips"*. IEEE Micro, v.22(5), 2002, pp. 36-45.

[9] Millberg, M.; Nilsson, E.; Thid, R.; Jantsch, A. *"Guaranteed Bandwidth Using Looped Containers in Temporally Disjoint Networks Within the NOSTRUM Network on Chip"*. In: DATE, 2004, pp. 890-895.

[10] Wiklund, D.; Liu D. *"SoCBUS: Switched Network on Chip for Hard Real Time Systems"*. In: IPDPS, 2003, 8p.

[11] Bolotin, E; Cidon, I.; Ginosar R.; Kolodny A. *"QNoC: QoS Architecture and Design Process for Network on Chip"*. Journal of Systems Architecture, v.50(2-3), 2004, pp 105-128.

[12] Véstias, M.; Neto, H. *"A Reconfigurable SoC Platform Based on a Network on Chip Architecture with QoS"*. In: XX DCIS, 2005 , 6 p.

[13] Mello, A.; Tedesco, L.; Calazans, N.; Moraes, F. *"Evaluation of Current QoS Mechanisms in Networks on Chip"*. In: International Symposium on System-on-Chip, v.1, 2006, pp. 115-118.

[14] Dally, W.J.; Towles, B. *"Principles and Practices of Interconnection Networks"*. Morgan Kaufmann Publishers, 2004, 550p.

[15] Andreasson, D.; Kumar, S. *"Improving BE Traffic QoS Using GT Slack in NoC Systems"*. In: NORCHIP, 2005, pp. 44-47.

[16] Kumar, S.; Andreasson, D. *"Slack-Time Aware Routing in NoC Systems"*. In: ISCAS, 2005, pp. 2353-2356.

[17] Giroux, N.; Ganti, S. *"Quality of Service in ATM Networks: State-of-Art Traffic Management"*. Prentice Hall, 1998, 252p.

[18] Moraes, F.; Calazans, N.; Mello, A.; Möller, L.; Ost, L. *"Hermes: an Infrastructure for Low Area Overhead Packet-switching Networks on Chip"*. Integration the VLSI Journal, v.38(1), Oct. 2004, pp. 69-93.

Parallelized Radix-2 Scalable Montgomery Multiplier

Nan Jiang and David Harris

Harvey Mudd College

301 E. Twelfth St. Claremont, CA 91711

{Nan_Jiang, David_Harris}@hmc.edu

Abstract- **This paper describes the FPGA implementation of a parallelized scalable radix-2 Montgomery multiplier. It improves upon previous designs by rearranging previously sequential calculations to take place in parallel. On a Virtex-II FPGA, this design can perform 1024-bit modular exponentiation in 6.3 ms using 6006 lookup tables, a 17% speed improvement over the previously fastest scalable radix-2 Montgomery multiplier.**

I. INTRODUCTION

Modular exponentiation is widely used in modern cryptography algorithms such as RSA and digital signatures. However, the operands of these algorithms usually involve 256 to 2048-bit numbers and the process is time-consuming due to the long divisions necessary in the modulo calculation. Montgomery multiplication transforms this difficult division into a simple bit shift, and is therefore very attractive for hardware implementation.

Since the advent of Montgomery's algorithm in [1], there have been many designs of the Montgomery multiplier that falls into different categories based on their radix. In a simple radix-2 design, an $n \times n$-bit multiplication is performed using n steps. The multiplier kernel contains processing elements (PEs) that acts on one bit of the multiplier and all n bits of the multiplicand. These designs are hardwired to support only one choice of n. A *scalable* radix-2 design described in [2, 3] breaks the n-bit multiplicand into w-bit words. The kernel of the designs contain p PEs organized in a systolic array. Each PE handles one bit of the multiplier and w bits of the multiplicand at a time. The kernel iterates until the entire multiplication completes. These designs are highly flexible; they can be configured to handle any n. The overall advantage of a radix-2 Montgomery multiplier design is its hardware simplicity. The $1 \times n$-bit or $1 \times w$-bit multiplication involved in these designs can be accomplished with an n-bit AND gate or a multiplexer, and the number of registers required to store intermediate values is minimal. However, the hardware efficiency comes at the cost of large number of iterations through the kernel.

The critical path in a standard Montgomery multiplier involves two dependent multiplications steps. Orup showed how to reorder the steps so that the multiplications can take place in parallel [7]. We have successfully applied the technique to very high radix Montgomery multipliers [4, 5, 6] to shorten the critical path. Our goal in this paper is to apply

the parallel modification of the Montgomery algorithm to the improved scalable radix-2 multiplier implemented in [3]. By parallelizing the existing radix-2 designs, we achieve a significant performance boost without increasing hardware cost.

II. BACKGROUND

The basic Montgomery Multiplication algorithm is

$$Z = (XYR^{-1}) \bmod M \qquad (1)$$

with the notations

X:	n-bit multiplier
Y:	n-bit multiplicand
M:	n-bit odd modulus, typically prime
R:	2^n
R^{-1}:	modular multiplicative inverse of R satisfying $(RR^{-1}) \bmod M = 1$
M':	n-bit integer satisfying $RR^{-1} - MM' = 1$

Montgomery in [1] showed how to perform this multiplication without dividing by M:

Multiply:	$Z = X \times Y$
Quotient:	$Q = Z \times M' \bmod R$
Result :	$Z = [Z + Q \times M] / R$

The Q term has the property such that it forces the numerator of the Result step to be divisible by R, simplifying the division to a shift.

A. Improved Radix-2 Design

The radix-2 design in [3] will be the basis of this paper's radix-2 implementation. The design follows Tenca-Koç's multiple word radix-2 Montgomery multiplication algorithm from [2] shown in Figure 1.

n:	size of operands
R:	2^n
M:	n-bit odd modulus
R^{-1}:	modular multiplicative inverse of R satisfying (RR^{-1}) mod $M = 1$

M': n-bit integer satisfying $(RR^{-1}-MM') = 1$
X: n-bit muliplier
Y: n-bit multiplicand

$Z = 0$
for $i = 0$ to n-1
    $Z = Z + X^i \times Y$
    if Z is *odd* then $Z = Z + M$
    $Z = Z/2$

Figure 1: Tenca-Koç's radix-2 Montgomery multiplication algorithm

Each iteration of the i loop is called a kernel cycle, and the clock delay between successive PEs is called a PE cycle. The Tenca-Koç design [2] requires a delay of two clock cycles between PEs, because at the end of a PE the resulting word of Z needs to be right shifted to account of the division by 2. Before this shift can occur, the next word of Z needs to be calculated so that the least significant bit of the next word of Z can be shifted to become the most significant bit of the current word of Z. Our improved design [3] is able to avoid right shifting of Z by left shifting Y and M, as shown Figure 2. Each PE no longer has to wait for the next word of Z to be computed before producing the correct Z output. Thus, the latency between PEs drops to only one clock cycle.

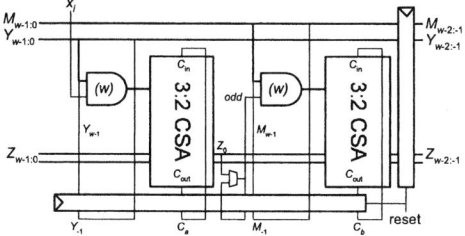

Figure 2: PE diagram for improved radix-2 design from [3]

B. Parallelized Algorithm

In [5], an alternative implementation of the Montgomery algorithm based on [7] is discussed. The parallel algorithm reorganizes the original algorithm to produce a new pre-calculated value $\hat{M}$ that allows the original algorithm's Multiply and Result steps' multiplications to occur simultaneously. After parallelizing, the basic algorithm is shown in Figure 3

$\hat{M}$: $((M' \bmod 2)M +1)/2$

$Z = 0$
for $i = 0$ to n
    $Q = Z \bmod 2$
    $Z = Z/2 + Q \times \hat{M} + X^i \times Y$

Figure 3: Parallelized Montgomery algorithm

There are two side effects caused by parallelizing the original algorithm. The first is that the result, Z, has increased

in size by 1 bit. This is because division by 2 is no longer the final operation of the algorithm. This side effect requires additional calculations to normalize the result back to the expected range. However, normalization can be performed at the end of a modular exponentiation after many iterations of Montgomery multiplication. The second side effect is that this design requires one more kernel cycle than the non-parallel algorithm to compensate for the loop reordering.

III. PARALLELIZED RADIX-2

The Montgomery algorithm implemented for this paper is a hybrid between the improved radix-2 and the parallel algorithm. The basic algorithm is nearly identical to the parallel very high radix algorithm in Figure 3. The features of the improved radix-2 algorithm are introduced by left shifting $\hat{M}$ and Y at each PE. The resulting algorithm takes advantage of both simultaneous multiplication and one-cycle latency between PEs.

The general algorithm can be recast in scalable form by splitting $\hat{M}$ and Y into w-bit words. Each PE has to run multiple times during a kernel cycle to process all bits of $\hat{M}$ and Y. Thus, an inner for loop iterates over the n/w words of $\hat{M}$ and Y. In addition, another iteration of the inner loop is necessary to process the left shifted bits of $\hat{M}$ and Y. Furthermore, as mentioned previously, one side effect of the parallel algorithm is that the result could be larger than expected. Thus the inner loop iterations is increase by one to account for the additional PE iteration required. The resulting scalable algorithm is shown in Figure 4.

w: multiplicand word length
e: $\lceil n/w \rceil$ + 2 PE iterations per kernel
C: 1-bit carry digit

$Z = 0$
$Q = 0$
for $i = 0$ to n
    $C = 0$
    $Q = Z^0 \bmod 2$
    for $j = 0$ to e-1
        $(C, Z^{j+1}) = Z^j + Q \times \hat{M}^j +$
                $X^i \times Y^j + C$

Figure 4: Scalable parallel Radix-2 algorithm

IV. HARDWARE IMPLEMENTATION

The overall hardware architecture of the parallel radix-2 multiplier is similar to those presented in [3, 4, 5, 6]. Figure 5 provides the overview architecture of a scalable Montgomery multiplier using p PEs. Every PE receives one bit of X and Q, and w bits of $\hat{M}$, Y, and Z on each step. In one kernel cycle, p digits of X are processed against all n bits of $\hat{M}$ and Y'. Hence, $k = \lceil n/p \rceil + 1$ full kernel cycles are necessary to process all the bits of X and satisfy the additional kernel cycle requirement of parallel algorithm. As results emerge from the last PE of the kernel, they are either stored in a FIFO until the first PE has finished its kernel cycle or bypassed directly to the input of the first PE.

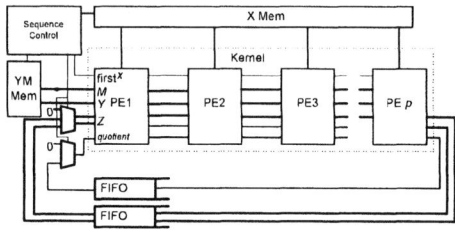

Figure 5: Parallel radix-2 hardware diagram

A. Processing Element

Figure 6 shows a processing element for the parallel radix 2 design. Compared to Figure 2, the two AND-multipliers are placed in parallel and the 2 input multiplexer is eliminated. The left shifting of $\hat{M}$ and Y are achieved using the two delay registers in the design. At each PE, the most significant bit of a word of $\hat{M}$ and Y are delayed to become the least significant bit of the next word of $\hat{M}$ and Y. After $\hat{M}$ and Y have passed through p PEs, they are in effect left shifted by p bits. The PE diagrams also show that the change from improved radix-2 design to parallel radix-2 design eliminates an AND and a multiplexer from the critical path without increasing the hardware cost.

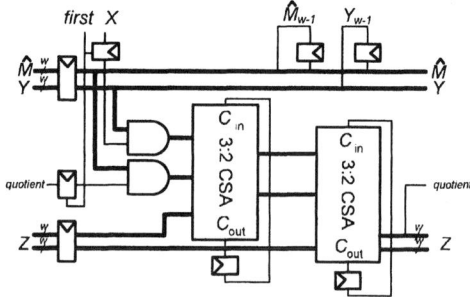

Figure 6: Parallel Radix-2 PE diagram

B. Latency

The latency of the parallelized radix-2 design is similar to that of [3]. One PE cycle consists of only one clock cycle due to the left shifting of $\hat{M}$ and Y. Each PE multiplies one bit of X with w bits of Y and $\hat{M}$. When a PE has processed all the bits of $\hat{M}$ and Y, a kernel cycle has completed. It will then wait for a new set of X bits to start the cycle all over again. Modification introduced by the parallel algorithm shows no visible effect on the latency graph in Figure 7. However, e and k are increased by 1.

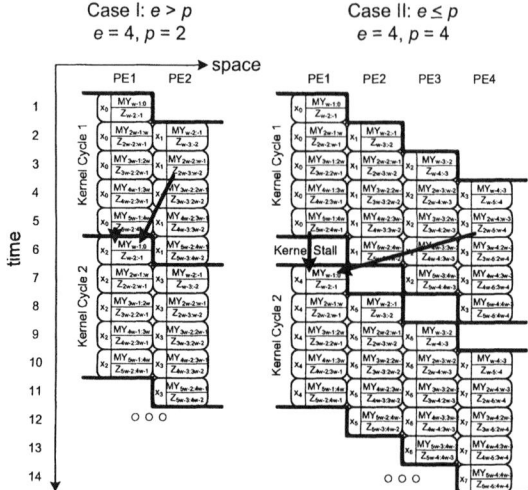

Figure 7: Parallel radix-2 kernel latency

An entire multiplication using p PEs takes k kernel cycles to complete. When the first PE has finished a kernel cycle, it cannot begin the next kernel cycle until the last PE has completed the first word of Z. The latency of a kernel cycle depends on e and p. Case I corresponds to a large number of PE cycles, e, relative to the number of processing elements, p. In this situation, when the first PE has finished its kernel cycle, the first word of Z from the last PE is already waiting in the FIFO, there is no stall between kernel cycles, and the kernel hardware is used with maximal efficiency. Case II corresponds to a large number of processing elements relative to the number of PE cycles. As shown in Figure 7, the first PE must wait until the last PE finishes calculating the first word of Z. Therefore, Case I occurs when $e > p$ and Case II occurs when $e < p+1$.

Case I: The first PE is used continuously e times per kernel cycle for k full kernel cycles. Therefore the total delay is

$$D_I = ke \qquad (2)$$

Case II: Each kernel cycle takes $p+p/w$ clock cycles until the first word of Z is ready, plus 1 to bypass the result back to the first PE. The p/w term is caused by the left shifting of $\hat{M}$ and Y. After passing p PEs, p zeros have been shifted into the least significant bits of $\hat{M}$ and Y which are ignored, causing p/w cycles of delay. Therefore the total delay Case II is

$$D_{II} = k(p+p/w+1) \qquad (3)$$

Rewriting these delays in terms of the design parameters n, w, and p, and assuming integer divisibility, we obtain

TABLE 1: SYNTHESIS PERFORMANCE COMPARISON OF VARIOUS
PROCESSING ELEMENTS

Architecture	Reference	w	4-input LUTs / PE	Registers / PE	Critical Path	PE Clock Speed (MHz)
Parallel Scalable Radix 2	This work	16	94	72	AND + 2CSA + REG	403
Improved Scalable Radix 2	[3]	16	97	72	2AND + 2CSA + BUF + MUX + REG	318
Tenca-Koç Scalable Radix 2	[2]	16	97	72	2AND + 2CSA + BUF + MUX + REG	318

$$D_I = \frac{n^2}{pw} + \frac{n}{p} + \frac{n}{w} \qquad (4)$$

$$D_{II} = n + p + \frac{n+p}{w} + \frac{n}{p} + 1 \qquad (5)$$

V. RESULTS

The parallel radix-2 Montgomery multiplier design described previously was implemented using Verilog. The design was synthesized using Synplify Pro and compared to the synthesis result of previous designs. All synthesis results were produced by targeting the Xilinx Virtext II XC2V2000 speed grade -6 FPGA with "sequential optimizations" disabled [8] to prevent flip-flops from being turned into shift registers. Table 1 shows the synthesis result of a single process element for several radix-2 designs. The result matches our expectation that the parallel algorithm removed an AND and a multiplexer from the PE critical path. As a result, the PE of a parallel radix-2 design achieved a 26% clock speed increase. In addition, because several gates were removed from the parallel radix-2 PE, there is a hardware decrease when compared to previous designs. Thus, the parallel radix-2 design is both faster and smaller than previous designs.

Table 2 compares overall system performance of the new parallel design to the other scalable radix-2 designs. For the kernel synthesis, the frequency of the kernels is significantly lower than the PE synthesis shown in Tables 1. This difference is caused by the interconnect delay estimated by Synplify Pro which can be minimized in a real implementation of the Montgomery multipliers through datapath floorplanning.

fastest radix-2 multiplier. In the 1024-bit multiplication using 16 PEs, it can perform a multiplication 17% faster than the improved radix-2 design and significantly faster than the improved radix-2 design. In addition, the hardware requirement of the parallel radix-2 multiplier is approximately the same as other designs, justifying our claim that this design provides a performance increase without additional hardware costs.

It is interesting to note that as the number of PEs in the kernel increase, the performance benefit of the parallel radix-2 design decrease. This effect is caused by the fact parallel algorithm requires an addition kernel cycle than traditional radix-2 Montgomery multipliers. Thus, as the number of PE increases, this performance overhead for parallel radix-2 is increased as well. It is possible to increase the number of PEs so much that the parallel radix-2 design would actually become slower than the traditional designs despite of running at higher clock frequencies.

VI. CONCLUSION

In this paper, we have demonstrated a novel approach to radix-2 scalable Montgomery multipliers by reordering the steps to perform multiplications in parallel. The design is both faster and smaller than previous radix-2 Montgomery multipliers. It provides a significant cycle time improvement at the cost of a small increase in cycle count. In simulation, the parallel radix-2 design was able to provide a 17% speed increase over the previous designs for a 1024-bit multiplication using 16 PEs.

TABLE 2: SYNTHESIS PERFORMANCE COMPARISON OF VARIOUS
MONTGOMERY MULTIPLIERS

From Table 2, we see that the parallel radix-2 design is the

ACKNOWLEDGEMENT

Description	Ref	Tech	w	v	p	LUTs	REGs	16×16 MULT	N	T_{mult} (ms)
Parallel Scalable radix 2	This work	Xilinx XC2V2000-06	16	1	16	1575	1189	0	256	0.41
									1024	21.8
					64	6006	4597	0	256	0.52
									1024	6.3
Improved scalable radix 2	[3]	Xilinx XC2V2000-06	16	1	16	1564	1202	0	256	0.49
									1024	27.2
					64	5932	4705	0	256	0.56
									1024	7.6
Tenca-Koç Scalable radix 2	[2]	0.5 µm CMOS	8	1	40	28 kgates		0	256	1.6
									1024	37
		Xilinx XC2V2000-06	8	1	40	3902	2937	0	256	1.0
									1024	15

The authors would like to thank the Clay-Wolkin Family Foundation fellowship as well as Intel Circuit Research Lab for funding this research project.

REFERENCES

[1] P. Montgomery, "Modular multiplication without trial division," *Math. Of Computation*, vol. 44, no. 170, pp. 519-521, April 1985.

[2] A. Tenca and Ç. Koç, "A scalable architecture for modular multiplication based on Montgomery's algorithm," *IEEE Trans. Computers*, vol. 52, no.9, pp. 1215-1221, Sept. 2003.

[3] D. Harris *et al.*, "An improved unified scalable radix-2 Montgomery multiplier", *IEEE Symp. Computer Arithmetic*, pp. 172-178, 2005.

[4] N. Jiang and D. Harris, "Quotient piplelined very high radix scalable Montgomery multipliers", *Proc. Asilomar Conf. Signal, Systems , and Computers 2006*

[5] K. Kelly and D. Harris, "Parallelized very high radix scalable Montgomery multipliers," *Proc. Asilomar Conf. Signals, Systems, and Computers*, pp. 1196-1200, 2005.

[6] K. Kelley and D. Harris, "Very high radix scalable Montgomery multipliers", *IEEE IWSOC Conference*, pp. 400-404, July 2005.

[7] H. Orup, "Simplifying quotient determination in high-radix modular multiplication," *Proc. 12th IEEE Symp. Computer Arithmetic*, pp. 193-199, 1995.

[8] Xilinx, Virtex-II Pro and Virtex-II Pro X Platform FPGAs Datasheet, June 30, 2004, www.xilinx.com

An efficient Heterogeneous Reconfigurable Functional Unit for an Adaptive Dynamic Extensible Processor

Arash Mehdizadeh   Behnam Ghavami   Morteza Saheb Zamani   Hossein Pedram   Farhad Mehdipour*

{ a_mehdizadeh, ghavami, szamani, pedram}@ aut.ac.ir, *farhad@c.csce.kyushu-u.ac.jp

Computer Engineering Department, Amirkabir University of Technology (Tehran Polytechnic)

424 Hafez Ave, Tehran 15785, Iran

*Computing and Communication Center, Kyushu University

3-8-33-309 Momochihama, Sawara-ku, Fukuoka 814-0001, Japan

ABSTRACT

Replacing functional units of an extensible processor with reconfigurable functional units enhances performance and flexibility of processors to execute custom instructions. That is due to the ability of reconfigurable functional units to perform computations in hardware to increase performance, while retaining much of the flexibility of a software solution. In this paper, we develop a heterogeneous architecture for the reconfigurable functional unit of an extensible processor. To verify the efficiency of our architecture, we applied it to 8 applications of Mibench. Our experiments show that compared to the similar architectures, ours supports a wide range of custom instructions. In addition, use of the new architecture improves execution time of custom instructions by 20% to 30% on average. Moreover, compared with the previous architecture, area is reduced by 15%.

Keywords: Custom Instruction, Extensible Processor, Reconfigurable Functional Unit.

1. Introduction

Embedded systems, having proven their abilities in a wide range of applications, are extensively used in communications and consumer products. Regarding prevalence of these systems, different methods, such as general purpose processors (GPPs) and application specific integrated circuits (ASIC), have been adapted to implement them. Although the use of GPPs, as a usual approach of implementing embedded systems results in high flexibility, because of their inefficiency in performance and power consumption, they are not widely applicable. As a result, ASICs have been proposed. Deep submicron issues of interconnect delay and signal integrity have significantly increased design costs of ASICs both due to the higher engineering costs resulting from longer design cycles and increasing cost of design tools [1]. Another recent approach of embedded systems implementation, and in fact a way to fill the gap between GPP and ASIC era, is the use of custom hardware in special applications. In such way, even instructions could be customized. Application-specific instruction-set processors (ASIPs) have been an important design and implementation methodology for system-on-chip processors in the last decade. Compared to GPPs, ASIPs have more potential to meet high-performance demands of embedded applications. However, synthesis of ASIPs traditionally involved the generation of a complete instruction set architecture for the target application. On the other hand, GPPs are very flexible but may not offer the necessary performance. Hence, as a complement to the approach of ASIPs, processors with extensible instruction sets have been introduced. The important motivation toward specialization of existing processors versus the design of complete ASIPs is to avoid the complexity of a complete processor and toolset development. In these systems, a core collaborates with a reconfigurable functional unit (which can be implemented either coarsely or finely). Even after the design and implementation of the instruction set architecture of such systems, custom instructions (CIs) can be added to the system. These instructions are extracted regarding hot basic blocks (HBB). A basic block is a sequence of instructions which is ended with a control instruction. HBBs are referred to the basic blocks which are repeated more than a threshold number of times during the execution of a certain program. With such definition, critical sections of programs are extracted as data flow graphs (DFG), mapped and executed on a hardware accelerator or a functional unit (FU) bound to the main core.

In this paper, A new tightly coupled fast-interconnected reconfigurable functional unit (RFU) is presented for the previously introduced Adaptive dynaMic extensiBlE processoR (AMBER) [2]. Enhancing AMBER's functionality, we reduced critical path delay of the RFU by replacing collections of individual identical FUs with some other non-identical ones.

The rest of paper is organized as follows: In Section 2 a background of systems with reconfigurable functional units will be given. In Section 3 and Section 4 AMBER processor, which is used as a basis for our implementations, is introduced continued with a proposed structure for the RFU in Section 5. The mechanics of DFG clustering and their mapping on the RFU are presented in Section 6. In Section 7, binding of the RFU to the main processor is discussed. The issue of configuration memory is discussed in Section 8. In Section 9, experimental results are reported and finally, the paper is concluded with some proposals for the future works.

2. Related Work

Recently design and implementation of extensible processors FUs has been much of concern in numerous papers. Programmable accelerators augmenting to a base processor fall in two categories based on the granularity of their structure, fine grain and coarse grain. Fine grain accelerators are suitable for very flexible computations. However, long latency and slow reconfiguration time are two of the most important drawbacks associated with these systems. They also need a large amount of memory for storing the configuration bits. To compensate the computational inefficiency and configuration latency most of them deal with very large sub-graphs. Some of the fine grained hardware accelerators are introduced in [3][12].

151

978-1-4244-1709-4/07/$25.00 © 2007 IEEE

Chimaera [4], OneChip [5] and XiRisc [6] are some instances of fine grain programmable hardware integrated with GPPs. ADRES [7] is a counterpart of the formers with a coarse grain structure.

The number of inputs/outputs and integration method of accelerator and base processor differ for each design. For example, PRISC uses an RFU with two inputs and one output, while RFU of Chimaera has nine inputs and one output.

Accelerators are divided into two general categories as loosely coupled and tightly coupled. A loosely coupled accelerator plays the role of a co-processor which helps balancing of the load on the main processor and itself. Use of these accelerators calls for exclusive compilers and refinement of portions of the opcode [10][11]. In loosely coupled systems like MorphoSys [8] and Garp [9] there is an overhead for transferring data between the base processor and the coprocessor. In contrast, use of tightly coupled accelerators does not require any overhead in information transfer. Further more, there is no need to worry about an individual compiler or refinement of opcode.

In [2][3] an extensible processor named AMBER is introduced which utilizes a tightly coupled coarse grain RFU. In AMBER, there is no need for a new programming model, compiler, opcode for new instructions, source code modification or recompilation. The user just runs the applications on the base processor then generation of custom instructions and handling their execution are done transparently and automatically. The main concern in [2][3] was to cover as much CIs as possible or in other words has a coverage percentile as close to as 100%. We further enhanced this structure by introducing a new heterogeneous architecture which reduces critical path delay and configuration bits while increasing CI coverage with no penalty in area or total wire length. AMBER architecture is introduced in the following section.

3. Overview of AMBER

AMBER is an extensible processor which can be utilized in different applications of embedded systems. It consists of a microprocessor, profiler, RFU, and a scheduler. The base processor is a 4-issue in-order RISC processor that supports MIPS instruction set. Figure 1 demonstrates AMBER components that will be elaborated in the following sections.

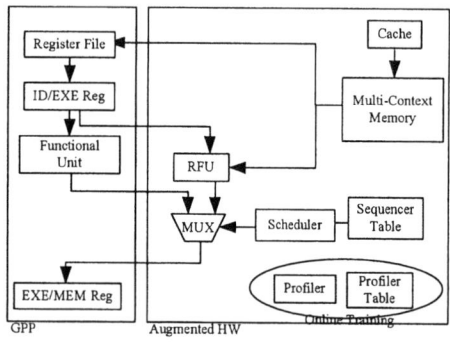

Figure 1. Components of AMBER

3.1 Profiler

AMBER has two modes of operation: Training and Normal. In the training mode, applications are profiled to extract HBBs. Then the object code is used to extract the configuration bits. Training can be done either dynamically or statically. In the

former case, there is a sheer need for extra hardware to perform profiling (profiler). In addition, all elaboration functions such as HBB recognition and CI generation are done on the main processor. On the other hand, in the latter case, a host computer simulates and profiles the programs prior to their execution on AMBER. The host computer works independently from AMBER. However, dynamic profiling can be done during intermittent idle periods of the main processor. In this case, generation of CIs does not interfere with the main tasks of the processor.

Profiler contains the following components: 1) two registers one for the previous program counter (PC) and the other for current PC, 2) a comparator to compare values of the two registers; and 3) a table to store the start addresses of HBBs and their execution frequency. In every clock cycle, the profiler compares values of the two aforementioned registers. If the difference of these two values is not equal to the instruction length, a taken branch or jump has occurred. The profiler has a table with a counter for each entry that keeps the execution frequency of basic blocks. In the case of a taken branch/jump, the profiler's table is checked. If the target address (the current PC) is in the table, the corresponding counter is incremented, otherwise current PC is added as a new entry and its counter is initialized to one. Using the profiler's table and a predefined threshold value, the start addresses of HBBs are detected according to their frequency of occurrence [3].

3.2 Reconfigurable Functional Unit (RFU)

Portions of applications suitable for acceleration are the ones which are executed frequently. These portions can be executed on a reconfigurable core in AMBER that is the RFU. The RFU is a matrix of FUs. As it has been also mentioned in [15], according to the processor's size of data, a matrix of FUs seems an efficient and reasonable hardware for accelerating sub-dataflow graphs as CIs. Exploiting such core can increase the execution speed dramatically [2]. Hence, promisingly, increasing the number of CIs (mappable on the RFU) can decrease the application runtime.

3.3 Scheduler

Scheduler is responsible to decide whether operations to be executed on the microprocessor or on the RFU. This unit contains a table in which the starting address of each CI (in the reconfigurable memory unit) along with the required clock cycles for execution of the instruction is stored. This table is given the values based on the starting address of each CI in the object code. During the execution, as soon as the scheduler observes the PC equal to one of the entries in this table, FU of the microprocessor halts and the RFU takes the responsibility to execute the CI. In this situation, the scheduler waits for completion of the CI, and then it sets value of the PC according to the length of the recently executed CI.

4. An Architecture for the RFU: A Quantitative Approach

The quantitative flow in Figure 2 was applied to 20 applications of Mibench [13] to identify suitable CIs. To do it, Simplescalar [14] is used as the simulation tool and it is modified to keep track of taken branches and jumps. The trace file is employed as input by the profiler to detect beginning of HBBs [3]. Then a DFG is generated for each HBB and passed to the CI generator tool. The CI generator makes CIs. Mapping tool receives CIs and maps them on the RFU. Results of the mapping tool lead to the RFU architecture. To reduce implementation overhead and increase efficiency, two primary constraints are considered for CIs: a) supporting only fixed-point instructions excluding multiply, divide and load, b) including at most one store and at most one control instruction in a CI. Multiply and divide were excluded due to their low execution frequency and large area occupancy.

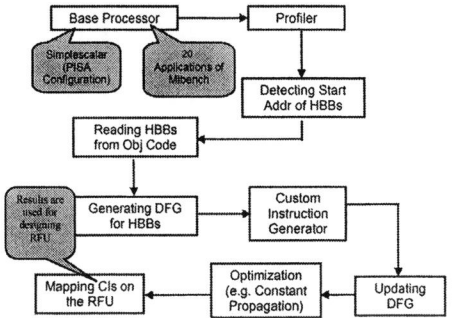

Figure 2. Quantitative approach flow

In order to justify the use of heterogeneous functional units, in this section an overview of the previous work in [3] is presented firstly then our proposed architecture is introduced.

4.1 Homogeneous Architecture of RFU

To determine the proper number of inputs and outputs, first it is assumed that all DFGs corresponding to the extracted CIs can be mapped on the architecture. Based on the analyses of CIs in 20 test applications of Mibench, concerning higher mapping rate as well as less consumption of resources, proper numbers of inputs and outputs were proved to be 8 and 6 respectively [3] (Figure 3). This means that having an RFU with 8 inputs and 6 outputs, nearly 100% of CIs with 8 inputs and 6 outputs are mappable on such structure (mapping rate is almost equal to 100%).

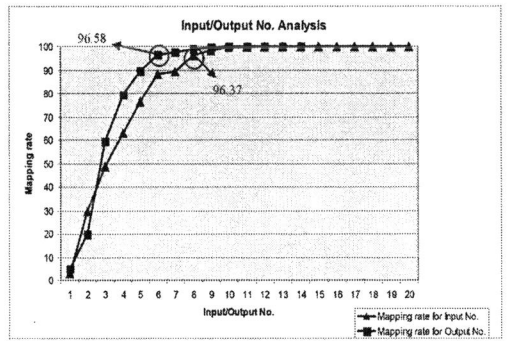

Figure 3. Mapping rate for different numbers of I/O

In the next step there should be an appraisal on the total number of Functional Units (FUs) in the RFU. This is to acquire a high mapping rate using as few FUs as possible. Based on the observations (Figure 4), provided that the limitation put on the number of inputs and outputs is considered, mapping rate curve levels out around the number 16. Hence the minimum proper number of FUs is 16.

Number of inputs, outputs and FUs being determined, to preserve the high mapping rate it is assumed that all the FUs are identical and each is able to implement an individual function per configuration. Based on this assumption other analyses were performed to determine the dept and number of FUs for every row.

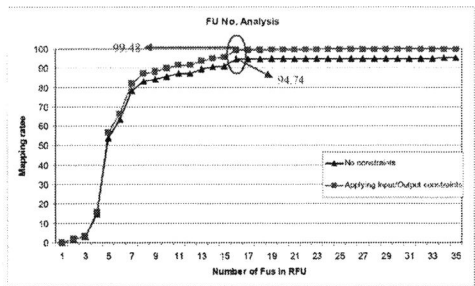

Figure 4. Mapping rate for different numbers of FUs

Aforementioned observations resulted in the architecture depicted in Figure 5. In this architecture, there are 16 identical FUs which implement a single function per configuration based on their configuration bits. Interconnection of these 16 units is established through multiplexers programmed by configuration bits other than the ones associated with the configuration of FUs.

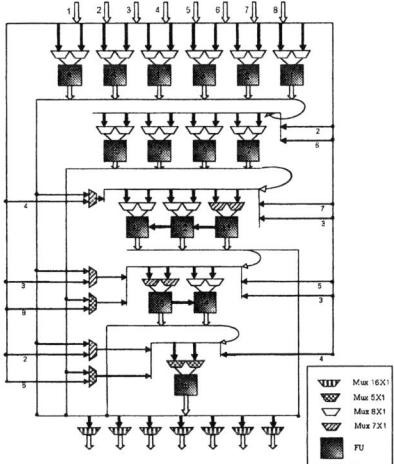

Figure 5. Proposed architecture of RFU using identical FUs

5. New Heterogeneous Architecture of RFU

One of the most important aspects of RFU construction which did not receive much of concern in the similar works is the length of critical path and its effect on performance. Critical path is referred to the path incorporating maximum number of active sequential FUs from one input to one output. Frequent occurrence of CIs inherently calls for reduction of CIs execution time. Consider a mapped CI which requires two clock cycles to be accomplished on the RFU. CI execution time is equal to the integer number of clock cycles multiplied by the length of a cycle. This time directly depends on the critical path delay. If the specified CI is to be repeated for 2000 times in the course of the whole program execution, reducing its run time to one clock cycle will make the whole program execution time 2000 cycles shorter which is a considerable gain in comparison to the former case. Shortening CIs execution time calls for reduction of critical path delay.

Based on these premises, experiments have been conducted on the relativity of delay of different components of RFU to the overall RFU critical path delay. Results show that delay associated with the multiplexers being used for interconnection of FUs constitutes a grate portion of the overall delay. This is shown in Table 1 (RFU's circuit is synthesized with TSMC 0.18u technology). Intuitively, reducing number of multiplexers affiliated with the structure of a mapped CI on the RFU will reduce the critical path dramatically.

Table 1. Critical path delays of different component of RFU in TSMC 0.18u technology.

Unit	delay of critical path (ns)	Unit	delay of critical path (ns)
mux3-1	1.16	mux7-1	1.7
mux 4-1	1.24	mux8-1	1.86
mux5-1	1.47	FU	6.94
mux6-1	1.52		

Based on this assumption, in order to reduce the number of multiplexers in the critical path without affecting the mapping rate, we proposed an RFU comprising non-identical FUs. This means, FUs of this new architecture are able to execute multiple instructions (one, two or three) with every regular consecution in a DFG structure (based on the FU's structure). We define a regular DFG as the one in which output of every node is not connected to more than one node. Analyses over 20 applications of Mibench show that almost 97% of DFGs have this attribute. Considering these regular DFGs structures, we proposed FUs which can implement every consecutive two and three-instruction sets representing sub-data flow graphs (sub-DFG) depicted in Figure 6. We referred these FUs as bi- and tri-instruction FUs respectively.

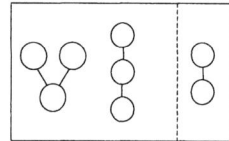

Figure 6. Regular executable sub-DFGs on tri and bi-instruction FUs (left to right)

By replacing a number of previous uni-FUs and multiplexers of RFU with bi/tri-FUs a considerable reduction in critical path delay of mapped CIs can be resulted as is shown in Table 2.

Table 2. Replacing a number of uni-FUs and multiplexers with bi/tri-FUs (Synthesized in TSMC 0.18u technology)

Units	Critical path delay (ns)	Replaced Unit	Critical path delay (ns)
2 uni-FU and 1 Mux	14.47	bi-FU	10.7
3 uni-FU and 2 Mux	23.8	tri-FU	15.9

To determine the structure of RFU, first we must indicate the proper numbers for RFU inputs and outputs. We inspected mapping rate of generated CIs of 20 applications of Mibench on the RFU. Then, we mapped our generated CIs on the RFU without considering any constraints. By examining the mapping rate for different numbers of inputs and outputs we tried to choose proper numbers. According to the results depicted in Figure 3, eight and six are good candidates for input and output numbers, respectively.

We also conducted an analysis to determine the number of uni/bi and tri-instruction FUs. According to Figure 3, putting constraints on the number of inputs and outputs, 85% of CIs contain less than 9 nodes. Hence, to have a more reliable analysis, we observed precisely all the regular DFGs consisting less than 9 nodes with different topologies (multiplicities of these DFGs are shown in Table 3). Moreover for the analysis

of CIs with more than 8 instructions we used mapping rate frequency by which we mean the percentage of generated CIs for 20 applications of Mibench that can be mapped on the RFU.

Table 3. Possible numbers of regular DFGs with different numbers of nodes

Num of DFG nodes	Num of regular DFGs	Num of DFG nodes	Num of regular DFGs
1	1	5	6
2	1	6	11
3	2	7	23
4	3	8	46

Many different architectures and configurations considering the mapping rate results were examined. Based on the conducted analyses the RFU architecture depicted in Figure7 is introduced.

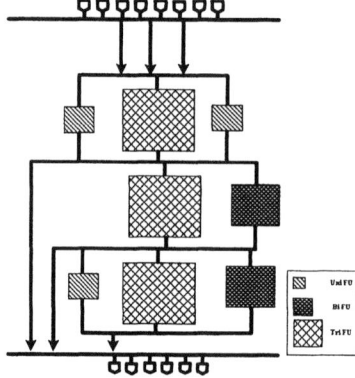

Figure 7. The first proposed heterogeneous architecture of RFU

In this architecture when an input data is needed by FUs located in rows other than first row or when the output of one row is used by FUs placed in a non-subsequent row, move instruction are mapped on the intermediate FUs to pass over the data. Assuming these limitations the mapping rate decreases to 84.72%. To improve the mapping rate, many different architectures and configurations considering the mapping rate results were examined. We reached to the architecture illustrated in Figure 8.

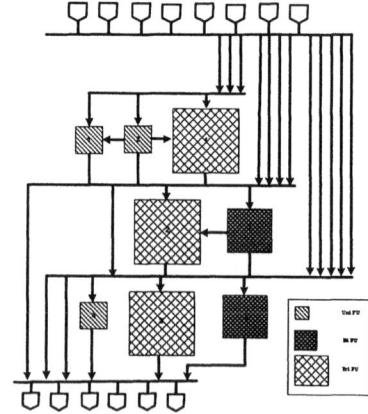

Figure 8. The proposed fast interconnected heterogeneous architecture of RFU

To facilitate data accesses for FUs and reduce the inserted move instructions (which occupy FUs), besides the connections that exist from outputs of each row to the inputs of subsequent row, ten other longer connections were added. One longer connection is: from outputs of row 1 to inputs of row 3 and other connections are: from main inputs of RFU to the inputs of row 2 and 3. In rows 1 and 2 unidirectional connections to the neighbor FUs were added. These three connections support those long CIs that do not have much parallelism but their operations are very dependent. Using these connections CIs with lengths less than 13 nodes can be supported by the RFU.

This RFU is able to map all CIs consisting of less than 9 nodes and 78% of CIs 9 to 16 nodes long. As a result the mapping rate becomes 95.31%. Experiments show that each FU of the RFU does not need to support all of microprocessor supported instructions. We defined three types of instructions: logical (type 1), add/sub/compare (type 2) and shift operations (type 3). Distribution and multiplicity of each type for each row are given in Table 4.

Table 4. Type of functions for each FU

Row Number	uni-FU	uni or bi-FU	tri-FU
1	Type 1,2	Type 3	Type 1,2,3
2	-	Type 2	Type 1,2,3
3	Type 1,2	Type 1,2,3	Type1,2

Enforcing all of constraints, the mapping rate becomes 94.86% which is almost 6% better than the previous architecture [2]. In addition regarding the homogeneous architecture of the RFU proposed in [2] (Figure 5), DFGs longer than 8 nodes are not mappable. This limitation is improved to DFGs 12 nodes long in our proposed architecture. Each CI configuration needs 287 bits for storing control signals and 201 bits for immediate values. Therefore, configuration of each CI on the RFU requires 488 bits totally compared to 512 bits for the RFU proposed in [3].

6. DFG Clustering

Regarding difference of FUs and their varying ability to implement certain sub-DFGs, clustering of DFGs requires certain considerations to be mappable on the RFU. For optimum utilization we used a greedy algorithm to cluster CIs, thus first the biggest branch of the graph is clustered, then a new sub-graph is established. The same step will be repeated on the sub-graph based on the remaining resources in the RFU. Optimum clustering is the most of concern in this algorithm. As efficiency of clustering is affected by the limit on the number of resources, architecture of the RFU is designed based on the mostly repeated patterns in DFGs. In other words, for establishment of interconnections, the structures which are more frequent are noticed. As a result our structure will inherently be optimum for our way of clustering. On the other hand, for less frequent structures, clustering will be affected by the limitations of interconnection network which will deviate the result from optimality. Figure 9 illustrates an example of DFG clustering and its corresponding CI mapping on the RFU.

In order to obtain better results we applied a new mechanism called "Structure Profiling" to map CIs less than 9 nodes long. In our survey, all regular structures of DFGs with less than 9 nodes were identified then traversed bottom up. So for all of these DFGs we generated a configuration profile stored in the configuration memory. During CI mapping every new DFG is traversed then compared with the previously profiled ones. If one of the stored profiles is identical to the corresponding

DFG, it will be used to map the CI on the RFU. Every configuration profile is stored in the mapping tool.

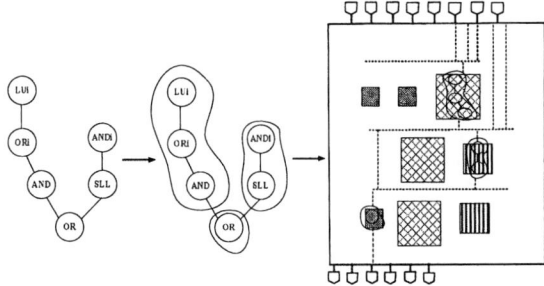

Figure 9. An example of DFG clustering and CI mapping on the RFU

7. RFU and Main Processor Connection

The core processor of AMBER is a 4-issue in-order RISC microprocessor. Connection of this processor to the RFU is shown in Figure 10.

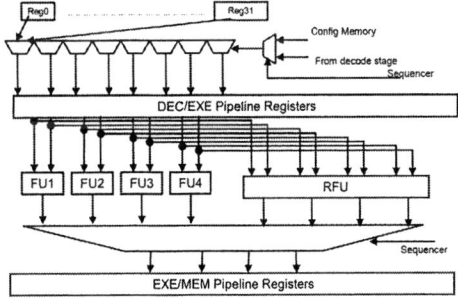

Figure 10. Integrating the RFU and the core processor

As it is shown, in/output ports of the microprocessor and the RFU are common. This way of connection eliminates the need for extra lines to separate intputs/outputs while eliminating parallelism of these two units as well. As a large portion of application code is executed on the RFU, inherently there is no opportunity for parallelism which in turn reduces the need for accommodating exclusive in/output lines. As stated before, the RFU has 6 outputs; hence, two registers are added to the RFU structure which will store extra outputs (more than 4). Contents of these registers are written to the register file in the next cycle.

8. Configuration Memory

As computed in Section 6, each CI configuration needs 488 bits. As a result for an application such as rijndael with 117 CIs we need around 6.96 KBs to keep the configuration data. However, experiments show that similar CIs provide good opportunities to reduce the configuration memory by merging their configuration data to one. The problem is that in most cases, the configuration bits which are related to functions and connections are the same but the control bits for inputs, outputs or immediate values differ. In order to reduce the size of configuration memory, CI configuration data is divided to four parts. In other words RFU is provided with partial reconfiguration. 138 bits for configuring intermediate connections and selecting functions of FUs (P1), 90 bits for selecting inputs (P2), 60 bits for outputs (P3) and 196 bits for immediate values (P4).

2007 IFIP International Conference on Very Large Scale Integration (VLSI-SoC 2007)

We added another stage to our tool flow in which we receive generated CIs as inputs and look for similar CIs and their subsets and then merge their configuration data into one. Two CIs are similar if their P1 are the same and a CI is a subset of another CI if its P1 is a subset of the bigger one. Although we generate one P1 for a CI and its subsets, as their P2, P3 and P4 are different they will generate the desired results. For inputs (P2), outputs (P3) and immediate values (P4), we just look for equal configuration bits and generate one configuration data for them. In the next step to make the configuration memory smaller, we try to merge P1 of small CIs into one configuration. Using these two techniques and using less intermediate multiplexers we were able to reduce the configuration memory. The other advantage of using these two techniques is that the number of context switching will be decreased due to fewer configurations.

9. Experimental Results

Average execution times of CIs derived from 8 applications of Mibench on both homogeneous and heterogeneous architectures are given in Figure 11. As it is shown, owing the fact that the critical path is reduced in our newly proposed architecture, CI execution times are 20-30% shorter than the previous architecture. Moreover as in the new architecture the number of multiplexers is reduced, configuration bits of multiplexers are reduces which in turn results in less memory and power consumption. All these improvements are besides the reduction in the area which proved to be reduced by 15%, according to the reports of MAGMA's layout synthesis tool.

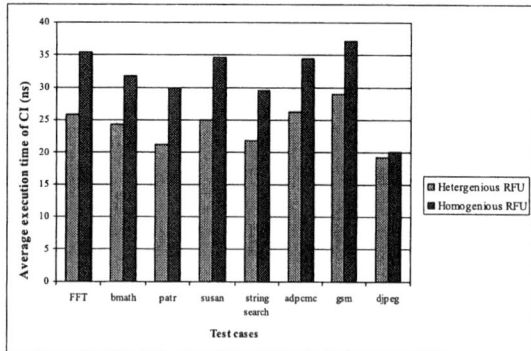

Figure 11. Comparing average execution time of CIs on previous and new architecture

10. Conclusion

Exploiting non-identical FUs in a heterogeneous RFU can improve the runtime and mapping rate of CIs compared to their homogeneous counterparts. In this paper we proposed a heterogeneous architecture for RFU of an extensible processor named AMBER, based on a quantitative and analytical approach. The mapping rate of CIs on this architecture was 94.86%. In addition, the CI execution speed on this architecture was improved drastically compared to the previous one. Our RFU consists of 8 inputs, 6 outpus, 3 tri-instruction FUs, 2 bi-instruction FUs and 3 uni-instruction FUs.

As a continuation of this work and to improve the overall runtime of applications, one can work on the architecture and mapping mechanisms to support floating point operations. In addition, it would be worth trying to work on some methods to reduce overall power consumption of the unit.

Reference

[1] Keutzer. K., S.Malik, A.R.Newton, J.M.Rabaey and A.Sangiovanni-Vincentelli. 2000. "System-Level Design: Orthogonalization of Concerns & Platform-based Design". In *IEEE Transactions on CAD of Integrated Circuits and Systems*, 19, No.12, 1523-1543.

[2] Hamid Noori, Farhad Mehdipour, Kazuaki Murakami, Koji Inoue and Morteza Saheb Zamani, "A Reconfigurable Functional Unit for an Adaptive Dynamic Extensible Processor," IEEE International Conference on Field Programmable Logic and Applications (FPL'06), Spain, 2006.

[3] H. Noori, K. Murakami, and K. Inoue, "A General Overview of an Adaptive Dynamic Extensible Procesor", in Proc. Workshop on Introspective Architecture, 2006.

[4] S. Hauck, T. Fry, M. Hosler, and J. Kao, "The Chimaera reconfigurable functional unit," in Proc. IEEE Symp. FPGAs for Custom Computing Machines, pp. 87–96, Apr.1997.

[5] J. E. Carrillo, and P. Chow, "The effect of reconfigurable units in superscalar processors," in Proc.of the 2001 ACM/SIGDA FPGA, pp. 141–150, 2002

[6] A. Lodi, M. Toma, F. Campi, A. Cappelli, R. Canegallo, and R. Guerrieri , "A VLIW Processor with Reconfigurable Instruction Set for Embedded Applications," IEEE Journal of Solid-State Circuits, vol. 38, no. 11, pp. 1876–1886, 2003.

[7] B. Mei, S. Vernalde, D. Verkest, and R. Lauwereinsg, "Design Methodolody for a Tightly Coupled VLIW/Reconfigurable Matrix Architecture: A Case Study," in Proc. Design, Automation and Test in Europe, 2004.

[8] M. H. Lee, H. Singh, G. Lu, N. Bagherzadeh, and F. J. Kurdahi , "Design and implementation of the MorphoSys Reconfigurable Computing Processor," Journal of VLSI and Signal Processing-Systems for Signal, Image and Video Technology, Mar. 2000.

[9] J. R. Hauser, and J. Wawrzynek, "GARP: A MIPS processor with a reconfigurable processor," in IEEE Symp. On FPGAs for Custom Computing Machines, Apr. 1997.

[10] K. Compton, and S. Hauck, "Reconfigurable Computing: A Survey of Systems and Software," ACM Computing Surveys, vol. 34, no. 2, pp. 171–210, 2002.

[11] F. Barat, and R. Lauwereins, "Reconfigurable Instruction Set Processors: A Survey," International Workshop on Rapid System Prototyping, 2000.

[12] S. Vassiliadis, and et al., "The MOLEN Polymorphic Processor," IEEE Transactions on Computers, vol. 53, no. 11, Nov. 2004, pp. 1363–1375.

[13] www.eecs.umich.edu/mibench

[14] www.Simplescalar.com

[15] N. Clark, M. Kudlur, H. Park, S. Mahlke and K. Flautner, "Application-Specific Processing on a General-Purpose Core via Transparent Instruction Set Customization", in Proc. MICRO-37, 2004.

Simulation of Hybrid Computer Architectures: Simulators, Methodologies and Recommendations

Pranav Vaidya and Jaehwan John Lee
Department of Electrical and Computer Engineering
Purdue School of Engineering and Technology
Indiana University-Purdue University Indianapolis

Abstract— In the future, high performance computing systems may consist of multiple multicore processors and reconfigurable logic coprocessors. Industry trends indicate that such coprocessors will be socket compatible to microprocessors and will be integrated on existing multiprocessor motherboards without any glue logic. Due to these trends, it is likely that such hybrid computing machines will be a breakthrough for various High Performance Computing (HPC) applications. It is essential to investigate the computer architecture of such hybrid computing machines that utilize reconfigurable logic coprocessors as application accelerators in a HPC system. Simulation can be used to aid this architectural research and guide design space exploration. In this paper, we first present a representative architecture for future hybrid computing machines. Next we present a survey of existing simulators and simulation methodologies for simulation of components of hybrid computing systems. Finally, we present some of the challenges and recommendations to encourage research in hybrid computing machines and their simulators.

Index Terms— Simulation, modeling of hybrid computer architectures, simulation of multiprocessor systems, simulation of FPGAs

I. INTRODUCTION

Two major trends are evident in the computing industry. Firstly, physical limitations of frequency scaling has led to major microprocessor manufacturers pushing for integration of multiple processor cores in a single chip. Secondly, novel computing fabrics such as reconfigurable devices are prominently being used for application acceleration. It is quite likely that these two trends will merge and hybrid computing machines made up of several processors and Reconfigurable Logic (RL) coprocessors will become commonplace. Commodity multiprocessor server platforms containing multiple processor cores and reconfigurable coprocessors [1]–[3] are indications of this trend. These machines offer high performance computation beyond the limitations of Von Neumann machines.

It is imperative to investigate the system architectures of such hybrid computing machines and understand any associated issues with design of such machines. Such investigation can be undertaken by using computer architecture simulation. Computer architects have long utilized simulators to guide the design space exploration and validate the efficacy of proposed architectural enhancements. In addition to traditional challenges such as trade-offs between simulation fidelity and speed, hybrid computing simulators face unique challenges in the form of lack of open source architectures, lack of open source synthesis, configuration and debugging tools. Furthermore, the variation in the reconfigurable logic coprocessor architectures make the design space exploration of hybrid computing architectures truly challenging.

Here, we first define several terms that will be used in this paper. We define a simulator designer as an individual responsible for designing the simulator. We also define a simulation designer/performer as an individual that leverages the simulator to perform simulation. Additionally, we follow the definition of simulation techniques and methodologies as described in [4] by Yi and Lilja. They define simulation methodology as a general term to describe how the simulator is constructed and simulation technique as the approach used by the simulation designer/performer to perform simulation such as using reduced input sets and microbenchmarks. The design decisions associated with the simulation methodology are usually made by the simulator designer while the design decisions associated with the simulation techniques are usually made by the simulation designer/performer.

The design decisions associated with the simulation methodology have direct consequences on the speed and fidelity of simulation. Here, fidelity of the simulation refers to the degree to which the simulated system models the real system. Any design decision associated with the simulation methodology should ensure that the simulation methodology is:

1) Efficient: The simulation methodology should be able to utilize greatly, if not completely, the capabilities of the simulation host. In this case, a simulation host refers to the computing system used to perform the simulation. Dynamic binary translation and parallel simulation are some of the examples of increasing the efficiency and the speed of simulation.

2) Elegant: The chosen simulation methodology should be easily understandable and extensible. This typically involves choices such as choosing an existing simulation language and/or a well validated simulation kernel. Hardware designers exercise this choice frequently where hardware designs are typically simulated using languages such as VHDL [5] and Verilog [6]. Recently, SystemC [7] has also become a popular option in hardware simulation.

3) Deterministic and Reproducible: The simulation should be able to produce identical results given identical initial conditions. Popular simulation language kernels are Sequential Discrete Event Simulators (SDES) because it is relatively easy to ensure determinism in SDES. Simulation kernels like SystemC ensure determinism by modeling concurrent activities in the simulation as user-level threads managed via cooperative multitasking. If concurrent activities are modeled as kernel-level threads, then non-determinism is introduced into the simulation as scheduling of kernel-level threads is seldom available to applications such as the simulation kernels.

Similarly, the design decisions associated with the simulation techniques have direct consequences on the accuracy and validity of simulation. Accuracy and validity of simulation refers to the degree to which the workload used during simulation reflects the true workload of the real system. Yi and Lilja [4] cite several simulation techniques such as reduced input set simulation techniques, truncated execution simulation techniques, processor warm-up approaches and sampling simulation techniques as the popular simulation techniques. Due to space limitation, we concentrate more on the simulation methodology that can be useful in simulation of hybrid computing machines.

Hybrid computing machines consist of multiprocessors and RL coprocessors. Hence, it is essential to identify the simulators, simulation methodologies and techniques used for simulating multiprocessors and RL coprocessors. This work surveys these facets of a hybrid computing system. The remainder of the paper is structured as follows. In section II, we present a representative architecture of future hybrid computing machines. This enables us to identify the main components that the simulators should simulate to a certain degree of fidelity. In section III, we present an overview of existing simulators, simulation methodologies and approaches that may be useful in simulating the hybrid computing system. Section IV presents the challenges and limitations of current simulation methodologies, and section V presents some recommendations for improving research in hybrid computing architectures and simulators.

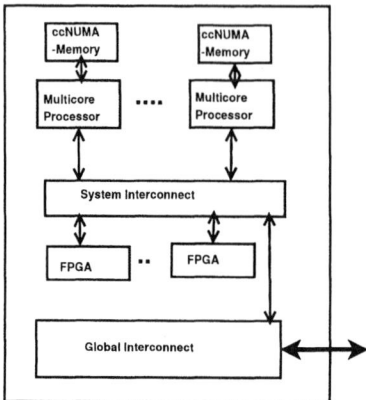

Fig. 1. A Single Node In A High Performance Hybrid Computing System.

II. A REPRESENTATIVE COMPUTER ARCHITECTURE FOR HYBRID COMPUTING MACHINES

In this section, we present a representative computer architecture for hybrid computing machines that we believe will be common in a High Performance Computing (HPC) environment. Figure 1 shows the most likely system architecture of a single node in a high performance hybrid computing system.

As shown in Figure 1, a single node of the hybrid HPC system will consist of several complex out-of-order issue RISC/CISC multicore processors and Reconfigurable Logic (RL) coprocessors. These coprocessors will be socket compatible to processors and hence will be integrated on existing motherboards without any glue logic. The processors and coprocessors will be interconnected through uniform chip-to-chip and board-to-board interconnects like Hypertransport [8]. To ensure scale-up as well as speed-up, it is quite likely that the most prevalent memory architectures of a single node in these hybrid computing machines will be cache-coherent Non-Uniform Memory Access (ccNUMA) [9]. The machines will have multiple levels of caches and main memory sizes of several gigabytes, if not terabytes [10].

Field Programmable Gate Arrays (FPGAs) [11]–[15] will be the most commonly used RL coprocessors. These FPGAs will be made up of hundreds of thousands of simple logic blocks such as Configurable Logic Blocks (CLBs). It is quite likely that with better fabrication processes, such FPGAs will have millions of CLBs. Other variations of FPGAs such as coarse-grained FPGAs [16] may also be used to reduce configuration times. Furthermore, these devices will support Run-Time Reconfiguration (RTR) and Run-Time Partial Reconfiguration (RTPR) so that the reconfigurable coprocessors can be used as multiplexed shared resources. We consider the ability of processors to configure and control these custom coprocessors as a distinguishing characteristic of these hybrid computing machines as compared to System-on-Chip (SoC). In SoC designs, the hardware modules are pre-configured to perform a specified function. In a hybrid computing machine, the RL coprocessor is used as either a shared or dedicated resource to perform several functions in hardware.

The DS2004 system from DRC Computer Corp. [2] is reviewed here as an example of the suggested representative architecture. This system is based on a Tyan Thunder K8QSD (S4882) 4-way motherboard with four processor sockets. It supports up to four AMD Opteron Model 875 dual core processors, 12GB ECC DDR, an Nvidia 7300GT PCI Express video card, one 160GB SATA

hard drive and one or two DRC Reconfigurable Processor Units (RPUs). The DRC RPU provides a tightly coupled RL coprocessor with direct access to DDR memory and any adjacent Opteron processor at full HyperTransport [8] bandwidth and low latency. The RPU is controlled via an RPU manager, which allows FPGA configuration over HyperTransport. This system is capable of hosting ccNUMA operating system namely Linux (64-bit) Ubuntu 6.x. It is an indication of the growing trend towards integration of RL coprocessors with multicore processors in the industry. Other competitive vendors [1], [3] offer similar platforms.

For the aforementioned representative architecture, it is crucial to note that any simulator for such architecture should be able to simulate the following components:

1) Multicore processors and caches.
2) Reconfigurable logic coprocessors.
3) System interconnects and global interconnects: It is crucial to model system interconnects to a certain degree of fidelity. This is essential as any HPC involves both computation and communication.
4) Run-Time Reconfiguration (RTR) and Run-Time Partial Reconfiguration (RTPR): A hybrid computing machine simulator should be able to model RTR and RTPR to simulate the RL coprocessor as a shared resource.
5) Memory modules: It is quite likely that in the future, each node of a hybrid HPC system will have ccNUMA memory access architecture.

III. SIMULATION AND CO-SIMULATION RESEARCH WORK

In this section, we present an overview of the popular simulators and simulation methodologies that will be useful for modeling and simulating the components of hybrid computing machines.

A. Simulators for Chip Multiprocessors

As can be seen in Figure 1, a node in most of the future hybrid computing systems will contain multiple multicore processors. Hence, we only survey the popular simulators that simulate multiprocessors and chip-multiprocessors. Popular uniprocessor simulators such as SimpleScalar [17] are not reviewed here as they do not model such multiprocessing systems.

1) RSIM: Rice Simulator for ILP Multiprocessors: RSIM is the Rice Simulator for ccNUMA, ILP multiprocessors. It was developed and released to public in 1997 [18], [27]. Key RSIM features include support for out-of-order issue, register renaming, branch prediction and nonblocking caches. RSIM also supports user-configurable parameters such as cache sizes and latencies, flit size and delay, as well as instruction window size [4], [18]. RSIM was different from most other simulators in that it modeled the ILP features of a multiprocessor system. RSIM's research showed that disregarding the ILP-level features of a multiprocessor system resulted in the overestimation of the execution time by as much as 132 percent.

RSIM's simulation methodology was derived from YAC-SIM [19]. YACSIM is a process-oriented, discrete-event simulator developed as part of Rice Parallel Processing Testbed. YACSIM supported user-level multithreading to represent multiple processes. Thus, each process in RSIM runs in a user-level thread and the simulation kernel manages the scheduling of these threads. As a result, RSIM does not take advantage of multiprocessing simulation hosts. RSIM utilizes execution driven simulation techniques to simulate applications compiled and linked for Sparc/V9/Solaris. RSIM uses standard Sparc compilers and linkers at all optimization levels. However, it lacks support for 64-bit integer and quad-precision floating-point operations. Furthermore, it lacks support for standard libraries and applications that rely on conventional Sparc traps. To overcome this limitation, RSIM provides standard C library to support applications.

2) Virtutech Simics - A Full System Simulation Environment: Virtutech Simics is a commercial full-system simulator that can simulate multiprocessor systems with enough accuracy to boot unmodified operating systems [38]. Simics executes unmodified binaries from an ISA perspective and provides a timing interface to user modules. For example, instruction fetch by the simulator is forwarded to the cache modules to stall the execution of instruction for an arbitrary number of cycles [20], [21].

As of Simics 2.0, Simics supports a Micro-Architectural Interface (MAI). Using MAI, the user module can determine when an instruction passes through the microprocessor pipeline such as fetch, decode, execute and commit phases. Using the timing interface provided by Simics, the user module can also support detailed timing modeling. Simics supports checkpointing as a useful simulation technique. This allows the user to run the application to a specific point of interest and save the state of the simulated machine to disk. This technique can reduce simulation time since application initialization phase is run only once. This has important consequences for commercial benchmarks where such initialization or warmup phase can require a significant amount of time, even requiring weeks of simulation [38], [43]

Simics is one of the most popular simulators used in the academia and industry to model entire computer systems and even distributed computing systems. Simics toolset has been used in the academia to develop the Wisconsin General Execution-driven Multiprocessor Simulator (GEMS) [20]. GEMS leverages the full-system functional simulation infrastructure of Simics to drive a set of timing simulator modules for modeling the timing of the memory system and microprocessors. Other projects which have used Simics are Vasa [21] and SimFlex [24]. VASA [21] is a highly configurable multiprocessor simulation package for Simics. Vasa includes models of multilevel caches, store buffers, interconnects, memory controllers and detailed complex out-of-order SMT/CMP processors. It also supports two additional, less detailed simulation modes which run up to 287 times faster than the detailed simulator. SimFlex [24] is a simulator package for Simics developed at the Carnegie Mellon University that leverages the statistical sampling of the inputs to reduce the simulation time of a chip multiprocessor

system.

Simics methodology involves simulating a multiprocessor system by simulating each processor in a round-robin fashion. Each processor is simulated for a given number of cycles controlled via a variable called *cpu-switch-time*. This variable allows the coarseness in thread interleaving to be scaled. However, adjusting *cpu-switch-time* to large value can have significant effect when simulating multithreaded applications with contended locks [21]. As a result, derived simulators such as VASA typically set the value of this variable to one. Other simulators in the academia use a similar round-robin simulation of each processor [22] to simulate a multiprocessor system. While Simics can be customized using the APIs that it provides, it does not expose its simulation methodology as it is a commercial software. On the other hand, simulators such as GxEmul [22], [23] are open source and simulate the processors at instruction set level.

3) Wisconsin Wind Tunnel II (WWT II): WWT II [25] differs from the above two simulators in that it is a parallel, discrete-event, direct-execution simulator that can be run across a wide range of platforms, such as desktop workstations, a SUN Enterprise server, a cluster of workstations, and a cluster of symmetric multiprocessing nodes.

WWT II simulates a parallel, ccNUMA system on various parallel systems connected using Myrinet [26]. It uses Synchronized Active Messages (SAM) to communicate between the host nodes for parallel simulation. Analytical modeling has been used to approximate the performance of WWT II for a variety of system sizes. WWT II uses direct execution and parallel hosts as the simulation methodology to speed up the execution. Direct execution executes an instruction of a target machine by directly executing it on the host system. Only operations unavailable on the host platform are simulated by the host platform. Direct execution typically runs orders of magnitude faster than pure interpreted software simulation [25]. Furthermore, WWT II performs parallel simulation by exploiting the parallelism inherent in the target parallel computer to achieve speed-ups of up to 5.8X. However, this approach does not allow changes in the processor models and other architectural parameters such as issue widths, speculative memory accesses and out-of-order execution [29].

WWT II uses SAM as its programming model for communication and synchronization operations. Since SAM runs only on the SPARC architecture, WWT II is not portable to other architectures.

4) Parallel Trace Driven Simulation approaches: Other approaches to parallel simulation of computer architectures include [30]–[35]. All of these approaches use parallel trace driven execution to speed-up simulation of benchmarks such as SPEC CPU 2000 [37]. A given benchmark application is executed concurrently on multiple instances of the simulator initialized with different configurations. Though such an approach increases throughput of simulation, it does not reduce the simulation time of a single simulation as demonstrated by [29].

5) Parallel Simulation of Chip-Multiprocessor Architectures: Research by Chidester *et al.* [29] targeted the simulation of Chip Multiprocessors (CMP) by performing parallel simulation of tightly coupled CMPs (which share L2 caches) on a distributed host system consisting of commercial-off-the-shelf (COTS) workstations. These workstations were connected by a high-speed network.

The simulation methodology used by Chidester *et al.* involves cycle-accurate simulation of the processors and L1 caches. They used the parallel, event-driven simulation built using the Message Passing Interface (MPI) [36] to model communication between L1 cache and the shared L2 cache. Using this approach, simulation speed-ups of up to 5X were obtained.

B. Simulators for Reconfigurable Logic

1) Levels of Abstraction in modeling custom logic: Due to the large complexity of hardware designs today, most simulations are done at various levels of abstraction. Gajski and Cai [28] explain the various levels of abstraction used in system models. They identified that system functionality/computation and communication can be developed independently of one another and refined at each subsequent stage.

As seen in Figure 2, Gajski and Cai have classified the following levels of abstraction:

i) Model I: Model I represents an untimed system architectural model. This model is typically used to specify the functionality and communication of the system and its subsystem without any attention paid to the timing of the interfaces. This model is used to verify the correct functioning of the system and system interconnects.

ii) Model II: Model II represents the Component Assembly Model (CAM). The CAM is used to integrate the empirical understanding of computational time into the model. However, data transfer between components is still untimed.

iii) Model III: Model III represents the Bus Arbitration Model (BAM) or transactional model. In this model, the information about each cycle of the bus is accurately modeled.

iv) Model IV: Model IV represents the Bus Functional Model (BFM). In this model, each signal transition of the bus is modeled as a single event. As a result, communications are timing accurate.

v) Model V: Model V represents the cycle accurate computation model. However, the timing is approximately timed. This model emphasizes communication at transaction level.

vi) Model VI: Model VI represents the register-transfer level model. In this case, both the communication and computation are modeled accurately. This model closely represents the actual hardware and is typically used for automatic synthesis to gates.

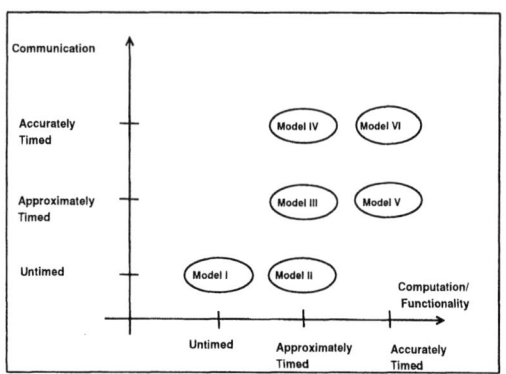

Fig. 2.   Abstraction Level Of Models. Courtesy of Cai and Gajski [28]

2) Traditional Simulation/Co-Simulation Approaches and Limitations: Compton *et al.* provided an extensive survey of reconfigurable computing systems and software [39]. However, they did not consider the ability of processors to configure and partially reconfigure RL coprocessors as the defining characteristics of hybrid computing machines. We feel as described in section II that these abilities are key features of hybrid computing machines.

Typically, hardware designers have designed and validated hardware models (Models I-VI) using vendor-specific tools and hardware design languages such as VHDL, Verilog and SystemC.

Tools such as ModelSim [40]–[42] are used to perform functional simulation, static timing analysis and timing simulation of the hardware designs. These simulators use the knowledge of cell and routing primitives of the actual device to perform simulations.

Most FPGA design suites assume that hardware design simulated using behavioral and timing simulation will work in actual hardware as intended. However, hardware designs validated using timing simulation may not work on the actual device due to several problems. For example, third party implementation tools may have inferred, places and routed the designs differently than what was specified.

These design suites also assume that the hardware design being synthesized is the only design resident on the reconfigurable device. Such an assumption is valid for most embedded systems which use reconfigurable devices to implement SoC designs. However, these assumptions may not be valid for hybrid computing machines where the RL coprocessors may be multiplexed across multiple applications. As a result, the RL coprocessor may be configured to support several hardware functions. Hence partial reconfiguration is an essential characteristic of such machines. Most of these simulators do not have support for reconfigurable design concepts such as partial reconfiguration. As a result, simulation and co-simulation approaches using traditional hardware design flow is of limited use to the simulation of hybrid computing machines.

3) VTSim - A Virtex-II Device Simulator: VTSim [44], [45] was a discrete-event simulator written in Java that modeled all the hardware resources present in a Virtex-II FPGA [11]. VTSim provided a virtual FPGA device which was compatible to existing Xilinx tools. Using VTSim, the designers could access all the resource values in the virtual FPGA such as flip-flop and look-up table values or values on a routed wire. VTSim was a bitstream level simulator that took the bitstream file (.bit extension) generated from the Xilinx tool chain to simulate the hardware designs.

VTSim was useful in reconfigurable designs as it was able to read and modify bitstreams used to configure and reconfigure the virtual FPGA device. Furthermore, VTSim was integrated into the JHDLBits [47] design suite allowing simulation in Java Hardware Description Language (JHDL) or as a standalone tool.

Unfortunately, this simulator was never released to the public because the permission to release this simulator was never granted by Xilinx, the vendor for Virtex II devices.

4) VirtexDS - A Virtex Device Simulator: Virtex Device Simulator (VirtexDS) [46] was a device level simulator for Virtex-II Pro devices [11] from Xilinx. It was released as part of the Xilinx JBits 2.8 SDK [11]. This simulator was similar to VTSim that simulated Virtex FPGA devices. VirtexDS provided a software model of the FPGA device for the entire Virtex family of FPGAs. It supported run-time configuration and run-time partial reconfiguration that could be controlled through the JBits 2.8 environment. VirtexDS allowed for existing tools such as the BoardScope [48] debug tool to interface directly to the simulator without any modification.

Subsequently, Xilinx released JBits 3.0. However, it did not release a device level simulator. During our survey, we found no device level FPGA simulators available in the industry or academia for research purposes.

IV. CHALLENGES AND LIMITATIONS FOR HYBRID COMPUTING MACHINE SIMULATORS

From our survey, we found the following limitations and challenges that the hybrid computing simulators and machines face:

1) *Current limitations for simulating multicore processors*: Most simulators in the industry and academia such as RSIM and Simics are built using Sequential Discrete Event Simulators (SDES). Even hardware simulation languages and kernels use SDES for functional and timing simulation. These simulators do not take advantage of the parallel computation facilities that are becoming available even at the desktop computing level. With the advent of multicore processors, these kernels should use the parallel computational facilities that current simulation hosts offer. WWT II and other parallel discrete event simulation approaches show that speed-ups can be obtained from parallel simulation of computer architectures without compromising on the fidelity of the simulator. However, these simulators have been built using specialized programming models for distributed computing such as SAM and MPI. While SAM is not portable, MPI suffers serious performance degradation on multicore shared memory architectures as it maps each node of computation to an OS process. Hence, one of the main challenges in simulating multicore processors is balancing portability of the simulator with the ease of using and extending the simulator. This challenge can be solved by identifying and using a good programming model for multicore and cluster simulation hosts. The key idea behind such a programming model should be exploiting local multiprocessor as well as cluster computing power. We state such a computing model in section V.

2) *Challenges in Simulating Reconfigurable FPGAs*: The challenges in simulating reconfigurable logic devices are greater than that of traditional processors. Reconfigurable devices typically have closed architectures, closed bitstreams, and even more so there is a lack of open source development tools, compilation-to-gates tools, verification and synthesis tools. Furthermore, there are no standard APIs for configuring and communicating with these reconfigurable coprocessors in a hybrid computing machine. It is reasonable to understand that the industry would most likely not release open architectures and tools due to the inherent financial gains associated with such tools. In another aspect, as the granularity of FPGA devices increases towards fine-grained architectures, it would be extremely inefficient to simulate these devices using SDES.

3) *Challenges in Simulating the Hybrid Computing System*: Most simulation kernels do not support multiple models of computation in the simulator. Different models of computation may be advantageous to model the various components of the hybrid computing system. For example, while Synchronous Data Flow Graphs (SDFG) may be advantageous to model streaming devices such as DSPs, Parallel Discrete Event Simulation (PDES) may be advantageous to model multicore processors. Hence, research and further exploration of such multi-model simulation kernels [52] should be encouraged.

V. RECOMMENDATIONS

Based on the aforementioned observations, we make the following recommendations for fostering research in the area of hybrid computing systems and hybrid computing simulators.

Recommendation 1: Use of Parallel Simulation Techniques for Current Simulation Hosts

It is essential to note that as hybrid computing machines are growing more complex, the simulation hosts are also becoming more powerful. Over the last few years, even desktop computers with two or more processors/processor cores have become available

to the general public [50], [51]. Simulator designers should take note of this, and research simulators using Parallel Discrete Event Simulation (PDES) should be investigated.

As identified in the challenges, the choice of programming model is a key challenge for developing simulators for multicore simulation hosts. It is also essential that to ensure scalability of the simulation host, such programming model should be seamlessly extensible to a cluster of computers. Streaming programming models based on well established process calculi such as Communicating Sequential Processes [49], [53] may be the solution to this issue. These programming models may be more flexible and faster than SAM and MPI for both multicore and other cluster simulation hosts. Such programming models can model physical processes as nano-threads, user-level threads and kernel-level threads. Thus, the designer is flexible in choosing the appropriate granularity of threads according to the level of communication between the modeled components. However, we did not find any simulator that is built using streaming languages. Developing simulators using such programming paradigms should be pursued.

Recommendation 2: Open FPGA Architecture and Open Source FPGA Design Tools

FPGA industry is a multi-million dollar industry. Device vendors have invested greatly in their proprietary architectures and FPGA design suites. However, it would be beneficial to both the academia and industry if a consortium similar to OpenFPGA consortium [54] is established for the development of open source FPGA architectures and design tools for hybrid computing machines. This would be both financially and intellectually beneficial to the industry and academia. For example, traditionally it is assumed that the RL coprocessors are dedicated for a single application. It would be beneficial if these RL coprocessors are used as multitasking shared resources. To make this happen, the detailed layout of the application on the RL coprocessor should be known beforehand. Hence, with open source FPGA architectures, a detailed architectural model of the RL coprocessor can be used to perform intelligent compilation and synthesis for these shared coprocessors. Additionally, required research impetus can be accelerated using open source reconfigurable coprocessor architectures in HPC applications. Furthermore, such an endeavor can create the required engineers and scientists who are exposed to reconfigurable computing internals.

VI. SUMMARY

In this paper, we have summarized some of the simulators and simulation methodologies that are likely to be useful in the simulation of hybrid computing machines made up of multiple multicore processors and reconfigurable logic coprocessors. It would be beneficial if the simulator methodologies fully utilize the computational power offered by the simulation hosts. Research into developing simulators built around the concept of Parallel Discrete Event Simulation (PDES) and/or streaming language paradigms such as Communicating Sequential Processes (CSP) [53] should be encouraged.

There exists an inherent trade-off between simulation speed and simulation accuracy. However, many simulation approaches target simulation speed by compromising the fidelity of simulation. Such a trade-off is acceptable for development of systems software; however it can result in the overestimation of execution speeds in some cases. With the current industry trends towards chip multiprocessing, it is essential that simulators model such systems with sufficient fidelity. As a result, as part of our recommendations, we have suggested that further research into parallel simulation of chip multiprocessors be pursued.

2007 IFIP International Conference on Very Large Scale Integration (VLSI-SoC 2007)

In addition, to foster further development into Reconfigurable Logic (RL) coprocessors, we have suggested that both the industry and the academia join hands to come up with open source FPGA architectures and programming tools. Currently, there exist no open source simulators that support run-time reconfiguration and run-time partial reconfiguration. Research into device level FPGA simulators would be greatly useful to both academia and industry and thus should be pursued. We predict with high confidence that such research will provide great impetus in developing open source compilation and synthesis tools. This will further the integration of such RL coprocessors with applications spanning from embedded systems, general purpose computing to High Performance Computing (HPC).

REFERENCES

[1] Celoxica RCHTX System,
http://www.celoxica.com/products/rchtx/default.asp, visited Mar 2007.

[2] DRC Computer Corporation,
http://www.drccomputer.com/, visited Mar 2007.

[3] XtremeData Inc,
http://www.xtremedatainc.com/, visited Mar 2007.

[4] J. Yi and D. Lilja, "Simulation of Computer Architectures: Simulators, Benchmarks, Methodologies, and Recommendations," IEEE Trans. on Computers, vol. 55, no. 3, pp. 268-280, Mar. 2006.

[5] VASG: VHDL Analysis and Standardization Group,
http://www.eda.org/vhdl-200x/, visited Mar 2007.

[6] IEEE Verilog Standardization Group,
http://www.verilog.com/IEEEVerilog.html, visited Mar 2007.

[7] SystemC Community Website,
http://www.systemc.org/, visited Mar 2007.

[8] Hypertransport Consortium,
http://www.hypertransport.org/index.cfm, visited Jan 2007.

[9] NUMA, HyperTransport, 64-Bit Windows, and You
http://developer.amd.com/article_print.jsp?id=8, visited Dec 2006

[10] Performance Guidelines for AMD Athlon 64 and AMD Opteron ccNUMA Multiprocessor Systems,
http://www.amd.com/us-en/assets/content_type/white
_papers_and_tech_docs/40555.pdf, visited Dec 2006.

[11] Xilinx Corporation, http://www.xilinx.com/, visited Jan 2007.

[12] Altera Corporation, http://www.altera.com/, visited Jan 2007.

[13] Actel Corporation, http://www.actel.com/, visited Jan 2007.

[14] Lattice Semiconductor Corporation, http://www.latticesemi.com/, visited Jan 2007.

[15] QuickLogic Corporation, http://www.quicklogic.com/, visited Jan 2007.

[16] E. Mirsky and A. DeHon, "MATRIX: A reconfigurable computing architecture with configurable instruction distribution and deployable resources," IEEE Symposium on FPGAs for Custom Computing Machines, pp. 157-166, 1996

[17] T. Austin, E. Larson and D. Ernst, "SimpleScalar: An infrastructure for computer system modeling," Computer, vol. 35, no. 2, pp. 59-67, 2002.

[18] V. Pai, P. Ranganathan and S. Adve. "RSIM: An Execution-Driven Simulator for ILP-Based Shared-Memory Multiprocessors and Uniprocessors," In Proceedings of the Third Workshop on Computer Architecture Education, February 1997.

[19] J. Jump, YACSIM Reference Manual. Rice University, version 2.1.1 edition, 1993, www.owlnet.rice.edu/ elec428/yacsim/yacsim.man.ps, visited Mar 2007.

[20] M. Martin et al., "Multifacet's general execution-driven multiprocessor simulator (GEMS) toolset," SIGARCH Comput. Archit. News, pp. 92-99, 2005.

[21] D. Wallin, H. Zeffer, M. Karlsson and E. Hagersten, "VASA: A Simulator Infrastructure with Adjustable Fidelity," Parallel and Distributed Computing and Systems, 2005.

[22] P. Vaidya and J. Lee, "Design Space Exploration of Multiprocessor Systems with Multicontext Reconfigurable coprocessors," In Proceedings of Engineering of Reconfigurable Systems and Algorithms, ERSA'07, pp. 51-60, June 2007.

[23] GxEmul, http://gavare.se/gxemul/, visited Jan 2007.

[24] T. Wenisch et al., "SimFlex: Statistical Sampling of Computer Architecture Simulation," IEEE Micro special issue on Computer Architecture Simulation, vol. 26, no. 4, pp. 18-31, Jul/Aug 2006.

[25] S. Mukherjee et al., "Wisconsin Wind Tunnel II: A Fast and Portable Parallel Architecture Simulator," In Workshop on Performance Analysis and Its Impact on Design, June 1997.

[26] Myricom Page for Myrinet, http://www.myri.com/myrinet/overview/, visited Jan 2007.

[27] R Covington et al., "The Rice Parallel Processing Testbed," In Proceedings of the 1988 ACM SIGMETRICS Conference on Measurement and Modeling of Computer Systems, pp. 4-11, May 1988.

[28] L. Cai and D. Gajski, "Transaction Level Modeling: an overview," Hardware/Software Codesign and System Synthesis, pp. 19-24, 2003.

[29] M. Chidester and A. George, "Parallel Simulation of Chip-Multiprocessor Architectures," ACM Trans. on Modeling and Computer Simulation, vol. 12, no. 3, pp. 176-200, July 2002.

[30] L. Eeckhout and K. De Bosschere, "Efficient Simulation of Trace Samples on Parallel Machines," Parallel Computing, vol. 30, no. 3, pp. 317-335, Mar. 2004.

[31] B. Falsafi and D. Wood, "Modeling Cost/Performance of a Parallel Computer Simulator," ACM Trans. on Modeling and Computer Simulation, vol. 7, no. 1, pp. 104-130, Jan. 1997.

[32] G. Lauterbach, "Accelerating Architectural Simulation by Parallel Execution of Trace Samples," Sun Microsystems Laboratory Technical Report TR-93-22, 1993.

[33] A. Nguyen, P. Bose, K. Ekanadham, A. Nanda and M. Michael, "Accuracy and Speed-Up of Parallel Trace-Driven Architectural Simulation," In Proceedings of Int'l Parallel Processing Symp., 1997.

[34] D. Poulsen and P. Yew, "Execution-Driven Tools for Parallel Simulation of Parallel Architectures and Applications," In Proceedings of Supercomputing, pp. 860-869, 1993.

[35] W. Wang and J. Baer, "Efficient Trace-Driven Simulation Methods for Cache Performance Analysis," ACM Trans. on Computer Systems, vol. 9, no. 3, pp. 222-241, Aug. 1991

[36] MPI Homepage, http://www-unix.mcs.anl.gov/mpi/, visited Mar 2007.

[37] SPEC CPU 2000, http://www.spec.org/cpu/, visited Mar 2007.

[38] P. Magnusson et al., "Simics: A full system simulation platform," Computer, vol. 35, no. 2, pp. 50-58, 2002.

[39] K. Compton and S. Hauck, "Reconfigurable computing: a survey of systems and software," ACM Comput. Surv. 34, pp. 171-210, 2002.

[40] Mentor Graphics, ModelSim. http://www.mentor.com/modelsim.

[41] Mentor Graphics, Hardware/Software Co-Verification:Seamless. http://www.mentor.com/seamless/, visited Jan 2007.

[42] Mentor Graphiscs, Seamless FPGA,
http://www.mentor.com/products/fv/hwsw_coverification/seamless_fpga/, visited Jan 2007.

[43] W. Fu and K. Compton, "A Simulation Platform for Reconfigurable Computing Research," IEEE International Conference on Field Programmable Logic and Applications, Aug. 2006.

[44] J. Hunter, P. Athanas and C. Patterson, "VTsim: A Virtex-II Device Simulator," In Proceedings of Engineering of Reconfigurable Systems and Algorithms, ERSA'04, Jun 2004.

[45] J. Hunter, "A Device-Level FPGA Simulator," Masters Thesis, June 2004.

[46] S. McMillan, B. Blodget and S. Guccione, "VirtexDS: A Virtex device simulator," In Proceedings of SPIE, pp. 50-56, Oct 2000.

[47] A. Poetter, "JHDLBits: An Open-Source Model for FPGA Design Automation," Master's Thesis, Aug 2004.

[48] D. Levi and S. Guccione, "BoardScope: a debug tool for reconfigurable systems," In Proceedings of SPIE vol. 3526, pp. 239-246, Oct 1998.

[49] W. Thies, M. Karczmarek and S. Amarasinghe, "StreamIt: A Language for Streaming Applications," In Proceedings of the 2002 International Conference on Compiler Construction, Apr 2002.

[50] AMD Multicore Website,
http://multicore.amd.com/, visited Mar 2007.

[51] Intel Multicore Website,
http://www.intel.com/multi-core/, visited Mar 2007.

[52] J. Eker et al., "Taming heterogeneity–the Ptolemy approach" In Proceedings of the IEEE Special Issue on Modeling and Design of Embedded Software, vol. 91, pp. 127-144, Jan 2003.

[53] C. Hoare, "Communicating Sequential Processes," Prentice Hall International, 1985.

[54] OpenFPGA consortium,
http://www.openfpga.org/, visited Mar 2007.

New Parallel Programming Techniques for Hardware Design

Satnam Singh
Microsoft Research
7 JJ Thomson Ave
Cambridge CB3 0FB,
United Kingdom
+44 1223 479905

satnams@microsoft.com

ABSTRACT

We present some recent advances in concurrent and parallel programming which are promising candidates for the design and specific of hardware and in particular reconfigurable systems. We explore the relationship between parallel programming and hardware design and ask the question "are these two activities the same thing at an important level of abstraction"? In particular, we consider join patterns from join calculus, software transactional memory, futures and nested data parallel programming.

1. Introduction

Digital hardware design and parallel programming have been accidentally separated at birth and this paper argues to have both activities reunited. The current design techniques for digital hardware are largely based on writing descriptions that are inputs to an event-base simulator and then inferring the corresponding network of gates that implement the required behavior. Although this technique has been widely adopted it is a rather indirect way of designing hardware and a desire for a more productive design method is often expressed by designers. Furthermore, digital hardware design is likely to become a more mainstream activity as future many-core processors present regular programmers with the challenge of not only programming things that looks like processors but also trying to code for things like that look like GPUs and FPGAs on a SoC that is a general purpose computing platform. We propose that it is essential to present a feasible programming model for targeting all these SoC computing elements and one direction worth exploring is attempting to use established and emerging parallel and concurrent programming techniques to express calculations that can map into software, hardware or some mixture of the two. Another alternative is to try and come up with techniques that support many different kinds of programming models and then find a way to relate different models of computation. This is also an interesting approach which is being explored by other research groups e.g. the Ptolemy II project at Berkeley.

Until now the mainstream software development community has not had to deal with the challenges of concurrent or parallel programming. Concurrent programming has been confined to specialized domains with closed world assumptions e.g. operating system kernels. Parallel programming has been successfully performed in very specific domain e.g. scientific computing using

highly skilled developers often on special parallel systems (e.g. supercomputers). Even more specialized are FPGA-based systems which provide parallel hardware implementations of problems which can outperform supercomputers. However, the highly specialized skills one needs to produce these systems makes them inaccessible to the mainstream developer or user of computationally intensive systems.

All this is about to change as we approach an era where single core performance has leveled off and processor vendors have started to move to a model where the number of cores available in a single die will double with each silicon generation [1]. What are we going to do with all those processors? How will we program them? Will we have to learn how to write concurrent and parallel programs? Or can we produce some special technology to take our sequential programs and automatically parallelize them? Are there better ways of using all that silicon other than stamping out many instances of the same core and connecting them through a series of caches to the same shared memory?

We are now approaching an era where mainstream programmers may face some of the same challenges that have faced the parallel hardware and reconfigurable computing communities. Every challenge can also lead to an opportunity. Today we have a sad state of affairs at the point where hardware meets software. Hardware and software are developed using quite different models of computation. Systems that comprise a mixture of hardware and software are difficult to design because it is hard to relate C components to VHDL components. The author believes that the change in processor architectures will necessarily force a change in the way we write software and the underlying models that we use to understand our software. This is an opportunity to carefully design models which make it easier to relate software to hardware and to even convert software descriptions into hardware descriptions.

One important reason for being able to convert a software implementation of a calculation into a hardware implementation is the author's belief that future many-core architectures will be very heterogeneous. They will comprise not only many processors of different kinds but also of other components that can compute e.g. reconfigurable data-paths that are the evolution of today's GPUs and reconfigurable fabrics that are the evolution of today's FPGAs. These devices may no longer have a single coherent memory and instead provide local memory to tens or hundreds of processors which are connected together via an on-chip network.

163

978-1-4244-1709-4/07/$25.00 © 2007 IEEE

To allow applications to run effectively on mixtures of such "Metropolis" architectures (see Figure 1) will require the ability to migrate calculations from one kind of computation element to another e.g. from software to hardware.

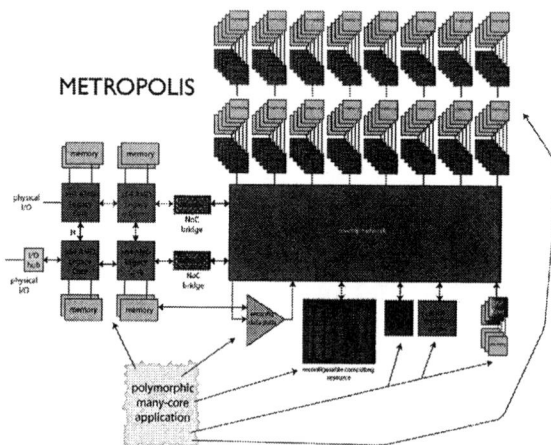

Figure 1. A potential many-core architecture.

The problem of how to model, write, integrate, test and verify such systems a daunting task but we have some good existing work to guide us. We absolutely need to think in terms of different models of computation for different parts of our application and then we need a way to compose or relate these models of computation. The Ptolemy II [5] system provides such a framework and has been successfully used to model embedded systems and it has also been used to generate hardware from certain models of computation. A successful future parallel systems programming model is likely to borrow some of the key ideas from Ptolemy II to help structure and compose the elements of heterogeneous parallel systems. For programming such architectures it is essential to allow the programmer to express relevant aspects of the parallel computation in a productive manner without saying too much about a particular parallel implementation architecture and here we have the example of work like the Chapel programming language [6] developed for high performance computing to help guide us in the programming language space. Chapel, along with languages like Fortress [10] presents a model that abstracts the number and kind of computational resources available to the programmer.

There has been much work on compiling C or C-like languages to hardware. With a few notable exceptions, there has been limited progress in this area. The author believes that starting from C one has already introduced so much unnecessary sequentiality that it is often impossible to recover the potential parallelism in the original algorithm. Furthermore, C does not have explicit constructs for expressing concurrency or parallelism and this makes it even more difficult to devise automatic parallelization flows.

Some have suggested compiling multi-threaded code into hardware or even explicitly implementing the notion of hardware threads. However, programs written explicitly in terms of threads

and locks are likely to be at the wrong level of abstraction for effective compilation into a variety of parallel computing elements especially FPGA-like fabrics. Threads and locks are systems programming abstractions that have proved effective for their original use but it is not clear that they are the appropriate abstractions for writing concurrent or parallel applications (hardware, software or a mixture) [17][16]. This programmer's perspective on how to write for parallel systems yearns for higher level constructs which capture the salient parallel operations and leave enough details unsaid to allow efficient compilation to many different kinds of parallel computing elements. Even modestly raising the concurrency abstraction can provide a better model for writing parallel systems and a better starting point for hardware synthesis tools. Here are some examples of such higher level concurrency abstractions:

- **Futures.** This technique implicitly introduces parallelism by identifying calculations whose values will be needed at a later stage and may be scheduled for parallel execution. There has been much research into compiling future based descriptions into software. But can future based descriptions also be synthesized into efficient hardware?

- **SHIM.** SHIM stands for Software/Hardware Integration Medium [11] and is an adaption of the C language which introduces deterministic concurrency through disciplined used of multi-way rendezvous. The design of this language combines imperative C-like semantics with the cycle-based VHDL/Verilog style semantics to yield a system which seems valuable for describing both hardware an software.

- **Cilk.** The Cilk programming language [2] adds three keywords to C which allows many interesting kinds of parallelism to expressed without the explicit mention of system level threads. These extensions can always be removed from a Cilk program to yield a valid C program. Very efficient compilation and run-time technology has been produced for Cilk. This may be a much viable starting point for a C to gates style flow than conventional methods.

- **Single assignment variables.** These are variables which can only be assigned once. If a parallel activity tries to read from an unassigned value it is blocked. This simple mechanism often provides a nice way to synchronize parallel activities and is effectively implemented in systems that have memory with full/empty bits. Can parallel systems described in terms of single assignment variables be translated into effective circuits which communicate and synchronize using appropriate hardware structures e.g. single element buffers?

- **Software transaction memory.** This technique [14] takes the notion of a transaction from the database world and applies it to a shared memory concurrent programming context. It allows multiple parallel activities to operate on shared information without taking explicit locks. This approach eliminates many kids of deadlocks and may also give rise to optimistic concurrency. Can this model be used to compose hardware and software blocks in a parallel system with shared memory? Preliminary work has shown that some other kinds of concurrency abstractions can be modeled using STM e.g. join patterns [19].

- **Join patterns**. Several research languages have experimented with the notion of using join patterns [11][13] for synchronization between parallel activities. An example is the COmega [3] language which nicely integrates joins into the object model of C# (similar work has been done with Java [15] and OCaml [8] and they have also been implemented as a library [7]. At least on research group has investigated compiling join pattern based descriptions into hardware and this approach merits further investigation as a way to help synchronize parallel activities in hardware and software. Join based descriptions raise the level of abstraction at which asynchronous message passing programming is performed and join patterns can effectively encode many other kinds of concurrency abstractions.

- **Data parallelism**. The approaches presented so far belong to the class of control oriented concurrency. By far the most effective way of producing parallel systems that map well to multiple processors or FPGA-like hardware is the use of data-parallel descriptions. These descriptions can often be compiled into very effective pipelined parallel hardware on reconfigurable platforms. Furthermore, data parallel programming is a much easier "sequential" model for the programmer than a thread based model which results in non-deterministic behavior. Challenges remain for the design of parallel systems built in a composable manner and there needs to be more work in the area of nested data parallelism.

- **Functional programming**. Programming languages that are inherently more amenable to parallelism are likely to play a more important role in the future, especially for data-parallel programming e.g. concurrent Haskell [18].

These are just a few of the techniques and approaches we can take to help us develop methodologies for programming parallel systems whether they are software, hardware or ambiguous.

Other challenges arise from the changing nature of the underlying fabric as we move beyond 65ns process technologies where we can no longer economically produce defect-free chips [4]. Can we produce abstractions (at the hardware or software level) which allows the developer of parallel systems to assume a defect-free computational fabric? Or will we have to architect future parallel systems (hardware and software) with the explicit assumption that there will be errors as we do with distributed systems programming?

2. Examples

We present two examples of concurrent and parallel programming techniques which we believe may be valuable for expressing calculations that can be compiled into effective hardware. In some cases research projects have already made some progress with some of these techniques e.g. there is work reported on compiling join patterns into hardware.

The first example we present is the notion of a join pattern. Several languages support join patterns as language extensions but here we shall show how join patterns can be incorporated into an existing language (C#) by defining a suitable library module [7]. These asynchronous join patterns can then be used to define highly efficient concurrent programs or they can be analyzed to produce digital circuits.

Join patterns express conditions over a collection of ports that will caused multiple values to be atomically consumed and an associated action to be executed in an asynchronous context. The C# code below shows how one can define a port and associate actions to be executed when a message arrives on the given port.

```
Port<int> pi = new Port<int>() ;
pi.post (42) ;
activate (pi.with(delegate (int i)
        { Console.WriteLine(i) ; }));
activate (!pi.with(delegate (int i)
        { Console.WriteLine(i) ; });
```

The port is created, has 42 written to it and then two handler functions are defined which compete to capture a value and write it out. The action to be performed is expressed in lambda expression style using C# 2.0 delegates which allows a function to be defined inline anonymously. The first activation defines a single shot handler which terminates after processing the message. The second handler is invoked with the ! operator which makes it re-activate itself after processing each message.

Non-deterministic choice is expressed using the | operator:

```
activate(p.with(MyIntHandler)
        |
        p.with(MyStringHandler);
```

which is similar in meaning to the Ada select statement code shown below:

```
select
  accept intPort (iv : in integer) do
    // int handler code
or
  accept stringPort (sp : in string) do
      // string handler code
end select ;
```

Note that the handlers are typed and the channel can be overloaded to carry different types. The type of the handler determines which messages it will process.

The code below shows an example of a join pattern:

```
activate(!join<int, int>(balance, deposit)
        .with(delegate(int b, int d)
            { balance.post(b + d); }));
```

This join pattern works by observing the channels for **balance** and **deposit**. When values are available in both channels the join pattern atomically consumes both values and binds them to b and d respectively and then runs the code specified in the delegate (which itself causes another asynchronous event to be triggered).

Using these simple building blocks it is possible to build high efficient concurrent programs that comprise of a large number of "threads" which can be multiplexed onto a small number of heavyweight operating system threads. One can also use this infrastructure to provide an alternative way to model large digital circuits which does not use an event-based kernel like System-C. One advantage of such a formulation is that it results in an automatic parallelization of the circuit descriptions on multi-core processors.

Taking this approach one step further leads us to consider these descriptions as input descriptions for synthesis to digital circuits. Such an approach is described in the work on Hardware Join Java [15]. One way of compiling such join patterns is to translate their basic semantics into synthesizable VHDL or Verilog. This approach has the advantage of raising the level abstraction for the circuit designer and at the same time exploiting mature synthesis technology to produce an efficient implementation.

The second example we focus on is on using functional languages for expressing implicitly parallel programs. The join-based approach still requires the programmer to express their program explicitly in terms of concurrent or parallel activities and then to explicitly use synchronization constructs to facilitate inter-thread communication. Implicit parallel programming techniques involve taking a sequential program with some annotations and then automatically parallelizing it. If this approach can be made to work then it has several advantages over explicit approach e.g. the model for writing and debugging code is much simpler since the programmer only has to think in terms of one program counter (although of course the implementation will create multiple threads which will each have their own program counter).

For our examples we pick the lazy functional programming language Haskell and show how one can write parallel annotations. As a starting point we use the following sequential code for the Fibonacci calculation:

```
nfib :: Int -> Int
nfib 0 = 1
nfib 1 = 1
nfib x = nfib (x-2) + nfib (x-1) + 1
```

The first line defines the type of the nfib function: it takes an integer and returns an integer. The next three lines defines the nfib function for three different cases by pattern matching on the argument. If the nfib function is called with 0 it returns 1, if it is called with 1 it returns 1 and if it is called with any other value then it makes the recursive calculation shown on the last line.

To make a parallel version of this code we can use the x `par` y annotation in Haskell which says that the calculation of x should be undertaken by another thread if possible because its value will be used later as part of the process of calculating y. This scheme has some similarities with futures in other languages except in the Haskell case due to the controlled nature of side-effects it is always safe to speculatively evaluate x in parallel with the parent thread. In addition to the par combinatory we need a x

`seq` y combinator to force the evaluation of expressions to ensure they are ready to be used at the appropriate time i.e. x is evaluated before computing y. This lets us now write the following parallel code for the Fibonacci calculation (with the 0 and 1 cases omitted):

```
parfib :: Int -> Int -> Int
parfib n
   = n1 `par` (n2 `seq` (n1 + n2 + 1))
     where
     n1 = parfib (n-1)
     n2 = parfib (n-2)
```

The parfib function requests that n1 be calculated in parallel with the calculation of the result of the function. If there is an thread available then the calculation of n1 i.e. the recursive case for parfib (n-1) will be calculated in parallel and be ready when the parent thread needs it. The parent thread uses a call to seq to ensure that the parallel task created to compute the other recursive call parfib (n-2) has completed and it then combined the two parallel sub-calculations to return the final result.

Using the multi-threaded GHC Haskell compiler on a eight core machine and adjusting the code above to perform thresh-holding yields nearly linear speedups. The run-time system will dynamically decide which calls to schedule for parallel execution and it will also perform load balancing. Furthermore, this parallelization has been achieved without explicitly forking off any threads or having to write any explicit synchronization code.

It is possible to build up higher order combinators that describe common patterns of parallel computations which can all be expressed in terms of par and seq (these are called *strategies*). For example, here is how we can define a parallel map operation:

```
parmap :: (a -> b) -> [a] -> [b]
parmap  f [] = []
parmap f (x:xs)
  = fx `par` pmxs `par` (fx:pmxs)
    where
    fx = f x
    pmxs = parmap f xs
```

This takes an operation f and a list (x:xs) where x is the head element and xs is the tail of the list and applies f in parallel to every element in the list. This function can then be used to define circuits which contain replicated elements which operate in parallel. We believe that this style of implicit deterministic parallel programming has much to offer developers of concurrent programmers and designers of digital circuits and SoCs.

3. Conclusions

This paper has briefly described a wide array of concurrent and parallel programming techniques that are candidates for the high level specification, modeling and implementation of digital circuits and SoCs. We have described in detail two contrasting approaches: one which uses explicitly created threads and explicit synchronization to yield efficient non-deterministic programs and the other which relies on annotations to yield deterministically parallel programs. Both approaches have already been used to generate circuits and we propose further research in the areas of compiling join patterns to hardware and for compiling data-parallel function descriptions to hardware. The join-base approach has the advantage of presenting an incremental approach which can be incorporated into an existing language. However, it is challenging to write explicitly parallel programs and to write in an explicit continuation passing style. The functional approach has the advantage of presenting a very simple model for implicit parallelism but it does require the system designer to learn a very different programming paradigm. Such approach are difficult to incorporate into mainstream languages due to the presence of side effects and the difficulty of performing programs transformations in the presence of side effects which makes it hard to implement techniques for nested data parallelism.

Although we describe just a few parallel programming approaches in their early stages of research for joint compilation to hardware and software we believe the direction of future many-core processor architectures will mean that much more research will need to be done to help create languages and models which allow us to describe calculations that can map efficiently to either hardware and software and the best way to do this is to re-unite hardware design and parallel programming into an interoperable set of models.

REFERENCES

[1] Krste Asanovic, Ras Bodik, Bryan Christopher Catanzaro, Joseph James Gebis, Parry Husbands, Kurt Keutzer, David A. Patterson, William Lester Plishker, John Shalf, Samuel Webb Williams and Katherine A. Yelick. *The Landscape of Parallel Computing Research: A View from Berkeley.* EECS Department University of California, Berkeley Technical Report No. UCB/EECS-2006-183, December 18, 2006.

[2] R.D. Blumofe, C.F. Joerg, B.C. Kuszmaul, C.E. Leiserson, K.H. Randall, and Y. Zhou, *Cilk: An efficient multithreaded runtime system*, Proceedings of the 5th ACM SIGPLAN Symposium on Principles and Practice of Parallel Programming, July 1995, Santa Barbara, CA, 207–216.

[3] Benton, N., Cardelli, L., Fournet, C. Modern Concurrency Abstractions for C#. ACM Transactions on Programming Languages and Systems (TOPLAS), Vol. 26, Issue 5, 2004.

[4] S. Borkar, *Designing Reliable Systems from Unrealiable Components: The Challenges of Transistor Variability and Degradation*, IEEE Micro, Nov.–Dec. 2005, pp. 10–16.

[5] C. Brooks, E. A. Lee, X. Liu, S. Neuendorffer, Y. Zhao, and H. Zheng. *Heterogeneous Concurrent Modeling and Design in Java*: Volume 1: Introduction to Ptolemy II. Technical Memorandum UCB/ERL M04/27, University of California, July 29 2004.

[6] D. Callahan, B.L. Chamberlain, and H.P. Zima. *The Cascade High Productivity Language*, in Proceedings of the 9th International Workshop on High-Level Parallel

Programming Models and Supportive Environments (HIPS 2004), IEEE Computer Society, Apr. 2004, pp. 52–60.

[7] Chrysanthakopoulos, G., Singh, S. *An Asynchronous Messaging Library for C#. Synchronization and Concurrency* in Object-Oriented Languages (SCOOL). October 2005.

[8] Conchon, S., Le Fessant, F. *JoCaml: Mobile agents for Objective-Caml.* In First International Symposium on Agent Systems and Applications. (ASA'99)/Third International Symposium on Mobile Agents (MA'99). IEEE Computer Society, 1999.

[9] Discolo, A., Harris, T., Marlow, M., Peyton Jones, S., Singh, S. *Lock Free Data Structures using STM Haskell.* Eigth International Symposium on Functional and Logic Programming (FLOPS 2006). April 2006.

[10] E. Allen, V. Luchango, J.-W. Maessen, S. Ryu, G. Steele, and S. Tobin-Hochstadt, *TheFortress Language Specification*, 2006. Available at http://research.sun.com/projects/plrg/

[11] S. A. Edwards and O. Tardieu, *SHIM: A Deterministic Model for Heterogeneous Embedded Systems.* In Proceedings of the ACM Conference on Embedded Software (Emsoft), Jersey City, NJ, September 2005.

[12] Fournet, C., Gonthier, G. *The reflexive chemical abstract machine and the join calculus.* In Proceedings of the 23rd ACM-SIGACT Symposium on Principles of Programming Languages. ACM, 1996.

[13] Fournet, C., Gonthier, G. *The join calculus: a language for distributed mobile programming.* In Proceedings of the Applied Semantics Summer School (APPSEM), Caminha, Sept. 2000, G. Barthe, P. Dybjer, , L. Pinto, J. Saraiva, Eds. Lecture Notes in Computer Science, vol. 2395. Springer-Verlag, 2000.

[14] Harris, T., Marlow, S., Jones, S. P., Herlihy, M. *Composable Memory Transactions.* PPoPP 2005.

[15] Itzstein, G. S, Kearney, D. *Join Java: An alternative concurrency semantics for Java.* Tech. Rep. ACRC-01-001, University of South Australia, 2001.

[16] Lee, E. A. *The Problem with Threads.* EECS Department, University of California, Berkeley, Technical Report No. UCB/EECS-2006-1. January 10, 2006.

[17] Ousterhout, J. *Why Threads Are A Bad Idea (for most purposes).* Presentation at USENIX Technical Conference. 1996.

[18] Peyton Jones, S., Gordon A., Finne S. *Concurrent Haskell.* In 23rd ACM Symposium on Principles of Programming Languages (POPL'96), pp. 295–308.

[19] Singh, S. *Higher Order Combinators for Join Patterns using STM.* TRANSACT 2006.

Efficient DSP Algorithm Development for FPGA and ASIC Technologies

Shiv Balakrishnan and Chris Eddington
Synplicity, Inc.

ABSTRACT

The use of Digital Signal Processing (DSP) in electronic products is increasing at a phenomenal rate. FPGAs, with their multi-million equivalent gate counts and DSP-centric features can offer dramatic performance increases over standard DSP chips. They also offer an attractive alternative for small and medium volume production FPGAs also make very powerful prototyping and verification vehicles for real-time emulation of DSP algorithms [1]. This paper discusses the challenges and requirements of creating portable algorithmic IP for FPGAs and ASICs and illustrates how an ESL synthesis methodology using Synplicity's Synplify DSP tool can significantly reduce the time and effort to implement either technology. The Synplify DSP tool automatically creates optimized logic implementations for both FPGAs and ASICs.

1. CHALLENGES IN PORTING RTL BETWEEN FPGA AND ASICS

The design team might ask: why is porting RTL between FPGAs and ASICs a problem? After all, is not RTL (synthesizable Verilog and VHDL) supposed to make the design portable? The answer can be quite lengthy and varies depending on the type of design. But for DSP algorithms, a general answer is that the RTL often specifies the exact mapping of key operations like multipliers, adders, and storage. Another way of saying this is that although the RTL is portable at the logic level, it is not at the architectural level. If synthesized to a different target, the same RTL will yield less than ideal results; in a different target technology, the result may be functionally correct but very sub-optimal.

Choosing an algorithm architecture involves the basic question of how much pipelining, parallelization, or serialization is needed to meet the sample rate and throughput requirements of the algorithm. In addition, fundamental DSP functions like FIR, FFT, sine, cosine, divide, etc. may have different optimal implementations depending on the target technologies. A good example is the direct form versus the transposed form of a FIR filter - one may be better for a particular FPGA device [2], and the other may be better for an ASIC technology.

Different architectures are usually required to get good results from an FPGA versus an ASIC. It's commonly known that FPGA devices tend to be more register-centric, and many ASIC-to-FPGA porting guidelines recommend lots of pipelining, registering of all ports, and breaking combinatorial logic into smaller portions. This results in an area increase if done in the ASIC, but might be required to meet timing in the FPGA [2].

For an ASIC target, the opposite is often desired. Register minimization is recommended to reduce area and power. Higher clock speeds can be exploited using time-multiplexing and resource sharing techniques to minimize multipliers and other expensive operations. Recent designs for the consumer and wireless markets balance these demands carefully.

One of the inevitable differences between ASIC RTL and FPGA RTL is the use of memory. In an FPGA, standard memory types are built into the device. Depending on the FPGA tool flow and vendor, specific coding styles are required to describe storage arrays and memories. High quality FPGA synthesis tools automatically infer memory use from the RTL. However, in the ASIC world, there are many different memory options available from IP and fab library vendors. Users select and compile memory for a particular configuration, and instantiate it into the RTL design.

There are many articles and resources describing the coding styles and porting techniques needed to move IP between FPGAs and ASICs [1], [3] and [4]. Such techniques include pipelining and building memory wrappers for handling resets, enables, and other differences between FPGA and ASIC memory interfaces. Suffice to say that it requires a significant amount of coding, verification, and expertise to move implementations between technology types.

Additional porting challenges arise in ASIC designs first prototyped in FPGAs. This occurs when real-time stimuli and at-speed verification are required. To support these requirements, it is necessary to maintain bit and sample accuracy between simulation models, in particular the FPGA implementation and the ASIC model. This can require a lot of effort especially if the implementations are different or changing rapidly, and the test harness must be manually modified, compared, and debugged.

2. AN ESL SYNTHESIS SOLUTION

Synplicity's Synplify DSP tool provides a powerful DSP synthesis methodology which overcomes many of the problems and issues described above. Synplicity's DSP synthesis concept is based on four key elements:

- Use of Electronic System Level (ESL) models with

978-1-4244-1709-4/07/$25.00 © 2007 IEEE

high levels of architectural and hardware abstraction

- Automatic optimizations based on user-specified sample rates

- User-selected target technologies

- Native support for multi-rate designs

With these elements, the DSP synthesis engine can synthesize different RTL implementations based on user constraints using system-wide, target-aware optimizations. The RTL, produced with optimized architecture and coding styles, can then be taken through the standard logic synthesis flow. Figure 1 shows this approach to highlevel design.

By using this ESL synthesis approach, designs are created and maintained at a high-level of abstraction, which increases portability, shortens development time, and improves the return on engineering effort. Instead of maintaining the IP at the RTL level, it can be done at the algorithm model level which increases portability and the return on the algorithm developer's efforts.

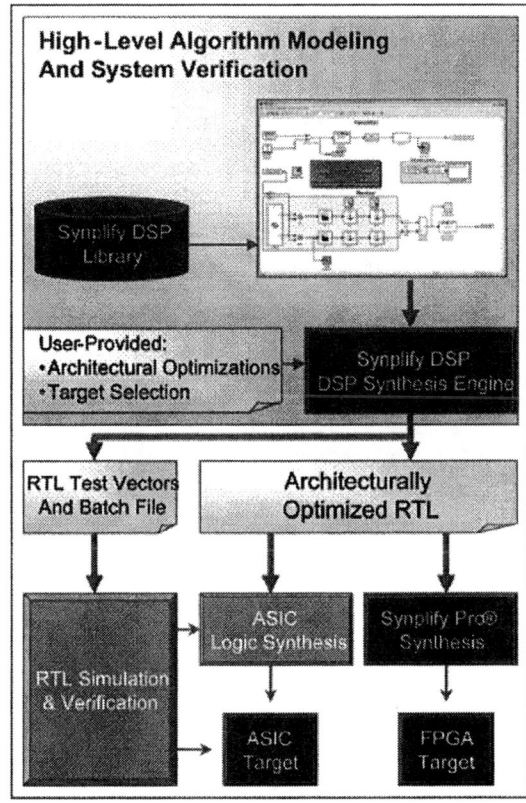

Figure 1: DSP Synthesis automates the generation of RTL optimized for a specific implementation target from a single ESL model.

DSP synthesis enables a user to quickly generate and explore a wide variety of implementations from a single algorithmic model as shown in Figure 2. A fully parallel and pipelined architecture can be used in the FPGA, or

a more serial, area efficient architecture can be used for an ASIC. Furthermore, bit and sample accuracy is automatically maintained across implementations as well as a complete verification path via standard RTL simulation tools. This contrasts with parameterized schematic entry or RTL methods which require users to commit to a specific architecture before being able to see its area-delay characteristics and require extensive modifications to port to a new implementation target.

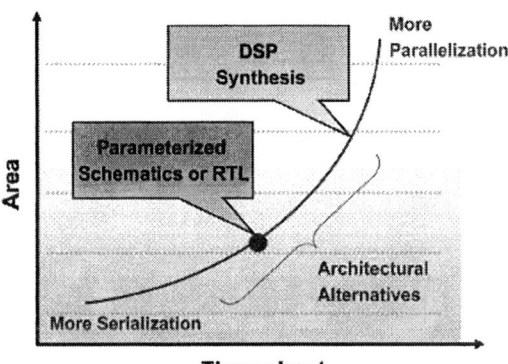

Figure 2: Rapid design exploration from a single ESL model

3. FEATURES FOR ASIC TARGETS

As discussed earlier, ASIC technologies and design flows are substantially different from those of FPGAs. As illustrated in Figure 1, there are some special features and capabilities required to support ASIC flows. Some of these features include:

- ASIC lithography performance characterization necessary for architectural optimizations

- Automatic memory extraction for flexible support of memory IP

- Choice of Synchronous/Asynchronous Resets

- RTL optimizations and suitable output files for smooth operation with ASIC logic synthesis flows

These ASIC-specific features, combined with complementary FPGA-specific features allow automatic porting of ASIC designs to FPGAs or vice versa replacing time-consuming manual coding and translation efforts.

4. DESIGN EXAMPLE

We illustrate these benefits with a simple FIR design example which highlights the power of high level modeling abstraction, architectural DSP synthesis, and target-aware optimizations. The first step is to create a model that includes fixed-point and sample rate behavior. The Synplify DSP library leverages the powerful quantization and multi-rate features in the Simulink environment to simplify fixed-point design capture and verification. The following figure shows a Synplify DSP model of a basic 16-tap FIR filter with fixed coefficients.

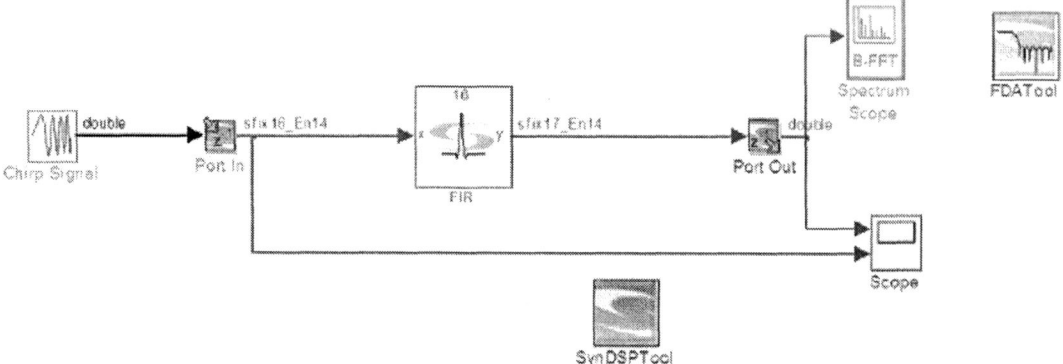

Figure 3: Simulink [5] block diagram of a 16-tap FIR. Also shown are the filter design tool, input waveform, and output analysis in time and frequency domains.

If we were to examine the specification of the filter itself, some of the key parameters are shown below in Figure 4. Note that the block inherits the fixed-point data type at the input and automatically propagates the data type according to user selected settings and the functionality of the internal calculations. During simulation, the exact quantized, discrete-time behavior can be verified. In addition, powerful analysis tools such as floating-point override and overflow logging can be used to explore the impact of word length and precision on the algorithm performance.

5. ALGORITHM IMPLEMENTATION USING DSP SYNTHESIS

Given this algorithmic model we can show the benefits of automatic DSP synthesis and architectural optimizations. An important detail to note is that the Synplify DSP algorithm model is vendor, technology, and architecture independent, that is, the simulation behavior is independent of these implementation choices. Not until the DSP synthesis step does the user define the target device and select architectural optimization choices. The Synplify DSP tool then synthesizes an optimized RTL implementation from the model. Figure 5 shows an example selection for an ASIC target.

Of particular note are the Retiming and Folding options. The Retiming option allows the Synplify DSP tool to modify the architecture to use pipelining and other techniques to get to the desired performance goal, at the expense of latency at the output. The Folding option allows the design to share hardware, at the expense of lower throughput (i.e. trade off maximum sample rate for resource utilization).

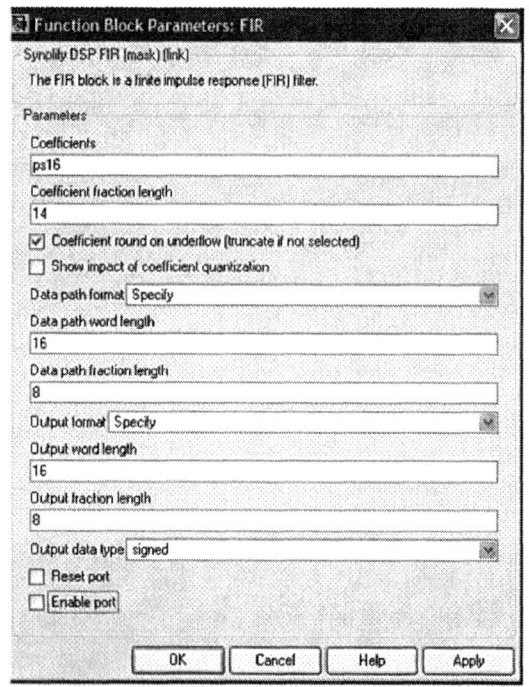

Figure 4: Parameters of the Synplify DSP FIR filter block. Note specified data formats.

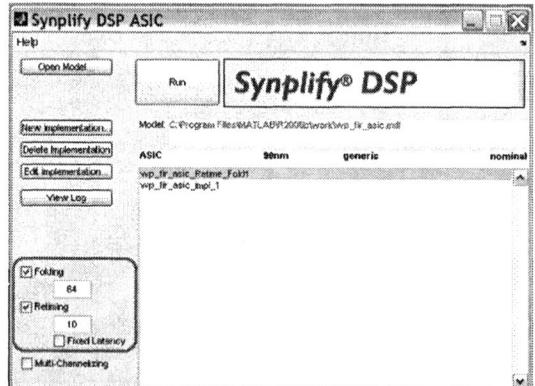

Figure 5: Defining parameters for architectural optimization to an ASIC target.

Table 1: Serialization and hardware sharing allows implementing the 65-tap FIR filter in half the area

Architecture	Extracted clock (MHZ)	Logic Area (gates)[1] 2.8 sq nm	# of 18x18 bit Multiplies	Memory (equiv. gates)[2] 2.8 * (sq nm)	Total core area (sq nm)
90nm Baseline (parallel)	10	53000	64[3]	0	53000
90nm Folding = 64 (serial)	640 MHz	11000	1	9800[4]	20800

6. ARCHITECTURAL EXPLORATION

The power of an automatic DSP synthesis engine is the ability to quickly explore a variety of architectures and target technologies. This design-space exploration process can lead to a significantly more optimal solution, especially if one is able to consider a variety of FPGA and ASIC technologies.

In this example, we will explore the retiming and folding optimizations and show how they offer significant tradeoffs in speed and area. First, we generate 4 cases of the 10 MHz 64-tap FIR filter in a Virtex-4 FPGA: a baseline and three different folding factors for area reduction. The results are generated by logic synthesis of the Synplify DSP RTL.

Table 2: Impact of automatically synthesized folding optimizations on filter throughput and HW sharing for Virtex-4 FPGA

Folding Factor	Est. Frequency	Max. Throughput	DSP48 Slices
None	414.5 MHz	¿ 400 Ms/s	16
4	346.7 MHz	86.7 Ms/s	4
8	352.0 MHz	44 Ms/s	2
16	338.5 MHz	21.2 Ms/s	1

A similar analysis for an ASIC implementation of the same design is shown below in Table 1, which shows the area difference in a 90nm technology for the two extremes i.e. fully parallel vs. fully serial implementation.

What we observe from Table 1 is that DSP synthesis provides area improvements automatically when lower sample rates permit shared hardware. Alternatively, the powerful ESL capability permits easy mapping to different technologies while exploiting higher clock frequencies. At the same time, working from a single algorithmic model eliminates the need to change or re-verify the model.

7. CONCLUSION

The simple FIR example above shows that the Synplify DSP software provides a rapid and efficient means of making architectural tradeoffs based on accurate simulation of their relative performance and area. This enables the user to explore multiple architectural possibilities, including important implementation details such as fixed point considerations, while efficiently obtaining useful cost versus performance data. The result is optimal FPGA and ASIC implementation of high level algorithms while minimizing design time.

The EDA industry seems to be evolving towards delivering on early promises of ESL design [6], both in an integrated

design flow for hardware prototypes as well as shipping systems [7]. Improvements to tools such as Synplify DSP will be crucial to further this advance in performance-sensitive applications.

8. REFERENCES

[1] M. Serughetti. System integration and testing before first hardware availability? it's possible! In *CoWare, Inc., SoC Central*, 2007.

[2] Xilinx Inc. Dsp: Designing for optimal results. In *Advanced Design Guide*, 2005.

[3] C. Eddington. Efficient development of wireless ip with high level modeling and synthesis. 2006.

[4] A. Haines. Software-intensive asics/assps demand integrated prototyping solutions. In *Synplicity Inc., EDA DesignLine*, 2007.

[5] MathWorks. Matlab r2006b. In *Version 7.3.0.267*, 2006.

[6] S.Bloch. Focusing on primary esl design solutions. In *http://www.chipdesignmag.com/display.php?articleId=879*.

[7] Design community trend survey reveals hidden market for esl and fpga. In *www.embeddedstar.com/press/content/2005/9/embedded18796.html*.

Incremental Placement for Structured ASICs using the Transportation Problem

Andrew C. Ling

Department Electrical and
Computer Engineering
University of Toronto
Toronto, Canada
e-mail: aling@eecg.toronto.edu

Deshanand P. Singh

Altera Corporation
Toronto Technology Centre
Toronto, Canada
e-mail: dsingh@altera.com

Stephen D. Brown

Altera Corporation
Toronto Technology Centre
Toronto, Canada
e-mail: sbrown@altera.com

Abstract— While physically driven synthesis techniques have proven to be an effective method to meet tight timing constraints required by a design, the incremental placement step during physically driven synthesis has emerged as the primary bottleneck. As a solution, this paper introduces a scalable incremental placement algorithm based upon the well known *transportation problem*. This method has an average speedup of 2x and a 30% reduction in memory usage when compared against a commercial incremental placer without any impact on area or speed of the final placed circuit. Furthermore, this method is scalable for structured ASICs.

I. INTRODUCTION

As the complexity of circuits increases exponentially, traditional synthesis methods often fall short of finding a timing-optimal solution. The primary reason for this is that delays of an unplaced circuit are hard to estimate, thus the synthesis tool lacks the knowledge required to optimize for timing. A solution to this is to resynthesize the circuit incrementally after placement. The benefit of doing this is that delays can be modeled accurately after placement since cell positions are known. This incremental process is known as physically-driven synthesis which we will refer to as physical synthesis [12]. The

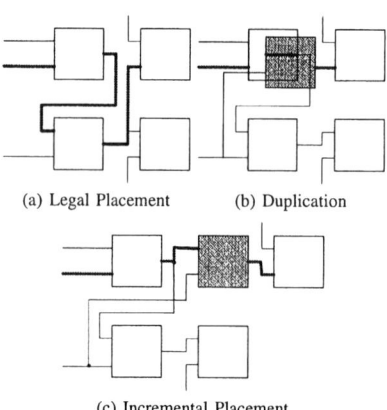

(a) Legal Placement   (b) Duplication

(c) Incremental Placement

Fig. 1. Physical synthesis illegality example.

optimizations that occur during physical synthesis will usually lead to a placement that is illegal. For example, consider Fig. 1. Fig. 1a is an initial placement of three logic cells where the highlighted connections show the critical path. The length of a wire is often proportional to the delay of that wire, thus most timing optimizations attempt to shorten the critical path. One method to do this is through duplication [3]. Nodes feeding a critical connection are duplicated

and each duplicated node is moved to its critical fanout connection as shown in Fig. 1b. A problem with this optimization is that congestion often occurs after the duplication, which needs to be legalized. Physical synthesis deals with these illegalities with an incremental placement (ICP) step. During ICP, all illegalities are resolved while maintaining the quality of the optimizations done by physical synthesis. Quality is maintained by minimizing the disruption of the original placement. Going back to the previous duplication example, ICP is shown in Fig. 1c. As we will describe shortly, the main problem with current incremental placement algorithms is their localized approach to remove illegalities in the placement. This has proven to be very effective for FPGAs where the number of illegalities incurred after physical synthesis is small [11]. However, as physical synthesis algorithms are migrated to larger devices, such as structured ASICs, current ICP techniques will not scale. As a solution, we present a novel ICP algorithm for structured ASICs using the *transportation problem* [7]. The benefit of this method is that it solves illegalities on a global scale. Thus, more illegalities can be resolved at once in a very fast and efficient manner which outperforms a commercial incremental placer both in terms of runtime and memory use. The novelty of our ICP algorithm is identified by three major heuristics which have proven pivotal in creating a fast incremental placer without any degradation to circuit performance. This includes the guided movement of logic cells to remove illegalities, the use of the transportation problem to create guide paths for logic cell movement, and an ordering scheme that ensures congested regions get placed before uncongested regions.

II. BACKGROUND AND MOTIVATION

A. Structured ASICs

Structured ASICs differ from full-custom chips since logic cells are predefined to form basic building blocks of the chip. These building blocks are used to form more complex cells which are uniform in height, but can vary in width. The goal of placement is to arrange cells such that no overlaps occur, while optimizing the circuit timing and area. From a conceptual point of view, valid placement regions form a grid where each point of the grid is given a planar coordinate (x,y). A single cell can occupy multiple adjacent grid coordinates due to their varying widths. Thus, a legal cell placement will have at most one cell covering each grid coordinate. Fig. 2 shows an example legal placement.

B. Previous Work

In general, placement algorithms for ASICs are tackled from a global perspective. This is necessary to ensure the placer runs quickly. In [8], the authors present an incremental placer based on floorplan

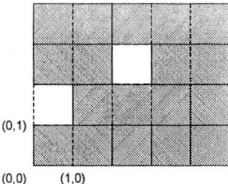

Fig. 2. Legal placement of cells on a grid.

management. They treat ICP as a resource allocation problem where congested areas need much more space resources than non-congested areas. In order to solve this problem, the authors propose a method to move space resources from non-congested areas to congested areas. This method guarantees that the relative position of cells remains the same, thus not affecting the original placement quality. A problem of this approach is that it assumes that the placement area is homogeneous. For heterogeneously structured ASICs, some regions are reserved for specialized logic, such as RAM blocks; thus this ICP technique is not applicable to heterogeneously structured ASICs.

In [13], an algorithm called FastPlace is described. In this work, the authors use a cell shifting technique to fix any overlaps created by a quadratic placer. This greatly speeds up the convergence of the quadratic placer to a final legal solution. Since cell shifting must be interleaved with a quadratic placement algorithm, all cells will be disturbed, including those cells that are already legally placed. This is not desirable in ICP since ICP should move only those cells that are illegal to maintain the quality of the original placement.

In [4], the authors present a placement tool called Domino. Like FastPlace, this work uses a quadratic placer [6] to get a loose placement of the cells. This is followed by detailed placement where the placement is legalized using network flow. However, as in [13], the detailed placement step disrupts all cells and thus cannot be applied to ICP. An improvement to this is presented in [5]. Here, a network flow approach is also used to legalize a globally optimized placement, however, they use critical path delay information and focus on optimizing critical paths.

In [2], a transportation based algorithm is used to place cells. Their algorithm is based on partitioning to spread and place cells. This is also not suitable for incremental placement since it causes too much disruption to the existing placement picture. In later sections, we will show how we can adapt previous placement techniques to ICP while minimizing the disruption of the original placement.

1) Localized ICP: Finally, we review the incremental placement algorithm used in Altera's QuartusII. An early description of this tool was published in [11]. This algorithm has proven very successful for FPGAs and has recently been adapted to Altera's structured ASIC family. The primary goal of ICP is to resolve the architectural violations created when the physical synthesis modifications are integrated into the existing placement. In modern FPGAs [1] [14], architectural constraints are found in the clustered logic blocks (also known as LABs) including: a limit on the number of logic elements within the cluster; a limit on the number of distinct inputs to the cluster; and a limit on the number of distinct control signals (e.g. clock, reset) that can be used within the cluster. Structured ASICs have much simpler constraints where the primary constraint is overlaps. This is analogous to a limit on the number of logic elements in a cluster, where in structured ASICs only one cell can occupy a given location. Thus, the ICP algorithm described here primarily needs to remove overlaps between neighbouring cells and ignores other constraints specific to FPGAs.

The ICP algorithm uses an iterative improvement strategy where

cells are moved according to a cost function. This cost function consists of three components:

- *Legality Cost.* Each cell is penalized if it contains any architectural violations such as overlaps. The cost is proportional to the total number of constraints violated.
- *Timing Cost.* The timing cost is used to ensure that critical cells are not moved into locations that would significantly increase the critical path delay.
- *Wire-length Cost.* Wire-length estimation is used to ensure that the circuit is easily routable after the cell moves.

The total cost is a weighted sum of these components:

$$C = K_L \cdot Legality + K_T \cdot Timing + K_W \cdot Wire-length \quad (1)$$

The weighting coefficients (K_i) are used to normalize the contribution of each component so that the components contribute equally when considering a move.

There are several types of moves that can be generated for each candidate cell, x, to be moved:

- *Move to fanin.* Attempt to move x near a location that contains a fanin of x.
- *Move to fanout.* Attempt to move x near a location that contains a fanout of x.
- *Move to sibling.* A sibling of x is a cell driven by a fanin of x. We attempt to move x near a location containing one of its siblings.
- *Move to neighbor.* Attempt to move to an adjacent location.
- *Move to space.* Attempt to move to a randomly selected empty cell location in the device.
- *Move in direction of critical vector.* Compute a *critical vector* for x and move to a location in this direction.

The moves to fanins, fanouts, and siblings are essential to ensure that wire-length is not degraded. The critical vector for a cell x is shown in Fig. 3. The direction of the critical vector is computed by summing the directions of all the critical connections attached to x. A move in the direction of the critical vector helps correct any mistakes when unexpected paths become critical as a result of moves performed in preceding iterations. Note that the critical vector move is similar to the move types attempted by iterative force-directed placement algorithms.

Although the selection of move type is random, the proposed moves are biased in the direction of free space. For example, if the target device had a large amount of free space close to the top edge of the chip, then moves that move cell closer to the top edge are selected more frequently.

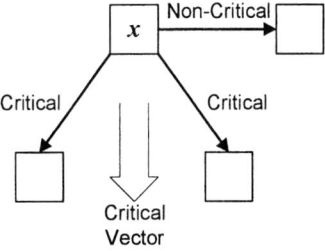

Fig. 3. A critical vector for a cell x.

The ICP algorithm makes cost-lowering moves until no further illegalities exist in the placement. This greedy strategy can easily become stuck in a local minima where none of the proposed moves seem to improve the cost. For example, consider the situation illustrated in Fig. 4. Assume that the current solution is x and that it still contains illegalities that need to be resolved. All moves in the neighborhood of x increase the cost of the solution and ICP is unlikely to find the global minima, y, through a process of randomized local moves. However, through *basin filling* we can change the shape of the cost function so that ICP is forced to move out of the local minima. Basin filling is achieved by slowly increasing the weight for those illegalities that are consistently present in the local solution space, thereby forcing ICP to select moves that resolve illegality at the cost of increases in timing and/or wire-length. Our basin filling approach is similar to the notion of negotiated congestion in the PathFinder [9] algorithm used for FPGA routing.

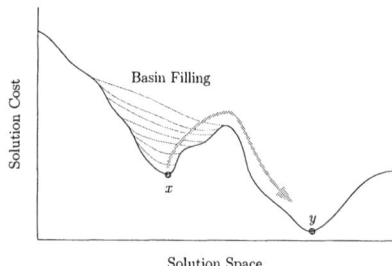

Fig. 4. Using basin filling to escape local minima.

Localized ICP has proven very successful to remove illegalities without harming the placement quality. However, the algorithm described previously can consume more than half the runtime of physical synthesis due to its localized nature. Furthermore, a large amount of data is required during localized ICP to maintain a history of the basin filling and cost function. Thus, this method is not scalable to large chips that contain millions of cells. In our approach, we resolve this problem in ICP by adapting techniques from both the localized ICP method and analytical placers described previously.

C. Transportation Problem

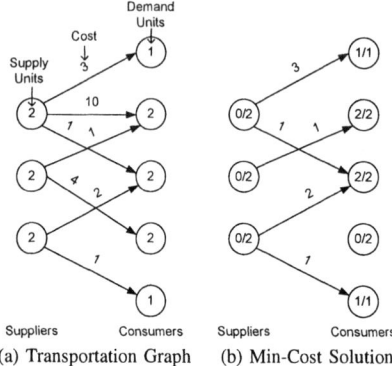

Fig. 5. Illustration of the transportation problem.

The incremental placer we have developed is based on a guided shifting technique where guide paths are calculated using the transportation problem: given a set of suppliers and consumers at fixed locations where there exists a set of directed paths from suppliers to consumers with an associated cost, find a subset of paths that fulfills the demands of all the consumers or exhausts the supply of all the

suppliers such that the total cost of the used paths is minimized. The directed paths are also known as edges. Edges leave the suppliers and enter the consumers, thus the suppliers are the edge *tails* and the consumers are the edge *heads*. Fig. 5 illustrates the transportation problem. Fig. 5a illustrates a possible set of consumers and suppliers connected by a set of paths, and Fig. 5b illustrates a minimum cost path solution such that the supply of all suppliers is exhausted. There has been much work that has been done to solve the transportation problem quickly [7], [10] and is not discussed here.

III. INCREMENTAL PLACEMENT

A. High-Level Algorithm

1	EMPTYQUEUE()
2	**foreach** $l \leftarrow IllegalLocation$
3	PUSHQUEUE(l)
4	**end foreach**
5	$s \leftarrow$ SIZEOFQUEUE()
6	$c \leftarrow 0$
7	**while** NOTEMPTYQUEUE()AND($c <$ CUTOFFPOINT(s))
8	$l \leftarrow$ POPQUEUE()
9	$c \leftarrow c + 1$
10	$R \leftarrow$ CREATESUBREGION(l)
11	REMOVECELLS(R)
12	$L \leftarrow$ PLACECELLS(R)
13	PUSHQUEUE(L)
14	**end while**
15	**if** NOTEMPTYQUEUE()
16	CALLLOCALICP()
17	**end if**

Fig. 6. High-level overview of transportation-based incremental placer.

A high-level overview of our incremental placer is shown in Fig. 6. The algorithm starts off by identifying all illegal positions found in the placement solution and placing the locations in a queue (line 1-4). These locations identify where overlaps occur in the cell grid. After the illegality queue is formed, each illegal location is processed sequentially (line 7). A subregion is created around the current illegal location. Within this subregion, all cells are removed from the region creating an empty area. The removed cells are then re-placed in the subregion one by one in such a way that most, if not all, illegalities in the subregion are removed. This is the main area of speedup when compared against the localized ICP algorithm, which in contrast moves cells one by one. It is possible for new illegalities to be created during the re-placement process (line 12, set L). If so, these are added to the back of the illegality queue (line 13). This ensures that the original illegalities are processed before newly formed illegalities.

The original size of the illegality queue is used to form a cutoff point (line 7). If the number of illegal locations processed reaches this cutoff point before the placement is legalized, the incremental placer falls back on a localized algorithm similar to that described in Section II-B.1. This finishes up any fine illegalities that could not be resolved with the transportation-based incremental placer (line 15-17).

B. Re-Placement in Subregion

After cells have been lifted in the subregion, they must be re-placed in a manner that removes illegalities. We use a shifting technique to accomplish this. The shifting works by first placing a cell at its original location (i.e. where it was lifted from). If that location is occupied by another cell, it is shifted until a legal position is found. If no legal position can be found within a fixed radius from its original location, it is placed back at its original location creating new illegalities. The shifting is guided by a direction vector which is calculated using the transportation problem.

C. Setting Up Transportation Problem

Formulating the transportation problem involves creating a set of suppliers and consumers and adding a set of relevant paths with associated costs between these points to form a transportation graph. In ICP, the goal is to move cells to valid locations such that all illegalities are removed. In order to represent this problem as a transportation problem, we need to introduce subcells. A subcell can be thought of as any part of the cell that occupies one grid location on the structured ASIC. Since the width of a cell spans multiples of x units on the placement grid and has a unit height, the number of subcells in a cell is equivalent to its width. Breaking up cells into subcells allows the cells to be thought of as suppliers who need to ship their subcells to a set of location consumers, where the demand of each location is at most one subcell. This is illustrated in Fig. 7. The heuristic used for determining the costs of paths between a cell and location will be described in the next section. Another consideration

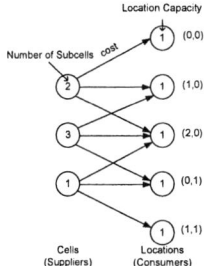

Fig. 7.   Representing incremental placement as a transportation problem.

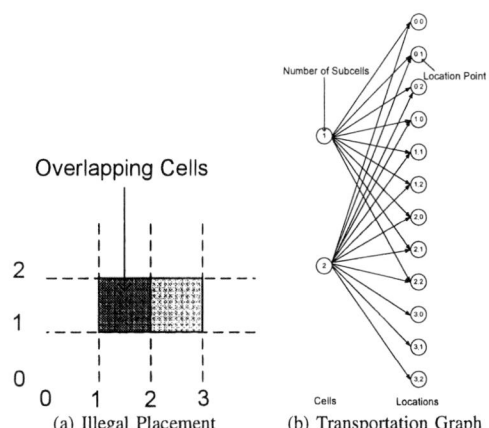

(a) Illegal Placement   (b) Transportation Graph

Fig. 8.   Incremental placement example.

Also, when splitting up cells into subcells, we assume that subcells will share the same path costs as the single cell the subcells were derived from.

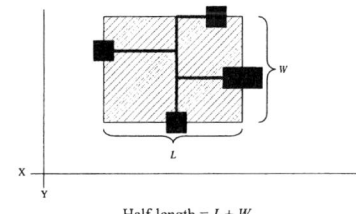

Half-length = $L + W$

Fig. 9.   Half-length perimeter of a net.

when forming the transportation graph is determining which locations will connect to which cells. One approach is to connect all locations to every cell as in [4]. The problem with this is that it creates a very dense transportation graph. Since transportation solver runtimes are proportional to the number of edges in the graph, a dense transportation graph will take much longer to solve than a sparse graph. Furthermore, ICP should minimize disruption of the original placement. Thus, we connect cells only to locations that are relatively close to the original location of the cell. For example, consider the illegal placement shown in Fig. 8a. Fig. 8b shows the transportation graph created using the illegal placement where edge costs have been ignored for simplicity. Notice that paths only exist between cells and locations that are fairly close to the original location of the connecting cell. After the transportation solver is run on the graph, a set of paths will be selected. The locations found at the heads of the selected paths are used to determine a suggested location for the cell found at the tail of the path.

1) Transportation Problem Path Costs: Since the transportation problem is a cost minimization problem, the edge costs in the transportation graph must ensure that the transportation solver will find cell locations that reduce the overall delay between cells. One estimate of delay is wire-length where delay is proportional to wire-length. Thus, using a path cost that represents wire-length will ensure that the overall wire-length, and hence delay, will be minimized. The metric we use to estimate the wire-length of a net is the half-length perimeter [4]. The half-length perimeter is the width plus height of the bounding box containing all the pins of a net as illustrated in Fig. 9. We assume that the cell pins are at the center of the cell.

The final cost of a path between a cell and a location is the sum of the half-lengths for all nets connected to the cell when placed at the given location as shown in Fig. 10. When costing the edges this way, we are assuming that the net cells connected to the current cell are stationary. For example, referring back to Fig. 10, while costing the edge costs for the center cell, we are assuming all the outer cells will not move. Although this is not true during the ICP process, we can make this assumption since we know that during ICP, cells will only move slightly since the majority of cells are already legally placed. In contrast, if cells were to move dramatically, such as during the initial stages of cell placement, our edge costs will be invalid leading to very poor placement solutions. This is a similar problem to physical synthesis optimizations since they use delay values of the current placement to guide their optimizations, even though the final placement after physical synthesis will be different due to ICP. Thus, even though some cells will move during ICP, the initial delay values used in physical synthesis optimizations are sufficient approximations.

D. Directed Replacement of Cells

Although the subcells of a single cell will usually be placed close to each other because they share the same edges and costs, there is no guarantee that subcells of a single cell will be placed adjacent to each other by the transportation solver. In order to consolidate the subcells, their center of gravity is used as the suggested location of the cell. This suggested location is used to create the directed vector needed by the guided shifting of cells as illustrated in Fig. 11a. Fig. 11b illustrates the shifting process if the original cell location is occupied.

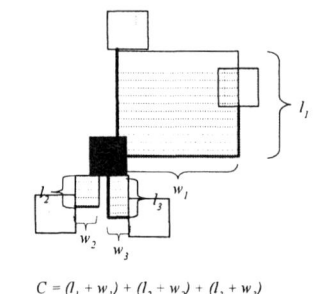

$$C = (l_1 + w_1) + (l_2 + w_2) + (l_3 + w_3)$$

Fig. 10.   Path cost calculation.

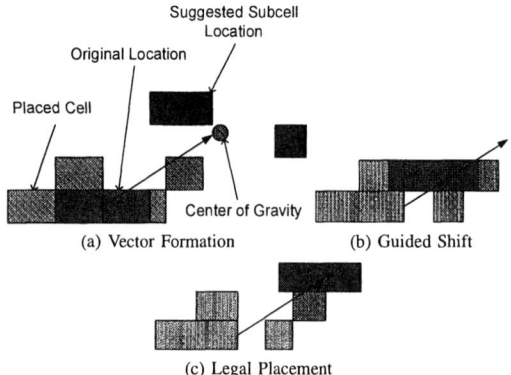

(a) Vector Formation   (b) Guided Shift

(c) Legal Placement

Fig. 11.   Example of replacement using suggested location from transportation problem.

This will continue until a final valid location is found as shown in Fig. 11c. In cases where the region is extremely crowded, some cells may be pushed fairly far. We account for this by creating a radius limit for placement. If a cell cannot be legally placed by the time it reaches the radius limit from its original location, it will be placed back at its original location, creating some overlap illegalities. Any new overlap locations will be placed at the end of the illegality queue for reprocessing later on.

E. Order of Re-Placement

After the subregion is emptied and all cells have been assigned a guided vector, re-placement of cells can occur. Cells are placed one at a time. The order of re-placement is extremely important in determining the quality of the final solution. Crowded regions should be placed before uncrowded regions since crowded regions have much less freedom for cell placement. Furthermore, cells on critical paths should be placed before other cells to ensure they will not be shifted far from their original location. Placing cells in crowded regions first is achieved by starting re-placement at illegal locations and working outwards. If there are more than one illegal location in the area, the illegal locations that contain cells on critical paths are placed first. This process is illustrated in Fig. 12. Fig. 12a illustrates the initial subregion and identifies the locations where illegalities occur. Note that one of the illegal locations is given higher priority since it contains critical cells. Next, placement regions are created (Fig. 12b). Placement will occur starting at each illegal location, then continuing in each vertical direction of the placement region (Fig. 12c). Once placement in the vertical regions completes (Fig. 12d), the placement regions are expanded in the horizontal directions (Fig. 12e). Placement continues in the vertical direction of

the newly expanded placement regions (Fig. 12f,g) and the process continues until the entire region is placed (Fig. 12h). Once the entire subregion has been placed, the next illegal location in the illegality queue can be processed.

IV. RESULTS

We implemented our version of ICP within a commercial placer called QuartusII and compared it against the original ICP algorithm described in section II-B.1. Once implemented, we ran over 60 industrial benchmark circuits through physical synthesis where several designs contained more than 100K gates (designs are proprietary and cannot be disclosed). Table. IV shows our summarized results for our tests.

Percent Reduction		
Runtime	Memory	Clock Frequency
52%	35%	0.2%

TABLE I

GEOMETRIC MEAN REDUCTION IN RUNTIME AND MEMORY USE OF OUR ALGORITHM WHEN COMPARED AGAINST QUARTUSII (RAN THROUGH 60 INDUSTRIAL DESIGNS).

Table IV shows the percent reduction in runtime when comparing the transportation-based ICP versus the localized ICP, with a geometric mean reduction of 52% or 2x speedup. Since it is known that ICP dominates the runtime during physical synthesis, this is a significant reduction in the overall physical synthesis runtime. Also shown is the percent reduction in memory usage, with a geometric mean reduction of 32%. The reason for a reduction in memory is because the transportation problem is only run on small subregions of the chip, thus a much smaller portion of data needs to be created at any given time. The localized ICP, however, must contain data for all locations that may contain a cell due to the basin filling and complex cost function described in section II-B.1. Note that the improvements shown in Tab. IV had almost no impact on the maximum clock frequency of the circuits tested where we saw an average reduction of only 0.2%.

One concern that could be raised is related to the convergence of our algorithm. Since our algorithm may create new illegalities during our re-placement process, convergence will be a problem if a large number of illegalities exist at the start of our algorithm. Also, if all the chip illegalities are clustered in a small region, we will need several iterations to shift cells in surrounding regions to make space to remove the overlaps in the densely packed area. However, from our experience, we have noticed that less than 4% of chip locations will contain illegalities after physical synthesis. Furthermore, the placement overlaps tend to be distributed throughout the chip. Thus, under these conditions convergence is not a problem, and our global ICP algorithm should be able to remove most overlaps found after physical synthesis optimizations. Our 2x speedup supports this claim and implies that our algorithm converges quite quickly.

V. CONCLUSIONS

We have developed a novel incremental placer based upon guided shifts. This method works well for large designs where there only exists one constraint for placement. This method has proven to be much faster than the existing ICP found in an existing commercial placement tool. More importantly, it is scalable to structured ASICs both in terms of runtime and memory usage.

REFERENCES

[1] Altera. Component selector guide ver 14.0, 2004.

[2] U. Brenner and M. Struzyna. Faster and better global placement by a new transportation algorithm. In *DAC '05: Proceedings of the 42nd annual conference on Design automation*, pages 591–596, 2005.

[3] G. Chen and J. Cong. Simultaneous timing-driven placement and duplication. In *FPGA '05: Proceedings of the 2005 ACM/SIGDA 13th international symposium on Field-programmable gate arrays*, pages 51–59, New York, NY, USA, 2005. ACM Press.

[4] K. Doll, F. Johannes, , and G. Sigl. Domino: Deterministic placement improvement with hill-climbing capabilities. In *VLSI*, pages 3b.1.1–3b.1.10, 1991.

[5] S. Dutt, H. Ren, F. Yuan, and V. Suthar. A network-flow approach to timing-driven incremental placement for asics. In *ICCAD '06: Proceedings of the 2006 IEEE/ACM international conference on Computer-aided design*, pages 375–382, 2006.

[6] J. Kleinhans, G. Sigl, F. Johannes, and K. Antreich. Gordian: VLSI placement by quadratic programming and slicing optimization. *IEEE Transactions on Computer-Aided Design of Integrated Circuits and Systems*, pages 356–365, 1991.

[7] D. Klingman and R. Russel. Solving constrained transportation problems. In *Operations Research*, pages 91–106, 1975.

[8] C. Li, C.-K. Koh, and P. H. Madden. Floorplan management: Incremental placement for gate sizing and buffer insertion. In *Asia and South Pacific Design Automation Conference (ASP-DAC)*, pages 349–354, 2005.

[9] L. McMurchie and C. Ebeling. Pathfinder: a negotiation-based performance-driven router for FPGAs. In *FPGA '95: Proceedings of the 1995 ACM third international symposium on Field-programmable gate arrays*, pages 111–117, New York, NY, USA, 1995. ACM Press.

[10] K. G. Ramakrishnan. Solving two-commodity transportation problems with coupling constraints. *J. ACM*, 27(4):736–757, 1980.

[11] D. P. Singh and S. D. Brown. Incremental placement for layout driven optimizations on FPGAs. In *ICCAD '02: Proceedings of the 2002 IEEE/ACM international conference on Computer-aided design*, pages 752–759, New York, NY, USA, 2002. ACM Press.

[12] D. P. Singh, V. Manohararajah, and S. D. Brown. Two-stage physical synthesis for FPGAs. In *CICC '05: Proceedings of the 2005 IEEE Custom Integrated Circuit Conference*, 2005.

[13] N. Viswanathan and C. C.-N. Chu. Fastplace: efficient analytical placement using cell shifting, iterative local refinement and a hybrid net model. In *ISPD '04: Proceedings of the 2004 international symposium on Physical design*, pages 26–33, New York, NY, USA, 2004. ACM Press.

[14] Xilinx. Virtex-ii complete data sheet ver 3.3, 2004.

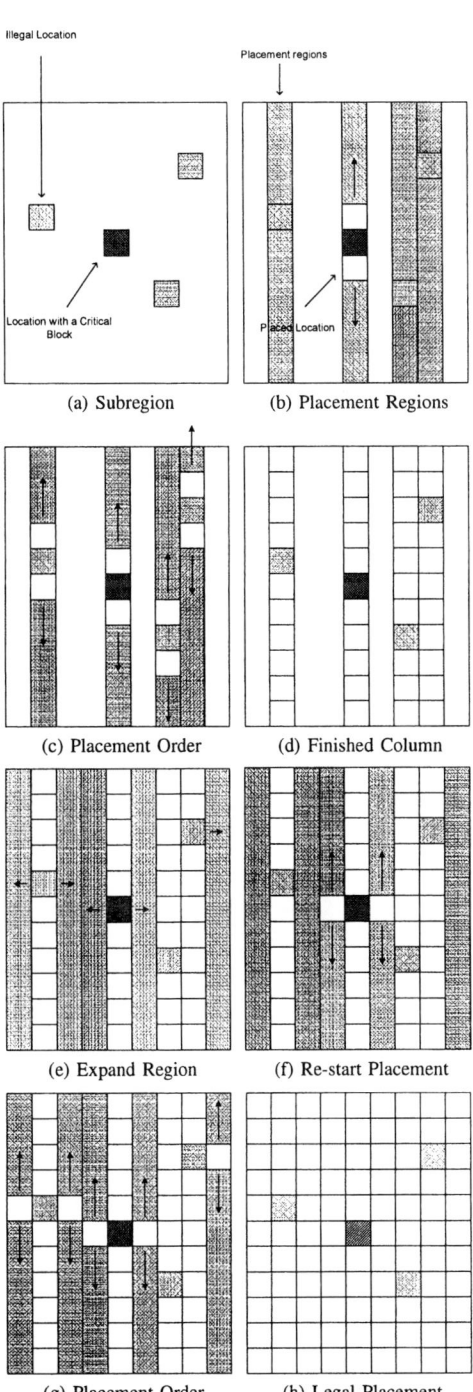

(a) Subregion (b) Placement Regions

(c) Placement Order (d) Finished Column

(e) Expand Region (f) Re-start Placement

(g) Placement Order (h) Legal Placement

Fig. 12.   Order of re-placement in illegal subregion.

Test Data Compression and TAM Design

Julien DALMASSO, Marie-Lise FLOTTES, Bruno ROUZEYRE
LIRMM, Univ. Montpellier II/CNRS
161 rue Ada, 34932 Montpellier cedex 5, France
{dalmasso, flottes, rouzeyre}@lirmm.fr, tel: (33)467418525, fax: 33)467418500

Abstract

Test Data Compression (TDC) techniques have been developed for reducing requirements in terms of Automatic Test Equipment resources. These techniques generally deal with stand alone circuits. In this paper, we explore the benefits of using TDC techniques in the context of core-based SoCs. TDC is used to reduce the test time by improving the parallelism of core tests without the expense of additional ATE channels. We first detail the constraints on test architectures and on the design flow inferred by the use of TDC. We propose a method for seeking an optimal architecture in terms of total test application time. The method is independent of the compression scheme used for reduction of core test data. The gain in terms of test application time for the SoC is over 50% compared to a test scheme without compression.

Keywords: System-on-Chip test, test resource partitioning, test data compression

1. Introduction

Testing a SoC mainly consists in testing each core in the system. In order to provide accessibility to cores, the SoC architecture is completed by a Test Access Mechanism (TAM) and wrappers interfacing cores with the TAM (IEEE 1500 standard [1]). The TAM is generally a bus whose bitwidth fits the number of SoC test IOs. The TAM and the wrappers must be co-designed in order to reduce the global Test Application Time (TAT) according to the available Automatic Test Equipment (ATE) channels and to the cores test parameters such as the number of internal scan chains. The best trade-off between number of buses, bus bitwidth, wrappers size in terms of I/Os number and test parallelism must be established. This can be formulated as an optimization problem and several methods have been proposed to solve it (e.g. [2], [3], [4], [5], [6]).

In the meanwhile, as the complexity of SoCs designs keeps on growing, testing becomes more and more expensive with regard to test time and test pins requirement. While increasing the number of scan chains helps to reduce the test time of a given core, it also increases the bitwidth of the interfaces between wrappers and TAM. As a consequence, it also increases the bitwidth of the interface between the SoC and ATE or decreases test parallelism.

Several Test Data Compression (TDC) techniques aiming at reducing the number of visible scan chains have been developed for stand alone cores. Concerning test pattern compression (also called horizontal compression), they consist in compressing test patterns off line (i.e. reducing their bitwidth), storing the compressed test data in the ATE, and decompressing test data on-chip for restoring initial test patterns; Input-data compression schemes rely on the fact that test patterns originally contain don't-care bits. These don't care bits do not have to be stored into ATE but can be supplied on-chip in some other ways. LFSRs [7][8], Xor networks [9] [10], ring generator [11], RAM [12] [13], arithmetic units [14] and test pattern broadcasting among multiple scan chains [15] [16] [17] constitute a range of solutions for minimizing the number of data to be stored into ATE. All these methods reduce therefore the number W_{ATE} of necessary ATE channels required to test a stand alone core including N scan chains with $N > W_{ATE}$.

Considering a parallel interface where $W_{ATE} = N$, increasing the number of internal scan chains and the interface bitwidth allows reducing the test time of a core since the resulting test scheme requires fewer scan-in clock cycles. However, for a fixed number N of scan chains, the reduction of W_{ATE} with the help of a compression scheme ($W_{ATE} < N$), may result in additional test time compared to a solution where the number of ATE channels equals the number of scan chains ($W_{ATE} = N$). In fact, in this case, there is no more a one-to-one mapping between the ATE channels and the scan chains. Indeed, no matter the TDC technique is used, the TAT may increase because compressing the N-bits vectors on W_{ATE}-bits words stored into ATE is not always possible. For keeping the fault coverage obtained with the original non compressed test sequence, it is then necessary to serialize the non-compressible patterns and to add a decompressor bypass mechanism or to look for additional compressible test patterns. In any case, a side-effect is an increase in TAT.

Concerning test responses, several methods have been proposed (e.g. [18], [19]). Conversely to TDC, those test responses compaction techniques do not impact TAT and are independent of the core netlist and of the test responses sequence. Thus, they can be directly employed in the framework of SoCs design. In the remainder of this paper, we focus on test pattern compression only.

Only few studies on the use of TDC at system level have been published. In [20], [21], [22] TAM architectures using TDC have been presented but they rely on specific TDC techniques. Moreover, all architectural solutions are not considered since these techniques essentially target TAM architectures with a single decompressor for all cores or architectures with a dedicated decompressor per core (or connected to duplicated versions of the same core).

We propose here a method for exploring all TAM/TDC architectures including dedicated and shared decompressors. The final goal is to generate test architectures and test schedules that minimize the system TAT. The proposed technique is independent of the adopted compression scheme.

Section 2 discusses the implication of TDC insertion at SoC level. The problem formulation as well as notations are given is Section 3. The algorithm is detailed is Section 4 whereas experimental results are reported in Section 5. Finally, Section 6 draws some conclusions.

2. SoC test architecture and compression

A SoC test architecture is proposed by the IEEE 1500 Standard. It mainly consists of a TAM bus and wrappers around cores. The TAM links the SoC's test IOs and the cores. Each core wrapper interfaces the core and the TAM bus. The test time of a core depends, among other things, on the size of its wrapper in terms I/Os interfaces with the TAM.

As in [2] and [3], we assume a TAM architecture organized around a partitioned test bus, each core being connected to one sub-bus, as depicted in Figure 1 in which the TAM is split into three sub-buses. Cores connected to the same sub-bus are tested serially (e.g. C1, C2, C3), cores assigned to different TAMs can be tested in parallel (e.g. C1 and C4 or C1 and C5). We do not make any assumptions about the wrappers of the cores: either the wrappers of the cores have been already designed (wrapper1500-ready cores), or they have to be designed when building the test infrastructure.

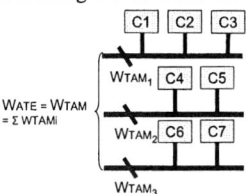

Figure 1: TAM architecture

In the rest of this paper, W_{TAM} denotes the TAM bitwidth, and W_{TAM_i} the bitwidth of each sub-bus.

Without changing W_{ATE}, the TAM bitwidth can be increased thanks to the use of TDC techniques. The tests parallelism can be therefore increased resulting in a shorter test time.

However, as explained in the introduction, TDC may also increase test time of individual cores. More precisely, for a fixed number N of scan chains (or equivalently for a fixed wrapper size), the test time of a core increases when the number of bits W_{ATE} at the input of the decompressor gets smaller. For instance, using the TDC technique presented in [14], the test of the S38417 benchmarks circuit with N=16 scan chains needs 21451 clock cycles when $W_{ATE} = 10$ and 38867 clocks cycles when $W_{ATE} = 3$ (see for instance, results given in Figure 7). In the remaining, the ratio W_{ATE}/N is denoted by ρ.

In this paper, we question the benefit of using TDC in the context of the design of test infrastructure for SoCs.

Let's recall that, under the chosen TAM model, building the test infrastructure mainly consists in: 1) finding a partition of the bus into p sub-buses and determining their bitwidth, 2) assigning the cores to the p sub-buses, and designing their wrappers 3) deriving a test schedule so that the total test time is minimized. An underlying data of these tasks is the test times of cores.

The test time of a core also depends on the size and the structure of its wrapper (size means here the number of visible scan chains). The wrapper size must be narrower than the sub-bus to which the core is connected to.

The use of TDC impacts the building of the test infrastructure in two aspects:

1) Since TDC modifies the test times of individual cores, the decompression ratios ρ must be established *during* the design of the test infrastructure and not after.

2) Decompressor sharing can be envisaged in two ways. Either a decompressor feeds several sub-buses (Figure 2.a) or feeds a single sub-bus (in Figure 2.b).

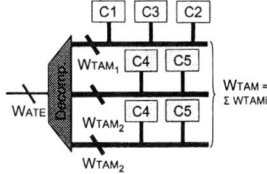

a) One decompressor for several sub-buses

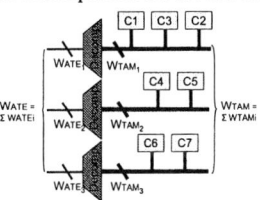

b) A single decompressor per sub-bus

Figure 2: TAM/decompressors architectures

The first architecture style leads to prohibitive CPU time when exploring architectural solutions. Indeed the evaluation of a solution (its test time) requires to define bus partitioning, core assignment, test parallelism, test sequence definition and finally compression of this sequence. Figure 2.a for instance depicts only one bus partitioning and core assignment possibility. It includes several test parallelism solutions (e.g. either C1 and C4 tested in parallel or C1 and C5). In turn, each one necessitates building the actual test sequence by concatenating the test sequences of the cores tested in parallel. Finally the resulting test sequence has to be compressed in order to obtain the actual test time. Another way of dealing with this model is 1) to build the optimal test infrastructure and related test schedule without looking at compression 2) derive the whole SoC test sequence and compress it. Doing so, there is no chance to obtain an optimal solution: the test infrastructure (without decompressor) is built given the test times of cores which are modified by the compression. We did such an experiment with the example given in section 5. Doing so, the obtained test time is 65699 cycles while a solution with 57941 cycles has been obtained using the method we propose here.

Conversely, the cores connected downstream a decompressor in the second architecture style are tested one after the other. The test sequences to compress are simply those of the cores and not issued from the concatenation of several ones. The compression of the test sequences can therefore be done independently of the test infrastructure building process. This alleviates the problems raised by the first model.

So in the remaining, we consider the second architecture style and we propose a method for *conjunctly* building up the TAM, the wrappers (if needed) and the decompressors.

Each path for the ATE channels, through a decompressor, up to sub-bus is called a *line*. The architecture depicted in Figure 2.b is composed of three lines. It must be noted that within this architectural model, and in the absence of additional constraints such as power limit for instance, the test scheduling is trivial (as without compression). The test time on a line is simply the sum of the individual test times of the cores since there is no test parallelism on the line. The total TAT is the maximal test times over the lines.

3. Problem statement and notations

We state the problem of building the test infrastructure with decompressors as an optimization problem. Given the number of available ATE channels, the bitwidth of the TAM, and the test patterns, we want to determine the best partition of the test infrastructure into p lines and the interconnection of the cores to the sub-buses so that the TAT is minimized.

In the remaining, we will use the following notations. Let n be the number of cores under test, w_c the number of visible scan chains for every core $c=1\ldots n$. Let p be the number of lines, and let W_{ATE_i} and W_{TAM_i}, $i=1,\ldots,p$ be respectively the number of ATE channels and the bitwitdh of sub-bus on line i. Let $\rho_i = W_{ATE\_i}/W_{TAM\_i}$ be the decompression ratio of line i Let's $t^c_{w_c,\rho}$ denote the test time of core c with a w_c bits wrapper for a ratio ρ.

The problem is to determine:
- the line number p
- the bitwidths W_{ATE_i} and W_{TAM_i} for $i=1,\ldots,p$;
- an assignment of the cores to the lines;
- optionally, the wrapper size w_c of each core,
- and a test schedule so that TAT is minimal;

The following constraints must be obeyed:

$$W_{ATE} = \sum_{i=1,..,p} W_{ATE\_i} \qquad \text{cons.1}$$

$$W_{TAM} \geq \sum_{i=1,..,p} W_{TAM\_i} \qquad \text{cons.2}$$

$$W_{TAM\_i} \geq W_{ATE\_i}, \ i=1,..,p \qquad \text{cons.3}$$

$$w_c \leq W_{TAM\_i} \text{ if c is connected to line i.} \qquad \text{cons.4}$$

Variables to be determined are depicted in bold in Figure 3.

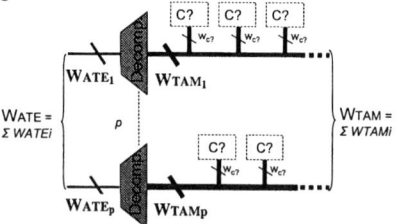

Figure 3: Problem statement

Concerning core wrappers, there are two cases: either the cores are wrapper-ready or their wrappers have to be designed. In the later case, it must be noted first that $1 \leq w_c \leq \max(\#PIs,\#POs)+ \#scan$ chains. Secondly, once w_c is determined, designing the wrapper so that the test time

of the core is minimized resumes simply to balance the lengths of the visible scan chains. This won't be detailed in the remaining.

In the scenario where the wrappers are already fixed, if a core is assigned to a line i for which W_{TAM_i} is strictly greater than the wrapper size w_c, only w_c bits of TAM_i are connected to the wrapper, the test time of the core is considered to be the same as if TAM_i was w_c bits wide. For instance, and for a core c with $w_c =4$, its test time $t^c_{w_c,\rho}$ is the same whether it is assigned to a line with $W_{ATE} = 2$ and $W_{TAM} =6$, or to a line with $W_{ATE} = 2$ and $W_{TAM} =4$, i.e. $\rho = 2/4$.

The test time $t^c_{w_c,\rho}$ of a core c must be pre-computed for all possible values of ρ (from 1 to $1/w_c$). (cf. section 4 to see how this process can be speeded up). For examining the benefit of using TDC when designing the test infrastructure of a SoC, we developed the heuristic presented hereafter.

4. Algorithm

The general flow chart of the method is depicted in Figure 4. First, all the possible combinations of lines are explored (line 1 and 2). The ATE channels partition can be easily determined knowing the total W_{ATE} width and the number of lines p by applying the formula of the partition of integer numbers. Namely, the number $X(n,p)$ of partitions of a set of n elements into p subsets can be computed as:

$$X(n,p) = \sum_{k=1}^{p} X(n-p,k) \text{ with } X(n,n) = X(n,1) = 1$$

and $X(n,p)=0$ if $p > n$

For instance, 10 ATE channels can be partitioned into p=3 subsets in X(10,3)=8 different ways (1+1+8, 1+2+7, 1+3+6, etc...).

Then for each ATE channels partition, all the compatible partitions of the TAM are calculated. A partition of the TAM is said to be compatible with a partition of the ATE channels if cons.3 is verified for all p lines. Furthermore, if cores are wrapper-ready i.e. w_c are fixed, the number of TAM partitions to be explored can be further reduced by considering cons.4. In other words, the narrowest TAM must be large enough to support the narrowest wrapper. It must be noticed that if $W_{TAM\_i}=W_{ATE\_i}$, no decompressor is present on this line. For a pair of partition, (ATE channels partition and TAM partition), cores must be assigned and the scheduling performed to obtain the TAT of this architecture (line 3 in Figure 4).

---

1. For all ATE channels partitions into p parts
2. For each compatible TAM partition into p parts
3. Find the best assignment of the cores to the p lines (that minimize TAT) ->cf Fig 5.
If this assignment reduces the global TAT, memorize this assignment and ATE/TAM architecture

---

Figure 4: Partition algorithm

Seeking for the assignment of cores to lines that minimizes TAT is an NP-complete problem. So we developed the heuristic given in Figure 5.

```
// Initial Solution
  –  Sort cores by decreasing test data volume
  –  Assign each core to the largest bus so that
     TAT increases as few as possible.
// Improvement of the solution
  •  While TAT is reduced
  –  Find the line i with the highest TAT_i
  –  For each core c assigned to i,
     •  For all other lines k ( k ≠ i )
        –  Move core c from i to k
        –  Compute newTAT and memorize i, k, c
           and newTAT
        –  Move back core c from k to i
  –  Move core c from i to k such that:
     1) the smallest TAT has been obtained
     2) the number of useless bits on k is minimized
     3) the standard deviation between TAT_i of all
        lines is maximized
```

Figure 5: Assignment algorithm

The first step determines an initial solution of the architecture, i.e an initial assignment of cores to the TAMs. Each core is positioned on the largest possible TAM i.e. and its wrapper size is set according to (cons.4). If the core is wrapper-ready, it is assigned to the smallest bus i.e. respecting cons.4. For instance, in case of 3 TAMs having resp. 5, 7 and 10 bits, a core with a 6-bits wrapper will be assigned to the 7 bits TAM. The first bus is not large enough to be connected to the core's wrapper (cons.4). The second bus is preferred to the third one since, a priori, it is beneficial to reserve the larger one for cores with larger wrappers.

The second step consists in improving this initial solution. For this, the cores are moved to other lines to reduce the global TAT.

The principle is to move a core from the line with the highest TAT to another line so that the global TAT gets reduced as much as possible. For that, all cores of the line are virtually shifted to other lines and TATs are computed accordingly. The move that gives the highest benefit is chosen. In case of equality, the algorithm chooses (Core c, Line i) such that the number of useless bits on the line is minimized i.e. $W_{TAM\_i} - w_c$ is minimal. This is done for getting more room to move cores with larger wrappers to large buses, in next steps. Similarly, a third order criterion is used to unbalance test times over lines.

Let's recall that the computation of TAT is straightforward (TAT_i denotes the test application time on line i):

$$TAT = \max\left(TAT_i, i = 1, ..., p\right) \text{ and } TAT_i = \sum_{\text{cores assigned to } i} t^c_{w_c, \rho_i}$$

Note that the test times $t^c_{w_c, \rho}$ for all cores and for all compression ratios (W_{ATE_i}/w_c) are inputs of the proposed algorithm. These data are necessary to compute the system TAT (i.e. schedule the tests). Thus, as a pre-process, the compression algorithm must be performed for all compression ratios, for all cores and all wrapper sizes. This can be very CPU expensive depending on the compression technique used. We propose here an alternative to the exhaustive computation

First, when the wrapper size is questioned, let's recall that as reported by many authors, the test time of a core, in the absence of compression i.e. ρ=1, is a stepwise decreasing function of the wrapper size. Furthermore it depends on the number of test vectors and not on the vectors themselves. Figure 6 reports the test time versus w_c for the 10th core of the D695 ITC'02 benchmark. In general the number of steps is small. Only 15 optimal values of w_c have to be considered for this core.

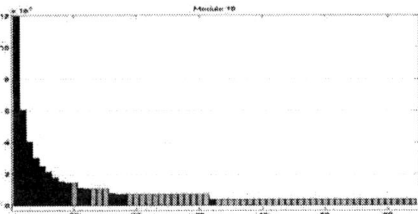

Figure 6: ITC'02 d695 benchmark (core 10) Test time vs wrapper size

Secondly, whatever the TDC technique is used, the same behavior of the test time of cores versus decompression ratio can be observed (for a given wrapper size w_c). It can be identified to the function:

$$(3) \qquad t^c_{w_c, \rho} = \frac{\alpha}{\rho} + \beta$$

Only two values of t for one core are sufficient to identify α and β. The estimated values of $t^c_{w_c, \rho}$ for several decompression ratios are thus obtained from only two measured values instead of w_c computations. In order to improve the precision of the estimation, the

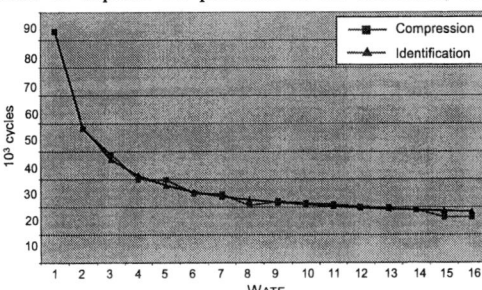

Figure 7: S38417 ($w_c = 16$) computed/estimated test times

compression algorithm is performed with the first and last decompression ratio values.

This property has been validated with the TDC method [14]. This compression scheme is applicable with intellectual property cores and it is Test Suite independent, i.e. it does not required specific test generation or fault simulation.

The measured and estimated $t_{c, \rho}$ values are reported on Figure 7 for the ISCAS'89 s38417 benchmark (16-bits wrapper). The maximum error between measured and estimated values is smaller than 1%. Similar results have been obtained for all ISCAS'89 benchmarks and several configurations of wrappers.

As a final remark let's note that the proposed heuristic can be very easily adapted to additional constraints such as power limit, precedence constraints, etc...

5. Results

The first SoC used for experiments is the one described in [9][20] and depicted in Figure 8. It is composed of 16 ISCAS'89 benchmark circuits used as cores (i.e. with wrappers).

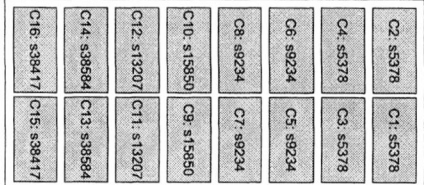

Figure 8: SoC example from [20]

In a first series of experiments, we assume that the wrappers are already designed. Wrappers sizes are equal to the number of scan chains. The test sequences of the circuits have been obtained with the Synopsis ATPG tool TETRAMAX [23] and compressed with our TDC technique described in [14]. The characteristics of the cores are given in table 1.

Core number	#scan chains (wc)	test cycles
1 to 4	5	9331
5 to 8	6	9030
9, 10	10	8804
11,12	12	16048
13,14	14	19845
15,16	16	45760

Table 1: Characteristics of the cores

In the experiments, we have set the number of ATE channels to 32 and the maximal total TAM bitwidth to 64. The algorithm has been applied with a number of lines ranging from 2 to 6. Results are reported in Table 2.

p	# conf.	TAT	Lines' parameters (W_{ATE_i} / W_{TAM_i})	#bits used on TAM
2	522	127413	(16,16) / (16, 48)	30
3	44639	90457	(8,9,15) / (14,16,34)	42
4	1345142	68361	(5,7,8,12) / (7,14,16,27)	53
5	18605924	57941	(5,5,7,7,8) / (6,12,14,16,16)	64
6	142238520	57941	(1,4,5,7,7,8) / (1,5,12,14,16,16)	63

Table 2: Architectures exploration results

The number of lines is given in col.1. Col.2 indicates the number of architectural configurations that have been explored while col.3 gives the TAT of the elected architecture. The details of the test infrastructure are given in col.4. The last column indicates the actual number of TAM bits.

For instance, for an architecture with 3 lines, 44639 configurations have been explored. The optimal one leads to a TAT equal to 90457 test cycles. The architecture contains 3 decompressors such that (W_{ATE_1}, W_{TAM_1}) = (8,14), (W_{ATE_2}, W_{TAM_2}) = (9,16), and (W_{ATE_3}, W_{TAM_3}) = (15,34).

From this table, some observations can be done:

- All potential test infrastructures are explored including those that do not contain decompressors. For instance, for the 2 lines configuration, the optimal architecture does not include a decompressor on the first bus $W_{ATE\_1} = W_{TAM\_1} = 16$.

- While a budget of a 64 bits TAM has been given, all those bits are not necessarily connected to cores (and thus are useless). This is the case for p=2, 3, 4, 6. This is mainly due to the wrapper sizes chosen for the cores. This means that the actual bitwidth of the TAM is smaller than 64 bits.

Among all experiments, the best TAT is obtained with p=5 lines. The corresponding test schedule and architecture are given in Figure 9 and Figure 10. Test parallelism cannot be fully exploited with smaller values of p since at most p cores can be tested in parallel. For larger values of p (6, 7, ...), the relative sizes of the wrappers to the possible TAM_i widths act as a brake on parallelisation.

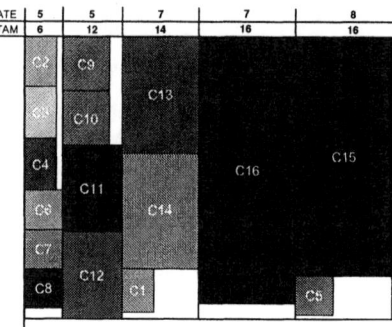

Figure 9: Test schedule for a 32 → 64 bits decompression with 5 lines. TAT = 57941 cycles

We measured the benefit of using compressors in SoCs test architectures by comparing them to standard architectures i.e. without using compression, while setting the same environmental constraints. In the first case, we assumed the same limit on the numbers of available ATE channels (32 bits and thus a TAM of 32 bits), in a second case, the same area budget for building the TAM (64 bits wide and thus 64 ATE channels).

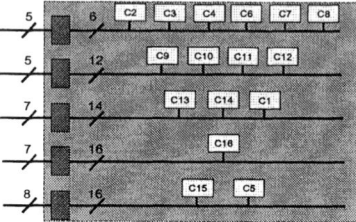

Figure 10: Final architecture

For the first case, the TAT is 127413 cycles for a standard TAM architecture when a number of sub-buses p ranges from 2 to 4 and 131210 cycles when p equals 5 or 6. These results have to be compared with the 57941 cycles when compression is used. Thus, the use of TDC technique in the context of SoC infrastructure design leads to a gain of 54.5% in terms of TAT for this example (at the expense of area overhead: larger TAM, decompressors).

In the second case, i.e. a TAM of 64 bits (which means 64 ATE channels for a standard architecture vs 32 ATE channels with compression), comparative results are reported in Table 3. At the evidence, TDC has allowed to divide by two the number of ATE channels at the expense of only a 4% increase on TAT.

Proposed architecture: W_{ATE} =32, W_{TAM}=64			Standard architecture: W_{ATE} = 64, W_{TAM}=64	
# lines	TAT	actual TAM bitwidth	# lines	TAT
2	127413	30	2	127413
3	90457	42	3	90457
4	68361	53	4	68361
5	57941	64	5	57941
6	57941	63	6	55738

Table 3: Architectures Comparison (fixed wrappers)

We did the same experiments, but without assuming fixed wrappers size i.e. letting the method determines the most adequate wrappers structures. TAT are reported in Table 4. It can be fist noted that since wrappers structures are questioned, bus width can be better utilized leading to shorter TAT. Secondly, as in the previous case, the use of TDC leads to a large TAT improvement.

p	32→64 Decomp. Architecture	Standard 64 bits Architecture	Standard 32 bits Architecture
2	66596	52953	97216
3	61277	49814	96624
4	57337	49129	96736
5	55101	48592	96624
6	54140	48517	96563

Table 4: Architectures Comparison

The same kind of experiments has been performed on the g1023 ITC'02 benchmark. Unfortunately, in the ITC'02 suite, neither cores netlists nor test patterns are provided, all information necessary to perform compression. Thus we got random patterns including 80% of don't care bits (many authors report a percentage of more than 90% of don't care bits on industrial circuits). Comparative results are given in table 5.

	Decomp. Architecture		Standard Architecture		
p	32→64	16→32	64 bits	32 bits	16 bits
2	17492	26256	15153	19633	33952
3	14185	23084	11274	17892	33718
4	12996	21409	11274	17235	33824
5	12399	20719	11274	17215	33824
6	12138	20667	11274	17235	33824

Table 5: g1023 Comparison results

6. Conclusion

In this paper, we explored the benefits of horizontal test data compression techniques in the context of the design of SoC test infrastructures. The increase in parallelism allowed by compression is fully exploited to reduce the test application time of the SoC. We propose a method that explores all architectural solutions from one single decompressor for all cores to architectures with a dedicated decompressor per core. Results obtained on a SoC based on ISCAS'89 benchmarks circuits have confirmed this TAT reduction with a ratio of more than 50%. While the experiments have been performed using

a particular TDC technique, the method is independent of the used TDC.

Presently, this method is geared to minimize the test time. Area overhead induced by decompressors and TAM is not taken into account. Seeking the best trade-off is a direction for future research.

7. References

[1] IEEE standard for embedded core test – IEEE Std. 1500-2004.

[2] V. Iyengar et al.. "Test wrapper and test access mechanism co-optimization for system-on-a-chip". J. Electronic Testing, vol. 18, no. 2, pp. 213-230, April 2002

[3] V. Iyengar et al., "Efficient Wrapper/TAM Co-Optimization for Large SOCs", DATE'02, pp: 491-497.

[4] V. Iyengar et al. "Wrapper/TAM co-optimization, constraint-driven test scheduling, and tester data volume reduction for SOCs", DAC '02. pp. 685-690.

[5] S.K. Goel, E.J. Marinissen, "Effective and Efficient Test Architecture Design for SOCs", ITC'02, p: 529- 535.

[6] G. Zeng, H. Ito, "Concurrent core test for SOC using shared test set and scan chain disable", DATE'06, pp: 1045-1050.

[7] A. Jas, B. Pouya, N.A. Touba, "Virtual Scan Chains: a means for reducing scan length in cores", VTS'00, pp: 73-78.

[8] L-T Wang et al., "VirtualScan: a new compressed scan technology for test cost reduction", ITC'04, pp: 916-924.

[9] I. Bayraktaroglu, A. Orailoglu, "Test volume application time reduction through scan chain concealment", DAC'01, pp: 151-155.

[10] K.J. Balakrishman, N.A. Touba, "Reconfigurable linear decompressor using symbolic Gaussian elimination", DATE'05, pp: 1130-1135.

[11] J. Rajski et al., "Embedded deterministic test for low cost manufacturing Test", ITC'02, pp: 916-922.

[12] L. Li, K. Chakrabarty, N. A. Touba "Test data compression using dictionaries with selective entries and fixed-length indices", ACM TODAES, Vol. 8, No. 4, October 2003, pp: 470-490.

[13] A. Würtenberger, C.S.Tautermann, S.Hellebrand, "Data compression for multiple scan chains using dictionaries with corrections", ITC'04, pp: 926-935.

[14] J. Dalmasso, M.L. Flottes, B. Rouzeyre, "Fitting ATE Channels with Scan Chains: a Comparison between a Test Data Compression Technique and Serial Loading of Scan Chains", DELTA'06, pp: 295-300.

[15] N. Sitchinava et al., "Changing the scan enable during shift", VTS'04, pp: 73-78.

[16] H. Tang, S.M. Reddy, I. Pomeranz, "On reducing test data volume and test application time for multiple scan chain designs", ITC'03, pp: 1079-1088.

[17] B. Arslan, A. Orailoglu, "CircularScan: a scan architecture for test cost reduction", DATE'04, pp: 1290-1295.

[18] S. Mitra, K.S. Kim, "X-compact, an efficient response compaction technique for test cost reduction", ITC 02, pp: 311-320

[19] J. Rajski,et al., "Finite memory test response compactors for embedded test applications", IEEE Trans. on CAD, April 2005, Vol. 24-4, pp: 622- 634

[20] V. Iyengar, A. Chandra, "A Unified SOC Test Approach Based on Test Data Compression and TAM Design", Proc. IEEE DFT'03, pp: 511-518

[21] P.T. Gonciari, B.M. Al-Hashimi, "A Compression-Driven Test Access Mechanism Design Approach", ETS'04, pp: 100-105.

[22] P.T. Gonciari, B.M. Al-Hashimi, N. Nicolici, "Integrated Test Data Decompression and Core Wrapper Design for Low-Cost System-on-a-Chip Testing", ITC'02, p: 64-70.

[23] www.synopsys.com/products/test/tetramax_ds.html

Dynamic Gates with Hysteresis and Configurable Noise Tolerance

Krishna Santhanam   Kenneth S. Stevens

Electrical and Computer Engineering

University of Utah

krishna.santhanam@utah.edu   kstevens@ece.utah.edu

Abstract

Dynamic logic can provide significant performance and power benefit compared to implementations using static gates. Unfortunately dynamic gates have traditionally suffered from low noise margins, which limits their reliability. A new logic family, called complementary dynamic logic *(CDL), is presented. CDL replaces the standard keeper logic with a dual dynamic keeper gate that is applicable to all dynamic gate structures. CDL provides dynamic gates with two novel characteristics: hysteresis and arbitrarily configurable noise margins. However, these two benefits come at the cost of reducing the gain and increasing the energy of the dynamic gate. This paper compares the noise, energy, performance, gain, and total transistor width tradeoffs of CDL and three other logic families applied to a 65nm cell library consisting of 23 functions. The results show that the performance advantages of dynamic domino gates can be maintained while providing significantly enhanced noise margins using CDL structures.*

1. Introduction

Deep submicron designs consist of a number of competing critical design tradeoffs. Performance has traditionally been the most important design metric. Others such as power and noise have become increasingly important due to scaling. Indeed, the impact of wires in our design is having an enormous effect on our architectures and circuits by increasing the delay, noise, and power of our designs [9, 13].

The circuit family used in our designs maintains a direct relationship to the performance, power, noise tolerance, and time to market of a design. The robustness and ease of mapping combinational functions to static logic are significant advantages that keep this logic family at the forefront of our design world. However, other logic families hold distinct advantages in terms of power and performance over traditional static logic design. For example, a domino implementation of a six-gate two-input NAND pipeline is 40% faster with 21% less peak switching energy than a static implementation driving an identical load.

A dynamic gate owes its significant performance advantages to its unique logic structure. Dynamic gates implement the *state change* of a function, then act as *latches*. Hence the transistor logic is only implemented to effectuate the change in a function from high to low (or low to high). Otherwise the gate is left in a high impedance state. This results in a very efficient gate. For instance, a traditional 2-input domino NAND gate has a logical effort [14] *less* than that of an inverter[1] – giving an input-to-output gain

greater than an inverter in traditional CMOS processes. The high gain of these gates are the primary reason for both performance and power advantages of dynamic gates, and their latching property can create design advantages.

However, these structures also have a serious drawback. The dynamic latched output states that are not covered by the set-reset function are sensitive to noise. Noise is one of the primary reasons that dynamic gates are not exploited in more designs for their performance and power advantages.

This work presents a new dynamic gate structure, called *complementary dynamic logic*, or CDL, that can provide dynamic gates with hysteresis and a configurable noise margin. Hysteresis provides the gates with a high switching threshold when the output is low, and a low switching threshold when the output is high. The configurable noise margin allows propagation and coupling noise effects to be mitigated – even to the point that noise sensitivity is less than that of a comparable static gate.

The noise tolerance comes at the cost of gain – increasing the delay and power of the gate. Therefore CDL gates will be sized to optimize gain based on the specific noise requirements of the interconnect.

In this paper, we report the results of our characterization of these logic gate as we trade off performance, power, gain, and total transistor width to increase the noise margin of this novel circuit family. We characterize and compare CDL gates to static and traditional dynamic domino gates with a weak feedback keeper.

2. Dynamic Gate Noise Reduction

Several gate structures have been used to increase the noise margin of dynamic gates while retaining speed and energy advantages compared to static gates. These techniques fall under two categories.

The first category of circuits were developed to reduce the propagated noise of dynamic gates by (a) dynamically increasing the switching threshold and (b) precharging intermediate nodes to increase the body effect [15, 1, 5, 3, 4]. These methods do nothing to improve coupling noise, and have a relatively limited improvement on propagated noise, but do reduce the leakage of these gates. We classify these as uncompetitive alternatives.

A second more effective method is to employ *keeper* structures that retain the output voltage in dynamic states of the gate. The circuits in this category are effective against both propagation noise and crosstalk noise. The most successful keeper design has been to implement a *jam latch*, or back-to-back inverters, on the output of the dynamic gate[2].

---

1   Logical effort is a metric of the gain of a gate that takes into account the complexity of logic required to switch a gate. An inverter is traditionally assumed to be the highest gain gate.

---

2   When dynamic gates are used inefficiently in clocked pipelines by connecting the precharge to the clock the pull-down structure of the second inverter in the jam latch can be removed because the gate will never be in a dynamic state with the output low.

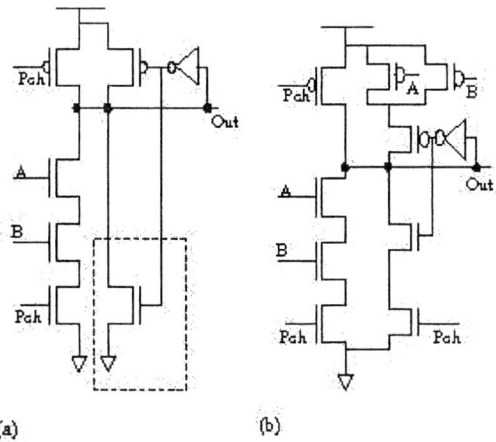

Figure 1. Footed Domino 2-input NAND gates with (a) jam latch, (b) CDL

This latch, shown in Figure 1(a) provides a pull-up or pull-down path for the output at all times. However it also has one deleterious property: The jam latch keeper logic will *always* oppose the output transition in the dynamic gate. This has several drawbacks: (a) This increases the power dissipation of the gate due to the short circuit current between power and ground when the output switches. (b) The switching delay of the gate is increased due to the fight between the keeper and dynamic gate. (c) The keeper becomes a ratioed gate and must be sized properly or the gate will not function. If the keeper is too large, the gate will not switch. This limits the ability to create a dynamic gate with a large noise tolerance. (d) The feedback inverter of the dynamic gate must be sized to switch the keeper logic quickly to reduce short circuit current. This increases the load on the output of the dynamic gate.

Stronger noise margins are needed than can be provided by the standard jam latch keeper. CDL gates provide the needed ability to obtain the necessary noise tolerance to allow usage of domino logic further into deep submicron technologies. CDL requires a second dynamic gate that is the dual of the dynamic set-reset gate. The outputs of these two gates are tied together as shown in Figure 1(b). The complementary CDL gate provides noise tolerance to the high impedance states of a dynamic gate. This secondary dynamic gate will *never* switch the output of the gate and does not fight the gate's transition. Instead, it's solitary purpose is to provide noise tolerance. The complementary dual gate can be arbitrarily sized to achieve any necessary noise margin – even to the point where it is stronger than the gate that toggles the output. CDL gates can therefore drive long-distance communication wires. However, in most applications dynamic gates will continue to drive short local wires. In such cases the complementary gate will be very small, having a minimal impact on the gain of the dynamic gate.

There are two main advantages to CDL logic. It is the only set-reset dynamic logic family that can have an effectively controllable noise margin. The second key advantage of CDL is hysteresis. This is due to the dual gate which continues to provide current to retain the previous state until the inputs have fully switched or the output has toggled.

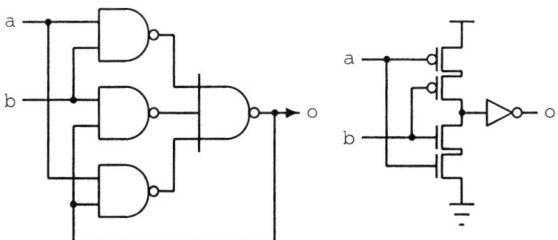

Figure 2. Static and dynamic C-Elements

3. Dynamic Gate Architectures

The logic of a dynamic gate is only intended to toggle the output. Therefore dynamic gates are most efficient when technology mapped from *production rules* [7] or other set-reset synthesis methodology. Such an approach is employed by asynchronous synthesis CAD [2, 16]. This is in stark contrast to the wasteful approach typically used in clocked designs where the precharge input is tied to the clock.

A simple C-Element, or rendezvous, will be used as an example to illustrate the design and benefits of the CDL gate using set-reset synthesis. The production rules for a C-Element are given as:

$$a\uparrow \cdot b\uparrow \;\mapsto\; o\uparrow$$
$$a\downarrow \cdot b\downarrow \;\mapsto\; o\downarrow$$

The implementation mapped to a static gate versus a dynamic gate are shown in Figure 2 and Table 1. The dynamic gate implementation is 40% faster with a 20% reduction in energy compared to the static implementation when driving the same output load. Hence a design methodology that efficiently uses dynamic gates has a two fold benefit (1) there can be a substantial reduction in overall logic when exploiting the dynamic states, and (2) the gates themselves have higher gain, particularly when the set and reset functions have disjoint input conditions.

The design of the complementary dual gate is illustrated based on the dynamic C-Element. The KMap of the dual gate is shown in Table 2. All dynamic states from Table 1 are specified as 0 or 1 based on the value of the output o. All other states are don't care but must be covered with either a zero or one based on the values of the dynamic gate. The full CDL logic is shown in Figure 3 where the (a) is the set-reset gate, (b) is an inverter to provide proper polarity to the CDL gate (c).

The primary disadvantage of CDL gates is the complexity of the keeper logic. First, since the two gates are duals, the keeper gate will produce additional load on the inputs, reducing the gain of the set-reset gate. Second, since the dual keeper gate does not switch the output, it will always have the output feeding back as an input, creating a latching structure. The keeper gate will therefore have more transistors than the set-reset dynamic gate. Third, the complementary nature of the gate results in inefficient series structures when the set-reset gate has wide OR functionality.

However, the complexity of the dual CDL gate may not be as significant as one initially expects. The complexity is mitigated by two factors. First, mapping functions into set-reset logic can have a significant overall reduction in the

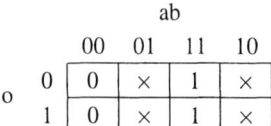

ab

		00	01	11	10
O	0	0	×	1	×
	1	0	×	1	×

Table 1. KMap for C-Element: × states are dynamic. Set, reset coverings $= \mathrm{ab}, \overline{\mathrm{ab}}$

ab

		00	01	11	10
O	0	$-_0$	0	$-_1$	0
	1	$-_0$	1	$-_1$	1

Table 2. Dynamic C-Element dual: high, low coverings $= (\mathrm{ao} + \mathrm{bo}), (\overline{\mathrm{ao}} + \overline{\mathrm{bo}})$

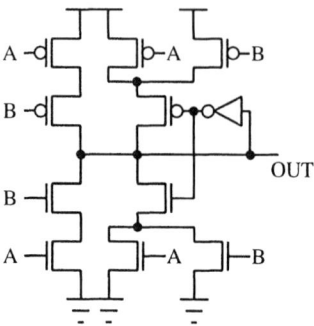

Figure 3. CDL Inverting C-Element

logic, particularly when mapping to sequential functions. For example, 18 transistors are required for the static implementation of the C-Element, versus 12 for the CDL and 8 for a dynamic gate with a jam latch keeper. Secondly, what really matters in a design from a power and performance perspective is the the *size* in terms of total transistor widths, not the number of transistors[3]. Unless very high noise margins are required, the total transistor width in a dynamic gate, including the more complicated CDL Logic, can be quite small when compared to a traditional static gate. For example, the total transistor widths of the standard keeper and CDL C-Elements are 23% and 27% respectively of the static gate for an identical load where the keepers are sized to drive 20% of the switching current of the devices. Note that the CDL implementation is only 18% larger than a gate using a traditional keeper. This highlights two important factors to remember: (a) the increased gain and reduction in logic of dynamic gates can result in a significant reduction in total transistor widths, and (b) the extra transistor width for the dual gate in CDL logic is very small unless high noise margins or hysteresis are needed.

The CDL gate structure has in the past been arbitrarily applied to a few dynamic gates. The CDL C-Element shown here was used in the design of the Post Office [12] and is compared against other C-Element circuit structures in [10]. However, a general approach to the design and sizing of a complementary keeper to control noise is novel.

Because of the complex relationship between the noise advantages and gain costs of the CDL, a rigorous evaluation of these gates has been carried out and will be reported in Section 6.

4. Circuit Comparison

The circuit comparison has been carried out using the functions of a complete 65nm cell library. This library contains 23 independent functions [11]. We have mapped four gate families to these 23 functions: two dynamic cell libraries and a complex gate static library. The dynamic libraries consist of traditional dynamic domino with a jam latch keeper and domino CDL gates. Only a fraction of this

---

3 Many deep submicron libraries now split single logical transistors into many smaller transistors through "*legging*" to reduce variation, etc.

data can be presented here due to space limitations and the scope of the study. We will therefore use a representative example and summarize the full data results here. The complete data set is contained in [8].

Noise Margin: A domino gate will fail if it flips state or produces non-monotonic output changes. Therefore an aggressive definition of failure under noise is adopted: any change of V_{th} on the output is deemed a failure. Noise immunity curves are reported because they show how noise margins scale compared to the size of the keeper logic and provide a timing perspective which is absent in a DC analysis.

Performance: Performance is measured as the delay between 50% change in the input to a 50% change in the output. Transistors are sized in this study by setting the PMOS to NMOS ratio to 2:1.

The current drive of the equivalent static gate is used as the baseline. Dynamic gates are sized to have an equivalent switching current and load as the static gate shown in Figure 4. This results in similar delay, but greatly underestimates the gain advantage of dynamic gates. An alternative approach is to match the input loads of the gate which takes the gain into account [6]. We opted to use identical drive size to create the worst case scenario for the dynamic gates and to mitigate variations based on our sizing of the keeper transistors.

Switching Power: The sum of the energy to drive both the inputs and the outputs are reported.

Gain: Gain of the gate is calculated as the ratio of output load to the input load C_{out}/C_{in}. The gain is reported as C_{in} in this study because the output load remains constant.

Transistor Width: The total transistor width is the best first order metric of the cost of the circuit in terms of leakage and transistor area (but perhaps not layout area).

Hysteresis: Measured by the DC switching points.

5. Experimental Setup

Noise margins are modified by changing the size of the keepers. Our experiments consisted of measuring the seven design metrics in the previous section upon varying the size of the keepers. The circuits are evaluated under both input noise propagation and aggressor or crosstalk noise coupling on the output of the gate.

5.1. Keeper Sizing Parameters

Three parameters are used to modify the gate sizes in the keeper structures.

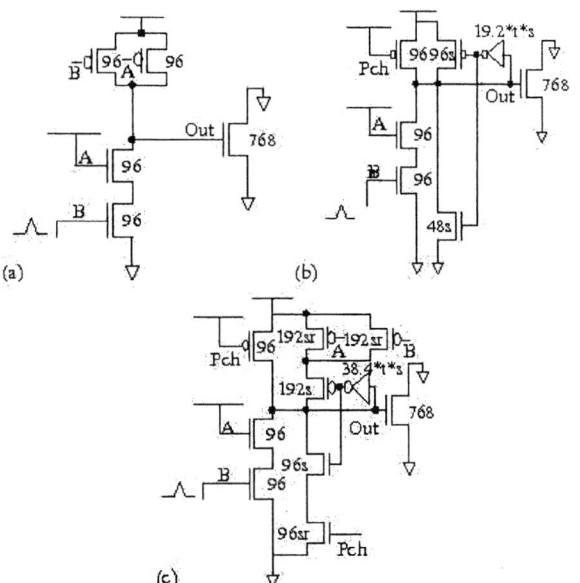

Figure 4. Propagation noise setup

The keeper logic is sized in relation to the set and reset function of the gate using the parameter s. When $s = 1$ the keepers will drive nearly the same current as the set-reset logic function as shown in Figure 4. When $s = 0$ the gate is fully dynamic. This is the primary keeper scaling parameter used to control noise immunity. Most of our graphs show results of keeper sizing with s ranging from 0 to 1.

Parameter r is used to optimize the sizing of the keeper in the CDL gate by keeping the noise margin the same while reducing the load on the input pins. This optimization technique improves gain at the cost of slight increased delay and switching power of the gate.

Parameter t is used to optimize the size of the first inverter in the feedback path to the keeper as shown in Figure 4. The fanout load of the inverter was varied using this parameter while measuring the delay, power, and noise margin. The optimal value of t for the domino gate with regular keeper is approximately a fan out of four, whereas in the case of the CDL gate structure it is a fan out of 10. The optimal values have been used in all the simulations while varying the other parameters.

5.2. Noise Modeling

Propagated noise: The setup used to measure noise propagation is shown in Figure 4 using 2-input NAND gates as an example. All simulations switch the output closer to ground. The transition to the input is a ramp that saturates at approximately $V_{cc}/2$.

Dynamic noise immunity curves are reported varying the s parameter that dictates the size of the keeper transistors. Spice simulations step the parameter s, sweeping the duration of the input noise pulse until the propagated noise changes by V_{th}. The results plot input duration versus parameter s. The traditional design range in size for a jam latch is from 0.1–0.2 times that of the dynamic gate. However, we have plotted the graphs across a much larger dynamic range, varying s from zero to one.

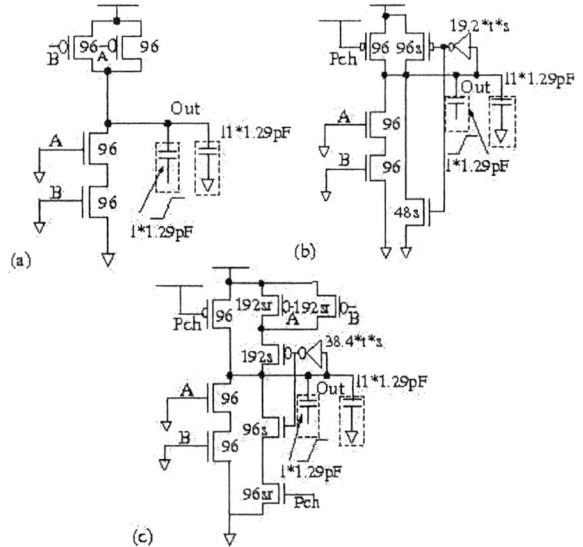

Figure 5. crosstalk noise configuration

Crosstalk noise: Parameter l is introduced to model aggressor noise as shown in Figure 5. This parameter specifies the percentage of the effective load on the gate that can be associated with a noise source, and ranges from 0 to 1. The total capacitance on the output node remains constant, but as l increases more of the total cap is attributed to cross-coupled wires. Therefore this is a figure of merit that can be used to determine the maximum wire length that can be safely driven for a given keeper size. The aggressor signals are ramps that saturate at V_{cc}. Dynamic noise graphs are created by incrementing the keeper size s, and sweeping l until the maximum noise on the output changes by a threshold V_{th}.

6. Simulation Results and Comparisons

All values are taken from spice simulations of the gates in a 180nm process with a power supply of 1.8V and threshold voltage of 0.4V. All signal ramps for propagated and crosstalk noise use ramps with a 150ps rise time that is equal to FO4 values.

Dynamic noise graphs are plotted in Figures 6 and 8. These plot footed domino NAND and NOR structures that range from two to four inputs using jam latches and CDL keepers. The NAND and NOR structures give the best intuition for the scaling and cost, since we cannot show the results of all 23 gate functions. The more complicated AOI gates exhibit an additive combination of the characteristics of these structures.

All values in these graphs are normalized to the values of a static gate. A value of 2 on the vertical axis is twice as good as the static gate, and 0.5 is half as good. The horizontal axis scales parameter s. Changes in parameter s have an effect on the keeper structures but there is no change in the static transistor's sizes. When zero, the gate is fully dynamic. When $s = 1$, the keeper logic has approximately the same drive as the static gate.

The devices are sized pessimisticly for the dynamic gates by matching the drive strengths and loads. Therefore the

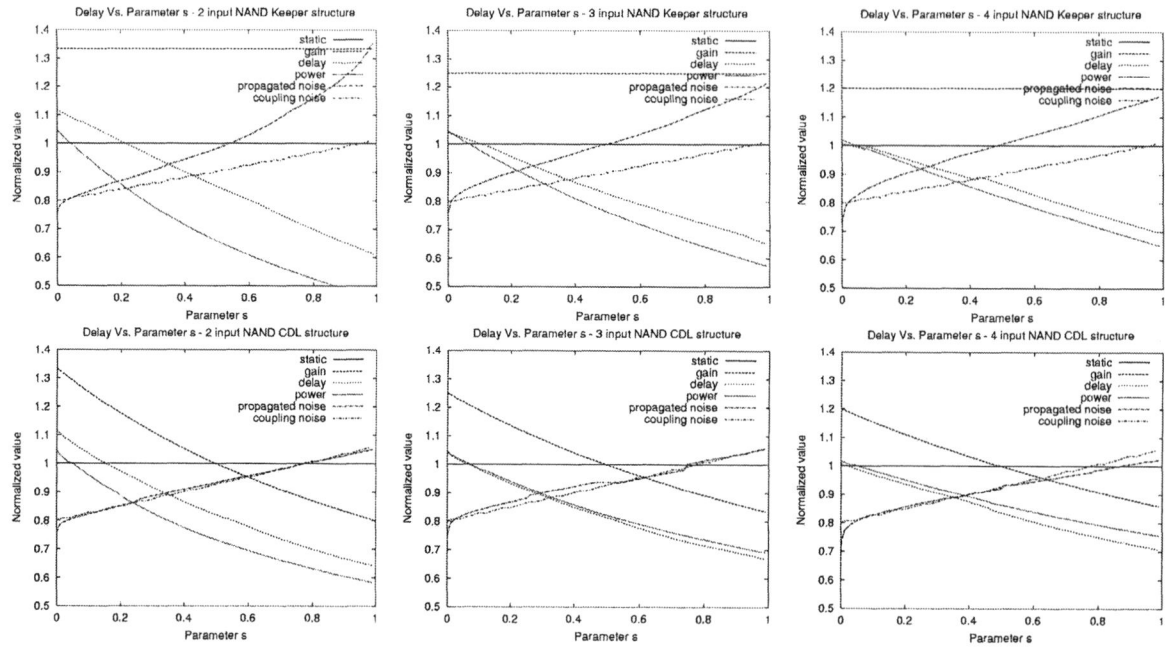

Figure 6. Dynamic Noise Graphs of NAND structures

performance and power of the dynamic gates are similar to the static gate. However, the gain for these circuits is considerably better than for the static gate. This implies that from a system perspective, substantial performance and power improvements are possible beyond what is reported here.

The arcs in the graphs align so that on the horizontal axis, the order of the arcs are gain, delay and power which are an improvement over the static gate. These values degrade as the keeper size is enlarged. The exception is the the gain of the domino with a week keeper which remains constant because the inputs are independent of the keeper logic. Next the coupling and propagated noise appear, each with a worse value than the static gate. Noise immunity improves as the size of the keeper is increased in all gates. The dynamic gates are identical with a zero sized keeper for gain, performance, power and noise margin. They begin to diverge based on the particular properties of the keeper.

The CDL gate scales better in NAND structures than the jam latch for performance, power, and coupled noise. The CDL noise immunity is equal to the static gate with a keeper size about 80% the size of the dynamic gate. The CDL gate doesn't scale as well for propagated noise. The performance/noise tradeoff improves compared to the static gate with the deeper NAND structures. A CDL gate has a 20% performance penalty when the same noise margin as the static gate is achieved for the 4-input NAND. The jam latch cannot reliably reach this point.

There is a substantially larger improvement in performance and power for the NOR structures in Figure 8 compared to the static gate with small keepers even given equivalent switching transistor sizes. Performance improves by a factor over 2× for the 3-input NORs. However, the propagated noise (the lowest line in the graphs) is substantially worse than the static gate. The CDL gate shows better noise margin scaling for large keeper structures. (The jam latch failed a little over 80% the size of the dynamic gate in the 2-

input NAND case.) The CDL gate achieves approximately the same performance as a static gate for an identical coupling noise margin.

Figure 7 shows the DC analysis of a 2-input NAND gate as the CDL keeper gate scales. For large keepers a substantial hysteresis differential is created. This was consistent across all of the gates in our 23 function library.

7. Conclusion and Future Work

A new gate structure (CDL) has been designed and evaluated showing configurable noise tolerance for dynamic gates that improves robustness and range of application. While there is no advantage for small keeper logic over the standard weak inverter implementation, the CDL gates show advantages in both NOR and NAND structures when increased noise tolerance is required and large keeper structures are used. The results show that the CDL gate can

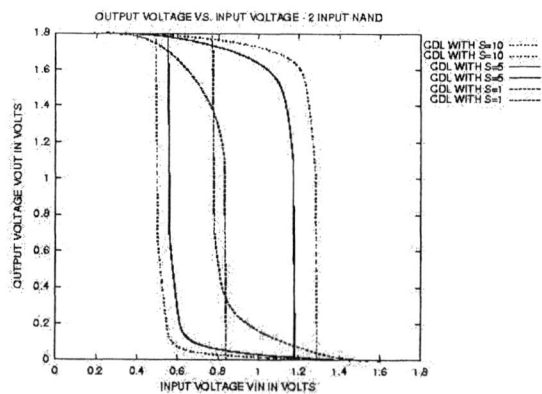

Figure 7. CDL Hysteresis of 2-input NAND

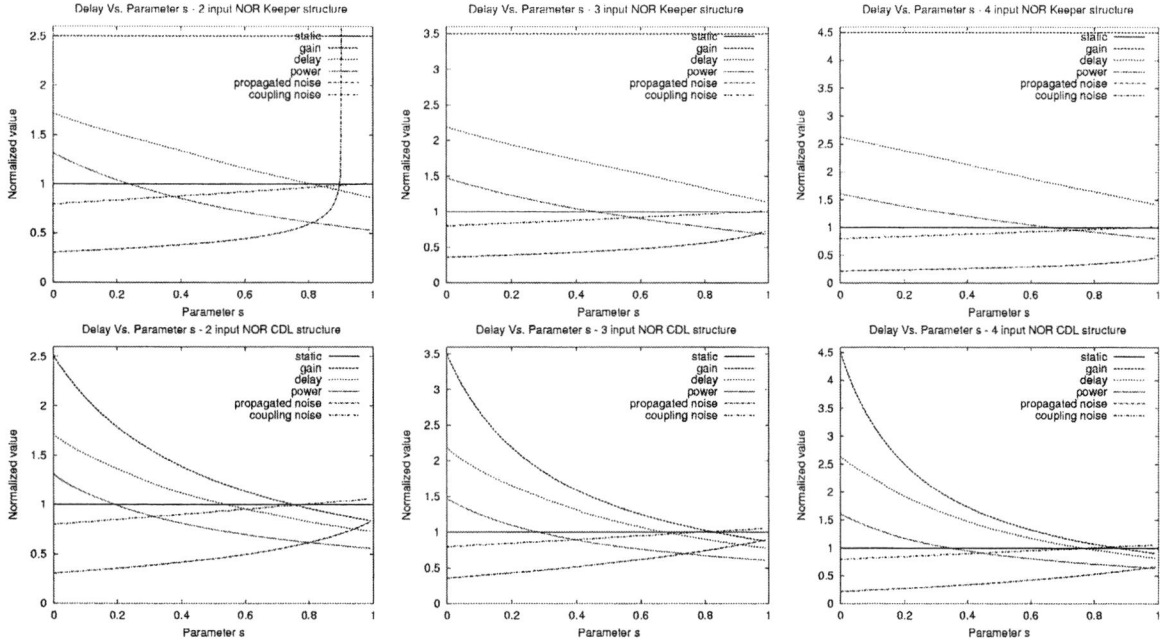

Figure 8. Dynamic Noise Graphs of NOR structures

achieve the same coupled noise margin as a static NOR gate with approximately the same performance and an improved gain. Dynamic gates were also shown to have over 600mV of hysteresis in a 1.8V process with large CDL structures.

There is significant cost in transistor width and decrease in gain for the NOR structures in a CDL gate. We briefly studied applying ratioed gates the the PMOS NOR structures in a CDL gate. This showed a potential for substantial improvements in gain and transistor width at a cost of some contention when the gate switches. We are also investigating datapath implementations using CDL logic, automatic sizing of the keeper gate for noise and wire lengths, and comparing CDL against static and traditional keeper designs in a test chip.

References

[1] G. Balamurugan and N. R. Shanbhag. The twin-transistor noise-tolerant dynamic circuit technique. *IEEE Journal of Solid-State Circuits*, 36(2):273–280, Feb 2001.

[2] J. Cortadella, M. Kishinevsky, A. Kondratyev, L. Lavagno, and A. Yakovlev. Petrify: a tool for manipulating concurrent specifications and synthesis of asynchronous controllers. *IEICE Transactions on Information and Systems*, E80-D(3):315–325, 1997.

[3] J. J. Covino. Dynamic CMOS circuits with noise immunity. Technical Report US Patent 5650733, IBM, Jul 1997.

[4] G. P. D'Souza. Dynamic logic circuit with reduced charge leakage. Technical Report US Patent 5483181, Sun Microsystems, Inc., Jan 1996.

[5] S. Goel, T. Darwish, and M. Bayoumi. A novel technique for noise-tolerance in dynamic circuits. In *IEEE Computer Society Symposium on VLSI*, pages 203–206, Feb 2003.

[6] D. Harris, G. Breed, M. Erler, and D. Diaz. Comparison of Noise Tolerant Precharge to Conventional Feedback Keepers for Dynamic Logic. In *Great Lakes Symposium on VLSI*, pages 261–264, 2003.

[7] A. Martin. Compiling Communicating Processes into Delay-Insensitive VLSI Circuits. *Distributed Computing*, 1(1):226–234, 1986.

[8] K. Santhanam. Novel Dynamic Gate Structure with Configurable Noise Tolerance. Master's thesis, University of Utah, May 2007.

[9] P. Saxena, N. Menezes, P. Cocchini, and D. A. Kirkpatrick. Repeater scaling and its impact on CAD. *IEEE Transactions on Computer-Aided Design of Integrated Circuits and Systems*, 23(4):451–463, April 2004.

[10] M. Shams, J. C. Ebergen, and M. I. Elmasry. Modeling and Comparing CMOS Implementations of the C-Element. *IEEE Transactions on VLSI Systems*, 6(4):563–567, December 1998.

[11] K. S. Stevens and F. Dartu. Algorithms for MIS Vector Generation and Pruning. In *International Conference on Computer-Aided Design (ICCAD-06)*, pages 408–414. IEEE Computer Society, Nov. 2006.

[12] K. S. Stevens, A. L. Davis, and W. S. Coates. The Post Office Experience: Designing a Large Asynchronous Chip. In *Proceedings of the 26th Hawaii International Conference on System Sciences*, pages 409–418, January 1993.

[13] R. Suaya, R. Escovar, S. Ortiz, K. Banerjee, and N. Srivastava. Modeling and extraction of nanometer scale interconnects: Challenges and opportunities. In *Advanced Metallization Conferennce (AMC-2006)*, pages 17–27. Materials Research Society, Sept. 2006.

[14] I. Sutherland, B. Sproull, and D. Harris. *Logical Effort: Designing Fast CMOS Circuits*. Morgan Kaufmann Publishers, Inc., San Francisco, 1999.

[15] L. Wang and N. R. Shanbhag. Noise tolerant dynamic circuit design. In *International Symposium on Circuits and Systems (ISCAS)*, pages 549–552. IEEE, Jun 1999.

[16] K. Y. Yun and D. L. Dill. Automatic Synthesis of 3D Asynchronous Finite-State Machines. In *International Conference on Computer Aided Design, ICCAD-92*, pages 576–580, Los Alamitos, Calif., November 1992. IEEE Computer Science Press.

2007 IFIP International Conference on Very Large Scale Integration (VLSI-SoC 2007)

A Low-Power Deblocking Filter Architecture for H.264 Advanced Video Coding

Jaemoon Kim, Sangkown Na and Chong-Min Kyung
Department of EECS, KAIST
373-1, Yuseong-gu, Guseong-dong
Daejeon, Republic of Korea
Email: {jmkim, skna}@vslab.kaist.ac.kr, kyung@ee.kaist.ac.kr

Abstract—In this paper, a low-power deblocking filter architecture for H.264/AVC is proposed. A hybrid filtering order has been adopted to boost the speed of the deblocking filter process up to 208 clock cycles per 16x16 macroblock. The processing order of the filter is optimized to reduce power consumption and filter size and this is done by reducing memory access and raising the reusability of register blocks. A hardware implementation, under Samsung 0.18 μm standard cell library, consumes 18.34K gates at a clock frequency of 125MHz. Comparing to some state-of-the-art designs, the proposed architecture delivers the lowest level of power consumption while achieving similar speed of performance.

I. INTRODUCTION

H.264/AVC is an emerging coding standard [1] [2]. The H.264/AVC offers high coding efficiency in comparison to previous video standards like MPEG-2 or MPEG-4. The coding efficiency is gained from heterogeneous video coding standard algorithms such as incorporation of the inter/intra predictions, which use smaller block size (4x4 block), integer discrete cosine transform (DCT), context-adaptive variable length coding (CAVLC), and deblocking filtering. H.264/AVC is a block-based video coding standard algorithm, in which video is encoded and decoded block by block through intra/inter predictions, DCT, and CAVLC. Deblocking filter is an efficient tool to reduce the blocking artifacts between the blocks generated by block-based tools listed ahead. Despite of the fact that deblocking filter is one of the most important tools of H.264/AVC, it is difficult to implement into software to perform real-time decoding or encoding of high resolution of video sequence due to the large computational complexity. Therefore, it is required to have an adaptive hardware for deblocking filter.

H.264/AVC has an advantage of transferring good-quality video in a low bit-rate. This video standard is used widely nowadays for the video-transfer in wireless communication and is the standard for Digital Multimedia Broadcasting (DMB) in Korea [3]. It is important to realize low-power video decoder because mobile applications have a limited battery capacity.

Ever since the announcing of the new video coding standard H.264/AVC, many researches have conducted designing deblocking filter architecture. Most architecture have been able to operate in a frequency above 100MHz using current

fabrication technology. Furthermore, recently the filtering performance has improved to deliver a full HDTV (1920×1080) real-time video sequence. However, no work has been reported in realm of applying low-power deblocking filter in mobile applications. In this paper, it is proposed that a low-power architecture for the H.264/AVC deblocking filter which supports not only high resolution video such as full HDTV but also supports low-power mobile applications.

The rest of this paper is organized as follows: Section II describes the memory organization of the deblocking filter architecture. Section III reveals the detail of the proposed architecture. The result of the synthesis and the comparison with other architectures are discussed in section IV. Finally, conclusions are drawn in section V.

II. MEMORY ARCHITECTURE AND ORGANIZATION

The performance of deblocking filter depends on the memory architecture and organization. Well-organized memory architecture enables the implementation of the deblocking filter in the pipelined architecture of H.264/AVC decoder system with an enhanced performance. Memory architecture also affects power consumption where internal or external memory access acts as a significant factor in determining the power consumption.

A. Reconstructed memory

Our H.264/AVC decoder system was carefully designed to implement 4x4 sub-block level pipelined architecture. It is difficult to establish a direct connection with the deblocking filter with 4x4 sub-block level pipelined architecture because the reconstruction unit of the deblocking filter is a 16x16 macroblock. The reconstructed memory (Fig. 1) acts as a kind of bridge between 4x4 level and macroblock level pipeline. A 96x32 bit SRAM contains reconstructed pixel data of one macroblock, which are the results of inter/intra prediction are added with residual data from IDCT(Inverse DCT). Reconstructed memory consists of two single-port 96x32 bit SRAMs; one is for reading by deblocking filter and the other is for writing by residual adder. This double buffering enables the entire decoder architecture to simultaneously operate the 4x4 sub-block level pipeline while the deblocking filter running under the 16x16 macroblock level pipeline.

978-1-4244-1709-4/07/$25.00 © 2007 IEEE

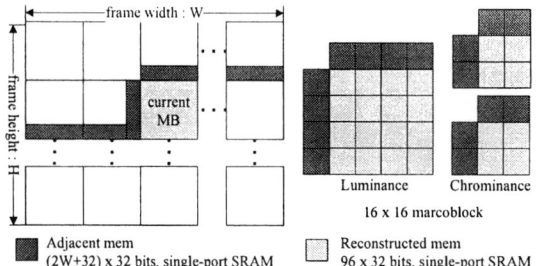

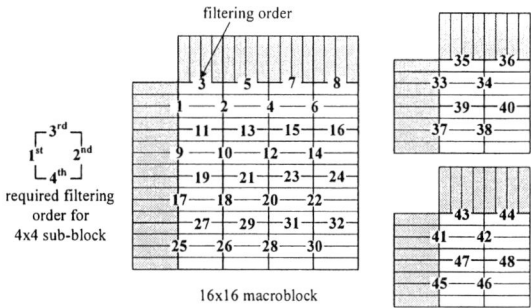

Fig. 1. Memory architecture and organization for deblocking filter

Fig. 2. Filtering sequence of hybrid filteirng

B. Adjacent memory

Deblocking filter not only requires the reconstructed pixels from the reconstructed memory but also requires of ready-filtered pixels when filtering the boundaries of macroblock. Fig. 1 represents these ready-filtered pixels, which is located adjacent to the current macroblock. As they are filtered pixels, the adjacent pixels should be delivered from the frame buffer through the data bus. However, the access to external memory decreases the performance of the deblocking filter and consumes a considerable amount of power. Employing a suitable size of SRAM could be the solution to this problem as SRAM can store the adjacent pixels [7] [8]. We used the term 'adjacent memory' to specify this particular SRAM for the rest of this paper.

The frame width, 'W', determines the size of the adjacent memory. The derivation of the equation used in Fig. 1 '(2W+32)×32' is as follows: '2W' is used to cover the luminance and chrominance. The multiplying coefficient '32' comes from the total number of bits which are divided into 4 pixels (8 bits each) vertically. The addition coefficient '32' is for the memory pixels stored in the left side adjacent memory. For example, to decode a CIF(352×288) size video sequence, a decoder requires a (2×352+32)×32 bit SRAM.

III. PROPOSED ARCHITECTURE

The objective of this paper is to achieve the low-power operation of the deblocking filter architecture by followings: 1)enhancing the speed of filtering per macroblock, 2)reducing resources which consume large amount of power and 3)reducing the switching of register.

A. Hybrid Filtering Sequence

Filtering sequence defined in the standard of H.264/AVC [2] is not appropriate for recycling data because all horizontal edges should be filtered only after all vertical edges are filtered. It is known that many researches have excelled the architecture of recycling data by using novel filtering sequence. We adopted a hybrid filtering sequence which is introduced as '2-D processing order' in [6]. Fig. 2 shows the filtering sequence of hybrid filtering. To begin with, vertical edge 1 is horizontally filtered before the filtering of vertical edge 2 occurs. Horizontal filtering and vertical filtering are then repeated alternatively for

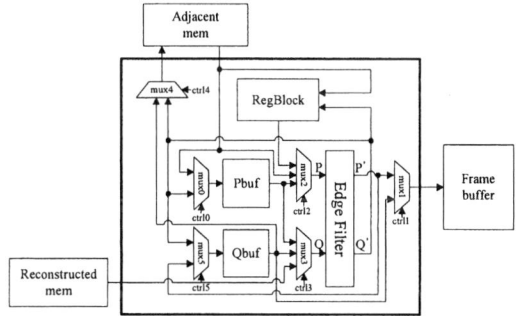

Fig. 3. An architecture for the low-power deblocking filter

edge 3, 4, 5, 6 and 7. Horizontal edge 8 is filtered vertically by following the alternating sequence. This filtering sequence satisfies the required filtering condition defined in the standard of H.264/AVC, which states that the order of filtering 4x4 sub-block should follow the edge sequence of vertical-left, vertical-right, horizontal-upper and horizontal-bottom as well as enabling the recycle of data.

The architecture described in [6], with several internal SRAMs, requires a frequent access to the SRAMs causing a large power consumption. Another disadvantage of this architecture is that it requires a large chip area for deblocking filter. In order to solve this problem, a new architecture is proposed which reduces excessive power consumption while maintaining the same filtering sequence.

B. Deblocking Filter Architecture

The implemented hardware architecture for the deblocking filter is shown in Fig. 3. In Fig. 3, the bold line indicates deblocking filter and the memories outside of the bold line indicates SRAMs and frame buffer. Frame buffer becomes an external memory connected via the 32bit data bus. The proposed architecture consists of a 1-D FIR filter (Edge Filter), two 4x4 transpose arrays (Pbuf, Qbuf) to transpose the data for filtering and one register block (RegBlock) which contains four 4x4 buffers for a temporal store of the filtered data.

Fig. 4 is a detailed architecture of RegBlock, Pbuf and Qbuf. RegBlock is a temporal buffer for pixels which require one more filtering. One of the register blocks of RegBlock,

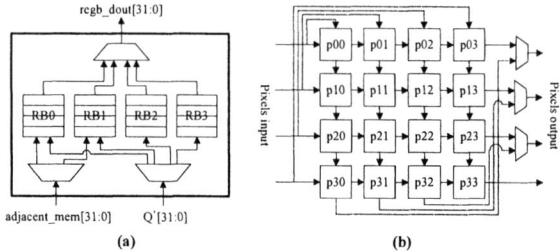

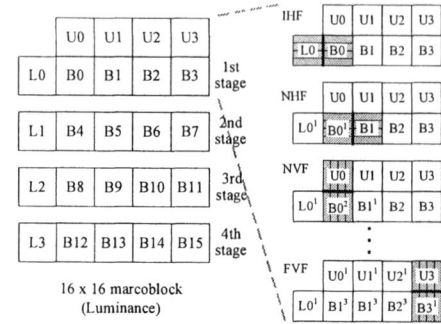

Fig. 4. The architecture for (a)RegBlock, (b)Pbuf and Qbuf

RB0-3, is stimulated to transfer data at one edge while the other register blocks keep their data. This means that no data transfer occurs in these register blocks. In other words, register transfer switching occurs at only one register block, giving an opportunity to reduce power consumption by clock gating scheme. Pbuf and Qbuf are 4x4 sub-block transpose matrices. These matrices transpose row-aligned pixels to column-aligned pixels or vice versa.

Fig. 5 shows the hybrid filtering sequence described earlier, where the macroblock is partitioned to reduce the number of registers and the number of access to SRAM(Adjacent memory). Each stage of partitioned macroblock consist of 8 edges, and the edges are filtered using four filtering modes - initial horizontal filtering (IHF), normal horizontal filtering (NHF), normal vertical filtering (NVF), and final vertical filtering (FVF).

In IHF mode to start with, left side of the edge block L0 comes from the adjacent memory and the right side of edge block B0 comes from the reconstructed memory. In one clock cycle, each of the four pixels of L0 and B0 are transferred as well as filtered by 'Edge Filter'. Pixel transposing is not a necessary requirement to start IHF mode because L0 in adjacent memory is already aligned as a row of pixels as shown in Fig. 2. After completing the IHF mode, the filtered L0 and B0 are expressed as $L0^1$ and $B0^1$ respectively, which the superscript 1 implies that the indicated block has been filtered once. Since L0 had undergone filtering for three times, it left only with one more filtering. As $L0^1$ is one step advanced filtered form than L0, it is transferred to frame buffer. Similarly, $B0^1$ gets stored in Pbuf and waits for the next filtering.

In NHF mode, B1 comes from the reconstructed memory and $B0^1$ comes from Pbuf. After completing NHF mode, $B0^2$ is stored into Qbuf, where $B1^1$ is stored in Pbuf. In NVF mode, $B0^2$ comes from Qbuf with a transposed form, and U0 comes from the adjacent memory. As shown in Fig. 2, U0 in the adjacent memory is already aligned in column, and hence NVF mode does not require any redundant clock cycle. After the NVF mode, $U0^1$ is transferred to frame buffer and B03 is stored into RegBlock. The NHF and NVF mode are then alternate twice. Lastly in FVF mode, U3 comes from the adjacent memory and $B3^1$ comes from Pbuf with a transposed form.

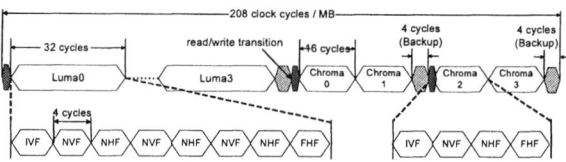

Fig. 5. The partitioned MB and each mode for hybrid filtering sequence.

Fig. 6. Total filtering clock cycles of deblocking filter.

The filtering processes described above are applied to other stages in the same way. Filtered blocks $\{B3^3, B7^3, B11^3, B12^3, B13^3, B14^3, B15^2\}$ are stored to the adjacent memory at the edges $\{17, 25, 26, 27, 29, 31, 32\}$. Block $B12^3$, $B13^3$, $B14^3$ and $B15^2$ are stored into the adjacent memory directly while $B3^3$, $B7^3$ and $B11^3$ require four clock cycles to be transposed from column-aligned pixels into row-aligned pixels, so that $B3^3$ and $B7^3$ can be stored at the edges $\{17, 25, 32\}$ respectively. At edge 32, the filtered block $B11^2$ and $B15^2$ is required to be stored into the adjacent memory at the same time. This is because we need the additional four clock cycles to store the block $B15^2$ into the adjacent memory.

Chrominance filtering is almost as the same as luminance filtering. We need 192 (i.e. 4×48) clock cycles to filter all of the horizontal and vertical edges in a macroblock, 12 clock cycles for the edges $\{32, 40, 48\}$ for the data congestion and 4 extra clock cycles for transition of SRAM to read and write. Thus, there are a total of 208 clock cycles to filter a macroblock(Fig. 6). The access to the reconstructed memory, adjacent memory and frame buffer occurs 96, 64, and 96 times respectively in the process of one macroblock. When comparing to the other architectures referred in [4]- [9], our architecture offers minimum number of memory access. Considering that the SRAM access requires more power consumption than that of register switching, the proposed architecture reduces the power consumption.

IV. IMPLEMENTATION RESULTS

In order to test the performance of our architecture, we described the architecture into synthesizable verilog HDL and synthesized it using Synopsys Design Compiler with Samsung 0.18 μm cell library. The clock frequency is set to 125MHz and the result of synthesis shows that our architecture takes

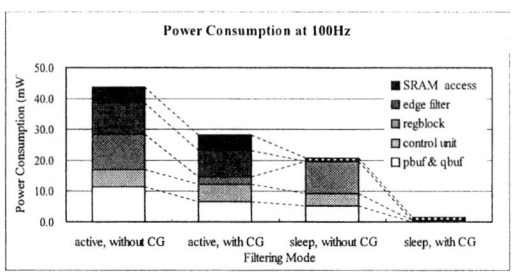

Fig. 7. Power Consumption at 100MHz - Comparison of Clock Gating and without Clock Gating at Active and Sleep Mode.

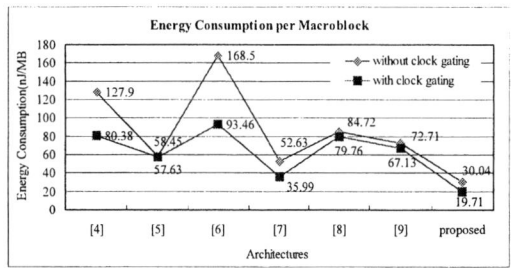

Fig. 8. Architecture Comparison - Energy Consumption per a Macroblock.

only 18.34K gate count. Synthesized design is verified with reference software [10].

Table I shows the comparison of our architecture with those in of [4]- [9]. The proposed architecture takes only 208 clock cycles per a macroblock while [4], [5] and [6] takes more than twice the clock cycles compared to the proposed architecture. This architecture shows a satisfactory performance, as 192 clock cycles is considered to be the optimal performance for deblocking filter with one 1-D FIR filter [8].

The proposed deblocking filter aims to achieve low-power consumption. In order to test for the power consumption, we extracted switching information of the synthesized architecture from simulations with several test sequences. We then simulated the power consumption using Synopsys Prime Power with the data from the switching information gathered previously. Fig. 7 shows our architecture drastically reduces power consumption using clock gating scheme. In 'active' mode, the power consumption of the clock-gated architecture is reduced by about 35%. In 'sleep' mode, the power consumption of the clock-gated architecture is further reduced by up to 92%. The drastic power reduction is the result of the decreased number of access to SRAM, the optimized register block architecture and the filtering process.

Fig. 8 clearly shows that the proposed architecture consumes the lowest power consumption. Due to the fact that the reference architectures [4]- [9] have different clock cycles to perform the deblocking filtering, the energy consumption of filtering a macroblock is chosen for comparison. The proposed

architecture is simulated with synthesized architecture, while other references are estimated based on the register switching and SRAM access using Samsung $0.18\mu m$ standard cell library. The power consumption of 1-D FIR filter and the control unit are ignored because the reference architecture does not offer detailed architecture and switching information. Fig. 8 shows that the proposed architecture consumes the smallest energy per one macroblock.

V. CONCLUSION

We have proposed a low-power deblocking filter architecture for H.264/AVC. The design is implemented in synthesizable verilog HDL and the synthesized results take only 18.34K gate at 125MHz clock frequency with Samsung 0.18 μm standard cell library. The proposed architecture has shown the lowest power consumption without any degradation of filtering performance compared to the state-of-the-art architecture of deblocking filter.

REFERENCES

[1] T. Wiegand, G. J. Sullivan, G. Bjntegaard, and A. Luthra, "Overview of the H.264/AVC Video Coding Standard," *IEEE Transaction on Circuits and Systems for Video Technology*, vol. 13, 2003, pp. 560-576
[2] "Draft ITU-T recommendation and final draft international standard of joint video specification (ITU-T Rec. H264 — ISO/IEC 14496-10 AVC)," JVT G050 2003.
[3] "Digital Multimedia Broadcasting," Telecommunications Technology Association in Korea, 2003SG05.02-046, 2003.
[4] Y. W. Huang, T. W. Chen, B. Y. Hsieh, T. C. Wang, T. H. Chang, and L. G. Chen, "Architecture Design for Deblocking Filter in H.264/JVT/AVC," *IEEE International Conference on Multimedia and Expo*, 2003.
[5] S. C. Chang, W. H. Peng, S. H. Wang, and T. Chiang, "A platform based bus-interleaved architecture for deblocking filter in H.264/MPEG-4 AVC," *IEEE Transaction on Consumer Electronics*, vol. 51, 2005, pp. 249-255
[6] B. Sheng, W. Gao, and D. Wu, "An Implemented Architecture of Deblocking Filter for H.264/AVC." *IEEE International Conference on Image Processing*, 2004, vol. 1, pp. 665-668
[7] T. M. Liu, W. P. Lee, T. A. Lin, and C. Y. Lee, "A Memory-Efficient Deblocking Filter for H.264/AVC Video Coding", *IEEE International Symposium on Circuit and Systems*, 2005
[8] G. Khurana, A. A. Kassim, T. P. Chua, and M. B. Mi, "A Pipelined Hardware Implementation of In-loop Deblocking Filter in H.264/AVC," *IEEE Transactions on Consumer Electronics*, vol. 52, No. 2, May 2006, pp. 536-540
[9] S. Y. Shih, C. R. Charng, and Y. L. Lin, "A Near Optimal Deblocking Filter for H.264 Advanced Video Coding," *Asia South Pacific Design Automation Conference*, 2006, pp. 170-175
[10] JVT H.264/AVC Reference Software JM 10.2

TABLE I
SYNTHESIS RESULTS

Architecture	Cycle /MB	# of 4×4 (Arrays)	Clock (MHz)	Technology (μm)	Gate Count(K)
Proposed	208	6	125	0.18	18.34
[4]	614	4	100	0.25	20.66
[5]	50-342	2	100	0.18	11.8
[6]	446	8	100	0.25	24.0
[7]	250	4	100	0.18	19.64
[8]	192	2	200	0.13	7.5
[9]	214	2	100	0.18	20.9

The Hazard-Free Superscalar Pipeline Fast Fourier Transform Algorithm and Architecture

Bassam Jamil Mohd   Adnan Aziz   Earl E. Swartzlander, Jr.

Abstract—This paper examines the superscalar pipeline Fast Fourier Transform algorithm and architecture. The algorithm presents a memory management scheme to prevent memory contention throughout the pipeline stages. The fundamental algorithm, a switch-based FFT pipeline architecture and an example 64-point FFT pipeline are presented. The proposed superscalar architecture substantially improves the FFT processing. The pipeline consists of $\log_2 N$ stages, where N is number of FFT points. Each stage can have M Processing Elements (PEs.) As a result, the architecture speed up is $M*\log_2 N$. The pipeline algorithm is configurable to any $M > 1$.

Index Terms— Discrete Fourier Transforms, Pipeline Processing, Memory Management

I. INTRODUCTION

THE FAST FOURIER TRANSFORM ALGORIRTHM, developed by [1], is a standard method for computing the Discrete Fourier Transform (DFT). The FFT algorithm consists of $\log_2 N$ loops; each loop executes N/2 complex operations. The operations in loop i depend on the results from loop i-1 creating potential data dependencies between algorithm loops. This is referred to as inter-stage dependency. The dependencies can be mitigated by use of temporal parallelism in the pipeline architecture where each algorithm loop is mapped to a pipeline stage. The performance of the machine can be further enhanced by exploiting spatial parallelism: processing operations within each stage in parallel. The challenge is to arrange the data in the pipeline memories. A simple mapping to memories results in multiple data elements residing in the same memory and creating structural hazards and pipeline-stalls. This dependency is referred to as an intra-stage dependency. The solution is to rearrange the data in the memory.

A variety of architectures and hardware implementations have been proposed to enhance speed, reduce power and resolve memory contention. One of the earliest implementations of a pipeline FFT was described in [2]. A variety of pipeline FFTs have been surveyed in [11]. Most pipeline FFT realizations use delay lines for data reordering between the processing elements. Although this gives a simple data flow architecture, it causes high power consumption. A memory address generation scheme was proposed by Cohen in [3], which allows parallel organization of memory so that the date used at any instant reside in different memories. The address generation is based on a counter, shifters and rotators.

In [4], Pease proposed dividing the memory into sub-memories for overlapping the access. He observed that the operand addresses differ only in the (n-i)-th bit for the butterfly operand pair in stage i, where n is number of address bits. A multi-bank memory address assignment for a radix-r FFT was developed in [5]. The memory assignment minimizes the memory size and allows conflict-free simultaneous memory access. Reference [6] developed a fast address generation scheme with hardware cost comparable to the address generation scheme in [3]. Ma and Wanhammar proposed an address generation scheme in [7] to reduce the hardware complexity and power consumption. Power is reduced by activating only half of the memory during memory access and by minimizing the number of memory accesses. Reference [10] proposed using cache-memory architecture to reduce communication energy between FFT processor and memory.

This paper proposes a superscalar pipeline architecture to achieve maximum speed for FFT processing. A switch fabric controls and connects single-port memories and processing elements (PEs). A memory management algorithm resolves any memory access contention. Rearranging data in the memories requires tracking them throughout the pipeline to process the right pair of data for FFT computations. The ordering of data elements is used to calculate the twiddle factors and other important indices. The algorithm provides an implicit method to track data. The superscalar pipeline achieves a speed up of $M*\log_2 N$.

The paper is organized as follows. Section-II explains the pipeline architecture and analyzes pipeline speedup hazards and optimizations. Section-III discusses hazard conditions and resolutions. It provides a pseudo code of the pipeline memory management algorithm. Section-IV details the design of a 64-point FFT with emphasis on the data movement and storage in the pipeline and memories. Section-V compares the proposed design with other pipeline FFTs.

II. ARCHITECTURE

This section discusses the proposed superscalar pipeline architecture for a radix-2 FFT.

A. Superscalar Pipeline Architecture

The pipeline architecture of an N-point FFT consists of $\log_2(N)$ stages. Figure 1 shows the block diagram of pipeline stage. Stage i of the pipeline executes the i-th loop of the

Radix-2 decimation-in-frequency FFT algorithm.

Each stage consists of:

1) A switch fabric that connects PEs and memories.

2) PEs which have three inputs (a, b, w) and two outputs (c, d) and perform the radix-2 FFT butterfly operation:

$$c = a + b$$
$$d = (a - b) * w \qquad (1)$$

(a, b) are inputs, w is the twiddle factor and (c, d) are outputs. There are M PEs per stage, where

- $N/2 \geq M \geq 2$
- $M = 2^p$, where p is an integer $p > 1$.

3) Memories that store intermediate results. There are $4*M$ single-port memories per stage, the size of each memory is equal to $N/(2*M)$. Memories can be implemented as RAM, caches, register files or flip-flops, based on the size of the memory and cost constraints. One half of the input memories will be active per cycle, while the other half will be active in the following cycle.

4) Memories to store twiddle factors. Since the twiddle factors do not change, twiddle factor memories can be implemented as ROMs. There are M ROMs per stage, each with size equal to $N/(2*M)$ words.

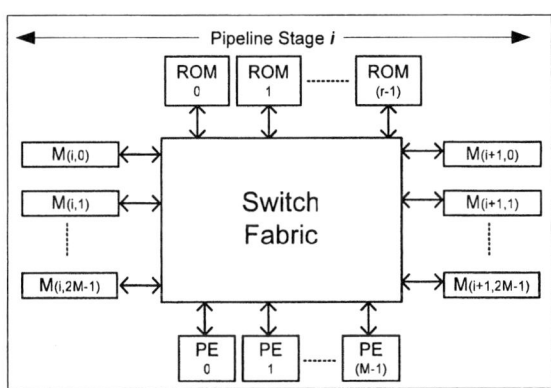

Fig. 1. Block Diagram of the Switch-Based Pipeline Stage

Figure 2 shows an overview of pipeline architecture. Each stage is capable of calculating M radix-2 butterfly results. Using the Instruction Level Parallelism (ILP) classification from [8], the architecture is a superscalar machine with Instruction Parallelism (IP) equal to M. It is also a super-pipeline where each cycle has $N/(2*M)$ minor-cycles. The architecture applies to the decimation-in-time FFT as well, where the specifications of stage i in the decimation-in-time algorithm is the same as that of stage $\log_2(N)$–i in the decimation-in-frequency algorithm. A scalar machine takes $(N/2)*\log_2(N)$ steps to execute an N-point radix-2 FFT algorithm. The architecture consists of $\log_2(N)$ stages, where each stage executes M operations. Therefore, the pipeline speedup can be expressed as: $M*\log_2(N)$. The maximum pipeline speedup is $(N/2)*\log_2(N)$, when M = N/2. In this case memories are reduced to registers, and the switch fabric

connects each any register to any PE. Clearly, while this case provides the most speed up, its hardware is expensive. The practical value of M is decided by design parameters: speed, area and power.

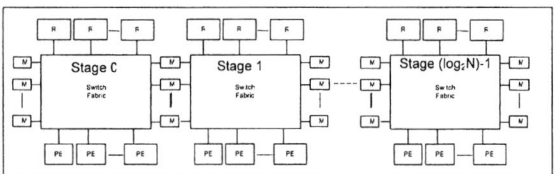

Fig. 2. Overview of Pipeline Architecture

B. Pipeline Design Optimizations

Upon close examination of the FFT algorithm, it is clear that not all twiddle factors are used in all stages. Also, the algorithm allows PEs to have identical twiddle factors in some stages, and therefore, not all the ROMs are required. In fact, the number and size of ROMs per stage can be reduced as outlined in Table 1.

TABLE 1: NUMBER AND SIZE OF ROMS PER STAGE

Stage "i"	Number of ROMs	Size of ROM
0	M	$N/(2*M)$
$\log_2 M \geq i \geq 0$	M	$N/(M* 2^i)$
$i > \log_2 M$	$M/2^{(i-\log_2 M)}$	1

If the pipeline is designed for a specific value of N, where N is static, the pipeline connectivity and twiddle factors are static. As a result, the design implementation can be optimized since the connectivity of each stage is predetermined. Figure 3 illustrates the connectivity of 16-point 2-PE pipeline. Furthermore, in many computations the value of twiddle factor is one. A twiddle factor of one reduces the PE computation to add/subtract operation. Also, several PEs executes specific sets of twiddle factors, which can lead to design simplification.

Fig. 3. Pipeline Hazard Example

As indicated earlier, the speed up of the pipeline depends on two factors: the number of PEs/stage (i.e., M) and the number of stages ($\log_2(N)$) since Speedup = $M*\log_2(N)$. One might ask this question: *"Given fixed target speedup (e.g., S), which factor should be increased to achieve more efficient design: the number-of-stages" or the number-of-PEs/stage?"* Consider a pipeline with a speedup of S with two designs: Design A and design B, as shown in Table 2. Design A has one PE per stage, while design B has one stage. Clearly,

- Design B requires less memory than design A since the design A total memory is proportional to S.
- Design A switch fabric is simpler than that of design B. The complexity of the design B switch fabric is quadratically proportional to S.

TABLE 2: ANALYZING SPEEDUP FACTORS

Parameter	Design A	Design B
Number of Stages	S	1
Number of PEs per Stage	1	S
Memory Size	N/2	N/(2*S)
Number of Memories	4*(S+1)	2*S
Total Memory	2*N*(S+1)	N
Switch Complexity	2*2	S*S

The main disadvantage of the increasing the number of stages is the increase in total memory. On the other hand, increasing the number of PEs per stage increases the complexity of the switch fabric. Hence, the tradeoffs between the two factors depend on the constraints on the total memory and the maximum complexity of the switch. Only specific design goals and technology processes can determine the optimum solution.

C. Pipeline Hazards

The main source of hazards in the pipeline is memory contention. Memory contention occurs when one or more PEs requests two or more accesses to a given memory at the same time. Memory contention results in stalling the pipeline and reduces the system speed. In the decimation-in-frequency FFT, memory contention does not occur in the early stages, it occurs from stage $\log_2(M)+1$ to the last stage. In the decimation-in-time FFT, the contention affects stage 0 to stage $\log_2(N)-\log_2(M)-1$.

Figure 3 shows an example of memory contention for N=16 and M=2. It is clear that stage 0 and stage 1 have no contention. However, contention occurs in stage 2 and stage 3.

Observe the following:

- In stage 2 the inputs for the top PE are $x_2(0)$ and $x_2(2)$, both of which reside in MEM0.
- In stage 3 the inputs for the top PE are $x_3(0)$ and $x_3(1)$, both of which reside in MEM0.

One solution for memory contention is to use a multi-port memory. However, multi-port memories are expensive and can slow down the system performance. In addition, the later stages of the pipeline have higher degree of contention which requires more ports in the memory. Eventually, it becomes impractical to implement the required multi-port memory. Moreover, the number of memory ports varies in the memory hierarchy. Register files usually have more ports than caches and SRAMs. Requiring a certain number of memory ports restricts where the intermediate results can be saved in the memory system. Another solution to resolve memory contention is to employ a memory management mechanism to mitigate the hazard, as discussed in the next section.

III. HAZARD FREE PIPELINE ALGORITHM

The main idea of the algorithm is resolve memory contention in the early stages of the pipeline. First, the condition that causes contention is described and then the hazard free algorithm is described.

A. Detecting Pipeline Hazard

From Figure 3, in stage 0, x(0) and x(8) are go to PE_0. Similarly, x(1) and x(9) go to PE_1,..., etc. Define stage distance as the index delta in each stage. The stage distance for a 16-point pipeline FFT is shown in Table 3.

TABLE 3: STAGE DISTANCE FOR 16-POINT PIPELINE FFT

Stage	Stage Distance	
	Decimation-In-Frequency	Decimation-in-Time
0	8	1
1	4	2
2	2	4
3	1	8

In general, for an N-point pipeline FFT, the stage distance for stage i is equal to $N/2^{(i+1)}$. Memory contention occurs when the stage distance falls in a single memory space. From Section II, the memory size is equal to N/(2*M). Hence, memory contention occurs in stage i if the following condition is satisfied:

$$N / 2^{(i+1)} \leq N /(2^M) \qquad (2)$$
$$i \geq \log_2(M)$$

A stage that satisfies condition (2) will be referred to as a hazard stage; the rest of the stages are safe stages. For instance, in Figure 3, stage 2 and stage 3 are hazard stages. Define memory pair $(i, j)_t$ as memory location x(i) and x(j) for stage *t*. In stage 2, the following memory pairs are hazard pairs: $(0, 2)_2$, $(1, 3)_2$, $(4, 6)_2$, $(5, 7)_2$. Other pairs will be referred to as safe pairs, for instance $(3, 5)_2$. The stage distance can be represented in binary form:

Stage-3 distance = 001

Define pair $(i, j)_t$ as a hazard pair if and only if:

1) *t* is a hazard stage
2) The bit wise Exclusive-OR of addresses i and j is equal to the stage t distance.

For example, the address pair $(5, 7)_2$ is a hazard pair since:

Stage-2 distance = 2_{10}

$5_{10} \oplus 7_{10} = 101_2 \oplus 111_2 = 010_2 = Stage\text{-}2\ distance$

On the other hand, address pair $(3, 5)_2$ is a safe pair because:

$3_{10} \oplus 5_{10} = 011_2 \oplus 101_2 = 110_2\ != \ Stage\text{-}2\ distance$

B. Memory Management Operations

Let $x_i(t)$ and $x_j(t)$ be the i-th and j-th elements in stage t and $i<j$. Define the memory management operations as follows (see Figure 4):

- **Normal Operation**: Input $x_i(t)$ and $x_j(t)$ are provided to the first and second inputs of the PE: a, b. The results (c and d) are saved in $x_i(t+1)$ and $x_j(t+1)$.

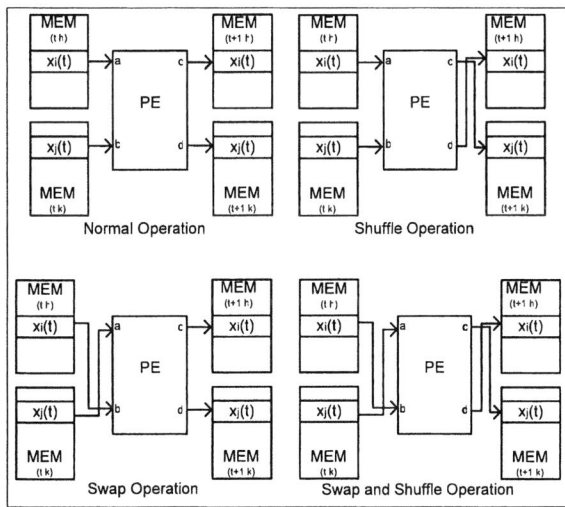

Fig. 4. Pipeline Memory Management Operations

- **Shuffle Operation** affects how PE results are saved back in memory. In shuffle operation, the results c and d are saved in $x_j(t+1)$ and $x_i(t+1)$
- **Swap Operation**: The swap operation affects the order of PE inputs. In swap operation, $x_i(t)$ is provided to b (instead of a) and $x_j(t)$ is provided to a (instead of b). The reason for the swap operation is because the PE is an asymmetric unit and the memory management algorithm changes the normal order of data in the memory. If the algorithm detects a case when inputs are incorrect, the swap operation is performed. A PE operation can have both swap and shuffle memory operations at the same time.

C. Pipeline Algorithm

The main idea of the pipeline algorithm is to identify hazard pairs in early stages and perform memory management operations to resolve the hazard. Because data is rearranged in memory, the algorithm has to track where data is. One idea to track the movement of data is to use a separate memory to store the data indexes (i.e., pointers), as shown in Figure 5. This approach provides a great flexibility in moving data in

the memory. It also simplifies the reordering logic of the final stage hardware. The downside of this approach is it increases memory size. Also, it increases loading the operands in the PE by one cycle to retrieve pointers from memory.

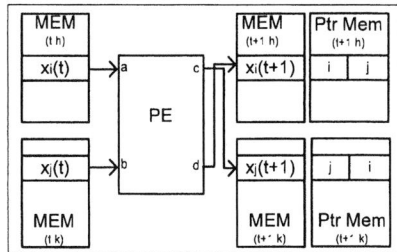

Fig. 5. Tracking Shuffled Data

Another (less flexible) solution is to move data in memory in a methodic way to simplify data tracking in the pipeline. This approach resolves hazards for next stage only. The algorithm can be summarized as follows. For each PE operation:

- If data has been reversed in memory, the PE input is swapped.
- If present data pair will create hazard in the next pipeline stage, the PE results are shuffled.

As a result of reordering data in the pipeline, results from the last stage in the pipeline should be reordered. Below is a detailed pseudo code of the algorithm for swap/shuffle operations.

```
// Preparation Step
Number_Of_Stages   = log2NUMBER_OF_FFT_POINTS
Cycles_Per_Stage   = N/(2*NUMBER_OF_PE)
Memory_Size        = N/2^(NUMBER_OF_PE+1)
Safe_Stage         = log2NUMBER_OF_PE
// Start main nester loops
for Current_Stage=0 to (Number_Of_Stages -1)
 Group_Size = N/2^(Current_Stage+1)
 for Current_Stage_Cycle=0 to (Cycles_Per_Stage -1)
  for Current_Cycle_Operation=0 to (NUMBER_OF_PE -1)

   // Calculate Operation Indices
   Horizontal_op_index = Cycles_Per_Stage *
                     Current_Cycle_Operation
                     + Current_Stage_Cycle
   Vertical_op_index   = NUMBER_OF_PE * Current_Stage_Cycle
                     + Current_Cycle_Operation
   Current_Stage_Rev = Number_Of_Stages - Current_Stage - 1
   Current_Group     = floor(Horizontal_op_index/
                     2^Current_Stage_Rev)
   Current_Operation = Horizontal_op_index mod 2^Current_Stage_Rev

   // Calculate Memory Address
   M0_addr = Current_Stage_Cycle
   If Current_Stage <= Safe_Stage
     M1_addr = M0_addr
   Else
     K = Safe_Stage +1
     L = Current_Stage
     M1_Addr = Reverse M0_Addr0 bits between K to L bits
   End

   // Calculate Memory Select
   If Current_Stage <= Safe_Stage
     Group_Offset = Current_Group * N /2^Current_Stage
     Group_Count  = Horizontal_op_index mod Group_Size
     Memory_Count = floor (Group_Count / Memory_Size)
     Offset       = Memory_Count * Memory_Size
     M0_Select    = Offset + Group_Offset
```

```
  M1_Select    = Offset + Group_Offset + Group_Size
Else
  Memory_Count = Vertical_op_index mod NUMBER_OF_PE
  Offset       = 2 * Memory_Count * Memory_Size
  M0_Select = Offset;
  M1_Select = Offset + 2 * Memory_SiZe
End
M0_data = Memory(Current_Stage, M0_Select0) [ M0_addr ]
M1_data = Memory(Current_Stage, M1_Select1) [ M0_addr ]

// Determine if swap operation is required
If  Current_Group is even
    AND Current_Sage <= Safe_Stage
  // Read data with no swap
  M0_data = Memory(Current_Stage, M0_Select) [ M0_addr ]
  M1_data = Memory(Current_Stage, M1_Select) [ M1_addr ]
Else
  // Read Data and perform Swap
  M1_data = Memory(Current_Stage, M0_Select) [ M0_addr ]
  M0_data = Memory(Current_Stage, M1_Select) [ M1_addr ]
End

// Read Twiddle
ROM_SELECT  = Current_Cycle_Operation
ROM_Address = Current_Operation * 2^Current_Stage
W           = ROM(Current_Stage, ROM_SELECT) [ROM_Address ]

// Enable PE to perform FFT butterfly operation
[Result1, Result0] =
    PE_Current_Cycle_Operation(M0_data, M1_data, W);

// Perform shuffle operation
Shuffle_Bit = log2 NUMBER_OF_FFT_POINTS
            - Current_Stage - 2
Shuffle_Flag = Horizontal_op_index [Shuffle_Bit]
If  Current_Stage >= Sage_Stage  AND
    Shuffle_Flag == 1
  // Shuffle ResultsShuffle = 1
  Memory(Current_Stage+1, M0_Select) [ M0_addr ] = Result1
  Memory(Current_Stage+1, M1_Select) [ M1_addr ] = Result0
Else
  // No Shuffling
  Memory(Current_Stage+1, M0_Select) [ M0_addr ] = Result0
  Memory(Current_Stage+1, M1_Select) [ M1_addr ] = Result1
End

 end // Current_Cycle_Operation
 end // Current_Stage_Cycle loop
 end // Current_Stage loop
```

IV. 64-POINT PIPELINE FFT DESIGN

This section explains a 64-point pipeline FFT design using four PEs per stage. Therefore, although there are 16 memories per stage, only eight memories will be active memory at any time. The memory size is eight words. There are four ROMs per stage, each with size of eight words. The pipeline speed up equals 6*4=24. The following tables detail the operation of the pipeline PEs and illustrate the memory contents.

Table 4 gives the PE operand pairs for Stage 0. The rows give the operand pairs for PE_0, PE_1, PE_2 and PE_3. The columns give the pairs for each micro-cycle in Stage 0 cycles. There are eight micro-cycles per stage. For example, at micro-cycle 0:

- PE_0 input operands will be MEM[0] and MEM[32]
- PE_1 input operands will be MEM[8] and MEM[40]
- PE_2 input operands will be MEM[16] and MEM[48]
- PE_3 input operands will be MEM[24] and MEM[56]

Tables 5-9 give the PE operand pairs for Stages 1-5. Underlined pairs indicate shuffle operation. Since Stages 0-2 are safe stages, the first shuffle operation starts in Stage 2 to prevent hazards in stage 3. Table 10 lists the memory contents for pipeline stages. For example, the output of stage 2 has the memory contents for Memory 0 as follows: 0, 1, 2, 3, 12, 13, 14, and 15.

TABLE 4: PIPELINE STAGE-0 PE OPERAND PAIRS

PE	Stage-0 Cycles							
	0	1	2	3	4	5	6	7
0	0,32	1,33	2,34	3,35	4,36	5,37	6,38	7,38
1	8,40	9,41	10,42	11,43	12,44	13,45	14,46	15,47
2	16,48	17,49	18,50	19,51	20,52	21,53	22,54	23,55
3	24,56	25,57	26,58	27,59	28,60	29,61	30,61	31,63

TABLE 5: PIPELINE STAGE-1 PE OPERAND PAIRS

PE	Stage-1 Cycles							
	0	1	2	3	4	5	6	7
0	0,16	1,17	2,18	3,19	4,20	5,21	6,22	7,23
1	8,24	9,25	10,26	11,27	12,28	13,29	14,30	15,31
2	32,48	33,49	34,50	35,51	36,52	37,53	38,54	39,55
3	40,56	41,57	42,58	43,59	44,60	45,61	46,62	47,63

TABLE 6: PIPELINE STAGE-2 PE OPERAND PAIRS

PE	Stage-2 Cycles							
	0	1	2	3	4	5	6	7
0	0,8	1,9	2,10	3,11	4,12	5,13	6,14	7,15
1	16,24	17,25	18,26	19,27	20,28	21,29	22,30	23,31
2	32,40	33,41	34,42	35,42	36,44	37,45	38,46	39,47
3	48,56	49,57	50,58	51,59	52,60	53,61	54,62	55,63

TABLE 7: PIPELINE STAGE-3 PE OPERAND PAIRS

PE	Stage-3 Cycles							
	0	1	2	3	4	5	6	7
0	0,4	1,5	2,6	3,7	12,8	13,9	14,10	15,11
1	16,20	17,21	18,22	19,23	28,24	29,25	30,26	31,27
2	32,36	33,37	34,38	35,39	44,40	45,41	46,42	47,43
3	48,52	49,53	50,54	51,55	60,56	61,57	62,58	63,59

TABLE 8: PIPELINE STAGE-4 PE OPERAND PAIRS

PE	Stage-4 Cycles							
	0	1	2	3	4	5	6	7
0	0,2	1,3	6,4	7,5	12,14	13,15	10,8	11,9
1	16,18	17,19	22,20	23,21	28,30	29,31	26,2	27,25
2	32,34	33,35	38,36	39,37	44,46	45,47	42,40	43,41
3	48,50	49,51	54,52	55,53	60,62	61,63	58,56	59,57

TABLE 9: PIPELINE STAGE-5 PE OPERAND PAIRS

PE	Stage-5 Cycles							
	0	1	2	3	4	5	6	7
0	0,1	3,2	6,7	5,4	12,13	15,14	10,11	9,8
1	16,17	19,18	22,23	21,20	28,29	31,30	26,27	25,25
2	32,33	35,34	38,39	37,36	44,45	47,46	42,43	41,40
3	48,49	51,50	54,55	53,52	60,61	63,62	58,59	57,56

TABLE 10: PIPELINE MEMORY CONTENTS

MEM	Input	0	1	2	3	4	5
0	0	0	0	0	0	0	0
	1	1	1	1	1	3	3
	2	2	2	2	6	6	6
	3	3	3	3	7	5	5
	4	4	4	12	12	12	12
	5	5	5	13	13	15	15
	6	6	6	14	10	10	10
	7	7	7	15	11	9	9
1	8	8	8	8	8	8	8
	9	9	9	9	9	11	11
	10	10	10	10	14	14	14
	11	11	11	11	15	13	13
	12	12	12	4	4	4	4
	13	13	13	5	5	7	7
	14	14	14	6	2	2	2
	15	15	15	7	3	1	1
2	16	16	16	16	16	16	16
	17	17	17	17	17	19	19
	18	18	18	18	22	22	22
	19	19	19	19	23	21	21
	20	20	20	28	28	28	28
	21	21	21	29	29	31	31
	22	22	22	30	26	26	26
	23	23	23	31	27	25	25
3	24	24	24	24	24	24	24
	25	25	25	25	25	27	27
	26	26	26	26	30	30	30
	27	27	27	27	31	29	29
	28	28	28	20	20	20	20
	29	29	29	21	21	23	23
	30	30	30	22	18	18	18
	31	31	31	23	19	17	17
4	32	32	32	32	32	32	32
	33	33	33	33	33	35	35
	34	34	34	34	38	38	38
	35	35	35	35	35	37	37
	36	36	36	44	44	44	44
	37	37	37	45	45	47	47
	38	38	38	46	42	42	42
	39	39	39	47	43	41	41
5	40	40	40	40	40	40	40
	41	41	41	41	41	43	43
	42	42	42	42	46	46	46
	43	43	43	43	47	45	45
	44	44	44	36	36	36	36
	45	45	45	37	37	39	39
	46	46	46	38	34	34	34
	47	47	47	39	35	33	33
6	48	48	48	48	48	48	48
	49	49	49	49	49	51	51
	50	50	50	50	54	54	54
	51	51	51	51	55	53	53
	52	52	52	60	60	60	60
	53	53	53	61	61	63	63
	54	54	54	62	58	58	58
	55	55	55	63	59	57	57
7	56	56	56	56	56	56	56
	57	57	57	57	57	59	59
	58	58	58	58	62	62	62
	59	59	59	59	63	61	61
	60	60	60	52	52	52	52
	61	61	61	53	53	55	55
	62	62	62	54	50	50	50
	63	63	63	55	51	49	49

V. COMPARISON WITH OTHER FFT PIPELINES

Table 11 summarizes features of FFT pipeline architectures discussed in reference [11] and the switch based architecture (shown in the last row of the table.) The other pipeline architectures require delay elements in the pipeline implementation. Delays are implemented by registers (which dissipate high dynamic power) or by RAMs with additional address generation hardware (which increases design complexity). The switch-based pipeline uses SRAM caches, which consume less power than registers and is easier to implement. Moreover, the throughputs of the other pipelines are limited to one (single-path) or a few (multi-path), while the switch based implementation has a throughput of M. Unfortunately, the switch based pipeline requires larger memorys and more hardware in the data path.

TABLE 11: FFT PIPELINE ARCHITECTURES

FFT Pipeline	Multiplier #	Adder #	Memory Size	Speed up
Radix-2 Multi-path Delay Commutator	$2(\log_4 N - 1)$	$4 \log_4 N$	$3N/2 - 2$	$\log_2 N$
Radix-2 Single-path Delay Feedback	$2(\log_4 N - 1)$	$4 \log_4 N$	$N - 1$	$\log_2 N$
Radix-4 Single-path Delay Feedback	$\log_4 N - 1$	$8 \log_4 N$	$N - 1$	$\log_2 N$
Radix-4 Multi-path Delay Commutator	$3(\log_4 N - 1)$	$8 \log_4 N$	$5N/2 - 4$	$\log_2 N$
Radix-4 Single-path Delay Commutator	$\log_4 N - 1$	$3 \log_4 N$	$2N/2 - 2$	$\log_2 N$
Radix-2^2 Single-path Delay feedback	$\log_4 N - 1$	$4 \log_4 N$	$N - 1$	$\log_2 N$
Switch-Based Pipeline	$M*2(\log_4 N - 1)$	$M*4 \log_4 N$	$2*N* (1+\log_2 N)$	$M* \log_2 N$

VI. CONCLUSION AND FUTURE WORK

In this paper we have proposed switch-based architecture for FFT engine implementation. We have also presented an algorithm to predict and resolve memory contentions. As a result the pipeline speedup is M*log₂N, where N is number of points and M is number of processing elements. An implementation of a 64-point FFT machine using the proposed architecture was presented. The proposed architecture was compared to other FFT pipelines. Future research should focus on reducing power consumption of the FFT pipeline and extending the work done in [7], [9] and [10].

REFERENCES

[1] J. W. Cooley and J. W. Tukey, "An algorithm for the machine calculation of complex Fourier series," *Math. Comput.,* vol. 19, pp. 297-301, 1965.

[2] H. L. Groginsky and G. A. Works, "A pipelined fast Fourier transform," *IEEE Transactions* on *Computers,* vol. C-19. pp. 1015-1019, 1970.

[3] D. Cohen, "Simplified control of FFT hardware," *IEEE Transactions on Acoustics, Speech, and Signal Processing*, vol. ASSP-24, pp. 577-579, 1976.

[4] M. C. Pease, "Organization of large scale Fourier processors," *JACM,* vol. 16, pp. 474-482, 1969.

[5] L. G. Johnson, "Conflict free memory addressing for dedicated FFT hardware," *IEEE Transactions on Circuits and Systems, II,* vol. 39, pp. 312-316, 1992.

[6] Y. Ma, "An effective memory addressing scheme for FFT processors," *IEEE Transactions on Signal Processing*, vol. 47, pp. 907-911, 1999.

[7] Y. Ma and L. Wanhammar, "A hardware efficient control of memory addressing for high-performance FFT processors," *IEEE Transactions on Signal Processing*, vol. 48, pp. 917-921, 2000.

[8] J Shen and M Lipasti, *Modern Processor Design: Fundamentals of Superscalar Processors*, New York: McGraw-Hill, 2005, pp. 27-32.

[9] A. El-Khashab and E. Swartzlander, "The Modular Pipeline Fast Fourier Transform Algorithm and Architecture," *Proc. of the Thirty-Seventh Asilomar Conference on Signals, Systems, and Computers*, November 9-12, 2003, Pacific Grove, CA, pp. 1463-1467.

[10] B. M. Baas, "A low-power, high-performance 1024-point FFT processor," *IEEE Journal of Solid-State Circuits*, vol. 34, pp. 380-387, March 1999.

[11] S. He and M. Torkelson, "Designing pipeline FFT processor for OFDM (de)modulation," *Proc. of URSI Int. Symp. on Signals, Systems, and Electronics*, 1998, pp. 257-262.

AN EFFICIENT H.264 INTRA FRAME CODER SYSTEM DESIGN

Ilker Hamzaoglu, Ozgur Tasdizen, Esra Sahin

Faculty of Engineering and Natural Sciences, Sabanci University

34956, Tuzla, Istanbul, Turkey

hamzaoglu@sabanciuniv.edu, tasdizen@su.sabanciuniv.edu, esra@su.sabanciuniv.edu

ABSTRACT

In this paper, we present an efficient H.264 / MPEG4 Part 10 Intra Frame Coder System. The system achieves real-time performance for portable applications with low hardware cost, and it includes a novel intra prediction hardware design. The proposed hardware is implemented in Verilog HDL. The Verilog RTL code works at 71 MHz in a Xilinx Virtex II FPGA and it code 35 CIF frames (352x288) per second. The system also includes a software running on an Arm926EJS processor for implementing pre-processing and post-processing functions. The H.264 Intra Frame Coder hardware and software are demonstrated to work together on an Arm Versatile Platform development board.

1. INTRODUCTION

Video compression systems are used in many commercial products, from consumer electronic devices such as digital camcorders, cellular phones to video teleconferencing systems. These applications make the video compression systems an inevitable part of many commercial products. To improve the performance of video compression systems, recently, H.264 / MPEG4 Part 10 video compression standard, offering significantly better video compression efficiency than previous standards, is developed with the collobaration of ITU and ISO standardization organizations.

The video compression efficiency achieved in H.264 standard is not a result of any single feature but rather a combination of a number of encoding tools. As it is shown in the top-level block diagram of an H.264 encoder in Figure 1, one of these tools is the intra prediction algorithm used in the baseline profile of H.264 standard [1, 2, 3]. Intra prediction algorithm generates a prediction for a Macroblock (MB) based on spatial redundancy. H.264 intra prediction algorithm achieves better coding results than the intra prediction algorithms used in the previous video compression standards. However, this coding gain comes with an increase in encoding complexity which makes it an exciting challenge to have a real-time implementation of H.264 intra prediction algorithm.

H.264 Intra Frame Coder is a video encoder which uses H.264 intra prediction algorithm for generating predictions for each MB [4, 5]. H.264 intra frame coder is a competitive alternative to JPEG2000 for still image compression, in terms of both coding efficiency and computational complexity. H.264 intra frame coder is also shown to be superior to Motion-JPEG2000, especially at lower resolutions, for motion picture production, editing and archiving, where video frames are coded as I-frames only to allow for random access to each individual picture.

In this paper, we present an efficient H.264 Intra Frame Coder System. The system achieves real-time performance for portable applications with low hardware cost, and it includes a novel intra

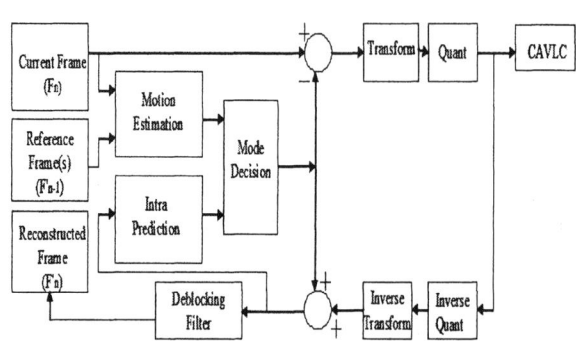

Figure 1. H.264 Encoder Block Diagram

prediction hardware design. The proposed hardware is implemented in Verilog HDL. The Verilog RTL code works at 71 MHz in a Xilinx Virtex II FPGA and it code 35 CIF frames (352x288) per second. The system also includes a software running on an Arm926EJS processor for implementing pre-processing and post-processing functions. The H.264 Intra Frame Coder hardware and software are demonstrated to work together on an Arm Versatile Platform PB926EJ-S development board.

An H.264 Intra Frame Coder hardware is presented in [4, 5]. This hardware achieves higher performance than our hardware design at the expense of a much higher hardware cost. Our hardware design is a more cost-effective solution for portable applications. They use four reconfigurable datapaths, which include 12 adders, 16 multiplexers, 4 shifters and 4 clippers, in their intra prediction hardware design. They use additional adders and multiplexers for preprocessing in 16x16 plane mode and 8x8 plane mode. On the other hand, we use three reconfigurable datapaths, which include 6 adders, 12 multiplexers, 6 shifters and 2 clippers, in our intra prediction hardware design. We don't use any aditional hardware resources for 16x16 plane mode and 8x8 plane mode.

The rest of the paper is organized as follows. Section 2 explains the H.264 intra frame coder algorithm. Section 3 describes the proposed architecture in detail. The implementation results are given in Section 4. Finally, Section 5 presents the conclusions.

2. H.264 INTRA FRAME CODER ALGORITHM

The top-level block diagram of an H.264 Intra Frame Coder is shown in Figure 2. An H.264 Intra Frame Coder has a forward path and a reconstruction path [1, 2, 3, 4, 5]. The forward path is used to encode a video frame and create the bitstream. The reconstruction path is used to decode the encoded frame and reconstruct the decoded frame. Since a decoder never gets original images, but

200

978-1-4244-1709-4/07/$25.00 © 2007 IEEE

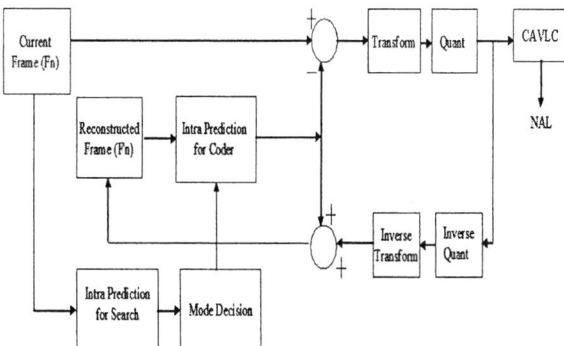

Figure 2. H.264 Intra Frame Coder Block Diagram

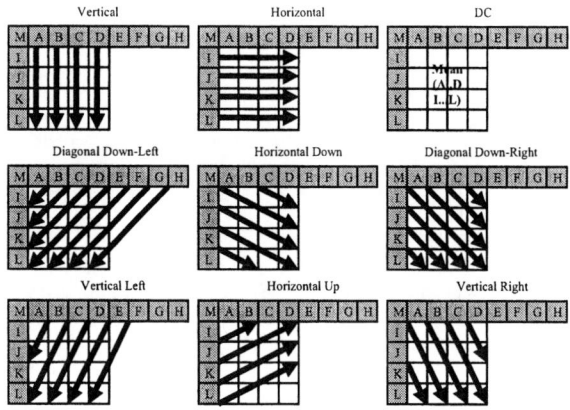

Figure 3. 4x4 Luma Prediction Modes

$$pred[0, 0] = A + 2B + C + 2 >> 2$$
$$pred[0, 1] = B + 2C + D + 2 >> 2$$
$$pred[0, 2] = C + 2D + E + 2 >> 2$$
$$pred[0, 3] = D + 2E + F + 2 >> 2$$
$$pred[1, 0] = B + 2C + D + 2 >> 2$$
$$pred[1, 1] = C + 2D + E + 2 >> 2$$
$$pred[1, 2] = D + 2E + F + 2 >> 2$$
$$pred[1, 3] = E + 2F + G + 2 >> 2$$
$$pred[2, 0] = C + 2D + E + 2 >> 2$$
$$pred[2, 1] = D + 2E + F + 2 >> 2$$
$$pred[2, 2] = E + 2F + G + 2 >> 2$$
$$pred[2, 3] = F + 2G + H + 2 >> 2$$
$$pred[3, 0] = D + 2E + F + 2 >> 2$$
$$pred[3, 1] = E + 2F + G + 2 >> 2$$
$$pred[3, 2] = F + 2G + H + 2 >> 2$$
$$pred[3, 3] = G + 3H + 2 >> 2$$

Figure 4. Prediction Equations for 4x4 Diagonal Down-Left Mode

rather works on decoded frames, reconstruction path in the encoder ensures that both encoder and decoder use identical reference frames for intra prediction. This avoids possible encoder – decoder mismatches.

Intra prediction algorithm predicts the pixels in a MB using the pixels in the available neighboring blocks [1, 2, 3]. For the luma component of a MB, a 16x16 predicted luma block is formed by performing intra predictions for each 4x4 luma block in the MB and by performing intra prediction for the 16x16 MB. There are nine prediction modes for each 4x4 luma block and four prediction modes for a 16x16 luma block. A mode decision algorithm is then used to compare the 4x4 and 16x16 predictions and select the best luma prediction mode for the MB. 4x4 prediction modes are generally selected for highly textured regions while 16x16 prediction modes are selected for flat regions.

There are nine 4x4 luma prediction modes designed in a directional manner. Each 4x4 luma prediction mode generates 16 predicted pixel values using some or all of the neighboring pixels A to M as shown in Figure 3. The pixels A to M belong to the neighboring blocks and are assumed to be already encoded and reconstructed and are therefore available in the encoder and decoder to generate a prediction for the current block. The arrows indicate the direction of prediction in each mode. The predicted pixels are calculated by a weighted average of the neighboring pixels A-M for each mode except Vertical, Horizontal and DC modes. The prediction equations used in 4x4 Diagonal Down-Left prediction mode are shown in Figure 4 where [y,x] denotes the position of the pixel in a 4x4 block (the top left, top right, bottom left, and bottom right positions of a 4x4 block are denoted as [0, 0], [0, 3], [3, 0], and [3, 3], respectively) and pred[y,x] is the prediction for the pixel in the position [y,x].

There are four 16x16 luma prediction modes designed in a directional manner. Vertical, Horizontal and DC modes are similar to 4x4 luma prediction modes. Plane mode is an approximation of bilinear transform with only integer arithmetic.

For the chroma components of a MB, a predicted 8x8 chroma block is formed for each 8x8 chroma component by performing intra prediction for the MB. There are four 8x8 chroma prediction modes. Vertical, Horizontal, DC and Plane modes are similar to 16x16 luma prediction modes. A mode decision algorithm is used to compare the 8x8 predictions and select the best chroma prediction mode for chroma components of the MB. Both chroma components of a MB always use the same prediction mode.

The predicted MB is subtracted from the current MB to generate the residual MB. Residual MB is transformed using forward transform algorithm [1, 2, 3]. Transform algorithm is based on a 4x4 integer transform which only uses integer addition and binary shift operations. Transform coefficients are then quantized and re-ordered in a zig-zag scan order [1, 2, 3]. The quantization algorithm uses a non-uniform quantizer and it requires an integer multiplication. Quantization parameter can take a value between 0-51 and an increment of 1 in quantization parameter results in 12.2% increment in quantization step size. The reordered quantized transform coefficients are entropy encoded using context adaptive variable length coding (CAVLC) algorithm [1, 2, 3]. CAVLC uses multiple tables for a syntax element and it adapts to the current context by selecting one of these tables for a given syntax element based on the already transmitted syntax elements.

The quantized transform coefficients are also reconstructed. The quantized transform coefficients are inverse quantized and inverse transformed to generate the reconstructed residual data. Since quantization is a lossy process, inverse quantized and inverse transformed coefficients are not identical to the original residual data. The reconstructed residual data are added to the predicted pixels in order to create the reconstructed frame.

2007 IFIP International Conference on Very Large Scale Integration (VLSI-SoC 2007)

3. PROPOSED HARDWARE ARCHITECTURE

The proposed H.264 intra frame coder hardware, as shown in Figure 5, includes a search & mode decision hardware and a coder hardware that work in a pipelined manner. After the first MB of the input frame is loaded to the input register file, search & mode decision hardware starts to work on determining the best mode for coding this MB. After search & mode decision hardware determines the best mode for the first MB, coder hardware starts to code the first MB using the selected best mode and search & mode decision hardware starts to work on the second MB. The entire frame is processed MB by MB in this order.

This is achieved by performing intra prediction in the search & mode decision hardware using the pixels in the current frame rather than the pixels in the reconstructed frame. However, intra prediction in the coder hardware is performed using the pixels in the reconstructed frame in order to be compliant with H.264 standard. This makes the MB pipelining possible at the expense of a small PSNR loss in the video quality [4, 5].

3.1 Proposed Search & Mode Decision Hardware

The proposed search & mode decision hardware, as shown in Figure 6, includes Intra Prediction, Residue, Hadamard Transform and Mode Decision modules. The efficient intra prediction hardware design presented in [6] is used in the proposed hardware. In the proposed hardware, there are two parts operating in parallel in order to complete the search & mode decision process faster. The upper part is used for finding the best 16x16 luma prediction mode for the luma component of a MB and the best 8x8 chroma prediction mode for the chroma components of a MB. The lower part is used for finding the best 4x4 luma prediction mode for each 4x4 block in the luma component of a MB.

Top level scheduling for the upper part of the search & mode decision hardware for 16x16 luma predictions is shown in Figure 7. First, the neighboring buffers in the intra prediction hardware are loaded with the corresponding neighboring pixels from the current MB register. Then, the intra prediction hardware generates the pixel predictions for the luma component of the current MB using the first available 16x16 luma mode and writes the predicted pixels to the prediction buffer. The Residue hardware, then, calculates the difference between the corresponding luma pixels in the current MB and the predicted MB. As the residue data associated with the first pixel position in a MB is calculated, Hadamard Transform module starts to calculate the Sum of Absolute Transformed Difference (SATD) for that mode using the residue data. So, Residue and Hadamard Transform modules are overlapped.

Hadamard transform for SATD calculations of 16x16 luma prediction modes requires storing DC coefficients in a register, because Hadamard transform has to be applied to these coefficients again. The multiplexer before Hadamard Transform module selects between DC coefficients and coefficients from the residue block.

After Hadamard Transform module finishes calculating SATD for an available 16x16 luma prediction mode of a MB, it decides whether it is the mode with lowest cost or not. After each available 16x16 luma prediction mode for a MB is searched, the prediction mode with the lowest cost and its cost information are sent to the Top Level Mode Decision hardware.

When the upper part of the search & mode decision hardware finishes with available 16x16 luma modes of a MB for luma samples, it starts to work with 8x8 chroma modes of the same MB for chroma samples. Top level scheduling for chroma samples is similar to that of luma samples.

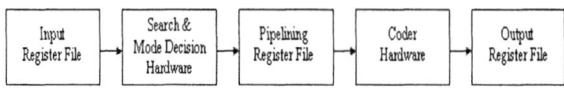

Figure 5. H.264 Intra Frame Coder Block Diagram

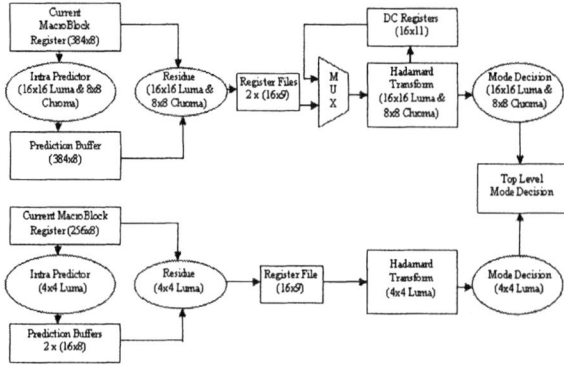

Figure 6. Search & Mode Decision Hardware Block Diagram

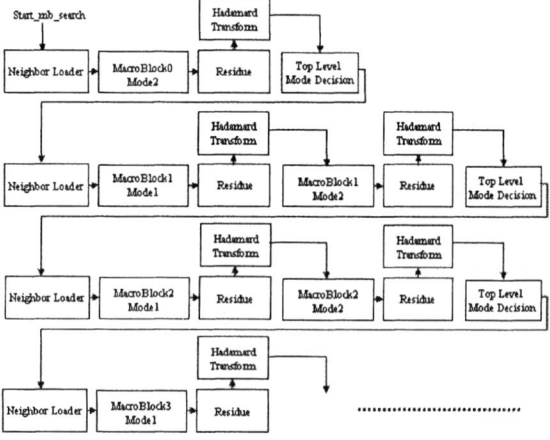

Figure 7. Schedule for 16x16 Luma Prediction Modes

Table 1. Latencies of the Modules in the Upper Part of the Search & Mode Decision Hardware

Module	Latency
Neighbor Loader	256 clock cycles
Hadamard Transform	288 clock cycles
Residue Module	256 clock cycles
Intra Prediction – Mode0	257 clock cycles
Intra Prediction – Mode1	257 clock cycles
Intra Prediction – Mode2	273 clock cycles
Intra Prediction – Mode3	340 clock cycles

The latencies of the modules in the upper part of the search & mode decision hardware are given in Table 1. In the worst case, when all 16x16 prediction modes are available, intra search for a MB takes 256*4 (Neighbor Loader) + 1127 (Intra Prediction) + 288*4 (Hadamard Transform) + 1*4 (Top Level Mode Decision) = 3307 clock cycles.

Top level scheduling for the lower part of the search & mode decision hardware is shown in Figure 8. Before intra prediction hardware for the first available mode of a 4x4 luma block starts, the corresponding entries of the neighboring buffers for that 4x4 block in the prediction hardware are loaded with the neighboring pixels from the current MB register file. After generating pixel predictions of a 4x4 luma block using an available 4x4 luma prediction mode, the difference (residue) between the current 4x4 luma block and the predicted 4x4 luma block is calculated by Residue module. When the Residue module finishes the calculation of residue data for a 4x4 luma block for the current available 4x4 luma prediction mode, Hadamard Transform module starts to calculate SATD for that mode using the residue data. After Hadamard Transform finishes to calculate SATD for a 4x4 luma prediction mode, mode decision hardware for 4x4 luma blocks determines whether this prediction mode is the mode with lowest cost or not.

Intra prediction module is overlapped with Residue and Hadamard Transform modules. As the Residue and Hadamard Transform modules are working on the current available 4x4 luma prediction mode for a 4x4 luma block, intra prediction module starts to generate the prediction for the next available 4x4 luma prediction mode for the same 4x4 luma block if the current available 4x4 prediction mode is not the last available mode for the current 4x4 luma block.

If the current available 4x4 prediction mode is the last available 4x4 luma prediction mode for the current 4x4 luma block, Neighbor Loader module starts to load the corresponding neighboring pixels for the next 4x4 luma block from the current MB register file as the Residue module is working on the current 4x4 luma block. The Residue module is again followed by Hadamard Transform module. After Neighbor Loader finishes loading the neighboring pixels of the next 4x4 luma block, intra prediction module starts to generate the prediction for the first available 4x4 luma prediction mode for the next 4x4 luma block. All the 4x4 luma blocks in a MB are processed in this order.

After intra prediction for all 4x4 blocks in a MB is finished, most probable mode calculation module determines the number of selected modes which are not the most probable mode for each 4x4 block in a MB and uses this information to calculate the cost of using intra 4x4 prediction for a MB (for each 4x4 block, $Cost_{4x4}$ = SATD + $4\lambda R$, where R=0 when selected mode is the most probable mode and R=1 otherwise). Most probable mode calculation module has vertical and horizontal buffers that are used for storing the most probable mode information of the 4x4 blocks in the MB boundaries.

Finally, Top Level Mode Decision module uses the results produced by the individual mode decision modules of the lower and upper parts of search & mode decision hardware to determine the prediction modes with lowest cost for a MB (one mode for luma samples and one mode for chroma samples) and sends this information to the coder hardware. The mode decision algorithm implemented in the proposed mode decision hardware is the same as the algorithm implemented in the H.264 Joint Model (JM) reference software encoder when there is no Rate-Distortion optimization.

In order to complete the SATD operations faster, the high speed Hadamard transform hardware shown in Figure 9 is designed. The proposed hardware finishes SATD operations of a 4x4 block in 18 clock cycles.

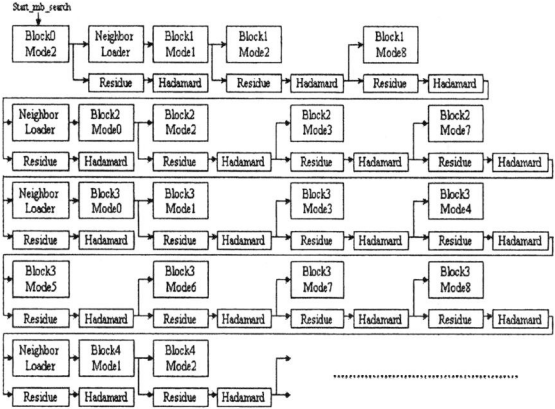

Figure 8. Schedule for 4x4 Luma Prediction Modes

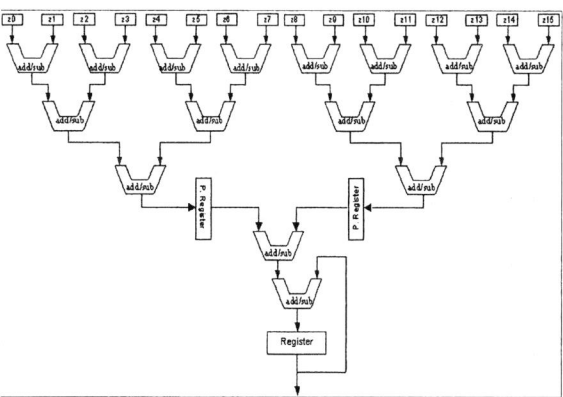

Figure 9. Hadamard Transform Hardware

Table 2. Latencies of the Modules in the Lower Part of the Search & Mode Decision Hardware

Module	Latency
Neighbor Loader	16 clock cycles
Hadamard Transform	18 clock cycles
Residue	18 clock cycles
Intra Prediction – Preprocessing	8 clock cycles
Intra Prediction – Mode0	17 clock cycles
Intra Prediction – Mode1	17 clock cycles
Intra Prediction – Mode2	19 clock cycles
Intra Prediction – Mode3	18 clock cycles
Intra Prediction – Mode4	18 clock cycles
Intra Prediction – Mode5	17 clock cycles
Intra Prediction – Mode6	17 clock cycles
Intra Prediction – Mode7	17 clock cycles
Intra Prediction – Mode8	17 clock cycles

The latencies of the modules in the lower part of the search & mode decision hardware are given in Table 2. After the prediction for each mode is generated (hadamard transform is overlapped), it takes 16 cycles for the Residue module to generate the residue block for that mode (loading neighbors is overlapped). 1 extra cycle is required after the Hadamard Transform module before starting intra prediction for the next available mode of the same 4x4 block. So, in the worst case when all 4x4 modes are available, it takes 165 (Intra Prediction) + 16*9 (Residue) + 1*9 = 318 clock cycles for performing intra search for a 4x4 luma block. After intra search for all 4x4 luma blocks in a MB is done, total cost for the selected modes for each 4x4 luma block in a MB is calculated in 18 clock cycles. Most probable mode calculation for 4x4 blocks in a MB is, then, started and this calculation takes 36 clock cycles. Finally, cost comparison between 16x16 and 4x4 intra search is initiated and it takes 9 clock cycles. Since the upper part of the search & mode decision hardware always finishes before the lower part, the lower part is the bottleneck. Therefore, intra search for a MB takes (16*318) + 18 + 36 + 9 = 5151 clock cycles.

3.2 Proposed Coder Hardware

The proposed coder hardware, as shown in Figure 10, includes Intra Prediction, Residue, Transform, Quant, Inverse Transform, Inverse Quant, Hadamard Transform, Reconstruction, and Entropy Coder modules. The efficient intra prediction, forward and inverse transform, forward and inverse quantization, and context-adaptive variable length coding hardware designs presented in [6, 7, 8] are used in the proposed hardware.

After the search & mode decision hardware determines the best modes for luma and chroma components of a MB, the MB is loaded to the current MB register file in the coder hardware. As soon as this loading operation finishes, intra prediction hardware generates the predicted MB using the selected best mode. Then, the Residue module creates the residual data by taking the difference between the current MB and the predicted MB and it loads the residual data to the input register file of the Transform-Quant hardware. Reconstruction module adds the results of Inverse Transform module which is stored in a 16x16 register file and the corresponding intra predicted data from the predicted MB register and clips the result to the [0-255] range. The results obtained from the reconstruction process are loaded to the neighboring pixel buffers in the intra prediction hardware and the reconstructed MB register file.

The scheduling of the Coder Hardware for a MB that will be coded with 4x4 luma prediction modes is shown in Figure 11. In the worst case, it takes 2676 clock cycles to code a MB that will be coded with 4x4 luma prediction modes. First, intra prediction hardware generates all pixel predictions for a MB based on the selected mode information for each 4x4 luma block and writes these results to the predicted MB register file. Then, the Residue block subtracts the predicted MB from the current MB. When the residual data for the first 4x4 luma block is available, Transform-Quant module starts to generate the quantized transform coefficients and loads these coefficients to the input register file of CAVLC hardware. After the quantized transform coefficients of the first 4x4 block are loaded, CAVLC and inverse Transform – Quant modules start to work. The bitstream generated by CAVLC module is stored in the output register file of CAVLC hardware. After Transform – Quant module finishes inverse quant and inverse transform operations for the first 4x4 block, reconstruction block starts to work. After the first 4x4 block of a MB is coded and reconstructed, the coder hardware starts to work on the second 4x4 block. In this way, all 4x4 blocks in a MB are coded and reconstructed.

The scheduling of the Coder Hardware for a MB that will be coded with a 16x16 luma prediction mode is shown in Figure 12. In the worst case, it takes 3680 clock cycles to code a MB that will be coded with a 16x16 luma prediction mode. Hadamard Transform has to be applied to DC coefficients after 4x4 integer transforms. Therefore, inverse quant, inverse transform, CAVLC and reconstruction operations for the MB can only start after the Hadamard transform finishes.

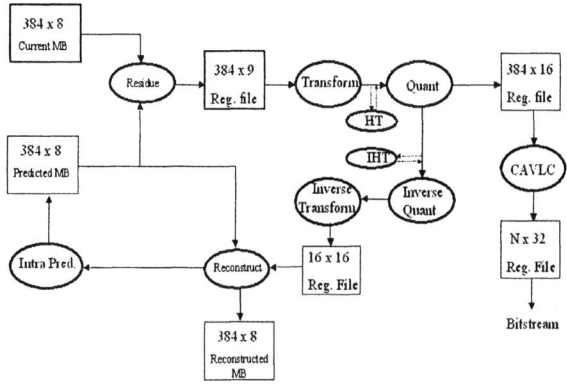

Figure 10. Coder Hardware Block Diagram

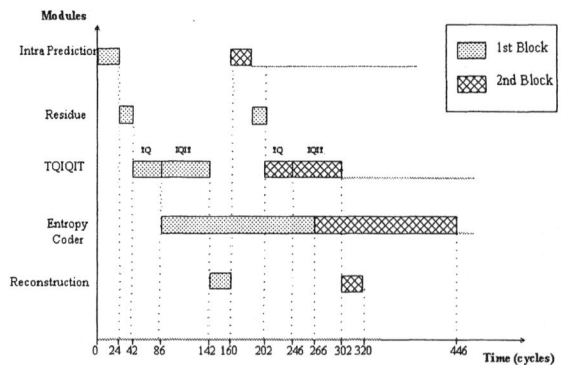

Figure 11. Coder Hardware Schedule for 4x4 Intra Modes

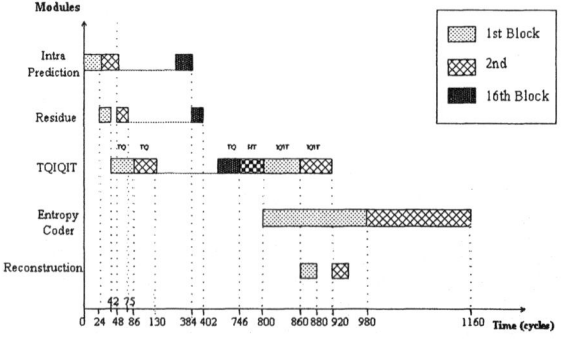

Figure 12. Coder Hardware Schedule for 16x16 Intra Modes

4. IMPLEMENTATION RESULTS

The proposed architecture is implemented in Verilog HDL. The implementation is verified with RTL simulations using Mentor Graphics ModelSim SE. The Verilog RTL is then synthesized to a 2V8000ff1152 Xilinx Virtex II FPGA with speed grade 5 using Mentor Graphics Leonardo Spectrum. The resulting netlist is placed and routed to the same FPGA using Xilinx ISE Series 7.1i. The FPGA implementation is verified to work at 71 MHz on a Xilinx Virtex II FPGA on an ARM Versatile PB926EJ-S development board shown in Figure 13.

As shown in Figure 14, an AHB bus Master interface is designed and integrated into H.264 intra frame coder hardware in order to communicate with ARM processor and SRAM through AHB bus and the H.264 intra frame coder hardware is integrated into the Xilinx Virtex II FPGA on the logic tile of the ARM Versatile PB926EJ-S development board as a master of the AHB S bus. The H.264 intra frame coder hardware is verified to work correctly on this board. The verification includes first capturing an RGB image, converting it into YCbCr format, partitioning it into MBs and writing it into an SRAM using the software running on ARM9EJ-S processor. Then, the intra frame coder hardware mapped to the Xilinx Virtex II FPGA reads the input image from the SRAM using AHB bus protocol, encodes the image and reconstructs it, and writes the reconstructed image to the SRAM using the AHB bus protocol. The conversion of reconstructed image into raster scan order and RGB color domain is then performed by software running on ARM9EJ-S processor. The reconstructed image is then displayed on a color LCD panel for visual verification.

The H.264 intra frame coder hardware is also verified to be compliant with H.264 standard. The bitstream generated by the H.264 intra frame coder hardware for an input frame is successfully decoded by H.264 Joint Model (JM) reference software decoder and the decoded frame is displayed using a YUV Player tool for visual verification.

The proposed H.264 intra frame coder hardware includes a search & mode decision hardware and a coder hardware that work in a pipelined manner. Since, in the worst case, the search & mode decision hardware takes 5151 clock cycles for a MB and the coder hardware takes 3680 clock cycles for a MB, the intra frame coder hardware takes 5151 clock cycles for a MB. Therefore, the FPGA implementation can process a CIF frame in 396 MB * 5151 clock cycles per MB * 14 ns clock cycle = 28.5 msec. Therefore, it can process 1000/28.5 = 35 CIF (352x288) frames per second.

The FPGA implementation including input, output and internal RAMs and register files uses the following FPGA resources; 19589 Function Generators, 9795 CLB Slices, 3698 DFFs, and 1 Block Multiplier, i.e. %21.02 of Function Generators, %21.02 of CLB Slices, %3.83 of DFFs, and %0.6 of Block Multipliers.

5. CONCLUSION

In this paper, we presented an efficient H.264 Intra Frame Coder System. The system achieves real-time performance for portable applications with low hardware cost, and it includes a novel intra prediction hardware design. The proposed hardware is implemented in Verilog HDL. The Verilog RTL code works at 71 MHz in a Xilinx Virtex II FPGA and it code 35 CIF frames (352x288) per second. The system also includes a software running on an Arm926EJS processor for implementing pre-processing and post-processing functions. The H.264 Intra Frame Coder hardware and software are demonstrated to work together on an Arm Versatile Platform development board.

Figure 13. Arm Versatile PB926EJ-S Development Board

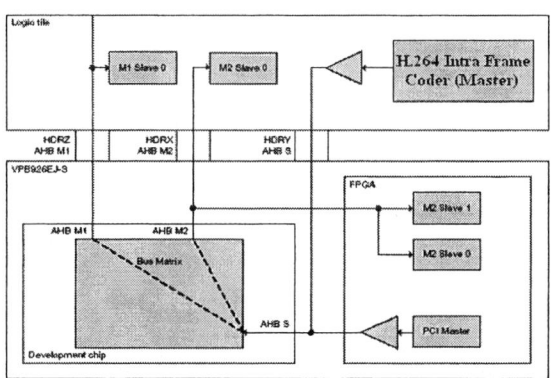

Figure 14. Integration of H.264 Intra Frame Coder Hardware into Arm Versatile PB926EJ-S Development Board

6. REFERENCES

[1] T. Wiegand, G. J. Sullivan, G. Bjøntegaard, and A. Luthra, "Overview of the H.264/AVC Video Coding Standard", IEEE Trans. on Circuits and Systems for Video Technology, vol. 13, no. 7, pp. 560–576, July 2003.

[2] I. G. Richardson, H.264 and MPEG-4 Video Compression, Wiley, 2003.

[3] Joint Video Team (JVT) of ITU-T VCEG and ISO/IEC MPEG, Draft ITU-T Recommendation and Final Draft International Standard of Joint Video Specification, ITU-T Rec. H.264 and ISO/IEC 14496-10 AVC, May 2003.

[4] Y. Huang, B. Hsieh, T. Chen, and L. Chen, "Hardware Architecture Design for H.264/AVC Intra Frame Coder", Proc. of IEEE ISCAS, pp. 269-272, April 2004.

[5] Y. Huang, B. Hsieh, T. Chen, and L. Chen, "Analysis, Fast Algorithm and VLSI Architecture Design for H.264/AVC Intra Frame Coder", IEEE Trans. on CAS for Video Technology, March 2005.

[6] E. Sahin and I. Hamzaoglu, "An Efficient Hardware Architecture for H.264 Intra Prediction Algorithm", Design, Automation and Test in Europe (DATE) Conference, April 2007.

[7] O. Tasdizen and I. Hamzaoglu, "A High Performance and Low Cost Hardware Architecture for H.264 Transform and Quantization Algorithms", European Signal Processing Conf., September 2005.

[8] E. Sahin and I. Hamzaoglu, "A High Performance and Low Power Hardware Architecture for H.264 CAVLC Algorithm", European Signal Processing Conf., September 2005.

Qualification of Behavioral Level Design Validation for AMS & RF SoCs

Yves JOANNON[1,2], Vincent BEROULLE[1], Chantal ROBACH[1], Smail TEDJINI[1], Jean-Louis CARBONERO[2]
LCIS-ESISAR (INPG), Valence, France[1], ST MICROELECTRONICS, Crolles, France[2]
1: firstname.name@esisar.inpg.fr, 2: firstname.name@st.com

Abstract

The expansion of Wireless Systems-on-Chip leads to a rapid development of design and manufacturing methods. In this paper, the test vectors used for design validation of AMS & RF SoCs are evaluated and optimized. This qualification is based on a fault injection method. A fault model based on variation of behavioral parameters and a related qualification metric are proposed. This approach is used in the receiver's design of a WCDMA transceiver. A test set defined by verification engineers during the validation of this system is qualified and optimized. Then, this test set is compared with a second test set automatically generated by a developed tool.

Keys words: Test Qualification, characterization, design validation, AMS & RF SoCs, VHDL-AMS, behavioral modeling, fault injection.

1 Introduction

Telecommunication and multimedia electronic products integrate more and more complex Integrated Circuits (IC). Nowadays these complex ICs often called Systems-on-Chip (SoCs) embed both Analog and Mixed Signal (AMS) and RadioFrequency (RF) components. This increasing complexity and the hybrid nature of the AMS & RF SoCs involve the use of Top-Down design and verification flow. This approach permits to achieve first time right design of SoC. A challenge for the system design verification is to ensure that all system functionalities are verified. Further in the flow, all electrical parameters of embedded analog components must also be verified. In addition, hybrid nature of IC imposes SoC designers to use methods adapted for both digital, analog and radiofrequencies systems.

Design validation and verification concern different abstraction levels of the design flow. [1] proposes an approach to generate and to validate the architecture of an RF transceiver described at functional level. In [1], the authors use known classical validation test benches (ie BER, Bit Error Rate, or EVM, Error Vector Magnitude...), as specified by telecommunication standards, which are assumed to be exhaustive enough for the complete validation of the design. The problem of the qualification of these test benches with low level abstraction descriptions is that it involves long time to simulate the system. In addition, the reuse of this approach for manufacturing test is very expensive.

A system verification with transistor level descriptions is presented in [2]. In this article, a formal verification method based on the extraction of parameters is developed. The verification is realized comparing the behavior of an extracted model with the specifications. This method is efficient because formal proof allows an exhaustive validation but the extraction of formal model from structural description is complex for the validation of large systems. [3] presents a verification method using behavioral level descriptions. The system is modeled with an HDL-A language and a code coverage metric is used for the qualification of the test set. This paper shows that code coverage metric is not enough for the verification and adds a frequency coverage metric. This method is interesting for the verification of the systems which could be described with a transfer function but it becomes too much complex for the validation of SoC. In [4], faults are injected on process parameters and then the induced behavioral parameters are extracted. In [5], faults are directly injected on behavioral parameters. Then, efficiency of multi tones signal to distinguish faulty and fault-free descriptions is evaluated. In this work, system specifications (i.e. expected output signal for a specified input signal) are considered as a reference. Relationships between specifications and measurements are computed and estimated. Comparison between behavioral simulation results and results computed from regressive estimations (MARS) allows to determine few optimized multi tone signals (magnitude and frequency). Both approaches [4, 5] generate vectors by using behavioral level descriptions, but they become too much time consuming when applied on complex systems.

In this article, a new qualification tool is proposed to qualify the validation test sets used for AMS and RF SoC. Our PLAtform for the qualification of Systems with Mixed and Analog signals (PLASMA) is first detailed in section 2 and 3. Then, the receiver part of WCDMA system we used to carry out experimental results is presented in section 4. Then, the fault models are discussed in section 5. The principle of fault injection is developed in part 6. Experimental results on WCDMA receiver are exposed in the last section with a reduction of validation test sets and a comparison with an automatically generated test set.

978-1-4244-1709-4/07/$25.00 © 2007 IEEE

2 Qualification of AMS & RF SoC Verification

In Top-Down design flow, the system is first described at functional level. Then, several successive refinement steps allow modeling the system at lower abstraction levels. At each step, simulations performed with test vectors defined by designers allow to verify the new description. This validation is achieved by comparing the simulation results of the next lower-level description with previously developed higher-level descriptions. Most of design faults due to bad choices of architecture or wrong parameters definitions should be detected using these simulations.

Some stimuli developed for the validation of high-level descriptions could not be applied for the validation of a lower level description because simulations become too time consuming. For example, a BER simulation using a lot of random input data cannot always be performed on a transistor level SoC description. Thus, efforts are required to minimize and to simplify these test cases. But how to be sure that all possible design faults are always detected after simplification?

Our approach for qualifying vectors is based on the classical fault injection technique [6, 7, 8]. Several faulty versions of the original description of the system under verification are created. Each faulty description contains only one fault. Test cases are used to stimulate these faulty descriptions with the goal of distinguishing the faulty programs from the original program. A faulty description is detected, when a test vector activates and propagates the induced fault toward an observable output. However, it is important to note that this test cases analysis does not achieve system verification, but only evaluates the quality of test vectors.

The choice of the abstraction level of the original description must be adapted for the verification of AMS and RF SoCs. In one hand, the use of a functional description is not possible to qualify test sets because this description does not make use of electrical parameters and so does not guarantee the validation of these parameters. On the other hand, the use of component level SoC description is impossible because the simulation time to qualify the test set is in this case too large. A trade off between these two abstraction levels is to use descriptions at the behavioral level. A few studies on behavioral modeling of RF transceiver have already been developing. For example, in [9], a BPSK transceiver has been modeling with the VHDL-AMS language. The accuracy of this model has been validated by comparing its simulation results with component level simulation results.

3 WCDMA transceiver

Our test sets qualification and optimization tool (PLASMA) has been experimented on an AMS and RF SoC. This SoC has been designed and manufactured by ST MICROELECTRONICS [10]. It is an integrated WCDMA (Wideband Code Division Multiple Access) transceiver. This device has been developed in a 0.25μm BiCMOS technology. The complexity of this SoC is representative of current industrial realizations. As the simulation of the complete system at the transistor level is very time consuming; an approach based on the simulations of behavioral level descriptions is needed.

3.1 Brief presentation of WCDMA

The studied system is the receiver part of a WCDMA (Wideband Code Divided Multiple Access) transceiver. WCDMA is a technology used for third-generation cellular systems (3G). The frequency range down-link (Bases Station to User Equipment) or receiver part is [2110-2170MHz]. The modulation defined in the WCDMA standard is an IQ modulation based on two signals: I "in-phase" component of the waveform, and "Q" represents the quadrature component. WCDMA standard specifies several parameters: maximal and minimal output power, maximal power out of frequency band, ACLR (Adjacent Chanel Leakage Ratio)…

3.2 Architecture definition

In this section, the architecture of the receiver part is presented. This architecture and the related blocks specifications will allow us to define the fault models.

The system has been designed using a Top-Down flow [10]. First, the architecture has been chosen. Then, the system has been divided into several blocks. The parameters of each block have been fixed in order to fit the system specifications. During these steps, the system has been modeled at different abstraction levels: functional, behavioral, and structural levels. In the following, only the behavioral level is used to qualify the test sets.

The architecture of the receiver part is a Zero Intermediate Frequency (Zero-IF) Wireless Radio architecture (*Fig. 1*). It only employs one stage to down-convert the RF signal directly to the desired base-band signal.

The figure 1 presents the architecture of the receiver part (Rx). It is a classical architecture made of Low Noise Amplifier (LNA), external Surface Acoustic Wave (SAW) RF filter, mixers, base-band Voltage Gain Amplifiers (VGA) and internal filters. Digital registers (not illustrated in *Fig. l*Extrenahtrol the LNA and VGA gains. These registers permit to control the receiver parameters; they can be used to control the system during validation.

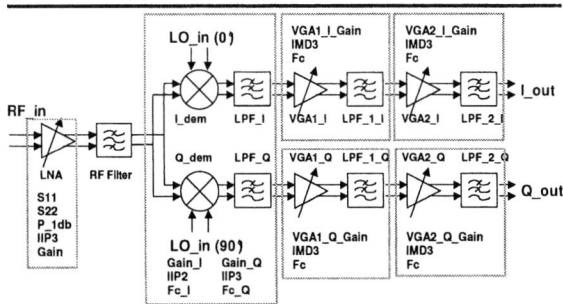

Fig. 1: WCDMA Receiver Part (RX)

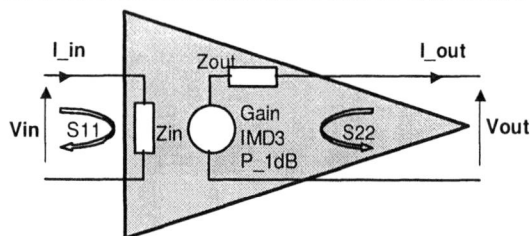

Fig. 2: Amplifier model.

3.3 Amplifier behavioral model

The amplifier behavioral model is presented in order to illustrate with one example this level of modeling. Amplifier functional and electrical parameters are also defined.

The amplifier model is made of one functional parameter, the power gain (Gain), and several electrical parameters: input and output impedances (Zin, Zout), S parameters (S11, S22), compression point at 1dB (P_1dB), third-order intermodulation distortion (IMD3). Figure 2 presents the model of the amplifier with its functional and electrical parameters. The table 1 specifies the limits of several parameters. Each parameter is defined by a typical value and with one or two worst-case values (minimum and/or maximum admitted values). In Tab. 1, the gain is specified by two limits because both limits are critical; the other parameters have only one critical limit.

The other blocks of the system are modeled in the same way plus additional parameters: IIP2 (Second-Order Intermodulation Distortion), DC offset, and cut off frequency. The figure 1 gives the modeled parameters of each block. Finally, the receiver part is modeled by 23 functional and electrical parameters (*Fig. 1*). These parameters are critical parameters in the system design. Obviously, our qualification process could be applied on a description involving additional parameters. This case study with 23 parameters will allow us to estimate the computation time required by this method. The WCDMA transceiver is modeled with VHDL-AMS an hardware description language [10]. The simulations are realized with the Mentor Graphics simulator's ADvance MS RF. The Mentor Graphics behavioral VHDL-AMS library's CommLib RF [11] is used for the modeling.

4 Fault model definition

The definition of the fault model is a crucial point of our approach because it is directly linked to the accuracy of the

test set qualification. Due to the choice of mutation-based fault injection, data contained in the original description or choice of modeled parameters are essential. In our fault model, only small modifications of the original description are considered. That means that we assume that wrong parameters values when they exist are not too far to their tolerance ranges. Moreover, we assume that if the test set is able to detect these small parameters variations then it is also able to detect larger variations. In addition, it is also assumed that multiple faults would be more detectable than single ones so only single faults are injected. During the Top-Down design flown, the behavioral level description is developed before transistor level description. The physical level does not already exist and correlations could not yet be extracted. Therefore, correlations between behavioral parameters are not taking into account in this paper.

During the development of the SoC, designers specify blocks parameters (local parameters) with typical, maximum and minimum values (*Tab. 1*). Then, the aim of the design validation is to verify that these local specifications satisfy global system specifications. Thus, we will verify that validation test set detects the variation of local parameters outside of their specified ranges. So, it seems possible to set the local parameter value just across its limits and then to compare simulation results with specifications of the SoC. But due to the natural desired analog system robustness, such small variations should have no effect on the global specification of the SoC and then, the closer the parameter value is to the limit; the more difficult is the mutant detection.

The mutants are created by translating one parameter of the original description to a value outside of its specifications. In [4, 12], faults are injected on physical parameters and their faulty values are fixed at a percentage of the typical value: for example in [12], 50% of the typical value. This approach may mislead the test qualification because this fault could be hard to detect in some cases or easy to detect in other cases. Hence, in our approach, to avoid this problem, the value of the faulty parameter is computed starting from tolerance ranges. The limit value of detection is obtained with the use of a dichotomy algorithm (*Fig. 3*). The initial faulty parameter value is computed using tolerance ranges (i.e. minimal and maximal specified values); an exploration range is first specified. The maximal accepted value P_{max_fault} is equals to 25 times the tolerance range: $P_{max\_fault} = P_{max} + 25.(P_{max} - P_{typ})$ where P_{max} is the high limit value (*Fig. 3*). The initial faulty parameter value of a mutant is specified at the middle of [P_{max}, P_{max_fault}] range. This faulty value depends on both its tolerance range ($P_{max}-P_{typ}$) and its maximal value P_{max_fault}. If this mutant is detected, the next faulty value is brought closer to P_{max}. On the other case, it is moved away. The detection limit is

Parameters	Typical Value	Minimum	Maximum
S11	-12dB	Not Specified	-10dB
S22	-10dB	Not Specified	-8dB
Gain	15dB	14dB	16dB
IMD3	1dB	-1dB	Not Specified
P_1dB	-9dB	-12dB	Not Specified

Tab. 1: Limits of a few amplifier parameters

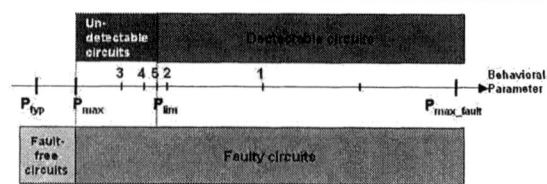

Fig. 3: Definition of faulty value

iteratively computed by simulating several faulty parameter values. The dichotomy algorithm is iterated 5 times and determines a good estimation of the detection limit P_{lim}. The injected faulty value is linked to the test quality: the higher the test quality is, the closer to the limits induced faults must be detected.

This detection limit is computed for every mutated parameter and every qualified vector. Relative Parametric Coverage (RPC) qualifies the detection limit; RPC is defined in Eq. 1.

$$RPC_{PL1} = 1 - \frac{(P_{fault_{PL1}} - P_{max_{PL1}})}{P_{max\_fault_{PL1}} - P_{max_{PL1}}} \quad (1)$$

When the detection limit is equal to the specification limit, the RPC is 100%. The farer this detection limit is from the specification limit, the lower is the RPC. The test set optimization is realized by saving the vector that leads the highest RPC.

The total relative parametric coverage RPC is the average of all behavioral parameters relative parametric coverage RPC_{PLx}.

5 Fault injection and simulation

Fault injection and fault simulation are used to qualify test set. The principle of the fault simulation relies on the comparison of the simulation results (*Fig. 4*). Faulty and fault-free models are simulated with the same input vectors. When the comparison of simulation results involves a difference, the fault has been both activated and observed.

Fault-free and faulty circuits are simulated with pre-defined test sets. The faulty values of detected and undetected mutants are modified as presented in part 4 and circuits are re-simulated with the same test sets. The simulation is realized 5 times and permits to determine a faulty value close to the limit of detection. Then, the total RPC is computed in order to qualify the test set. In addition, the test set optimization is also possible. This optimization can be done by keeping only vectors that make observable at least one fault and induced the maximal RPC.

In Fig. 4, the simulation results of faulty models are compared with several fault-free models simulation results. The use of several fault-free simulation results permits to

obtain a simulation results range representative of fault-free models. The choice of several fault-free models will be justified in part 5.2.

5.1 Faulty models

Faulty models or mutants are created from fault-free model. The mutant is generated by modifying the value of a behavioral parameter. The number of faulty models is imposed by the number of behavioral parameters. When a parameter is specified by two limits (for example, the gain in Tab. 1), two mutants are generated: a first for the low limit and a second for the high limit; when it is specified by one limit (ex: IMD3), only one mutant is generated. Due to this consideration, the receiver part of WCDMA SoC modeled by 23 behavioral parameters involves the generation of 36 faulty models. All faulty models simulation results are compared with fault-free models simulation results. This comparison determines the ability of the test vectors to detect the behavioral faults.

5.2 Computation of the fault-free models

Faults detection is based on the comparison between faulty models and fault-free models. In fact, analog circuit fault simulations could not be compared to only one "typical model". Typical model means a fault-free model made of behavioral parameters fixed at their typical value. Although this model has the best probability to be manufactured, other fault-free models instanced with parameters inside of their specification ranges also have to be considered in order to obtain the realistic ranges of fault-free simulation results. Finally, these measurement ranges are the references for the mutant detection.

Obviously, it is impossible to simulate all fault-free models because the variations of each parameter involve an infinite number of possible fault-free models. Therefore, a direct and simple approach to define fault-free ranges is to sample the fault-free population. The samples set must represent an accurate estimation of the real fault-free models population. The number of samples increases the simulation time for the vectors qualification. Thus, a trade off between accuracy and number of fault-free models has to be made. The fault-free sample set is generally generated with two main approaches:

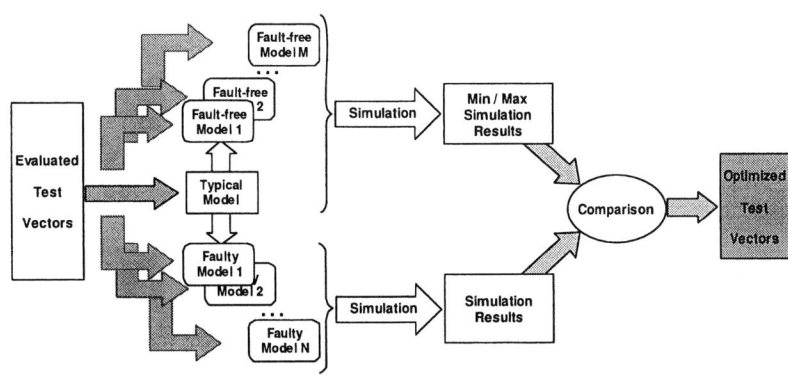

Fig.4: Test vector qualification & optimization

- *Random sampling* approach as in the Monte Carlo analysis [12, 13]. Instances of the fault-free circuit are generated randomly choosing parameters in the specifications ranges.
- *Worst-case min-max* approach [14]. Only worst-case fault-free models are selected. All parameters are fixed to their minimal or maximal values.

Our test vector qualification method has been developed for the validation of complex systems. The number of fault-free models required by the random sampling approach depends on the total number of parameters and the desired accuracy. In our case, the RX part modeled at behavioral level could be described with 23 parameters. Thus, this random approach leads to the use of an unacceptable number of fault-free models. With the second method, the number of worst-case possibilities is $2^{23} \approx 8$ millions. So, it is also impossible to simulate all worst-case combinations.

In our case, fault-free models are defined using the statistical distributions of behavioral parameters. Then, the statistical parameters of the simulation results can be extracted. Generally, Gaussian estimation is done thanks to Monte Carlo analysis but this method requires the simulation of many samples to be accurate enough. In PLASMA, the number of fault free simulated models is decreased by analyzing the statistical characteristics of the results: i.e. the average (μ) and the standard deviation (σ) of fault-free simulation results. The number of fault free simulated models is increased as long as μ and σ variations are not negligible. Then, the fault free measurement ranges can be determined by defining the limits at 6σ.

6 Experimental Results

This approach has been applied to the WCDMA receiver part. A list of test vectors has been manually generated to validate the design in its operating range. PLASMA tool will identify the useful vectors and reject redundant vectors. First, fault-free models and mutants are simulated with the pre-defined vectors. Only the vectors that detect the mutated parameters are saved. Then, the RPC obtain with this initial test set is compared with a second RPC obtained by the simulation of test set automatically generated.

	Frequency	Power
RF	2113.4 MHz	-60dBm
RX LO	2112.4 MHz	-7dBm

Tab. 2 A: Two tones input signal parameters

LNA Gain	-15dB
VGA1 Gain I	-10dB
VGA1 Gain Q	-10dB
VGA2 Gain I	23dB
VGA2 Gain I	23dB

Tab. 2 B: WCDMA Receiver part configuration

6.1 Test sets

During the design of the WCDMA system, a list of test vectors has been defined for its validation. These vectors aim to validate system specifications within different configurations (Gain, Offset) and with different input signals (frequency, power). They should verify the values of the programmable gains, the IMD3, the IMD2... The validation test set is made of vectors with single tone or bi-tone signals on RF and RX Local Oscillator (LO) inputs; one example of test vector parameters is given in Tab. 2. Tab. 2A describes RF input signal whereas Tab. 2B gives receiver gains configuration controlled thanks to digital registered inputs. In a first stage, designers define simple vectors to validate the design (mainly single tone). If the validation is not achieved, more complex vectors will be used. In this paper, we verify behavioral parameters with only single tone vectors.

In the following experimental results, a test set made of 80 single tone vectors are be evaluated. PLASMA will analyze these test benches and are identify the useful vectors and reject redundant vectors.

In a second part, a test set automatically generated by PLASMA is qualified. During the generation of this second test set, only single tone vectors are generated and the values of frequency and magnitude are randomly defined. These values are selected into the model validity domain. The test sets are evaluated by comparing total relative parametric coverage.

6.2 Simulation results

80 vectors have been evaluated with 37 fault-free models and 36 faulty models. These vectors manually defined aim to measure the RX Gains for the numerous possible gain configurations. These vectors have been defined to validate the RX behavioral model in its operating range only.

The simulation has been realized with ADMS RF from Mentor Graphics on 3 GHz Pentium-4, with 1 GB RAM, running a Linux Operating System. Over 36 mutated parameters, 28 have been detected during the entire dichotomy process. The total computation time is about 15 hours. For this sample of test data, the total Relative Parametric Coverage FC previously defined in section 4 is 68% (Tab. 3).

Mutants which have not been detected are mutants with faults injected on non-linearity parameters (compression point, IIP2, IIP3). In fact, the non-linearity parameters cannot be detected by using small power input signals specified by the operating range. These parameters can have an effect only when a block works in its saturation range. In addition, only a few variations on filters' cut off frequencies are detected. Indeed, for all vectors, the deviation between RF signal frequency and RX LO signal frequency is lower than filters' cut off frequencies. So, the increase of cut off frequency value can never be detected. However, decrease of cut off frequency value can be detected but only for high variations.

Each vector of the 80 vectors detects a least one behavioral faulty parameter but after compaction, only 4 vectors are kept to achieve the same RPC. After compaction, the number of stimuli is divided by 20 (Tab. 3).

	Validation test set manually defined	Automatically generated test set
Number of initial vectors	80	20
Mutation score	28/36=0.77	34/36=0.94
Relative Parametric Coverage	0.68	0.90

Tab. 3: Qualification results

In a second step, the test set automatically generated has been evaluated. Over 36 mutants, 34 mutants have been detected. It takes only about 4 hours to generate these test vectors and to perform their evaluation. This duration is lower than the previous one because there are only 20 vectors to evaluate. The relative parametric coverage of this test set is 90.1%. This FC is better than the previous one because the input signal parameters are selected in a large range corresponding to the model validity domain. In opposition with previous test vectors included in the operating range, with this large range, mutants induced on cut off frequencies and non-linearity parameters can be detected. However, the simulation of the system in its saturation range can be problematic because some mutants could be masked. In fact, in that case, variations induced on LNA parameters as coefficient reflection (S_{22}) and variations induced on amplifier non-linearity are not detected because their impact on system behavior is hidden.

7 Conclusion

A method and a tool (PLASMA) for the qualification of test vectors have been presented in this paper. The method is based on faults injection and simulation. A behavioral fault model relying on the mutation approach has been defined. The different descriptions (faulty and fault-free models) are automatically instantiated by PLASMA. The classification of vectors is realized by comparing simulation results of faulty descriptions with fault-free descriptions. The main advantage of our method is to evaluate complex systems by considering behavioral faults. A relative parametric coverage metric based on faulty parameters value gives a measure of test quality. This measure allows designers to complete their test vectors.

First, manually predefined test set chosen for the design validation of AMS & RF SoCs is qualified. In a second step, a test set automatically generated by the developed tool is also evaluated and compared with previous results. Simulation results show that an optimized test set can be rapidly generated and reach better performances than the initial manually generated test bench.

8 References:

[1] S. Vitali, D. Laurentiis, N. Albertazzi, G. Agnelli, F. Rovatti, "Multi-standard simulation of WLAN/UMTS/GSM transceivers for analog front-end validation and design", ISWCS'04, pp 16- 20, 2004.

[2] A. Ghosh, R. Vemuri, "Formal Verification of Synthesized Analog Designs," IEEE International Conference on Computer Design (ICCD'99), p. 40, 1999.

[3] Yuan-Bin Sha; Mu-Shun Lee; Chien-Nan Jimmy Liu, "On code coverage measurement for Verilog-A" High-Level Design Validation and Test Workshop, pp 115 – 120, 2004.

[4] F. Liu, S. Ozev, "Fast Hierarchical Process Variability Analysis and Parametric Test Development for Analog/RF Circuits", International Conference on Computer Design (ICCD '05), pp. 161-170, 2005.

[5] A. Halder, S. Bhattacharya, A. Chatterjee, "Automatic Multitone Alternate Test Generation For RF Circuits Using Behavioral Models," International Test Conference 2003 itc03, p. 665, 2003.

[6]G. Al-Hayek, C. Robach, "From Design Validation to Hardware Testing: a Unified Approach", Journal of Electronic Testing : theory and application 14, pp 133-140, 1999.

[7]A.J. Offutt, R.H. Untch, "Mutation 2000: Uniting the Orthogonal", Mutation 2000: Mutation Testing in the Twentieth and the Twenty First Centuries, pages 45--55, San Jose, CA, October 2000.

[8] B.Charlot, S.Mir, E.F.Cota, M.Lubaszewski, B.Courtois, "Fault modeling of suspended thermal MEMS" IEEE International Test Conference (ITC'99), USA, 28-30 , pp. 319-328, 1999.

[9] E. Normark, L. Yang, C. Wakayama, P. Nikitin, R. Shi, "VHDL-AMS behavioral modeling and simulation of a Pi/4 DQPSK transceiver system", Behavioral Modeling and Simulation Conference, 2004. BMAS 2004. Proceedings of the 2004 IEEE International Issue pp 119 - 124, 2004.

[10] Y. Joannon, V. Beroulle, R. Khouri, C. Robach, S. Tedjini, J-L. Carbonero, "Behavioral modeling of WCDMA transceiver with VHDL-AMS language", Design and Diagnostics of Electronic Circuits and Systems (DDECS'06), pp. 113-118, 2006.

[11] MENTOR GRAPHICS, "CommLib RF VHDL-AMS Library", Manual, October 2005.

[12] A. Khouas, A. Derieux, "Methodology for Fast and Accurate Analog Production Test Optimization", 5th IEEE International Mixed Signal Testing Workshop (IMSTW'99), pp. 215-219, 1999.

[13] K. Saab, N. Ben-Hamida, B. Kaminska, "Parametric Fault Simulation and Test Vector Generation", Design Automation and Test in Europe (DATE '00), p. 650, 2000.

[14] M.W. Tian and C. J. Shi, "Worst Case Tolerance Analysis of Linear Analog Circuits using Sensitivity Bands", IEEE Transactions on Circuits and Systems - I: Fundamental Theory and Applications, vol. 47, n. 8, pp. 1138–1145, 2000.

Evaluating Memory Sharing Data Size and TCP Connections in the Performance of a Reconfigurable Hardware-based Architecture for TCP/IP Stack

Jean Carlo Hamerski, Everton Reckziegel, Fernanda Lima Kastensmidt

Universidade Federal do Rio Grande do Sul
Instituto de Informática – PPGC
Porto Alegre - Brazil
{jchamerski, ereckziegel, fglima}@inf.ufrgs.br

Abstract – **The TCP/IP (Transmission Control Protocol/Internet Protocol) Stack processing based on software becomes a bottleneck for the explosive growth of data transmission rate on the Internet. Software-based TCP/IP is not able to process the packets at the same rate of transmission lines, which has been pushing the TCP/IP processing implementation into hardware. The use of dedicated hardware for TCP/IP stack processing aims reducing the Central Unit Processing (CPU) load and increase as possible the throughput for Internet services that need a large bandwidth. In this way, a reconfigurable hardware-based architecture to transport and network layers protocols processing is proposed. The effect of shared memory data size and the number of TCP connections were evaluated in terms of area and packets computation performance.**

I. INTRODUCTION

Due to the advance of semiconductors technology, the *Ethernet* transmission lines have increased from 3 Mbps to 10 Gbps, and the processor speed has increased at the same rate. But the capability of software-based TCP/IP stack protocols processing has not increased at the same rate [1]. Some researches [2, 3] indicate that the biggest overhead of TCP/IP processing is because of I/O management and buffers of operating systems. When a high transmission rate is needed, and so processing, these tasks (I/O) can need a very big CPU time in order to process TCP/IP packets. Consequently, other applications that are being executed on the same platform could not have CPU time to execute theirs tasks.

By implementing the hardware-based TCP/IP stack in high-speed computing environments, administrators can help relieve network bottlenecks and improve application performance [17].

In this paper, we propose dedicated hardware architecture for the TCP/IP stack processing for reconfigurable devices. A set of implementation versions based on different memory sharing data size and the number of TCP connections were synthesized into Xilinx Virtex II-Pro FPGA. Results were analyzed in terms of

area and performance evaluating the influence of data size and TCP connections. Moreover, a network traffic functional simulation was developed over both implementations in order to analyze the performance in relation to the packets computing time.

The proposed architecture aims providing a friendly interface between the TCP/IP processing in hardware and the application layer processing in software. The goal of this approach is surpass traditional TCP/IP processing limitations in software and minimize the system resources needed to this processing.

The paper is divided as follows. Section 2 presents the related works that aim accelerate the TCP/IP stack processing in software and in hardware and how this work is inserted on the researches done. Section 3 introduces the TCP/IP model used on the implementation of the proposed architecture on this work. Section 4 presents proposed architecture, implementations, and the simulation environment of the network flow used for testing proposes. Section 5 shows a preliminary communication interface between the TCP/IP Module in hardware and the application layer in software. In the end, the section six shows the work conclusions.

II. RELATED WORKS

The works which aim to reduce the overhead of processing of TCP/IP stack can be divided in two groups: software solutions – they identify the processing bottlenecks that can be optimized, such as kernel operations and device drivers; hardware solutions – they have the advantage of hardware processing speed, which solves communication problems.

The mainly bottlenecks of software processing are: interruption operations, data copy, etc. Some works try to solve these questions. In [4], the bottleneck is identified on memory traffic that exists between the user area and the kernel area, beyond the traffic between the kernel area and the buffers of network devices. So the goal is to develop network devices architecture together with a communication interface with the operating system kernel, where the checksum and packets storage tasks are done in a dedicated hardware.

One software solution that has been proposed in [5] shows an optimized algorithm on TCP packets processing. The most frequently used method is reducing the overhead through checksum calculation of packets by a dedicated hardware. In [6], a technique proposed combine copy and checksum operations to reduce the checksum computing cost.

Other proposed technique is a "*zero-pass checksum*" [7], where the packets are moved straight from system memory to network devices. Another scheme is implementing the Internet checksum directly in the network devices [8].

A TCP/IP Core for reconfigurable logic presented in [9] is able to process TCP/IP stack packets fully in dedicated hardware. Other works [10,11] also follow the same idea.

The referenced works use several techniques to decrease the processing overhead. However, they are not parameterized for different application requirements. The proposed architecture can be customized for the number of TCP connections, ARP (Address Resolution Protocol) table size and shared memory size. In addition, the authors implemented two versions of the TCP/IP stack using two word sizes for the shared memory in order to analyze its influence in the performance of the protocol processing. Beyond this, it provides an interface that makes possible the services implementation which uses the stack to the "application-to-application" communication.

A comparison between the proposed TCP/IP Module and related works will be done in future works, after its validation on real network environment.

III. INTERNET COMMUNICATION MODEL

Before explain architecture details, it is important to show some models that have defined the communication form of Internet.

A standard model that makes possible understanding and projecting network architectures is known as OSI model (Open Systems Interconnection). This model defined a theoretic standard based in seven layers (Fig. 1a). It has just served as a reference to a good understanding of the communications processes of computer networks.

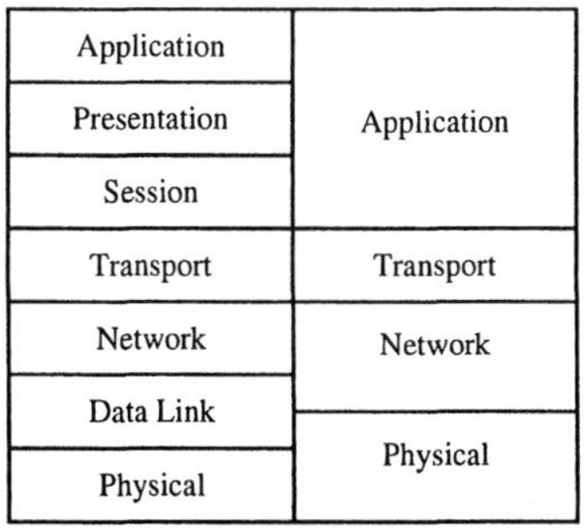

Application	
Presentation	Application
Session	
Transport	Transport
Network	Network
Data Link	Physical
Physical	

a) OSI Model   b) TCP/IP Model

Figure 1 – OSI Model and TCP/IP Model

A given layer in the OSI model generally communicates with three other OSI layers: the layer directly above it, the layer directly below it, and its peer layer in other networked computer systems. For any sent information, at each transition level, a header is inserted in order to allow the peer layer to understand what it means and to know what to do with the receiving data. On the receiving, after the treatment, these headers are discarded and the packet is sent to the above layer. When the packet arrives at the receiver application layer, it is a comprehensive data. The application layer handles messages, the transport layer, segments, the network layer, datagram, the data link layer, frames and the physics layer, bits.

The TCP/IP Model (Fig. 1b) is a hybrid model originated from the OSI model. It joins the three higher level layers in only one application layer. The Fig. 1 shows the difference between the models.

Thus, the proposed architecture is based on the four layers of TCP/IP Model. The physics layer is implemented by the existing one in a network device, for example, Ethernet. The network and transport layers are implemented by the TCP/IP Module in hardware. The software platform is responsible by the tasks execution of the application layer.

IV. TCP/IP MODULE IN HARDWARE

A. Architecture

The preliminary architecture proposed on this paper is based on the idea of a shared store space between the network interface, TCP/IP hardware module and the application which is running on a software platform. The Fig. 2 exposes the differences between both the traditional and proposed approaches.

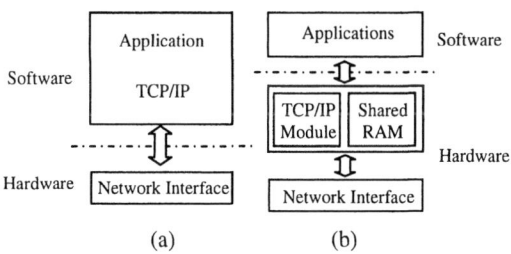

Figure 2 (a) Software TCP/IP and (b) Hardware Architecture

The network interface is responsible for assembling packets which come from physics layer and store them on the shared memory. An important observation is that the TCP/IP hardware module is independent of network interface which is being used, e. g.: Ethernet, Wireless, etc.

The TCP/IP Module implemented in hardware is responsible for the IP (Internet Protocol), ICMP (Internet Control Message Protocol), ARP (Address Resolution Protocol) and TCP (Transmission Control Protocol) processing protocols. Tasks such as connection establishment, data reception or transformation, checksum and connections management are done by the TCP/IP

Module. The use of dedicated hardware for this processing aims to increase the throughput and to reduce the CPU charge, comparing with the method which all TCP/IP processing is performed by software [17].

In the case of hardware implementation, the entire Internet data stream management is done by TCP/IP Module, while higher level instructions, more precisely application layer services, are performed at software level.

The modular characteristic of TCP/IP model protocols stack composition allowed the proposed architecture division in modules according to the protocol type. Thus, it can be configured according with the service that is aiming to run on TCP/IP stack. For example, those services that do not need functions implemented by the ICMP protocol can dispense the correspondent module, so the designer could just extract the ICMP module in order to configure it to the specific application.

The TCP/IP Module, on its internal architecture, is composed by the sub-modules described on Fig. 3. The Main Control Module (MCM) is responsible for the TCP/IP Module coordination and effective operation. It makes the first data checking, like IP header checksum, protocol version checking and, if it is a valid packet, it identifies the protocol type and redirects the packet to the corresponding sub-modules (or, on the other case, it discards the packet).

The other modules (ICMP, ARP and TCP) are responsible for processing their own protocols. Particularly, the ARP Module has a table, which maps the physical address (MAC) to the logical address (IP). The TCP Module keeps the connections states into internal register to module. The ICMP module builds the ICMP echo response packet to be retransmitted to the network.

The sub-modules are implemented using state machines that handle the sending and receiving packets process and the operations which are done during the machine states evolution.

The TCP/IP Module architecture was described in VHDL language. Some configuration parameters must to be set before of the synthesis, which characterize the feature of architecture customizing. The configurations parameters are:
➢ The number of simultaneous TCP connections;
➢ Shared memory size that stores the packet in processing;
➢ ARP table size which does the mapping between physical address (MAC) and logical address (IP);
➢ IP address, *gateway*, physical address, and host sub-net mask configuration.

Concerning of the connections number that TCP module should control: this number is set by the architecture user before its synthesis. The connection states are stored in internal registers. Therefore, the simultaneous connections number is proportional to the area of the TCP Module. Likewise, the ARP module area is proportional to the entrances number in the ARP table.

B. TCP/IP Hardware Implementations

The network interface is responsible for assemble packets coming from physics layer and store them in shared memory. The manner how packets are stored in memory, in relation to the word size, will define the state machine structure of the TCP/IP Module architecture. So, two different TCP/IP Module implementations were developed. The difference between them is the word size written or read from memory (16 and 32 bits) and consequently state machine descriptions. Both implementations were analyzed altering the number of simultaneous TCP connections which the implementation is able to manage.

The number of TCP connection has an influence only in the TCP module since this module is the one responsible to store and handle all the number of TCP information. The main effect is in the number of registers used to store the TCP connections.

These implementations were all described in VHDL and synthesized using the *Xilinx ISE 8.1* tool for the Xilinx Virtex II-Pro XC2VP30-5ff896 FPGA. The number of lookup tables (LUTs), which are the minimum configurable logic gates used to implement the user combinational logic inside the FPGA, and the number of flip-flops (FFs) were counted to evaluate the used area for the proposed implementations. Also the estimated frequency results of each sub-module and of the entire TCP/IP Module were also evaluated. Results are presented in Tab. 1.

By analyzing the algorithm that implements the TCP/IP stack, one can observe that the majority of data fetched on memory has a 32 bits word size, such as the network address (32 bits). Consequently, when 32-bit data in the shared memory is considered, the complexity of the state machine can be reduced because now all the process computation is performed in 32-bit. However, the problem is that 32-bit operators, such as 32 bits adders, decrease the frequency compared with the 16 bits implementations. So, even the reduction in the state machine complexity in terms of the number of states can not compensate the increase in the delay of the operators.

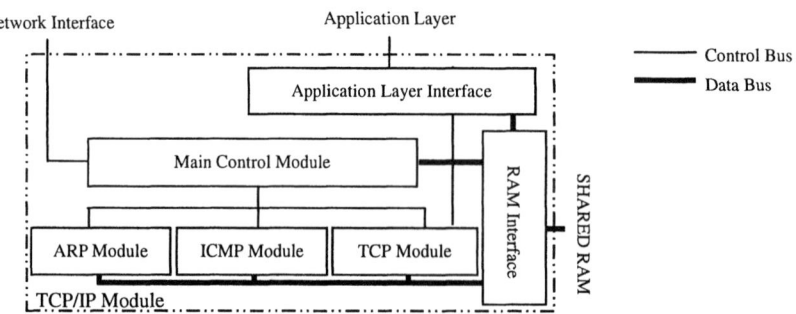

Figure 3 – Internal architecture of TCP/IP Module

214   *2007 IFIP International Conference on Very Large Scale Integration (VLSI-SoC 2007)*

Table 1: Synthesis results of two architecture implementations

TCP Connections	Factor	16 bits					32 bits				
		MCM	ICMP	ARP	TCP	TOTAL	MCM	ICMP	ARP	TCP	TOTAL
10	LUTs	758	208	836	6035	7936	728	165	672	5873	7551
	FFs	248	113	385	2556	3346	181	56	302	2491	3015
	Fmax (MHz)	77,5	207,7	222,8	70,9	71,2	59,8	414,3	159,3	58,7	58,7
5	LUTs	758	208	836	4565	6459	728	165	672	4074	5735
	FFs	248	113	385	1578	2368	181	56	302	1527	2047
	Fmax (MHz)	77,5	207,7	222,8	70,9	71,2	59,8	414,3	159,3	58,7	58,7
1	LUTs	758	208	836	2555	4480	728	165	672	2278	3968
	FFs	248	113	385	758	1547	181	56	302	648	1172
	Fmax (MHz)	77,5	207,7	222,8	70-9	71,2	59,8	414,3	159,3	58,7	58,7

One can see that the bottleneck in terms of frequency is always the TCP module and this result does not change with the number of TCP connections (10, 5 or 1). The reason of this low frequency is due to the complexity of the checksum operation that is performed in modules MCM and TCP. The number of TCP connections influences only the area of the TCP module as seen in Tab 1.

So, the entire TCP/IP module must work at the frequency limited by he TCP module. In summary, the 16-bits version operates at a frequency of 71,2 MHz, while the 32-bit version operates at 58,7 MHz.

C. Experiments

We developed a network flow simulation environment in order to perform a functional simulation of the TCP/IP Module to verify the implementations performance in case-study applications. This environment is composed by the TCP/IP Module presented on section three, besides a packet generator module and a third module which represents the shared memory area where the processing packet is stored. The Fig. 4 shows the diagram blocks of this environment.

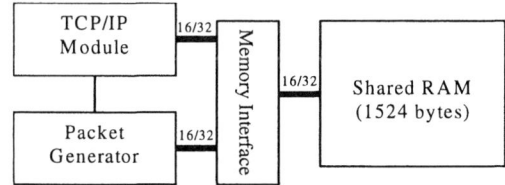

Figure 4 – Blocks diagram of the simulation environment

On this simulation environment, the Packet Generator has a set of packets with well-known behavior. The network flow simulation begins when the first packet is generated by the Packet Generator and stored in shared memory by itself. The TCP/IP Module is set in motion by the Packet Generator, informing that there is a packet to be processed. The TCP/IP Module will do a set of verifications on the packet, like destiny network address, packet version and checksum. At the end of processing, if a packet was generated by TCP/IP Module and should be dispatched, it will emit a command signal with this information. On this environment the packets sending is not showed and its implementation inside a real network environment is easy.

The Packet generator is able to generate a set of four packets, which define a little packets flow that happens in a real network environment:

➢ Arp_Reply (60 bytes) – ARP packet response to an ARP_Request;
➢ ICMP_Echo (74 bytes) – ICMP solicitation packet;
➢ Syn (62 bytes) – Connection solicitation (sync) to a TCP service;
➢ Syn_ack (60 bytes) – Confirmation (sync + ack) of a connection solicitation to a TCP service;

The proposed architecture implementations were analyzed in relation to the computing time of the referred packets. The Tab. 2 shows the results obtained from the packets flow simulation.

As it can be observed on Tab. 2, the results of 16 bits description are better than the 32 bits description in relation to the computing time of the four packets that simulate a real network flow. The Syn packet processing was the only flow which had the computing time affected by the number of TCP connections. This occurs because the first packet used for establishing the connection, the Syn packet, needs to analyze every one registers that store the connection states. Consequently, the computation time of this packet will be proportional to the number of TCP connections.

A comparative analyzes of obtained results on two experiments detailed on Tab. 1 and 2 reveals that the occupied area of the generated circuits from the 16 bits description is 12% greater, but, the computing time is 18% minor than the 32 bits description (these rates were calculated to the three configurations of TCP connections number). Based on these results, the chosen architecture will depend on the restrictions and requirements of the application. Consequently, if the application restriction is the circuit area, so the best option will be the 32 bits description. Otherwise, if there is a performance requirement, the 16 bits description will be the best choice.

Table 2 – Results of Packets Computing Time (μs)

TCP connections	16 bits				32 bits			
	ARP_Reply	ICMP_Echo	Syn	Syn_ack	ARP_Reply	ICMP_Echo	Syn	Syn_ack
10	0,2031	0,5252	1,3515	0,7073	0,2706	0,6061	1,6234	0,7251
5	0,2031	0,5252	1,2815	0,7073	0,2706	0,6061	1,5152	0,7251
1	0,2031	0,5252	1,2254	0,7073	0,2706	0,6061	1,4286	0,7251

V. THE HARDWARE/SOFTWARE COMMUNICATION INTERFACE

The TCP/IP Module architecture, as showed on section 4, has a communication interface with the application layer. Some information about TCP connection should be passed to the service that is being executed on the application layer. In this way, the TCP/IP model makes available through this interface some data like the port number (which will sue the respectively service), and the pointer to the data area of TCP packet.

The software layer also initializes the TCP/IP Module, moreover it synchronizes the incoming and outgoing packets, and sue the correspondent service.

For example, in a *Web Server*, the communication between the TCP/IP model and the service that is executing on microprocessor is done when there is a page request (*get http*). When it happens, the service has access to the packet data area. And this packet will be stored on the shared memory. The software service will transfer the requested page content and will return the control to the TCP/IP model which will fill the header packet and calculate the checksum for the packet sending. Once finished the processing, the TCP/IP model will sue the device *driver* to allow the sending packet through the network interface.

For the implementation of the application layer in software and the synchronization with the TCP/IP model in hardware, it is used the *Xilinx Virtex II-Pro* device [12]. This device has an embedded Power-PC which can be used on this kind of board. The HW/SW architecture projected uses the device drivers available by communication API of *Xilinx* environment to manage the incoming and outgoing packets. The *EDK 8.1 Xilinx* tool [13] was used to modules integration in hardware and in software. It uses some characteristics on development board, such as device drivers, link layer managing modules insertion, among others functionalities.

The HW/SW architecture, with all used platform components is showed on Fig. 5.

The BRAM_BLOCK of demonstrated architecture on Fig. 4 is the shared memory block used to store the packet.

The used platform allows the addition of an interface to an external SRAM memory. We could use this characteristic to implement a buffer to store packets in a queue, for example. On the current version this functionality is not implemented and the used memory block has size enough to store at the most an Ethernet maximum size packet (1500 bytes). Some studies demonstrate that Jumbo packets (9000 bytes) utilization increases the processing throughput of a network channel [14]. This kind of packet can be used on the proposed architecture. For this, before the architecture description synthesis, we have to configure the parameter corresponding to the shared memory size that stores the packet.

Finally, if we think on embedded systems, the used platform is similar to the embedded systems platforms. So, the proposed architecture could be easily adjustable for use in embedded systems. These systems need high CPU load to execute its applications and don't desire that CPU be occupied with TCP/IP stack processing. It increases the throughput and allows the Internet specialized services use, like iSCSI [15] and RDMA [16]. Other interesting characteristic of the proposed architecture is that it's not necessary a lot of memory to store the packet. This is interesting because embedded systems have memory restrictions.

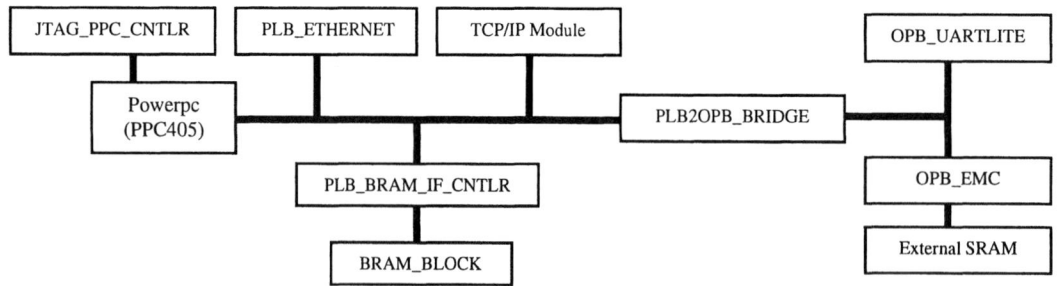

Figure 5 – The HW/SW architecture on Xilinx Virtex II-Pro platform

VI. CONCLUSIONS AND FUTURE WORKS

This work has analyzed the influence of memory sharing data size and number of TCP connections in the performance of Hardware TCP/IP architectures for high-throughput internet services in reconfigurable architectures. The proposed architecture is composed by a specific module to process packets from network and transport layers. The dedicated hardware accomplishes all processing of these protocols. It gets better the throughput and frees the CPU of the TCP/IP stack processing that would be done in software.

Two versions of the architecture, one with 16 bits and other with 32 bits were implemented and synthesized according to the state machine structure which defines the sub-modules behavior, and three different configurations in relation to the TCP connections number. The obtained results were analyzed. The choice of which description will be used will depend on the restrictions and requirements of the goal application.

The hardware/software preliminary interface to the specialized services implementation on TCP/IP stack was presented.

As future work, it is defined a complete validation of the proposed architecture and its integration with an Ethernet network interface in order to validate a real network environment. After the validation of the TCP/IP Module on real network environment, a study about the current approach gain in relation to the processing in software will be done. So we will be able to prove the advantages of the presented architecture of this work.

REFERENCES

[1] WANG, W., WANG, J., LI, J., Study on Enhanced Strategies for TCP/IP Offload Engines, ICPADS ' 05: Proceedings of the 11th International Conference on Parallel and Distributed Systems, pp. 398-404, 2005.

[2] CLARK, D., JACOBSON, V., et al., An analysis of TCP processing overhead, IEEE Comm., vol. 27, no. 6, pp. 23-29, 1989.

[3] MARKATOS, E., Speeding up TCP/IP: faster processors are not enough, in 21th IEEE Int. Perf., Comput., and Comm. Conf., pp. 341-345, 2002.

[4] STEENKISTE, P., Design, implementation and evaluation of a single-copy protocol stack, Software - Practice and Experience, vol. 28, no. 7, pp. 749-772, 1998.

[5] JACOBSON, V., 4BSD header prediction, ACM Comput Comm. Rev., vol. 20, no. 1, pp. 13-15, 1990.

[6] CLARK, D., Modularity and efficiency in protocol implementation, RFC 817, 1982.

[7] FINN, G., HOTZ, S., METER, R. V., The impact of a zero-scan Internet checksumming mechanism, ACM Comput. Comm. Rev., vol. 26, no. 5, pp. 27-39, 1996.

[8] KLEINPASTE, K., STEENKISTE, P., ZILL, B., Software support for outboard buffering and checksumming, in Proc. of the ACM SIGCOMM' 95 Conf. on App., Tech., Archit., and Protocols for Comput. Comm., pp. 87-98, 1995.

[9] DOLLAS, A., ERMIS, I., KOIDIS, I., ZISIS, I., KACHRIS, C., An Open TCP/IP Core for Reconfigurable Logic, proceedings of the 13th Annual IEEE Symposium on Field-Programmable Custom Computing Machines (FCCM' 05), p. 297-298, 2005.

[10] BOKAI, Z., CHENGYE, Y., TCP/IP Offload Engine (TOE) for an SOC System, Institute of Computer & Communication Engineering, National Cheng Kung University, 2005.

[11] KANT, K., TCP offload performance for front-end servers, Global Telecommunications Conference, vol. 6, p. 3242 – 3247, 2003.

[12] Virtex II Pro Platform FPGA Handbook, http://www.xilinx.com.

[13] Platform Studio and EDK, http://www.xilinx.com.

[14] DYKSTRA, P., Gigabit Ethernet Jumbo Frames, White Paper, WareOnEarth Communications, 1999.

[15] Clark, T., IP SANs, A Guide to iSCSI, iFCP and FCIP Protocols for Storage Area Networks, Addison-Wesley, 2001.

[16] Romanow, A., Bailey, S., An Overview of RDMA over IP, The Internet Society, 2002.

[17] Senapathi, S., Hernandez, R., Introduction to TCP Offload Engines, Network and Communications, Power Solutions, 2004.

Impact of Hardware Emulation
on the Verification Quality Improvement

Youssef Serrestou, Vincent Beroulle, Chantal Robach

LCIS-INPG, 50 rue Barthélémy de Laffemas, 26902 cedex, Valence

firstname.name@esisar.inpg.fr

Abstract— Software simulation remains the most used method for VHDL RTL functional verification. The functional verification process essentially consists of two parts. The first one is the functional qualification; the second one is the qualification-driven stimuli generation. Currently, the qualification and the generation tasks are iterative processes based on VHDL simulation which is dramatically time consuming. The simulation time increases with the circuits' size and the required level of quality. In our previous works, we have proposed some approaches based on the mutation testing technique to evaluate and to improve functional validation quality. Now, to reduce this simulation time, we propose in this paper a new approach based on FPGA emulation. So, an hardware-software platform called "Meta-Mutant Testbench" is used to emulate mutants. Experimental results for some ITC'99 benchmark circuits show that our mutation emulator is about 20 times faster than classical software simulators; this speedup increases with the circuits' size.

Index Terms— Functional verification, qualification, mutation testing, emulation, FPGA.

I. INTRODUCTION

Mutation-based test technique gives an efficient estimation of the design validation quality [1-6]. This technique based on a fault model involves the creation of numerous faulty circuits' descriptions (called mutants) from the original description of the *Design Under Verification* (DUV). In addition, this technique requires the application of a given validation test set to the DUV's inputs as well as to every mutant's inputs. Then, if at least one mutant's output differs from one DUV's output, this mutant is detected (we say "*killed*"). At the end of the simulation, the *Mutation Score* (MS) which is the ratio of killed mutants over all injected mutants [4-6] measures the functional verification quality. Hence, MS gives a high level of confidence on the design verification process. However, the simulation of numerous mutants with numerous test vectors is very time consuming. In fact, in current industrial RTL descriptions, this number of mutants can reach 10,000 and the number of test vectors can exceed 100,000. So, the computational effort required to measure the quality of a significant number of stimuli applied to a large design, becomes a bottleneck in the verification process. Thus, as new generations of reprogrammable circuits includes millions of gates, and embeds processors and memories, the cost of analyzing a large number of mutants can be reduced thanks to the use of emulation techniques. In fact, these modern reprogrammable devices present high capabilities and interesting characteristics allowing emulating large designs.

In current works and since many years [7, 8, 9], the FPGAs (Field Programmable Gate Array) have been used as hardware platforms for the emulation of test approaches originally based on fault simulation. Obviously, this emulation (or prototyping) is used in order to accelerate fault simulations. Thanks to the FPGA capabilities for fast hardware emulation, the cost of simulations can be greatly reduced.

For RTL design qualification, FPGA can also be used as an emulation platform. The main FPGA characteristics that justify their use for functional verification qualification are the following:

❖ FPGAs include millions of equivalent gates which allow emulating numerous mutants associated to complex RTL descriptions.

❖ FPGAs include large SRAM memories which allow FPGAs to load great number of stimuli.

❖ FPGAs include processor which allows FPGA to control the emulation process. In addition, this embedded processor with its on chip available peripherals provides many ways to communicate (RS232, Ethernet, USB) with a host computer.

In this paper, we present a new technique allowing high speed design verification qualification. This technique is based on mutation testing and exploits the FPGA technological improvements. We propose in this paper a mutation testing technique based on a synthesizable fault model which allows hardware acceleration. A complete testbench allowing reducing both hardware resources and synthesis time is also presented. Mainly, we aim at showing the feasibility and the advantages of this approach, and describing this developed hardware-software emulation platform for functional verification qualification.

The efficiency of this approach has been validated on two descriptions from the ITC'99 VHDL benchmarks. The used FPGA is a Xilinx Virtex-II Pro including a PowerPC 405 microprocessor.

The rest of the paper is organized as follows: the next section summarizes previous works concerning mutation-based test and describes our new Meta Mutation approach. Section 3 presents our hardware acceleration platform. The fourth section shows, analyses and discusses the obtained experimental results. Finally, section 5 draws some conclusions and lays the bases for future work.

II. MUTANTS INJECTION

A. Mutation Testing

Mutation testing was originally intended for software testing to locate and to expose weaknesses in test data [2]. It was also proposed as a generation technique for unit software testing. In these works, the aim of mutation testing is to measure the efficiency of a test set to exercise the different functions of a program.

Thanks to its efficiency and advantages, this technique was adopted as a functional qualification metrics for hardware validation [1, 4, 5]. In this case, a mutant is a faulty induced version of an RTL-level description which differs from the original description by a single error. A mutation operator is a function which is applied to the original description to generate a mutant. An example of mutation operation is the Arithmetic Operator Replacement (AOR), which replaces each arithmetic operator in the original description with another arithmetic operator. A subset of mutation operators has been defined in [5]; this subset is summarized in Table 1.

Operator	Description	Examples
AOR	Arithmetic operator replacement	X <=Y+Z→X <=Y-Z
LOR	Logical operator replacement	X<=Y and Z→X <=Y or Z
ROR	Relational operator replacement	X<Y → X>Y
UOR	Unary operator replacement	X<=Y → X<= -Y X<=Y → X<= not Y
CR	Constant replacement	X <= «01» →X <= «10»
CVR	Constant for variable replacement	X <=Y+Z→ X <=2+Z
VCR	variable for Constant replacement	X <=Y+1→ X <=Y+Z
VR	Variable replacement	X <=Y+Z→ X <=W+Z

Table 1: Examples of Mutation Operators

This technique is a fault oriented test data evaluation. In addition, the quality of the verification process can be measured respectively to the following three metrics:

- *Fault activation (or weak mutation [4])*: verification data are evaluated by their capacity to generate a mismatch between the fault free description and the faulty descriptions. This mismatch is directly observed after the line of the fault location.
- *Fault propagation (or firm mutation [4])*: verification data are evaluated by their capacity to activate a mismatch and to propagate it from its original location (the faulty line) towards an output of the faulty block; this faulty block being the process (i.e. a VHDL group of sequential instructions) concerned by fault injection.
- *Error detection (or hard mutation [4])*: fault propagation is observable by the way of at least one circuit primary output.

Thanks to mutants' simulations, this approach leads to a metrics called the Mutation Score (*MS*). This metrics qualifies the validation stimuli (*VS*) for a *Design Under Verification* (*DUV*). Before defining this MS, let us first more precisely define *detected mutants* and *equivalent mutants*.

A detected mutant is a mutant that can be distinguished from the original DUV because there is at least one sequence in *VS* that, when applied on the inputs of the original DUV and the mutant, generates differences on the outputs. An equivalent mutant can not be distinguished from the original program whatever the simulated input data. The mutation score MS relatively to a test set TS and a program P is computed as in Eq. 1:

$$MS(VS, DUV) = 100 * \frac{D}{(M - E)} \qquad (1)$$

where M is the number of generated mutants, D the number of detected mutants and E the number of equivalent mutants [10].

B. Mutant Injection for simulation

For fault simulation purposes, our mutant generator analyses the DUV and then applies suitable mutation operators, creating one concurrent description for every mutant. Figure 1 illustrates this structure. This scheme generates as many VHDL files or descriptions as mutants. In fact, each fault is inserted in one replicated description called concurrent mutant. Global Testbench architecture is then created including all mutants, the DUV and comparators. To measure MS, output signals of all mutants are compared with the DUV's outputs. To measure the quality of the validation data relatively to weak and firm mutations the RTL-codes of the DUV and mutants are instrumented in such a way that we can track the evolution of internal signals during simulation [4]. This method allows the activation and propagation of the injected faults to be measured. This architecture is memory and compilation time consuming, however the simulation times are optimal.

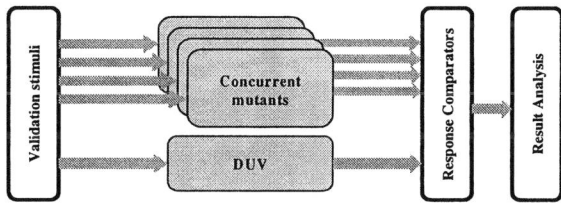

Figure 1: Testbench for Mutants injection for simulation

C. Mutant injection for emulation

The adopted strategy for mutation injection in the case of simulation is not suitable for emulation. For example, let L be the ratio of logic needed for the DUV on the FPGA, if we inject N_{mut} concurrent mutants, the ratio of required logic is around $N_{mut} *L$. It means that for a design corresponding to 1% of the available logic, we are limited to inject less than 100 mutants. The number of mutants increases with the DUV complexity more quickly than the FPGA available resources. The same problem also appears for the synthesis time: mutant synthesis times are cumulative. Thus, another strategy must be adopted to optimize hardware resources and compilation time for the emulation of mutants.

In order to inject mutants, we create from the original circuit description a single *Meta-Mutant* description containing all possible mutants. After analyzing the VHDL DUV description, we construct a VHDL package consisting of functions which models all mutation operators. These functions allow us to sequentially select, for the Meta-Mutant, the normal or a mutated operation behavior. Each statement in the original DUV, is replaced in the Meta-Mutant by its corresponding mutation function. This Meta-Mutant generation process is illustrated with an example in what following.

In Figure 3, the first code describes a function that models UOR (see Tab. 1) for a bit type operand. Then, Figure 2 shows a part of an original DUV code and its corresponding code in the generated Meta-Mutant. To select one mutant embedded in the Meta-Mutant description, the Meta-Mutant uses an additional input, with a length of $[log_2 (N_{mut}) + 1)]$ bits, with N_{mut} the number of mutants and [x] indicates the floor of x. If this input called *mutant selector* matches with the mutant selector input in the mutation function, this function returns the corresponding mutated behavior of the operation that it models, otherwise it returns the normal behavior. For example, in Figure 2, we replace unary logical operator "not", for *std_logic* type operand, by its mutation function *UORstd_mutation*; the parameters of this function are respectively the "original operand" (the signal *s* in this example), the "original operation" ("not" in this example), the "local mutant selector" (equal to 19) and the "global mutant selector" (mutant_selector). When the "global mutant selector" is equal to "local mutant selector" this function mutates the original operation and returns the signal *s* without inversion, otherwise this function inverts this signal.

```
if counter > RED_TONE then
    speaker <= not s;
    counter := 0;
else
        counter := counter + 1;
end if;
```

```
if CORInteger_mutation (counter,RED_TONE,sup,13, mutant_selector) then
    speaker <= UORstd_mutation(s,opnot,19, mutant_selector) ;
    counter := Integer_mutation(0,20, mutant_selector) ;
else
        counter:= AORInteger_mutation (counter,1,'+', 22, mutant_selector) ;
end if;
```

Figure 2: Meta-Mutant Example

In term of FPGA hardware resources requirement, we observe that the size of the Meta-Mutant is not proportional to N_{mut} but to $log_2(N_{mut})$. Each mutation function is associated to a selection operator that allows returning the normal behavior or one mutated behavior. The number of gates required to implement this selection operator depends on the number of induced mutants. In addition, the functional core of each mutation function after synthesis does not exceed five times the required logic for the original function. It means that if L is the ratio of required logic to synthesize the original DUV, then the required logic for its Meta-Mutant with N_{mut} injectable mutants is in the order of $F*L+ L*log_2(N_{mut})$, with F the factor of transformation (i.e. the ratio of logic required for a mutation function to the logic required for its original operation). For example, for an original circuit that occupies 2% of the available logic in the FPGA, its corresponding Meta-Mutant,

which allows to inject 1500 mutants, occupies 25% of the available logic.

Figure 4 and 5 show an inverter gate and the synthesis of the previous mutation function UORBIT_mutation for this simple logical unary operator. When the *mutant_selector* signal is '0' the mutation function performs the faulty-free operation (a bit inversion in this case), and when the *mutant_selector* is '1' *the* mutation function performs the mutated operation (inversion in this example).

```
type ULOR_type is (opnot, noop);
-- opnotunary operator not,
-- noop no operator
function UORBIT_mutation
        (
        R              : bit;
        OP             : ULOR_type;
        Mutant_Ref     : integer;
        Mutant_selector : integer
        )
    return bit is
-----------------------------------------------
    Begin
    if (Mutant_Selector /= Mutant_Ref ) then
    -- Normal behavior
        case OP is
                when opnot   => return not R;
                when noop    => return R;
        end case;
    else
    -- Mutation
        case OP is
                when opnot => return R;
                when noop  => return not R;
        end case;
    end if;
end UORBIT_mutation;
```

Figure 3: Mutation function for the Bit Unary Operator Replacement

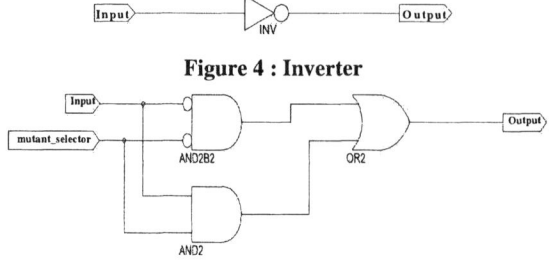

Figure 4 : Inverter

Figure 5: Inverter mutation function synthesis

III. HARDWARE PLATFORM FOR IP FUNCTIONAL QUALIFICATION

A. *Qualification Platform Architecture*

Our previous software tools targeted both functional verification qualification and test data improvement [4, 5]. This tool relies on iterative simulations. For complex circuits the time required to reach a high validation quality becomes too high. To reduce this cost, we plan to replace the software simulator engine by a hardware emulator. As described in the last section, we transform the original VHDL DUV description into one VHDL description containing all mutants called

Meta-Mutant. To create the engine part of the emulation platform, we assemble in the same synthesizable description the original DUV, this Meta-Mutant, and several comparators.

This Meta-Mutant testbench is integrated into the FPGA as a peripheral of a complete software architecture including memories, microprocessor and communication modules. In addition, this FPGA board communicates with a host computer performing stimuli generation and analysis. Figure 6 presents the whole implemented platform and it details the *Meta-Mutant structure*.

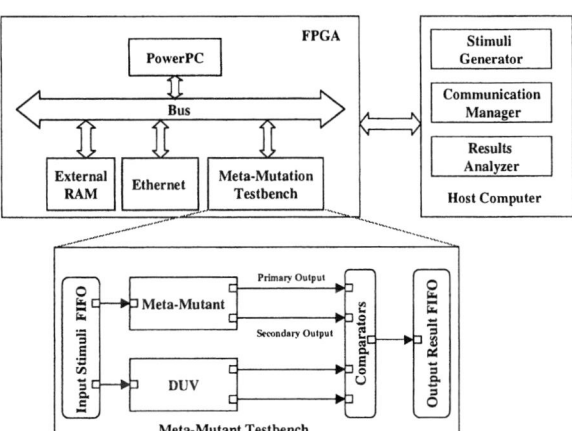

Figure 6: FPGA-based Mutation Emulator

The used FPGA board contains one Xilinx Virtex-II Pro SRAM-based FPGA embedding a 405 PowerPC (PPC) [10]. This embedding PowerPC is a 32-bits microprocessor running at a 100MHz frequency.

The whole system is composed of the following modules and functionalities:

❖ Embedded PPC 405 microprocessor: The PPC manages communications. It downloads the generated stimuli from the host computer into the board DDRAM memory. It sends these stimuli toward the Meta-Mutation testbench input FIFO and uploads the output FIFO results. These results are stored into the DDRAM FPGA board memory, and then they are transmitted to the host computer for analysis.

❖ External RAM: the 512Mo DDRAM FPGA board memories used to store stimuli and results.

❖ High-speed Ethernet IP to exchange data between the host computer and the FPGA board. The average measured transmission speed is about 2 Mbyte/s.

❖ *Meta-Mutant testbench:* The main module that allows us to qualify validation test data. This module consists of the following sub-modules, as it is described in Fig. 6.

  • Original DUV: Design Under Verification
  • Meta-Mutant: This sub-module contains all injectable mutants. A mutant can be selected to be emulated and compared to DUV by the input pin for mutant selection (cf. section 2.C).

• Comparators: When a mutant is selected, its outputs are compared to the DUV's outputs. The mutant is said detected if there is at least one difference between the outputs values. The internals signals are also compared, the mutant is said activated, or propagated if there is a difference between DUV and mutant internals signals.

• Input and Output FIFOs: These two FIFOs permit to synchronize stimuli reception and results upload.

❖ Stimuli Generator: It generates valid stimuli for the DUV. These stimuli are transmitted to the FPGA external RAM.

❖ Result Analyzer: Emulation results are received from the FPGA and analyzed to determine the quality of the generated stimuli relatively to *Hard Mutation Score* and *Firm Mutation Score*.

To program the FPGA or to load the executable program into the PPC, we use the JTAG interface. The communication channel, used for the validation data download or the results upload, is the High-speed Ethernet cable.

B. Emulation flow

After generating and downloading the bits stream implementing the whole system, summarized in Figure 6, into the FPGA, we run the emulation process. Firstly, we download the stimuli generated in the host computer into the FPGA memory. For each mutant the PPC reads these data validation sequences from the board memory and sends these stimuli to the Meta-Mutant testbench input FIFO, with the corresponding mutant selector signal. The results of the emulation process for each mutant are first stored into the board memory. When all mutants have been emulated, the host computer recovers these results and analyzes them in order to compute the quality of stimuli. The entire flow of our approach is described in figure 7.

```
//Host computer
Generating valid Stimuli for the target DUV
Loading Stimuli to FPGA external memory
// PowerPC
mutant_emulation ()
{
  for each mutant
  {
    mutant_selector = mutant_number
    for each stimuli in memory
    {
      Reading Stimuli from memory
      Sending stimuli to Meta-Mutant Testbench
      Receiving Emulation Result
      Storing Result into Memory
    }
  }
}
Sending Emulation Results to Host computer
//Host computer
Analyzing Results
```

Figure 7: Emulation process Flow

IV. EXPERIMENTAL RESULT

To compare simulation and emulation time results, we use two circuits, B12 and B14, from ITC'99 benchmark. The VHDL description characteristics of these circuits are summarized in Tab. 2: complexity: number of primary inputs (PI), primary outputs (PO), lines, processes and number of induced mutants. The available resources in our Xilinx Virtex-II Pro FPGA are 27392 flip-flops, 27392 Look-Up Table (LUTs) and 13696 Slices [10]. To perform software simulation, we use *ModelSim* from Mentor Graphics, installed on a 1.6 GHz Pentium-4, with 1 GB RAM, running a Linux Operating System. The hardware resources needed to implement into the FPGA our emulation platform are reported in the tables 3 and 4 for B12 and B14 circuits.

Circuits	Characteristics				
	#PI	#PO	#Lines	#Process	#Mutants
B12	7	6	567	4	1500
B14	34	54	509	1	600

Table 2: B12 and B14 Benchmark Circuits Characteristics

	B12	Meta-Mutant	Global System	Available
#FFs	132	172	4232	27392
	0.5%	0.62 %	15.44 %	100 %
#LUTs	762	6572	11774	27392
	2.78 %	23.99 %	42.98 %	100 %
#Slices	403	3526	7734	13696
	2.94 %	25.74 %	55.25 %	100 %

Table 3: Hardware Resources used for B12 Meta Mutation System

	B14	Meta-Mutant	Global System	Available
#FFs	635	635	5281	27392
	2.31 %	2.31 %	19.27 %	100 %
#LUTs	2422	14662	21774	27392
	8.84 %	53.52 %	79.49 %	100 %
#Slices	1268	7719	13683	13696
	9.25 %	56.35 %	99.9 %	100 %

Table 4: Hardware Resources used for B14 Meta Mutation System

In the figure 4, we observe that emulation reduces the computation time by 10 to 20 times. For B12, the average time required to simulate one vector for only one mutant is $T_S = 59.98\mu s$. By emulation this time is $T_E = 3.16\ \mu s$. So, we have $T_S/T_E = 19$. For B14, the same average times are $T_S = 67.14\ \mu s$ and $T_E = 4.61\ \mu s$, so we have $T_S/T_E = 14.56$.

The simulation time depends on both the number of mutants and the complexity of the DUV VHDL description. At contrary, emulation time almost only depends on the number of mutants. In fact, in a first order analysis, the impact of DUV complexity on the maximal clock frequency can be neglected. So, the simulation time T_{Sim} and the emulation time T_{Emul} can be estimated as follows:

$$T_{Sim} = N_m * N_s * T_S \text{ and } T_{Emul} = T_{syn} + N_m * N_s * T_E$$

When N_m be the number of injectable mutants, N_s the number of test vectors, T_s be the average time required to simulate the

	#Mutants	#Stimuli	$T_{Emulation}[s]$	$T_{Simulation}[s]$	T_S/T_E
B12	1500	1,000	5.34	101.76	19.05
		5,000	22.21	433.37	19.51
		10,000	47.56	946.5	19.90
		20,000	94.98	1536.2	16.17
		100,000	446.35	8028.4	17.98
B14	600	1,000	2.67	44.71	16.74
		5,000	7.2	100.64	13.97
		10,000	27.18	473.15	17.4
		20,000	53.15	894.2	16.82
		100,000	286.02	4471.2	16.63

Table 5: Simulation vs. Emulation Time Cost

DUV for only one vector, and T_E be the average time required to emulate the DUV for again one vector, and T_{Syn} be the time required to implement into the FPGA the whole system. In the studied example the time Tsyn required to generate the programming bitstream is less than 10 minutes.

Obviously, our emulation approach is suitable for complex circuits, because for simple circuits, both the number of mutants and the number of test vectors are low. So, in this case, the time to implement the emulation system T_{syn} is higher than the time required for performing the simulation. On the contrary, for a complex circuit, N_m, N_s and T_S dramatically increase and the emulation greatly reduces the cost of simulation as it has been shown in the reported results for B12 and B14 circuits.

V. CONCLUSIONS

In this paper we propose to use a hardware acceleration or emulation technique for verification qualification purpose. A complete hardware-software emulation platform is described and evaluated. The performances of this platform are compared to software simulation results. Experimental results show that this emulator allows the simulation speed to be increased by more than 10 times. In this paper, we only focus on the optimization of the time to achieve verification qualification and not on the improvement of validation stimuli quality. In the next work, we will explain how to adapt a stimuli generator to this emulation platform in order to improve the verification quality.

REFERENCES

[1] G. Alhayek and C. Robach. From specification validation to hardware testing: A unified method. *In International Test Conference, pages 885–893, October 1996.*

[2] R. De Millo and A. Offutt, "Constraint-based Automatic Test Data Generation", IEEE Transactions on computers, Vol. 17, No. 9, pp. 900-910, 1991.

[3] A.J. Offutt, R.H. Untch, "Mutation 2000: Uniting the Orthogonal", Mutation 2000: Mutation Testing in the Twentieth and the Twenty First Centuries, pages 45--55, San Jose, CA, October 2000.

[4] Y. Serrestou, V. Beroulle, C. Robach, "IP validation using genetic algorithm guided by mutation testing", *DCIS'06, XXI Conference on Design of Circuits and Integrated Systems, Barcelona, November 2006.*

[5] Y. Serrestou, V. Beroulle, C. Robach, "How to improve a set of design validation data", DDECS'06", *9th IEEE Workshop on Design and Diagnostics of Electronic Circuits and Systems, Prague, April 2006.*

[6] A.J. Offut and J. Pan "Detecting equivalent mutants and the feasible path problem". The Journal of Software Testing, verification. and Reliability, vol. 7, pp. 165-192Septembre 1997.

[7] C.Lopez-Ongil, M. Garcia-Valederas, M. Portel-Garcia, L. Entrena, "Autonomous Fault Emulation: A New FPGA-Based Accelarator System for Hardeness Evaluation", *IEEE Transaction   On Nuclear Science vol. 54, NO. 1, February 2007.*

[8] M. Sonza Reorda, L.Sterpone, M. Violante, C.Lopez-Ongil, M. Portel-Garcia, L. Entrena, « Fault inection-based Reliability Evaluation of SoPCs », IEEE European Test Symposium, 2006.

[9] S.Hwang, J.Hong, C.Wu, "Sequential Circuit Fault Simulation Using Logic Emulation", *IEEE Transactions On Computer-Aided Design of Integrated Circuits and Systemes, vol. 17 NO. 8, August 1998.*

[10] Xilinx Product Specification, "Virtex-II Pro and Virtex-II Pro X Platforme FPGA: Complete Data Sheet", *DS083 v4.6, Mars 05, 2007*

Fast Estimation of Software Energy Consumption Using IPI(Inter-Prefetch Interval) Energy Model

Jungsoo Kim   Kyungsu Kang   Heejun Shim   Woong Hwangbo   Chong-Min Kyung

Dept. of EECS at KAIST
373-1, Guseong-dong Yuseong-gu, Daejeon 305-701
Republic of Korea

{jskim,kskang,shimy,woonghb}@vslab.kaist.ac.kr   kyung@ee.kaist.ac.kr

ABSTRACT

In this paper, we present the way of fast and accurate estimation of software energy consumption in off-the-shelf processor using IPI(Inter-Prefetch Interval) energy model. In our previous work[1], we proposed a new energy estimation method, and presented the way to characterize IPI energy model. However, there were some drawbacks in our previous work. First, the previous IPI energy model was only able to cover basic block. Second, it took too much time to estimate energy consumption of software due to the absence of energy estimator. To tackle these problems, in this paper, we propose modified IPI energy model, and present the way to implement IPI energy estimator using off-the-shelf system-level simulator. By using this, we can analyze such dynamic effect as address jump due to branch instructions and stalls due to hazard fast and accurately. The major challenging issue when implementing IPI energy estimator is cycle-by-cycle behavioral discrepancy between physical chip and ISS(Instruction Set Simulator) in system-level simulator. In order to overcome this problem, we propose training method to reduce the discrepancy. To verify our method, we applied our idea to ARM1136JF-S real chip. By the result of this work, the accuracy of energy estimation is more than 90% compared to measured data on prototyping board.

Categories and Subject Descriptors

I.6.5 [**Computing Methodology**]: Simulation and Modeling—*Model Development*; C.0 [**Computer Systems Organization**]: General—*Modeling of Computer Architecture*

General Terms

Algorithm

Keywords

Black-box Processor, Software Energy Estimation, IPI Energy Model, System-Level simulator

1. INTRODUCTION

With increasing software complexity drastically[2], software energy optimization has become one of the most important issues in SoC(System-On-Chip) design. To optimize software, fast and accurate analysis on energy consumption of target software must be preceded in software design step.

In general, the flow for developing energy model consists of three steps - measurement(or simulation), characterization and estimation. In measurement(or simulation) step, we measure energy consumption using some test vectors. By using the data, we extract parameters which affect large portion of energy consumption, and develop energy model by LUT-based[3] or equation-based[4] approach in characterization step. And in the estimation step, we implement energy estimator reporting energy consumption fast and accurately with considering dynamic effects such as stall and address jump.

Techniques for developing energy model have been studied so far. They can be largely divided into two categories -white-box[1] [5],[6],[7],[8],[9],[10] and black-box[11],[12],[13] approach. Compared to black-box[2] energy model, white-box energy model is more accurate. In real situation, however, to obtain implementation data of advanced off-the-shelf processor is almost impossible. So, black-box approach has been considered more importantly than white-box approach in real situation. In black-box approach, it only requires limited information, such as ISA (Instruction Set Architecture), event monitoring counter information or architectural information of target processor, to model the energy consumption of target processor. The details of the previous energy models are explained well in [1].

In the previous works on energy model for black-box processor, there is a fatal defect: they can only report total energy consumption of application software executed in a target processor. This make it less useful when optimizing software energy consumption. When optimizing software code, information on energy hotspot or instantaneous energy consumption is needed. In order to overcome this problem, we proposed new energy model called IPI(Inter-Prefetch Interval) energy model in [1]. Although IPI energy model cannot report energy profile in each cycle, it can report energy profile in each prefetch interval. It is enough

---

[1]Cycle-by-cycle behavior of target processor can be analyzed from implementation data such as RTL code, gate-level and post-layout netlist or cycle-accurate ISS.
[2]Physical chip is available, but cycle-by-cycle behavior can not be known.

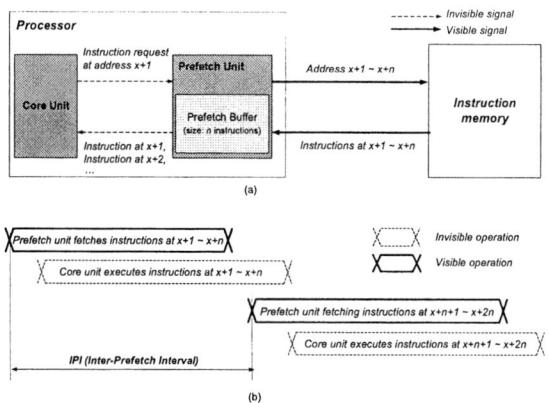

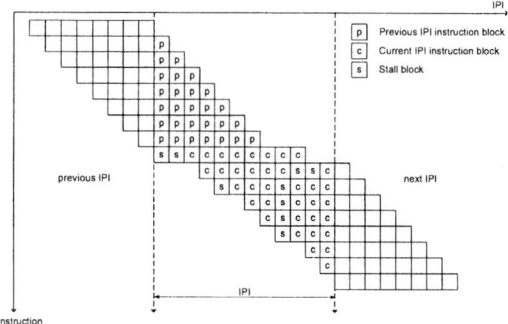

Figure 1: Relation between IPI(Inter-Prefetch Interval) and instruction prefetch (a) Instruction prefetch process (b) The operation of processor using instruction prefetch

Figure 2: The IPI energy is composed of three components: energy due to current IPI instruction(E_C^I), energy due to previous IPI instruction(E_P^I) and energy due to stalls(E_S^I)

to find energy hotspot, then it can be helpful for software optimization.

However, there were some defects in our previous work[1]. First, it was only able to cover basic blocks. That means, there was no way to consider dynamic effects such as address jump. Second, due to the absence of of IPI energy estimator, it took lots of time to estimate energy consumption of software. To tackle above two issues, in this paper, we propose advanced IPI energy model which considers dynamic effects. Also, we implement IPI energy estimator using off-the-shelf system-level simulator. The major challenging issue when implementing IPI energy estimator is how to reduce the cycle-by-cycle behavioral discrepancy between physical chip and ISS of target processor.

The rest of this paper organized as follows. The next section briefly overviews the previous work on IPI energy model. Section 3 elaborates advanced IPI energy model with considering dynamic effects. Section 4 elaborates on how to implement IPI energy estimator using off-the-shelf system-level simulator. Section 5 gives out our experimental setup and result. Finally, section 6 concludes the paper.

2. PREVIOUS WORK

2.1 Concept of IPI

IPI means Inter-Prefetch Interval. That is, IPI is the time interval between two consecutive instruction prefetches[3], as shown in Figure1. In each prefetch phase, microprocessor generates addresses to access data in external memories. By triggering a start signal of prefetch phase, we can divide IPI time unit in an application software. The reason for which we use prefetch information is to report not only total but also instantaneous energy consumption of software energy consumption.

2.2 IPI Energy Model

---

[3]A technique which attempts to minimize the waiting time for instructions to be loaded from memory.

Basically, IPI energy model is based on instruction-based energy model[12][13]. To overcome the drawbacks of instruction-based model, however, we introduce additional parameters called energy weight. The meaning of energy weight is how many portions of instructions, fetched current IPI, are executed in the current IPI time unit.

In each IPI, there are three energy components: energy consumption due to current IPI instructions(E_C^I), energy consumption due to instructions launched in the previous IPI(E_P^I) and energy due to stalls in the current IPI (E_S^I) as shown in Figure2. We divide an instruction into the number of pipeline stages in target processor, which is called instruction block.

First, energy consumption due to current IPI instructions(E_C^I) can be expressed as a weighted sum of energy of each instruction launched for execution during the current IPI;

$$E_C^I = \sum_{i=1}^{N_{fetch}} a_i \cdot E_i \qquad (1)$$

where N_{fetch} is the number of fetched instructions in each prefetch phase. At first, we assumed that N_{fetch} was fixed to simplify IPI energy model. E_i is energy of the i-th instruction, and a_i is the energy weight given as the time fraction of execution of the i-th instruction in the current IPI,

Second, energy consumption due to instructions launched in the previous IPI(E_P^I) is given as weighted sum of energy consumption of instructions launched in the previous IPI and finished in the current IPI.

$$E_P^I = \sum_{i=1}^{N_{fetch}} (1 - a_i') \cdot E_i' \qquad (2)$$

where a_i' is the energy weight, and E_i' is the instruction energy of the i-th instruction launched in the previous IPI. The weight $(1 - a_i')$ denotes the time proportion of the i-th instruction(launched in the previous IPI) executed in the current IPI.

The final term energy consumption due to stall(E_S^I) is expressed as multiplying stall cost by the number of additional stalls in the current IPI.

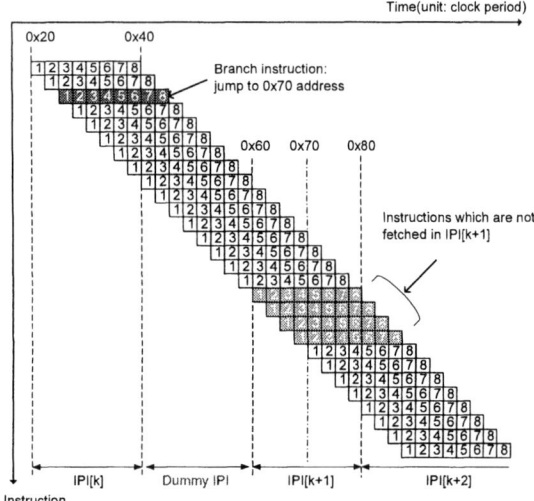

Figure 3: Effects due to branch instructions in case of N_{fetch} is eight: dummy IPI and variable number of fetched instructions

$$E_S^I = k \cdot e_s = (k_m - k_i) \cdot e_s \qquad (3)$$

where k is the number of stalls in the current IPI, e_s is the energy consumption when a stall occurs, k_m is the total number of stalls in the current IPI, and k_i is the number of ideal stalls when no hazard. By subtracting ideal cycle count from total cycle count of the IPI, we can calculate how many additional stalls has been occurred in the current IPI.

Based on Eqn(1),(2), and (3), the proposed model for the total energy consumption(E_T^I) during an IPI becomes

$$
\begin{aligned}
E_T^I &= E_C^I + E_P^I + E_S^I \\
&= \sum_{i=1}^{N_{fetch}} \{ a_i \cdot E_i + (1 - a_i') \cdot E_{ins\_i}' \} + k \times e_s
\end{aligned} \qquad (4)
$$

3. ADVANCED IPI ENERGY MODEL

3.1 Branch Consideration

In the previous IPI model, it was only able to cover basic blocks because we assumed that the number of fetched instruction was fixed. When branch is taken, however, and branch destination address is at the address other than times of the $4 \times N_{fetch}$[4], we cannot hold the assumption anymore. For example, N_{fetch} is eight in ARM1136JF-S, and branch instruction is stored at the address 0x4, and it jumps to instruction at the address 0x50. In this case, only four not eight instructions are fetched in the next prefetch, because the start address of the next IPI is 0x60. Furthermore, there are IPI where instructions are fetched but not executed. We call it *dummy IPI*. These concepts are shown in Figure3

3.1.1 Dummy IPI

---
[4]N_{fetch} is determined in each architecture, e.g.,N_{fetch} is eight in case of ARM1136JF-S processor

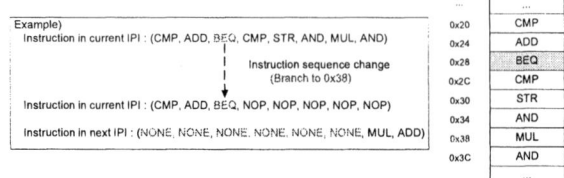

Figure 4: Example of variable number of fetched instructions when N_{fetche} is eight: when branch is taken, instructions belonging to the same IPI with the branch are considered as NOP, and instructions belonging to the same IPI with branch destination instruction are considered as NONE.

The dummy IPI is the IPI where instructions are fetched, but not executed because of branch instruction. Figure3 shows the example of dummy IPI. In a dummy IPI, it consumes energy when it fetches instructions from external memory to prefetch buffer, but these fetched instructions are not entered to execution datapath. So, we can handle this as same as *NOP* instruction, because instructions in dummy IPI region have the same characteristic as NOP instruction in that they do not change any processor state.

3.1.2 Variable number of instructions

As I mentioned above, the number of instructions which is executed in an IPI is not fixed value anymore due to branch instruction. Figure4 shows the example of the effect. Branch instruction is stored at physical address 0x28. If the branch is taken, address jumps to the instruction at physical address 0x38. From the address, another prefetch is generated. That is, the next prefetch is started from instruction at address 0x38, and finished at the address 0x3C, when N_{fetch} is eight. That is, there are only two not eight instructions in the IPI. To model such that circumstances, we introduce virtual instructions called NONE(No Operation No Energy consumption) and NOP(No OPeration) instruction.

First, NONE instruction models the instructions belonging to the same IPI but stored before the address of branch destination instruction, which are not fetched at all. Second, NOP instruction models the remaining instructions stored after the address of branch instruction, which are fetched but not executed like instructions in dummy IPI. Figure4 shows the example of how to use virtual instructions.

By introducing the virtual instruction, NONE and NOP, we can still use fixed $N_{fetched}$ value obtained from architecture information.

4. IPI ENERGY ESTIMATOR

Building energy model of black-box processor consists of characterization and dynamic analysis based energy estimation as shown in Figure5. Characterization step of IPI energy model is explained in [1]. So, in this paper, we only focus on dynamic analysis based estimation using system-level simulator for fast estimation of software energy consumption.

IPI energy estimator is implemented in estimation step to report energy consumption of an software application fast and accurately. In an application software, many dynamic effects such as branch and stalls occurs frequently. To an-

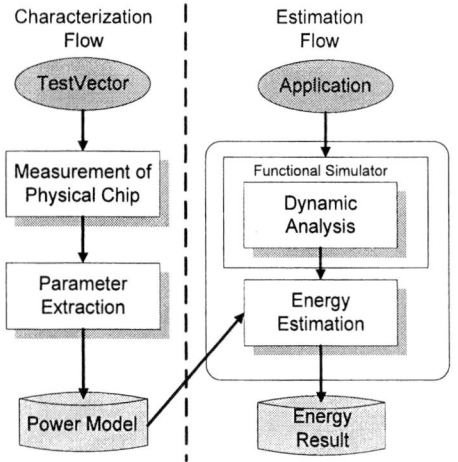

Figure 5: Energy Estimation Flow : It can be divided into characterization and estimation.

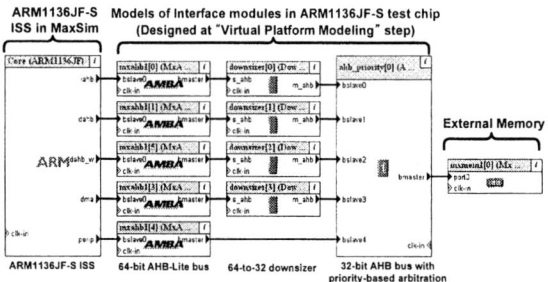

Figure 6: Virtual platform for the simulation of ARM1136JF-S test chip model using MaxSim, a commercial system-level simulator. The virtual platform consists of ISS model, model of the interface modules, and external memory.

alyze the effect fast and accurately, we analyze the effect using system-level simulator to reduce analysis time.

However, in most cases, ISS of target processor in the system-level simulator does not have exactly the same cycle-by-cycle behavior as the real chip, because ISS, for the sake of simulation speed, does not perfectly reflect the cycle behavior of real chip. Also, there may be some interface modules in real chip. To use the data extracted from system-level simulator, we have to modify the discrepancy between ISS and real chip of target processor. We call the modification process *preprocessing*. Preprocessing consists of two steps, virtual platform modeling and ISS error compensation.

4.1 Virtual Platform Modeling

There are several components for the discrepancy. The first component is due to interface moduels employed for chip testability. As the simulation model of interface modules in a real chip is not included in such commercial system-level simulator as MaxSim[14] and ConvergenSC[15], the behavior of the interface modules must be modeled in the system-level simulator to close up the discrepancy. As internal information of the interface modules is not available either, we need to model the interface modules using bus signals appearing when processor is accessing external memories. Figure 6 shows the example of so-called virtual platform of ARM1136JF-S test chip using MaxSim simulator. Although this process based on bus signal observation still does not reproduce exact cycle-by-cycle behavior of the interface modules, the remaining error can be reduced with the adjustment process called ISS error compensation explained below.

4.2 ISS Error Compensation

The second component for the behavior discrepancy is behavior implication(e.g., transaction-level) in the system-level simulator to improve simulation speed, or some intentional modifications in each chip vendor when fabricating real chip to improve performance they want to meet. To reduce the effect, we have to reflect the discrepancy onto the interface modules, because, in most cases, it is not available

to modify the behavior of ISS in the system-level simulator.

To do this, we train platform with some test vectors which are classified by the number and position of load/store instructions in the current and the previous IPI. If we trained platform for all the cases, it could produce the best solution. However to reduce training time, we only focus on the load/store instruction because they produce the most critical discrepancy between real chip and virtual platform. To solve this, we construct an equation as follows:

$$C_m^I = C_e^I + \Delta C_P + \Delta C_C \qquad (5)$$

where,

C_m	:	IPI cycle count measured in chip
C_e	:	IPI cycle count estimated in ISS
ΔC_C	:	Number of clock cycles to be added to compensate for error in the simulator due to current IPI instruction sequence
ΔC_P	:	Number of clock cycles to be added to compensate for error in the simulator due to previous IPI instruction sequence

At first, we calculate the value of the C_C. This is calculated by excluding effect of previous IPI. To do this, we apply test vector which is filled with NOP instruction in the previous IPI and some instructions of which we want to know the amount of behavioral discrepancy in current IPI like test vector for finding energy coefficient weight[1]. After obtaining the value of the C_C, we can obtain the value of the C_P with test vector which is composed of combinations of instruction sequences in previous and current IPI. After this process, we can closely estimate the number of clock cycles of real chip by summing the ISS error compensation parameters to the estimated cycle count in ISS.

4.3 Energy Estimation Method

After the above adjustment, the discrepancy between the behavior of a real chip and that of ISS is reduced such that the extracted data from the virtual platform including the ISS can be used for IPI energy estimation. Figure 7 shows the method to estimate the IPI energy consumption using simulation result on the virtual platform. In IPI energy estimation, we need two kinds of data extracted from the func-

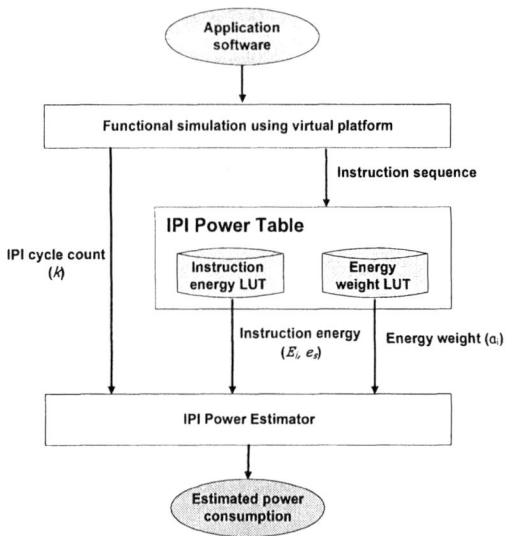

Figure 7: IPI Energy Estimator : IPI energy estimator uses dynamic analysis data and characterized energy table to estimate energy consumption

tional simulation using virtual platform to provide the values needed in Equation(4) to estimate the power consumption. First, instruction sequence in IPI is necessary to find the instruction energy of the constituent instructions and the energy weights in IPI, which are obtained in the characterization path and stored in the IPI power table. Second, cycle count of each IPI is necessary to calculate the stall energy.

5. EXPERIMENTAL RESULTS

To evaluate IPI energy model, we implemented IPI energy estimator of ARM1136JF-s[17] processor in MaxSim5.2[14]. At first, We implemented measurement environment like Figure7 . We measured energy consumption of ARM1136JF-S physical chip using current mirror method described in [16]. By generating trigger signal at the start of and the end of application, we could measure the amount of current which was drawn into the target processor by measuring output node of current mirror using oscilloscope. Then, we constructed virtual platform in commercial system-level simulator, MaxSim[14], as shown in Figure6. As you can see in Figure6, there are such interface modules as AHB-priority bus and downsizer in ARM1136JF-S testchip.

At first, each column of Table1 shows the result of the error for the number of IPI clock cycles and energy estimation error compared to measurement data, respectively. As shown in Table1, the average error for the estimated energy is less than 10%. The error is acceptable to optimize software in system-level. There are some reasons for error. The first error component is the error due to energy model without considering inter-instruction energy. The next error component is behavioral discrepancy between real chip and ISS cycle-by-cycle behavior. We tried to close up the discrepancy by the training flow proposed in Section4, but error amounting to less than 5% was still existed. This error affects the stall energy in the IPI. Because stall energy is less than 50% compared to other instruction cost, the discrep-

Testbench	IPI cycle count count error(%)	IPI energy error.(%)
Fibonacci	0	5.09
wordcpy	4.17	4.59
strcpy	6.23	5.79
Dhrystone	10.25	9.99

Table 1: Error of IPI Cycle Count : discrepancy between ISS and test chip behavior is less than 10%, and estimation error is also less than 10% compared to measured value.

ancy affect less than5% to the total energy consumption

Table2 shows the effect of our proposed training method reducing behavioral discrepancy between real chip and virtual platform.

	IPI cycle count error(%)	IPI energy error(%)
Type1	59.36	58.65
Type2	13.62	14.00
Type3	10.25	9.99

Table 2: Improvement on discrepancy between real chip and virtual platform for Dhrystone Application : error between virtual platform and real chip is drastically reduced using training method

Type1 is the simulation result using ISS without any preprocessing. Type2 is the simulation result using ISS with model for interface logic. And Type3 is the simulation result using ISS with model for interface logic modules and their latency.

As shown in Table1, the error is reduced 82.7% and 82.9% in respect to cycle count and energy, respectively. It proves that our training method is effective if ISS has different behavior to the real chip.

At last, we compare estimated and measured energy profile in the each IPI. In energy estimation, energy tracking is one of the most important metric to evaluate the energy estimator. Because our objective is to estimate energy consumption in system-level, relative value is important when software energy optimization. The Figure10 shows this ex-

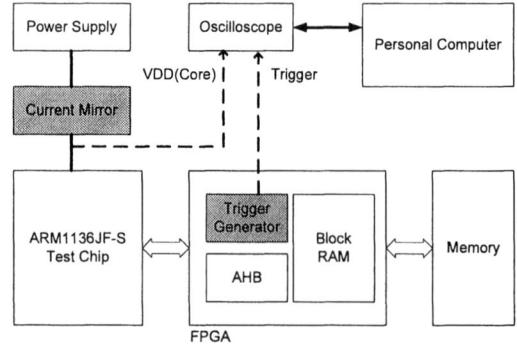

Figure 8: Prototyping platform : The amount of current drawn into ARM1136JF-S testchip is measured by oscilloscope.

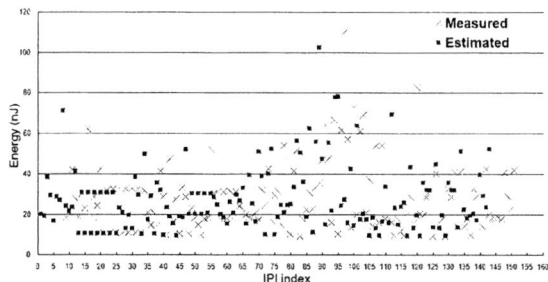

Figure 9: Estimated energy vs. Measured energy for Dhrystone application : Estimated energy tracks very closely to measured energy.

perimental result.

As you can see in Figure10, the estimated energy is well tracked to measured energy.

6. CONCLUSION AND FUTURE WORK

The IPI energy model is the unique model which can consider energy-profile in advanced black-box processor. That is, the IPI energy model is very effective when information of target processor's behavior is not available, and ISS of target processor is inaccurate. Also, the IPI energy model is better than instruction-based model in respect to having capability of reporting energy profile for an application. By reporting energy profile, it has the possibility to be used in finding energy hotspot, so that the energy model can be used in energy optimization. Upon our experimental result, our proposed model can give more than 90% accuracy on average. Also, by implementing IPI estimator, we can estimate energy consumption of an application software fast and accurately. Our proposed idea has been proved in commercial embedded processor ARM1136JF-S

7. REFERENCES

[1] K.Kang, J.Kim.H.Shim, and C.-M.Kyung. Software Power Estimation using IPI(Inter-Prefetch Interval) Power Model for Advanced Off-the-Shelf Processor. GLSVLSI, 2007.

[2] MEADEA+. Design automation roadmap 2005. http://www.medeaplus.org.

[3] S.Gupta and N.Najm. Power macromodeling for high level power estimation. DAC, 1997.

[4] S.Gupta and F.N.Najm. Analytical models for rtl power estimation of combinational and sequential circuits.

[5] W. Ye, N. Vijaykrishnan, M. Kandemir, and I. M.J. The design and use of SimplePower: A cycle-accurate energy estimation tool. DAC, 2000.

[6] N. Chang, K. Kim, and H. G. Lee. Cycle-accurate energy consumption measurement and analysis: Case study of ARM7TDMI. ISLPED, 2000.

[7] N. Chang, K. Kim, and H. Lee. Cycle-accurate energy measurement and characterization with a case study of the ARM7TDMI. *IEEE Transaction on Very Large Scale Integration(VLSI) Systems*, 10(2), April 2002.

[8] H. Kim, S. Kim, I. Lee, S. Yoo, E.-Y. Chung, K.-M. Choi, J.-T. Kong, and S.-K. Eo. An industrial case study of the ARM926EJ-S power modeling. ISOCC, 2005.

[9] D. Brooks, V. Tiwari, and M. Martonosi. Wattch: A framework for archituecutral-level power analysis and optimizations. ISCA, 2000.

[10] A. Sinha and A. P. Chandrakasan. Jouletrack - a web based tool for software energy profiling. DAC, 2001.

[11] R. Joseph and M. Martonosi. Run-time power estimation in high performance microprocessors. ISLPED, 2001.

[12] V. Tiwari, S. Malik, and A. Wolfe. Power analysis of embedded software : A first step towards software power minimization. *IEEE Transactions on Very Large Scale Integration Systems*, 2(4), December 1994.

[13] N. Kavvadias, P. Neofotistos, C. Kosmatopoulos, and T. Laopoulos. Measurement analysis of the software-related power consumption in microprocessors.

[14] ARM. MaxSim. http://www.arm.com.

[15] CoWare. ConvergenSC. http://www.coware.com.

[16] T. Laopoulos, P. Neofotistos, C. Kosmatopoulos, and S. Nikolaidis. Measurement of current variations for the estimation of software-related power consumption.

[17] ARM. ARM1136JF-S and ARM1136J-S technical reference manual. http://www.arm.com.

Power Optimization for Conditional Task Graphs in DVS Enabled Multiprocessor Systems

Parth Malani, Prakash Mukre, Qinru Qiu

Department of Electrical and Computer Engineering, Binghamton University
Binghamton, NY 13902
{parth, pmukre1, qqiu} @binghamton.edu

Abstract — **In this paper, we focus on power optimization of real-time applications with conditional execution running on a dynamic voltage scaling (DVS) enabled multiprocessor system. The targeted system consists of heterogeneous processing elements with non-negligible inter-processor communication delay and energy. Given a conditional task graph (CTG), we have developed novel online and offline algorithms that perform simultaneous task mapping and ordering followed by task stretching. Both algorithms minimize the mathematical expectation of energy dissipation of non-deterministic applications by considering the probabilistic distribution of branch selection. Compared with existing CTG scheduling algorithms, our online and offline scheduling algorithms reduce energy by 28% and 39% in average, respectively.**

I. INTRODUCTION

Multiprocessor System-on-Chip (MPSoC) is becoming a major system design platform for general purpose and real-time applications, due to its advantages in low design cost and high performance. Minimizing the power consumption is one of the major issues in designing battery operated MPSoC. One of the widely used power reduction technique is *Dynamic Voltage Scaling (DVS)*, which allows the processor to dynamically alter its speed and voltage at run time to trade power for performance.

In a multiprocessor system, the mapping and ordering of tasks changes the task slack time, i.e. the intervals when a processing element (PE) is idle, and hence have a significant impact on the efficiency of DVS. As the system complexity grows, the latency and energy of inter-processor communication increases. A holistic technique must be developed for task mapping, ordering and stretching to reduce both communication and computation energy.

Many of the real-time applications are non-deterministic. The application is divided into several tasks. Some tasks are activated only if certain conditions evaluated by previously executed tasks are true. A conditional task graph (CTG) [4]~[7] captures such relation and hence enables us to model more general application.

Although conditional branch prediction is a common practice in high performance processors, the prediction will not be perfect. Furthermore, the task graphs that we are working with are high level descriptions of large applications. Their selection of conditional branches depends mostly on the input data, which are random. Techniques that dynamically assign confidence levels [8] or probabilities [9] to the conditional branches have been proposed by previous research works.

In this work, we propose a set of *communication aware and profile-based* (CAP) scheduling algorithms for CTG on a DVS enabled multiprocessor system. The targeted system has a set of heterogeneous PEs, such as DSPs, FPGAs or ASICs, that are connected by interconnect network. The energy and delay for inter-PE communication is not negligible. Each PE has DVS capability. We assume that the branch probabilities are available through static or dynamic branch profiling.

Online and offline CAP scheduling algorithms are presented in this paper. They consider task mapping, task ordering and task stretching altogether to minimize energy dissipation and also satisfy the performance constraint. The task mapping and task ordering are performed simultaneously and their goal is to minimize the inter-processor communication and maximize the task slack. The task stretching algorithm finds the best speed and starting time for each task so that the computing energy is minimized. The algorithms consider the branch probabilities and they minimize the mathematical expectation of energy dissipation.

The offline CAP algorithm formulates and solves the task stretching problem as a *linear programming* (LP) problem which is time consuming. The online CAP algorithm replaces the LP based algorithm with a heuristic algorithm. Experimental results show that compared with the offline algorithm, the online CAP heuristic is 120,000X faster and almost as effective as offline CAP.

Many techniques have been proposed that consider the task mapping and ordering for DVS [1]~[3]. However, these algorithms only consider traditional data-flow graph without conditional execution. One of the major characteristic of CTG is that some tasks are mutually exclusive. These tasks can be mapped to the same PE at the same time. Reference [4] and [5] consider scheduling and mapping for CTG, however, they do not minimize energy dissipation. Wu et al. [6] proposed an algorithm for task ordering and stretching of CTGs running on a DVS enabled system. They search for the optimal task mapping using genetic algorithm (GA). The proposed algorithm provides a complete solution for power optimization of CTGs. However, it does not consider the branch probabilities. An implied assumption of this technique is that all the conditional branches will be selected with equal probability. Furthermore, the GA based task mapping algorithm has high complexity because the inner loop of this algorithm needs to perform the task ordering and stretching of the entire CTG. Shin et al. [7] proposed an algorithm for task ordering and stretching of CTG which considers the run-time behavior. They refer this approach as condition aware scheduling. Under condition aware scheduling, a task has different start time and speed for different combinations of possible branch selections. Therefore, a large table is needed to store the scheduling result. The probabilistic distribution of the branch selections is considered only during task stretching. Another limitation of this algorithm is that it takes task mapping as a fixed input so that the communication overhead cannot be considered.

The characteristics of the proposed CAP algorithms are described as follows.

1. The proposed algorithms consider task mapping, ordering and task stretching altogether for energy reduction.

2. We consider the application with conditional execution as a random procedure. The algorithm explores the fact that the conditional branches will be selected with different probabilities. The algorithm utilizes the probabilistic information that is collected through static or dynamic branch profiling. Its objective is to minimize the mathematical expectation of energy dissipation.

3. Offline and online versions of the CAP algorithm are proposed. The offline algorithm is a compile time scheduling algorithm while the online algorithm has very low complexity so that it can be used with runtime branch prediction for dynamic scheduling.

The experimental results show that, comparing with the scheduling algorithms presented in [7], the offline and online CAP algorithms provide an average of 39% and 28% energy saving respectively.

The rest of this paper is organized as follows. Section II introduces the application and hardware architecture models. Section III provides detailed introduction of our scheduling algorithm. Sections IV and V present the experimental results and conclusions.

II. APPLICATION AND ARCHITECTURE MODELING

The CTG that we are using is similar as the one specified in [7]. A CTG is an acyclic graph $<V, E>$. Each vertex $\tau \in V$ represents a task. An edge $e=(\tau_i, \tau_j)$ in the graph represents that the task τ_i must complete before τ_j can start. A conditional edge e is associated with a condition $C(e)$. We use $prob(e)$ to denote the probability that the condition $C(e)$ is true. The node with output conditional edge is a *branch fork node*.

A node can be either *and-node* or *or-node*. An and-node is activated when all its predecessor nodes are completed and the conditions of the corresponding edges are satisfied. On the other hand, an or-node is activated when one or more predecessors are completed and the conditions of the corresponding edges are satisfied.

The condition that the task τ is activated is denoted as $X(\tau)$. The condition of an and-node τ_i can be written as $\wedge_{\tau_k} (C(\tau_k, \tau_i) \wedge X(\tau_k))$, where τ_k is the predecessor of τ_i. The condition of an or-node τ_j can be written as $\vee_{\tau_k} (C(\tau_k, \tau_j) \wedge X(\tau_k))$, where τ_k is the predecessor of τ_j. A *minterm* m is a possible combination of all conditions of the CTG. We use M to denote the set of all possible minterms of a CTG. A task τ is associated with a minterm m if $m \subseteq X(\tau)$. In another word, a task τ is associated with a minterm m if $X(\tau)$ will be true when m is evaluated to be 1. The set of minterms with which τ is associated is denoted as $\Gamma(\tau)$. Two tasks τ_i and τ_j are mutually exclusive if they cannot be activated at the same time, i.e. $X(\tau_i) \oplus X(\tau_j)=0$. To simplify the implementation and discussion, we refer the condition "1" (i.e. always true) as one of the minterms as well.

The volume of data that pass from one task to another is also captured by the CTG. Each edge (τ_i, τ_j) in the CTG associates with a value Comm(τ_i, τ_j) which gives the communication volume in the unit of Kbytes. Finally, we assume a periodic graph and use a common deadline for the entire CTG.

Example 1: Figure 1 shows an example of a CTG. All nodes except node τ_8 are and-nodes. The edges coming out from τ_3 and τ_5 are conditional edges. The symbol marked beside a conditional edge gives the condition under which the edge will be activated. For example $C(\tau_3, \tau_4) = a_1$. There are total of 4 minterms in the CTG and $M=\{1, a_1, a_2b_1, a_2b_2\}$. We have $\Gamma(\tau_1) = \Gamma(\tau_2) = \Gamma(\tau_3) = \{1\}$, $\Gamma(\tau_4) = \{a_1\}$, $\Gamma(\tau_5) = \{a_2\}$, $\Gamma(\tau_6) = \{a_2b_1\}$, $\Gamma(\tau_7) = \{a_2b_2\}$ and $\Gamma(\tau_8) = \{1, a_1\}$. The execution profile and communication volume are given beside the CTG. The fact that τ_8 is an or-node indicates that if condition a_1 is true then τ_8 cannot start until τ_2 and τ_4 finish and if condition a_1 is false then τ_8 does not have to wait for τ_4. Note that, in reality, we do not know weather a_1 is true or false until τ_3 finishes. Therefore, in any case, τ_8 must wait until both τ_2 and τ_3 finish. This example shows an implied dependency between an or-node and the branch fork node. More detailed discussion will be provided in the next section.

The following models the architecture of an MPSoC:

- The set of PEs, $P = \{p_1, p_2, ..., p_n\}$

- The energy $E(\tau_i, p_j)$ and worst case execution time $WCET(\tau_i, p_j)$, $\forall \tau_i \in V$ and $\forall p_j \in P$. These values give the energy and delay of each task when it is running on different PEs at the nominal V_{DD}.

- The bandwidth $B(p_i, p_j)$ and transmission energy $E_{tr}(p_i, p_j)$, $\forall p_i, p_j \in P$. These values specify the bandwidth as well as the transmission energy per byte of the communication link between p_i and p_j. We modeled a point-to-point communication link for our interconnect network and dedicated communication resource for each PE. We also assume that the voltage scaling cannot be applied to the communication tasks.

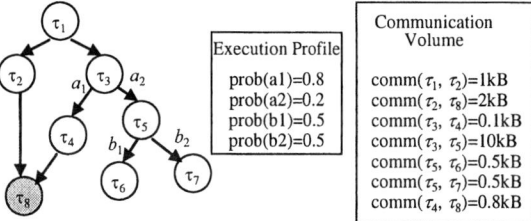

Figure 1 An example of CTG.

III. PROPOSED SCHEDULING ALGORITHM

A. Offline CAP algorithm

This section provides an insight into the offline version of the communication aware profile-based (CAP) scheduling algorithm. The algorithm can be divided into two steps. The first step finds task mapping and ordering while the second step finds the task starting time and task speed.

1) Task mapping and ordering

Both online and offline CAP algorithms use the same task mapping and ordering algorithm. It is based on *Dynamic Level based Scheduling* (DLS) proposed by [10]. The DLS algorithm is a list scheduling algorithm. It considers computation scheduling and communication scheduling altogether. The *ready list* is a list of tasks whose predecessors have been scheduled and mapped. For each task τ_i in the candidate list, the dynamic level $DL(\tau_i, p_j)$ is calculated using the following formula:

$$DL(\tau_i, p_j) = SL(\tau_i) - \max[DA(\tau_i, p_j), TF(p_j)], \quad (1)$$

where p_j is one of the processing elements, $SL(\tau_i)$ is the static level of task τ_i, which is equal to the longest distance from node τ_i to any of the end nodes in the task graph, $DA(\tau_i, p_j)$ is the earliest time that all data required by node τ_i is available at the jth PE with the consideration of both computation and communication delay, and $TF(p_j)$ is the time that the last task assigned to the jth PE finishes its execution. The pair of (τ_i, p_j) which gives the maximum dynamic level will be selected and the mapping is performed accordingly. The task is scheduled to be started at the time $\max[DA(\tau_i, p_j), TF(p_j)]$. After that, the candidate list is updated and the dynamic level of each task in the candidate list is re-calculated.

In this work, we modified the DLS algorithm to consider the mutual exclusiveness among conditional tasks and also consider the probabilistic distribution of branch selection.

The static level $SL(\tau_i)$ is calculated using a dynamic program. The algorithm starts calculating SL of end nodes first and traversing whole graph upwards by updating SL of each node. Since we assume heterogeneous processor environment, we take the average $WCET$

(denoted by *WCET) for each task to account for variability in execution time on different processors.

The static level of a non-branching node is the maximum static level of its successors plus the *WCET of itself. Let $S(\tau_i)$ be the set of successor nodes of τ_i, equation (2) calculates the static level of a non-branching node.

$$SL(\tau_i) = *WCET(\tau_i) + \max SL(\tau_j), \tau_j \in S(\tau_i) \qquad (2)$$

The static level of a branch fork node is the mean of the static level of all its successors plus the *WCET of itself. Let c_{ij} denote the condition of edge (τ_i, τ_j), equation (3) calculates the static level of a branch fork node.

$$SL(\tau_i) = *WCET(\tau_i) + \sum_j prob(c_{ij}) * SL(\tau_j), \ \tau_j \in S(\tau_i) \qquad (3)$$

The main idea of our mapping and ordering algorithm is to find the most critical path in terms of execution cycles for each node while considering probability of execution for each path. The SL remains constant for each node once calculated. We also modified the calculation of Dynamic Level (DL) to account for the mutual exclusiveness among conditional tasks. The dynamic level of task processor pair (τ_i, p_j) can be calculated as the following.

$$DL(\tau_i, p_j) = SL(\tau_i) - AT(\tau_i, p_j) + \delta(\tau_i, p_j) \qquad (4)$$

The term $\delta(\tau_i, p_j)$ is the difference between *$WCET(\tau_i)$ and $WCET(\tau_i, p_j)$ which accounts for heterogeneous processor architecture. Adding this offset ensures correct evaluation of a task's DL for different processors since SL is computed using average $WCET$. $AT(\tau_i, p_j)$ is the earliest time that task τ_i can start on processor p_j. It must satisfy the following two conditions:

- At time $AT(\tau_i, p_j)$ all the data required by τ_i is available at p_j, i.e. $AT(\tau_i, p_j) \geq DA(\tau_i, p_j)$.

- If task τ_j is scheduled during the interval $[AT(\tau_i, p_j), AT(\tau_i, p_j) + WCET(\tau_i, p_j)]$, then τ_j and τ_i are mutually exclusive. This condition allows two mutually exclusive tasks to share the same processor at the same time, and thus making the schedule more efficient.

Computations and communications could be overlapped considering the availability of dedicated communication resource. Multiple data transfers from same node to different nodes are serialized provided the data values are different.

Our mutual exclusion detection procedure for each task is based on branch labeling method discussed in [5]. Considering example CTG of Figure 1 our algorithm detects tasks τ_6 and τ_7 to be mutually exclusive. Some other combinations are not mutually exclusive. For example, when condition a_2 is evaluated to be true, both τ_8 and τ_5 will be executed, and thus are not mutually exclusive. Our algorithm also detects the mutual exclusiveness among data transfers.

Figure 2 shows the flow diagram of our task ordering algorithm. The algorithm begins with the generation of initial ready list which has all start nodes. For each possible (τ_i, p_j), where τ_i is one of the tasks in the ready list, the algorithm calculates the static level and then finds the best pair that has the highest DL given by (4). Task τ_i is then scheduled on p_j at time $AT(\tau_i, p_j)$. Since the schedule of τ_i imposes new precedence order between τ_i and other tasks that are scheduled on the same processor, we also update the CTG to reflect this change. After that, the ready list will be updated and the above mentioned procedure will repeat until the ready list is empty.

Our algorithm that searches for $AT(\tau_i, p_j)$ is similar to the Find_AvailableTime() routine in [7]. However, in the Find_AvailableTime() routine, τ_i will be scheduled immediately after the first available time is found and the CTG will be updated. Our

algorithm simply returns the first available time without modifying the CTG because not all τ_i will be scheduled to its first available time. We will update the CTG after the best pair of (τ_i, p_j) is selected and scheduled. The term $AT(\tau_i, p_j)$ also captures communication scheduling. For example, in Figure 1 if τ_3 is scheduled before τ_2, comm(τ_1, τ_3) will be scheduled before comm(τ_1, τ_2) as the data transfers from τ_1 are serialized.

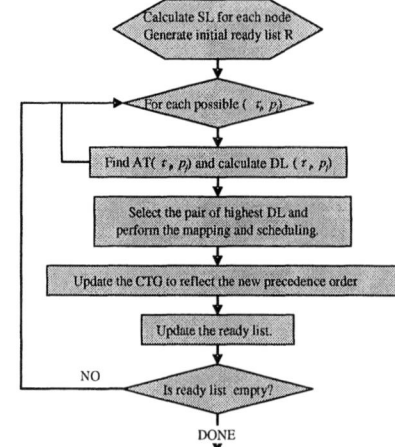

Figure 2 Task ordering algorithm flow.

Unlike the algorithm presented in [7], our task mapping and ordering algorithm is not condition aware. The mapping and ordering of each task will not change at runtime even if some conditional branch has been selected. The benefits of using a condition unaware scheduling algorithm are low computation complexity and less storage requirement, both of which are essential for online scheduling. The offline CAP algorithm is overall condition aware because it uses a condition aware task stretching algorithm. However, as we will show later, the online algorithm is condition unaware. Only one speed will be selected for each task and hence the storage of the scheduling results is simplified.

Another major difference between the proposed algorithm and the previous works [6][7] is that the proposed algorithm utilizes the profiled information of the branch probabilities and it reduces the average schedule length instead of the worst case schedule length.

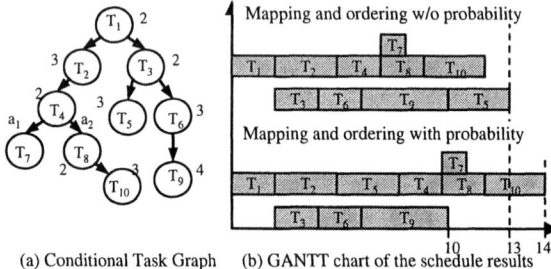

(a) Conditional Task Graph   (b) GANTT chart of the schedule results

Figure 3 Considering branching probabilities in task ordering

Example 2: Consider the CTG given in Figure 3 (a). The *WCET* of each task is given beside the circle. The branches a_1 and a_2 are taken with probability 0.9 and 0.1. Assume that the system has 2 PEs and the latency for inter-processor communication is negligible. Without considering the branching probability, T_4 has higher DL than T_5 and it will be mapped to PE1 and T_5 will not be scheduled until the very end. No matter which branch is taken, the overall schedule length is 13. With the branching probability, T$_5$ has higher DL than T$_4$. Therefore, it is mapped to PE1 before T4. The schedule length is 10 or 14 depending on whether the branch a_1 or a_2 is taken. Therefore, the average schedule length is 10.4. The reduced schedule length will be transformed to energy saving using task stretching. Table 3 (b) shows the GANTT chart of the schedule results with or without considering the profile information.

2) Task stretching

The task stretching routine finds the best starting time and speed of each task that minimizes energy while meeting the performance constraints. The problem is formulated and solved as a constrained linear program. The task stretching algorithm consists of three steps:

Step1: Preprocess the CTG to capture the implied control dependency between the or-node and its related branch fork nodes. As example 1 shows, the precedence requirements of an or-node is unknown until all of its related branch fork nodes have finished. Therefore, for an or-node τ_i and a condition c, if c is one of the literals in $\Gamma(\tau_i)$ and τ_j is the branch fork node of c, then an edge (τ_j, τ_i) must be inserted into the CTG.

Step 2: Duplicate the task graph. For each minterm $m \in M$, a task graph $G_m = (V_m, E_m)$ is created based on the CTG. A task $\tau_i \in V_m$, if τ_i is activated when m is true, i.e. $m \oplus X(\tau_i) \neq 0$. For two nodes τ_i and τ_j in V_m, if there is an edge (τ_i, τ_j) in the original CTG, then the same edge will be added in G_m.

Step 3: Formulate and solve the task stretching problem as a linear program (LP). For each task τ and a minterm $m \in \Gamma(\tau)$, two variables $\sigma_\tau(m)$ and $f_\tau(m)$ are defined. They represent the start time and speed of τ respectively when minterm m is true. The task stretching problem can be formulated into the following linear program:

$$\min \sum_{m \in M, m \neq 1} prob(m) \sum_{\tau \in V_m} \frac{E(\tau, p_\tau)}{F_\tau(m)^2} \quad \text{s.t.} \quad (5)$$

$$\sigma_{\tau_i}(m) + \frac{WCET(\tau_i, p_{\tau_i})}{F_{\tau_i}(m)} + \frac{Comm(\tau_i, \tau_j)}{B(p_{\tau_i}, p_{\tau_j})} \leq \sigma_{\tau_j}(m),$$

$$\forall m \in M, m \neq 1, \forall \tau_i, \tau_j \in V_m \text{ and } (\tau_i, \tau_j) \in E_m \quad (6)$$

$$\sigma_{\tau_i}(m) + \frac{WCET(\tau_i, p_{\tau_i})}{F_{\tau_i}(m)} \leq deadline,$$

$$\forall m \in M, m \neq 1, \forall \tau_i \in V_m. \quad (7)$$

In the above equations, p_τ is the processor that task τ is mapped to. $prob(m)$ is the probability that minterm m is true. It can be calculated based on the branch probability. The symbol $F_\tau(m)$ is $f_\tau(m)$, if $m \in \Gamma(\tau)$; otherwise, it is $f_\tau(m')$, where $m' \in \Gamma(\tau)$ and $m' \oplus m \neq 0$.

Equation (5) specifies that the objective of the LP is to minimize the mathematical expectation of the energy dissipation. Equation (6) and (7) specify the precedence constraints and deadline constraints of the tasks.

Example 2: Consider the CTG given in Figure 1. Because $\Gamma(\tau_8) = \{1, a_1\}$ and the branch fork node of condition a_1 is τ_3, the edge (τ_3, τ_8) is inserted to represent the control dependency between the or-node τ_8 and the branch fork node τ_3. Then for each minterm in M a task graph is generated, which consists of only the nodes that will be activated in the corresponding minterm. Figure 4 (a)~(c) show the

new task graphs. The probabilities of minterms are: $prob(a_1)=0.8$, $prob(a_2b_1)=0.1$, and $prob(a_2b_2)=0.1$.

The LP has 18 variables:

- $\sigma_1(1)$, $f_1(1)$, $\sigma_2(1)$, $f_2(1)$, $\sigma_3(1)$ and $f_3(1)$ are the start time and speed of task $\tau_1 \sim \tau_3$. They do not depend on which condition branch is selected.

- $\sigma_4(a_1)$ and $f_4(a_1)$ are the start time and speed of task τ_4 when branch a_1 is selected.

- $\sigma_5(a_2)$ and $f_5(a_2)$ are the start time and speed of task τ_5 when branch a_2 is selected.

- $\sigma_6(a_2b_1)$ and $f_6(a_2b_1)$ are the start time and speed of task τ_6 when branch a_2 and b_1 are both selected.

- $\sigma_7(a_2b_2)$ and $f_7(a_2b_2)$ are the start time and speed of task τ_7 when branch a_2 and b_2 are both selected.

- $\sigma_8(a_1)$ and $f_8(a_1)$ are the start time and speed of task τ_8 when branch a_1 is selected while $\sigma_8(1)$ and $f_8(1)$ are the start time and speed of task τ_8 when branch a_2 is selected.

Based on the definition of $F_\tau(m)$, the symbol $F_{\tau 8}(a_1)$, in equation (5)~(6) will be replaced by $f_{\tau 8}(a_1)$, while $F_{\tau 8}(a_2b_1)$ and $F_{\tau 8}(a_2b_2)$ will be replaced by $f_{\tau 8}(1)$.

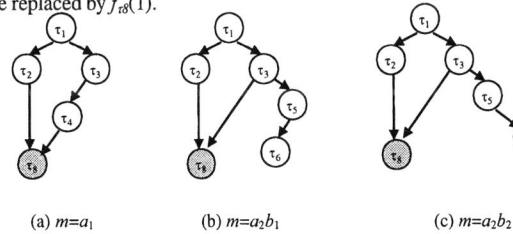

| (a) $m=a_1$ | (b) $m=a_2b_1$ | (c) $m=a_2b_2$ |

Figure 4 Task graph duplication.

B. Online CAP algorithm

Although solving a constrained linear program is a tractable problem, it is still time consuming. As we show in our experimental results, the runtime of the algorithm is significantly high especially for the graphs with many nodes.

As an alternative solution to the stretching problem we suggest a heuristic which performs task stretching with negligible runtime at the reasonable loss of energy saving compared to offline CAP algorithm. The heuristic is based on critical path based slack distribution algorithm considering conditional execution and is similar to the one proposed in [6]. However, there are few noteworthy changes. 1) The DVS techniques suggested in [6] and [7] have multiple speeds for a single task corresponding to different minterms. Therefore, the algorithms have a high complexity and a large schedule table is needed to store the stretching information. 2) The technique in [6] applies same stretching ratio at a time to all PEs, which is suboptimal. 3) The task stretching algorithm in [6] does not consider the branch probabilities. Unlike these techniques, our heuristic algorithm is an online approach in that it does not necessitate a schedule table to store the stretching information and use it at runtime. It is a profile-based approach considering branch probabilities. It calculates only single speed for each task and it can facilitate different scaling ratio for different PEs. Since the heuristic has very low complexity, the whole online CAP algorithm including task ordering and stretching can be called iteratively during runtime when branch probability changes significantly.

Once the CTG is updated, all possible paths in CTG are calculated using Breadth First Search (BFS) algorithm. Also associated with each path p is the slack and delay which are denoted as *slk(p)* and *delay(p)* respectively. Associated with each task τ on path p, there is a probability *prob(p, τ)*, which gives the probability of path p given the condition that task τ is started. *prob(p, τ)* is calculated as the joint probability of all the conditional branches lying on the path after node τ. For example, consider the example in Figure 1, the probability *prob(τ_1-τ_3-τ_5-τ_6, τ_5)=prob(b_1)*=0.5 because the only conditional branch along the path τ_1-τ_3-τ_5-τ_6 after node τ_5 is b_1.

Online CAP task stretching heuristic for CTG G

1. *Process initial schedule generated by task ordering algorithm shown in* Figure 2
2. *Calculate possible paths in CTG using BFS;*
3. *For each task τ_i {*
4.   *CalculateSlack (τ_i);*
5.   *Stretch τ_i, lock its schedule and speed;*
6.   *Update the delay and slack of all paths spanning τ_i ;*
7.   *Update the schedule for CTG G;*
    }

CalculateSlack (τ_i)

1. *For each minterm $m \in \Gamma(\tau_i)$ {*
2. *For all paths of $m \in \Gamma(\tau_i)$ that span node τ_i {*
3.   *Find the critical path p_{worst} where $prob(p_{worst}, \tau_i) \neq 1$*
4.   *$slk1 += prob(p_{worst}, \tau_i) * wcet(\tau_i) *(slk(p_{worst}) / delay(p_{worst})) * prob(\tau_i);$*
    }
  }
5. *For each path of $m \in \Gamma(\tau_i)$ where $prob(m) = 1$*
6. *Find the critical path t_{worst} ;*
7. *$slk2 = wcet(\tau_i) * (slk(t_{worst}) / delay(t_{worst})) * prob(\tau_i);$*
8. *$slk(\tau_i) = min [slk1, slk2];$*
9. *If there is a path p that spans node τ_i and $slk(\tau_i)>deadline-delay(p)$*
10. *$slk(\tau_i)=deadline-delay(p);$*

Figure 5 Online CAP task stretching heuristic.

For each task, step4 in Figure 5 determines the available slack by calling CalculateSlack(τ_i) routine. This routine finds the most critical path that has a minimum slack applicable to task τ_i. In case of multiple paths pertaining to different minterms with probabilities less than 1, first the critical path that has the lowest distributable slack ratio ($slk(p)/delay(p)$) and has the probability less than 1 is identified for each minterm. After that the initial slack of τ_i is taken as a probability weighted sum of all these critical path slacks corresponding to each minterm $m \in \Gamma(\tau_i)$ as shown in step 4 of routine. Note that the weight for each path is $prob(p, \tau_i)$, which is the probability of path p given the condition that task τ_i is started. One more slack value is calculated as shown in step 7 for critical path with probability equal to 1. It is noteworthy that both slack values are further weighted by the activation probability of node τ_i. More slack will be allocated to the task that has higher probability to be activated. The slack of τ_i is now minimum of these two slack values. Because the slack is the average for all possible minterms, at the end of the routine, we need to check for each path that the deadline can be met otherwise, the slack will be adjusted.

Once slack is calculated for a task, the task is stretched and its schedule and speed are locked. Next, all paths that span this task are updated in terms of their respective delay and slack. Updating these variables dynamically alters the criticality of paths for different nodes and subsequently releasing the tasks, that have been stretched, from consideration. The online CAP algorithm then updates CTG and repeats the above mentioned procedure for another task following the task order generated by offline CAP.

Given a CTG with total nodes |V| and edges |E|, the time complexity of the online stretching heuristic (derivation not shown here due to space limitation) is O ($2|V|^3 + C|V| + |E|$) where constant C is an upper bound on number of outgoing edges from a node. This low complexity enables the algorithm to be used for dynamic scheduling in a system with the capability of runtime branch prediction.

IV. EXPERIMENTAL RESULTS

Simulations have been carried out to evaluate the efficiency of the proposed algorithm. Six scheduling algorithms are evaluated in the experiments. They are: offline CAP, CAP w/o PM (offline CAP without profile-based mapping), CAP w/o PS (offline CAP without profile-based stretching), online CAP w/o PM, online CAP and Reference algorithm. We implemented the scheduling algorithm in [7] according to the best of our ability and denote it as Reference algorithm. The Reference scheduling does not consider profile information in task ordering and it assumes fixed task mapping. Table 1 summarizes the characteristics of the 6 scheduling algorithms. For all the experiments, we consider the execution of the CTG as a random process and we report the mathematical expectation of energy dissipation.

Table 1 Difference of the evaluated scheduling algorithms.

Algorithm	Task ordering		Task stretching		Flexible mapping
	Profile based	Condition aware	Profile based	Condition aware	
Offline CAP	X		X	X	X
CAP w/o PM			X	X	X
CAP w/o PS	X			X	X
Online CAP w/o PM			X		X
Online CAP	X		X		X
Reference		X	X	X	

Firstly, a real life example of a vehicle cruise controller system [12] is experimented for different algorithms. This application is modeled as a CTG with 32 tasks and two branching nodes, each of them forking two conditions. The original cruise control model in [12] did not have probability information, so it is assigned randomly in our experiment. We tested this CTG with fix mapping and no communication between tasks. The application is mapped on five different PEs. The schedule length of 124ms reported by CAP was same as Reference algorithm and the best length reported in [12]. The energy consumption is 102 mJ for both CAP offline and online and same for the Reference algorithm. The results favor the CAP online approach for this application due to its low complexity and equal energy compared to other algorithms.

Next, five test cases are randomly created with different CTGs and different MPSoC architecture. The CTGs are modified from the random task graphs generated by TGFF [11]. The MPSoC architecture consists of either 3 or 4 PEs.

Table 2 Fixed mapping and zero communication cost.

CTG	a/b/c	Offline CAP	CAP w/o PS	Online CAP	Reference
1	25/3/3	596	1414	723	874
2	16/3/1	591	769	641	568
3	15/4/2	223	490	292	215
4	15/4/1	386	630	497	386
5	25/4/3	285	1075	498	501

The first experiment focuses on demonstrating the effectiveness of our task ordering algorithm. The same fixed task mapping is used in both Reference and CAP algorithms. The communication cost is also set to be 0. Table 2 shows the energy dissipation of different scheduling algorithms. Second column of Table 2 displays the characteristics of the CTGs we used. We use a triplet (*a/b/c*) to characterize a test case where *a* represents the number of nodes in the CTG, *b* represents the number of PEs in the MPSoC and *c* represents the number of conditional branching nodes in the CTG. These five

CTG IDs shown in column 1 are used to report results for all experiments in rest of this paper. In average, the CAP has 13% energy reduction over the reference scheduling and 49% energy reduction over the CAP w/o PS. The online CAP has 11% more energy than the Reference algorithm. This is expected since the potential of the DLS based algorithm is known to unfold with flexible mapping.

The second experiment focuses on demonstrating the effectiveness of our task mapping algorithm. In this experiment, the CAP based algorithms perform task mapping together with task ordering. The communication cost is again set to 0. Table 3 shows the energy dissipation of different scheduling algorithms. As we can see, with flexible task mapping, the offline CAP gives more than 43% energy reduction over the Reference algorithm. The improvement is more significant than the first experiment. The difference between offline CAP versus CAP w/o PS is now 46%. The online CAP has an average of 39% of energy reduction over the Reference algorithm. Compared to offline CAP, the online CAP has 8% more energy dissipation. Average energy dissipation of offline CAP and CAP w/o PM is almost same while online CAP has 7% less energy than online CAP w/o PM. This is because offline CAP tries to allocate slack based on critical path while online CAP stretching heuristic calculates slack based on average path length.

Table 3 Flexible mapping and zero communication cost.

CTG	Offline CAP	CAP w/o PS	CAP w/o PM	Online CAP w/o PM	Online CAP	Reference
1	389	1014	382	459	449	874
2	365	498	365	393	393	568
3	158	339	144	187	166	215
4	252	353	260	325	277	386
5	167	422	160	183	173	501

Table 4 shows the comparison of runtime between the offline CAP and online CAP. Note that the unit of the runtime for offline CAP is second while the unit for the online CAP is millisecond. The average speedup of the online CAP is 120,000X. Although the time reported here includes task ordering, mapping and stretching, because both the online and offline algorithms use the same task ordering/mapping routine, the difference in runtime comes only from task stretching step.

Table 4 Comparison of runtime.

CTG	1	2	3	4	5
Offline CAP (sec)	72.69	66.41	37.06	24.11	147.28
Online CAP (msec)	1.26	0.59	0.37	0.31	0.58

Table 5 Flexible mapping and non-zero communication cost.

CTG	Offline CAP	CAP w/o PS	CAP w/o PM	Online CAP w/o PM	Online CAP	Reference
1	422	1146	487	534	508	1396
2	374	522	441	469	404	1206
3	194	592	237	225	211	356
4	413	615	455	546	469	805
5	218	620	254	250	242	777

The last experiment focuses on demonstrating the communication aware capability of our algorithm. In this experiment, we generate the communication volume from a node to its successor based on Computation to Communication ratio (CCR).For this experiment, we chose a value of CCR=1 with some variance (10%) to model data transfer from a node to multiple nodes. Table 5 shows the energy dissipation of different scheduling algorithms. Since the communication cost is non-zero, we can see that the energy dissipation for all test cases increases. However, the energy reduction

of the offline CAP scheduling over the Reference scheduling also increases compared with previous two experiments. The average energy reduction is now 60%. The differences between offline CAP versus CAP w/o PS increases to 51% in this case. The online CAP has an average of 56% energy reduction than the Reference algorithm. Compared to offline CAP, the online CAP has 12% more energy dissipation. The offline CAP has now 14% less energy than CAP w/o PM and the improvement of online CAP over online CAP w/o PM increases to 8%.

V. CONCLUSIONS

Online and offline algorithms are proposed that perform simultaneous task mapping and ordering followed by task stretching of a conditional task graph (CTG). The algorithms minimize the mathematical expectation of energy dissipation of non-deterministic applications with random branch selection by utilizing the task execution profile. Both communication and computing energy are reduced in the scheduled result. The experimental results show that, comparing with the previous scheduling algorithm, our offline algorithms give more than 39% energy reduction in average. The online algorithm gives more than 28% energy reduction in average with a speed up of 120,000X over offline algorithm. Our future efforts target a development of an adaptive version of CAP online algorithm that can fit the changing runtime system conditions and utilize the low complexity of CAP. Considering contentions inside communication network would also be an interesting analysis.

REFERENCES

[1] J. Luo and N. K. Jha, "Static and Dynamic Variable Voltage Scheduling Algorithms for Real-time Heterogeneous Distributed Embedded Systems," *Proceeding Of International Conference on VLSI Design*, pp.719-726, 2002.

[2] Y. Zhang, X. Hu, and D. Z. Chen, "Task Scheduling and Voltage Selection for Energy Minimization," *In Proc. Of Design Automation Conference*, pp.183-188, 2002.

[3] J. Hu and R. Marculescu, "Energy-Aware Communication and Task Scheduling for Network-on-Chip Architectures under Real-Time Constraints," *Proceeding of Conference and Exhibition on Design, Automation and Test in Europe*, 2004.

[4] P. Eles, K. Kuchcinski, Z. Peng, A. Doboli, and P. Pop, "Scheduling of Conditional Process graphs for the Synthesis of Embedded Systems," *Proceedings of Design, Automation and Test in Europe*, 1998.

[5] Y. Xie and W. Wolf, "Allocation and Scheduling of Conditional Task Graph in Hardware/Software Co-synthesis," *Proceedings of Conference and Exhibition on Design, Automation and Test in Europe*, 2001.

[6] D. Wu, B.M. Al-Hashimi and P. Eles, "Scheduling and Mapping of Conditional Task Graph for the Synthesis of Low Power embedded Systems," *IEE Proceedings of Computers and Digital Techniques*, Volume 150, Issue 5, pp. 262-273, Sept. 2003.

[7] D. Shin and J. Kim, "Power-Aware Scheduling of Conditional Task Graphs in Real-Time Multiprocessor Systems," *Proceedings of International Symposium on Low Power Electronics and Design*, 2003.

[8] E. Jacobsen, E. Rotenberg, and J.E. Smith, "Assigning confidence to conditional branch predictions," *Proceedings of the 29th Annual International Symposium on Microarchitecture*, Nov. 1996.

[9] A. K. Uht and V. Sindagi, "Disjoint Eager Execution: An Optimal Form of Speculative Execution," *Proceedings of the 28th Annual International Symposium on Microarchitecture, Nov. 1995*.

[10] G.C. Sih and E.A. Lee. "A Compile Time Scheduling Heuristic for Interconnection-Constrained Heterogeneous Processor Architecture," *IEEE Transactions on Parallel and Distributed Systems*, Volume 4, Issue 2, Page(s):175 – 187, Feb. 1993.

[11] R. P. Dick, D. L. Rhodes, and W. Wolf, "TGFF: Task graphs for free," *Proc. of Int. Workshop Hardware/Software Codesign*, Mar. 1998.

[12] Paul Pop, "Scheduling and communication synthesis for distributed real-time systems", *Ph.D. thesis, Linkopings University,2000*.

A Minimum-Latency Block-Serial Architecture of a Decoder for IEEE 802.11n LDPC Codes

Massimo Rovini, Giuseppe Gentile, Francesco Rossi and Luca Fanucci

Department of Information Engineering - University of Pisa

Via G. Caruso, I-56122 Pisa - Italy

{*massimo.rovini, giuseppe.gentile, francesco.rossi, luca.fanucci*}@iet.unipi.it

Abstract—This paper describes a scalable architecture of a decoder for IEEE 802.11n low-density parity-check (LDPC) codes. The decoder runs the layered decoding algorithm and its architecture is arranged in clusters of serial functional units, which are configured to process all codes in the standard. The decoder works in pipeline, and a very effective technique to re-arrange the sequence of its elaborations is proposed in order to minimize the iteration latency; this relates to the order of the messages input and output by the processing units, as well as the sequence of layers followed for decoding. Moreover, memory optimization techniques have been applied to get a very efficient partitioning, allowing the pipeline of the operations. The synthesis on 65 nm CMOS technology with low-power standard-cell library, shows that the proposed design is suitable for portable devices, the throughput ranging from 136 to 355 Mbps, and the power consumption being below 185 mW.

I. INTRODUCTION

After their recent re-discovery by MacKay and Neal [1], low-density parity-check (LDPC) codes [2] have quickly gained the momentum of the scientific and industrial community, due to their remarkable performance (even at low signal-to-noise ratios (SNR) and for small codeword sizes), to the flexibility in the design of the code parameters, and to the relatively simple decoding algorithm, suitable for parallelization. For such reasons, the physical layer interface of modern high-throughput communication standards is more and more based on LDPC codes for advanced forward error correction.

In line with the trend started by DVB-S2 [3], and followed by wireless metropolitan area networks (WMAN) [4], wireless local area networks (WLAN) are also converging on *structured* LDPC codes for high-throughput applications, desirably reaching data-rate up to 100 Mbps. Although the standard is still being defined by the IEEE 802.11n working group, draft amendments have recently been issued [5], and preliminary designs of the related decoder are already available [6], [7].

This paper describes a very-efficient architecture of a decoder for IEEE 802.11n LDPC codes. As mentioned above, *structured*, or architecture-*aware* codes (AA-LDPC [8]) are used by all standards above as an effective means to reduce the area and power consumption of the related decoder, to improve its memory efficiency, and to get increased decoder scalability; furthermore, as shown in [8], AA-codes perform very close to random codes.

In this work we describe a scalable, semi-parallel or block-serial architecture of a layered LDPC decoder [9], based on serial processing units. In power and area constrained applications, such as for mobile devices, a block-serial architecture that serially processes *blocks* of parity-check constraints, is preferred to fully parallel implementations [10]; moreover, serial processing units are very flexible in the support of different and irregular codes, such as the 802.11n ones.

As a drawback, block-serial implementations might dramatically increase the decoding latency, so that are typically operated in pipeline. However, in a pipelined layered decoder, the access to the memory units must be properly scheduled, and the dependence between elaborations carefully followed; as an improvement to similar state–of–the–art works [6], [11], [12], this paper proposes a systematic technique to schedule the order of the messages, by acting on the sequence of block-rows updated by the functional units, as well as on the (input and output) order of block-columns within block-rows.

II. LOW-COMPLEXITY LAYERED DECODING

Decoding of LDPC codes is achieved by applying the maximum *a posteriori* (MAP) algorithm, commonly arranged in the form of *belief propagation* (BP). BP is optimal when the code does not contain cycles, but it is still used and considered as a reference for practical codes with cycles. In this case, the sequence of the elaborations, also referred to as *schedule*, highly affects the performance in terms of convergence speed and (although less) error correction.

Following the work of Mansour and Shanbhag [8], BP has recently been re-formulated in a very effective schedule, known as *shuffled* or *layered* decoding [9], [13]–[15]. The underlying principle is to decode a code as a sequence of several super-codes, or *layers*, which exchange *a posteriori* reliability messages, as in a common MAP algorithm. Layered decoding is preferable to the original *flooding* schedule (FS) [14] because of the increased convergence speed, which is almost doubled, both for codes with cycles and cycle-free [16]. This is achieved through the early propagation of intermediate, *a posteriori* messages, made available to the next layers immediately after their computation, and not only at next iteration as in FS.

Although layers could be arranged both by check-nodes (CNs) and variable-nodes (VNs) (see [13] and [14]), CN-*centric* or *horizontal* solutions are desirable since they allow serial, flexible and simpler architectures of the check-node unit (CNU). Horizontal layered decoding (HLD) iteratively updates the parity-checks in a layer according to:

$$-\,\mathrm{sign}\left(\epsilon_{m,n}^{(q+1)}\right)=\prod_{j\in\mathcal{N}(m)\backslash n}-\,\mathrm{sign}\left(y_j-\epsilon_{m,j}^{(q)}\right)\quad(1)$$

$$\left|\epsilon_{m,n}^{(q+1)}\right|=\Phi^{-1}\left\{\sum_{j\in\mathcal{N}(m)\backslash n}\Phi\left(\left|y_j-\epsilon_{m,j}^{(q)}\right|\right)\right\}\quad(2)$$

with y_n, $n=0,1,...,N-1$ the *total information* or *soft-output* (SO) associated to the bits in the codeword, $\mathcal{N}(m)$ the sets of the bits involved in parity-check m, $\epsilon_{m,n}$ the check-to–variable (c2v) reliability message from CN m to VN n, $\Phi(x)=\Phi^{-1}(x)=-\log(\tanh(x/2))$, and q the iteration index. Using the results of (1) and (2), all involved SOs are updated by summing the input variable–to–check (v2c) message and the new c2v message, i.e., $y_n=(y_n-\epsilon_{m,n}^{(q)})+\epsilon_{m,n}^{(q+1)}$, with $n\in\mathcal{N}(m)$; then, SOs are made available to the next layers, and faster convergence is achieved. The algorithm is initialized with the channel *a priori* log-likelihood ratios (LLRs) $y_n=\lambda_n$, $n=0,1,...,N-1$.

Magnitude update in (2) can be recursively reworked in several low-complexity approximations, very suitable for serial implementations. A good trade-off between complexity and performance was found in the *2-output* M-min* algorithm described in [17] with $P=1$. In this case, only two values are computed, namely:

$$\begin{cases}|\epsilon_{m,j}^{(q+1)}|=\text{M-min}^*_{j\neq j_{\min}}\left\{\left|y_j-\epsilon_{m,j}^{(q)}\right|\right\}\doteq\epsilon'_m & \text{if } n=j_{\min}\\ |\epsilon_{m,j}^{(q+1)}|=\text{M-min}^*\left(\epsilon'_m,\left|y_{j_{\min}}-\epsilon_{m,j_{\min}}^{(q)}\right|\right)\doteq\epsilon_m & \text{otherwise}\end{cases}$$

where $j_{\min}$ is the index of the v2c message $y_j-\epsilon_{m,j}^{(q)}$ with the smallest reliability, which is updated with ϵ'_m; then, one common metric ϵ_m is propagated to the remaining VNs ($n\neq j_{\min}$).

III. IEEE 802.11N AA-LDPC CODES

IEEE 802.11n LDPC codes proposed in [5] are designed for three codeword lengths, $N_0=648$, $N_1=1296$ and $N_2=1944$, referred to as short, middle and long codeword in this paper. Then, for each length, four code rates are specified, $1/2$, $2/3$, $3/4$ and $5/6$, for a total of 12 different codes. These are AA-LDPC codes, block-wise partitioned into smaller $B_i\times B_i$ sub-matrices, or *blocks*, with $B_0=27$, $B_1=54$ and $B_2=81$ for the short, middle and long codeword, respectively.

Particularly, the parity-check matrix $\mathbf{H}$ is arranged in $n=N_i/B_i=24$ block-columns and $m_i=(1-r)\cdot n$ block-rows, with r the code rate; each sub-matrix is either a rotation of the identity matrix or an all-zero matrix. This allows the very compact description of $\mathbf{H}$, known as *prototype* matrix, by only specifying the locations and rotations of the non-zero blocks. An example of prototype matrix is shown in Fig. 1.

IV. DECODING PIPELINING & IDLING

To work properly, practical implementations of HLD must respect dependence between layers, and always use the most updated total information. However, since the elaborations of a

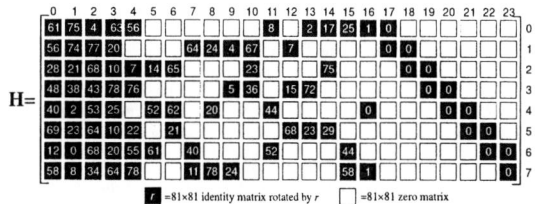

Fig. 1. Parity-check prototype matrix for codeword size $N_2=1944$ and rate $r=2/3$. Black squares correspond to rotations of the identity matrix, also indicated in the square, while empty squares correspond to all-zero matrices.

serial check-node unit CNU are typically pipelined, a SO could be accessed before being updated, thus spoiling performance. This problem can be solved with the insertion of *null* or *idle* cycles between two consecutive layers [12], where the CNU is halted to wait for updated data. In opposition to this route, rearrangements of HLD are also viable, as shown in [11] and [18], but extra complexity is brought into the design.

As a matter of fact, the number of idle cycles heavily impacts the decoder performance in terms of latency, and then throughput; hence, it is highly desirable to minimize its value. The time spent in idling strictly relates to the number of SOs shared between consecutive layers, and the use of an ad-hoc sequence of layers could help to relax the problem.

In this work, we propose a very efficient technique to arrange the order of the messages updated in a layer, and the sequence of layers followed by the decoder, in order to minimize the overall number of idle cycles, and accordingly, the decoding latency.

A. Unconstrained Output Processor

Focusing on serial processing units, one of the main factors increasing the number of idle cycles in a HLD decoder, is the presence of any type of correlation between the order used to input and output SO messages. Typically, they are delivered with the same order they were acquired, but better results are expected if this constraint is removed and the input and output sequences are uncorrelated.

To formalize the problem, let us define:

i. $\mathcal{S}_k$ the set of SOs updated on layer k, with size $d_k=|\mathcal{S}_k|$;

ii. $\mathcal{A}_k=\mathcal{S}_k\cap\mathcal{S}_{k-1}$, the subset of SOs updated on layer k in common with layer $k-1$, with $\alpha_k=|\mathcal{A}_k|$;

iii. $\mathcal{B}_k=\mathcal{S}_k\cap\mathcal{S}_{k+1}$ the subset of SOs of layer k in common with layer $k+1$, with $\beta_k=|\mathcal{B}_k|$;

iv. $\mathcal{E}_k=\{\mathcal{S}_k\cap\mathcal{S}_{k-2}\}\backslash\mathcal{S}_{k-1}$ the messages common to layer $k-2$ but *not* to layer $k-1$, with $\epsilon_k=|\mathcal{E}_k|$;

v. $\mathcal{F}_k=\{\mathcal{S}_k\cap\mathcal{S}_{k+2}\}\backslash\mathcal{S}_{k+1}$ the messages common to layer $k+2$ but *not* to layer $k+1$, with $\phi_k=|\mathcal{F}_k|$;

The I/O data streams of an *unconstrained* output processor (UOP) are shown in Fig. 2, where L_{SO} is the latency of the SO data path, measured in clock cycles[1]. The CNU is designed to take the messages in $\mathcal{A}_k$ later, and, to deliver the messages

---

[1]As shown in Fig. 4, L_{SO} includes the delay of the CNU, the shifting network and the two accesses to memory. Typical values are 4-6 clock cycles.

in $\mathcal{B}_k$ earlier. In this way, more time is given to the previous layer to complete its update, while the next layer receives its messages earlier.

Although this strategy solves the dependence between two consecutive layers, the same issue may arise between three consecutive layers. Indeed, if the set $\mathcal{F}_k$ is output soon after $\mathcal{B}_k$, as shown in Fig. 2, and at acquisition time, $\mathcal{E}_k$ is taken just before $\mathcal{A}_k$, then there will be no chance for layer $k+2$ to get not-updated messages. To summarize, the number of idle cycles $\mathcal{I}'_k$ needed between layers k and $k+1$ for a fully pipelined decoding, must be such that:

$$\mathcal{I}'_k + |S_{k+1}\backslash\mathcal{A}_{k+1}| \geq L_{\text{SO}} \tag{3}$$

and the minimum number of idle cycles is:

$$\mathcal{I}'_k = L_{\text{SO}} - (d_{k+1} - \alpha_{k+1}) \tag{4}$$

Note that messages in $\mathcal{A}_k$ and $\mathcal{E}_k$ must be acquired with the same order they were written in memory by previous layers, i.e., the order used to output $\mathcal{B}_{k-1}$ and $\mathcal{F}_{k-2}$, respectively.

Furthermore, a serial CNU can not process consecutive layers with decreasing degrees: in this case, the outputs of the two layers would overlap in time, with corruption of internal data. Again, idle cycles in the number of $\Delta d_k = d_k - d_{k+1}$ might be inserted between the two layers to solve the issue; thus, the overall number of idle cycles is:

$$\mathcal{I}_k = \max\{\mathcal{I}'_k, \Delta d_k, 0\} \tag{5}$$

where the result of (4) is further constrained to be non-negative.

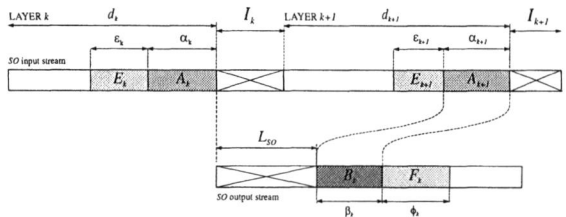

Fig. 2.   Input and output data streams of a CNU with UOP.

B. Optimum Sequence of Layers

The system was then optimized in terms of the actual sequence of layers followed by the decoder, to minimize the idle time. To this aim, let us define the following cost function:

$$C(p) = \sum_{k=0}^{m_i-1} \mathcal{I}_k \tag{6}$$

where m_i is the number of block-rows or layers in the code, and $\mathcal{I}_k$ is computed according to (5). The optimum sequence of layers is the one minimizing (6), i.e., $\hat{p} = arg\min_{p\in\mathcal{P}} C(p)$, with $\mathcal{P}$ the set of all possible permutations of m_i layers.

Note that, in view of different sequences of layers, different error rates and convergence speeds should be expected; nevertheless, simulations showed only negligible variations in performance.

V. DECODER ARCHITECTURE

As structured codes, 802.11n LDPC codes allow a considerable reduction of the decoding latency through the parallel update of at most B_i parity-checks, grouped in a layer. Moreover, since $B_0 = 27$ is the greatest common divisor between the three block sizes B_0, B_1 and B_2, the decoder can be conveniently arranged into several data-paths, managing an array of B_0 data each. Figure 3 shows a scalable decoder architecture made of K independent data-paths, merged in a multi-size circular-shifting (MS-CS) network to perform the parallel rotation over an array of B_i messages. To cope with all block sizes, at least three data-paths ($K = 3$) are needed; however, aiming at high throughput implementations, the number of data-paths could be increased to decode several codewords in parallel. More in detail, when a burst of codewords with block size B_i is received, then $F_i = \lfloor K \cdot B_0/B_i \rfloor$ frames can be decoded in parallel, thus using $F_i \cdot (B_i/B_0)$ data-paths.

The decoder interface is designed to receive $B_{\text{IO}} = B_0$ LLR data per clock cycle and, the I/O unit (input buffer FSM in Fig. 3) properly feeds the data-paths according to the actual code parameters. Finally, the control unit stores the code descriptors and supervises all decoding operations.

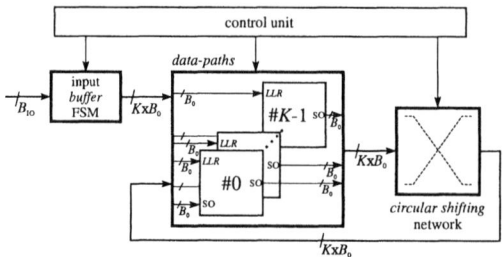

Fig. 3.   Scalable top-level architecture of the LDPC decoder.

The architecture of a single data-path, shown in Fig. 4, is composed of a cluster of $B_0 = 27$ CNUs plus dedicated logic and memories. An array of B_0 c2v messages, directly output by the CNUs, is stored in a row of the related c2v memory, while a similar array of B_0 SOs, computed in the SO adder array and shuffled by the MS-CS network, is not immediately stored but is first marginalized through the V2C adder array; in this way, similarly to [6], the array of v2c related to the next update of the current block-column is computed and stored in the related v2c memory, ready for use. As a result, the c2v memory stores up to $Q \cdot B_0$ messages with $Q = 88$ the maximum number of non-null sub-matrices among the 12 codes, while the v2c RAM always contains $(N_i/B_i)\cdot B_0 = 648$ data.

Figure 4 also shows that a pipelined implementation of the decoder needs three concurrent accesses to the v2c memory. In particular, two reading operations are needed to feed the CNUs and the SO adder array, and one writing is performed to update the v2c memory. Some memory compilers for modern ASIC technologies are already able to generate three-port memories, but, in this work, the v2c memory was built upon standard

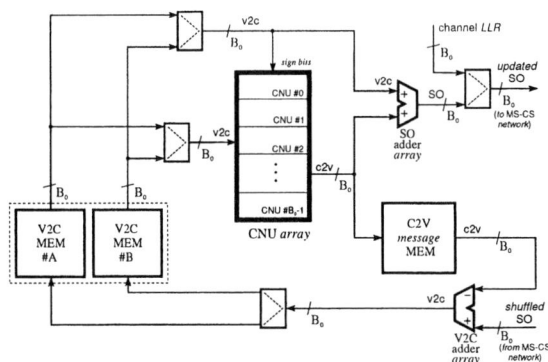

Fig. 4.   Architecture of the single data-path.

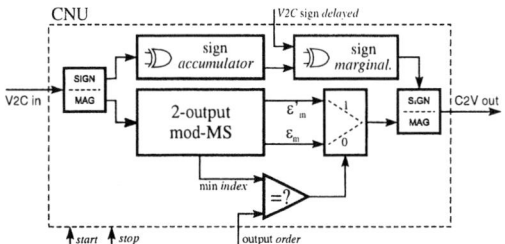

Fig. 5.   Architecture of the serial CNU for the UOP strategy.

dual-port memories, thus keeping a high level of portability between different technologies, from FPGA to almost every standard-cells library.

More in detail, the memory should be partitioned in a proper number of dual-port banks, such that two concurrent reading operations always retrieve data from different banks. It follows that the greater is the number of banks, the higher is the overhead due to the additional logic (control and selection of the banks), as well as the possible waste of memory cells during the technology mapping onto physical cuts with different size.

As an improvement of [11], where the memory is just mirrored, a good trade-off was found in partitioning the v2c memory into two banks (labeled A and B in Fig. 4) and in allowing, if necessary, redundancy of data: some messages are stored in both banks in order to solve those unavoidable reading conflicts. To minimize the redundancy, an optimization procedure based on a cycle-true model of the decoder was run, resulting in a maximum overhead of 33%.

A. Pipelined Check Node Unit Architecture with UOP

The architecture of the CNU is shown in Fig. 5. This is a serial unit working in pipeline and implementing the *2-output* mod-MS algorithm of Sect. II. Once the two c2v magnitudes and the index of the least reliable input are available, the output stream is re-built according to the UOP strategy by providing to the CNU the actual order of the output data (*output order*) and the sequence of signs for marginalization (*V2C delayed sign*), taken from the v2c memory.

Idle cycles are managed with two synchronization strobes, named *start* and *stop*. The former notifies that the first data of a layer is available on the input of the CNU, while the latter is asserted when all input data have been acquired. The clock cycles between *stop* and *start* are then idle cycles and during this time, the acquisition of new data is suspended.

B. Multi-Size Circular-Shifting Network

The MS-CS network allows the scalability of the decoder architecture and provides the means to manage the three different block sizes with K independent data-paths of size

B_0 each. In a multi-frame architecture, the MS-CS network is not only capable of rotating a single vector of messages with variable size B_i, but it can also do it on F_i vectors in parallel.

The network is arranged into K barrel shifters, feeding an *adaptation* network (AN). Each shifter works independently and shuffles an array of B_0 data with a cascade of $\rho = \lceil \log_2(B_0) \rceil$ stages of B_0 2-to-1 *multiplexers*; a single stage can either work as a short circuit or rotate data by 2^r, with $r = 0, 1, \ldots, \rho - 1$ the index of the stage. This turns out in a very simple control law, requiring only one control bit per stage.

The AN is in charge of further shuffling the data output from the shifters and is composed of a single stage of $K \cdot B_0$ K-to-1 *multiplexers*; the control of the AN is crucial, and is based on information about the block size B_i and the current rotation R_i to properly configure all the *multiplexers* and implement the multi-frame and multi-length functionality.

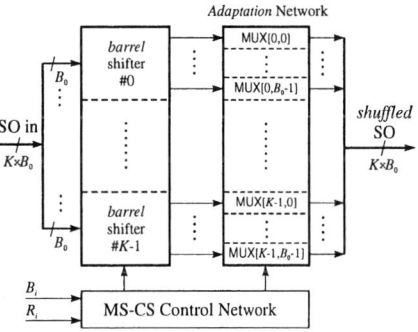

Fig. 6.   MS-CS network architecture.

VI. IMPLEMENTATION RESULTS

A. Throughput and Decoding Latency

Table I shows the throughput in information bits and the decoding latency achieved with the UOP strategy and 12 iterations. The total number of idle cycles and the related decoding latency in clock cycles are reported both before ($\mathcal{I}_{orig}$, $L_{dec,orig}$) and after ($\mathcal{I}$, L_{dec}) the optimization with the UOP strategy. As a result, a remarkable saving in the decoding time, in the range of 20-46%, is achieved.

Unlike the decoding latency, the throughput also depends on the architectural parallelism K and roughly scales linearly

TABLE I

LDPC DECODER PERFORMANCE WITH UOP, $L_{SO} = 5$ AND 12 ITERATIONS AT THE CLOCK FREQUENCY OF 240 MHZ.

r	$N = 648$				$N = 1296$				$N = 1944$			
	1/2	2/3	3/4	5/6	1/2	2/3	3/4	5/6	1/2	2/3	3/4	5/6
L_{dec}	1312	1220	1247	1384	1191	1172	1199	1264	1263	1220	1198	1168
$\mathcal{I}$	95	0	24	153	47	0	24	117	71	0	11	47
$\mathcal{I}(\%)$	7.2	0	1.9	11.1	4	0	2	9.3	5.6	0	1	4
$L_{dec,orig}$	2453	1922	1912	1517	2353	1874	2033	1711	2349	1990	1943	1706
$\mathcal{I}_{orig}$	1236	702	689	286	1209	702	858	564	1157	770	756	585
$\mathcal{I}_{orig}(\%)$	50.4	36.5	36	18.9	51.4	37.5	42.2	33	49.3	38.7	38.9	34.3
$L\,(\mu s)$	5.67	5.65	5.98	5.68	5.9	5.65	5.97	5.96	5.75	5.87	5.95	5.85
Γ (Mbps) $(K = 4)$	246	354	390	389	262	354	390	411	196	271	311	355
Γ (Mbps) $(K = 3)$	188	271	298	297	136	185	203	214	196	271	311	355

with the number of frames F_i decoded in parallel. Aiming at a low-complexity design, K is set to 3, thus boosting the throughput of the short codeword codes only. Anyway, for the sake of comparison, Tab. I reports the throughput Γ both for $K = 3$ and $K = 4$ at the clock frequency of 240 MHz.

B. Error Correction Performance

The fixed-point precision analysis of the decoder was carefully performed to achieve a good trade-off between error correction performance and design complexity. Given N_{c2v} bits to represent *c2v* messages, *SO*s need $N_{SO} = N_{c2v} + \lceil \log_2 (d_{v,\max}) \rceil - p$ bits, where $d_{v,\max}$ is the maximum VN degree and p bits are dropped for a reduced complexity. Preliminarily, the impact on performance of the *c2v* precision was evaluated by assuming $p = 0$. Simulations showed that $N_{c2v} = 6$ results in no implementation loss (IL) w.r.t. the floating-point reference model, while the use of $N_{c2v} = 5$ yields about 0.15 dB IL at the frame error rate (FER) of 10^{-4}. Since the design complexity roughly scales linearly with N_{c2v}, we considered $N_{c2v} = 5$ for implementation.

Figure 7 shows the FER performance of the long codeword codes with $N_{c2v} = 5$ and p ranging from 0 to 3; targeting the FER to 10^{-4}, a good trade-off for all codes was found in using 7 bits for *SO* messages (see curves 5-7), which corresponds to $p = 2$ for $r = 1/2$, $p = 1$ for $r = 2/3$ and $3/4$, and $p = 0$ for $r = 5/6$. Finally, the precision of input LLR (N_{LLR}) and *v2c* messages (N_{v2c}) is strictly related to that of *c2v* and *SO* respectively, and we set $N_{LLR} = N_{c2v}$ and $N_{v2c} = N_{SO}$.

C. Synthesis Results

The scalable LDPC decoder was synthesized on 65 nm CMOS technology, using low-power, 1.20 V standard-cell library at the clock frequency of 240 MHz and Tab. II shows the complexity and power consumption breakdown for logic and memories. Here, the power refers to the rate 1/2, $N = 648$ code, simulated at three different SNRs.

The LDPC decoder makes extensive use of configuration ROMs to store information about all 802.11n codes, to properly manage the two banks of the *v2c* memory and to re-order the CNU output sequences according to the UOP strategy; as a matter of fact, they count for 10% and 11% of the area and power consumption of the whole design, respectively.

Finally, as expected, RAM memories play the biggest role in the overall complexity (58%) and power consumption (61%) of the design.

TABLE II

COMPLEXITY/POWER BREAKDOWN WITH $K = 3$, $N_{c2v} = 5$, $N_{v2c} = 7$.

logic	#	Area (μm^2)	Power (mW)
CNU & adders	81	$81 \times 1,414$	81×0.412
MS-CS	1	22,752	9.495
Control Unit	1	21,022	3.506
tot logic		0.158 mm^2	46.37
Logic gates		75.96K	
ROM	$[\# \times \text{(cut)}]$ *bits*	Area (μm^2)	
Degree/Idle	$2 \times (90 \times 8)$	$2 \times 4,300$	2×0.96
Rotation	$1 \times (2362 \times 7)$	9,000	4.18
Parity Read	$1 \times (1037 \times 5)$	6,100	3.14
Parity Write	$1 \times (1325 \times 6)$	9,353	3.20
*c2v*Read	$1 \times (1037 \times 7)$	8,540	3.81
UOP	$1 \times (1037 \times 5)$	6,100	3.14
tot ROM	43,553 *bits*	47,693	19.39
RAM	$[\# \times \text{(cut)}]$ *bits*	Area (μm^2)	
v2c#1	$3 \times (16 \times (27 \cdot 7))$	$3 \times 30,857$	3×12.14
v2c#2	$3 \times (16 \times (27 \cdot 7))$	$3 \times 30,857$	3×12.41
c2v	$3 \times (88 \times (27 \cdot 5))$	$3 \times 31,200$	3×9.74
tot RAM	53,784 *bits*	278,742	102.87
Decoder		0.48 mm^2	168.63

VII. CONCLUSION AND REMARKS

This paper has presented an efficient implementation of a scalable LDPC decoder for IEEE 802.11n codes. The decoder implements the horizontal layered decoding (HLD) algorithm and idle cycles are used as a means to solve the dependance between layers. The number of idle cycles was effectively minimized through the off-line re-ordering of both the sequence of layers followed for decoding and the input/output data stream of the processors, according to the unconstrained output processor (UOP) strategy. As a remarkable result, only from 0 to 11% of the iteration time is spent in idling.

Moreover, a memory partitioning technique was used to map a *three*-port memory onto two *dual*-port banks, with a minimum redundancy to cope with unavoidable conflicts.

Overall, the decoder complexity is about 0.48 mm^2 on 65 nm CMOS technology, and its power consumption is estimated to range from 168 to 185 mW. Table III shows that

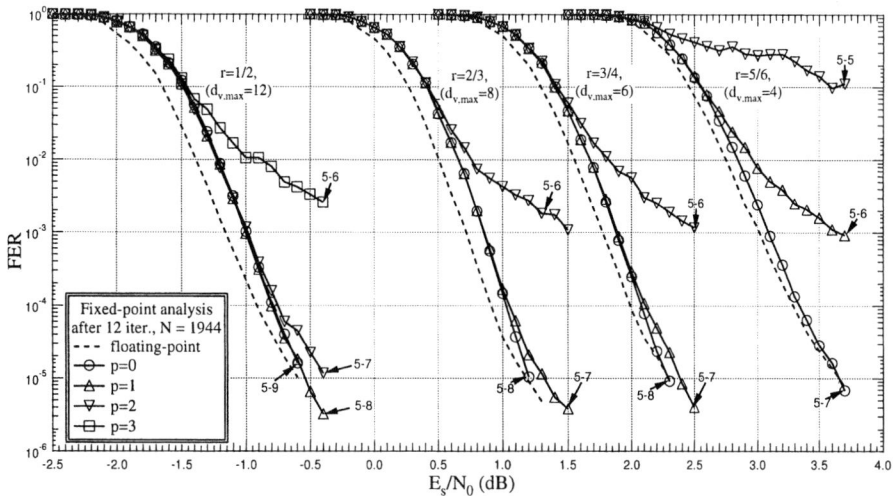

Fig. 7. FER performance of IEEE 802.11n LDPC codes with $N = 1944$ and $c2v$ messages on 5 bits.

the proposed design compares favorably with similar state-of-the art implementations [6], [7] in terms of area and power consumption. Aiming at a fair comparison, we can define the architectural efficiency η as the number of clock cycles to perform an iteration, inclusive of data acquisition and flushing. This can be computed as $\eta = T_{dec}/(t_{clk} \cdot N_{it}) = rN \cdot f_{clk}/(\Gamma \cdot N_{it})$.

As shown in Tab. III, the proposed design not only saves complexity and power w.r.t. similar works, but its efficiency η is from $13,9\%$ to $23,8\%$ higher than [6] and from $4,9\%$ to $10,4\%$ higher than [7]. Furthermore, the multi-frame approach we adopted is neither found in the cited implementations, nor in similar block-LDPC decoders, and this allows us to boost the throughput of the smaller block size (see Tab. I), with the same computational power ($K = 3$) of [6] and [7].

TABLE III
IEEE 802.11n LDPC DECODER IMPLEMENTATIONS.

	HLD	[6]	[7]
CMOS process	65 nm LP	0.13 μm	65 nm
f_{clk} (MHz)	240	500	400
nand-2 (μm^2)	2.08	6.58	2.08
area (mm^2)	0.48	1.85	1.02
area (Kgates)	231	281	490
RAM (bits)	53,784	55,344	n/a
power (mW)	168-185	238	n/a
Γ (Mbps)	136-355	541-1618	54-281
iterations (N_{it})	12	5	25-20
η	91.3-103.4	119.8-120.1	96-115.3

REFERENCES

[1] D. MacKay and R. Neal, "Good codes based on very sparse matrices," in *5th IMA Conference on Cryptography and Coding*, Springer-Verlag, Ed., 1995, pp. 100–111.

[2] R. Gallager, "Low-Density Parity-Check Codes," Ph.D. dissertation, Massachusetts Institutes of Technology, 1960.

[3] "Digital video broadcasting (DVB); second generation framing structure, channel coding and modulation systems for broadcasting, interactive services, news gathering and other broadband satellite applications," ETSI, June 2004.

[4] "Air Interface for Fixed and Mobile Broadband Wirelss Access Systems," IEEE Computer Society, Feb. 2006, IEEE Std 802.16e™-2005.

[5] "IEEE P802.11n™/D1.06," Draft amendment to Standard for high throughput, Nov. 2006, 802.11 Working Group.

[6] K. Gunnam, G. Choi, W. Wang, and M. Yeary, "Multi-rate layered decoder architecture for block LDPC codes of the IEEE 802.11n wireless standard," in *IEEE Intern. Symp. on Circuits and Systems, ISCAS*, May 2007, pp. 1645–1648.

[7] T. Brack, M. Alles, T. Lehnigk-Emden, F. Kienle, N. Wehn, N. L'Insalata, F. Rossi, and M. Rovini, "Low Complexity LDPC Code Decoders for Next Generation Standards," in *Design Automation and Test in Europe, DATE 07*, Apr 2007.

[8] M. Mansour and N. Shanbhag, "High-throughput LDPC decoders," *IEEE Trans. VLSI Syst.*, vol. 11, no. 6, pp. 976–996, Dec 2003.

[9] D. Hocevar, "A Reduced Complexity Decoder Architecture via Layered Decoding of LDPC Codes," in *IEEE Workshop on Signal Processing Systems, SISP 2004*, 2004, pp. 107–112.

[10] A. Blanksby and C. Howland, "A 690-mW 1-Gb/s 1024-b, rate-1/2 low-density parity-check code decoder," *IEEE J. Solid-State Circuits*, vol. 37, no. 3, pp. 404–412, Mar 2002.

[11] T. Bhatt, V. Sundaramurthy, V. Stolpman, and D. McCain, "Pipelined block-serial decoder architecture for structured LDPC codes," in *Proc. IEEE Conf. on Acoustics, Speech and Signal Proc., ICASSP*, Apr. 2006.

[12] Y. Sun, M. Karkooti, and J. Cavallaro, "High throughput, parallel, scalable LDPC encoder/decoder architecture for OFDM systems," in *Proc. IEEE Work. Circuits and Systems*, Oct. 2006, pp. 225–228.

[13] E. Sharon, S. Litsyn, and J. Goldberger, "An Efficient Message-Passing Schedule for LDPC Decoding," in *23rd IEEE Convention of Electrical and Electronics Engineering in Israel*, Sep 2004, pp. 223–226.

[14] F. Guilloud, E. Boutillon, J. Tousch, and J.-L. Danger, "Generic description and synthesis of LDPC decoders," *IEEE Trans. Commun.*, 2006.

[15] J. Zhang and M. P. C. Fossorier, "Shuffled Iterative Decoding," in *IEEE Trans. Commun.*, vol. 53, no. 2, Feb 2005, pp. 209–213.

[16] H. Kfir and I. Kanter, "Parallel versus sequential updating for belief propagation decoding," *Physica A Statistical Mechanics and its Applications*, vol. 330, pp. 259–270, Dec. 2003.

[17] M. Rovini, F. Rossi, N. L'Insalata, and L. Fanucci, "High-precision ldpc codes decoding at the lowest complexity," in *14th European Signal Processing Conference, EUSIPCO 2006*, Sep 2006.

[18] M. Rovini, F. Rossi, P. Ciao, N. L'Insalata, and L. Fanucci, "Layered Decoding of Non-Layered LDPC Codes," in *9th Euromicro Conference on Digital System Design (DSD)*, Aug-Sep 2006.

Full Custom Design of a Three-Stage Amplifier with 5500MHz·pF/mW Performance in 0.18μm CMOS

Run Chen[†], Liyuan Liu[*], Dongmei Li[*], Zhihua Wang[†]

[†] Institute of Microelectronics, [*] Electronic Engineering Department

Tsinghua University, Beijing, P.R.China

chen-r02@mails.tsinghua.edu.cn

Abstract—**A full custom design of a three-stage amplifier is described in this paper. A feedback transconductance stage and a feedforward stage combined with two Miller compensation capacitors are used for frequency compensation. The circuit is designed in 0.18μm CMOS process with a 1.8V supply voltage. When driving a 150pF capacitive load, the amplifier achieves over 100dB dc gain, 2.24MHz gain-bandwidth product (GBW), 62° phase margin (PM), 1.2V/μs slew rate (SR) and 61μW power dissipation. Compared to conventional multistage amplifiers, this work provides improvement in both GBW and SR, and also shows a significant improvement in MHz·pF/mW performance.**

I. INTRODUCTION

The supply voltage of the VLSI continuously scales down with advanced deep sub-micrometer technologies. Low supply voltage and low power electronic system becomes the main design challenge for many applications. The operational amplifier (OPA), which is needed in almost all analog and mixed-signal systems, has to keep up with the fast advances in present-day process. However, single-stage or two-stage amplifiers based on cascoding transistors are no longer suitable for low supply voltages. In order to achieve high gain and comparatively higher output voltage swing, a multistage amplifier is widely used by increasing the number of gain stages horizontally. Nonetheless, all multistage amplifiers suffer from the closed-loop stability problem since high-resistance nodes between the stages generate poles and zeros with the parasitic capacitances, which can greatly interfere with frequency response. Thus a multistage amplifier must be frequency compensated in order to either cancel the redundant poles and zeros or shift them to higher frequency range.

Several frequency compensation topologies for multistage amplifiers are reported [1]-[6]. The nested Miller compensation (NMC) is well known to split the poles so the nondominant poles can be shifted to higher frequency than unity-gain frequency [1]. The structure of NMC is depicted in Fig. 1. Parameters g_{mi}, r_i and C_{pi} are transconductance, output resistance and parasitic capacitance for the i-th stage, respectively. C_L stands for the load capacitance, and C_{m1} and C_{m2} are compensation capacitors. Careful scrutiny on this topology reveals that the Miller capacitor C_{m2} is a serious source of instability since the phase shift reaches 180° with

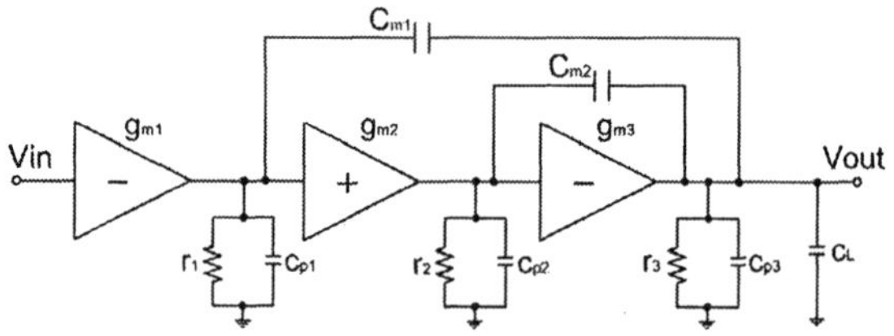

Fig. 1. NMC frequency compensation tolopogy.

the increase of frequency. As a result, extra power is required to ensure stability. Since the first and second nondominant pole p_{nd1} and p_{nd2} is often required to be above 3 and 5 times of Gain-Bandwidth Product, g_{m2} and g_{m3} are given by

$$g_{m2} \geq 6\pi \cdot \text{GBW} \cdot C_{m2},\qquad(1)$$

$$g_{m3} \geq 10\pi \cdot \text{GBW} \cdot C_L.\qquad(2)$$

The NMC topology is not suitable for low power applications since the required transconductance is 5 times the transconductance for single stage amplifier.

Many frequency compensation topologies are based on NMC structure, such as multipath nested Miller compensation (MNMC) [1], nested G_m-C compensation (NGCC) [2], and NMC with feedforward transconductance stage and nulling resistor (NMCFNR) [3]. Some other advanced topologies have been proposed to improve bandwidth, such as Damping-factor control frequency compensation topology (DFCFC) [4], positive-feedback compensation topology (PFC) [5], active-feedback frequency compensation (AFFC) [6] and transconductance with capacitances feedback compensation (TCFC) [7].

In this paper, a full custom design of a three-stage CMOS amplifier based on TCFC is described with high figure of merit (FOM) defined as

$$\begin{cases} \text{FOM}_S = \dfrac{\text{GBW} \cdot C_L}{\text{power}} \\[3mm] \text{FOM}_L = \dfrac{\text{SR} \cdot C_L}{\text{power}} \end{cases}.\qquad(3)$$

The frequency compensation structure includes a transistor to provide feedback transconductance and a feedforward stage. The feedback transistor attenuates the feedforward

---

This work was supported by National High Technology Research and Development Program of China (No. 2004AA1Z1100).

242

978-1-4244-1709-4/07/$25.00 © 2007 IEEE

signal, and the feedforward stage is used to form a Class-AB output stage so as to provide large slew rate. In addition, this amplifier can be used to drive large capacitance load and it occupies low die area.

II. Adopted Compensation Technique

A. Topology and Transfer Function

The adopted frequency compensation structure based on TCFC is depicted in Fig. 2. Compensation exploits C_{m1}, C_{m2}, a feedback transistor Ma and an active transconductance stage. The parameters shown in Fig. 2 are the same as defined before. Here g_{mf} represents the transconductance of the feedforward stage, and $1/g_{ma}$ is output resistance of transistor Ma. For simplification, we assume the dc gain of each stage is much larger than 1, and C_{m1} and C_{m2} are much larger than parasitic capacitors C_{pi}, and C_L is much larger than C_{m1} and C_{m2}. Thus, the small-signal transfer function of the open-loop gain can be expressed by (4). From (4), we can get dc gain A_{dc}, the dominant pole p_{-3dB} and GBW as

$$A_{dc} = g_{m1}g_{m2}g_{m3}r_1r_2r_3 \qquad (5)$$

$$p_{-3dB} = -\frac{1}{2\pi C_{m1}g_{m2}g_{m3}r_1r_2r_3} \qquad (6)$$

$$\text{GBW} = A_{dc} \cdot \left| p_{-3dB} \right| = \frac{g_{m1}}{2\pi C_{m1}} . \qquad (7)$$

Notice that g_{mf} does not appear in the denominator of the transfer function. Thus the feedforward stage does not introduce any poles and it helps to improve slew rate of the amplifier.

B. Stability Analysis and GBW

The stability condition of the adopted structure can be studied by first neglecting the zeros in (4). Actually, there is a left half plane (LHP) zero. The close-loop transfer function $A_{v\_close}(s)$ of the amplifier connected as in unity-gain feedback configuration can be derived as (8). As the order of the numerator in (8) is less than that of the denominator, the stability is determined by the denominator. Routh stability criterion can be used to evaluate system stability conditions [2], and it reveals that the amplifier is stable if

$$2\pi\text{GBW} < \frac{C_{m2}}{C_{p2}} \cdot \frac{g_{m3}}{C_L} . \qquad (9)$$

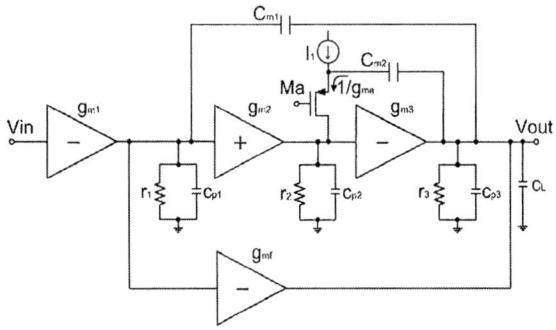

Fig. 2. Adopted compensation sturcture for three-stage amplifier.

According to (9), the achievable GBW depends on the ratio of C_{m2} and C_{p2} for a specific transconductance g_{m3} and a given load C_L. Apparently, the ratio C_{m2}/C_{p2} can be made large since C_{p2} is a lumped parasitic capacitor. Therefore, the GBW can be dramatically extended with respect to that of the NMC structure.

C. Dimension Conditions

The stability of the adopted structure can also be achieved when the denominator of the close-loop transfer function (8) has a fourth-order Butterworth polynomial B(s) with a cutoff frequency of ω_0 [8], which is given by

$$B(s) = 1 + 2.613\frac{s}{\omega_0} + 3.414\frac{s^2}{\omega_0^2} + 2.613\frac{s^3}{\omega_0^3} + \frac{s^4}{\omega_0^4} . \qquad (10)$$

By comparing the coefficients of the denominators of (8) with those of (10), we can get

$$\frac{C_{m1}}{g_{m1}} = \frac{2.613}{\omega_0} \qquad (11)$$

$$\frac{C_{m1}C_{m2}}{g_{m1}g_{m2}}\left(1+\frac{g_{m2}}{g_{ma}}\right) = \frac{3.414}{\omega_0^2} \qquad (12)$$

$$\frac{C_{m1}C_{p2}C_L}{g_{m1}g_{m2}g_{m3}} = \frac{2.613}{\omega_0^3} \qquad (13)$$

$$\frac{C_{m1}C_{m2}C_{p2}C_L}{g_{m1}g_{m2}g_{m3}g_{ma}} = \frac{1}{\omega_0^4} . \qquad (14)$$

The dimension conditions of C_{m2} can be deduced from (12)–(14) and eliminating ω_0. This can be written as

$$A_v(s) = \frac{r_1r_2r_3g_{m1}g_{m2}g_{m3}\left[1+\dfrac{C_{m2}}{g_{ma}}s-\dfrac{C_{m1}C_{p2}+C_{m2}C_{p2}g_{mf}/g_{ma}}{g_{m2}g_{m3}}s^2-\dfrac{C_{m1}C_{m2}C_{p2}}{g_{m2}g_{m3}g_{ma}}s^3\right]}{1+\left(r_1r_2r_3C_{m1}g_{m2}g_{m3}\right)s+\left[r_1r_2r_3C_{m1}C_{m2}g_{m3}\left(1+\dfrac{g_{m2}}{g_{ma}}\right)\right]s^2+\left(r_1r_2r_3C_{m1}C_{p2}C_L\right)s^3+\dfrac{r_1r_2r_3C_{m1}C_{m2}C_{p2}C_L}{g_{ma}}s^4} . \qquad (4)$$

$$A_{v\_close}(s) = \frac{1}{1+\dfrac{C_{m1}}{g_{m1}}s+\left[\dfrac{C_{m1}C_{m2}}{g_{m1}g_{m2}}\left(1+\dfrac{g_{m2}}{g_{ma}}\right)\right]s^2+\dfrac{C_{m1}C_{p2}C_L}{g_{m1}g_{m2}g_{m3}}s^3+\dfrac{C_{m1}C_{m2}C_{p2}C_L}{g_{m1}g_{m2}g_{m3}g_{ma}}s^4} . \qquad (8)$$

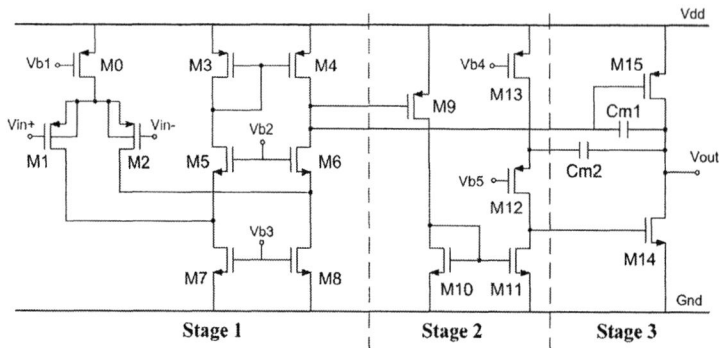

Fig. 3. Circuit schematic of a three-stage amplifier.

$$C_{m2} = \sqrt{\frac{C_{p2}C_L g_{ma}}{2g_{m3}(1+\frac{g_{m2}}{g_{ma}})}}$$

$$= \sqrt{\frac{C_{p2}g_{ma}g_{m3}}{50g_{m1}^2(1+\frac{g_{m2}}{g_{ma}})C_L} \cdot \frac{5g_{m1}}{g_{m3}}C_L}, \quad (15)$$

$$= \frac{1}{N}C_{m1(\text{NMC})}$$

where

$$N = \sqrt{\frac{50g_{m1}^2(1+\frac{g_{m2}}{g_{ma}})C_L}{C_{p2}g_{ma}g_{m3}}}, \quad (16)$$

and

$$C_{m1(\text{NMC})} = \frac{5g_{m1}}{g_{m3}}C_L. \quad (17)$$

$C_{m1(\text{NMC})}$ stands for the compensation capacitor C_{m1} in the NMC structure. Since C_L is much larger than C_{p2}, the size of C_{m2} in the adopted structure is much smaller than $C_{m1(\text{NMC})}$ as $N \gg 1$. Also, the dimension conditions of C_{m1} can be deduced from (11)–(12) as

$$C_{m1} = \frac{2g_{m1}C_{m2}(g_{m2}+g_{ma})}{g_{m2}g_{ma}}. \quad (18)$$

(18) reveals that C_{m1} has similar size as C_{m2}, and it is much smaller than $C_{m1(\text{NMC})}$. Thus, the adopted topology occupies much less die area than NMC structure because the compensation capacitors need not to be very large.

D. Phase Margin

From the open-loop transfer function in (4), the first nondominant pole p_{nd1} and the LHP zero z_1 are given by

$$p_{nd1} = -\frac{g_{ma}g_{m2}}{(g_{m2}+g_{ma})C_{m2}} \quad (19)$$

$$z_1 = -\frac{g_{ma}}{C_{m2}} = \frac{g_{ma}+g_{m2}}{g_{m2}}p_{nd1}. \quad (20)$$

Both p_{nd1} and zero z_1 are independent of imprecise parasitic capacitors. Other nondominant poles and zeros are much larger so their effects on phase margin can be neglected. The overall phase margin (PM) is given by

$$\text{PM} \approx 90° - \arctan\frac{2\pi\text{GBW}}{|p_{nd1}|} + \arctan\frac{2\pi\text{GBW}}{|z_1|}. \quad (21)$$

Since z_1 is slightly larger than p_{nd1}, the phase margin can be improved by the LHP zero.

E. Slew Rate

With the feedforward stage g_{mf}, the last stage of the amplifier becomes a Class-AB output stage which slews fast in both directions. Thus the first stage which drives the C_{m1} becomes the dominant limitation of the overall SR, which is given by

$$\text{SR} = \frac{I_{stage1}}{C_{m1}}, \quad (22)$$

where I_{stage1} stands for the current of the first stage.

F. Low-Power Design Considerations

The stability conditions require the NMC amplifier has large transconductance of the last stage, i.e. $g_{m3} \gg g_{m1}$, thus it is not suitable for low-power applications. The adopted structure is desirable for low-power design because the transconductance of the output stage need not be too large. This can be deduced from (9) as

$$g_{m3} > \frac{2\pi \cdot \text{GBW} \cdot C_{p2}C_L}{C_{m2}} = \frac{C_{p2}}{C_{m2}}g_{m(single)}, \quad (23)$$

where $g_{m(single)}$ represents for the transconductance of a single stage amplifier. g_{m2} can be small since C_{p2} is much smaller than C_{m2}.

III. CIRCUIT IMPLEMENTATION

The circuit schematic of a three-stage amplifier is depicted in Fig. 3. The first stage is a classical folded cascode OTA. It consists of transistors M0-M8, which ensure that the common mode input range can reach low rail voltage Gnd. The second noninverting gain stage includes transistors M9-M13 and the last stage includes M14-M15. The feedback transconductance g_{ma} is generated by M12, which also acts as a cascoding stage in the second stage. The feedforward stage g_{mf} is generated by M15, which is part of the last stage. C_{m1} and C_{m2} are compensation capacitors.

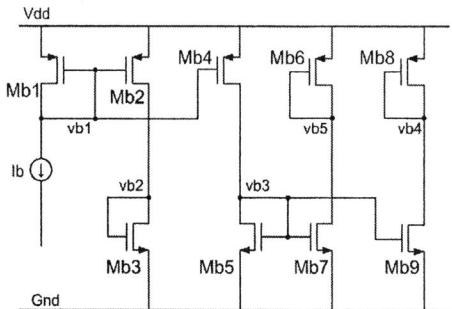

Fig. 4. Biasing network schematic.

TABLE I. PARAMETERS OF THE AMPLIFIER

Parameter	g_{m1}	g_{m2}	g_{m3}	g_{ma}	g_{mf}
Value (designed)	30μS	150μS	150μS	160μS	50μS
Value (Simulation)	29.1μS	146.0μS	156.3μS	161.0μS	48.8μS
Parameter	C_{m1}	C_{m2}	C_L	Vdd	Ib
value	2.1pF	6.9pF	150pF	1.8V	30μA

The biasing network schematic is shown in Fig. 4. Ib is the input reference current which generates the biasing voltages. Transistors Mb1-Mb8 work at saturation region to generate vb1-vb5.

The transistor size is designed to achieve following specifications: Open-loop gain above 120dB, SR larger than 1V/μs, power less than 80uW, GBW more than 1.5MHz, PM more than 50°. These specifications reveal that the amplifier has a FOM_S above 2813MHz·pF/mW, and FOM_L above 1875V·pF/μs·mW. This shows that the amplifier has high performance since it has high FOM value. Much design effort is put on further optimizing the FOM value.

The circuit parameters of the amplifier are shown in Table I. The amplifier is designed in a 0.18μm CMOS process and the supply voltage is 1.8V. The capacitors used to compensate frequency response are implemented by metal-insulator-metal (MIM) capacitors. This process is compatible with most VLSI systems. In addition, the stability is insensitive to the absolute value of the compensation capacitors.

IV. SIMULATION RESULT AND LAYOUT

The three-stage amplifier is simulated with a 150pF capacitive load. The frequency response of the amplifier has been tested with input common-mode voltage of 800mV, and the transient response together with SR have been tested when the amplifier is connected in unity-gain with a 1V step input. The simulation result of frequency and transient response are shown in Fig. 5 and Fig. 6, respectively. From Fig. 5, the amplifier achieves a GBW of 2.241MHz, a dc Gain of 134.7dB and a PM of 62°. From Fig. 6, the SR₊ and SR₋ are 1.16V/μs and 1.23V/μs, respectively. The results are summarized in Table II. The design specifications are well satisfied.

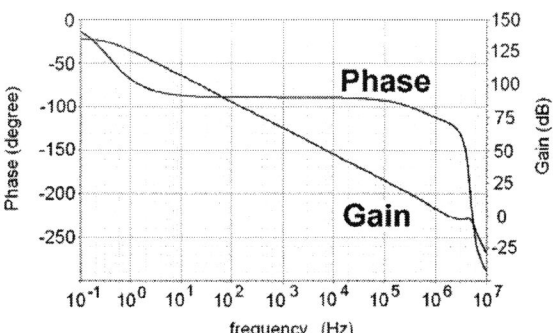

Fig. 5. Frequency response of the three-stage amplifier.

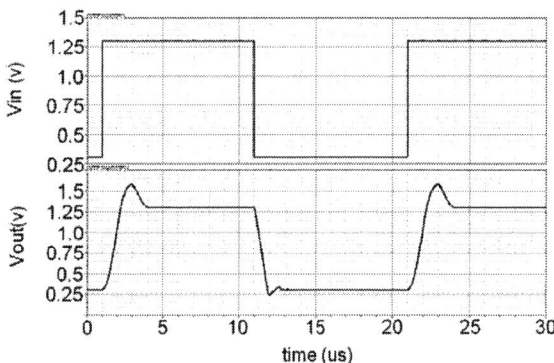

Fig. 6. Transient response of the amplifier with 1V step input.

TABLE II. SUMMARIZED SIMULATION RESULTS

Parameter	Specifications	Simulation Value
DC Gain	>100 dB	134.7 dB
Gain-bandwidth Product	>1.5MHz	2.241 MHz
Phase Margin	>50°	62°
Vdd	1.8 V	1.8 V
Idd	<45 μA	33.8 μA
Power	<80 μW	60.9 μW
Slew Rate +/-	>1 V/μs	1.16/1.23 V/μs
Settling Time +/- (to 1%)	–	2.93/1.72 μs

A. Circuit Robustness Analysis

In actual electronic systems, nonideal factors may deteriorate the circuit performance especially for the analog system. In this design, we mainly consider how the fluctuation of supply voltage Vdd and biasing current Ib affect the FOM_S value, which reflects the performance of frequency response and power dissipation. The relationship among FOM_S, Vdd, and Ib are illustrated in Fig. 7 within different environment temperature. Assuming that Vdd fluctuats between 1.6V and 2.1V, Ib fluctuates between 25μA and 35μA. In addition, different technology corners are considered during the simulation, including typical NMOS with typical PMOS (TT); fast NMOS with fast PMOS (FF); slow NMOS with slow PMOS (SS); slow

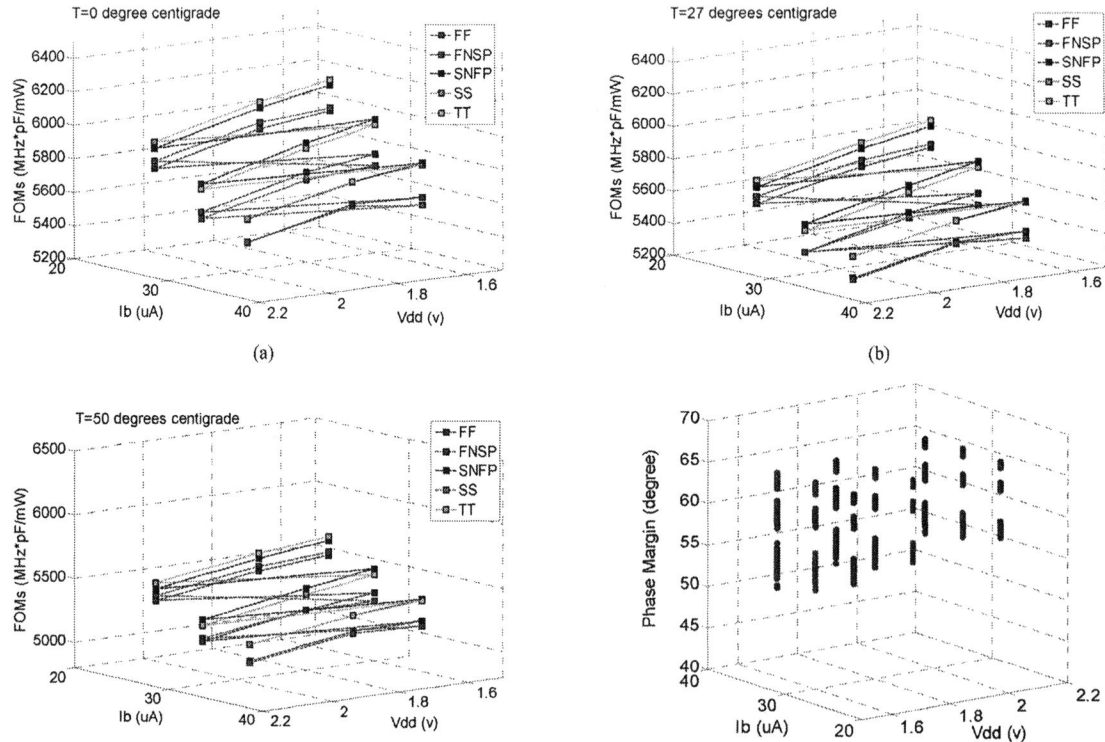

Fig. 7. The relationship among FOMs, Vdd, and Ib

(a) T = 0℃   (b) T = 27℃   (c) T = 50℃.

Fig. 8. Phase Margin in different conditions (different Vdd, Ib, temperature, technology corner, input common mode voltage are considered).

TABLE III.   COMPARISON OF DIFFERENT MULTISTAGE AMPLIFIERS

Parameter	NMC [1]	MNMC [1]	NGCC [2]	NMCFNR [3]	DFCFC [4]	PFC [5]	AFFC [6]	TCFC [7]	This work
C_L (pF)	100	100	20	100	100	130	120	150	150
DC Gain (dB)	100	100	100	>100	>100	>100	>100	>100	135
GBW (MHz)	60	100	0.61	1.8	2.6	2.7	4.5	2.85	2.24
Phase Margin (degree)	70	70	60	51	43	52	65	58.6	62
Vdd (V)	8.0	8.0	2.0	2.0	2.0	1.5	2.0	1.5	1.8
Power (mW)	76	76	0.68	0.41	0.42	0.275	0.40	0.045	0.061
Slew Rate (V/µs)	20	35	2.5	0.79	1.32	1.0	1.49	1.035	1.20
FOMS (MHz·pF/mW)	79	132	18	443	619	1276	1350	9500	5500†
FMOL (V·pF/µs·mW)	26	46	74	195	314	473	447	3450	2950†
Technology	3GHz f_t BJT	3GHz f_t BJT	2µm CMOS	0.8µm CMOS	0.8µm CMOS	0.35µm CMOS	0.8µm CMOS	0.35µm CMOS	0.18µm CMOS

† Typical value @ TT Corner, 27℃.

NMOS with fast PMOS (SNFP) and fast NMOS with slow PMOS (FNSP). The simulation results reveal that the FOM$_S$ value is insensitive to Vdd and Ib fluctuation and also the temperature, and it keeps well above 5000 MHz·pF/mW, which is a desirable value. The same analysis can be applied to FOM$_L$, which also reflects the robustness of the amplifier. From Fig. 7 (a)-Fig. 7 (c), the FOM$_S$ value increases when Vdd and Ib decreases, or the temperature goes down within

a certain range. Fig. 8 shows the phase margin range in different Vdd, Ib, technology corner and input common mode voltage conditions, where input common mode voltage ranges from 0.2V to 0.8V. From Fig. 8, the PM stays well between 51° and 66°, which guarantees stability of the amplifier. The comparison results of this work and previous reports are given in Table III. The simulation result prevails over most previous topologies.

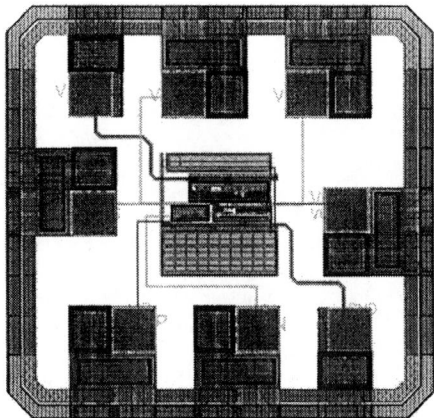

Fig. 9. Layout of the amplifier with pads.

B. Layout

The layout of the amplifier including gain stages and biasing network is depicted in Fig. 9. By using 0.18μm CMOS process, the area of the amplifier can be reduced effectively. The active area for the amplifier is about 0.02mm². In order to reduce mismatch, several dummy transistors are also included in the layout.

V. CONCLUSION

A full custom design of a three-stage amplifier is presented in this paper. We adopt a frequency compensation topology with a feedback transconductance stage and a feedforward stage combined with two Miller compensation capacitors. This topology improves the GBW and SR performance. In addition, the compensation capacitors are small compared to conventional topologies; thus the die area can be reduced. The amplifier achieves an overall FOM_S value of 5500MHz·pF/mW. It is desirable for low supply voltage and low power applications and is compatible with VLSI systems.

VI. ACKNOWLEDGEMENT

The authors would like to thank Fule Li, Institute of Microelectronics, Tsinghua University, and Tao Zhou, RDA Microelectronics, for their helpful suggestions.

REFERENCES

[1] J. H. Huijsing and D. Linebarger, "Low-voltage operational amplifier with rail-to-rail input and output stages," *IEEE J. Solid-State Circuits*, vol. SC-20, No. 6, pp. 1144–1150, Dec. 1985.

[2] F. You, S. H. K. Embabi, and E. Sánchez-Sinencio, "Multistage amplifier topologies with nested Gm-C compensation," *IEEE J. Solid-State Circuits*, vol. 32, No. 12, pp. 2000–2011, Dec. 1997.

[3] K. N. Leung and P. K. T. Mok, "Nested Miller compensation in lowpower CMOS design," *IEEE Trans. Circuits Syst. II: Analog Digital. Signal Process.*, vol. 48, No. 4, pp. 388–394, Apr. 2001.

[4] K. N. Leung, P. K. T. Mok, W. H. Ki, and J. K. O. Sin, "Three-stage large capacitive load amplifier with damping-factor-control frequency compensation," *IEEE J. Solid-State Circuits*, vol. 35, No. 2, pp. 221–230, Feb. 2000.

[5] J. Ramos, and M. Steyaert, "Positive feedback frequency compensation for low-voltage low-power three-stage amplifier." *IEEE. Trans. Circuits Syst .I: Regular paper.*, vol. 51, No. 10, pp. 1967–1974, Oct. 2004.

[6] H. Lee and P. K. T. Mok, "Active-feedback frequency-compensation technique for low-power multistage amplifiers," *IEEE J. Solid-State Circuits*, vol. 38, No. 3, pp. 511–520, Mar. 2003.

[7] X. Peng and W. Sansen, "Transconductance with capacitances feedback compensation for multistage amplifiers," *IEEE J. Solid-State Circuits*, vol. 40, No. 7, pp. 1514–1520, July. 2005.

[8] A.V. Oppenheim and R.W. Schafer, *Discrete-time signal processing*, 2nd edition, Prentice Hall, 1998.

A 128dB dynamic range 1kHz bandwidth stereo ADC with 114dB THD

YuQing Yang[1,2], Terry Sculley[1],
Texas Instruments Inc., Austin Texas, USA[1]

Jacob Abraham[2]
University of Texas, Austin Texas, USA[2]

Abstract -- *A high performance, single die stereo delta sigma ADC is designed for high precision measurement applications. A single loop, fifth-order, thirty-three level delta-sigma analog modulator with positive and negative feedforward path is implemented. An interpolated multilevel quantizer with unevenly weighted quantization levels replaces a conventional 5-bit flash type quantizer in this design. These new techniques suppress signal dependent energy inside the delta sigma loop, reduce internal channel coupling and power consumption. Integrated with an on-chip bandgap reference circuit and decimation filter, the ADC achieves 128dB dynamic range, −114dB THD over 1kHz bandwidth. Power consumption is less than 140mW per channel.*

I INTRODUCTION

High precision measurement applications have lately shown increased demand for higher level of linearity and dynamic range with lower cost. The reference voltages of high precision ADCs are normally supplied by off chip low noise circuits. This approach relaxes design requirements and power consumptions for the ADCs but increases overall system level cost and power dissipations. Multi-channel ADC is preferred to further reduce circuit board area and layout complexity.

Multibit delta sigma topology becomes popular nowadays for high performance converter design mainly due to its ability to reduce integrator output swings, which relaxes analog circuit design requirements and reduces power consumption. Such benefit becomes more critical for high performance A/D converter design. A multibit delta sigma topology with a large number of quantization levels also significantly lowers the out-of-band quantization noise energy, which greatly relaxes the decimation filter design requirements. Since the quantizer is inside the delta sigma loop, it can only tolerate minimum latency. A flash type A/D converter is normally adopted for the quantizer due to its low latency. However, each additional bit of resolution for the quantizer doubles its power consumption and die area. The power consumption increase of the quantizer quickly outweighs the benefit and power saving from low integrator output voltage swings. A large number of comparators also inject a substantial amount of signal dependant noise energy into the substrate, which causes degradation of converter

performance and severe interference between channels in the case of multi-channel converter design.

Clock jitter sensitivity is another major concern for high performance converter design. It is important to ensure that the converter clock jitter requirement is reasonable to achieve in a real world environment. A switched capacitor topology is widely used for analog circuit implementation due to its better immunity to clock jitter. However, charge injection from the switched capacitor input sampling network tends to degrade the performance of the input anti-alias filter, which limits the achievable performance of the converter. Although increasing the oversampling ratio decreases the sampling capacitor size, this translates to less settling time for integrator opamp and digital switching noise, which makes the converter sensitive to digital coupling noise.

This paper presents a newly developed single die, high precision stereo delta sigma ADC. A dynamic element matching (DEM) circuit is included inside the loop to remove tones and nonlinearity caused by feedback digital to analog converter (DAC) capacitor mismatches. Integrated with a decimation filter and on-chip reference circuits, this stereo ADC achieves 128dB dynamic range, -114dB THD over 1kHz bandwidth. Total core die area is around 14.8 mm^2 .

II MODULATOR ARCHITECTURE

Quantization noise and tones from the first stage, low order delta sigma loop of a multistage converter tends to leak to the output due to stage-to-stage mismatch [1] [2]. The delta sigma analog modulator used in this design is shown in Fig. 1. It is a modified 5th order feedforward modulator with thirty-three quantization levels. The signal feedforward path 1 prevents most signal energy from leaking into the delta sigma loop, which limits the output swings of all integrators. Since the quantizer gain is not constant across input range, there are residual signal energy leaks into the delta sigma modulator, which diminishes the advantage of adopting a multilevel topology. Residual signal energy leaks into the modulator are mostly accumulated at the last integrator output. The input signal is added negatively to the input of the last integrator, which is easy to implement in a fully differential structure, to cancel the remaining signal energy. This approach limits the output swings of all integrators with minimum increase of system and circuit level complexity. This architecture also helps to improve modulator stability and prevent overload for full scale input signals. The maximum modulation index exceeds 0.95 for this design. Clocked at 4.096MHz, the theoretical SNR of this $\Delta\Sigma$ analog modulator is 146dB over a 1kHz bandwidth.

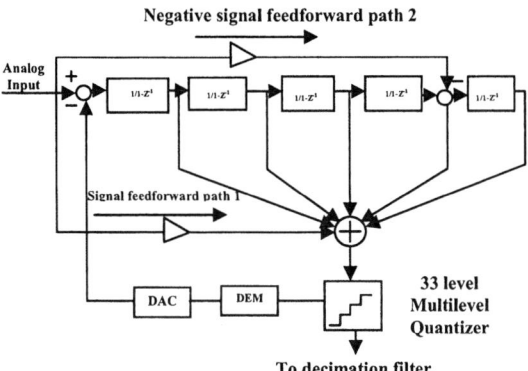

Fig.1 Delta-sigma modulator architecture

III ANALOG DESIGN

A. Multilevel Quantizer Design

A conventional flash type quantizer doubles the power dissipation and area of the quantizer for each additional bit of resolution. Furthermore, having many quantization levels decreases the reference voltage step substantially, which increases design and layout requirements of individual comparators. A large number of comparators also create large "kick-back" on the quantizer reference line, which potentially introduce excessive errors. Tracking quantizers [3] have been reported to deal with the issue. However, they suffer from several major shortcomings such as requiring a separate circuit to set up the initial loop operating point, additional digital to analog converter (DAC) circuits, potential modulator instability due to high frequency input signal leakage etc.

The multilevel quantizer topology applied in this design is shown in Fig. 2. The quantization levels are divided into two sections, which are coarse and fine. The quantizer changes one quantization level in the fine section, while it changes two quantization levels in the coarse section. In order to utilize existing dynamic element matching (DEM) logic circuits [4], the middle quantization output bit is interpolated by ANDing two adjacent comparator outputs in the coarse quantization section. Since the multilevel quantizer is inside the loop, nonlinearity introduced by this approach is suppressed by the delta sigma modulator. The fine quantization section, which is from level 8 to level 24, is further divided into two phases, a low resolution phase and high resolution phase. There are four comparators in the low resolution phase, and three reference voltage levels interleave

between two adjacent comparators. The high resolution phase consists of three comparators. During the low resolution phase, if adjacent comparators outputs are the same, then comparison operations between these bits are skipped and the outputs are "interpolated" by simple digital logic circuits. Switches S1, S2 and S3 are controlled by an XOR of adjacent low resolution comparator outputs. For example, if the level 20 output is low and level 16 output is high, the XOR of these two outputs is high. This operation turns on the switches S1, S2 and S3, which steers the high resolution comparators to detect level 17 to level 19. On the other hand, if both level 20 and level 16 are high or low, this suggests that the input level is either higher than level 20 or lower than level 16, and quantization operations between level 17 to level 19 can be skipped.

For large input signals, the ADC output performance is dominated by nonlinearity components such as even and odd harmonics. This new scheme, which increases quantization noise slightly for large input signals, virtually has no impact on large input signal ADC performance. For small input signals, when nonlinearity components such as even and odd harmonics don't exist, the fine quantization level is applied to achieve the lowest noise floor. The boundary between the coarse and fine quantization is defined by modulator behavioral model simulations.

The multilevel quantizer clock diagram and its output for sine wave input are shown in Fig. 3. The low resolution quantization occurs in the middle of the integration phase. This arrangement creates a time slot to switch in the proper reference voltage and turn on comparators for the high resolution quantization phase. This operation also divides quantization into two phases, which effectively increases the settling time for comparison glitches prior to the critical sampling clock edge. The low resolution phase comparators also set up the initial DC operational points for the $\Delta\Sigma$ loop. Ideally, an anti-alias analog filter in front of this modulator filter out all high frequency signals. However, in real world applications, there are cases where high frequency energy is leaked to the A/D converter input that forces the quantizer to change multiple levels during one operation. In such cases, low resolution phase comparators are also able to track the input changes and maintain the stability of the delta sigma loop.

Compared with conventional approaches, this design reduces the number of comparators from thirty two to twelve in the low resolution phase and three in the high resolution phase, which significantly lowers the "kick back" on the quantizer reference line especially for the critical high resolution quantization phase. Furthermore, comparators in the rough phase only need to distinguish every two or four quantization levels, which relax the design requirements for individual comparators.

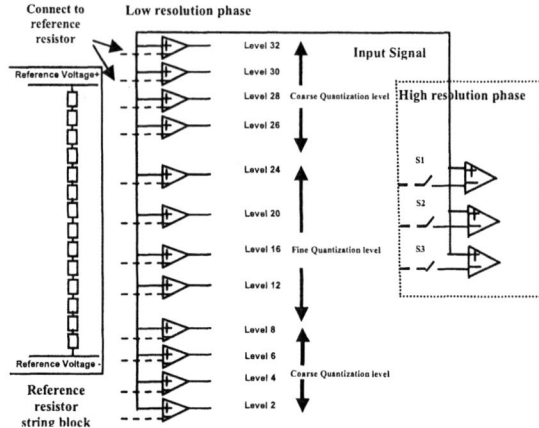

Fig.2 Multilevel quantizer scheme

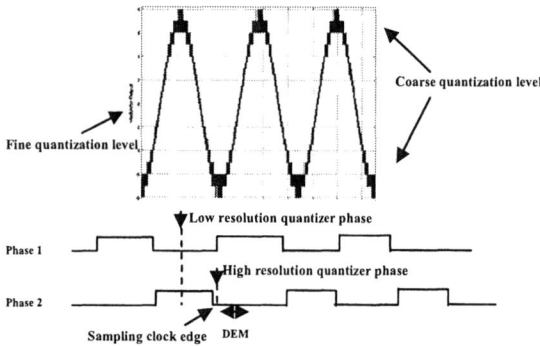

Phase 1 and Phase 2 are non-overlapped clock phase.

Fig.3 Quantizer clock diagram and output waveform

B. Integrator and Feedback DAC Design

Although chopper stabilization is able to remove 1/f noise [5], it may also demodulate high frequency quantization noise back to baseband. During chopper stabilization operation, digital switching noise can also be coupled into Integrator 1. In order to maximize the digital switching noise immunity, chopper stabilization is purposely avoided in this design.

Designing a low noise on-chip reference circuit is also a challenge at this performance level. The implementation of the reference circuit has large impact on overall stereo ADC power consumption and die area. On the other hand, multiple approaches have been proposed to limit charge injection from the switched capacitor signal sampling network, including boot strap, front unity gain opamp, slew boost switching opamp, etc. However, these approaches increase circuit complexity, die area, and power consumption considerably.

The converter kT/C noise is defined as:

$$\text{Total } kT/C \text{ noise} = kT/Cin(1+ Cfb/Cin) \qquad (1)$$

Cin: input sampling capacitor, Cfb: feedback DAC capacitor
The ratio of Cfb and Cin is limited by modulation index.

$$\text{Modulation index} \quad MI = (Vin*Cin)/(Vref*Cfb) \qquad (2)$$
Vin: input signal magnitude, Vref: Reference voltage
Insert equation (2) to (1)

$$\text{Total } kT/C \text{ noise} = kT/Cin(1+Vin/(MI*Vref)) \qquad (3)$$
The noise contribution from the reference circuit is:

$$\text{Reference noise contribution}=Cfb/(Cin+Cfb) \qquad (4)$$
Insert equation (2) to (4)

$$\text{Reference noise contribution}=1/((MI*Vref/Vin)+1) \qquad (5)$$

Equation 3 indicates, for a given input signal, a high modulation index and large reference voltage minimizes the kT/C noise contribution from the feedback DAC capacitors, thus reducing the input sampling capacitor size for a given design target. Equation 5 also suggests that, for a given input signal, a high modulation index and large reference voltage minimizes both thermal and flicker noise contribution from the reference circuit.

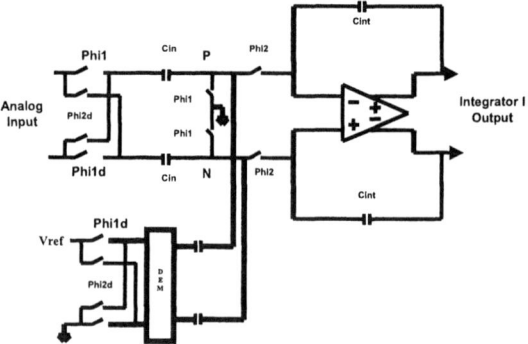

Fig. 4 Integrator 1 and feedback DAC

A double sampling scheme is used for both input and reference feedback network. A modulation index of 0.9 is used in this design and the single ended reference voltage is set at 4V. An on-chip reference circuit also provides better PSRR than sampling the power supply [5]. Reducing the input sampling capacitor size leads to smaller sampling switches and less charge injection. Since the modified feedforward modulator eliminates most signal energy leakage into the delta sigma loop and reduces integrator output swing, design requirements for Integrator 1 opamp are greatly relaxed. Transistors of the Integrator 1 opamp are carefully sized to suppress flicker noise. A folded cascode, single stage opamp with PMOS input pair is selected for the first integrator opamp. Simulated DC gain is 85dB and unity gain bandwidth is 90MHz.

IV DIGITAL DESIGN

Decimation Filter Design

Besides attenuating out-of-band quantization noise energy folded back into input signal band, the decimation filter needs to maintain linear phase and minimize passband ripple for high precision measurement applications. For single die A/D converter design, it is also important to suppress the switching noise of the decimation filter to achieve the ADC performance targets. The adopting of thirty three quantization levels for the analog modulator reduces out-of-band quantization noise energy substantially, which helps to relax stop band attenuation requirement for the decimation filter. An on-chip high pass filter further processes the decimation filter output to remove the DC offset from the analog modulator.

In conventional multistage comb filter design, an additional circuit is needed to compensate for the passband droop. Halfband filters are used for this design. Since a halfband filter has minimum passband ripple, the passband droop compensation circuit is eliminated. A transposed form structure is applied to implement the half band decimation filter to reduce memory requirements. The last two stages of the decimation filtering and the high pass filtering are implemented by a fixed point MAC engine. Digital circuit synthesis is optimized to achieve optimum trade off between minimizing worse case delay and digital die area. This linear phase filter is simulated with 100dB stopband attenuation and 0.00015dB passband ripple.

V MEASUREMENT RESULTS & CONCLUSION

The die photo is shown in Fig. 7. All measurement results were taken with a 5V analog and 1.8V digital power supply. The A/D converter achieves 128dB dynamic range, –114dB THD over 1kHz bandwidth. FFT plots for -60dBFS and -1dBFS 100Hz input signals are shown in Fig. 5 and Fig. 6. The chip is manufactured in a 0.35μm double poly, triple metal CMOS process. The core die area, which includes the reference voltage circuit and decimation filter is 14.8 mm^2. Power consumption is less than 140mW per channel.

REFERENCE

[1] I. Fujimori, K. Koyama, D. Trager, F. Tam, and L. Longo, "A 5V single-chip Delta-Sigma audio A/D converter with 111 dB dynamic range," Proc. 1996 IEEE CICC, May 1996.

[2] I. Fujimori, L. Longo, A. Hairapetian, K. Seiyama, S. Kosic, J. Cao, and S. Chan, "A 90-dB SNR 2.5-MHz output rate ADC using cascaded multibit delta sigma modulation 8x oversampling ratio," ISSCC2000.

[3] Lukas, et al, "3mW, 74dB SNR 2MHz Continuous time Delta-Sigma ADC with a tracking ADC Quantizer in 0.13um CMOS", ISSCC 2005.

[4] Steven R, et.al, "Delta-Sigma Data Converters, Theory Design and Simulation", IEEE press 1997, pp.262.

[5] Yuqing Yang et al, "A 114dB 68mW chopper stabilized stereo multibit audio A/D converter in 5.62 mm^2", ISSCC 2003.

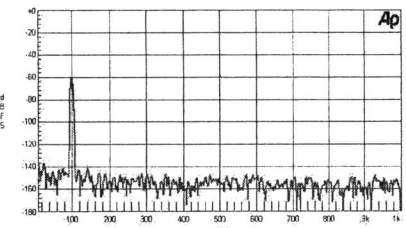

Fig.5 FFT plot for -60dBFS 100Hz input signals

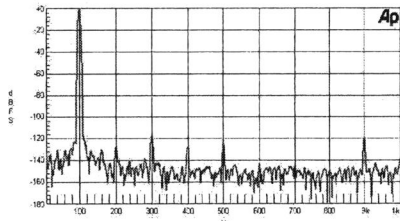

Fig.6 FFT plot for -1dBFS 100Hz input signals

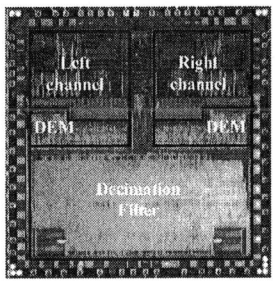

Fig. 7 Die photo

A Bit-sliced, Scalable and Unified Montgomery Multiplier Architecture for RSA and ECC

M. Sudhakar, R.V. Kamala, M.B. Srinivas

Center for VLSI and Embedded Systems Technologies (CVEST)

International Institute of Information Technology (IIIT) - Hyderabad
Gachibowli, Hyderabad - 500032, Andhra Pradesh, INDIA
sudhakarmaddi@research.iiit.ac.in, rv.kamala@gmail.com, srinivas@iiit.ac.in

Abstract— This paper presents a reconfigurable, bit-sliced, scalable Montgomery multiplier architecture which can operate in both prime and binary fields, that is, GF(p) and $GF(2^n)$. It can be configured for any bit length thus making it applicable for emerging elliptic curve cryptography (ECC) as well as widely used RSA cryptosystems. Existing word-based, scalable multiplier architectures perform well for key sizes in RSA (but not ECC) as they result in higher computational time. Limited utility of word-based architectures for ECC precisions, which are in general not equal to an integer multiple of word-size, is discussed and a new bit-sliced architecture to improve the performance in terms of delay is proposed. The new bit-sliced, scalable architecture computes the Montgomery multiplication with fewer clock cycles compared to existing architectures by configuring them at bit-level rather than at word-level, without compromising on the performance. Synthesis results (Mentor Graphic's Leonardo Spectrum) are compared with that of other scalable architectures and discussed.

I. INTRODUCTION

Elliptic curve cryptography (ECC) [1,2] is increasingly becoming an attractive alternative to traditional RSA systems [3], as it offers same level of security with smaller key sizes [4]. For example, it has been proved that a 163-bit ECC key size provides same level of security as an equivalent 1024-bit RSA key [5]. However, since RSA is still being used in most of the applications, it is attractive to design hardware architectures that can operate for both ECC and RSA, efficiently.

Modular multiplication is the basic operation for both RSA and ECC cryptosystems but uses expensive division operations. To avoid this, Montgomery [6] proposed an algorithm that replaces division operations with simple shift operations and comparatively easy to perform in hardware. Many algorithms and hardware implementations based on Montgomery multiplication have been developed for a fixed precision of the operands [7-9]. Since RSA algorithm operates on a precision of 1024-bits or even more for reasonable security, it requires more area when realized in hardware. To limit this area overhead, various scalable architectures have been proposed [10-15]. Hardware designs are said to be scalable if there is an ability to reuse the same design in both space and time until the desired result is obtained. A unified architecture is one which can perform arithmetic in both GF(p) and $GF(2^n)$ fields and various unified and scalable architectures have been proposed in literature [11-15].

Großschädl [12] proposed a bit-serial, scalable and unified multiplier architecture, in which modular multiplication was based on

MSB-first shift-and-add method. Except [12], all other architectures [11, 13-15] are based on processing multiple bits, in general powers of two, at a time. These architectures perform well for RSA precisions by processing 'w' bits at a time, where 'w' represents the word-size because RSA key sizes (256, 512, 1024, 2048 or even more) are exact multiples of word-size (w = 8, 16 or 32). But most of the recommended ECC key sizes, especially for binary field, are not integer multiples of word-size as shown in Table I [16, 17]. As explained in section V, use of existing word-based scalable multiplier architectures for ECC key-sizes requires significant number of extra clock cycles (compared to the architecture being proposed in this work) and result in more computational time and thus extra power consumption. The proposed scalable and unified architecture eliminates these extra clock cycles by configuring the hardware at bit-level rather than at word-level, while still maintaining the same performance (area-time product) for RSA precisions.

Rest of the paper is organized as follows: Section II discusses some basic concepts related to Montgomery multiplication and carry save arithmetic (CSA). Section III introduces the scalable and unified Montgomery multiplication algorithm and its performance estimate. The overall organization of the bit-sliced, scalable multiplier architecture and its main functional blocks are shown and described in section IV. Experimental results are shown in section V, followed by conclusions in section V.

TABLE I.   ECC VERSUS RSA KEY SIZES

ECC Key sizes		RSA Key sizes
GF (p)	$GF(2^n)$	
192	163	1024
224	233	2048
256	283	3072
384	409	7680
521	571	15360

II. MONTGOMERY MULTIPLICATION & CARRY SAVE ARITHMETIC

Given the two M-residues a, b and the modulus p of length n bits, Montgomery multiplication in GF(p) is defined as MonPro (a, b) = a × b × r^{-1} (mod p), where $r = 2^n$ and p is an integer in the range $2^{n-1} < p < 2^n$ such that gcd(r, p) = 1. Major bottleneck to perform arithmetic operations in GF(p) is carry propagation during addition/subtraction operations. To circumvent this problem, Kim et al. [18] used carry save adders which consist of two levels of carry save logic (CSL). Later, Bunimov et al. [19] improved this by replacing one level of CSL with a look-up table. An important limitation of these designs is the conversion between input and output formats for applications where repeated multiplications are required like in RSA exponentiation. To overcome this conversion problem, McIvor et al.

252

978-1-4244-1709-4/07/$25.00 © 2007 IEEE

[20] proposed two algorithms using five-to-two CSA (three levels of CSL) and a four-to-two CSA (two levels of CSL). Whereas in GF(2^n), given input polynomials a(x), b(x) and modulus p(x), Montgomery multiplication is defined as MonPro [a(x), b(x)] = a(x).b(x).x^{-n} (mod p(x)). Since binary extension field is free from carry propagation, the addition operations can be replaced by simple xor operations.

Savas et al. [11] discussed the importance of designing a unified architecture which can operate in both GF(p) and GF(2^n). To date, considerable amount of work has been done in realizing unified Montgomery multiplier architectures [11-15, 21, 22] in hardware. Sudhakar et al. [21] proposed an efficient, unified and reconfigurable multiplier that uses four-to-two CSA (figure 1) which requires fewer clock cycles for n-bit multiplication (n $\leq$ N), where N is the maximum precision that the multiplier can support. However, with increasing operand precision as in RSA, the design requires more area, precisely more number of bit-slices. To overcome this area overhead, in this paper authors propose first bit-sliced, scalable multiplication algorithm and its hardware architecture using CSA.

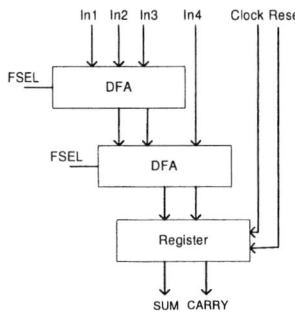

Figure 1. Four-to-two CSA

III. A NEW SCALABLE AND UNIFIED MONTGOMERY MULTIPLICATION ALGORITHM

The modified Montgomery multiplication algorithm is presented as algorithm 1, which can operate in both prime and binary fields. It can perform multiplication by configuring the architecture at bit-level rather than word-level, which is more attractive for ECC key-lengths. The inputs to the algorithm are (A1, A2), (B1, B2), p and FSEL, where (X, Y) is dual-field carry-save representation (DCSR) of a number Z such that Z = X + Y, operator '+' representing addition in Galois field. Since B1, B2 and p do not change throughout the multiplication, C1 and C2 can be precomputed. While operating the algorithm in GF(2^n) mode, B1, A1, C1 and R1 are always zero [21]. Signal 'FSEL' is used to select the mode of operation (FSEL = 1 for GF(p) mode and FSEL = 0 for GF(2^n) mode). When FSEL = 1, carry is computed whereas when FSEL = 0, carry is always forced to zero which is the desired property of GF(2^n) addition.

This algorithm reuses the hardware e = $\lceil n / N \rceil$ number of times per each process cycle, where n and N represent input operand precision and number of bit-slices in hardware, respectively. In one process cycle, one bit of A (A_i) will generate partial values of R1 and R2 by computing modular multiplication on n bits of B1, B2, p, C1 and C2. The A_i value is computed by adding the least significant bits of the A1 and A2 input operands using a dual field adder (DFA). Once the current A_i value has been used, A1 and A2 operands are barrel-shifted by one bit so that the next A_i value can be determined.

Inputs   A1, A2, B1, B2, p, FSEL

Outputs $R1^{(n)}, R2^{(n)}$

$C1, C2 = DCSR(B1 + B2 + p + 0)$

$R1_0^{(0)} = 0, \ R2_0^{(0)} = 0,$

$e = \lceil n / N \rceil$

for i = 1 to n

   $q_i = (R1_0^{(i-1)} + R2_0^{(i-1)}) + (A_i \times (B1_0 + B2_0)) \bmod 2$

   for j = 1 to e

$$m = \begin{cases} N, & j \neq e \\ n - ((e - 1) \times N), & j = e \end{cases}$$

     for k = (1 + (j - 1) \times N) to (m + (j - 1) \times N)

      if $A_i = 0$ and $q_i = 0$ then

        $R1_k^{(i)}, R2_k^{(i)} = DCSR(R1_k^{(i-1)} + R2_k^{(i-1)} + 0 + 0) >> 1$

      elseif $A_i = 1$ and $q_i = 0$ then

        $R1_k^{(i)}, R2_k^{(i)} = DCSR(R1_k^{(i-1)} + R2_k^{(i-1)} + B1_k + B2_k) >> 1$

      elseif $A_i = 0$ and $q_i = 1$ then

        $R1_k^{(i)}, R2_k^{(i)} = DCSR(R1_k^{(i-1)} + R2_k^{(i-1)} + p_k + 0) >> 1$

      else

        $R1_k^{(i)}, R2_k^{(i)} = DCSR(R1_k^{(i-1)} + R2_k^{(i-1)} + C1_k + C2_k) >> 1$

   end for loop

  end for loop

end for loop

return $R1^{(n)}, R2^{(n)}$

Algorithm1: Proposed Scalable and Unified Montgomery Multiplication Algorithm

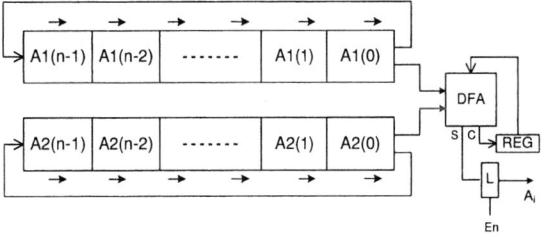

Figure 2. Barrel-shifter

This A_i value is latched until next process cycle starts as shown in figure 2. It does not add to the critical path delay of the proposed algorithm, as only one bit dual field addition is required per each processing cycle, which can be computed in parallel with the dual field CSAs. Thus, no extra clock cycles are required to compute A_i value. In the representation $X_Z^{(Y)}$, superscript (Y) represents the process cycle whereas subscript Z represents the bit number. In algorithm 1, outer-most loop represents the process cycle, whereas the inner loop computes modular multiplication on 'm' bits for every clock cycle, which is determined by the control unit. For example, if $n = 7$ and $N = 4$ ($n > N$), the hardware is reused 2 times per each process cycle. In the first clock cycle, this algorithm performs Montgomery multiplication on LSB 4-bits of B1, B2, p, C1 and C2. For the next clock cycle, it performs on the remaining MSB 3-bits, which is determined by m value as shown in algorithm 1. Since n and N values do not change throughout the multiplication process, these m values can be precomputed. For $n \leq N$, $e = 1$ and $m = n$, which means for each clock cycle it computes Montgomery multiplication on n-bits of B1, B2, p, C1 and C2. The remaining $(N - n)$ bits are inactive, which means that the corresponding slices are not used during the actual multiplication operation.

Figure 3 represents the dependency graph for the proposed algorithm for different values of n and N. Each column is computed by a separate bit-slice. In contrast to conventional word-based pipeline architectures, bit-slices in the proposed architecture perform the computation parallel to each other. The inactive slices are also shown in figure 3. These features distinguish the proposed design from existing scalable multipliers which can not be configured at bit-level.

A. Performance Estimate

For a given N value, the total computation time T in terms of number of clock cycles to compute the Montgomery multiplication with n-bits of precision is

$$T = \begin{cases} n + 2 & n \leq N \\ \lceil n/N \rceil \times n + 2 > & n \quad N \end{cases} \quad (1)$$

The first case represents the situation where the input operand precision 'n' is less than or equal to the number of bit slices in the hardware, N. In this case, the proposed architecture computes n-bit multiplication in just n+2 clock cycles as described in [21]. The second case represents the situation where sufficient bit-slices are not present in the hardware. Then for each bit of A, the hardware should be reused e $\left(\approx \lceil n/N \rceil \right)$ times to complete one process cycle. The extra two clock cycles are required to load the input operands and precomputed values to the corresponding input registers. As the N value increases, there is a reduction in the total computation time up to a lower bound as shown in (1). The best value of N depends on the execution time and area   trade-offs.

IV. PROPOSED SCALABLE AND UNIFIED MULTIPLIER ARCHITECTURE

Figure 4 shows the top level organization of the proposed bit-sliced, scalable and unified multiplier architecture. The main functional blocks are processing unit (PU), memory elements, barrel shifter and control unit. An N-bit PU consists of N single bit-slices. For example, a 3-bit PU is shown in figure 5. The control unit function can be inferred from the algorithm description that is provided, combined with other data manipulation tasks that must be done to transfer data between the multiplier and the host system. Here the different input operands are considered to be in CSA form, which can be precomputed and stored in RAM module.

A. Memory Organization

Prior to begin the multiplication process, the RAM should be loaded with the input operands A1, A2, B1, B2, p, C1 and C2. Upon reset, the input registers are loaded with the corresponding operand values and output registers are initialized with zeroes. For $n \leq N$, these registers are loaded only once for each multiplication process, and also there is no need for queuing the intermediate results except for final result conversion. But for $n > N$, they need to be loaded with corresponding input operands for each clock cycle and also queue is required to store the intermediate results. This overhead can be reduced by properly choosing the N value. The maximum length of the queue (Q_{max}) depends on the maximum number of iterations (e_{max}) per each process cycle and the N value. This length is determined as:

$$Q_{max} = \begin{cases} e_{max} \times N & \text{for } n > N \\ 1 & \text{for } n \leq N \end{cases} \quad (2)$$

B. Register Controller

For $n > N$, one extra clock cycle is required to read and write the intermediate results from/to the queue. To eliminate the delay due to this extra clock cycle, two sets of output registers are considered which can be inter-operable by using multiplexer based switch control logic. As the multiplication process is going on, the intermediate results generated in the present clock cycle get stored in one set of output registers, while another set of registers are already loaded with the previous cycle values.

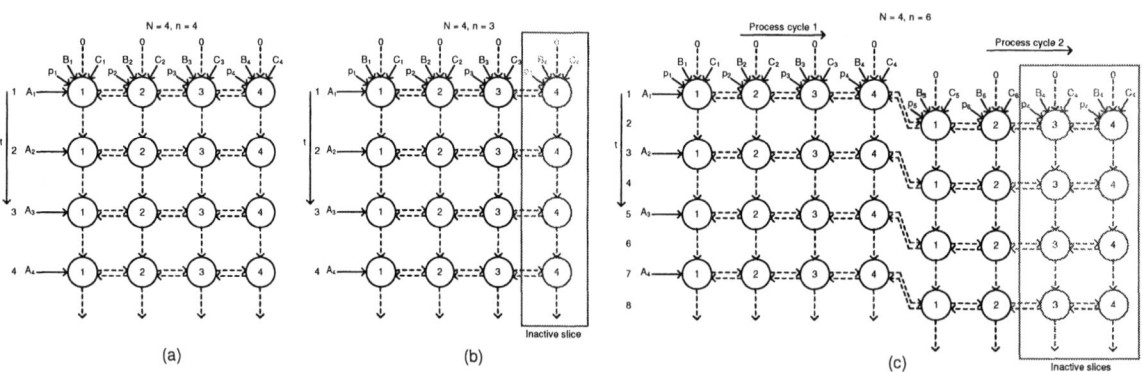

Figure 3.   Dependency graph for (a) N = 4, n = 4 (b) N = 4, n = 3 and (c) N = 4, n = 6

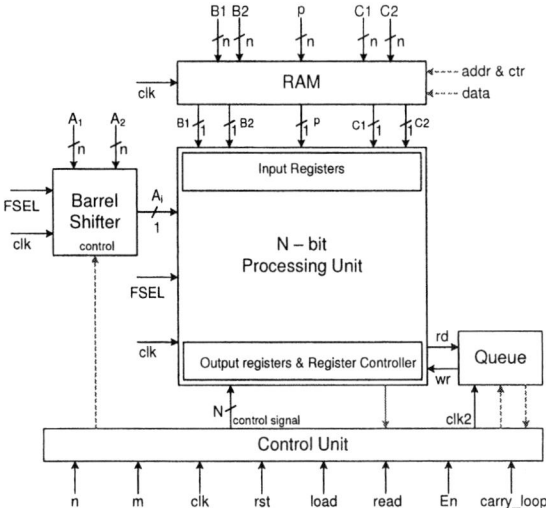

Figure 4. Proposed Scalable and Unified Multiplier Architecture

For the next clock cycle, the latter set of registers attach to the corresponding bit-slices, while the intermediate results present in the former set of registers are loaded to the queue. For this purpose, another clock (clk2) with twice the external clock frequency is required for queue. This can be easily achieved by using Digital Clock Managers (DCMs) [23], which are present internally in all modern Virtex series FPGAs. A comprehensive description of DCMs is provided in [24, pp 355 - 361]. This process adds a little delay for the normal flow of execution which is however very less compared to the critical path delay of one bit-slice.

C. 3-bit Processing Unit (PU)

The internal design of the 3-bit PU is shown in figure 5, which has been modified to the one presented in [21] to provide scalability. These modifications include register controller and some extra hardware and control signals. Here all inputs and outputs are in carry-save representation except the modulus, p. Each bit-slice consists of one 2:1 multiplexer, two 4:1 multiplexers, and two dual field adders along with few registers. The control signal generator, which is internal to the control unit (figure 4), generates control signals based on the precomputed 'm' values. These control signals are responsible to achieve reconfigurability at bit-level, which is a distinguishing feature of the proposed bit-sliced architecture. The value of control signal for i^{th} slice is defined as:

$$control[i] = \begin{cases} 1 & \text{if } i \le m \\ 0 & \text{otherwise} \end{cases} \qquad (3)$$

Signal 'control[i]' is used as selection line for the 2:1 multiplexer for i^{th} slice as shown in figure 5. When selection line is zero, C[i] is stored in the output register R2[i], else S[i+1] is stored, where C[i] and S[i+1] are the carry bit out of the first dual field adder (DFA^1) of ith slice and sum bit out of the second dual field adder (DFA^2) of $(i+1)^{th}$ slice respectively. The ability to configure the design at bit-level has the advantage to carry out the multiplication in fewer clock cycles compared to the existing architectures for ECC key-lengths. When FSEL is active high, it works as a scalable multiplier in prime field, GF(p) and when active low it works as a scalable multiplier in binary extension field, GF(2^n). Also the control signals can be used as gated

signals to avoid unnecessary transitions in the inactive slices, which can result in reduced power consumption.

V. PERFORMANCE RESULTS AND DISCUSSION

The functionality of the proposed architecture has been verified using verilog HDL for different values of n and N. It has been synthesized using Mentor Graphic's Leonardo Spectrum targeting Xilinx VirtexII Pro 2VP100ff1704 (speed grade-7) FPGA [23]. For 1024-bit Montgomery multiplication, a maximum clock frequency of 140MHz has been obtained. The design has however not been synthesized on an ASIC. From the experimental results obtained, it has been estimated that an N-bit PU requires approximately 6N flip-flops and (5N+3) 4-input lookup tables (LUTs). RAMs for the input and output operands and queue depend on the operand precision, n and the number of bit-slices presented in the hardware, N. But a rough estimate indicates that '6n' bits of storage for input operands is required along with [e x (2N-1)] bits for the queue. For example with N=256, a 256-bit multiplication requires another 132 flip-flops and 278 4-input LUTs. Putting all together, the complete 256-bit Montgomery multiplier contains 1668 flip-flops and 1594 LUTs. As mentioned earlier, it is more attractive to design hardware architectures that can support both ECC and RSA cryptosystems in a single unit. In the following sub-sections, authors compare latency (clock cycles) of the proposed scalable architecture with existing scalable ones for different ECC and RSA key-lengths.

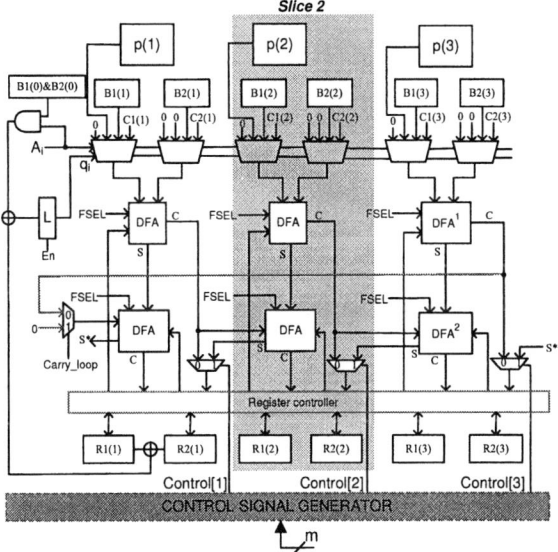

Figure 5. 3-bit Processing Unit (PU)

A. Comparison for different ECC key-lengths

Table II lists the total computational time in number of clock cycles and percentage gain of the proposed scalable architecture over previous architectures for recommended ECC key-lengths. Großschädl [12] proposed a bit-serial unified multiplier architecture, in which modular multiplication is based on MSB-first shift-and-add method.

TABLE II.   SCALABLE MULTIPLIER LATENCIES (CLOCK CYCLES) FOR ECC KEY-LENGTHS

ECC key lengths		[12]		18 PE x 32-bit = 576 (% Gain)								Proposed N = 576
				[11]		[13]		[14]		[15]		
Binary Field	163	163	(-1)	333	(50)	175	(6)	374	(56)	201	(18)	165
	233	233	(-0.8)	477	(51)	249	(6)	487	(52)	262	(11)	235
	283	283	(-0.7)	578	(51)	300	(5)	599	(52)	321	(11)	285
	409	409	(-0.5)	838	(51)	434	(5)	862	(52)	462	(11)	411
	571	571	(-0.4)	1177	(51)	599	(4)	1200	(52)	643	(11)	573
Prime Field	192	288	(33)	398	(51)	391	(50)	411	(53)	220	(12)	194
	224	336	(33)	457	(51)	456	(50)	486	(53)	260	(13)	226
	256	384	(33)	528	(51)	529	(51)	561	(53)	300	(14)	258
	384	576	(33)	782	(51)	783	(51)	824	(53)	441	(13)	386
	521	782	(33)	1060	(51)	1062	(51)	1088	(52)	584	(10)	523

While it computes n-bit multiplication in 'n' clock cycles in GF(p), it requires approximately 1.5n clock cycles in GF(2^n) which is 33% more compared to that of the proposed one. Since the architectures presented in [11, 13-15] are word-based architectures, authors consider word-size as 32 in this paper for the sake of best comparison. Note that the computational time decreases as the word-size increases. Also since the maximum key-length in ECC is 571 bits, it is assumed that the number of bit-slices present in the hardware ('wp' for word-based architectures and N for the proposed one) is 576.

Savas et al. [11] proposed a radix-2 based scalable architecture using the design methodology proposed in [10], which can operate in either field, GF(p) and GF(2^n). They also proposed [13] a dual-radix multiplier architecture which operates in radix-2 in GF(p) mode and in radix-4 in GF(2^n) mode. Thus, they achieved almost 50% reduction in number of clock cycles in GF(2^n) mode compared to GF(p) mode. Later Tenca et al. [14] described in detail the design trade offs in terms of area and time for different word sizes. Recently, Harris et al. [15] improved Tenca-Koc multiplication algorithm [14] to eliminate the two-clock cycle latency from one PE to the next which results in fewer clock cycles. The drawback of the existing word-based scalable multipliers is their inefficiency for input precisions which are not exact integer multiple of word-size. Because as described in [14], for n = 5, w = 1 and e = 6, two processing elements (p = 2) require 23 clock cycles to compute the modular multiplication. Whereas the proposed architecture requires only 17 clock cycles with N = 2, because of the ability to configure the architecture at bit-level rather than at word-level. This effect significantly predominates as the word-size increases as shown in Table II for w = 32. And it is clear that the proposed design requires fewer clock cycles compared to [11-15] for ECC key-lengths.

B. Comparison for different RSA key-lengths

Since RSA is the most widely used public-key cryptosystem, authors also discuss the performance of the proposed scalable architecture for RSA key-lengths. Even though the architectures proposed in [11, 12, 13] are scalable, they are not flexible in selecting the number of processing elements according to the input operand precision and the number of bit-cells present in the hardware. This is important when operating on larger precisions to get better time and area trade-offs. Hence in Table III and Table IV, authors compared proposed bit-sliced, scalable multiplier architecture with that

presented in [14, 15]. It has shown that for operand precisions 'n' up to the number of bit cells 'wp', [15] is about twice as fast as that in [14]. For larger operand lengths, the performance of the two designs is comparable. However, being a bit-sliced architecture with parallel execution (shown in figure 3), it requires much fewer clock cycles (15%~2%) to compute Montgomery multiplication than [15] as shown in Table III for RSA precisions also.

TABLE III.   SCALABLE MULTIPLIER LATENCIES (CLOCK CYCLES) FOR RSA KEY-LENGTHS

Bit cells	n	w = 32 (% Gain)				Proposed
		[14]		[15]		
256	256	550	(53)	303	(15)	258
	512	1102	(6.8)	1111	(7.7)	1026
	1024	4238	(3.3)	4263	(3.9)	4098
	2048	16654	(1.6)	16711	(1.9)	16386
512	256	534	(52)	287	(10)	258
	512	1070	(52)	575	(10.6)	514
	1024	2142	(4.3)	2159	(5)	2050
	2048	8350	(1.9)	8399	(2.4)	8194
1024	256	526	(51)	279	(7.5)	258
	512	1054	(51)	559	(8)	514
	1024	2110	(51)	1119	(8.3)	1026
	2048	4222	(2.9)	4255	(3.7)	4098

In Table IV, a comparison of proposed scalable architecture with existing ones [14, 15] in terms of area, maximum frequency of operation and the time to execute 256 and 1024-bit modular exponentiation is presented. Here it is assumed that n-bit exponentiation requires at most (2n+2) modular multiplications including the conversion to and from p-residues [15]. An FPGA implementation of 1024-bit RSA modular exponentiation using the proposed bit-sliced architecture performed in 15 ms, which is slightly better than (6%) [15]. At the same time, there is a slight area overhead (5%) than [15], which imply that the overall area-time product is almost same for both architectures for RSA precisions.

TABLE IV.   COMPARISON OF DIFFERENT SCALABL ARCHITECTURES FOR MODULAR EXPONENTIATION TIMES

Ref	Bit cells	Technology	Hardware	Clock speed	Scalable/ Unified	256-bit time (ms)	1024-bit time (ms)
Proposed	N = 1024	Xilinx VirtexII Pro	5892 LUTs	140 MHz	Yes / Yes	0.95	15
[15]	64 PEs x 16 bits	Xilinx Virtex Pro	5598 LUTs	144 MHz	Yes / Yes	1.0	16
[14]	40 PEs x 8 bits	0.5 μm CMOS	28K gates	80 MHz	Yes / Yes	3.8	88.2

VI.   CONCLUSION

In this paper, limitations of the existing word-based, scalable architectures are discussed for operand precisions which are not equal to exact integer multiple of word-size. To overcome this, authors proposed a bit-sliced, scalable Montgomery multiplication algorithm and its architecture which can be configured at bit-level. It has been shown that the proposed architecture requires fewer clock cycles to compute Montgomery multiplication compared to the existing ones, not only for ECC precisions but also for RSA precisions. The design trade-offs in terms of latency, speed and area have been discussed and compared with existing scalable architectures. The ability to configure the architecture at bit-level and inactivating the unused bit-slices to minimize the power clearly differentiates the proposed multiplier from the existing ones. Finally, to the best of authors' knowledge, this is the first scalable and unified multiplier architecture proposed that can be configured at bit-level.

REFERENCES

[1] V.S. Miller, "Use of elliptic curves in cryptography," in Proc. Adv Cryptolog. (Crypto'85), 1986, pp. 417-426.

[2] N. Koblitz, "Elliptic curve cryptosystems," Math. Computers., 1987, vol. 48, pp. 203-209.

[3] Rivest. R.L A. Shamir, and L. Adleman., "A method for obtaining Digital Signatures and Public-key cryptosystems", Comm. ACM, 1978, Vol. 21, No. 2, pp. 120-126.

[4] A.K. Lenstra and E.R. Verheul, Selecting cryptographic key sizes, In H. Imai and Y. Zheng (eds.), Public Key Cryptography – PKC 2000, Lecture Notes in Computer Science, Springer-Verlag, Berlin, Germany, 2000, vol. 1751, pp. 446-465.

[5] US National Institute of Standards and Technology (NIST), Cryptographic Toolkit Jun. 2002, pp. 83–84 [Online]. Available: http://csrc.nist.gov/CryptoToolkit/kms/guideline-1.pdf

[6] Montgomery. P.L, "Modular Multiplication without Trail Division", Math. Computers, 1985, Vol 44, No.70, pp. 519-521.

[7] A. Bernal and A. Guyot, "Design of a Modular Multiplier Based on Montgomery's Algorithm," Proc. 13th International Conf. Design of Circuits and Integrated Systems (DCIS '98), Nov. 1998.

[8] C.-C. Yang, T.S. Chang, and C.-W. Jen, "A New RSA Cryptosystem Hardware Design Based on Montgomery's Algorithm," IEEE Trans. on Circuits and Systems - II: Analog and Digital Signal Processing, vol. 45, no. 7, pp. 908-913, July 1998.

[9] C.-Y. Su, S.A. Hwang, P.-S. Chen, and C.-W. Wu, "An Improved Montgomery's Algorithm for High-Speed RSA Public-Key Cryptosystem," IEEE Trans. Very Large Scale Integration (VLSI) Systems, vol. 7, no. 2, pp. 280-284, June 1999.

[10] A.F. Tenca and C.K. Koc,, "A Scalable Architecture for Montgomery Multiplication," Proc. First Int'l Workshop Cryptographic Hardware and Embedded Systems—CHES '99, C, .K. Koc, and C. Paar, eds., pp. 94-108, Aug. 1999.

[11] Savas. E, A. F. Tenca and C. K. Koc., "A scalable and unified multiplier architecture for finite fields GF(p) and GF(2^m)", Proceedings of CHES, 2000, pp. 281-296.

[12] Johann Großschädl, "A bit-serial unified multiplier architecture for finite fields GF(p) and GF(2^m)", Proc. CHES 2001, August 2001, pp. 202-218.

[13] Savas. E, A.F. Tenca, M.E. Ciftcibasi and C. K. Koc., "Multiplier architectures for GF(p) and GF(2^n)," IEE Proc.-Comput. Digit. Tech., March 2004, Vol. 151, No. 2, pp. 147-160.

[14] A. Tenca, C. K. Koc, "A scalable architecture for modular multiplication based on Montgomery's algorithm", IEEE Trans. Computers, Sept 2003, Vol. 52, No. 9, pp. 1215-1221.

[15] David Harris, Ram Krishnamurthy, Mark Andres, Sanu Mathew, Steven Hsu., "An Improved Unified Scalable Radix-2 Montgomery Multiplier," IEEE. (ARITH-17), 2005, pp. 172-178.

[16] Certicom Corporation, The Basics of ECC 2006 [Online]. Available: http://www.certicom.com/index.php?action=res,ecc_faq

[17] NIST DSS Standard, Basicrypt – Elliptic Curve Cryptography (ECC) Benchmark Suite benchmark, http://csrc.nist.gov/publications/fips/fips186-2/fips186-2-change1.pdf

[18] Kim, Y.S., Kang, W.S., and Choi, J.R.: 'Implementation of 1024-bit modular processor for RSA cryptosystem'. http://www.ap-asic.org/2000/proceedings/10-4.pdf

[19] Bunimov, V., Schimmler, M., and Tolg, B.: 'A Complexity-Effective Version of Montgomery's Algorithm'. Presented at the Workshop on Complexity Effective Designs (WECD02), May 2002

[20] C. McIvor, M. McLoone and J.V. McCanny., "Modified Montgomery Modular Multiplication and RSA Exponentiation Techniques," IEE Proceedings – Computers & Digital Techniques, 2004, Vol. 151, No. 6, pp. 402-408.

[21] M. Sudhakar, R.V. Kamala, M.B. Srinivas: 'An Efficient, Reconfigurable and Unified Montgomery Multiplier Architecture', Proceedings. IEEE/ACM International Conference on VLSI Design Bangalore, India, January 6-10, 2007, pp. 750-755.

[22] T. Blum and C. Paar, "Montgomery Modular Exponentiation on Reconfigurable Hardware," Proc. 14th IEEE Symp. Computer Arithmetic, pp. 70-77, Apr. 1999.

[23] Virtex-II Pro and Virtex-II Pro X Platform FPGAs: Complete Data Sheet www.xilinx.xom/bvdocs/publications/ds083.pdf

[24] Xilinx Libraries Guide: http://toolbox.xilinx.com/docsan/xilinx7/books/docs/lib/lib.pdf

Low Power On-Chip Thermal Sensors based on Wires

Basab Datta and Wayne P. Burleson

Electrical and Computer Engineering Department

University of Massachusetts, Amherst, MA, U.S.A.

{bdatta, burleson}@ecs.umass.edu

ABSTRACT

Current thermal scaling trends in multilevel low-k interconnect structures suggest an increasing heat density as we move from substrate to higher metal levels. Thus, the deterioration of interconnect performance at extreme temperatures has the capability to offset the degradation in device performance when operating at higher-than-normal temperatures. Existing thermal sensing approaches rely heavily on devices (MOS/diodes). They are optimized for a low area and power overhead but continue to suffer from leakage and self-heating and also, tend to disregard the thermal impact on interconnects. We propose an alternate approach of using interconnects to perform the thermal sensing. With feature-size shrinking, metal layers are closer to the substrate suggesting a strong correlation between interconnect temperature and thermal profile of the underlying substrate. Thus, in addition to quantifying the temperature impact on interconnect signal delay; output of proposed sensors can be used to estimate substrate thermal status as well. The simplistic schemes proposed allow reuse of existing on-chip resources such as drivers and time-digitizers, have a low power requirement and are robust against variations in wire dimensions, non-uniform temperature distribution and supply noise.

1. INTRODUCTION

Aggressive technology scaling and an increasing demand for high performance VLSI circuits has resulted in higher current densities in the interconnect lines and increasingly higher power dissipation in the substrate [1]. Because a significant fraction of this power is converted to heat, an exponential rise in heat density is also experienced [2]. The higher die temperatures have a critical impact on CMOS circuit operation and reliability [6]. Different activities and sleep modes of the functional blocks in high performance chips cause significant temperature gradients on the substrate [3]. As we enter the GHz frequency regime, the magnitude of thermal gradients in the substrate can be expected to further increase. It has been recently reported that thermal gradients as large as $50^{\circ}C$ can exist across high performance microprocessor substrates [7]. To capture such significant temperature variations across the processor chip we need efficient but inexpensive (in terms of area and power) thermal sensing to optimize and localize thermal management schemes.

However, [4] predicts that in deep nanometer technologies, global metal lines are going to be much hotter than the substrate due to larger I_{rms} values and shows that even after considering densely embedded vias, the interconnect temperature is expected to increase significantly with scaling, due to increasing surface and grain boundary contributions to metal resistivity and decreasing ILD thermal conductivity. The massive increase in vertical

thermal gradients with technology scaling indicates that the deviation in interconnect performance at high temperatures can possibly override any performance improvement achieved though thermal sensing at the substrate level. Thus, efficient thermal sensing is needed not only to maintain an optimal power-performance envelope for the devices at the substrate level but also, to ensure accurate signal delay calculation at the interconnect level [8]. Therein lies the motivation for our work.

In this paper we outline a dual-objective for thermal sensing; one to get a relative measure of interconnect thermal status for both intra-layer and inter-layer gradients and two, to obtain an estimate of the substrate temperature profile. We propose schemes that make use of the deviation in interconnect behavior in a variable temperature environment to quantify the thermal susceptibility of interconnects in a particular region of the layer and to estimate the vertical gradients that exist between the same region in different layers. As the minimum feature size shrinks, the top most metal layers that carry the global signals get closer to the substrate [5] and hence, a strong correlation is bound to exist between the interconnect thermal profile and the underlying substrate temperature. In most existing on-chip thermal sensing approaches in literature, both the temperature sensing and digitization is done by devices (MOS/ diode). In deep sub-micron, MOS transistors in particular are plagued with high dynamic power consumption, leakage (enhanced at the higher temperatures), self-heating and thermally induced reliability issues. Thus our solution of using interconnects for thermal sensing will not only offer the obvious advantages of lesser power consumption, no leakage component and much lesser self-heating but also, will be a step towards addressing an emerging problem of dominance of metal temperatures over corresponding substrate values.

The remaining paper is organized as follows. In section 2 we discuss the temperature dependent signal delay model for a temperature dependent driver resistance and propose a 'thermal vulnerability' metric of wires. In section 3 we propose different techniques for temperature measurement using interconnects – both for horizontal and vertical gradients and outline the design concerns. In section 4, we present our analysis on thermal sensing using interconnects, the technology scaling trends, impact of variation in physical and environmental parameters and the impact of cross-talk and supply-noise. We draw our conclusions and present scope of future work in section 5.

2. TEMPERATURE DEPENDENT INTERCONNECT SIGNAL DELAY

We use the same model as the one proposed in [2] to explain signal behavior across interconnects in a variable temperature environment. The model considers an interconnect line of length

'L' and uniform width 'w' that is driven by a driver with on-resistance 'R_d', parasitic output capacitance 'C_p' and terminated by a load with capacitance 'C_{load}'. The line is partitioned into 'n' equal segments, each having a length 'Δx'. Using a distributed RC Elmore Delay model and assuming that 'n' tends to infinity; the delay 'D' of a signal passing through the line is given as: [2]

$$D = R_d.[C_p + C_{load} + \int_0^L C_0(x).dx] + \int_0^L r(x).(\int_x^L C_0(u).du + C_{load})dx$$

[1]

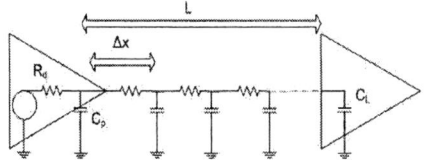

Fig.1 Distributed RC interconnect line

The capacitance per unit length (C_o) can be assumed to remain invariant to temperature variations across the length of interconnect. In [2], the delay equation was derived assuming a constant steady-state driver resistance and a simplistic model were used to illustrate the temperature dependence of the driver. Here, we will develop a more detailed model for 'R_d' and its temperature dependence. The equivalent resistance of the driver is given as: [11]

$$R_d \approx \frac{3}{4}.\frac{V_{dd}}{I_{dsat}}.(1 - \tfrac{7}{9}.\lambda V_{dd})$$

[2]

Wherein λ is the channel-length modulation constant and 'I_{dsat}' is given as [11]:

$$I_{dsat} = \mu(T).C_{ox}.\frac{W}{L}.[\{V_{dd} - V_t(T)\}.V_{dsat} - \frac{V_{dsat}^2}{2}]$$

[3]

The temperature dependent CMOS parameters are [12]:

The mobility of the channel carriers is modeled as:

$$\mu(T) = \mu(T_0).[\tfrac{T}{T_0}]^{-K_1}$$

[4]

And the threshold voltage variation with temperature is given as:

$$V_t(T) = V_t(T_0) - K_2(T - T_0)$$

[5]

Wherein 'T' is the absolute temperature, 'T_o' is the reference temperature, 'k_1' and 'k_2' are constants having values of (1.5-2) and (0.5-4 mV/K) respectively.

In terms of the temperature dependent parameters, 'R_d' can be written as:

$$R_d = \frac{3}{4}.\frac{V_{dd}.(1 - \tfrac{7}{9}\lambda V_{dd})}{\mu(T).C_{ox}.\frac{W}{2L}.[(V_{dd} - V_t(T))V_{dsat} - \frac{V_{dsat}^2}{2}]}$$

[6]

Though both mobility and the threshold voltage decrease with temperature, the mobility effect dominates for large overdrive voltages and hence, the drive resistance will increase with temperature.

Resistance of interconnect has a linear relationship with its temperature and can be written as: [2]

$$R(x) = R_o. (1 + \beta.T(x))$$

[7]

Where r_o is the resistance per unit length at the reference temperature and β is the temperature coefficient of resistance. Using [8] we can rewrite [1] as: [2]

$$D = D_0 + (C_0.L + C_{load})R_0.\beta \int_0^L T(x)dx - C_0.R_0.\beta. \int_0^L x.T(x)dx$$

[8]

Wherein $D_0 = R_d.(C_p + C_{load} + C_0.L) + C_0.R_0.\frac{L^2}{2} + R_0.L.C_{load}$

The value of R_d can be obtained from equations [6], [4] and [5]. For shorter wires, the delay component due to the driver might dominate over that of the wire and hence, any variation in 'R_d' due to temperature gradients will vary the signal-delay significantly. In such cases, R_d increase due to high temperatures of operation will override the increase in metal resistance of interconnect in terms of contribution to total signal delay. However, for long lines driven by large drivers, interconnect delay could dominate over driver delay and hence, metal resistance variation with temperature could become the deciding factor when computing the overall signal delay. Keeping the above in mind, we propose a metric called '*Thermal Vulnerability*' of a wire; we define it as:

Thermal Vulnerability (d) = D_{wire}/ D_{driver} [9]

Wherein D_{wire} is the interconnect-delay and D_{driver} is the driver-delay. Using [9] we can describe the component delays as:

$$D_{driver} = R_d.(C_p + C_{load} + C_0.L)$$

[10]

$$D_{wire} = C_0 R_0 \frac{L^2}{2} + R_0 L.C_{load} + (C_0.L + C_{load})R_0\beta \int_0^L T(x)dx - C_0.R_0\beta \int_0^L x.T(x)dx$$

[11]

The above metric can be used to gauge the susceptibility of a metal line to high interconnect temperatures. For a large, positive value of 'd', even a slight increase in the thermal gradients across the wire will result in significant error in the delay computation. For a small value of 'd' (much less than unity), temperature effect on interconnect delay will be offset by delay-variation of driver. The 'd' value for global metal lines will be rendered unnecessary because of their high capacitive component and relative insensitivity to temperature.

3. THERMAL SENSING SCHEMES USING INTERCONNECTS

Lower level interconnects (Metal layers 1/2/3) have a high resistive component which exhibits a linear dependence on temperature. This can be utilized for sensing temperature because the variable temperature gradients over a wire will reflect in the signal propagation delay across it. We use the term '*ThermoWire*' to describe a class of circuits that use interconnects for thermal sensing. The basic circuit as shown in Fig.2 is used to measure thermal gradients within a particular layer for a specific region in that layer. In our schemes, it has to be ensured that the 'thermal vulnerability' factor of the wires is much greater than unity so that the thermal status of the metal layers is captured accurately by the sensor output i.e. $D_{wire} > D_{driver.}$. The 1st scheme operates in the following manner.

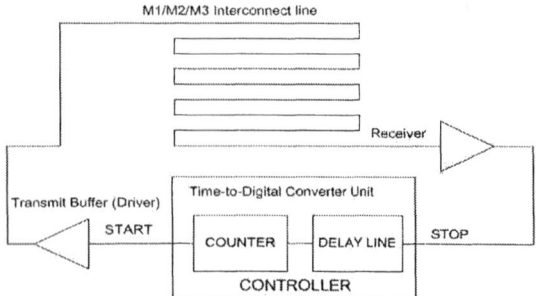

Fig.2 ThermoWire Scheme-1

The sensor-controller generates a 'Start' pulse that triggers an internal counter of the Time-to-Digital Converter (TDC) unit. A single, much larger-than-minimum-strength buffer drives the pulse through a long, winding, lower-level metal line (which spans the region for which temperature sensing is to be done but covers as small an area as possible) and ultimately terminates at a receiver buffer. The receiver output then becomes the 'stop' signal for the TDC. The start-to-stop delay or the TDC generated digital code will have a linear mapping with temperature. Since the thermal vulnerability factor of the wire used is much greater than unity, a linear increase in the delay output will map to a corresponding increase in the wire-temperature. Chip area can be saved by utilizing existing on-chip drivers when in non-functional mode. Because of the usage of lower-level metal lines; sensor reported temperature will be an indicator of underlying substrate temperature as well since for all technologies, the temperature differential up-to a vertical height of 0.3μm is very small [4]. A tight packing of the wire within a small area is necessary towards capturing local hotspots. A TDC akin to the one proposed in [14] can be used which is capable of providing 13ps single-shot resolution (in 0.35μ technology) with a temperature drift of less than 0.05ps/°C. The popular techniques of time-digitization are:

(i) Time-interval stretching followed by the counter

(ii) Time-to-amplitude followed by standard analogue-to-digital (A/D) conversions

(iii) Vernier method with two startable oscillators

(iv) TDC utilizing the tapped delay line

(v) Vernier method with a 'differential delay line' comprising of two tapped delay lines

TDCs implemented in ASICs such as the one proposed in [14] perform time-interval measurement by sampling the current states of the delay-line and counter and storing them 'on the fly', without interruption of the counting process. Precision, integrated TDC's with delay lines can be grouped in two categories, depending on the use of a PLL or DLL circuit [15]. The PLL contains a voltage-controlled-oscillator (VCO), whose frequency f_0 after optional dividing is compared with a reference frequency, f_r. The difference is detected, filtered, amplified, and used to adjust the frequency of the VCO to minimize the difference. In the DLL approach, the loop contains the voltage controlled delay line (VCDL). The delay of the line is varied to align the phases at the inputs of the phase detector. TDC's have been extensively researched for improvement of their resolution, dynamic range, conversion speed, calibration procedure, buffering and read-out interface. On-chip TDC's are generally used for jitter and drift-time measurements. Our thermal sensing approach proposes another utility for them, thus allowing reusability of on-chip resources.

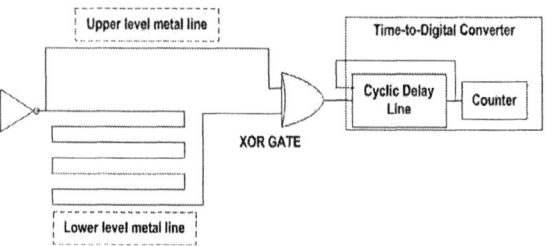

Fig.3 ThermoWire Scheme-2

A 2nd scheme, differential in nature can be used to estimate vertical thermal gradients. In this scheme, the driver buffer drives a signal across 2 wires, one belonging to the lower level and the other, an upper-level metal line. The lower-level metal line is around three times the length of the upper-level line and made to wound over a small area. The upper-level metal line although shorter in length is made to traverse the same region in its layer. Both terminate at a 'XOR' gate whose output will be a pulse having a width proportional to the temperature difference between the 2 layers. The basics of this scheme have been established for the delay-based sensor developed in [10], only this time we use interconnects to generate delays and use the same concept for inter-layer gradient estimation. The upper-level metal lines have a high capacitive and a low resistive component as a result of which their signal delays are relatively invariant to temperature although they are expected to reach extremely high temperatures. The signal traverses through the upper-level metal line with a relatively fixed delay while delay through the lower level metal line varies depending upon its current thermal status. Thus, the mismatch in the arrival times of the 2 signals at the input of the XOR gate will increase with decrease in temperature difference (metal temperature of lower-level line approaching that of upper level) and will cause the output pulse-width to increase accordingly. The time-to-digital converter (TDC) is of a specific type and proposed in [13]; it makes us of a pulse-shrinking mechanism, controlled by the in-homogeneity of inverters cascaded to form a delay line. The input pulse to TDC circulates through a cyclic delay line and is shrunk by a specific amount of width per cycle until it diminishes completely. A counter is used to count the circulation times of the input pulse and generates the corresponding digital output [13]. For both schemes, bulk of the power dissipation will be at the transmitter-receiver units and the units responsible for digitization of the output.

Some of the obvious design problems in these schemes include: achievable resolution of the TDC to be used impacts the minimum wire-length needed to produce commensurate delays, power requirement of the driver needed to drive the long-wires, packing the wire within a small area, power requirement of the TDC and the loss in spatial granularity due to ineffective wire-placement.

4. ANALYSIS & SIMULATION RESULTS

HSPICE™ (Synopsys) was the main simulation platform used in our analyses. The 45nm, 32nm and 22nm technology files were obtained from the Predictive Technology Model website [9].

Table 1 Scaled Local Interconnect Parameters

Tech. Node	Width (μm)	Spacing (μm)	Thickness (μm)	Height (μm)	Dielectric Constant (κ)
45nm	0.069252	0.069252	0.138504	0.138504	2.1
32nm	0.049244	0.049244	0.098488	0.098488	1.9
22nm	0.035394	0.035394	0.070789	0.070789	1.7

Interconnect parameters used were scaled versions of the one provided for 65nm by PTM. Most simulations were performed for 45nm technology node .The nominal V_t's used were (-0.218V, 0.266V) for pmos and nmos respectively. The nominal VDD used were 0.8V, 0.9V and 1.0V for 22nm, 32nm and 45nm respectively and the range of temperature sweep was 25-150°C. A 10-π distributed RLC model was used to model the wires. In our analyses, we ignore the TDC although it remains a key aspect of the design. We studied the effect of variation in different circuit and environmental parameters of the ThermoWire scheme-1 on the sensor-output-delay values to illustrate the feasibility of wire-based thermal sensing.

4.1 Driver strength and thermal vulnerability of wires

It is essential that the wires employed in our thermal sensing schemes have a thermal vulnerability factor much greater than unity. Experimental results suggest that for a given wire-length if the thermal vulnerability factor is greater than unity, it tends to gradually increase with temperature as the driver strength is increased. This validates our theoretical deduction in section 2. Furthermore, for the same driver strength, capable of driving a longer wire, thermal vulnerability is increased greatly for the longer wire (helpful to our cause of wire-based sensing). A large driver improves the thermal vulnerability of a wire-segment but at the cost of consuming more power. For a given wire-length, the power increment incurred for improving thermal vulnerability by upsizing the driver strength is however marginal and accounts for a small fraction of the overall power consumption of a sensor-system.

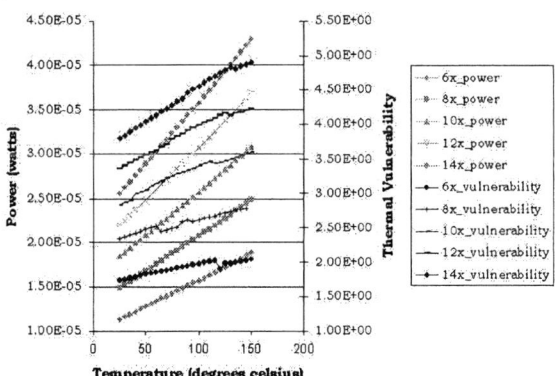

Fig.4 Variation of power values and thermal vulnerability with driver strength for a wire-length of 600μ

4.2 Sensitivity of delay to wire-length

We use the ThermoWire scheme-1 to find out the sensitivity of delay response to various wire-lengths for both upper and lower

level metal lines. It is ensured that the thermal vulnerability is greater than unity for all the setups. For lower-level metal lines, the delay is perfectly linear w.r.t. temperature and remains so at all lengths with gradually increasing slopes with increase in wire-length. In case of the upper-level metal lines (global layer) the delay-response is relatively invariant to temperature mainly because of a comparatively higher capacitive and smaller resistive component in these wires (larger cross-sectional area).

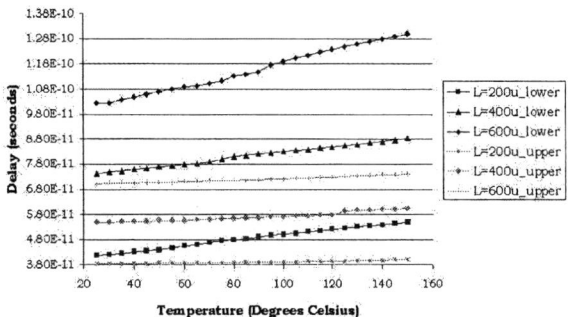

Fig.5 Length sensitivity of metal lines to temperature

Technology scaling trends suggest that for future generations, an interconnect based thermal sensor will not only provide a better resolution (steeper slope of sensor output) but also do so consuming much lesser power (both dynamic and leakage) making it a very power efficient sensing scheme. For this experiment, the wire-length used was 600μm and the driver-sizing was adjusted so that they all had the same thermal vulnerability factor (greater than unity). Moving from 45nm to 32nm, both the leakage and dynamic power undergo vast reduction of around 80%. Reducing the signal swing brings down the power values significantly (20-40%) while at the same time the sensor output delays increase (improving resolution). No level restoration is needed as long as the signal voltage levels remain discernible at the receiver.

4.3 Horizontal gradient estimation

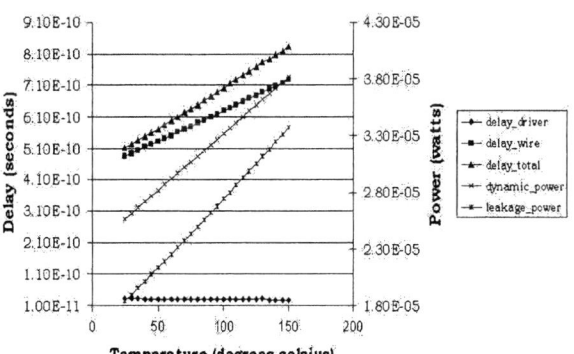

Fig.6 Operation of ThermoWire scheme-1

A wire-length of 2000μm is used for the lower-level metal line and the driver-strength (10x) is adjusted such that a thermal vulnerability factor of >40 can be achieved at all temperatures. This ensures the sensitivity of the wire to minute temperature variations and the total delay varies linearly with temperature (wire-delay accounting for more than 90%). For this setup, implemented in 45nm technology, both dynamic and leakage power are in the range of 10-38μW which is comparable to that of contemporary thermal sensors. A resolution of 5°C is achievable (2.7ps/°C) keeping in mind the resolution and precision of current time-to-digital converters. With minimum wire-spacing, a 2000μ long line (winded about in a small area) occupies approximately 250μ² of chip-area.

4.4 Vertical thermal gradient estimation

We simulated ThermoWire scheme-2 to illustrate measurement of vertical thermal gradients. We used a lower-level metal line of length 2000μm and an upper-level metal line of length 600μm (1/3rd approx.) with a relative driver-strength of 10x. The upper-level metal line was assumed to be at a uniform temperature of 160°C while the temperature of the lower-level line was swept from 50-150°C (gradually decreasing the temperature difference). The setup is such that the signal-delay across the upper-level is as small as possible (1st to reach XOR input) while delay across the lower level line varies depending on its thermal condition. As the temperature of the lower-level interconnect line approaches that of the upper-level, the signal delay across it increases and so does the difference in the arrival time of the 2 signals at the XOR input.

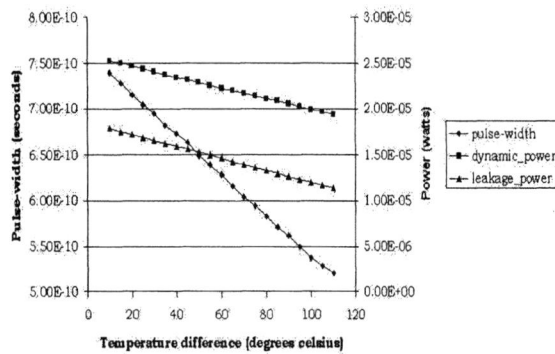

Fig.7 Operation of ThermoWire Scheme-2

This causes the pulse-width to increase as the temperature difference is decreased. Fig. 7 shows how the pulse width of the XOR gate output can be mapped to the temperature difference between the 2 layers. For this particular setup, the dynamic power consumption was 25-22μW while the leakage power was 18-12μW. A resolution of 6°C is achievable for this design (2.2ps/°C temperature difference).

4.5 Impact of variation in wire-dimensions

As the on-chip feature sizes shrink, the fundamental physical limits of traditional lithography impact design performance. Below 180nm, the size of on-chip elements fall below the 193nm wavelength of light used to print those elements on silicon. Distortion effects impact pattern fidelity and edge placement on silicon. Even slight pattern distortions can affect the fine-tuning of the design in terms of timing, power or crosstalk. We studied the effect of wire-width and wire-thickness variation in 45nm using ThermoWire scheme-1 with a lower-level wire of length 600μ reducing the width and thickness progressively by 5% to

15%. We define the measurement error as the error incurred when calibrating against a uniform width and thickness wire based sensor i.e. $error\ (\%) = (delay_{unif} - delay_{var\_})/delay_{unif\_} * 100$. In most practical designs the variation in wire-dimensions will be at-most 5% which will cause a measurement error of around 1%.

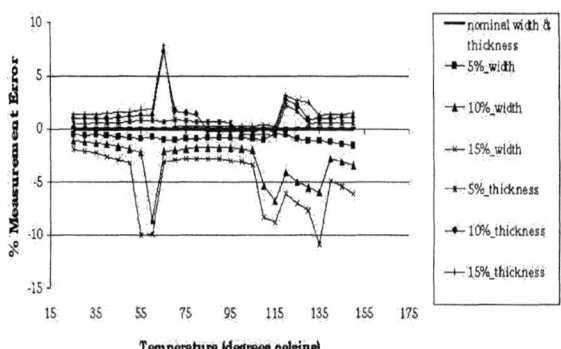

Fig.8 % Measurement error for different wire widths & thickness for ThermoWire scheme-1

4.6 Impact of non-uniform temperature distribution

In our basic design we assume exact wire-dimensions and uniform temperature distribution. However, the direction of thermal gradients could play an important role in determining the overall performance of an interconnect line (different portions of the wire in different temperature zones). We use ThermoWire scheme-1 and assume the maximum possible gradient of 50°C across the wire (over the base-circuit simulation temperature) in either direction from transmitter to receiver with a wire length of 600μm.

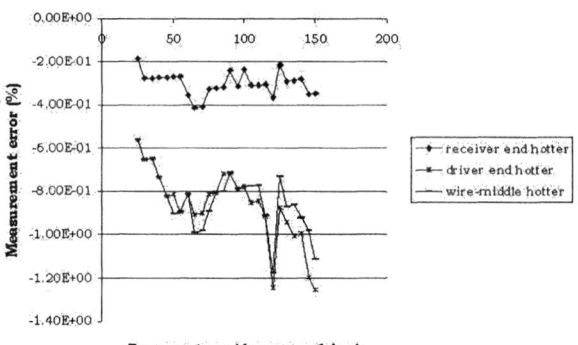

Fig. 9 % Measurement error for variable temperature distribution across wire

We compute the measurement error in a manner similar to that in section 4.5 (against uniform temperature distribution). In all cases, if different portions of the wire are above the base-simulation temperature of the circuit, the delay values increase. For a realistic temperature gradient of 50°C (worst-case), the delay deviation is very small as is suggested by the measurement error (%) values obtained (less than 1.5%). The direction of thermal profile has negligible effect for a gradient of this magnitude. Since, the wire is supposed to be looped in a small area; no significant thermal gradients are expected to exist.

4.7 Impact of cross-talk on sensor response

Table 2 % Measurement error due to crosstalk

Driver Strength	↓↑ –	↓↑↓
8x	21.4	49.2
10x	16.1	41.4
12x	12.8	35.6

Crosstalk has a detrimental impact on the signal delay particularly when both of the neighboring wires are switching in the opposite direction. We determine the measurement error incurred for the cases where one or both neighbor switch in the opposite direction. For a wire-length of 600μm used in ThermoWire scheme-1, the measurement error(%) due to crosstalk is tabulated above. The error becomes really high when both of the neighboring wires are switching in opposite directions. As the driver strength is increased, the % error reduces. The problem of crosstalk will mainly arise if we are attempting to use existing on-chip interconnects for thermal sensing purpose. Wire-placement plays a critical role in assessing the true impact of cross-talk on sensor response. The above values are misleading, in the sense that they assume uniformly straight wires, while in a practical implementation of a wire-based sensor, (in order to capture local hotspots) it will be imperative to place them winding about in a small area (keeping the minimum spacing rules intact). In such a case, self-coupling will be a bigger issue and self-capacitance will be the defining metric in the final sensor-delay-value. The popular design solutions to cope with RLC effects are: dedicated ground lines, differential signaling, buffer insertion, wire-splitting, matching termination and continuous power/ground planes; but these add to the sensor area and power consumption.

4.8 Impact of supply noise

The random variations in the power supply were modeled using a Normal distribution and Monte Carlo simulations were run with a maximum deviation of +/-10%. We studied the effect of supply noise on ThermoWire scheme-1 for a wire-length of 600μ with a driver-strength of 6x.The impact of supply noise on the temperature measurement was found to be negligibly small i.e. +/- 1.24% error for the entire temperature range.. This can be attributed to the small number of MOS transistors in the sensing device (supply noise afflicts MOS transistor operation more than signal propagation across wires).The above results suggest that the proposed sensing scheme offers a high degree of immunity to supply noise.

5. CONCLUSIONS

We highlight the potentially greater need of thermal sensing at the interconnect level than at the substrate level (keeping in mind the thermal scaling trends in interconnects), and propose low-power, wire-based thermal sensing approaches. We propose a 'thermal vulnerability' factor for lower-level metal wires that can be used as an indicator of their susceptibility to temperature variations. The schemes described can be used to measure temperature of the lower layers metal layers (which will closely reflect substrate temperature as well) and estimate vertical gradients w.r.t the upper layers. Technology scaling trends suggest an even better resolution at the cost of lesser power for future generations. The sensing schemes have a high degree of immunity to non-uniform

temperature distributions (realistic temperature gradients), variation in wire-dimensions and supply noise. There is a significant impact of cross-talk on sensor response and preventive measures are a necessity. However, self-coupling will have a more critical impact than cross-talk because of the particular placement style of the proposed wire-based sensor. The variation in sensor accuracy and area-consumption due to different wire placement styles is a possible scope of future work. The impact of the time-digitizer on the design in terms of area, power and accuracy must also to be evaluated.

6. ACKNOWLEDGEMENTS

This work has been funded by SRC Task 1415, AMD and Intel Corp.

7. REFERENCES

[1] F. Baez et al "Reducing power in high performance microprocessors", Proc. Design Automation Conf., 1998, pp.732-737

[2] A. Ajami, K. Banerjee and M. Pedram, "Modeling and analysis of non-uniform substrate temperature effects on global ULSI interconnects" In IEEE trans. CAD-ICS, vol.24, no.6, 2005

[3] A. Ajami, K. Banerjee and M. Pedram, "Analysis of substrate thermal gradient effects on optimal buffer insertion", In Proc. ICCAD, 2001

[4] S. Im, N. Srivastava, K. Banerjee and K. Goodson, "Thermal scaling analysis of multilevel Cu/Low-k interconnect structures in deep nanometer scale technologies" In Proc. VLSI multilevel interconnect conference, 2005, pp. 525-530

[5] S. Im and K. Banerjee, "Full chip thermal analysis of planar (2-D) and vertically integrated (3-D) high performance ICs", Tech Digest IEDM, 2000, pp.727-730

[6] K. Banerjee, A. Mehrotra, A. Vincentelli and C.Hu, "On thermal effects in deep sub-micron VLSI interconnects", In Proc. DAC, 1999, pp. 567-572

[7] S. Borkar et al "Parameter variation and impact on circuits and microarchitecture", In Proc. DAC, 2003, pp. 338-342

[8] A. Ajami, M. Pedram and K. Banerjee, "Effects of non-uniform substrate temperature on the clock signal integrity in high performance designs", In Proc. Custom Integrated Circuits Conference, 2001, pp. 233-236

[9] Predictive Technology Model, http://www.cas.asu.edu/~ptm/

[10] P. Chen, C.Chen et al "A time-to-digital converter based CMOS smart temperature sensor", In Proc. JSSC, 2005

[11] Rabeaey, Chandrakasan, Nikolic "Digital Integrated Circuits", 2nd edition, 2003

[12] A. Syal et al "Sensing temperature in CMOS circuits for thermal testing", In Proc. 22nd IEEE VTS, 2004

[13] P. Chen and S. Liu, "A cyclic CMOS time-to-digital converter with deep sub-nanosecond resolution", CICC, 1999

[14] J. Jansson, et al,"A delay line based CMOS time digitizer IC with 13ps single-shot resolution", ISCAS 2005

[15] J. Kalisz, "Review of methods for time-interval measurements with picosecond resolution", Institute of Physics Publishing, Metrologia, 2004, pp. 17-32

A Low-Power CAM using a 12-Transistor Design Cell

Saleh Abdel-Hafeez
Department of Computer Engineering
Jordan University of Science & Technology
Irbid, Jordan 21110
sabdel@just.edu.jo

Shadi M Harb & William R. Eisenstadt[*]
Electrical & Computer Engineering Department
University of Florida
Gainesville, FL 32611, USA
sharb@tec.ufl.edu, wre@tec.ufl.edu[*]

Abstract—A low-power CAM design using a 12-transistor cell is proposed. The CAM cell is based on the conventional 6T cross-coupled inverters used for storing data with an addition of two NMOS transistors for reading out. In addition, the CAM has another four transistors for mask comparison operation through classical pre-charge operation. The read-out port exploits a pre-charge reading mechanism in order to alleviate the drawback of power consumption generated from sensing amplifiers and all other related synchronization circuits which are structured in every column in the memory. Thus, the read and match features can have concurrent operations. An experimental CAM structure of storage size 64-bit x 128-bit is designed using 0.18-μm CMOS single poly and three layers of metals measuring a cell die area of 24.4375 μm² and a total silicon area of 0.269192 mm². The circuit works up to 200 MHz in simulation with total power consumption of 0.016 W at 1.8-V supply voltage

Keywords: CAM, low power, pre-charge, sense amplifier, 6T-cell, 8T-cell.

I. INTRODUCTION

A CAM memory is a parallel functional memory that contains large amounts of stored data for simultaneous comparison with input data. The match result of the CAM cell is the match data address. CAMs provide highly efficient architecture for high speed fully parallel data searching which are used for a wide range of applications such as high performance graphics, associative computing, data compressions, processor caches, lookup tables, TLBs, database accelerators, neural networks, image coding and IP classification [1]-[4]. In the standard CMOS technology, several 6T SRAM cells structures have been utilized for CAM memory cells [4]-[5] where the transistor count for the traditional CAM requires nine transistors which are comprised of six transistors for read/write and three transistors for comparison. However, most of these designs have either a reliability problem, high power consumption, or are not suitable to continue technology scaling. For example, the 9T traditional CAM inherits a stability problem while performing a disturb-read operation which affects the memory reliability and significantly decreases the static noise margin (SNR). Furthermore, most of these designs consume a considerable amount of power during the read operation due mainly to the I/O interfaced buffer with standby sensing amplifier currents. Furthermore, these designs normally lack the read-out and mask ports, where the concurrent read

and match operation is an important feature for testing procedure. In addition, the read function is important for data retrieval and refreshes purposes. On the other hand, several CAM designs with less than nine transistors have been proposed in the literature [6]-[7] which require special manufacturing technology [8]. For example, a 4T dynamic CAM presented in [6] achieves high memory densities but suffers from a match line coupling effect between adjacent cells and a write/match interference. Selective-precharge CAM is proposed in [9] to reduce the match line power consumption by doing a partial comparison first, and only if a partial match is obtained in a given row does the match line pre-charge. However, this approach has a time penalty for the worst case if all lines are precharged. Toggling Match line CAM [9] alternates between active high and active low match line to reduce its switching activities by half. However, this technique requires an additional active-high/active-low (AHAL) signal which incurs more hardware cost. Furthermore, it increases the hit match power consumption. Other work has been done based on a single bit line design with five-transistor D-latch and different comparison circuit topologies such as 7T, 9T and 10T CAMs [9]-[10]. To avoid the drawbacks of the dynamic circuit design such as noise margin, clock skew, and charge sharing, these designs exploit the static pseudo NMOS logic structure to reduce the switching activity in the match line. However, these designs suffer from considerable static power consumption. To realize fast access time, low voltage technology features, low power consumption and comparable silicon area with low area overhead, the 8T cell presented in [11]-[13] is adapted in our CAM cell which results in a new 12T CAM structure. In the 8T structure, adding two stacked nFETs to a 6T cell provides a read mechanism that doesn't disturb the internal node of the cell. This requires separate read and write word lines and can accommodate dual-port operation with separate single/multi-port read and write bit lines as shown in Figure 1. Not having a read-disturb issue will allow more scaling by lowering the Vth of the nFETs of the cells to the same level as the Vth of CMOS logic transistors can be lowered [13]. Furthermore, the dual-port 8T cell alleviates any stability problem which significantly provides larger SNR especially at low voltage and even provides a performance advantage over the 6T cell, if the pass-gates and read buffer are designed to be strong devices [13]. However, the area of the 8T cell is increased by 30% [12]-[13] compared to the 6T area, but it was shown in [13] that for the same speed, the complete SRAM module area is reduced by 15%. Although this area overhead is large in the field of memory, it is shown in [13] that the overall memory

978-1-4244-1709-4/07/$25.00 © 2007 IEEE

silicon area for the 8T cell is less for the same capacity and operating speed. This is due to the elimination of synchronization and sequential timing adjustments circuits used in the 6T cell SRAM design.

The paper is organized as follows: Section II discusses the CAM cell structure and layout, design topology is given in section III, simulation and results are given in section IV, conclusion is given in section V.

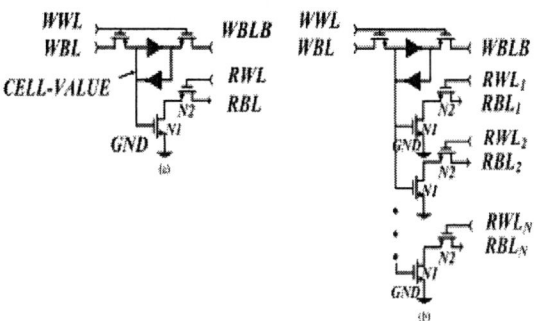

Figure 1. Single and Multi-Read Port 8T CMOS Cell

II.   12T CAM CELL

A.   Cell Structure

The proposed 12T CAM cell contains an eight-transistor memory structure for memory read/write operation and four additional transistors for data comparison. In the 8T memory structure, the conventional CMOS six-transistor cross-coupled inverters are employed since they are stable, compact, and more reliable than any other regenerative circuit design. Consequently, the 6T portion of the cell is only optimized in minimum sizes for a write operation, while the two additional stacked NMOS transistors are optimized in minimum sizes for reading out as shown in Figure 2. An added transistor (K1) controlled by a control signal called mask (MK) is used in order to prevent a path to ground which is generated during the pre-charge phase. This modification allows the match line to have a high flexibility of selecting the cells needed for comparisons. When the mask line is low, no comparison can take place and no activities occur at the match line. Once the mask signal is enabled, the stored data are compared against the incoming input data through the complementary bit lines. If the stored data and input data are matched in the masked cell, the match line is kept high; otherwise, the match line will be pulled down to the ground level value.

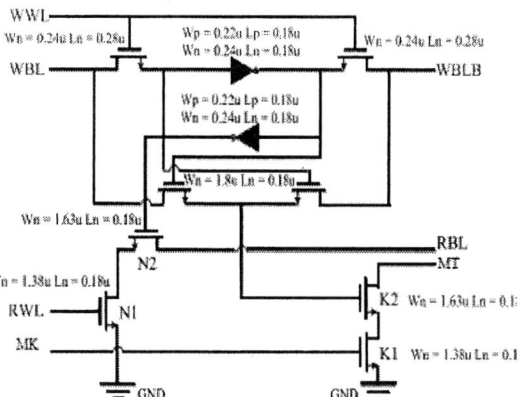

Figure 2. 12T CAM Cell with suggested transistor sizes using 0.18μm

B.   Cell Layout

The layout of the 12T cell is presented with an efficient layout geometry arrangement and small area as shown in Figure 3. The proposed cell is designed using 0.18-μm and three metal layers. The write operations are performed in 6T geometry of approximate size 2.2 x 2.2 μm2, where the total cell size is 5.75 x 4.25 μm2. The read bit line is accomplished with separate poly line which yields to an isolation between the read and write mechanisms. Furthermore, the sizes of the two NMOS read transistors provide fast and sufficient conducting capability. On the other hand, the sizes of transistors in the write portion of the cell are only optimized for the write operation since the read operation uses different portions of the cell. This makes the write portion of the 8T cell 6% smaller than that of the conventional 6T SRAM cell, which implies that the presented cell total area is only 30% larger than that of the 6T cell as presented in [12]-[13]. This area overhead is composed of not only the two added transistors but also of the contact areas of the VDD, GND, WWL, RWL, WBL, WBLB, and RBL for complete memory cells array overlap structure. This helps on routing for complete structure design through eliminating intermediate busses between cells.

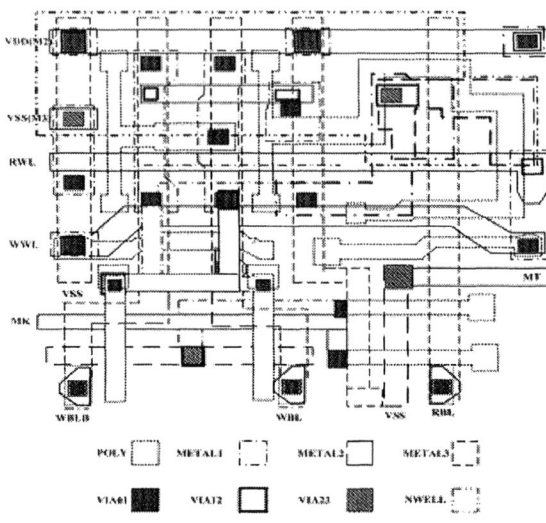

Figure 3. CAM Cell Layout

III. DESIGN TOPOLOGY

A. Architectural Overview

The high level CAM architecture for 64-bit (Height) X 128-bit (Width) is shown in Figure 4. The main functional parts are the CAM array which is organized as rows and columns with parallel search capability, read/write pre-decoder/decoder circuitry, and I/O buffers. The CAM structure is split into two main blocks, where the address decoders are passing in the middle in order to minimize the loading capacitances on the signal drivers and optimize the layout routing geometry.

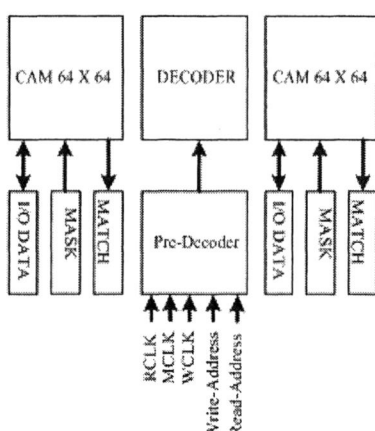

Figure 4. Architectural Overview

B. CAM Array Structure

The CAM array structure presented in Figure 5 consists of an array of the proposed 12T cell with read, write, and comparison capabilities. The read word line (RWL), write word line (WWL) and match line (MT) traverse the CAM array horizontally for each row using metal 2 (M2). On the other hand, the write bit line (WBL) and mask control signal (MK) traverse the CAM array vertically for each column using metal 3 (M3). A pull-up pre-charge pseudo PMOS transistor derived by the match clock signal (MCLK) is attached to each match line for every row. The match line (MT) capacitance is the transistor (K2) diffusion capacitance per cell with wire capacitance, where the read line capacitance is the transistor (N2) diffusion capacitance per cell with wire capacitance. The power supply line (VDD) runs horizontally using M2, where the ground line (GND) runs vertically using (M3).

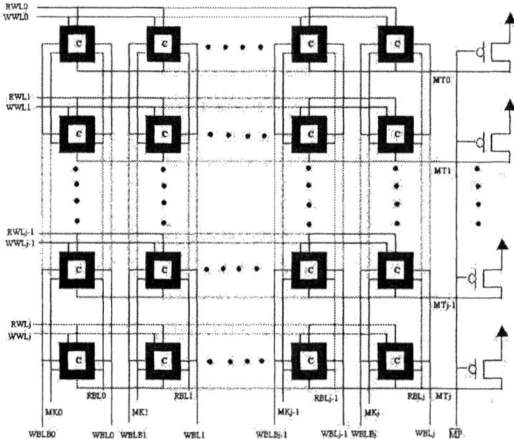

Figure 5: CAM Array Structure

C. Read/Write Pre-decoder/Decoder Circuitry

A standard pre-decoder/decoder circuitry with low power latches [14] is used to form the read and write address buses as shown in Figure 6. The timing specifications for read clock, write clock, data and address (i.e. setup, hold and accesses) are carefully designed by adjusting the sizes of pre-decoder/decoder circuits and input buffers. Thus, timing constraints are preserved without the need of a control timing sequential circuit. All simulations were verified at different technology corners with approximate metal layers wire models. The read pre-decoder circuit is a combination of latches that latch the read address during the falling edge of the clock concurrent with pre-charging the read-bit line. During the high edge of the RCLK, the read address enables the appropriate memory row cells that start to evaluate the data through RBLs. On the other hand, the write address line is structured with only a NAND gate pre-decoder and only an inverter gates decoder, where the input data are derived by a simple inverter gate structure.

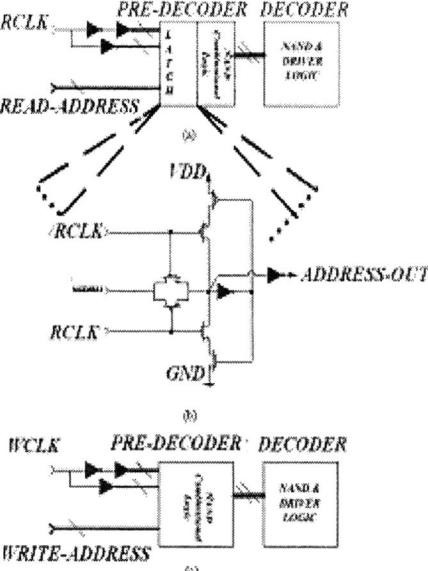

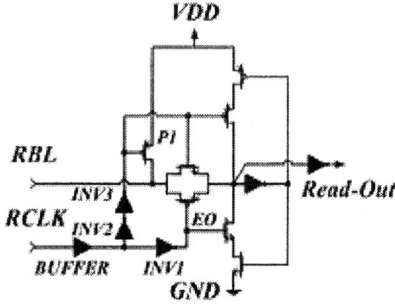

Figure 6. (a) Read organization, (b) Read address latch, (c) Write Organization

D. Output Buffer and Match Sense Circuitry

A low-power, low-voltage, and high speed sense buffer is designed to realize the pre-charge mechanism as well as sensing the output value on the read lines as shown in Figure 7. Similar to the sensing amplifier circuit with the 6T structure, the output buffer is implemented for every column. The PMOS (P1) transistor is used to pre-charge the RBL at the falling edge of the read clock, while the output latch holds the previous outcome. During the rising edge of the clock, the P1 transistor is disabled and the output latch updates the current RBL value. Timing intervals between P1, output latch, and RWL are preserved in order to optimize power dissipation reduction and provide high speed read access with a reliable outcome. The timing synchronization is simply accomplished through the use of INV1, INV2, and INV3.

VDD

Figure 7. Output Buffer and Match Sense Circuitry

E. Functional Overview

Since the write mechanism in the 8T cell is similar to the 6T cell conventional approach, the focus in this brief is only on the read and comparison operations. For the read operation as shown in Figure 8, the RBL is pre-charged during the asserted low of RCLK, implying

$$Tc2 = \frac{\alpha C_{Lm}}{\beta_p Vdd} = (\frac{\alpha Tox}{\mu\varepsilon})(\frac{L_p}{W_p}\frac{C_{Lm}}{Vdd}), \tag{1}$$

where $(\frac{\alpha Tox}{\mu\varepsilon})$ is the technology parameters of 0.18μm and Vdd is the power supply voltage (1.8V);

$\frac{L_p}{W_p}$ is the dimension of the pre-charge sense circuit transistor (P1), which is selected to have the value of L_p =0.18μm and W_p = 8μm; C_{Lm} is the RBL capacitance.

On the other hand, the RWL is enabled during the rising edge of RCLK, and the pre-charge sense buffer transistor (P1) is disabled. Thus, the data cells of the selected row are evaluated and latched through the sense buffers with an access time of Tdacc and held in the output buffer until the next rising edge. The worst case read access time is given by

$$Tdacc = (\frac{\alpha Tox}{\mu\varepsilon})(\frac{L_{neff} C_{Lm}}{W_{neff} Vdd}) + \text{Sense buffer delay time},$$

$$\tag{2}$$

where W_{neff} and L_{neff} are the effective sizes of the cell read transistors N1 and N2.

Furthermore, the power consumption consumed by a single RBL is simply measured by the equation

$$P = C_{Lm}V_{dd}^2 f, \tag{3}$$

where f is the maximum RCLK operating frequency.

Accordingly, the overall read bit lines power of a CAM block with N column lines is

$$P_{rt} = NC_{Lm}V_{dd}^2 f \tag{4}$$

In order to prevent any direct path power consumption between the supply connected by P1 through RBL and ground through the read buffer cell, timing constraints need to be carefully designed such that, the decoded row cells selected by RWL are disabled before RCLK enables the pre-charge on RBL by some during time (Tp). Following the same argument, pre-charge should be disabled before enabling the row cells by (Tp). In addition, the output circuit is enabled after the pre-charge is disabled on RBL by during Th. This holding time constraint prevents the pre-charge from sneaking in during the pre-charge phase. Hence, avoiding any small contention between the output latch and cell read buffer during the evaluation phase in order to optimize speed and read access. Lastly, the read bit line has full voltage swing between the supply voltage level and ground level which meets the requirement for low power consumption through low voltage level; and furthermore, eliminates the use of sensing amps which is a source of standby sensing currents.

Similarly, for the comparison operation, the match lines are pre-charged for each row at the falling edge of the MCLK similar to the pre-charge phase of RCLK. At evaluation, the pre-charge is disabled, and the mask lines are decoded which enables the masked cells for comparison. As long as all cells in the masked row match, the match line will be kept pre-charged, otherwise, it will be pulled down to ground. It is worthwhile mentioning that simultaneous read/write and comparison can be performed concurrently.

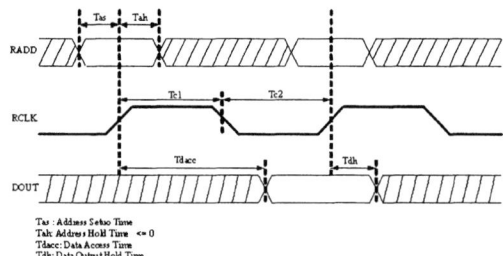

(a)

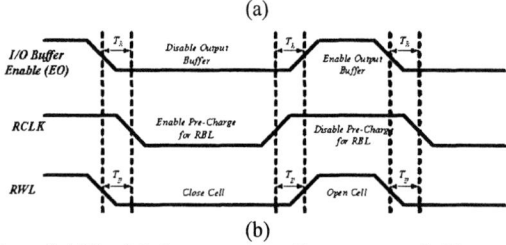

(b)

Figure 8. (a) Read timing constraints with respect to read address and read data, (b) Read pre-charge timing constraints with respect to output buffer and read cell

IV. SIMULATION AND RESULTS

A 64-bit x 128-bit CAM is designed using 0.18-μm and three metal layers. The CAM simulation results conducted by SpectreS simulator and Cadence tool are presented in Figures 9-11. As can be observed, the input data written into the memory cell is read out correctly and the memory is functioning correctly. Figure 11 shows the waveforms of the comparison operation. As shown, the input data is written in one cycle and compared against different input value in the next cycle which results in discharging the match line as soon as the mask line is enabled. The worst case output delay is measured to be 3.91 ns which implied that the CAM component can operate at 200 MHz for SS process corner, and can operate further at 250 MHz for FF process corner. In addition, the worst case power consumption was reported to be 0.016 W, while the reported 6T cell SRAM-based

CAM is 0.2936 W. This is due mainly to the standby current of the sensing amps that was generated for every column. Figure 12(a) shows a comparison of the maximum power consumption at different operating frequencies between 6T and 8T cells which shows that the 8T structure reduces the power consumption by an average rate of 34% for the same capacity and speed. On the other hand, Figure 12(b) shows the area cost of the 8T cell which is less than the 6T at sizes below 20k-bits by 15%~5% due to the area overhead induced by the synchronization circuitry in the 6T cell, while it is larger otherwise due to the large layout size of the 8T cell.

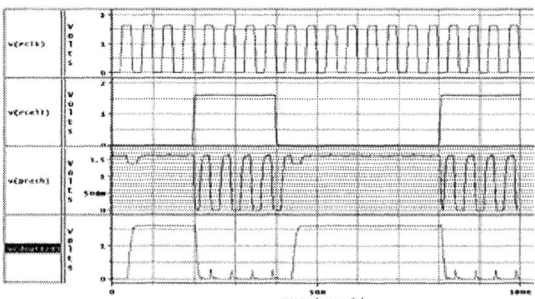

Figure 9. Simulation results of memory read outcome

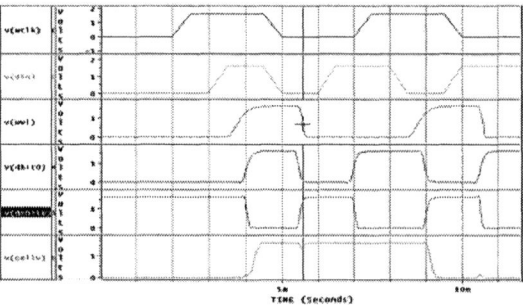

Figure 10. Simulation results of memory write outcome

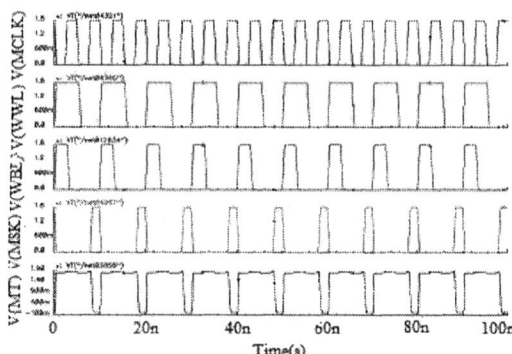

Figure 11. Simulation Results of memory comparison

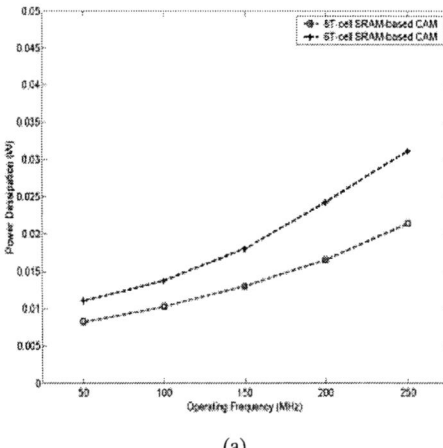

(a)

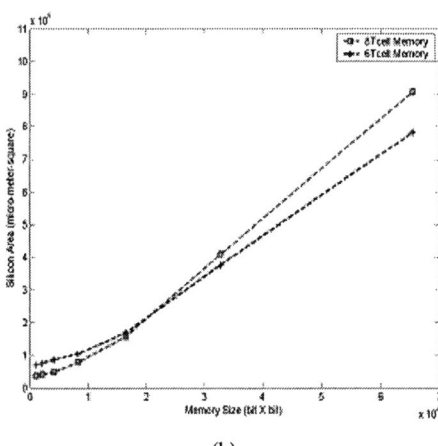

(b)

Figure 12. (a) Maximum power dissipation versus operating frequencies, (b) Total layout area in μm2 versus memory sizes in bit storage area

V. CONCLUSION

A novel 12T CAM cell for a low-power, low-supply voltage, and high density embedded CAM structure is proposed. The cell read and match portions provide high and fast conducting capability with concurrent operation. On the other hand, the proposed cell has separate read, write, and compare mechanisms, where all traditional sensing differential pair amplifiers are eliminated with all synchronization overhead circuitry. The proposed pre-charge and evaluate sense buffer is designed for low power consumption and high speed operation. Accordingly, the simulation results show that our CAM (64-bit x 128-bit) operates at 200 MHz and consumes less power consumption of about 34% than that the 6T SRAM-based

CAM structure. In addition, the overall silicon area is about 12% less than that of a similar CAM storage with 6T SRAM based cell that operates at the same speed.

REFERENCES

[1] T.B. Pei and C. Zukowski, "VLSI implementation of routing tables: tries and CAMs," in Proc. IEEE INFOCOM, vol. 2, 1991, pp 515-524.

[2] S. Panchanathan and M. Goldberg, "A content-addressable memomry architecture for image coding using vector quantization," IEEE Trans. Signal Process, vol. 39, no. 9, pp 2066-2078, Sep 1991.

[3] C.Y. Lee and R. Y Yang, "High-throughput data compresser designs using content addressable memory," IEE Proc. Circuits, Devices and Syst., vol 142, no 1, pp 69-73, Feb 1995.

[4] J.P. Wade andn C. G. Sodini, "A ternary content-addressable search engine, " IEEE J. Solid-State Circuits, vol. 24, pp 1003-1013, Aug. 1989.

[5] H. Miyatake, M. Tanaka, and Y. Mori, "A design for high –speed low-power CMOS fully parallel content-addressable memory macros," IEEE J. Solid-State Circuits, vol. 36, no. 6, pp. 956-968, June 2001

[6] Delgado-Frias, J.G.; Yu, A.; Nyathi, J.; "A dynamic content addressable memory using a 4-transistor cell," Design of Mixed-Mode Integrated Circuits and Applications, 1999. Third International Workshop on 26-28 July 1999 Page(s):110 - 113

[7] Lin, C,S.; Chang, J.C.; Liu, B.D.;, "Design for low-power, low-cost, and high reliability precomputation-based content-addreessable memory," Circuits and Systmes, 2002 APCCAS '02. 2002 Asia-Pacific Conference on. Volume 2, 28-31 Oct. 2002 Page(s): 319-324 Vol.2

[8] T. Miw, H. Yamada, Y. Hirota, T. Satoh, and H. Hara, "A 1MB 2-Tr/b nonvolatile CAM based on flash memory technologies", IEEE J. Solid-State Circuit, vol. 31, no 11, pp. 1601-1609, Nov., 1996

[9] Thirugnanam, G.; Vijaykrishnan, N.; Irwin, M.J.; "A novel low power CAM design," ASIC/SOC Conference, 2001. Proceedings. 14th Annual IEEE International, 12-15 Sept. 2001 Page(s):198 – 202

[10] Kuo-Hsig Cheng; Chia-Hung Wei; Yu-Wen Chen;, "Design of low-power content-addressable memory cell," Symposium on Circuits and Systems, 2003. MWSCAS '03. Proceedings of the 46th IEEE International Midwest. Volume 3, 27-30 Dec. 2003 Pages9s): 1447-1450 Vol3.

[11] S. M. Abdel-hafeez and S. P. Sribhashyam, "System and method for efficiently implementing double data rate memory architecture," US patent No. 6,356,509, Issued March 12, 2002.

[12] L. Chang et al., "Stable SRAM cell design for the 32 nm node and beyond," in Symp. VLSI Technology Dig., Jun. 2005, pp. 128-129.

[13] K. Takeda et al., "A Read-Static-Noise-Margin-Free SRAM Cell for Low-Vdd and High-Speed Applications," IEEE JSSC, Vol. 41, No. 1, January 2006, pp. 113-121.

[14] V. Stojanovic and V. G. Oklobdzija, " Comparative Analysis of Master-Salve Latches and Flip-Flops for High Pefrormance and low-Power Systems," IEEE J. Solid-State Circuits, Vol. 34, No. 4, pp 536-548, April 1999.

Improvement of dual rail logic as a countermeasure against DPA

A. Razafindraibe, M. Robert, P. Maurine
LIRMM/ CNRS/ University of Montpellier
161 rue Ada, 34392 Montpellier Cedex 5 - France

Abstract. Dual rail logic is considered as a relevant hardware countermeasure against Differential Power Analysis (DPA) by making power consumption data independent. In this paper, we deduce from a thorough analysis of the robustness of dual rail logic against DPA the design range in which it can be considered as effectively robust. Surprisingly this secure design range is quite narrow. We therefore propose the use of an improved logic, called Secure Triple Track Logic, as an alternative to more conventional dual rail logics. To validate the claimed benefits of the logic introduced herein, we have implemented a sensitive block of the Data Encryption Standard algorithm (DES) and carried out by simulation DPA attacks.

I-Introduction

It is now well recognized that the Achilles' heel of secure applications lies in their physical implementation. Among all the potential techniques to retrieve the secret key, one can mention side channel attacks. If there are many side channel attacks, DPA attack, is considered as one of the most efficient since it requires only less skills and materials, than others attacks such electromagnetic attacks, to be successfully implemented. Because of it dangerousness, many countermeasures have been proposed in former works [2, 3]. Recently, synchronous [4] or asynchronous dual rail logic [5, 6] has been identified as a promising solution to increase the robustness of secure applications. However some experiments have shown that the use of basic dual rail structures is not sufficient to warrant a high level of robustness against DPA. To overcome this problem, specific dual rail cells [4,7,8] and ad hoc place and route methods [9] have been developed. Goals of these countermeasures are to make the power consumption of logic gates independent of the manipulated data and to balance the wire capacitance of each differential pair during place & route steps.

Within this context, the first contribution of this paper is to analyze thoroughly the robustness of dual rail logic against DPA and to identify the secure design range. From the latter, we will identify the most sensitive parameters in secure dual rail design and will propose adequate countermeasures while staying as close to classical design flow as possible.

By looking closely this secure design range, it appears that it is too narrow. To address this problem, we propose the STTL (Secure Triple Track Logic) secure logic to implement key modules of ciphering algorithms. This is the second contribution of the paper.

The remainder of the paper is organized as follows: First, the basics of DPA are briefly summed up and the claimed benefits of DRL are reviewed. Then, the masked assumptions

supporting these claims are identified and their validity range evaluated by simulation on a 130nm process. After a discussion about this secure design range, the STTL is introduced as an adequate hardware countermeasure against DPA. The design features of this logic are also detailed. Before concluding, validations of the robustness of the proposed logic are presented.

II-Differential power analysis

DPA, first introduced in [1], succeeds in retrieving the secret key by exploiting the fact that the power consumption of cryptosystems is data dependent. Generally, DPA attack is executed in three phases: data collection, data sorting and data analysis.

Data collection consists in running a large number of cryptographic operations and recording the sampled corresponding power traces.

Data sorting consists in extracting, for all possible sub-secret keys, two sets of power traces from the whole power trace collection. These sets of power trace, $S_{.0'}$ and $S_{.1'}$, are built considering the expected value of the bit under attack according to both the guessed value of the sub-key and to the input data.

Data analysis consists in computing, for each possible guess of the secret key, the average power traces of $S_{.0'}$ and $S_{.1'}$ and in performing the difference between the averages. Finally, the secret key is usually disclosed by identifying the guess leading to the difference with the higher amplitude.

If this protocol is quite simple, one can wonder about the syndrome which is really captured by the DPA while applied on a dual rail circuit. In order to identify it, let us consider that a DPA is performed, with v vectors ($\in V$), on the output bit z of a logic block made of P gates. Among the v vectors applied to the cryptosystems, $t \in T$ of them forced z to the logic value '1', while the $f = v - t$ forced z to '0' ($f \in F$). With such definitions, the syndrome, S_{DPA}, captured by the DPA is:

$$S_{DPA}(Z) = \frac{1}{t} \cdot \sum_{u=1}^{t} I_u(t) - \frac{1}{f} \cdot \sum_{w=1}^{f} I_w(t) \qquad (1)$$

In the above expression, $I_u(t)$ and $I_w(t)$ are the current profiles of the whole block under attack while vectors $u \in V$ and $w \in W$ are applied on its inputs. These current profiles can be defined as the current consumes by all the gates of which is made up the block:

$$I_v(t) = \sum_{p=1}^{P} i_p(t) \quad I_w(t) = \sum_{p=1}^{P} i_p(t) \qquad (2)$$

Considering the definitions above and defining r_p^T, f_p^T (r_p^F et f_p^F) as the numbers of vectors of T (F) forcing the output of gate p to the a logic '1' and '0' respectively, it is then possible to deduce from (1) the following DPA signature expression :

$$S_{DPA}(Z) = \sum_{p=1}^{P-1} \left(\frac{f_p^F}{f} - \frac{f_p^T}{t} \right) \cdot \Delta i_p(t) + \Delta i_z(t) \qquad (3)$$

where $\Delta i_p(t)$ is the differential current profile of gate p, and $\Delta i_z(t)$ is the differential current profile of the gate driving the bit under attack. Here, we denote by differential switching current the waveform obtained by performing the difference between the currents provided by V_{DD} to the considered gate to settle respectively a logic '1' and a logic '0' on its output. Note, that if f_p^T/t and f_p^F/f are close one from the other, the expression (3) resumes to the differential current profile of the gate driving z, within its operating context. This highlights the great sensitivity of DPA.

II-Dual Rail Logic: a countermeasure against DPA

To secure cryptosystem against such an attack, the first action to be made is to break its assumptions in making power consumption independent of the manipulated data. Countermeasures have been proposed in [2] at all level of abstraction. Most of them aim at reducing the correlation between the data and leaking syndromes. Dual rail logic is one of these countermeasures.

The main advantage of dual rail Logic lies in the associated encoding used to present logic values. Indeed, for such an encoding, a rising transition on one of the two wires indicates that a bit is set to a valid logic '1' or '0', while a falling edge indicates that the bit returns to the invalid state which has no logical meaning. Consequently, the transmission of a valid logic '1' or '0' always requires switching a rail to V_{DD}. Therefore the differential current profiles of dual rail cells, and thus circuits, should be significantly lower than the ones of single ended gate. However, this claim holds if and only if the power consumption and the propagation delay of dual rail cells is data independent i.e. if the current waveform related to the settlement of logic '1' and '0' are rigorously the same.

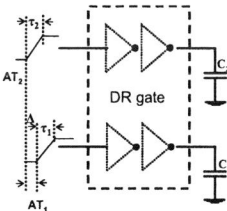

Fig.1: A Dual rail cell within its context

Since conventional dual rail cells, such as DCVSL or asynchronous DIMS logic [16] do not have perfectly balanced power consumption a lot of effort have been devoted in [4,7,8] to define secure dual rail cells. In its seminal paper [7], K. Tiri has introduced the Sense Amplifier Based Logic as logic with constant power consumption. Dynamic Current Mode Logic

has also been identified in [10] as an alternative to SABL while secure Dual Rail CMOS schematics are given in [4, 7].
Even if all these formerly proposed solutions appear efficient, at cell level, to counteract the DPA, they are all based on three crude assumptions. Indeed, in all these works, it is assumed that after place & route steps:

• Assumption n°1: each wire of each differential output is loaded by an identical capacitance value ($C_2=C_1$).
• Assumption n°2: all the inputs of the gate under consideration are controlled by identical drivers, i.e. that the transition times (labeled by τ in the remainder of the paper) of all the input signals have the same value ($\tau_T=\tau_F$),
• Assumption n°3: the switching process of the gate under consideration starts always at the same time ($AT_T=AT_F$).

Considering that both the power consumption and the timings of Dual Rail CMOS gates strongly depend on the transition time of the signals triggering the gate switching, and on the output capacitance switched, one can wonder about the validity domain of the three aforementioned assumptions.

III-Secure design range

To evaluate this validity domain, the modeling of the switching current waveform of CMOS dual rail gate is of prime importance. Considering that any single rail gate can be reduced to an equivalent inverter [11] or buffer (fig.1), we did model, at first order, the maximum amplitude Δi_{MAX} of the differential switching current profile of a dual rail gate loaded by unmatched capacitances, controlled by imbalanced transition time and finally triggered by imbalanced arrival time signals [27].
Considering I_{TH} as the smaller current imbalance that can be monitored with a given number N of current profiles measures according to the SNR definition:

$$SNR = \frac{I_{TH}}{\sigma} \cdot \sqrt{N} \qquad (4)$$

we deduced from the modelling of Δi_{MAX} [27], three criteria allowing to quickly estimate the robustness against DPA of a Dual Rail cell within its context. These criteria are the following:

$$\left. \frac{C_2}{C_1} \right|_{Crit} = max\left\{ \frac{1}{R_i} \cdot \frac{\frac{V_{DD}}{V_{DSAT}} - 1}{(1-\beta)} + 1 ; \left(\frac{V_{DSAT}}{\beta \cdot V_{DD}} \left(1 - \frac{1}{R_i} \right) \right)^{-1} \right\} \qquad (5)$$

$$\left. \frac{\tau_1}{\tau_2} \right|_{Crit} = 1 - \frac{(V_{DD} - V_T)}{V_{DD}} \cdot \frac{1}{R_i} \quad if \ \ I_{MAX} > I_{TH} \qquad (6)$$

$$\left. \frac{\Delta}{\tau} \right|_{Crit} = \frac{(V_{DD} - V_T)}{V_{DD}} \cdot \frac{1}{R_i} \quad if \ \ I_{MAX} > I_{TH} \qquad (7)$$

with

$$R_i = \frac{I_{MAX}}{I_{TH}} \qquad (8)$$

In the above expressions, V_{DD}, V_T and V_{DSAT} are the supply, threshold and saturation voltages of the considered transistor, β is the ratio of current provided by a transistor while its drain source voltage is respectively equal to V_{DSAT} and V_{DD}.

As shown, the first criterion allows evaluating the robustness of a dual rail cell in presence of unmatched loads. More precisely, for a given threshold of current I_{TH}, expression (9) provides the imbalance that can be tolerated between the outputs. In the same way, the second and third criteria allow evaluating the robustness of a Dual Rail cell in presence of imbalanced input transition and arrival times respectively.

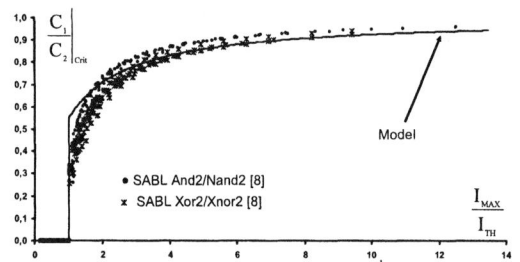

Fig. 2: Simulated and calculated values of $C_1/C_2|_{Crit}$ vs. $Ri=I_{MAX}/I_{TH}$ for two different SABL gates

One property of these criteria is that they do depend only on process parameters. This implies that, for a given cell topology, it is possible to obtained, by electrical simulation, characteristic curves of its robustness against DPA in presence of load, transition and arrival times imbalances. This provides a really interesting way to compare the robustness of different cell topologies regardless of their sizing provided to apply a unique gate sizing policy for all drives.

To demonstrate the validity of these first order criteria, we simulated and computed the critical load, transition and arrival time imbalance curves of SABL and2/nand2 and xor2/xnor2 gates. Fig 2, 3 and 4 report the obtained results.

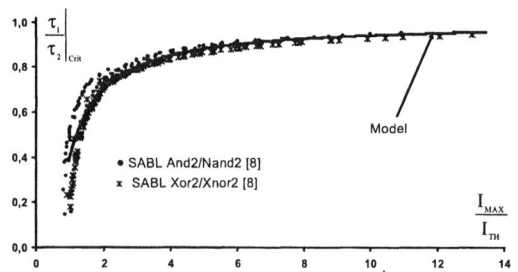

Fig. 3: Simulated and calculated values of $\tau_1/\tau_2|_{Crit}$ vs $Ri=I_{MAX}/I_{TH}$ for two different SABL gate

As shown, the accuracy of the proposed robustness criteria is satisfactory. However, a detailed interpretation of these characteristics provides more interesting results.

Let us consider that $R_i=I_{MAX}/I_{TH}$ is equal to 2 (100μA / 50μA). For such a R_i value, we may conclude that the two considered SABL gates remains robust against DPA if:

- the load imbalance C_1/C_2 is smaller than 0.7 ($C_1>C_2$), i.e. if C_2 remains smaller than 1.4 times C_1
- the transition time imbalance τ_1/τ_2 is smaller than 0.7,

- and the arrival time imbalance $|\Delta/\tau|$ is smaller than 0.2 ($\tau_1=\tau_2=\tau$), i.e. if all the signals triggering the gate arrive within a time window of width equal to 0.2 time the smaller input transition time τ. This is quite small considering that typical transition time values range between 20ps and 300ps for the 130nm process under consideration.

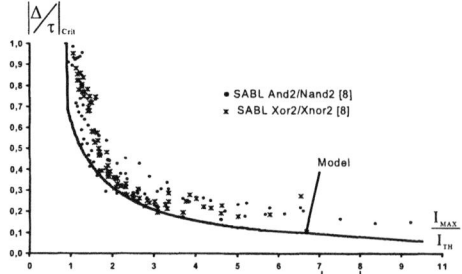

Fig. 4: Simulated and calculated values of $|\Delta/\tau|_{Crit}$ vs $Ri=I_{MAX}/I_{TH}$ for two different SABL gates [8]

This demonstrates that dual rail logic may be considered as robust against DPA in presence of significant load and transition time imbalances but does not suffer any significant arrival time imbalances. This is all the more true since arrival time imbalances may grow with the data path logic depth.

From the preceding expressions and results, it appears that there is effectively a design range in which dual rail logic can be considered as robust against DPA. However this secure design space is quite narrow since the tolerable arrival time imbalances are is quite small.

Based on the previous expressions, we have to make Ri (i.e. I_{MAX}) as small as possible to enlarge this secure design range. With this intention, naturally, one possible solution is to work with reduced V_{DD} values. However this imposes to manage properly the power versus timing trade off. Considering once again the narrowness of the secure design range, it appears that another alternative lies in the progressive development of dedicated CAD tools and/or design solutions to balance not only the parasitic capacitances introduced during the place & route as proposed in [9], but also the transition and arrival times. Within this context, expressions (5-8) constitute clever design criteria to evaluate the dangerousness of elementary cells within a secure dual rail circuit. However, as CAD tools will not be available in a near future, we therefore concentrate our effort on design solutions and more precisely on the structures of dual rail cells used to implement secure design.

IV-Secure Triple Track Logic

If the results obtained above demonstrate that the main benefit of the Dual Rail countermeasure lies in its ability in reducing the differential current profiles and thus the correlations between data and the power consumption, they also point out its main weakness: dual rail logic does not sufficiently reduce the correlation between data and computation times to constitute an extremely robust countermeasure.

To eliminate this remaining weakness, we developed a CMOS logic with data independent timing and power consumption called Secure Triple Track Logic (STTL in the rest of the paper). In fact, it is a variant of the dual rail logic.

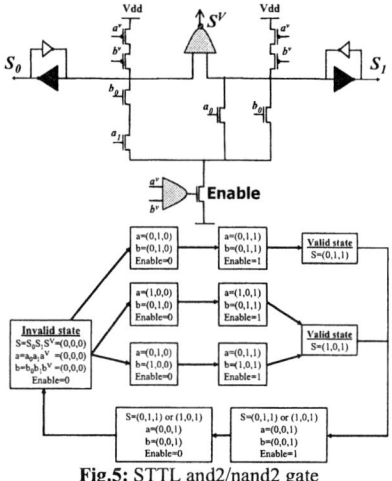

Fig.5: STTL and2/nand2 gate

To introduce the main characteristics of this logic style, an STTL and2 /nand2 gate is represented in Fig.5 as well as a graph illustrating its operation. As shown, instead of using two output wires to convey one logical value, STTL uses three. Indeed an additional output wire S^V is used to indicate whenever the output data S is valid or not. Similarly, two additional input wires a^v and b^v, indicating the validity of the incoming signals a and b are used. STTL operates thus according a kind of triple rail encoding of data (fig.6). Note that this is not the first time that the use of an additional wire to encode the validity of a signal is proposed. Indeed, in [25] an additional wire is used to obtain "efficient hardware implementations" but not to obtain secure designs or a data independent logic.

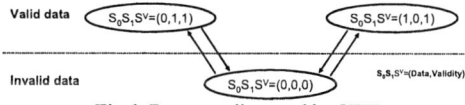

Fig.6: Data encoding used by STTL

As illustrated by Fig.6, the encoding of data is not a true triple rail encoding since the additional code value is redundant and does not convey any information about the bit value itself. This additional code value (and thus the power consumption of greyed gates on Fig. 5) is therefore uncorrelated with the value of input data. This property is extremely important. Indeed, one key design characteristic of STTL gate is that all validity signals such (a^v, b^v and S^V) are delivered by low switching current gate (greyed gates on fig.5), i.e. gates having a greater delays than high switching current gates (blackened gates on fig.5) in order to ensure that all input validity signals (a^v, b^v) settle after the data signals (a_0, a_1, b_0, b_1). This can be

obtained easily by sizing transistors of greyed cells smaller than those of blackened cells.

With such a Return to Zero encoding of data and specific gate design rules, the and2/ nand2 represented in Fig.5 operates as follows. Starting from the invalid state, data (a_0,a_1,b_0,b_1) settle first. In a second step, validity signals (a^v, b^v) rise forcing 'Enable' to '1' which allows in a fourth step the computation of the outputs. The return to the invalid state is performed in a similar way. First data (a_0,a_1,b_0,b_1) returns to '0'. Then the validity signals (a^v,b^v) are also forced to '0' by the environment to '0' allowing the gate to return to the invalid state.

If the use of an additional wire implies, at cell level, a data independent power overhead, estimated roughly to be within 10% to 30% compared to Dual Rail cells introduced in [16] depending on the complexity of the gate, it allows designing STTL gates having four interesting properties from security and design points of view:

First: avoid any internal cell activity while the data signals (a_0,a_1,b_0,b_1) are settling since no currents may flow if validity signals (a^v, b^v) are not true

Second: a quasi data independent power consumption as most of the proposed secure dual rail gates [4,7,8,10,16,22],

Third: a quasi data independent propagation delays, at block level, since the firing of gates will always be triggered by a data independent signals (Enable) computed from validity signals (a^v, b^v and S^V) which are also data independent.

Fourth: STTL gates are quite compact compared to other dual rail cells. As an illustration, Table I gives the number of transistors required to realize different basic functions in STTL and in others design styles

Table I: transistor count comparison

Gates	STTL	[21]	[4]	[8]
Nand2/Nor2And2/Or2	27	64	112	14
Nand3/Nor3And3/Or3	29	128	224	28
Xor2/Xnor2	29	68	80	18
Xor3/Xnor3	35	136	160	36
AO21/ AOI21	39	128	224	28
AO22/ AOI22	42	192	336	42

The third property aforementioned counterbalances the identified weakness (relative to the arrival time imbalances) of basic or secure dual rail gates introduced in former works. Indeed, the gate firings are independent of the data processed if the incertitude on the arrival time of all input signals, introduced by the place and route steps, is smaller than the time window Q that separates the settlements of the data (E_0, E_1) and the validity signal E^V (see Fig.7).

An important point here is that this time window Q can be tuned by sizing adequately the low switching current gates. In other words, the robustness of a STTL circuit can easily be managed by enlarging or reducing the width of this time window.

In order to evaluate the effectiveness of STTL, we have implemented a sensitive sub-module of DES algorithm namely the sbox1 which is driven by XORs gates. With this intention, we make use of our STTL library but also formerly introduced dual rail logics in order to perform comparison. Among these others dual rail libraries, we may distinguish the ones including secure Dual Rail gates [4, 16] but also the SABL gate [8] and finally AO222 based logic [18-21]. With such a simulation setup, the expected properties of the STTL were analysed and verified.

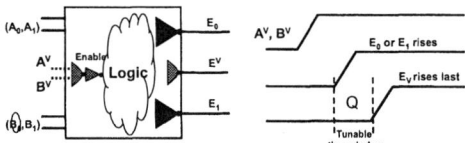

Fig.7: Timing behaviour of an STTL gate

In a first validation step, we have realized by simulation DPA attacks on the four output bits of the Sbox1 in order to identify precisely the impact of the routing on the robustness of the STTL. More precisely, to obtain a thorough evaluation of the robustness against DPA of the considered logic styles, all the simulations were first based on an ideal netlists (without parasitic capacitances) and subsequently on a back-annotated netlists. Note that to be fair, we have adopted the same sizing policy for all the Dual Rail cells but also the same parameters for the place & route steps done with Soc Encounter tool [20]. Fig.8 gives, for the 64 possible guesses of the secret key, the DPA signatures obtained during the attack of the third output (S_3) bit of the Sbox1 implemented with STTL gates. From this figure two conclusions may be drawn. First, performing DPA attacks on S_3, as for the three others, does not provide any information about the value of the secret key since its DPA curve is not distinguishable from the 63 others DPA signatures. Therefore STTL counteracts in this case the attack. Finally, the most important result that can be drawn considering Fig.8 is that STTL is, as expected, quasi insensitive to the load imbalances introduced by the place and route steps since the DPA signatures obtained with the ideal or back-annotated netlists are quasi identical.

In a second validation step, we wanted to demonstrate that STTL effectively leads to quasi-data independent propagation delay values at block level. We therefore extracted from electrical simulations of back-annotated netlists, the time spent by the signals to propagate from the inputs to the outputs. This was done for all possible input vectors considering STTL as well as the other logic styles introduced in [4, 8, 16]. Note that all input signals were assumed to be stable at t=0.

On Fig.9, we plotted the time spent by the signals to propagate from the inputs to the output S_3. More precisely, this figure gives the propagation delay distributions while S_3 settles logic '1' and '0' respectively for different Dual Rail Logic. Note, that we have also reported the average propagation delay values $<T_1>$ and $<T_0>$ spent by the circuit to settle a '1' and a '0' on output S_3. The obtained representation is interesting to evaluate the robustness of a logic block against DPA. Indeed, the more symmetrical are the distributions, the more data

independent is the considered logic and thus the more robust the physical implementation is.

As shown, depending on the logic, the gap ($<T_1>$ - $<T_0>$) between the average times spent to settle logic '1' and '0' can be quite small (few ps) or significant (several tenths of ps). Obviously, STTL exhibits a quasi data independent timing behaviour, while the ones introduced in [8, 16] do not. However the price to be paid is longer propagation delays due to the use of low switching current gates to control the validity signals.

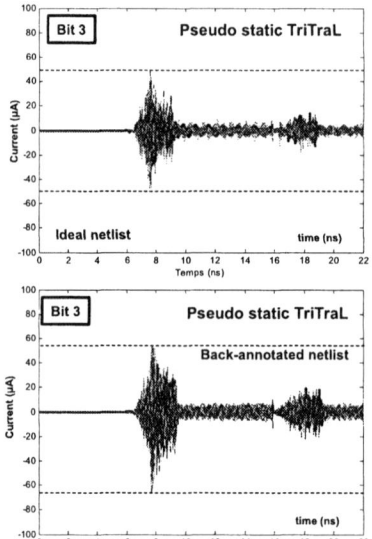

Fig.8: DPA signatures obtained considering and without considering routing capacitances.

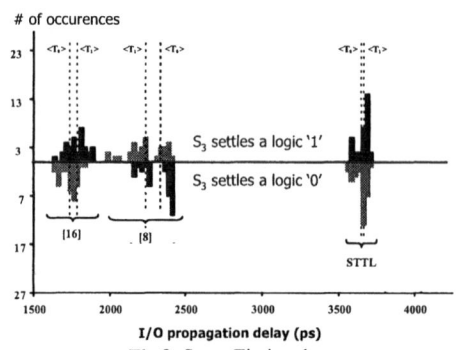

Fig.9: Some Timing data

In a final step, we compared the robustness against DPA of all the considered logic styles. We thus performed by simulation DPA attacks on all the outputs of the structure represented on Sbox1. The netlists considered during these simulations were back-annotated ones Fig. 10 reports some relevant results we have obtained. These results may be summarized as follows. First, for all the attacked output bits, DPA was unsuccessful while done on the STTL implementation. Second, these

274   *2007 IFIP International Conference on Very Large Scale Integration (VLSI-SoC 2007)*

attacks may be considered as successful while done on SABL [8], and on circuits implemented with gates introduced in [4, 18, 21]. However as shown on Fig.9, for [4], the revealed syndrome is quite small.

V-Conclusion

A thorough evaluation of the robustness of Dual Rail Logic has been carried out in this paper. This analysis has pointed out that Dual Rail Logic does not sufficiently reduce correlation between data and computation times to be a fully robust countermeasure against DPA. This observation has led to the proposal of an improved logic called STTL. The main characteristics of this logic that made of it a robust countermeasure against DPA are: quasi data independent power consumption and timing behaviour. The latter characteristic ensures that STTL is particularly robust to load, and arrival time imbalances introduced by the place and route steps while the resulting cells remain quite compact with respect to formerly introduced logic styles.

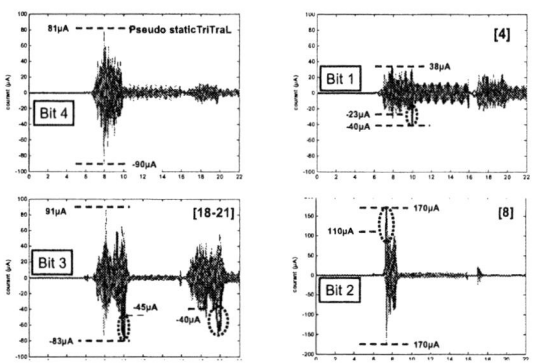

Fig.10: Simulated DPA signature of the Sbox1 outputs with back-annotated netlist (X-axis unit is ns)

References

[1] P. Kocher & al "Differential power analysis". CRYPTO'99, Lecture Notes in Comp. Science, vol. 1666, pp. 388-397.

[2] Suzuki & al, "Random Switching Logic: A Countermeasure against DPA based on Transition Probability", Cryptology ePrint Archive, report 2004/346

[3] A. Bystrov, Alex Yakovlev, Danil Sokolov, Julian Murphy, "Design and Analysis of Dual-Rail Circuits for Security Applications," IEEE Trans. on Computers, vol. 54, no. 4, pp. 449-460, April, 2005.

[4] S. Guilley & al "CMOS Structures Suitable for Secure Hardware". 2004 Design, Automation and Test in Europe Conf. and Exposition, (DATE 2004), 16-20 Feb 2004, France

[5] J.J. A. Fournier & al, "Security Evaluation of Asynchronous Circuits", Workshop on Cryptographic Hardware and Embedded Systems (CHES 2003), pp. 137 – 151, 2003.

[6] G.F. Bouesse & al "DPA on Quasi Delay Insensitive Asynchronous Circuits: Formalization and Improvement", 2005 Design, Automation and Test in Europe Conference and Exposition (DATE 2005), 7-11 March 2005, Munich, Germany.

[7] A. Razafindraibe & al "Secure structures for secure asynchronous QDI circuits", DCIS'04: 19th International Conference on Design of Circuits and Integrated Systems (DCIS'04), 2004.

[8] K. Tiri & al "Securing Encryption Algorithms against DPA at the Logic Level: Next Generation Smart Card Technology," Proc. of CHES 2003, LNCS 2779, pp. 125--136, Sept. 2003.

[9] K. Tiri & al, "A VLSI Design Flow for Secure Side-Channel Attack Resistant ICs" 2005 Design, Automation and Test in Europe Conference and Exposition (DATE 2005), 2005.

[10] F. Mace & al "A dynamic current mode logic to counteract power analysis attacks", DCIS'04: 19th International Conference on Design of Circuits and Integrated Systems (DCIS 2004).

[11] P. Maurine & al, "Transition time modeling in deep submicron CMOS", IEEE Trans. on CAD, vol.21, pp. 1352-1363, 2002.

[12] K.O. Jeppson. "Modeling the Influence of the Transistor Gain Ratio and the Input-to-Output Coupling Capacitance on the CMOS Inverter Delay", IEEE Journal of Solid State Circuits, Vol. 29, pp. 646-654, 1994.

[13] T. Tsividis and al, "Operation and Modeling of the Mos Transistor", Oxford University Press, 1999.

[14] T. Sakurai and al. "Alpha-power law MOSFET model and its applications to CMOS inverter delay and other formulas", IEEE J. Solid-State Circuits, vol. 25, pp. 584-594, April 1990.

[15] J. Sparso & al. "Principles of Asynchronous Circuit Design: A Systems Perspective", Kluwer Academic Publishers.

[16] A. Razafindraibe & al "Asynchronous Dual rail Cells to Secure Cryptosystem against Side Channel Attacks" SAME'2005

[17] T. S. Messerges & al, "Examining Smart-Card Security under the Threat pf Power Analysis Attacks", IEEE Trans. On Computer, Vol. 51, n°5, pp. 541-552, May 2002

[18] P. Maurine & al, 'Static Implementation of QDI Asynchronous Primitives'' 13th International Workshop on Power and Timing Modeling, Optimization and Simulation, 2003, (PATMOS'03), September 10-12, 2003, pp. 181-191

[19] A. Chaterzigeorgiou & al,"Collapsing the Transistor Chain to an Effective Single Equivalent Transistor", 1998 Design Automation and Test in Europe (DATE '98), February 23-26, 1998

[20] www.cadence.com/products/digital_ic/soc_encounter/index.aspx

[21] C. Piguet & al "Electrical Design of Dynamic and Static Speed Independent CMOS Circuits from Signal Transistion Graphs", 8th International Workshop on Power and Timing Modeling, Optimization and Simulation (PATMOS '98), Technical University of Denmark - October 7 - 9, 1998, pp. 357-366, 1998.

[22] K. J. Kulikowski & al "Delay Insensitive Encoding and Power Analysis: A Balancing Act", 11th IEEE International Symposium on Asynchronous Circuits and Systems (ASYNC 2005), March 13-16, 2005, New York City, USA, pp.116-125

[23] National Bureau of Standards, "Data Encryption Standard," Federal Information Processing Standards Publication 46, 1977.

[24] "Eldo User's Manual", Mentor Graphic's Corp, 1998

[25] T. H.-Y. Meng, & al, "Automatic Synthesis of Asynchronous Circuits from High-Level Specifications", IEEE Trans. On Computer Aided Design, vol. 8, No. 11, November 1989

[26] P. Kocher, R.Lee & al, "Security as a new dimension in embedded system design', Proceedings of the 41st annual conference on Design automation (DAC 2004), USA, pp. 753 – 760

[27] XXX for blind review, 'Evaluation of the robustness of dual rail logic against DPA', IEEE International Conference on Integrated Circuit Design and Technology, 2006. ICICDT '06. 2006, 24-26 May 2006

A VHDL Based Approach for Fast and Accurate Energy Consumption Estimations

César A. M. Marcon, Sérgio Johann Filho, Fabiano P. Hessel

PPGCC / FACIN / PUCRS – Av. Ipiranga, 6681, Porto Alegre, RS – Brazil

cesar.marcon@pucrs.br

Abstract

Efficient energy consumption became an important requirement and constraint to be considered in many systems implementations, mainly to the embedded ones. Accurate and efficient power estimation during the design phase is required, in order to meet the power specifications without a costly redesign. High abstraction levels descriptions enable fast energy consumption estimations, but hardly enable accurate estimations. It normally requires evaluations at low abstraction levels, such as electric ones. On the other hand, low abstraction levels require too much design effort and design time. In this sense, this work presents an approach for energy consumption estimation for systems written in synthesizable VHDLs. A VHDL cell library is the base of the methodology, which is characterized with some relevant energy consumption information according to foundry parameters. The use of this approach leads to high-quality energy consumption estimations and design time saving.

1. Introduction

The increases in chip density, operating frequency, and demand for portable applications have made power consumption a VLSI design major concern. Excessive power dissipation in integrated circuits not only discourages their use in a mobile environment, but also causes overheating, which degrades performance and reduces chip lifetime [1]. Therefore, designers need methods and tools to minimize not only the energy consumption, but also the high power density, avoiding reliability problems and expensive cooling systems.

It is imperative the use of tools for power consumption estimation, in order to guide the designer in solving these kind of problems. Unfortunately, accurate estimations can be achieved only with low description levels, which demand too much design and processing time. In addition, low abstraction level descriptions are more error prone, difficult to understand, to maintain, and to document the systems, when compared to high abstraction levels.

These drawbacks encourage researchers in finding new methods and/or models that leads to accurate estimations that overcome the problems. This paper proposes an approach for energy consumption estimation, named **EngyLib** (which stands for Energy Library), which allows the designer describing systems in high abstraction levels, with accuracy estimations of low abstraction levels.

EngyLib highlight is a VHDL cell library, which is characterized according to static power dissipation, dynamic energy consumption, and number of transistors switching. The approach consists of converting a high-level system description to an equivalent low-level one, which is a VHDL netlist composed by the characterized cell. Next, through a simulation step, the energy consumption is estimated according to the input stimuli.

Many levels and description languages may be used to describe the system under energy consumption estimation, like behavioral VHDL and structural Verilog. However, the high-level description has to be synthesizable to a VHDL composed by the characterized cells.

EngyLib is part of CAFES [2], which is a Java open-source framework developed targeting different sort of analyses and syntheses of embedded systems.

Results achieved by comparing electric simulation of small processors and embedded applications with the proposed approach attest the efficiency and efficacy of EngyLib. This approach reduces in average 38 times the estimation time, paying the cost of only 13 percent of energy consumption imprecision, when compared to energy consumption reference, accomplished by electric simulation.

2. EngyLib, a VHDL Based Approach for Energy Consumption Estimation

EngyLib is the approach proposed here, which uses the precision of electric levels to build a VHDL cell library. This one is timing and energy consumption characterized, enabling to estimate through VHDL simulation the static and dynamic energy consumption of circuits. EngyLib provides many benefits, as for instance, VHDL simulation is much faster than an electric one, speeding up the energy consumption estimation of the system under evaluation. In addition, although the approach proposed here synthesizes the system description to a low-level one, the designer may only work at high abstraction levels, since low-level description are seen as an intermediary language of the estimation flow.

Figure 1 illustrates that starting from a system description the designer may estimate the energy consumption, number of switching and execution timing, of the entire system or part of it by performing three steps: (*i*) *logic synthesis*, (*ii*) *connection analysis*, and (*iii*) *system simulation*.

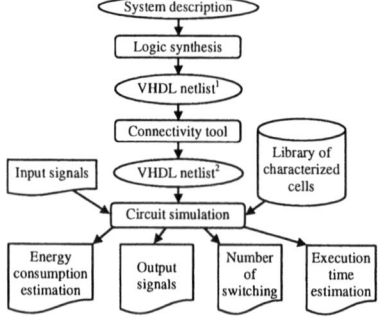

Figure 1 – Workflow for energy consumption estimation

Any abstraction level, which reduces the designer description effort and the error prone, may be used to perform the *system description*. The only restriction here is that this description has to be synthesizable, so that a *logic synthesis* tool may convert it to a netlist of VHDL cells. This synthesis is constrained to only the set of cells that are pre-characterized according to energy consumption parameters. This work uses Leonardo Spectrum [3] as the synthesis tool.

Figure 2 depicts a partial VHDL description of an accumulator system, which is a synthesizable one.

```
architecture accumulator of accumulator is
                                           ...
    begin
                                   ...
        soma <= opA + opB;
        process(clock, reset)
        begin
                if reset = '1' then
                    opA <= (others=>'0');
                elsif clock'event and clock = '1' then
                    opA <= opA + 1;
                end if;
        end process;
                                   ...
    end accumulator;
```

Figure 2 – Partial description of a synthesizable VHDL

The synthesis of the VHDL behavioral of Figure 2 generates a VHDL netlist description, which is partially illustrated in Figure 3, and named *VHDL netlist[1]* in Figure 1.

```
architecture accumulator of accumulator is
                                           ...
begin
                                   ...
opA_5: dffr port map(Q=>opA_5, QB=>nx210, D=>nx140, CLK=>nx248, R=>nx258);
ix216: nand02 port map(Y=>nx215, A0=>opA_3, A1=>nx50);
ix177: xor2 port map(Y=>saida_6, A0=>nx156, A1=>nx174);
ix157: mux21 port map(Y=>nx156, A0=>nx210, A1=>nx204, S0=>nx146);
                                   ...
end accumulator;
```

Figure 3 – Partial VHDL netlist of Figure 2 system

The *connection analysis* step enhances the synthesized VHDL netlist by adding the fan-out estimation of each output signal of each cell. A tool that searches all connectivity of output signals and knows the fan-in of each type of cell does it. This tool is implemented inside CAFES framework [2]. In addition, this tool may help the designer inserting special cells that represents metal lines. This is a partial manual step, which is normally performed when the system under evaluation have large transmission wires.

Figure 4 shows the *VHDL netlist[2]* generated by connection analysis tools. Here, all output signals have annotated a fan-out according to the associated load capacitance.

```
architecture accumulator of accumulator is
                                           ...
begin
                                   ...
opA_5: Entity work.DffReset port map(Q=>opA_5, FanOut_Q=>2,
        QB=>nx210, FanOut_QB=>3, D=>nx140, CLK=>nx248, R=>nx258);
ix216: Entity work.Nand2 port map(Y=>nx215, FanOut_Y=>2, A0=>opA_3,
        A1=>nx50);
ix177: Entity work.Xor2 port map(Y=>saida_6, FanOut_Y=>1, A0=>nx156,
        A1=>nx174);
ix157: Entity work.Mux2to1 port map(Y=>nx156, FanOut_Y=>2,
        A0=>nx210, A1=>nx204, S0=>nx146);
                                   ...
end accumulator;
```

Figure 4 – Partial VHDL netlist with fan-out annotation

```
architecture accumulatorTestBench of accumulatorTestBench is
                                           ...
begin
    PowerEstimation1: Entity work.PowerEstimation;
    UUT: Entity work.accumulator port map (
            clock => clock,
            reset => reset, ...)
                                   ...
end accumulatorTestBench;
```

Figure 5 – Partial test bench description

Any kind of commercial VHDL simulator can perform the *circuit simulation* step. The designer has just to insert the appropriated input stimuli to the inputs of the circuit under analysis and include the **PowerEstimation** entity inside the

VHDL source file. **PowerEstimation** is an EngyLib entity that contains static and dynamic energy consumption estimations, the sum of transistors switching of each transition, timing, and other information. Figure 5 depicts a partial VHDL containing the input file for accumulator system simulation with energy consumption estimation.

3. VHDL Library Characterization

The first step in VHDL library characterization is choosing an appropriate set of cells. If this propose, it was selected a large number of embedded applications to verify the most used cells, and the effect of discard the less used one. This process carried out to a basic library with a set of 15 cells composed by logic gates and flip-flops. Once the set was select, the cells were geometrically and electrically described according to a target technology (for instance, CMOS TSMC 0.35μm). A stimuli generation tool, which was implemented inside CAFES, generates all possible stimulus combinations of each cell. Figure 6 illustrates that having as inputs the target technological library, the input signals, and the electric library, containing the cell behavior and its electric characteristics, an electric simulator generates a set of outputs, reflecting logic, timing, and electric behaviors.

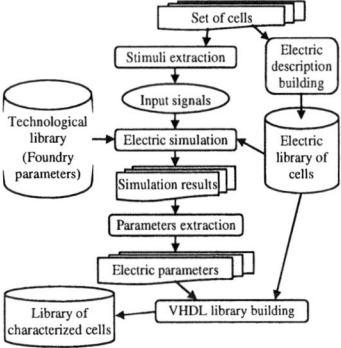

Figure 6 – VHDL library characterization

Each set of input stimulus produces different energy consumptions. Table 1(a) shows the energy consumption of all input combinations of a not gate that generate transistors switching. It considers a fan-out of one gate. Table 1(b) shows the static power dissipation, when inputs do not change.

While fan-out does not affect the static power dissipation, the load capacitance influences directly the dynamic energy consumption. To represent the fan-out influence, two solutions were evaluated: (i) to construct functions that allows the knowledge of energy consumption of a given fan-out by interpolation, and (ii) to annotate explicitly all fan-out possibilities. This work chose the last one, since this last leads to more accurate estimations.

Table 1 – Dynamic energy consumption and static power dissipation for all input combinations of a not gate

	Input transition	Energy consumption (J)
(a)	0 ⇒ 1	2.2993 E-13
	1 ⇒ 0	2.2258 E-13
	Input transition	**Power dissipation (W)**
(b)	0 ⇒ 0	3.6825 E-10
	1 ⇒ 1	7.1430 E-14

Figure 7 depicts a partial VHDL of the characterized not gate, where some energy and power parameters listed in Table 1 may be seen in lines 21, 22, 34 and 35.

```
1   use work.POWER_PCK.all;
2   entity Inv is
3     port(Y: out STD_LOGIC; FanOut_Y: in NATURAL := 3;
4          A: in STD_LOGIC);
5   end Inv;
6   architecture Inv of Inv is begin
7     process(A)
8       variable PreviousA: STD_LOGIC := '0';
9       variable staticPower, cellDynamicEnergy: REAL := 0.0;
10      variable dynamicEnergy: REAL;
11      variable switching: NATURAL := 0;
12      variable transition: STD_LOGIC_VECTOR(1 downto 0);
13    begin
14      transition := PreviousA & A;
15      dynamicEnergy := 0.0;
16      switching := switching + 2;
17      totalSwitching := totalSwitching + 2;
18      case transition is
19        when "01" =>
20          case FanOut_Y is
21            when 1 => dynamicEnergy := 2.2993E-13;
22            when 2 => dynamicEnergy := 3.1251E-13;
23                     ...
24        when "10" =>
25          case FanOut_Y is
26            when 1 => dynamicEnergy := 2.2258E-13;
27                     ...
28      end case;
29      cellDynamicEnergy := cellDynamicEnergy + dynamicEnergy;
30      totalDynamicEnergy := totalDynamicEnergy + dynamicEnergy;
31      totalStaticPower := totalStaticPower - staticPower;
32      transition := (A & A);
33      case transition is
34        when "00" => staticPower := 3.6825E-10;
35        when "11" => staticPower := 7.1430E-14;
36      end case;
37      totalStaticPower := totalStaticPower + staticPower;
38      Y <= not A;
39      PreviousA := A;
40    end process;
41  end Inv;
```

Figure 7 – Partial VHDL description of a not gate

In addition, Figure 7 shows some local and global variables used to compute static and dynamic energy consumption, and the number of transistors switching. For instance, **cellDynamicEnergy** is a local variable containing the total dynamic energy consumed by the cell during the simulation. The global variable **totalDynamicEnergy** contains the sum of all **cellDynamicEnergy** of the system. The local vector **transition** stores the previous input value and the new one. It is used to evaluate if occurred or not a transition in any input signal enabling to estimate both static and dynamic energy consumption. The global variable **totalStaticPower** contains the static energy consumption of all system cells. Each transition the static power value is recomputed to its new value, according to the latest input value. It preserves the consistence of the static energy consumption of the entire system, which is computed each time step of simulation.

VHDL allows global variable through declaring them as *shared* inside a global package. This is the way that each instance of a cell can add its influence in global estimation. Figure 8 illustrates some of the main *shared* variables inside POWER_PCK, which is the power package of EngyLib.

```
package POWER_PCK is
    shared variable totalDynamicEnergy: REAL := 0.0;
    shared variable totalStaticPower: REAL := 0.0;
    shared variable totalStaticEnergy: REAL := 0.0;
    shared variable totalSwitching: NATURAL := 0;
                    ...
end POWER_PCK;
```

Figure 8 –Partial description of the main EngyLib package

Although, this Section depicts a flow to characterize a VHDL library of a particular technology, a set of tools built inside CAFES framework allows to building new VHDL libraries according to others technologies, as long as it will be used the same set of selected cells.

4. Library Refinements and Verification

Estimations achieved with a netlist simulation, based on a cell library, may be imprecise due to many issues, like an inappropriate extraction of electric parameters and an unfitting abstraction level modeling. To overcome it Figure 9 shows the main steps applied to cell library refinement.

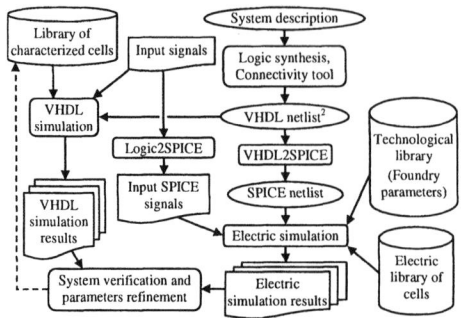

Figure 9 – Cell library refinements and verification

Innumerous circuit descriptions with different complexities may be evaluated with the depicted flow, including the precision of EngyLib. If the acquired precision is not satisfactory, a new energy model may be planned for a new evaluation.

The system description is converted by logic synthesis and the connectivity tool on the same VHDL netlist used in VHDL simulation of previous Section. Then, the VHDL2SPICE tool converts the VHDL netlist onto a SPICE one, in way of performing two equivalent estimations flows for future comparison. The values used as input of VHDL simulation are converted to electric ones with Logic2SPICE tool, according to some electric parameters, such as power supply, slope up, and slope down. Then, simulation results are compared in manner of verifying the accuracy of the proposed method and check the system behavior.

Table 2 – Energy consumption comparison between SPICE simulation and EngyLib of a chain of not gates

Interval (ns)	Energy consumption (J)		Imprecision (%)
	SPICE simulation	EngyLib approach	
2000	8.2274 E-11	8.2206 E-11	0.0827%

Table 2 illustrates the energy consumption achieved by SPICE simulation and EngyLib of a chain of not gates. A simulation during 2000 ns shows an imprecision of less than 0.1%. This imprecision practically maintains for larger evaluation.

Figure 10 illustrates **PowerEstimation,** which is the simulation core architecture. It performs the integration of all cells computation and some global calculi.

```
use work.POWER_PCK.all;
entity PowerEstimation is
end PowerEstimation;
architecture PowerEstimation of PowerEstimation is
    signal step: STD_LOGIC;
                    ...
begin
    process
    begin
        step <= '0', '1' after TREF / 2;
        wait for TREF;
    end process;
                    ...
    process(step)
        variable cycles: NATURAL := 0;
        variable totalEnergy: REAL;
    begin
        cycles:= cycles + 1;
        totalStaticEnergy := totalStaticEnergy +
                             totalStaticPower * TREF;
        totalEnergy := totalDynamicEnergy + totalStaticEnergy;
    end process;
end PowerEstimation;                    ...
```

Figure 10 – PowerEstimation architecture, which is the core of EngyLib approach

The parameter **TREF** of Figure 10 is a time reference, implying the precision of VHDL simulator, and affecting the total static energy computation.

5. Experimental Results

It was performed energy consumption estimation on a benchmark using electric simulation and EngyLib approach. Both were performed on a PC with Pentium 4, 3.2 GHz processors and 2 GB RAM. A router, two processors, and three embedded applications compose the benchmark. The selected router is a subcircuit of mesh NoC Hermes [4]. Plasma and 8051 processors are the ones selected. As embedded applications, it is evaluated a Fast Fourier Transform (FFT) and two image applications, one for object recognition and another for image encoding.

Table 3 – Benchmark characteristics

Characteristics		VHDL Code size (lines)	
Circuit	# of transistors	high-level	Synthesized
Router	52,376	630	2,985
8051	148,040	2057	8,967
Plasma	255,240	3089	16,124
FFT	66,550	697	3,699
Object recognition	54,774	612	2,578
Image encoding	55,732	734	2,903

Table 3 reports some characteristics of the benchmark. Column 2 reports the number of transistors for the technology mapped net lists (CMOS TSMC 0.35μm), while Columns 3 and 4 indicate the number of code lines in the original VHDL code and the one achieved by the synthesis tool.

For each circuit, it was firstly employed SPICE simulations to obtain accurate reports of the reference energy consumption. After that, with equivalent input stimuli, all benchmarks were simulated applying EngyLib approach. A random traffic provides input signals to energy consumption estimation of the router; two small programs were used as input of the processors, while embedded applications were stimulated by typical input traces. After that, the corresponding results generated by EngyLib were compared with the SPICE reference report. It produces absolute and relative comparison values depicted in Table 4 and Table 5.

Table 4 shows the elapsed time achieved by electric simulation and EngyLib for each circuit. As it can be observed, the approach proposed here reduces around of 34 times the simulation average time.

Table 4 – Time comparison between electric simulation and EngyLib approach

Simulation time (seconds)			
Circuit	SPICE	EngyLib	Reduction
Router	73,983	2,521	29.35
8051	235,473	5,778	40.75
Plasma	211,850	6,710	31.57
FFT	100,113	2,953	33.90
Object recognition	180,224	5,880	30.65
Image encoding	47,187	1,157	40.78
Average			34.50

Column 2 of Table 5 depicts the energy consumption achieve by each circuit during an electric simulation. Having these energy consumptions as references, Column 3 to Column 5 show the module percentage deviation of three energy models variations. Column 4 presents the results of the most accurate model, since it considerer static power

dissipation and the fan-out of each cell. It is also important to observe that energy consumption of each kind of circuit cannot be compared one with each other, since they are completely applications running for different times.

Table 5 – Accuracy of energy consumption estimation, regarding to different energy models

Circuit	SPICE energy consumption reference (mJ)	Energy consumption deviation from EngyLib to SPICE reference (%)		
		With static power dissipation		All fan-outs without static power dissipation
		Single fan-out	All fan-outs	
Router	1.13	5.07	3.11	3.12
8051	5.67	14.78	11.51	11.51
MIPS-like	11.19	31.23	25.12	25.13
FFT	1.47	9.35	8.49	8.49
Object recognition	0.90	8.42	5.03	5.04
Image encoding	1.02	21.05	16.94	16.95
Average		14.98	11.70	11.71

EngyLib approach has, in average, only 11.7% of deviation from SPICE energy consumption reference. It is well accepted when considering the benefits achieved by EngyLib approach, such as simulation time reduction. More deviation is achieved if the model considers only a single fan-out, even if it is achieved as an average fan-out.

Considering this technology the static energy consumption may be neglected when compared to the dynamic energy. On the other hand, the static energy has an important role in the estimation process of deep submicron technology, since it has the same magnitude order of the dynamic energy consumption.

6. Conclusions

Power estimation tools are required to manage the energy consumption of modern VLSI designs during the design phase, to avoid a costly redesign process. In this scenery, this paper proposed an approach based on a pre-characterized cell library, named EngyLib. It enables fast and accurate energy consumption estimations of systems described in high-level abstraction levels. The approach and auxiliary tools make part of a Java open-source framework, enabling the designer make any modification or enhancement into the library, tools, or the proposed approach. Experimental results show that EngyLib achieves high quality estimations, with expressive reduction of the computation time.

References

[1] Farid Najm. A Survey of Power Estimation Techniques in VLSI Circuits. **IEEE Transactions on VLSI Systems**, v. 2, n. 4, p. 446-455, Dec. 1994.

[2] C. Marcon et al. Modeling the Traffic Effect for the Application Cores Mapping Problem onto NoCs. **IFIP VLSI-SOC**, v. 1, p. 391-396, 2005.

[3] Mentor Graphics. Leonardo Spectrum Datasheet. Available at www.mentor.com/products/fpga_pld /synthesis/leonardo_spectrum, 2007.

[4] F. Moraes et al. HERMES: an infrastructure for low area overhead packet-switching networks on chip. **VLSI the Integration Journal**, v. 38, n. 1, p. 69-93, Oct. 2004.

CIRCUIT PROSPECTS OF DGFET: VARIABLE GAIN DIFFERENTIAL AMPLIFIER AND A SCHMITT TRIGGER WITH ADJUSTABLE HYSTERESIS

Srimoyee Sen[1], Urmimala Roy[1], Chaitanya Kshirsagar[2], Navakanta Bhat[2], Chandan Kumar Sarkar[1]
[1]Jadavpur University
[2]Indian Institute of Science

ABSTRACT: Double Gate(DG)FET is one of the most promising technologies for sub-50 nm transistor design. Various attempts have been made to exploit this newly emerging technology in electronic circuits[1, 2]. Threshold voltage of a double gate MOSFET can be varied by varying the bias on the back gate [1]. In this paper we propose a variable gain differential amplifier with current mirror load utilizing this property. Also it is shown that the saturation voltage of the differential amplifier varies with the varying back gate bias of the DGFETs. The variable gain differential amplifier has been used to propose a variable hysteresis voltage schmitt trigger, the hysteresis voltage of which can be varied without the help of external circuitry and only by varying backgate bias of the DGFETs. The nature of variation of hysteresis voltage with varying backgate bias of the transistors is also investigated.

INTRODUCTION

Double Gate MOSFET(DGFET)s are being widely studied because of their abilities to overcome several limitations of bulk-FETs like short channel effects[2]. The main idea of a double gate MOSFET is to have a Si channel of very small width and to control the Si channel by applying gate contacts to both sides of the channel. This helps to suppress short channel effects and leads to higher current as compared with a MOSFET having a single gate. It has been demonstrated that a DGFET can be scaled to a very short channel length (25 to 30 nm)[3, 4]. Double Gate MOSFETs are found to have some desirable features like dynamic threshold voltage control and transconductance modulation [1, 2]. Due to its greater ability in suppressing short channel effect, very lowly doped or even undoped channels can be used for DGFETs [3]. Moreover, DGFETs have good fabrication and integration features [3]. Generally in its common mode of operation the two gates of a DGFET are made to switch on simultaneously. But to control threshold voltage dynamically, only one gate is switched and a bias is applied in the other gate.

This device can be used for various purposes, one of which is variable gain amplifier as the gain of a single DGFET can be controlled by changing the back gate bias [1]. This feature can be utilized in a differential amplifier with currentmirror load.

In this paper a differential amplifier with current mirror load is presented where the Single Gate MOSFETs were replaced by DGFETs. The saturation voltage variation of this amplifier with backgate bias voltage has been investigated as well. Also an application of this feature,

which is a schmitt trigger with adjustable hysteresis voltage has been proposed.

Previous Work

Recently the application of independently controlled gate DGFETs in low power low voltage IC design and tuning the threshold voltage of the DGFETs varying backgate bias have been explored[1]. Operational amplifer operation using DGFETs with variable threshold voltage has also been demonstrated [1]. DGFETs have been used in wide variety of microwave circuit applications too [5]. The applications of DGFETs include automatic gain control amplifiers, frequency multipliers, mixers, phase shifters, switches and power dividers [5]. Inspite of being a versatile device, DGFET is not commonly used in microwave circuits due to a lack of accurate and efficient modeling technique [5].

Contribution of This Paper

The benefit of a schmitt trigger over a similar system with a single input threshold is that the schmitt trigger is more stable. With only one input threshold, a noisy input signal near that threshold could rapidly switch back and forth, causing the output to switch back and forth from low to high. With the schmitt trigger, a noisy input signal near one threshold could cause only one switch in output value, after which it would have to move to the other threshold in order to cause another switch.

Schmitt triggers are used in both analog and digital instrumentation and measurement system, on/off control, relaxation oscillators and reducing sensitivity to noise and disturbances [6, 7].

If we can achieve a schmitt trigger with adjustable hysteresis voltage, it can be used for different environments with different noise levels.

Schmitt triggers with fully or semi adjustable hysteresis voltages have been designed with the help of external circuitry, some of which used only MOSFETs [7]. But, here we propose a schmitt trigger with adjustable hysteresis voltage, achievable without the help of any external circuitry and only by varying the backgate bias of the DGFETs involved in the schmitt trigger circuit. Here we have shown that saturation voltage of a differential amplifier designed using DGFETs vary with the backgate bias voltage. We have made use of the feature of saturation voltage variation of the differential amplifier consisting of DGFETs. The hysteresis voltage of a schmitt trigger circuit depends on the saturation voltages of the differential amplifier used. As the saturation voltage is found to vary with the backgate bias, the hysteresis voltage should also vary with increasing backgate bias voltage.

In short, the proposed Schmitt trigger circuit has an adjustable hysteresis voltage where the adjustability has been achieved by utilizing intrinsic device characteristics and not by external circuitry.

DGFET DESIGN AND SIMULATION

The device topography and doping profiles are defined in MDRAW of ISE-TCAD package. The device structure used in this work is shown in figure 1.

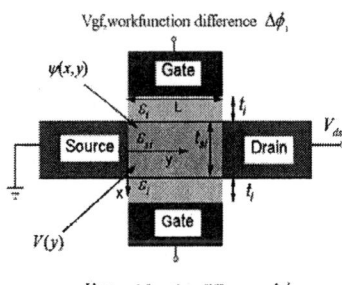

Figure 1. Schematic diagram of Double Gate MOSFET

Device Validation

The I_d-V_{fg} (drain current vs frontgate voltage) characteristics are validated against that of a fabricated device [2].

I_d-V_{fg} curves are simulated in DESSIS, a commercial device simulator of ISE-TCAD package, for the device dimensions like that of the fabricated device [2].

Experimentally obtained I_d-V_{fg} curves for the fabricated device is given in [2]

Corresponding curves from ISE-TCAD simulations are given in figure 2

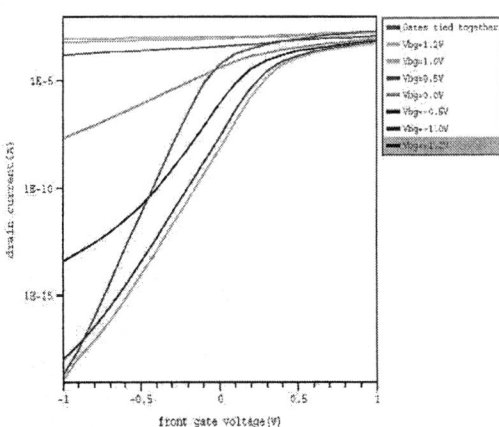

Figure 2. I_d − V_{fg} curves from ISE-TCAD simulations

This validation was done to obtain DGFETs with standard drain current vs frontgate voltage

characteristics with which we could design the differential amplifier and schmitt trigger circuit. This part is necessary to ensure the reliability of the results found in this paper.

DIFFERENTIAL AMPLIFIER DESIGN

The circuit is that of a differential amplifier with currentmirror load (p-channel DGFET), as shown in figure 3. The supply voltage used is 0.2 volts. The maximum gain is obtained with a backgate bias of -1.0 volts. The transistor supplying the current bias has a frontgate dc voltage of 0.15 volt and a backgate bias of - 0.6 volts. The maximum gain obtained by this configuration is 17.317 dB.

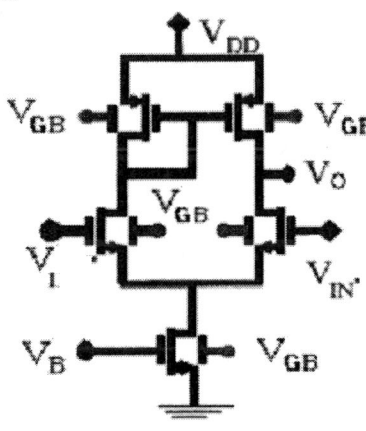

Figure 3. Circuit of the differential amplifier with current mirror load designed using DGFETs

It is noted that this differential amplifier amplifies at a very low supply voltage of 0.2 volts. This feature is also supported by the earlier findings. OPAMPs, designed using DGFETs, have been found to operate at a supply voltage as low as 0.5 volts [1].

Objective

The objective behind designing this amplifier is

a) To obtain a variable gain differential amplifier, the gain of which can be controlled by varying the backgate bias voltage of DGFETs used.

b) To observe the variations in the saturation voltage of the differential amplifier with changes in backgate bias voltage.

Results

Impact of Backgate Bias on Differential Amplifier Gain.
Gain vs backgate bias curves were simulated in DESSIS simulator, varying back gate bias in steps of 0.1 volt

from -0.5 volt to -1.1 volt. We get the maximum gain of 17.3 dB at a backgate bias voltage of -1.0 volts for n-channel DGFETs. If we increase backgate bias from this value the gain of the amplifier decreases. This much gain is obtained with a supply voltage of 0.2 volts.

Physical Insight Regarding Impact of Backgate Bias on Saturation Voltage of the Amplifier.

The threshold voltage of an n-channel DGFET is given by [2],

$$V_{Tn} = V_{TO} - kV_{BG} \quad \text{------- (1)}$$

where
V_{Tn} / V_{Tp} = threshold voltages of n-channel/p-channel DGFETs
V_{TO} = threshold voltage of the n-channel DGFET with no backgate voltage
V_{BG} = voltage applied at the backgate
k = an arbitrary constant

For the given circuit if implemented using bulk FETs, we can write for positive frontgate input voltage ie for the nMOSs operating in linear region and pMOSs in saturation

$$\frac{K_n[2(V_{GSn} - V_{Tn})V_D - \frac{V_D{}^2}{2}]}{2} = \frac{K_p[V_{GSp} - V_{Tp}]^2}{2} \quad \text{----(2)}$$

where
K_n / K_p = transconductance parameter for n-channel/p-channel DGFETs
V_{GSn} / V_{GSp} = gate to source voltage for n-channel/p-channel DGFETs
V_D = drain voltage

Therefore for DGFETs, applying eqn (1) in eqn (2) we get

$$2K_n(V_{GSn} + kV_{BG} - V_{TO})V_D - K_n\frac{V_D{}^2}{2} = K_p(V_{GSp} - V_{Tp})^2$$

or,

$$V_D{}^2 - 4V_D(V_{GSn} + kV_{BG} - V_{TO}) + 2\frac{K_p(V_{GSp} - V_{Tp})^2}{K_n} = 0$$

$$\text{----(3)}$$

or,

$$V_D = \frac{4(V_{GSn} + kV_{BG} - V_{TO}) + \sqrt{\begin{array}{l}4^2(V_{GSn} + kV_{BG} - V_{TO})^2 \\ - 8\frac{K_p}{K_n}(V_{GSp} - V_{Tp})^2\end{array}}}{2}$$

$$\text{-----(4)}$$

or,

$$V_D = 2(V_{GSn} + kV_{BG} - V_{TO}) + \sqrt{\begin{array}{l}4(V_{GSn} + kV_{BG} - V_{TO})^2 \\ - 2\frac{K_p}{K_n}(V_{GSp} - V_{Tp})^2\end{array}}$$

$$\text{------(5)}$$

From eqn 5 we can conclude that the drain voltage should increase with increasing backgate bias voltage for n-channel DGFETs for positive input voltages. Now, saturation voltage vs backgate bias voltage curves were simulated in Dessis simulator of ISE-TCAD package varying backgate bias voltage from -0.2V to 0.2V in steps of 0.1V. The maximum saturation voltages were obtained at a backgate bias of -0.2V. As the backgate bias was increased the output voltage, which is also the drain voltage of the DGFETs concerned was found to increase for positive frontgate voltages as expected from eqn 5 and the magnitude of saturation voltage was found to vary accordingly as shown in figure 4. No considerable effect on the output voltage was seen for negative frontgate voltages as we did not vary the backgate bias of the p-channel DGFETs involved.

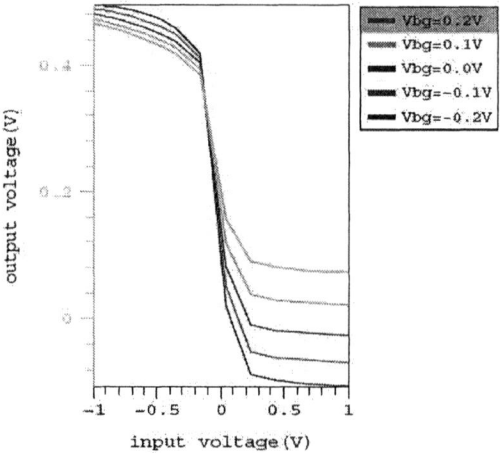

Figure 4. Impact of backgate bias voltage variation on saturation voltages of the diffamp

SCHMITT TRIGGER DESIGN

A schmitt trigger circuit was designed using the validated DGFETs, as shown in figure 5.The circuit used supply voltages of +0.5 volts and -0.5 volts as +Vdd and -Vss. The schmitt trigger was found to have maximum saturation voltages of +0.48 volts and -0.25 volts.

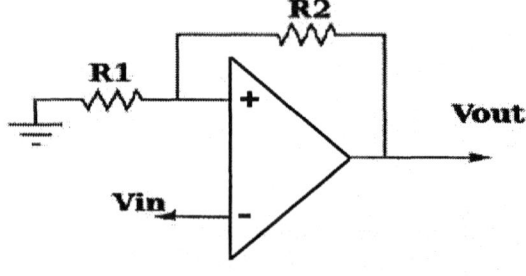

Figure 5. Circuit of the schmitt trigger using the differential amplifier designed

Impact of Backgate Bias Voltage on Hysteresis Voltage of Schmitt Trigger

Transfer curves of the designed schmitt trigger were plotted for different backgate bias voltages and the hysteresis voltage was found to be maximum at a backgate bias of -0.25 volts. The maximum value of the hysteresis voltage is 0.3 volts. As we increase the bias from this value the hysteresis voltage decreases as shown in figure 6.

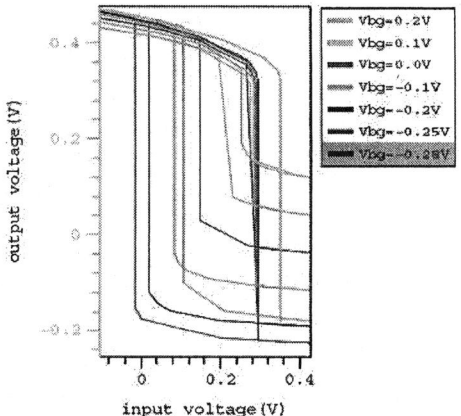

Figure 6. Impact of backgate bias on hysteresis voltage of schmitt trigger

TABLE 1.Hysteresis voltages obtained at different backgate bias voltages

Backgate bias (V)	Hysteresis voltage(V)
-0.25	0.3089
-0.2	0.2607
-0.1	0.2054
0.0	0.1378
0.1	0.0516
0.2	0.0233

Application

The hysteresis voltage of a schmitt trigger can be controlled by varying the backgate bias of the DGFETs used. Now, using the proposed schmitt trigger with adjustable hysteresis voltage we can obtain a schmitt trigger with adjustable noise tolerance. For low noise applications we can decrease the hysteresis voltage so as to make upper and lower threshold voltages close to each other, thereby increasing the precision of the comparator circuit and for higher noise applications, we have to compromise on the comparator quality to get a larger hysteresis voltage so as to increase the noise tolerance.

CONCLUSION

Here we have gained important physical insight regarding the characteristics of the device DGFET with independently controllable gates (MIGFET). The important property of DGFET of sensitivity of threshold voltage and saturation voltage to changing back gate bias is explored to propose unique circuit application of DGFET like schmitt trigger circuit with adjustable hysteresis voltage.

REFERENCES

[1] Arvind Kumar and Sandip Tiwari "A Power - Performance Adaptive Low Voltage Analog Circuit Design Using Independently Controlled Double Gate CMOS Technology"

[2] Weimin Zhang; Fossum, J.G.; Mathew, L.; Yang Du; "Physical insights regarding design and performance of independent-gate FinFETs", IEEE Transactions on Electron Devices, Volume 52, Issue 10, Oct. 2005 Page(s):2198 – 2206

[3] H. S. P.Wong "Beyond the conventional transistor"Solid State Electronics, vol. 49, pp. 755 – 762 (2005).

[4] D. Frank, S. Laux, and M. Fischetti, Monte Carlo Simulation of a 30nm Dual-Gate MOSFET: How Far Can Si Go?, IEDM Tech. Digest, p. 553 (1992).

[5] Licqurish, C.; Howes, M.J.; Snowden, C.M. Microwave Devices, Fundamentals and Applications, IEE Colloquium on volume,Issue,22 mar 1988 Page(s):2/1-2/7

[6] WANG, z., and GUGGENBUHL, w.: CMOS current Schmitt trigger with fully adjustable hysteresis , Electron.Lett., 1989, 25,(6), pp. 397-398

[7] Wang,z: CMOS adjustable Schmitt triggers,IEEE trans.Instrum.Meas.,1991,IM-40,pp.601-605

High Speed SOC Design for Blowfish Cryptographic Algorithm

Brian Cody[1] Justin Madigan[2] Spencer MacDonald[3] Kenneth W. Hsu[4*]

1, Brian.j.cody@gmail.com, Kulicke & Soffa Industries 2, Justin.madigan@gmail.com,3,
som0749@gamil.com, DAE Systems, 4, kwheec@rit.edu, Rochester Institute of Technology

* Correspondence author

Abstract—This paper seeks to implement the Blowfish algorithm in VHDL and provide a simple, robust implementation of Blowfish in hardware. As of today, the Blowfish algorithm has no known cryptanalysis. A hardware implementation of Blowfish would be a powerful tool for any mobile device or any technology requiring strong encryption. Our final design uses the core_slow library for worst-case scenario analysis and reaches an incredible encryption speed of 590 MBits/sec and a decryption speed of 559 MBits/sec. The area is 4996 standard cell,s and the power is a mere 63 mW. These results are very competitive and beat out the competition as far as speed. The overall design is an incredibly fast, efficient Blowfish implementation suitable for a plethora of applications. The speed can be drastically increased further at the expense of space and power by pipelining.

Index Terms—**Blowfish, crypto, encryption, high speed SOC**

I. INTRODUCTION

A network is a series of individual elements transmitting and receiving various data. Whenever sensitive or confidential information is transmitted, there is the possibility of an unauthorized third party "eavesdropping" on a transmission and learning the contents of the sensitive message. This possibility is unacceptable in many scenarios. Cryptography is the process of translating a message into a form which is unreadable to everyone except the intended recipient. This is typically done with the use of *keys*. A cryptographic key is roughly equivalent to the concept of a physical key which can unlock a lock. Many locks are made with many different keys, but only the correct key can unlock the correct lock. In cryptography, keys are used to encrypt a message into a format which would appear as unreadable random information to an unauthorized third party. Cryptography, then, is a required element of security for any sensitive communications.

Networks are susceptible to interception by their very nature. Wired networks can be "tapped" into by an eavesdropper with physical access to the network cabling. Messages transmitted across the internet are susceptible to eavesdropping attacks via any path along the transmission of a message. Wireless networks, which are quickly gaining popularity in all areas of technology, are particularly susceptible to interception since the message is transmitted by an antenna which broadcasts the message in all directions. Here cryptography is required to protect information from being intercepted and stolen by an unwanted third party.

The Blowfish encryption scheme was designed by Bruce Schneier in 1993 to replace Data Encryption Standard (DES), which was the Federal Information Processing Standard Cryptography (FIPS Crypto) [1]. The intent was to create a cryptographic algorithm which did not possess the limitations and issues common in other crypto algorithms and to provide an open, readily available crypto for users rather than the common patented or classified crypto algorithms being used contemporaneously. Due to its standing as a crypto algorithm, Blowfish is now part of the Linux kernel [3]. Blowfish continues to attain its lofty goals of secure, open encryption that is realizable in software and hardware.

The Blowfish algorithm is conceptually simple, but its actual implementation and use is complex. Blowfish has a fixed 64-bit block size. The key length of Blowfish is anywhere from 32 bits to 448 bits. [4] The cipher is a 16-round Feistel network and uses password-dependent S-boxes. A Feistel network is one that utilizes a structure which makes encryption and decryption very similar through the use of the following elements:

- P-boxes (permutation boxes; these perform bit shuffling)
- S-boxes (substitution boxes, simple nonlinear functions)
- XORing to achieve Linear Mixing

Blowfish encapsulates all these elements into an efficient and powerful algorithm. The action of Blowfish can be seen below.

284

978-1-4244-1709-4/07/$25.00 © 2007 IEEE

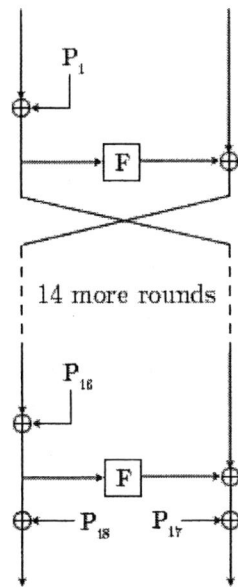

Figure 1: The Blowfish Algorithm [5]

The F function (the boxes containing an F in Figure 1) is the Feistel Function of Blowfish, the contents of which are shown below:

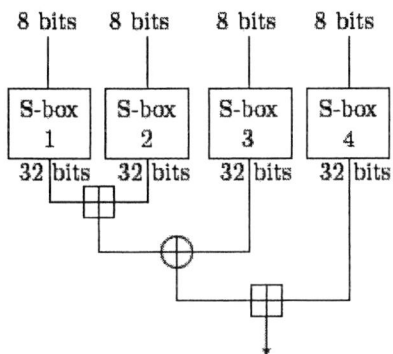

Figure 2: The Feistel Function of Blowfish [6]

Hence, Blowfish encrypts by splitting half the block (32 bits) into 8-bit chunks (quarters) and inputting this into the S-box. The results from the S-boxes then are added (with the carry dropped, resulting in mod 2^{32} addition) and XORed. Reverse Blowfish (decryption) is quite simple and accomplished by merely inverting the P17 and P18 cipher blocks and using the P entries in reverse. This makes Blowfish a very effective and conceptually simple algorithm with which to encrypt and decrypt.

Among Blowfish's main strengths is its complex key schedule. The S-boxes and P-boxes are initialized with values from the hex digits of pi. The variable-length user-input key is then XORed with the P-entries. Then a block of zeros is encrypted, and this result is used for the P1 and P2 entries. The

ciphertext resulting from the encryption of a zero block is then encrypted *again* and used for P3 and P4. This process continues until every P-box entry and S-box entry has been replaced, as shown in Figure 2, resulting in 521 successive key generations. This involves about 4 kB of data processing. This relatively complex key schedule makes Blowfish an effective and durable cryptographic algorithm.

Blowfish is among the fastest block ciphers available, according to an analysis done in [7]. The result of this analysis has been reproduced in Table 1.

Algorithm	Type	Clocks/round	# rounds	Clocks/byte of output
RC4	Stream cipher	n.a.	n.a.	7
SEAL	Stream cipher	n.a.	n.a.	4
Blowfish	Block cipher	9	16	18
Khufu/Khafre	Block cipher	5	32	20
RC5	Block cipher	12	16	23
DES	Block cipher	18	16	45
IDEA	Block cipher	50	8	50
Triple-DES	Block Cipher	18	48	108

Table 1: Block and Stream Cipher Speed Comparison

Clearly, Blowfish is among the fastest and yet remains cryptographically secure. Many of the other algorithms, it should be noted, have one or more known cryptanalysis methods, such as RC4, which routinely produces weak keys which can be used to crack the algorithm. As such, RC4 is not considered secure at this stage. The same can be said of RC5. These examples show the relative security and speed advantages Blowfish has over competitors in the block and stream cipher areas [8][9][10][11].

II. CRYPTANALYSIS OF BLOWFISH

After the initial proposal of the Blowfish cipher, Dr. Dobb's journal sponsored a cryptanalysis contest [12] in order to ascertain the security of Blowfish. The high hopes and large promises provided by the algorithm yielded much activity regarding a potential cryptanalysis. Given this fact, it is not surprising that many interesting results were proposed; however, none came close to actually successfully cracking or providing a cryptanalysis of Blowfish. Some of the most intriguing results of the last century will be presented here for completeness.

Only five results in total were submitted [13]. John Kesley could only break 3-round Blowfish (nowhere near the 16-round final version), and his cryptanalysis cannot be extended beyond 3 rounds [14]. Serge Vaudenay used an intentionally weakened version of Blowfish and found a known-plaintext attack requiring $2^{8r}+1$ (where r is the number of rounds) known plaintexts to break; however, this method is impractical in reality and does not work against the full 16-round Blowfish algorithm. The most promising attack was proposed in 1996 by Vincent Rijmen in his doctoral dissertation, but this attack can only break 4 rounds of Blowfish and no more [15]. The most recent work is from Dieter

Schmidt, who noted that the third and fourth subkeys are independent from the user's 64-bit key [16].

Given that these attempts are the only ones known thus far and that they are surprisingly weak in decrypting the actual Blowfish algorithm, the future of Blowfish as a secure algorithm is very promising indeed.

III. PROPOSED DESIGN

Our design utilizes the simplicity of Blowfish to create a relatively straightforward implementation. The S-box contains a table of non-linear data which maps inputs to outputs. This design severely disrupts any sort of correlation between the input and the output of the design. When the Blowfish design is started, the S-boxes and the P-boxes are initialized based on the input key and nonlinear data. The writing of S-box data to table is tied to a clock and is currently the slowest part of the design. The table lookup procedure is handled without a clock pulse, so a uniform clock is possible between initialization and runtime without sacrificing performance.

The design uses a state machine to initialize and perform its encryption/decryption functionality. The first state initializes the internal state counter, effectively preparing it for use in the next phase and throughout the encryption/decryption process. The second state initializes the P functions with nonlinear data derived from PI. This is an important step in assuring proper encryption, as the digits of PI have no currently known cryptographically exploitable pattern. The counter is incremented 1024 times as the P functions get XORed against the key. This step prepares the Blowfish algorithm for encryption and decryption by properly preparing the system for use. The next state rewrites the S-boxes with new information based on the key. A ready state is now entered where the system waits for an encryption or decryption signal from the user. Upon receiving this signal, it enters either an encryption or a decryption state which lasts 17-18 cycles. This design is compact, efficient, and very fast.

As you can see, our design aims for a slim-profile IC and a tightly-packed design. This can be seen in the schematic below, showing the final overall design. The synthesized chip results can be seen in Figure 3. The desired result as far as a size profile is clearly achieved.

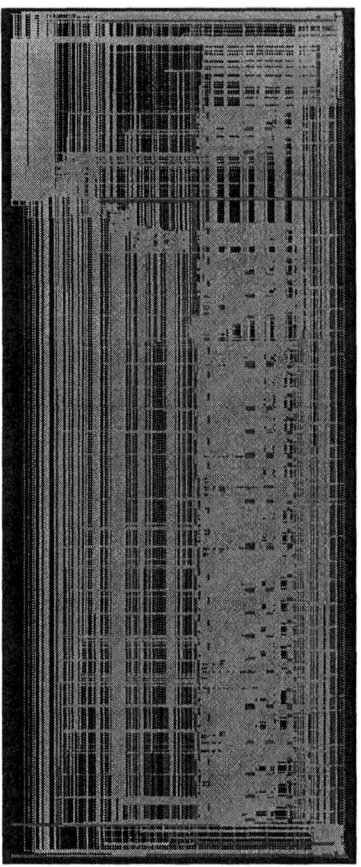

Figure 3: The Synthesized Blowfish Design

IV. RESULTS

The Blowfish implementation proposed here has been implemented, and several important test vectors were encrypted and decrypted to guarantee correct function. One simple test vector used was the hexadecimal value 12345678ABCDEF, which was encrypted and decrypted. The waveform from this process is shown below.

Our finalized Blowfish cipher implementation can perform block cipher encryption and decryption in only 18 and 19 cycles, respectively, as shown in Figure 4. Using the compile-ultra switch, the maximum clockspeed is 167 MHz (a 6 ns clock). The throughput of the design is 590 Mbits/sec for encryption, and 559 Mbits/sec for decryption. Power consumption is a mere 63 mW, and area is 4996 cells. A comparison of our design with the leading publicly available *high-speed* Blowfish implementations is shown in Table 2.

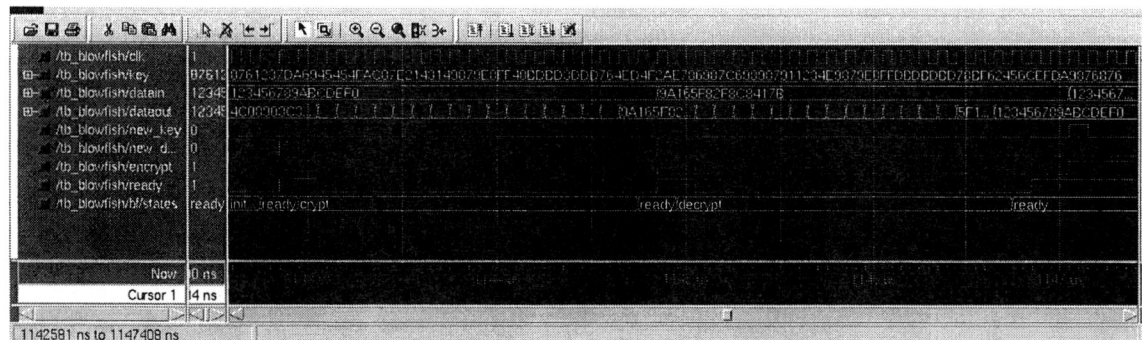

Figure 4: Encryption and Decryption of a Test Vector

Table 2: Comparison of High Speed Blowfish Hardware Implementations

Architecture	Through put	Size	Clock Frequency	Process (TSMC)
Salomao et al [10]	266 Mbits/sec	4620 std. cells	66 Mhz	0.7
Lin et al [11]	200 Mbits/sec	16k	50 Mhz	0.6
Lai et al [12]	288 Mbits/sec	13k	72 Mhz	0.35
Our design	**590 Mbits/sec**	4996 cells	167 Mhz	0.35
	(559 Mbps decrypt)			

Clearly, our implementation is the fastest of those surveyed. Although some sacrifices must be made in size and frequency, the overall result is an incredibly fast implementation of Blowfish. If our design were pipelined, the speed would increase to an incredible **10,666,666,624 bits/sec** (10.667 Gbits/sec). There was insufficient time to perform a full pipelining of the design; however, this could be achieved with relatively straightforward future enhancements to our design.

V. CONCLUSION

A high-speed implementation of the Blowfish cryptographic algorithm has been presented. Even without pipelining, which other high-speed methods utilize, our method was rated at 590 Mbits/sec maximal throughput, which is 204% as fast as the leading (pipelined) competitor. With pipelining, our design could reach 10.667 Gbits/sec throughput. The overall resulting design from our scheme was considerably large but only consumed 63 mW of power during operation. Our solution takes advantage of the conceptual simplicity of Blowfish and is optimized for high-speed encryption and decryption. The results show that our design objective is achieved, especially when compared to the leading competitors.

REFERENCES

[1] U.S. National Bureau of Standards, "Data encryption standard," U.S. Fed. Inform. Processing Standards Pub., FIPS PUB 46, January 1977.

[2] B. Schneier, "Applied Cryptography," 2nd ed. New York: , John Wiley & Sons, Inc., 1996.

[3] B. Schneier. The Blowfish Encryption Algorithm. Retrieved 12:04:58, July 27, 2007 from http://www.schneier.com/blowfish.html

[4] Ibid.

[5] B. Schneier, "Description of a New Variable-Length Key, 64-bit Block Cipher (Blowfish)," *Fast Software Encryption: Second International Workshop, Leuven, Belgium, December 1994, Proceedings*, Springer-Verlag, 1994, pp. 191-204.

[6] Ibid.

[7] B. Schneier, Speed Comparisons of Block Ciphers on a Pentium. Retrieved 12:04:58, July 27, 2007 from http://www.schneier.com/blowfish-speed.html

[8] B. Schneier, D. Whiting, Fast Software Encryption: Designing Encryption Algorithms for Optimal Software Speed on the Intel Pentium Processor, *Lecture Notes in Computer Science,* Volume 1267, January 1997, p. 242.

[9] S.L.C. Salomao, J.M.S. de Alcantara, V.C. Alves, and A.C.C. Vieira, "SCOB, a soft-core for the blowfish cryptographic algorithm," *in Proc. IEEE Int.. Conference on Integrated Circuit and System Design*, pp. 220-223, 1999.

[10] M.C.J. Lin and Y.L. Lin, "A VLSI implementation of the blowfish encryptioddecryption algorithm," in *PWC. ZEEE ASP-DAC*, pp. 1-2, 2000.

[11] Y.-K. Lai and Y.-C. Shu, "VLSI architecture design and implementation for BLOWFISH block cipher with secure modes of operation," *IEEE International Symposium on Circuits and Systems*, 2001, pp. 57-60.

[12] *Dr. Dobb's Journal*, September 1995.

[13] B. Schneier, "Blowfish: One Year Later," available online at http://www.schneier.com/paper-blowfish-oneyear.html, [Accessed: March 27, 2007].

[14] Ibid.

[15] V. Rijmen, "Cryptanalysis and design of iterated block ciphers," doctoral dissertation, Katholieke Universiteit Leuven (K.U. Leuven), October, 1997.

[16] D. Schmidt, "On the Key Schedule of Blowfish," Cryptology ePrint Archive, 2005.

Implementing Cellular Automata modeled Applications on Network-on-Chip Platforms

N. Zompakis, L. Papadopoulos, G. Sirakoulis, D. Soudris

VLSI Design and Testing Center, Department of Electrical and Computer Engineering,

Democritus Univ. Thrace, 67100 Xanthi, Greece

{nzompaki, lpapadop, gsirak, dsoudris}@ee.duth.gr

Nowadays, embedded consumer devices are expected to support demanding applications in terms of performance and energy consumption. For implementing such applications on Network-on-Chips (NoCs) a design methodology for performing exploration at system-level is needed, in order to select the optimal application-specific NoC architecture. In this paper we present a methodology for designing application-specific NoC platforms at system-level. The methodology is based on the exploration of different NoC aspects (e.g. topology, routing algorithms etc.) and is supported by a flexible NoC simulator. In this work we apply our methodology to applications modeled with Cellular Automata (CA).

I. INTRODUCTION

Today, complex parallel applications (e.g. network protocols, speech, image and video processing applications) can be implemented on embedded devices. The implementation of such sophisticated algorithms to portable devices is a difficult task, due to their limited resources and the stringent power constraints of such systems. Single processing platforms are not capable of providing the required performance for complex applications with Task-Level Parallelism (TLP) and hard real-time constraints. MPSoC platforms are expected to overcome computational power limitations.

With the contemporaneous technology scaling, the integration of billions of transistors on a chip is enabled. Current MPSoC platforms usually contain bus-based interconnection infrastructures, which suffer from limited scalability, poor performance for large systems and high energy consumption.

The computational power along with energy efficiency that modern applications require, cannot be provided by shared bus types of interconnections. During the last years, the communication structure of MPSoC has been proposed to be handled by Network-on-Chip (NoC). The NoC is an energy-efficient on-chip communication architecture, which, instead of using bus structures or dedicated wires, is composed of an on-chip, packet-switched network. It provides more predictability and better scalability compared to bus communication schemes [1].

We have developed a methodology of four steps for designing application-specific NoCs based on the exploration of NoC characteristics. The methodology is supported by a flexible NoC simulator which emphasizes in communication aspects (such as packet rates, buffer size etc.) and is based on an extension of Nostrum NoC Simulation Environment (NNSE) [2]. The NoC simulator is developed for implementing NoC exploration at system-level. The high abstraction level of the simulator and its simplicity allows easy exploration of NoC parameters and quick modifications. The input of the methodology is the communication task graph (CTG) of the application mapped onto the NoC architecture. The output is an optimal application-specific NoC architecture which meets the design constraints.

In this paper we apply the aforementioned methodology to the Cellular Automata (CA) domain because we want to explore applications with parallel characteristics in a high-performance parallel platform like NoC. In addition, parallel systems have the most suitable architecture for a CA algorithm. Specifically, we evaluate the optimal NoC topology according to performance, link utilization and communication energy consumption, for a number of different applications modeled with CA.

The rest of the paper is organized as follows. In Section 2, we describe some related work. In Section 3, we analyze the design methodology that is supported by the NoC simulator. In Section 4 we describe the applications which are modeled with CA and present the experimental results. Finally, in Section 5 we draw our conclusions.

II. RELATED WORK

NoC as a scalable communication architecture is described in [1]. The existing NoC research is presented in [3] and shows that NoC constitutes a unification of current trends of intra-chip communication.

Several research groups have proposed tools to simulate NoCs at different levels of abstraction. For instance, in PROTEO NoC model [4], VHDL was used to evaluate several features of virtual channels in mesh-based and hierarchical NoC topologies. Although the accuracy of the VHDL model is high, it suffers from the low simulation speed [5].

288

978-1-4244-1709-4/07/$25.00 © 2007 IEEE

For instance, SPIN network [6] implements a fat-tree topology using wormhole routing. CHAIN NoC [7] is implemented using asynchronous circuit techniques. MANGO [8] is a clockless NoC which provides both best-effort and guaranteed services.

However, the aforementioned approaches are not flexible enough, since they limit the design choices. Also, they do not focus on application-specific NoC design and therefore, they are not suitable for implementing exploration of NoC parameters according to the characteristics of the application under study. Therefore, a system-level and flexible simulator is needed for exploring interconnection characteristics.

Nostrum NoC simulator (NNSE) [2] focuses on grid-based, router-driven communication media for on-chip communication. As we show in the next section, the new tool based on Nostrum, which supports our methodology, adds new features to Nostrum and allows the easy exploration of several NoC parameters.

From the application perspective, Cellular Automata (CA) models have been implemented on a wide area of hardware. A hardware implementation of a CA algorithm for automated visual inspection, using hardware description language (VHDL), is presented in [13], in [16] are demonstrated the combination's benefits of parallel computing and CA. The [18] shows how an implementation of a CA based on artificial brain can be ported to MIT's cellular automata machine CAM-8. [17] presents a cellular automaton based on a neural network model which can be implemented in evolvable hardware. Generally, the CA have been applied mainly on single processor systems and on FPGA platforms, as shown in [14] [15].

In this work we implement CA on a NoC platform and evaluate the optimal NoC topology for each application modeled with CA.

III. SYSTEM-LEVEL NoC DESIGN METHODOLOGY

In this section we analyze step by step the flow of our methodology and describe the NoC simulator we developed to implement it.

A. Design Methodology Description

The first step of our methodology is the parallelization of the application in a number of tasks. This can be done with several methods, such as in [9].

The second step of the design flow is the insertion of C++ monitors into each task, to trace the inter-task communication. In our work, we used an in-house tool, which provides automatically the necessary information about the amount of data, exchanged between each task and the memory.

The following step is the exploration of the NoC characteristics. In this step, the designer can explore a number of different NoC parameters such as packet size, buffer size, topology etc. This is direct outcome of the flexibility provided by the simulator and allows the modification of the NoC

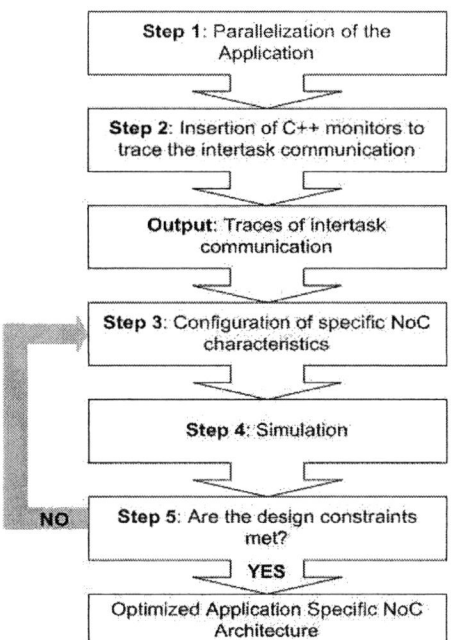

Fig. 1 Application-specific NoC design methodology.

communication infrastructure easily. Finally, after the simulation and the analysis of the evaluation results, if the constraints are satisfied the NoC architecture can be chosen otherwise the procedure restarts from the third step.

In this work we aim to prove that an optimal NoC platform can be achieved using a design methodology based on exploration. Although the tasks assigning process of application cores onto the NoC platform has been done manually, our purpose is to evaluate our methodology using CA models.

B. Simulator Description

The NoC simulator is an exploration tool for implementing NoC at system-level. The tool emphasizes in communication aspects (such as packet rates, buffer size etc.). Therefore, it abstracts away lower level aspects, such as the cache and other memory effects, in order to keep the complexity of the NoC model under control. The high abstraction level of the simulator and its simplicity allows the easy exploration and quick modifications. Additionally, the simulator is based on Nostrum, but provides more topologies and more evaluation metrics.

The simulator allows the construction of irregular topologies and routing can be done either using XY routing or routing tables. Also, provides more evaluation metrics such as performance, average performance, link utilization and communication energy consumption. Thus, allows an in-depth exploration of different NoC aspects at system-level.

The simulator is developed in SystemC and the resources, switches and channels are implemented as *sc_modules*. Application's threads are allocated in each resource which implement functions *read* and *write*. Resources, which are an abstraction of processing elements, provide the required interface for allocating application threads on them and the required network inter-face, to connect the specific resource to the network.

The resources communicate via the channels. The way the resources are connected and the number of channels used are defined by the designer. Thus, various topologies can be implemented. Each channel handles data independently, according to the bandwidth restriction. Every resource contains a switch, which implements the selected routing algorithm. By using adaptive routing, congestion avoidance can be implemented.

The simulator provides the essential evaluation metrics to explore NoC characteristics. Average packet delay refers to the time that a packet needs to be transferred from the source to the destination through the network. Energy consumption is calculated as described in [11] and is affected by the switch architecture and the number of hops passed by packets.

IV. EXPERIMENTAL RESULTS

In this work we take advantage of the parallelism of the CA applications to implement them on a NoC platform. More specifically, we chose to explore different topologies in order to evaluate optimal NoC platform for each CA application. The topologies we implemented using the NoC simulator are: 6x6 2D-Mesh and 6x6 Torus, Binary Tree and Fat Tree of 31 nodes. Other NoC characteristics such as buffer size, packet size and routing algorithm can also be explored. The cost factors we used to evaluate each NoC topology are: performance, link utilization and communication energy consumption. These evaluation metrics are provided by our simulator. The NoC simulation process was implemented using wormhole routing with flit size of 4 bytes (every packet contains 4 flits).

The benchmarks we used to evaluate our methodology are three applications modeled using CA. The first application is the model of a computer network. The second one is a popular sort algorithm implemented in parallel using CA namely Trend Bubble Sort .The third application is an image processing algorithm which surveys the quality and the eligibility of an object towards a prototype.

A. Computer Network Simulation modeled with CA

A computer network simulation using CA is proposed in [12]. This application uses the NaSch network model and is implemented as a 1D CA. The nodes, the packets and the routers are modeled as cells implementing different CA rules. As the simulation takes place, network characteristics such as performance and congestion can be observed.

The application was partitioned in 8 tasks. Then we inserted the C++ monitors to each task and extracted the inter-task communication. Each task was allocated on a different resource. Finally, the exploration on the NoC simulator took

place for each one of the different topologies we selected.

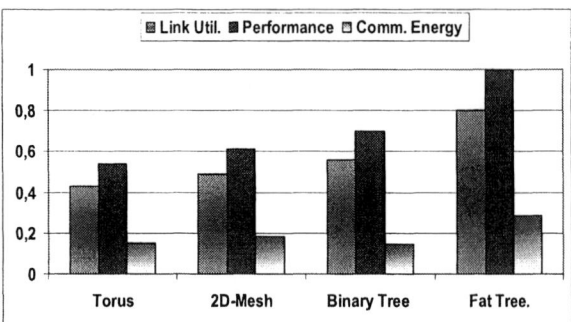

Fig. 2 Normalized link utilization, performance and communication energy consumption of the Network application.

From figure 2, the optimal topology in terms of performance is the Fat tree. In addition on the Fat Tree topology, optimal link utilization is achieved, but also communication energy consumption is increased at 33% compared to the Binary tree and Torus topology. It is up to designer to choose the best topology that satisfies the imposed design constraints of the NoC platform.

B. Trend Bubble Sort modeled with CA

Trend Bubble Sort is the CA implementation parallel of the common Bubble Sort algorithm [19], [20]. In this approach, the sorting process takes place in parallel. More specifically, every element sorts itself by comparing its value with the values of each neighbours.

We partitioned the application into 4 tasks. Then we inserted the C++ monitors, so as to extract the traces that describe the inter-task communication. Finally, we allocated each task on a different resource of the NoC simulator to perform exploration for different topologies. In figure 3 we demonstrate the experimental results. From figure 3, it can be seen that the optimal topology is the Torus topology. Although, this topology has the same energy consumption with others, performance and link utility are increased at 35-40%.

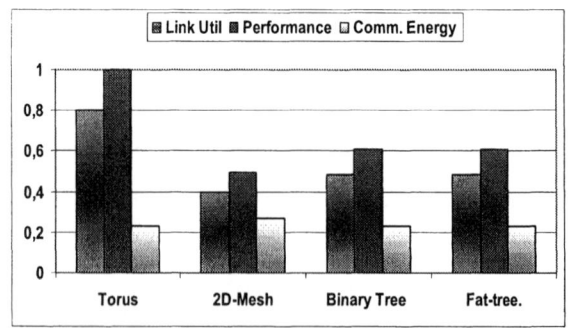

Fig.3 Normalized link utilization, performance and communication energy consumption of the Trend Bubble Sort application.

C. Image processing application modeled with CA.

Image processing application is a general CA purpose algorithm which can satisfy the simulation needs many physical systems without any specific scientific limitations. We will present only the general framework of this algorithm to avoid any unnecessary programming specifications. The implementation is described thoroughly in [13].

From figure 4, it is shown that the optimal topology in terms of performance the Fat tree. In this case, Fat Tree topology is clearly the optimal one, not only in terms of link utilization but also in communication energy, compared with the other topologies. In detail, performance is increased at 25-40%, link utilization is increased at 20-30% and communication energy consumption is decreased at 33% compared to the others topologies. Therefore, the application should be implemented on Fat Tree topology.

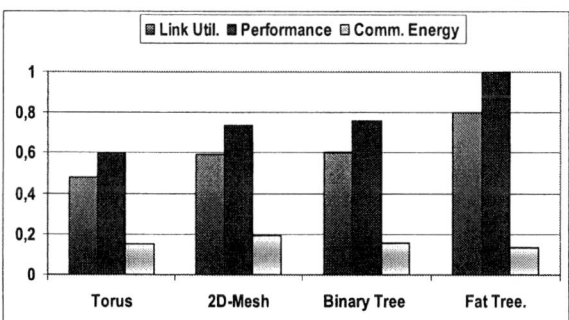

Fig. 4 Normalized link utilization, performance and communication energy consumption of the Image Processing application.

V. CONCLUSION

It is fact, that the efficient design of an embedded system supposes auspicious methodologies and useful tools. In this paper we have presented a system-level methodology using a flexible NoC simulator which gives us the ability to define a variety of critical parameters. We applied our methodology in three CA applications and explored the optimal communication architectures. The conclusions are based on experimental results which have presented in normalized graphs. The same process can be done for a number of CA applications. Our future work focus on the further automation of our procedure.

VI. ACKNOWLEDGEMENT

We would like to thank Christophe Poucet (NES Group, IMEC, Leuven, Belgium) for his help and support in the development of the profiling framework.

VII. REFERENCES

[1] L. Benini, G. De Micheli: Networks on Chips: A new SoC paradigm. IEEE Computer, 35(1), (2002).

[2] M. Millberg, E. Nilsson, R. Thid, S. Kumar, A. Jantsch: The Nostrum backbone-a communication protocol stack for networks on chip in Proc. VLSI Design, (2004).

[3] T. Bjerregaard, S. Mahadevan: A survey of research and practices of network-on- chip. ACM Computing Surveys (CSUR), 38(1), (2006).

[4] D.S.Tortosa *et al.*, "VHDL-based Simulation Environment for PROTEO NoC", In *Proc. of HLDVT Workshop*, (2002).

[5] Xi. Jinwen, Zhong Peixin "A Transaction-Level NoC Simulation Platform with Architecture-Level Dynamic and Leakage Energy Models", presented at the Great Lakes Symposium on VLSI archive (341 – 344) 2006.

[6] P. Guerrier, A. Greiner: A generic architecture for on-chip packet-switched interconnections. In Proc. DATE (250-256) 2000.

[7] J. Bainbridge, S. Furber: CHAIN: A delay-insensitive chip area interconnects. IEEE Micro 22, 5 (16-23) 2002.

[8] T. Bjerregaard: The MANGO clockless network-on-chip: Concepts and implementation. Ph.D. thesis, Information and Mathematical Modeling, Technical University of Denmark, Lyngby, Denmark.

[9] The Cadence Virtual Component Co-design (VCC), Available:http://www.cadence.com/company/pr/09_25_00vcc.ht.

[10] M. Meyer: Energy-aware task allocation for network-on-chip architectures. MSc thesis, Royal Institute of Technology, Stockholm, Sweden.

[11] T. Tao Ye, L. Benini, G. De Micheli,: Packetized on-chip interconnects communication analysis for MPSoC. in Proc. DATE (2003).

[12] Ren Zhiliang, Deng Zhidong, Sun Zengqi: Cellular automaton modeling of computer network.(243-251) 2002.

[13] .G. Ch. Sirakoulis, I. Karafyllidis, A. Thanailakis, and V. Mardiris, "A methodology for VLSI implementation of Cellular Automata algorithms using VHDL," Advances in Engineering Software, vol. 32, no. 3, pp. 189-202, (2001).

[14] Hugo de Garis, Michael Korkin, Felix Gers, Norberto Eiji Nawa, and Michael Hough. *Building an artificial brain using an FPGA based 'CAM-brain machine'.* Applied Mathematics and Computation Journal, Special Issue on Artificial Life and Robotics, Artificial Brain, Brain Computing and Brainware, (2000).

[15] Mathias Halbach, Rolf Hoffmann, Patrick Röder FPGA "Implementation of Cellular Automata Compared to Software Implementation" (2004).

[16] Domenico Talia ISI-CNRc/o DEIS, Università della Calabria, "Cellular Automata + Parallel Computing = Computational Simulation" (1997).

[17] Felix GERS (1), Hugo de GARIS (2), Michael KORKIN (3), "A cellular automata based neural net model simple enough to be implemented in evolvable hardware" Lecture Notes in Computer Science (1998).

[18] Felix GERS , Hugo de GARIS: Porting a cellular automata to MIT´s cellular automata machine CAM-8 SEAL'96 Conference Proceedings S7-3 , (1996).

[19] Antoine Spicher, Olivier Michel and Jean-Louis Giavitto, "A Topological Framework for the Specification and the Simulation of Discrete Dynamical Systems" Lecture Notes in Computer Science, Volume 3305/2004, pp.238-247.(Proceedings of 6th International Conference on Cellular Automata for Research and Industry, ACRI 2004, Amsterdam, The Netherlands) October 25-28,(2004)

[20] Hui-Hsien Chou, Wei Huang, James A. Reggia, "The Trend Cellular Automata Programming Environment," SIMULATION, Vol. 78, No. 2, 59-75 (2002).

Optimum IR Drop Models for Estimation of Metal Resource Requirements for Power Distribution Network

Rishi Bhooshan (r-bhooshan1@ti.com), Bindu P Rao (bindu@ti.com)
Texas Instruments India Ltd, Bangalore-560093, India

Abstract—**In this paper, we present closed form IR drop models for power distribution network in N-metal layer system for wire-bond and flip-chip packages, given design constraints such as chip power dissipation, total static IR drop budget and power supply voltage along with manufacturability and EM constraints as per the technology. The models proposed empowers designers to perform trade-off analysis for effective metal resource utilization in the power distribution network and meeting design specific signal routing needs in desired metal layers. The power distribution network designed using the proposed models have been verified for EM and IR drop on 90nm, 65nm and 45nm designs with industry standard EMIR analysis tool and are within 1-5% of error limit for both wire-bond and flip-chip designs.**

Index terms—**PG- Power and Ground, EM- Electro-migration, PDN- Power Distributions Network. SC- Bump Square cell**

I. INTRODUCTION

With scaling in technology, on-chip power dissipation of SOC designs increase, leading to higher current density in the power distribution network. The above, coupled with reduction in power supply voltage [1] causes increased power supply noise adversely affecting the chip performance [6] and reliability. Power supply noise is a combination of IR voltage drops in the resistive PG network and L di/dt noise [6], [2]. IR voltage drop degrades the drive capability of transistors and in severe cases can cause functional failure [6]. Resistance increase in narrow line width wires of DSM technologies caused by surface and grain boundary scattering has increased the need for more metal resource in the power distribution network to achieve tight IR drop constraints. This adversely affects the amount of metal resource available for signal and clock routing. Thus is it very critical to optimally plan the metal resource sharing between power and signal routing in initial phase of the design.

Today, designers design the initial power grid network with nominal set of wire widths and pitches for different metal layers, and the optimize power grid network to meet the allowed IR drop budget using circuit simulation techniques [8]. Most previous work in this field mainly focus on power network wire sizing using graph topology, linear or non-linear programming techniques, to meet allowed resistive IR voltage drop [7], [9-11]. These methods are computationally expensive and cannot be used in early stages of the design for power grid planning and optimal metal resource distribution for power and signal routing. Though, some of the works on power grid design [2-5] have proposed closed form models for rough estimation of global metal resource required for power network, they are based on two metal layers. Further, some of these models [2], [5] are derived specifically for isotropic grid (equal resistance for horizontal and vertical metal layers) and assume equal usage of both horizontal and vertical metal layers in PDN. However, the N metal layers used in the PDN may have different sheet resistances resulting in anisotropic power distribution network. Further, the metal layers offered by the technology need to be shared across power, signal and clock routing needs, warranting unequal usage of different metal layers in the PDN.

In this paper, we propose closed form models for power distribution network in N-Metal layer system for wire-bond and flip-chip packages, given design constraints such as power dissipation, power supply voltage and static IR drop budget along with reliability and manufacturability constraints. These models are an extension to the work presented in [12]. They are generalized to handle various aspect ratios of the chip and are more accurate. The proposed models are applicable to both isotropic and anisotropic power distribution network.

These models can be used effectively for the following:

- In early stages of the design, to determine the number of metal layers and their densities needed in the power distribution network for given design constraints.
- To perform trade-off analysis for optimal sharing of metal resources for power and signal routing requirements of the design.
- These closed form models are easy to implement for quick generation of initial power distribution network.

This paper is organized as follows. In section II and III, we propose the models for designing power distribution network for wire-bond and flip-chip packages respectively. In section IV, we design power distribution network using proposed models and tabulate EMIR analysis results for both wire-bond and flip-chip packages. The results are validated with industry standard tool, blast-rail [13] for 90nm, 65nm and 45nm designs. Finally, we conclude this paper with summary of the work presented in the last section.

II. MODELS FOR DESIGNING POWER DISTRIBUTION NETWORK FOR WIRE-BOND PACKAGES

A. Power distribution network model for wire-bond designs

For wire-bond designs, the commonly used power distribution network consists of horizontal and vertical power (V_{DD}) and ground (V_{SS}) metal strap pairs placed at regular pitch and connected by via array at intersection [12]. These power straps terminate on the core ring which are connected to the IO pads. In N-metal layer PDN, the lowest metal layer (eg. M1) is typically reserved for V_{DD}/V_{SS} cell row straps for feeding power to standard cells. Cell row metal strap width, pitch and hence density is fixed as per the standard cell

292

978-1-4244-1709-4/07/$25.00 © 2007 IEEE

library architecture for a particular technology. However, utilization of higher metal layers need to be planned by the designer for designing optimum power distribution network based on given design constraints.

We use following notations as needed.

V_{DD} Supply voltage
H   Height of chip
L   Length of chip
ΔV Total IR voltage drop on V_{DD} and V_{SS}
δ Percentage total IR drop budget $(\Delta V / V_{DD})$
N   Number of metal layers
M_n n^{th} metal layer
$M_{n,width}$ Width of metal strap of nth metal layer
$M_{n,pitch}$ Pitch of metal strap of nth metal layer
$R_{SH}(n)$ Sheet resistance of nth metal layer
$D(n)$ Metal density of nth metal layer
$G(n)$ Equivalent conductivity of nth metal layer
$d(n)$ Metal weightage parameter of nth metal layer
$g(n)$ Normalized conductivity of nth metal layer, $g(1)=1$
$I(n)$ Current distributed on the nth metal layer
$J_{EM}(n)$ EM current limit of nth metal layer
h   Height of a cell row
K   Number of cell rows for a given metal layer, K = H / h
I_R Current drawn per cell row
I_t Total current drawn by the PDN

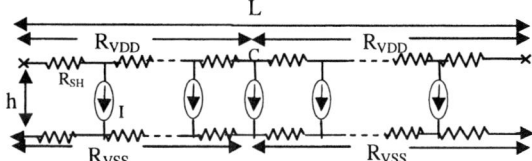

Fig-1: Single Cell Row Model of Power Distribution Network

In order to develop a model for the power distribution network, we first develop a model for single cell row for one metal layer of the power distribution network. We then extend the single cell row model to multiple cell rows and to the N-metal layers of the power distribution network. Consider a pair of core straps forming the single cell row of length L (chip length) and height h (where H= K*h) as shown in fig-1. The V_{DD} and V_{SS} power straps of this cell row model are connected to core rings at the periphery as shown with cross (power) and arrow (ground) in fig-1. The core ring is assumed to be *equipotential.* This assumption is valid if IO ring periphery has sufficient number of power and ground pads distributed uniformly in the IO ring. Let I_R be the current drawn by single cell row. Assuming uniform power distribution, the maximum IR drop will be at the centre point C as shown in fig-1. The metal straps of the single cell row with resistance $R=R_{[VDD|VSS]}$ is broken into m/2 metal squares with distributed resistance of R_{SH} (where $R_{SH}=R/(m/2)$). The standard cells placed in single cell row are modeled as distributed equivalent current sinks, I (where $I=I_R/m$, and m>>1) as shown in fig 1.

The total IR drop (V_{DD} and V_{SS}) will be the sum of IR drop across resistor segments.

$$\Delta V = 2 \times [I \times R_{SH} + 2I \times R_{SH} + \ldots + \frac{m}{2} I \times R_{SH}]$$

$$\Delta V = \frac{m^2}{4} \times I \times R_{SH} = \frac{(mI) \times (\frac{m}{2} R_{SH})}{2} = \frac{I_R \times R}{2} \quad (1)$$

We represent the current, I_R drawn by a single cell row model in the form of voltage drop and equivalent conductivity G of power distribution network using (1).

$$I_R = \frac{\Delta V \times 2 \times G}{m} \quad \text{where} \quad G = \frac{2}{R_{SH}} \quad (2)$$

Number of metal squares m, can be related to width and length of each metal strap as

$$m = \frac{L}{M_{n,width}} \Rightarrow L = m \times M_{n,width} \quad (3)$$

The equivalent metal density D of PDN, defined as the ratio of metal area to the total chip area is represented in terms of m, using (3) as follows

$$D = \frac{2 \times M_{n,width} \times L \times K}{L \times H} = \frac{2 \times K}{A \times m}, \text{Where}, \quad A = \frac{H}{L} \quad (4)$$

Where A, is the aspect ratio for a given metal layer, defined as the ratio of height and length of the chip, where height represents the side perpendicular to current flow and length represents the side parallel to the direction of current flow for a given metal layer. In a N-metal layer system, all metal layers in a given direction (horizontal/vertical) will have the same aspect ratio. In case of a square die A(n) = 1 for all metal layers.

Substituting the value of m (metal squares) from (4) in (2) I_R can be represented as

$$I_R = \frac{\Delta V \times G \times D \times A}{K} \quad (5)$$

Substituting IR drop ($\delta = \Delta V/V_{DD}$) in (5), the current I_R of the single cell row model is written as

$$I_R = \delta \times V_{DD} \times G \times D \times A \times \frac{1}{K} \quad (6)$$

Now, we can extend this single cell row model to K cell rows of the nth metal layer. The current distributed in n^{th} metal is written as

$$I(n) = K \times I_R = \delta \times V_{DD} \times D(n) \times G(n) \times A(n) \quad (7)$$

Where I(n) is the current carried on the nth metal layer. We can extend this model to other metal layers of N-metal layer PDN by considering each metal layer as multiple independent single cell rows, drawing its current from the *equipotential* core ring. Orthogonal metal straps of PDN are connected by via arrays. Since via resistance is negligible in comparison to metal segment resistance they can be considered as ideal connections [4]. Hence, voltage drop across via arrays is neglected in the derivation.

The total current carried by N-Metal layer PDN is the sum of current carried by each metal layer and written using (7) as

$$I_t = \sum_{n=1}^{N} I(n) = \delta \times V_{DD} \times \sum_{n=1}^{N} D(n) \times G(n) \times A(n) \quad (8)$$

Using (8), total power P ($P=V_{DD} \times I_t$) can be written as

$$P = \sum_{n=1}^{N} P(n) = \delta \times V_{DD}^2 \times \sum_{n=1}^{N} D(n) \times G(n) \times A(n) \quad (9)$$

In order to perform trade-off analysis for optimal sharing of metal resources, we define D(n) as a function of metal weightage factor d(n) and common metal density D as in (10)

$$D(n) = \frac{M_{n,width}}{M_{n,pitch}} = D \times d(n) \qquad (10)$$

Here the metal weightage parameter d(n) is the ratio in which metal densities are used in PDN. One can observe that by varying metal weightage parameter d(n) suitably, designers can perform trade-off analysis between utilization of different metal layers for power and signal routing needs of the design.

Conductivity G(n) of each metal layer can be represented in terms of equivalent conductivity G of fixed reference cell row metal layer and relative conductivity g(n) as

$$G(n) = \frac{2}{R_{SH}(n)} = \frac{R_{SH}(1)}{R_{SH}(n)} \times \frac{2}{R_{SH}(1)} = g(n) \times G \qquad (11)$$

Substituting D(n) and G(n) using (10-11) into (9), the common metal density D is calculated as

$$D = \frac{P}{\delta \times V_{DD}^2 \times \left[\sum_{n=1}^{N} d(n) \times g(n) \times A(n)\right] \times G} \qquad (12)$$

Once D is known using (12), the metal density D(n) of each metal layer is computed and represented in terms of width and pitch using (10) for N-metal power distribution network design and implementation.

Fig.-2 depicts the relationship between IR drop and metal density D(n) needed to design the power distribution network in 7 metal layer system for wire-bond packages in 90nm and 65nm technology nodes. As one can see, tighter the IR drop budget greater is the metal density needed in the power distribution network. Lower technologies needs more metal density in PDN due to increase in resistances of metal layers of narrow widths.

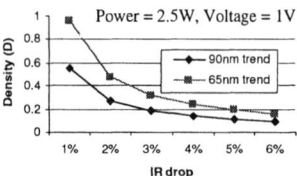

Fig-2 Power metal density vs. IR drop for wire-bond

B. Electromigration and Manufacturability constraints

Assuming the lowest metal layer width (eg M1, a part of standard cell) used satisfies EM requirement, the strap width $M_{n,width}$ of core strap is constrained w. r. t fixed cell row metal strap width $M_{1,width}$ as per below equation to satisfy EM and manufacturability constraint.

$$M_{n,width(max)} \geq \left[M_{n,width} \geq \frac{J_{EM}(1) \times G(n)}{J_{EM}(n) \times G(1)} \times M_{1,width}\right] \geq M_{n,width(min)} \qquad (13)$$

III. Models for Designing Power Distribution Network for Flip-Chip Packages

In flip-chip packages, power and ground bump pads are placed uniformly over the die as shown in fig-3(a) and are connected to top metal layer of the power distribution network. Fig-3(b) shows one bump square cell (BSC) area,

and this pattern is repeated all over the die. Ground bumps are also distributed in similar manner. For each BSC, aspect ratio A(n) is 1, considering equal bump pitch in horizontal and vertical direction.

Assuming uniform power distribution, the IR drop within each BSC is the same. Each BSC shows wire-bond like IR drop, maximum in the center and lowest at the edges of BSC and hence each metal layer in BSC is modeled on the basis of single cell row model as shown in fig-1. Power dissipation in one BSC area is considered to design power distribution network. For flip-chip packages IR drop is additive in nature, minimum at highest metal layer and maximum at lowest metal layer. Thus for each BSC, the total IR drop is sum of IR drop in all metal layer and given using (9) assuming A(n)=1 as

$$\delta = \sum_{n=1}^{N} \delta(n) = \frac{P}{N_{BUMP} \times V_{DD}^2 \times \frac{1}{N}\sum_{n=1}^{N} D(n) \times G(n)} \qquad (14)$$

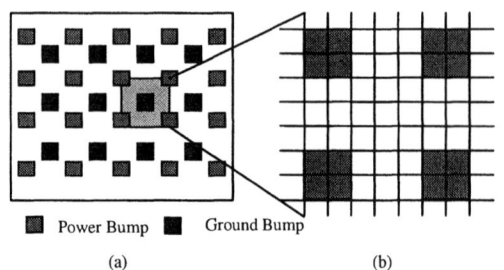

Fig-3 (a) Power and ground bumps for a Flip-chip (b) Bump square cell

Substituting D(n) and G(n) using (10-11) into (14), the effective metal density D is calculated as

$$D = \frac{P}{N_{BUMP} \times V_{DD}^2 \times \delta \times G \times \left[\frac{1}{N}\sum_{n=1}^{N} d(n) \times g(n)\right]} \qquad (15)$$

Here, N_{BUMP} is the total number of bumps pads of same net and N is the number of metal layers used in the PDN. Once D is known using (15) then metal density D(n) of each metal layer is computed and represented in terms of width and pitch using (10) for N-metal power distribution network design. The EM and manufacturability constraints (13) described in section II.B also need to be taken care of for determining the width of the core straps of power distribution network.

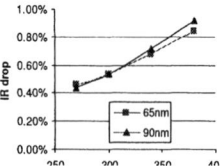

Fig-4 IR drop vs bump pitch with constant power per chip

Conventional power grid design for the flip-chip utilizes available packages with fixed bump pitches. PDN/package co-design can empower designers with additional parameters to design efficient and optimum power distribution network to meet stringent design constraints. Equation (15), indicates the inverse relation between IR drop and number of bump pads in a flip-chip design for given metal density. A graph of IR drop versus bump pitch for constant chip power and metal density for 7 metal layer systems in 90nm and 65nm technology nodes is plotted in Fig. 4. This graph depicts that in case of limited meal resource availability for power network stringent IR drop budget can be still be achieved by appropriate selection of bump pitch.

IV. VALIDATION OF THE PROPOSED MODELS ON VARIOUS TECHNOLOGY NODES

The first step in power network design is to define the design constraints such as number of metal stacks, power supply voltage, IR drop targets etc. Once the design constraints are set the models proposed in sec II and III can be used to design power distribution network and estimate the metal resource requirements. Keeping in mind the signal routing needs of the design metal weight age parameter d(n) can be varied accordingly to design optimal power distribution network. The designed power network can be implemented using physical design tool and analyzed for EMIR using analysis tools. Any possible metal resource crunch for signal and clock routing can be addressed at this stage by redesigning the power distribution network with modified design constraints such as metal layer system, IR drop budget etc.

Table-1: IR Voltage Drop Analysis for Various Tech. Nodes

Tech node	Power (W)	Die size (sq. mm)	Metal Density (%)	Target IR drop (%)	Computed IR drop (%)	Error (%)
WIRE BOND						
45nm	1W	4x4	M5=2.4 M6=16.8 M7=19.2	2.5%	2.6%	4%
65nm	2.5W	2x8	M5=21.2 M6=42.4 M7=63.6	3.5%	3.59%	2.57%
90nm	5W	9x9	M4=10 M5=21.2 M6=42.4 M7=56.5	5%	5.04%	0.8%
FLIP CHIP						
65nm	18W	8x8	M5=5.5 M6=11 M7=18.9	1%	1.02%	2%
90nm	5W	8x2	M4=5.28 M5=6.6 M6=13.2 M7=18.8	0.4%	0.82%	2.5%

The proposed models have been used to design power distribution network on various technology nodes and validated using Blast Rail[13] as EMIR analysis tool. The details of implementation and validation in terms of design constraints, targeted IR drop (summation of VDD and VSS drops) and the IR drop achieved as calculated using blast-rail for wire-bond and flip-chip packages is summarized in Table 1. All design test cases use seven metal layer system. M<N> in the density column of Table 1 indicates N^{th} metal layer and its corresponding density in the power distribution network. These metal densities are obtained by specifying appropriate d(n) for the metal layers. The metal layers with no entry for density in Table 1 are reserved solely for signal routing by setting d(n) to zero.

At early stages of the design before place and route blast rail assumes uniform power distribution in the core area for EMIR analysis which is similar to our assumption while deriving the models. The IR drop results of the power distribution network designed using proposed models correlates well with blast rail analysis and are within 5% of error limits. These models can therefore be used to optimally plan the metal resource utilization of N metal layers system for power and signal routing needs of the design.

V. CONCLUSION

In this paper, we proposed power grid models for designing optimum power distribution network for N metal layer system, given design constraints such as power, IR drop budget and supply voltage for both wire-bond and flip-chip packages. We presented and validated the results on 90nm, 65nm and 45nm designs. The models proposed empower designers to efficiently use optimal metal resources in the power distribution network, taking into account signal and clock routing needs of the design. These models can also be used to determine the number of metal layers that need to be supported for a given class of designs. They are thus very useful for predicting the metal layer system requirements to meet power, IR drop and EM constraints for future technologies.

References

[1] International Technology Roadmap for Semiconductors, Semiconductor Industry Association, 2005.

[2] Zarkesh-Ha P., Meindl J.D., "An integrated architecture for global interconnects in a gigascale system-on-a-chip (GSoC)," pp.149-152, ICM, 2000.

[3] Joyner J.W., Meindl J.D., "A compact model for projections of future power supply distribution network requirements," pp..376-380, ASIC/SOC Conference, Sept 2002.

[4] Shakeri K., Meindl J.D., "Compact physical IR-drop models for Chip/Package Co-Design of Gigascale Integration," vol 62, no. 6, IEEE Trans. Electron Device, June 2005.

[5] Zarkesh-Ha P., "Global Interconnect Modeling for a Gigascale System-on-a-Chip (GSoC)" A Ph. D. Thesis Presented to the Academic Faculty in Georgia Institute of Technology, Feb 2001.

[6] Lin S., Chang N., "Challenges in Power-Ground Integrity," pp. 651-654, ICCAD, 2001.

[7] Tan S.X.D., Shi C.J.R., "Efficient very large scale integration power/ground network sizing based on equivalent circuit modeling," vol.22, no.3, pp. 277-284, IEEE Trans. Computer-Aided Design of Integrated Circuits and Systems, Mar 2003

[8] Chen H. H., Ling D. D., "Power Supply Noise Analysis Methodology for Deep-Submicron VLSI Chip Design,"pp.638-643, Design Automation Conference, 1997.

[9] Liu C.W. et al, "Floorplan and Power/Ground Network Co-Synthesis for Fast Design Convergence", pp.86-93, ISPD, 2006.

[10] Singh J., Sapatnekar S.S., "Topology optimization of structured power/ground network", pp.116-123, ISPD, 2004.

[11] Xiaohai Wu et al, "Area minimization of power distribution network using efficient nonlinear programming techniques", pp.153-157, ICCAD, 2001.

[12] Bhooshan R., "Novel and Efficient IR-Drop Models for Designing Power Distribution Network for Sub-100nm Integrated Circuits", pp.287-292, ISQED, 2007.

[13] Blast-Rail User Manual, Magma Design Automation Inc.

Impact of Task Migration in NoC-based MPSoCs for Soft Real-time Applications

Eduardo Wenzel Brião, Daniel Barcelos, Fabio Wronski, Flávio Rech Wagner

Universidade Federal do Rio Grande do Sul - UFRGS
Instituto de Informática
Porto Alegre, RS, Brazil
{ewbriao, danielb, fwronski, flavio}@inf.ufrgs.br

Abstract— This work analyzes the impact of task migration in the context of multiprocessor systems-on-chip, where processors are interconnected by a network-on-chip (NoC) and the system must adapt itself to a dynamic workload, such that performance and real-time constraints are met and energy consumption is minimized. The task migration model assumes that the whole code and data of the tasks are transferred from an origin node to the chosen destination node. Experimental results show that, even with a higher overhead than other possible approaches, task migration may be applied in NoC-based architectures, since it pays off the costs involved in the transfer in terms of overall system performance and energy consumption and may help to improve the fulfillment of task deadlines in soft real-time systems.

I. INTRODUCTION

The complexity of electronic embedded systems design has been increasing due to the technological evolution that allows the integration of a complete system on a single chip (SoC – System-on-Chip). In order to cope with the corresponding design complexity and reduce design costs and time-to-market, systems are built by assembling pre-designed functional modules (processors, memories, dedicated hardware blocks, etc), called IP (Intellectual Property) cores. They can be reused from previous designs or acquired from third-party vendors.

An adequate communication architecture is required to interconnect a large number of IP cores. The most commonly used communication architectures are not suitable for the communication requirements of future SoCs, such as scalability and performance. Networks-on-Chip (NoC) arise as a solution to fulfill these requirements [1].

Today's state-of-the-art MPSoC platforms contain multiple heterogeneous processing elements (PEs). Several applications of same or different domains can be loaded and executed on an MPSoC platform, since it provides appropriate resources to allow the simultaneous execution of several applications. However, the system must know where and when allocating tasks that compose the application.

In distributed computing systems, the tasks must be reallocated as soon as they are available, such as to minimize both the impact of system degradation, due to a bad workload distribution, and the cost of system correction. With this in mind, there are on-line algorithms that may be combined to decrease the communication among tasks and to allocate tasks to processors such that timing constraints are met. Linear clusterization [2] may be applied such that clustered tasks are allocated to a single processor in order to minimize interprocessor communication. This approach may be combined with bin-packing algorithms [3], which apply heuristics to allocate tasks to processors, considering the processor utilization. Bin-packing algorithms may enable either load balancing or concentration of the system workload. These algorithms present a low overhead and may be efficiently applied in a dynamic context, in embedded systems. Various combinations of algorithms may explore different trade-offs between energy, power, and performance.

Mechanisms for task migration are needed to provide the infrastructure that allows dynamic load balancing or concentration. It may enable resource savings, since resources would be typically planned for worst-case conditions that rarely occur. Beyond traditional objectives, task migration may maximize energy savings, facilitate thermal chip management by moving tasks away from hot processing element, and balance the workload of parallel processing elements [4].

In this work, we present a model of task migration that can be used in the context of embedded systems based on network-on-chip architectures. The model can be classified as *copy* (or *total freezing*) [5], where the task is stalled and then all its code and data are transferred through network links. This model was chosen due to its simplicity and highest overhead. We want to show that, even using a model of task migration with higher overhead than other techniques found in the literature, task migration may be applied in embedded systems based on NoC architectures. The use of task migration is justified since it pays off the performance and energy costs involved in the system. Results show that task migration may greatly improve the fulfillment of task deadlines in soft real-time systems. In our work, task migration is triggered when the allocation heuristic is executed. When a new load of tasks is available, the allocation heuristic tries to balance the system on demand. Afterwards, task migration will be executed again only when a new application is loaded by the user. To our knowledge, this is the first work to show the impact of task migration in NoC-based systems in terms of energy consumption and deadline misses.

The remaining of this paper is organized as follows. Section II discusses related work, while Section III shows our energy and task models. Section IV presents our simulator and task migration support. Section V presents experimental results, and, finally, Section VI draws main conclusions and addresses future work.

II. RELATED WORK

Task assignment may be modeled as a bin-packing problem [3]. We concentrate on two bin-packing heuristics: Best-Fit (BF) and Worst-Fit (WF). The BF strategy places a new object in the bin whose remaining space will be the smallest one. The WF strategy, in turn, places an object in the bin whose remaining space will be the largest one. As a consequence, WF generates a distribution with load balancing, while BF generates a distribution that is concentrated in some bins.

Al Enawy and Aydin [6] use algorithms with bin-packing heuristics for allocating real-time tasks, applying rate monotonic scheduling. Among the heuristics, WF presents the largest gain (energy reduction), since the better the load balancing, the larger the gain with dynamic voltage scaling (DVS). BF leads to load concentration and offers a better opportunity for the application of dynamic power management (DPM), turning-off idle cores. However, this work did not consider inter-processor communication costs and used a bus-based communication architecture.

Few works cover task migration in the context of embedded systems. Nollet et al. [7] show the description of a NoC resource management heuristic that makes efficient use of reconfigurable hardware tiles and is closely linked to a run-time task migration mechanism. However, in this work, just performance data have been measured. Bertozzi et al. [8] propose a user-managed migration scheme based on code checkpointing [5] and a characterization methodology for task migration overhead in a shared memory. However, in this work, no figures for energy and power were measured. In addition, this work supports task migration only in a bus-based communication architecture. Wronski et al. [9] present a NoC-based simulator that executes clustering and bin-packing algorithms for task allocation. The authors conclude that the usage of a combination of load concentration with DPM may result in lower energy consumption. However, the usage of load balancing minimizes the number of deadline misses. This work did not consider the task migration costs.

Our work, in turn, considers task migration overhead in a dynamic environment and shows its impact in the context of NoC-based MPSoCs, in terms of energy, performance, and real-time constraints.

III. ENERGY AND TASK MODELS

The dynamic power consumption of the network routers and links is calculated with help of the Orion library [10] (the same used in the Xpipes NoC [11]). It implements an energy estimator for the crossbar and the buffers inside the routers. The components described in the library are the arbiter, buffers, and the crossbar. Regarding them, and similarly to [12], the energy spent by a data phit to be transferred between two routers is defined as:

$$E_{phit} = E_{wrt} + E_{arb} + E_{read} + E_{xb} + E_{link} \qquad [1]$$

where E_{wrt}, E_{arb}, E_{read}, E_{xb}, and E_{link} represent, respectively, the energy spent in writing the phit in the buffer, selecting the input channel (arbiter), reading the phit from buffer, crossbar, and output channel.

The static consumption of the memory is estimated by using the model from [13]:

$$E_{static} = \frac{V_{dd} \cdot I_{leak} \cdot k_{design} \cdot 6N_{bits}}{f} \eta \qquad [2]$$

The k_{design} constant is a design parameter, and N_{bits} is the number of bits in the memory.

Each application is a directed acyclic graph $T = G(K,A)$, where each node $k_i \in K$ is a periodic task and each arc $a_{i,j} \in A$ is a dependency and flow of messages between tasks k_i and k_j. The arc weight $a_{i,j}^w$ represents the amount of bits to be transferred between the tasks.

Each task $k_i \in K$ is a tuple $\{C,T,S,D,\alpha\}$, where C is the worst case execution time, T is the task period, S is the task size in bytes (including program size and data size), D is the task deadline, and α is the average number of gate switchings per cycle of the task in the core.

IV. SIMULATOR AND TASK MIGRATION

A. Serpens Simulator

The SystemC Serpens simulator has been developed to simulate the behavior of NoC-based systems that run sets of synthetic tasks, generated by TGFF [14], which are dynamically loaded. The simulator also executes clustering and bin-packing algorithms for task allocation and implements an on-line scheduling for tasks that are mapped to the same processor. DVS and DPM mechanisms are implemented in order to minimize energy consumption. The system is based on Femtojava processors [15], each one with its local, non-shared memory (see Figure 1), and on the SoCIN NoC [16], both of which have been developed in our research group. The pipelined FemtoJava microcontroller implements an execution engine for Java in hardware through a stack machine compatible with the Java Virtual Machine (JVM) specification. Each processor has its own EDF local scheduler. The simulator uses the Orion [10] library to evaluate the power consumption. The CACO-PS [17] tool is used to calibrate the processor power values.

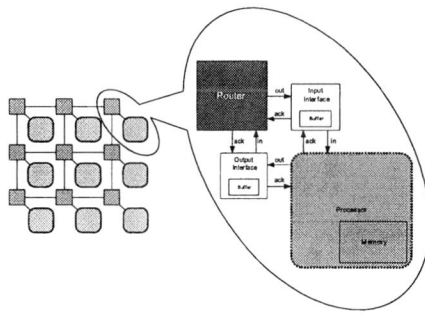

Figure 1. MPSoC with network interfaces and memory organisation.

The simulation model uses RaSoC routers [16], designed for the synthesis of low power and low area NoC-based embedded systems. The RaSoC architecture utilizes wormhole packet switching, XY routing algorithm, and handshake control flow. Its energy evaluation was also implemented with the Orion library. Each router has 5 bi-directional ports with input buffer size of 4 phits. The phit size is 4 bytes.

The DAR (Dynamic Average Rate) algorithm [18] was used for DVS. Each task is characterized with respective WCET and gate switchings per cycles. With these data, tasks are scheduled and their energy consumptions are evaluated. More information about the simulator can be found in [9].

B. Task migration infrastructure

The model for task migration adopted in this work, as previously mentioned, is based on a *copy* model. This model is very simple, with highest overhead, since the whole context (code, data, stack, and contents of internal registers) is migrated.

In each processor, there are mechanisms for interprocess communication based on messages (send/receive primitives). These primitives support two kinds of messages: inter-task communication and task migration. When a migration event occurs, the "send" primitive writes in the message header a service identification tag, identifying that the message is of migration type. The "receive" primitive executes by polling, trying to read a packet from the input link. If there is a packet in the input link, the primitive reads the header. If the service identification tag is of migration type, another field is read: the task id, which identifies the task that will be instantiated in the local processor. The task code and data are then read from the incoming messages, and the complete task becomes ready to execute in the destination processor, resuming from the stop point.

Figure 2 illustrates the scheduling sequence for a task migration between two cores P0 and P3. The origin core P0 is executing a task T0. An interrupt happens, and the system decides to migrate a task T1 from P0 to P3. The label "MS" in the scheduling indicates that P0 (origin core) is sending T1 to P3. Processor P3 is in idle until receiving the migration packets sent from P0 ("MR"). When the last packet was received by P3, the scheduler "S" in P3 monitors if there is some new task, and the new task is released through "R" ("R" inserts the new task in the ready queue of the core). The migrated task T1 then starts to run.

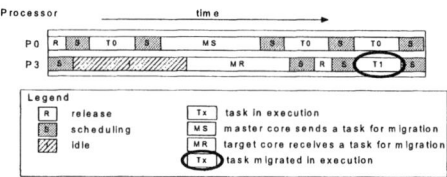

Figure 2. Scheduling in the processors of the system.

The migration overhead of our mechanism corresponds only to the data and code transfer overhead. Costs of task shut-off (deallocation of kernel-level data structures and user level-memory space) and task re-spawning (task creation system calls) are not taken into account. Of course they could bring an additional contribution to the migration cost.

It is assumed that the packet transmission time is much larger than the time of copying memory data. Besides, these activities are performed in pipeline. Therefore, in our migration scheme, we afford to neglect the time of copying memory data.

V. EXPERIMENTAL RESULTS

This section presents several experiments which demonstrate the feasibility and efficiency of the task migration mechanism.

A. Case studies

Experiments use two distinct applications, shown in Table I. The first one is a synthetic application, which has a small amount of communication, but a large number of tasks. In turn, the second one is part of the embedded system synthesis benchmark suite (E3S) [19] and comes from the telecom domain. The telecom application has a small number of tasks and a high amount of communication.

A second version of the telecom application (telecom-2) has 10 times more communication than the first one.

TABLE I. TASK GRAPHS USED IN THE EXPERIMENTS.

Application	# Task	# Edges	#Comm. in bytes
Synthetic	64	52	52
Telecom	30	18	85 Kb
Telecom 2	30	18	850 Kb

B. Simulation strategy

We implemented the WF bin-packing algorithm, based on the processor utilization, incremented with linear clusterization (LR) for minimizing the cost of communication among tasks. We chose the WF strategy to balance the workload system. According to Wronski et al. [9], this strategy, combined with LR, represents the best trade-off to minimize energy and deadline misses.

In both case studies, we first simulate the application with tasks allocated in an ad-hoc way. We then apply WF combined with LR. Both experiments are simulated during 200 ms. The processor frequency varies from 133 MHz to 266 MHz, and the voltage varies accordingly from 1.3 V to 2.0 V using DVS. The NoC has a mesh configuration, with a default size of 4x4, and runs at 266 MHz. The memory size of each core is 64 Kb.

C. Experiments

Tasks are considered to be periodic. They are scheduled and later on concluded, and then we count a *task finalization*, even if the task is terminated after its period. If this occurs, we count a *deadline miss*. In order to avoid counting non-terminated tasks, we calculate the average energy consumption per finalized task. This calculation already includes the overhead of task migration.

Figure 3 presents the average energy per task finalization as a function of the task context to be migrated for the two applications (telecom and synthetic). The synthetic application requires more energy, since it has a larger number of tasks to execute. The horizontal straight lines in the chart represent the energy consumption without task migration. We notice that, up to 100 KB of task context, the task migration does not influence the energy spent in the system. The straight liAs explainedWe n

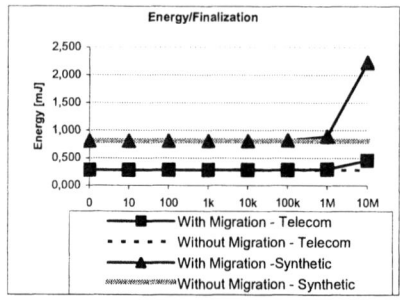

Figure 3. Impact of context size in task migration regarding energy per task finalization.

Figure 4 shows the impact of task migration on the number of deadline misses, measured as a percentage of task finalizations. In the telecom application, task migration slightly decreases the number of deadline misses up to context sizes of 10 KB, while it greatly increases the number of deadline misses for context sizes of 10 MB. For the synthetic application, however, task migration almost completely removes deadline misses for task contexts up to 100 KB. This happens because the synthetic application has smaller communication requirements, which are the bottleneck of the system. This is well explored by the worst-fit heuristic, improved with linear clustering, which can find a much better task distribution.

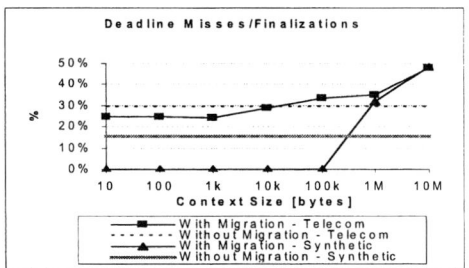

Figure 4. Impact of context size in task migration regarding deadline misses by finalization.

Figure 5 presents the energy spent along the simulation time, between 1 ms and 1000 ms, in four different scenarios. In the first one, task migration is applied, together with WF and LR for task allocation. In the second one, task allocation is implemented in an ad-hoc way and there is no task migration. These two experiments are performed in two sizes of network-on-chip: 16 (4x4) and 25 (5x5) cores. The telecom-2 application is used, since a higher communication among processors better amortizes the cost of task migration in terms of energy. It can be noticed that, for all times, the scenario with task migration consumes less energy than the scenario with ad-hoc allocation. Experiments show that, starting from 20 ms of execution time, for both NoC sizes, task migration shows less energy consumption when compared to the ad-hoc allocation. For a larger NoC, the reduction of energy consumption is larger when task migration is applied together with a task allocation based on WF and LR.

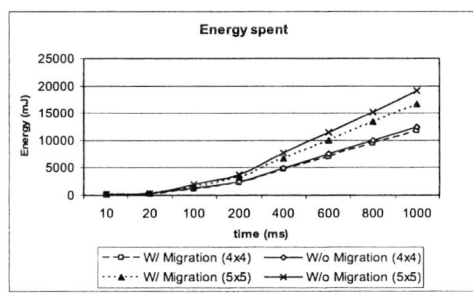

Figure 5. Energy spent for different execution times of the telecom-2 application, with two scenarios: ad-hoc allocation and task migration.

VI. CONCLUSIONS AND FUTURE WORK

This work demonstrated by experimental evidences that task migration may be applied in the context of MPSoCs, where processors are interconnected by a network-on-chip. Even considering the inherent overhead of the task migration, experiments show that task migration may help the fulfillment of soft real-time constraints, while avoiding a degradation in performance and energy consumption.

As future work, we will measure the overhead imposed by on-line monitoring resources running on the various processors. This monitoring will be basic to more advanced allocation algorithms.

REFERENCES

[1] L.Benini, G.DeMicheli. "Networks on Chips: a New SoC Paradigm". *IEEE Computer*, vol.35, n.1, 2002. pp. 70-78.

[2] A.Gerasoulis, T.Yang. "On the Granularity and Clustering of Directed Acyclic Task Graphs". *IEEE Trans. Parallel Distrib. Syst.*, v.4, n.6, 1993. pp. 686-701.

[3] D.Johnson. *Near-optimal Bin-packing Algorithms*. Cambridge, Mass. 1973.

[4] M.Kandemir, G.Chen. "Locality-Aware Process Scheduling for Embedded MPSoCs". In: *Design, Automation and Test in Europe Conference* (DATE), Munich, Germany, 2005, pp. 870-875.

[5] D.Milojicic, F.Douglis, Y.Paindaveine, R.Wheeler, S.Zhou. "Process Migration Survey". *ACM Computing Surveys*, September 2000.

[6] T.AlEnawy, H.Aydin. "Energy-aware Task Allocation for Rate Monotonic Scheduling". In: *IEEE Real-time and Embedded Technology and Applications Symposium*, San Francisco, USA, 2005, pp. 213-223.

[7] V.Nollet, T.Marescaux, P.Avasare, J-Y.Mignolet, D.Verkest. "Centralized Run-Time Resource Management in a Network-on-Chip Containing Reconfigurable Hardware Tiles". In: *Design, Automation and Test in Europe Conference* (DATE), Munich, Germany, 2005, pp 252-253.

[8] S.Bertozzi, A.Acquaviva, D.Bertozzi, A.Poggiali. "Supporting Task Migration in Multi-processor Systems-on-chip: a Feasibility Study". In: *Design, Automation and Test in Europe Conference* (DATE), Munich, Germany, 2006, pp. 15-20.

[9] F.Wronski, E.Brião, F.R.Wagner. "Evaluating Energy-Aware Task Allocation Strategies for MPSoCs". In: IFIP *Working Conference on Distributed and Parallel Embedded Systems* (DIPES), Braga, Portugal, 2006.

[10] H.Wang. "Orion: A Power-performance Simulator for Interconnection Networks". In: *ACM MICRO*, Istanbul, Turkey, 2002, pp. 294-305.

[11] A.Jalabert et al. "XpipesCompiler: A Tool for Instantiating Application Specific Networks on Chip". In: *Design, Automation and Test in Europe Conference* (DATE), Paris, France, 2004.

[12] T.Ye, G.DeMicheli, L.Benini. "Analysis of Power Consumption on Switch Fabrics in Network Routers". In: *Design Automation Conference* (DAC), New Orleans, USA, 2002, pp 524-529.

[13] J.Butts, G.Shoi. "A Static Power Model for Architects". In: *International Symposium on Microarchitecture*, Monterey, USA, 2000, pp 191-201.

[14] R.Dick, D.Rhodes, W.Wolf. "TGFF: Task Graphs for Free". In: *International Workshop on Hardware/Software Codesign*, Seattle, USA, 1998.

[15] S.A.Ito, L.Carro, R.P.Jacobi. "Making Java Work for Microcontroller Applications". *IEEE Design & Test of Computers*, vol 18, n.5, pp. 100-110. 2001.

[16] C.Zeferino, A.Susin. "SoCIN: a Parametric and Scalable Network-on-Chip". In: *Symposium on Integrated Circuits and Systems Design* (SBCCI), São Paulo, Brazil, 2003, pp.169–174.

[17] A.Beck, J.Mattos, F.R.Wagner; L.Carro. "CACO-PS: A General Purpose Cycle-Accurate Configurable Power Simulator". In: *Symposium On Integrated Circuits And Systems Design* (SBCCI), São Paulo, Brazil, 2003. pp.349-354.

[18] J. Zhuo, C. Chakrabarti. "An Efficient Dynamic Task Scheduling Algorithm for Battery Powered DVS Systems" In: *Asia and South Pacific Design Automation Conference* (ASP-DAC), Shanghai, China, 2005, pp. 846-849.

[19] EEMBC. The Embedded Microprocessor Benchmark Consortium. http://www.eembc.org, 2006.

A Flexible Design Flow for a Low Power RFID Tag

José C. S. Palma

PPGC - II - UFRGS - Av. Bento Gonçalves, 9500,
Porto Alegre, RS – Brazil
jcspalma@inf.ufrgs.br

César Marcon, Fabiano Hessel, Eduardo Bezerra,
Guilherme Rohde, Luciano Azevedo, Carlos Reif,
Carolina Metzler

PPGCC - FACIN – PUCRS - Av. Ipiranga, 6681,
Porto Alegre, RS – Brazil
{cesar.marcon, fabiano.hessel, eduardo.bezerra, grohde
}@inf.pucrs.br

Abstract

This paper describes the implementation of a passive RFID tag targeting low power implementation, which works on 915 MHz UHF frequency. The proposed architecture allows customizing the command sets implemented inside its digital block, according to the target application needs, saving area and reducing power consumption. A flexible design flow is proposed for the customization, verification and synthesis of the digital block, targeting low power requirements.

Keywords: RFID Systems, Tag customization, low power consumption.

1. Introduction

RFID (Radio Frequency IDentification) technology exists since the Second World War, but only nowadays, it is becoming attractive [1]. Technology advances and components costs reduction motivate it.

RFID systems can be used in applications such as supply chain management, tracking, and security. Several large companies have shown interest in this technology since Wall Mart's announce of start employing RFID.

RFID systems are composed of tags, readers and antennas. Tags exchange data with the reader (or interrogator) through radio frequency (RF) communication [2]. The reader has one or more antennas that send/receive radio waves to/from tags. The number of antennas depends on the communication area required by the application, since the use of more antennas implies larger communication area.

Tags can be classified as active or passive. Active tags have their own power supply, operating in high frequencies and usually are larger than passive tags. Passive tags are low power and need to be smaller, since they do not have internal power supply. They transform the communication RF signals from reader in energy to operate. As a result, passive tags are cheaper than the active ones [3].

This paper describes a passive RFID tag, which has the following features: low power, 915 MHz and communicates by means of UHF. The tag is class 1 generation 2, according to ISO 18000-6B and EPC standards. In addition, the paper also describes its flexible design flow, which allows command set customization according to application requirements. Tag area usage and power consumption can be optimized by selecting a customized and reduced command set. This work focuses only on the implementation, customization and validation of the tag's digital block.

This paper is organized as follows. Section 2 presents some works related to RFID systems. Section 3 presents the architecture of the proposed tag. Section 4 describes the gate-level energy flow used to estimate the tag power consumption. Section 5 discusses the flexible design flow used for tags building. Section 6 describes the entire system validation. Section 7 shows occupation area and power consumption resulted by synthesis step and, finally, Section 8 presents conclusions and ongoing work.

2. Related Work

Choi et al. [4] present a 13.56 MHz RFID reader for home security applications, which allows multi-tag recognition. Leong et al. [5] propose an approach to synchronize multiples RFID readers in order to enable successful dense RFID reader deployment, and minimize the reader collisions.

An important issue on RFID systems is the reading distance, typically limited to less than 1.2 m. Karthaus and Fisher [6] propose a passive UHF transponder with reading distance increased to 4.5 m by using a different voltage generator, a PSK in backward link, and a careful layout and antenna matching.

This work, similarly to [7], [8], [9] and [10], presents the design of a flexible low power RFID tag. The main contribution is the tag flexible design aiming to reduce area and power consumption.

3. The Proposed Architecture

Figure 1 shows the basic diagram of RFID tag architecture, which has three main blocks: External Antenna, Analog Block and Digital Block. The External Antenna interfaces the IC tag to the environment, and it is responsible for capturing data as electromagnetic waves. The tag power supply is implemented by the Analog Block that converts the electromagnetic waves into electric energy, which is stored in a bank of capacitors. In addition, this block is responsible for analog and digital signals demodulation/modulation. Finally, the Digital Block performs all tag operations, such as data decoding, commands interpreting, collision arbitration, sent data encoding and memory access. Six modules described next perform these functionalities.

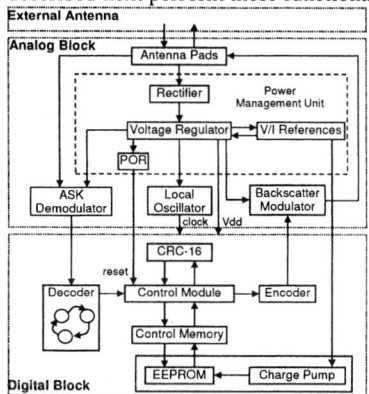

Figure 1 – RFID tag architecture diagram.

The Decoder receives from the ASK Demodulator a 40 Kbps or 160 Kbps Manchester encoded data, which is decoded to a binary format that is sent to the Control Module. Before starting data decoding, it is necessary to synchronize the input signal with a 640 KHz clock provided by the Local Oscillator module.

The Control Module identifies and handles all commands received from the Reader, sending signals and controlling the other digital modules. The CRC-16 module creates a 16-bit cyclic redundancy check of a new packet and checks the integrity of a packet received from the reader. When a CRC error is detected, all data are discarded and the frame is suspended. The Control Memory module is the interface between the Control Module and the EEPROM Memory. It is responsible for enabling the memory access and for returning the instruction results. The tag Digital Block has an internal 512 bits EEPROM, arranged in 8 rows of 4 words. The Encoder module codifies the output frame received from the Control Module, and that must be sent to the Backscatter Modulator module. The coding is done according to FM0 encoding method.

4. Gate-level Energy Flow

Simulations based on a library of gates provide the energy consumption estimation of the tag. The gates, which are described in VHDL, are pre-characterized according to the static and dynamic energy consumption.

Figure 2 presents the flow used to characterize the gate-level energy model and to obtain the energy consumption estimative of the circuit under analysis. The characterization begins with the selection of a set of gates, which are manually described in SPICE and VHDL. After that, the set of gates described in SPICE

are simulated at electrical level, considering the adopted technology library (CMOS TSMC 0.35 μm). Each gate is simulated in a SPICE simulator with all input signals combinations and different *fan-outs*, calculating the dynamic energy consumed by the gate due to input transitions.

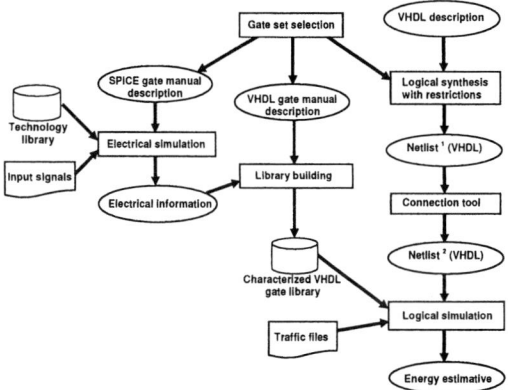

Figure 2 – Flow used to characterize the gate-level energy model and to obtain the energy consumption estimation.

Since leakage current grows dramatically as the feature size of CMOS circuit scaling down and it is the main parcel to static energy consumption [11]. It is also necessary to model the static energy of gates. This is performed by simulating the same gate described in SPICE during periods without transitions. In this case, it is not necessary to perform simulations with different *fan-outs*, since it does not interfere in the static energy consumption. The total static energy consumption is achieved when multiplying the total static energy by the execution time. The sum of static and dynamic parcels computes the total energy consumed by the whole system.

Performing electrical simulations enables to characterize all gate by inserting electrical information to the VHDL description of each gate, creating the characterized VHDL gate library. The next step is to perform the logical synthesis of the circuit under analyzes. It is executed restricting the synthesis tool to use only the set of gates previously characterized on the VHDL gate library. The synthesis tool generates a VHDL netlist, which is analyzed by a connection tool. This tool verifies all connections among gates, calculating the *fan-out* of each one. Such information is important to estimate more accurately the energy consumption of the circuit.

The connection tool generates a second netlist with *fan-out* information, which is simulated in a VHDL simulation tool. Such simulation uses the characterized VHDL gate library and input traffic files to estimate the energy consumption of the circuit under analysis.

5. Flexible Design Flow

The proposed architecture allows the selection of different command sets, according to the target application needs. This feature makes the RFID tag more flexible, when considering area, energy consumption and read/write range specifications.

Customization is performed over Digital Control module, which implements the set of commands supported by the tag. Since it is the most area and energy consuming module, as shows Table 2, the facility for performing different configurations in the command set will result in a significant impact on the overall tag costs.

Figure 3 shows the command select flow. It starts from the library of commands and an application specification file. Based on these inputs, the customer can select the set of commands to be implemented in the Digital Control module.

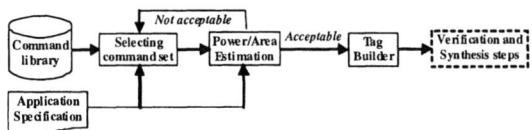

Figure 3 – Command select flow.

The next step is the area usage and energy consumption estimation, which is performed by some iteration. Each one selecting a new instruction set, in order to meet the expected area usage and energy consumption specified in the design requirements. This step corresponds to the right side of the flow presented in Figure 2 – starting from the VHDL description of the customized tag, until achieve its energy consumption estimation.

Table 1 shows area and energy consumption estimation of Digital Block of two different tag configurations. The first one is a complete tag, which implements all supported commands, according to ISO 18000-6B standard. This configuration corresponds to 28 commands, occupies 60060 transistors and consumes an average power of 258 µW. The second configuration is a tag, which implements only two commands: (*i*) *read* – reads one byte of data from the memory – and (*ii*) *write_four_byte* – writes four bytes of data in the memory.

Average power and energy estimative were obtained after the execution of the two commands supported for both tag configurations: *read* and *write_four_byte*. The reduction in the energy consumption was 25,4 % compared to the complete tag.

Table 1: Area and power estimations of Digital Block for two different tag configurations.

Version	Area (transistors)	Average Power (mW)	Energy (J)	%
Complete	60060	2.58e-1	5.2524e-6	100
Reduced	45432	1.924e-1	3.9256e-6	74.6

When all requirements are met, the next step concern the new tag manufacture, followed by validation and synthesis steps, described in the next section.

6. System Validation and Synthesis

This Section describes the system validation through functional and gate-level simulation. Before the validation stage, a tag was configured according to the command select flow, described in the last section. The configuration of a generic tag used, for instance in supply chain applications, has been chosen as a case study. Such tag supports all the commands determined by ISO 18000-6B standard, and some proprietary commands planed to accomplish supply chain purposes.

Figure 4 shows the system validation and synthesis flow, which is performed by functional and gate-level simulation.

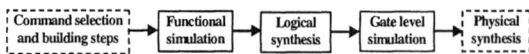

Figure 4 – System validation flow.

Four stages split the validation of the tag VHDL design using Modelsim. In the first three stages, it is performed VHDL functional simulation and, the last one is a gate-level simulation.

In the first stage, the digital modules were individually submitted to white box testing during the development, in order to verify their correct functioning. After this stage, the digital modules were interconnected, creating a new entity named Top Digital Block.

Second stage consists of a white box testing of Top Digital Block, which is performed using waveform analyses. The Top Digital Block has been simulated and its basic functions have been tested. The basic functions include receiving packets, decoding commands, checking CRC, building and sending responses. An auxiliary tool randomly generates frames containing commands, which are read from a text file.

In the third stage, the tests were automated, in order to perform massive verification of the Top Digital Block. First, a C program, named C-Tag, generates all Digital Block outputs according to its inputs. The C-Tag works executing the same stages of a real tag, which are data receiving, data decoding, command executing and response sending. In other words, it reproduces the Digital Block behavior under many conditions, following the ISO 18000-6B standard. This standard defines how this tag should work. Massive verification uses around of ten thousand commands per session, and each command have the same size and data of commands received by a real tag. The next step simulates both C-Tag and Top Digital Block. During C-Tag execution, all outputs and some internal debug signals are stored in a text file. In a similar way, signals from Digital Block are also monitored and stored in a second text file. After that, another external program analyses the data in the two text files and verifies if the behavior of Digital Block matches the C-Tag behavior. If no errors are detected, a new round of tests begins. Otherwise, the detected errors can be rechecked several times, simulating the same system state until the errors remain. Simulation status is stored in a third text file. In case of error detection, all debug signals are also stored in this file. The simulation stop criterion is a 100% of code coverage and no errors detection. Once simulation stops, the system is ready for logical synthesis. This process creates a Verilog netlist and delay files.

The last stage consists of an automated gate-level simulation with netlist and delay files. The simulation process of the third functional validation stage is performed again, but now with gates and interconnections details. The netlist file is also used to estimate power consumption, while delay files have all signals delays according to connection wires characteristics.

When the simulation reached the stop criteria, the system was sent to the foundry. In a first moment, the foundry will provide a small set of tags for exhaustive testing activities. Afterwards, the final tag design will be used to produce a commercial set of tags.

7. Synthesis Results

Table 2 presents area and power results for Digital Block and its internal modules. The first three columns present, respectively, the total cell area occupied by each module, the total number of cells and the number of sequential cells inside each module. The difference in the number of cells between the Top Digital Block and the sum of all internal modules is due to the insertion of buffers, used for power optimizations in the Top Digital Block. The fourth to seventh columns show the typical and worst case of power consumption of each module in a 3.0 V with 25°C and in a 3.3V with 70°C scenarios.

TABLE 2: AREA OCCUPATION AND POWER CONSUMPTIONS RESULTS FOR DIGITAL BLOCK AND ITS INTERNAL MODULES.

Digital Block	Area occupation			Power consumption (w)			
	Total Cell Area (µm²)	Total Cell	Sequential Cells (FFs)	3.0V and 25°C		3.3V and 70°C	
				Typical	Worst	Typical	Worst
Decoder	34143.2	270	48	7.0454e-03	7.9445e-03	9.1280e-03	1.0269e-02
Control	440203.4	4471	340	3.6410e-02	4.0963e-02	4.7220e-02	5.3013e-02
CRC	7425.6	38	16	1.7745e-03	1.9919e-03	2.3047e-03	2.5814e-03
Memory Control	108945.2	972	125	1.3791e-02	1.5470e-02	1.7896e-02	2.0030e-02
Encoder	9864.4	96	14	1.2958e-03	1.4648e-03	1.6803e-03	1.8925e-03
Test	13722.8	123	2	3.0224e-03	3.4164e-03	3.9190e-03	4.4184e-03
Top Digital	616652.4	5998	545	8.8119e-02	9.9060e-02	1.1264e-01	1.2637e-01

Figure 5 shows the final layout of RFID tag circuit, which uses 0.35µm standard CMOS technology. As previous described in Section 3, this work focusing only in the digital block, which occupies more than 60 % of the entire circuit.

8. Conclusions and Ongoing Work

This paper describes the implementation of a passive RFID tag, which has the following features: low power, 915 MHz and communicates by means of UHF. The main contributions are: (*i*) a new tag architecture, which has been written in VHDL, simulated and synthesized; and (*ii*) a flexible design flow, allowing the commands set customization, aiming efficient power consumption.

Different configurations for the tag's command set

have been tried, and a chip design for a typical application was chosen as a case study. The tag has been fully tested and validated at several simulation levels, and the synchronizing algorithm performed according to the expected. The resulting design has been sent to a foundry. Field tests are going to be performed on the actual chip, and a new and revalidated design version will be sent to the foundry for commercial production in large scale.

Figure 5 – Layout of the RFID tag.

9. References

[1] Want, R. The Magic of RFID Just How does Those Little Things Work Anyway? ACM Queue, v. 2, n. 7, Oct. 2004.

[2] Mamei, M. et al. *Making Tuple Spaces Physical with RFID Tags.* **Symposium on Applied Computing, ACM Press**, France, 2006.

[3] Weinstein, Ron. *RFID: A Technical Overview and Its Application to the Enterprise.* **IEEE Computer Society**, v. 7, n. 3, pp. 27-33, 2005.

[4] Choi, N-G. et al. *A 13.5 MHz RFID System.* Dec. 2005.

[5] Leong, K. et al. *Synchronization of RFID Readers for Dense RFID Reader Environments.* **International Symposium on Applications and the Internet Workshops**, Jan. 2006.

[6] Karthaus, U. and Fisher, M. *Fully Integrated Passive UHF RFID Transponder IC with 16.7 µW Minimum RF Input Power.* **IEEE Journal on Solid-State Circuits**, v. 38, n. 10, pp. 1602-1608, Oct. 2003.

[7] Leenaerts, D. *Low Power RF IC Design for Wireless Communication.* **International Symposium on Low Power Electronics and Design**. pp. 428-433, Aug. 2003.

[8] Vancherand, F. *New Technologies for Contactless Microsystems.* **ACM International Conference Proceeding Series**, v. 121, pp. 13-17, 2005.

[9] Porret, A-S. et al. *Tradeoffs and Design of an Ultra Low Power UHF Transceiver Integrated in a Standard Digital CMOS Process.* **IEEE Journal of Solid-State Circuits**, v. 35, n. 3, Mar. 2000.

[10] Jones, A. et al. *An automated, FPGA-based reconfigurable, low-power RFID tag.* **Proceedings of the 43rd Annual ACM IEEE Design Automation Conference**, pp. 131-136, 2006.

[11] X. Zhao, K. Wang, X. Cheng, D. Tong. *A Leakage Power Estimation Method for Standard Cell Based Design.* **IEEE Conference on Electron Devices and Solid-State Circuits**, p. 821-824, Dec. 2005.

Co-Synthesis of Custom On-Chip Bus and Memory for MPSoC Architectures

Sujan Pandey   Christian Genz   Rolf Drechsler

Department of Computer Science
University of Bremen, Germany
{pandey,genz,drechsle}@informatik.uni-bremen.de

Abstract— **The advancement in process technology has made it possible to integrate multiple processing modules on a single chip. As a result of this, there is a sharp increase of communication traffic on the communication bus architecture. In this case, the traditional single bus based architecture may fail to meet the real-time constraints. The major concern of the scaled technology is an effect of coupling capacitance due to the trend of shrinking pitches, i.e., the distance between two wires. Its consequence is higher crosstalk noise, which degrades the signal integrity and modifies the power consumption of the wires. This motivates the synthesis of a custom on-chip bus architecture, which is efficient in terms of power and performance. Further, the memory of a complex multiprocessor system has a significant contribution to power and delay.**

In this paper, we present a co-synthesis of on-chip buses and memories, which finds an optimal bus architecture, memory sizes, and the number of memories. The bus synthesis problem is formulated as an optimization problem as proposed in [11], [9]. Then it is solved efficiently using an optimization tool. The memory synthesis problem is based on the graph partitioning algorithm, which partitions a data dependency task graph into a set of sub graphs with the minimum number of data dependencies called *cut*. The experiments carried out on the real-life multimedia applications validate the proposed technique for the co-synthesis of bus architecture and memory.

I. INTRODUCTION

Due to the advancement in process technology and the increasing demand of performance requirements for the next generation multimedia, broadband, and network applications, it is expected that by year 2009 more than 4 billion transistors will be integrated on a single chip according to the ITRS'05 roadmap [1]. As a result of these, more and more functionalities are being integrated onto a single chip which, in turn, results in a sharp increase of overall on-chip communication traffic among the integrated modules. In such complex systems, on-chip communication is expected to become a major performance bottleneck. Traditional approaches are mainly based on the synthesis of a single shared bus based architecture, which may not meet the real time constraints.

The early works on communication bus synthesis are mainly based on the simulation of an entire Hw/Sw system, which takes huge amount of time while exploring a large design space. The first approach for synthesizing a single global bus was proposed in [5], which finds the minimum bus width in order to minimize the chip size. In [14] an automatic bus generation for an MPSoC architecture was proposed. The approach considers for three different types of buses, which can be generalized to a shared bus, point-to-point, and FIFO based architecture. A bus architecture for a given bus width is generated considering real-time constraints. In [13] a method of communication synthesis based on the library elements and constraints graph was presented, where the library elements are a collection of communication links and communication nodes. The approach focuses mainly on synthesizing a communication bus topology for a point-to-point communication architecture. In [17] a bus model for communication in embedded systems with arbitrary

topologies was proposed, where a point-to-point communication is a special case for the real-time application. Their algorithm selects the number of buses, the type of each bus, the message transferred on each bus, and schedules the communication bus. In [6] a synthesis flow which supports shared buses and point-to-point connection templates was presented. In [9], [10] an energy conscious on-chip bus synthesis technique was presented. All the above techniques are for synthesizing on-chip bus architectures, however, non of them synthesizes memories. Recently, in [12], an approach for co-synthesis of bus and memory was presented for MPSoC architectures. The technique synthesizes on-chip bus and memory, however, the synthesized bus architecture is not custom in terms of bus widths. It is rather based on the standard bus templetes provided by vendors.

In this paper, the bus synthesis algorithm is based on the approach proposed in [11], [9], which synthesizes a custom bus architecture. The synthesis problem is formulated as an optimization problem and finds the optimal bus widths and the number of buses. The memory synthesis is based on the graph partitioning algorithm, which clusters a set of tasks that have data dependencies. While partitioning a graph, the algorithm finds the minimum number of cuts among the clusters in order to minimize the communication via a bridge. This, in turn, results in the clusters of tasks, which seldom access the memory of another bus. The term cut is the number of edges connecting the clusters. It determines how many times an on-chip module accesses a memory using a bridge. Finally, the algorithm sums the data size of each cluster to find the memory size and maps each cluster onto a memory.

The reminder of this paper is organized as follows. Section II introduces a motivational example for co-synthesis of bus and memory. Section III gives a mathematical formulation and optimization techniques for on-chip bus synthesis problem. The memory synthesis algorithm based on graph partitioning is described in Section IV. In Section V, we present experimental results to validate our method of on-chip bus and memory synthesis and finally, in Section VI, we give the conclusion of this work.

II. MOTIVATIONAL EXAMPLE

In this section we give a motivation for co-synthesis of on-chip buses and memories and show that the synthesized buses and memories are optimal in terms of 1.) bus widths and the number of buses and 2.) memory sizes and the number of memories, respectively. We consider a partitioned and mapped Hw/Sw system. Based on the partitioned and mapped system, a communication task graph $G_C(C, \Pi)$ with nine communication tasks and their data dependencies is extracted as shown in Fig. 1(a). In the figure, tasks $\{c_4, c_5, c_7\}$, $\{c_8, c_9\}$, $\{c_1, c_2, c_3\}$, and $\{c_6\}$ are initiated by on-chip modules M1, M2, M3, and M4, respectively. After scheduling of tasks $c \in C$ as shown in Fig. 1(b), 16 and 24-bit buses are synthesized.

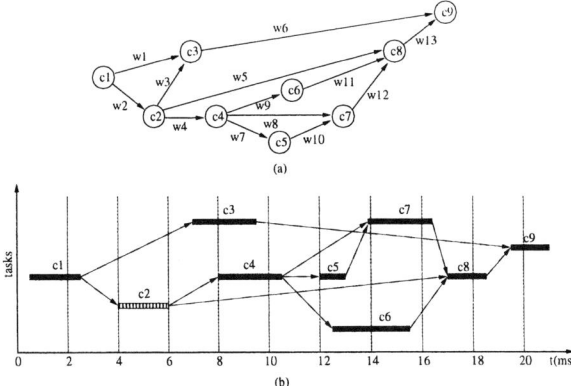

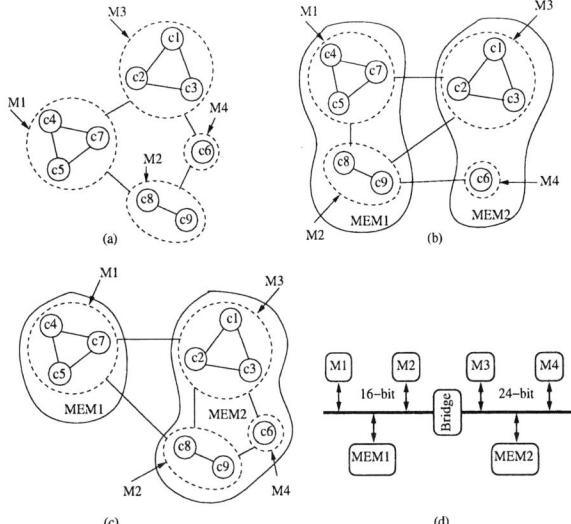

Fig. 1. Communication tasks and their schedule. (a) Example communication tasks. (b) The optimal schedule of tasks.

The tasks c_1, c_2, c_3, and c_6 are mapped to a 24-bit wide bus, while the tasks c_4, c_5, c_7, c_8, and c_9 are mapped to a 16-bit wide bus.

For the synthesis of memory sizes and the number of memories, we first extract an undirected data dependency tasks graph $G_D(C, \Pi)$ from the directed communication task graph $G_C(C, \Pi)$. In graph $G_D(C, \Pi)$, each vertex is a communication task c, while an edge between two vertices gives a data dependency between two communication tasks. From the data dependency tasks graph $G_D(C, \Pi)$, we find the maximum cliques with edges connecting each clique as shown in Fig. 2(a). The main aim is to cluster the data associated with tasks $c \in C$, which have data dependencies and map each cluster to a memory. From the given cliques of tasks $c \in C$ as shown in Fig. 2(a), we further cluster the cliques in order to find the minimum number of memories unless there is a memory access conflict (when two tasks access a memory at a same time). Fig. 2(b) and (c) show the synthesized memories and data associated with communication tasks. In the first figure, data of tasks $c \in C$, which are initiated by modules M1 and M2 are mapped to MEM1. While data of tasks initiated by modules M3 and M4 are mapped to MEM2. In the figure, there are three cuts, which mean that either on-chip modules M1 and M2 or M3 and M4 access memory MEM2 or MEM1 for three times using a bridge. Similarly, Fig. 2(c) depicts the synthesized memory sizes, number of memories, and the number of cuts. In the figure, the number of cuts is less than the synthesized memory of Fig. 2(b). This, in turn, results in less power and delay overhead due to communication via a bridge. Thus, the synthesis results of Fig. 2(c) give the optimal memory sizes and the number of memories with the minimum number of bridge accesses. The memory size is evaluated by summing all tasks $c \in C$ in a cluster. The synthesized on-chip buses and memories with their interconnection are shown in Fig. 2(d).

III. BUS SYNTHESIS

The on-chip bus synthesis problem is formulated as an optimization problem, which performs scheduling, allocation, and binding of tasks $c \in C$ and finds the optimal bus widths and the number of buses. During the scheduling, the slack of each task $c \in C$ is exploited to share the bus(es). The formulation of the optimization problem is given as follows [11], [9],

minimize:

$$\sum_{r \in R} Cost_r \cdot r_i \qquad (1)$$

Fig. 2. Co-synthesis of on-chip buses and memories. (a) Clique of data dependency tasks and their dependencies. (b) Synthesized memories with number of cuts = 3. (c) Synthesized memories with number of cuts = 2. (d) Synthesized bus architecture and memories with interconnection of on-chip modules and bridge.

subject to,

$$s_\tau + w_\tau \leq dl_\tau \ \forall \tau \in T \qquad (2)$$

$$s_\tau \geq s_{c'} + CLTI_{c',r} \cdot X_{c,t,r} \ \forall (c, c') \in \Pi \qquad (3)$$

$$t \cdot X_{c,t,r} \geq (s_{\tau'} + w_{\tau'} \cdot X_{c,t,r} \ \forall \ (c', c) \in \Pi \qquad (4)$$

$$(dl_c - s_c - CLTI_{c,r}) \cdot X_{c,r} \geq 0 \qquad (5)$$

$$w_\tau = \sum_{\tau \in T} NC_\tau \cdot T_d \qquad (6)$$

$$CLTI_{c,r} = \left\lceil \frac{NB_c}{b_r} \right\rceil \cdot T_d \qquad (7)$$

$$T_d = \frac{\kappa_3 \cdot V_{dd}}{[\kappa_1 \cdot V_{dd} + \kappa_2 \cdot V_{bs} - V_{th}]^\alpha} \qquad (8)$$

$$b_{r_{min}} \leq b_r \leq b_{r_{max}} \qquad (9)$$

The main objective is to minimize the communication bus cost (bus widths and the number of buses) as shown in Eq. (1), where $r_i \in R$ is a library of on-chip communication buses with different bus width. The $Cost_r$ of each bus r_i is expressed in terms of bus width. In Eq. (2), start time s_τ and execution time w_τ of each task τ should be less than or equal to its deadline dl_τ. Further, a task τ can start its execution only after its predecessor (communication task c) completes transferring data as shown in Eq. (3). A binary decision variable $X_{c,t,r} \in \{0, 1\}$, indicates scheduling of a communication task c at time $t \in \{0, \cdots, \lambda\}$, with bus width r. Eq. (4) gives a dependency between successor (communication task c) and predecessor (data processing task τ'). Since the delay $CLTI_{c,r}$ (communication lifetime interval) of a task c is a function of data size and bus width r (see Eq. (7)), Eq. (5) gives a delay constraint such that the overall delay of each task c must be less than or equal to deadline dl_c. The gate delay is calculated according to Eq. (8) [16], where V_{dd}, V_{bs}, and V_{th} are supply, body bias, and threshold voltages, respectively. Terms κ_1, κ_2,

κ_3, and α are the technology dependent parameters. While scheduling communication tasks for different bus widths r, Eq. (9) gives the constraint for bus widths. In the above formulation, the objective function is linear in optimization variable r_i and the delay constraint Eq. (7) is non-linear in the optimization variable r_i, thus, the above bus synthesis problem belongs to the convex quadratic optimization problem, which finds a global optimal solution in a polynomial time complexity [8].

IV. MEMORY SYNTHESIS

The memory synthesis algorithm presented in Algorithm 1, takes an undirected data dependency task graph $G_D(C, \Pi)$ as input, which is extracted from a communication task graph $G_C(C, \Pi)$. The algorithm finds the optimal memory sizes and the number of memories with the minimum number of cuts. The term cut is the number of edges that connect the clusters. It determines how many times an on-chip module accesses a memory using a bridge. Thus, the aim is to map the data of communication tasks to an individual memory in order to minimize the number of communications between a module and a memory via a bridge. This, in turn, reduces power and delay overhead due to the communication. The clustering of tasks is based on the graph partitioning problem called clique partitioning [15], which finds a set of cliques from the given data dependency task graph $G_C(C, \Pi)$.

At line 1 and 2, the algorithm reads the data dependency task graph and the synthesized number of buses. From line 4 to 13, a super graph $G'(S, E')$ is derived from the graph $G_D(C, \Pi)$. Each node $s_i \in S$ is a super node [15] that can contain a set of one or more vertices $c_i \in C$. Edge E' is identical to E except that the edges in E' link to super nodes in S. At line 4 and 5, the algorithm initializes sets S and E' to empty sets. From 6 to 9, each vertex $c_i \in C$ of $G_D(C, \Pi)$ is moved to a separate super node $s_i \in S$ of G'. In the graph $G'(S, E')$ each vertex $c_i \in C$ represents a communication task. An edge $e_{i,j} \in E$ between two vertices c_i and c_j represents a data dependency. From line 15 to 38 the algorithm finds the cliques of the data dependency graph. The algorithm stays in the while loop defined at line 15 until the set E' is not empty. In this loop the algorithm finds the super node of the graph, where each super node consists of all the nodes in connected nodes s_{Num1} and s_{Num2} with the maximum number of common nodes. By definition a super node $s_i \in S$ is a common node of the two super nodes s_j, $s_k \in S$ if there exist edges $e_{i,j}$, $e_{i,k} \in E'$. At line 21, the function COMMONNODE(G', s_i, s_j) returns the set of super nodes that are common nodes of s_i and s_j in G'. At line 22, if the returned number of common nodes is greater than the variable $MostCommons$ then the content of the variable is updated at line 24. When the $MostCommons$ nodes are found, from line 30 to 32, two super nodes are merged into a single super node, $s_{Num1Num2}$, which consists of all the vertices in s_{Num1} and s_{Num2}. From line 36 to 38, the algorithm adds edges among the super nodes. The variable $CommonSet$ consists of all the common nodes of s_{Num1} and s_{Num2}. At line 39 of the algorithm, a new while loop starts to synthesize memory sizes and the number of memories from the available cliques of a data dependency task graph $G_D(C, \Pi)$. The loop is repeated until the number of cuts is minimum and there is no more memory access conflict. The function $ClusterClique$ at line 41 clusters the cliques in order to minimize the number of memories. Each time the algorithm finds the maximum number of edges connecting two cliques and clusters them to make a new super node. The loop is repeated until there is no minimum number of cuts and there is no memory access conflict. After the completion of the while loop defined at line 39, the algorithm gives a set of super nodes, which are mapped to an

```
MEMORYSYNTHESIS()
 1   G_D(C, Π) ← GETDATADEPTASKS();
 2   n ← GETNUMOFBUSES();
 3   /*Create a super graph G'(S, E') */
 4   S ← ∅;
 5   E' ← ∅;
 6   for c_i ∈ C
 7   do
 8        s_i ← {c_i};
 9        S ← S ∪ {s_i};
10   endfor
11   for e_{i,j} ∈ E
12   do
13        E' ← E' ∪ {e_{i,j}};
14   endfor
15   while E' ≠ ∅
16   do
17        /*Find S_{Num1}, S_{Num2} having most common node*/
18        MostCommons ← −1;
19        for e'_{i,j} ∈ E'
20        do
21             c_{i,j} ← |COMMONNODE(G', s_i, s_j)|;
22             if c_{i,j} > MostCommons
23             then
24                  MostCommons ← c_{i,j};
25                  Num1 = i; Num2 = j;
26             endif
27        endfor
28        CommonSet ← COMMONNODE(G', S_{num1}, S_{num2});
29        /*Merge S_{Num1} and S_{Num2} into S_{Num1Num2}*/
30        S_{Num1Num2} ← S_{Num1} ∪ S_{Num2};
31        S ← S − S_{Num1} − S_{Num2};
32        S ← S ∪ {S_{Num1Num2}};
33        /*Add edge from S_{Num1Num2} to super nodes*/
34        for s_i ∈ CommonSet
35        do
36             E' ← E' ∪ {e'_{i,Num1Num2}};
37        endfor
38   endwhile
39   while cuts ≠ minimum ∧ MemAccessConflict ≠ true
40   do
41        cluster ← CLUSTERCLIQUE(S, E');
42   endwhile
43   for c_k ∈ cluster
44   do
45        memSize_i ← SUM(DataSize(c_k));
46   endfor
47   return
```

Algorithm 1: Memory synthesis algorithm.

individual memory. At line 45, the data size of all the communication tasks in a super node s_i is summed to find the memory size for each memory.

V. EXPERIMENTAL VALIDATION

We validate the effectiveness of the proposed technique using real-life multimedia applications, i.e., an audio decoder [2] and a speech recognition system [3]. The audio decoder includes four main decoding steps, which are inverse quantization, channel decoupling, reconstruct curve, and IMDCT. After manually partitioning and mapping of the decoder [11], the IMDCT was mapped to a single hardware and the rest of the functionalities were mapped to a processor. Furthermore, the raw audio data was mapped to a compact flash (CF) memory with a CF-interface and the extracted audio data was mapped to an audio buffer for streaming. Similarly, the second speech recognition system consists of three main components: front

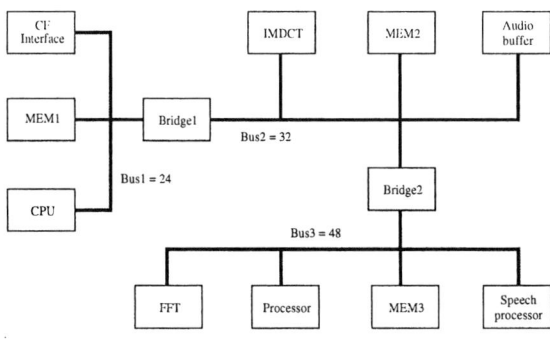

Fig. 3. Synthesized bus architecture with three buses and memory blocks.

Mem. Block	Mem. Size	No. of Cuts
MEM1	1.8 KB	(MEM1 and MEM2)=6
MEM2	2.4 KB	(MEM2 and MEM3)=10
MEM3	3.7 KB	(MEM1 and MEM3)=0

TABLE I

SYNTHESIZED MEMORIES, THEIR SIZE, AND NUMBER OF CUTS.

VI. CONCLUSION

Traditional on-chip bus synthesis approaches are mainly based on the simulation of an entire Hw/Sw system. After simulation, the communication requirement is mapped to a bus architecture provided by a vendor. However, the resulting synthesized bus architecture may not be efficient due to the under utilization of bus resources. In this paper, we proposed a custom on-chip bus architecture and memory co-synthesis techniques for a given application. The bus synthesis is formulated as a scheduling, allocation, and binding problems, and finds the optimal bus widths and the number of buses. The memory synthesis is formulated as a graph partitioning problem, which takes a data dependency task graph and synthesizes memory sizes and the number of memories. The experiments conducted on the real-life multimedia applications validate the effectiveness of the proposed technique.

REFERENCES

[1] http://www.itrs.net.
[2] http://www.xiph.org.
[3] http://www.speech.cs.cmu.edu/sphinx/.
[4] http://www.mosek.com/documentation.html#manuals.
[5] M. Gasteier and M. Glesner. Bus-based communication synthesis on system level. *In ACM Tran. of design automation electronic systems*, pages (1–11), 1999.
[6] D. Lyonnard, S. Yoo, A. Baghdadi, and A. A. Jerraya. Automatic generation of application specific architectures for heterogeneous mpsoc. In *proc. of Design Automation Conference (DAC)*, pages (518–523), 2001.
[7] Z. Ming. Architecture exploration for speech-feature-extraction acceleration. *Bachelor thesis, Institute of Microelectronics Systems, Darmstadt University of Technology, Darmstadt, Germany*, September 2005.
[8] Y. Nesterov and A. Nemirovskii. *Interior-point polynomial algorithms in convex programming*. Studies in Applied Mathematics, 1994.
[9] S. Pandey and M. Glesner. Statistical on-chip communication bus synthesis and voltage scaling under timing yield constraint. In *proc. of Design Automation Conference (DAC)*, pages (663–668), 2006.
[10] S. Pandey and M. Glesner. Simultaneous on-chip bus synthesis and voltage scaling under random on-chip data traffic. *In IEEE Trans. on Very Large Scale Integration (VLSI) Systems*, 2007.(accepted for publication).
[11] S. Pandey, M. Glesner, and M. Mühlhäuser. Performance aware on-chip communication synthesis and optimization for shared multi-bus based architecture. In *ACM proc. of SBCCI*, pages (230–235), 2005.
[12] S. Pasricha and N. Dutt. COSMECA: Application specific co-synthesis of memory and communication architectures for mpsoc. In *proc. of Design Automation Test in Europe (DATE)*, pages 700–705, 2006.
[13] A. Pinto, L. P. Carloni, and A. V. Singiovanni. Constraint driven communication synthesis. In *proc. of Design Automation Conference (DAC)*, pages (783–788), June 2002.
[14] K. K. Rye and V. MooneyIII. Automated bus generation for multiprocessor soc design. *IEEE Tran. of Computer-Aided Design (CAD) of Integrated Circuits and Systems*, Vol. 23(No. 11):(1531–1549), Nov. 2004.
[15] C. J. Tseng and D. P. Siewiorek. Automated synthesis of data paths on digital systems. *IEEE Tran. of Computer-Aided Design (CAD) of Integrated Circuits and Systems*, Vol. CAD-5(No. 3):(379–395), 1986.
[16] N. H. E. Weste and K. Eshraghian. *Principles of CMOS VLSI Design*. Addison wesley, 1994.
[17] T. Y. Yen and W. Wolf. Communication synthesis for distributed embedded systems. In *proc. of Int. Conf. on Computer-Aided Design (ICCAD)*, pages (288–294), 1995.

end, decoder, and linguist. The front end includes series of data processing tasks such as pre-emphasis, hamming window, FFT (fast Fourier transformation), mel frequency filter, IFFT, cepstral mean normalization, and feature extraction to generate the features from the speech. The training takes as input a large number of speech along with their transcriptions into phonemes to provide the speech models for the phonemes. The recognition is based on the HMM (hidden Markov model) to decode the speech. We used the American English lexicon consisting of 32 phonemes and a database of 17 different words (spelling out the names of the months, numbers, and digits) [7], [11]. The length and the number of phonemes in a speech varies from application to application. After partitioning of the speech recognition system, the front end was mapped to a dedicated hardware including FFT and filters. The task training and recognition were mapped to a PowerPC processor. Based on the partitioned and mapped system, communication tasks and their data size were extracted by profiling the Hw/Sw system [7].

The on-chip communication buses are given as a library of buses with different bus widths, which ranges from 16 to 128-bit wide. The bus synthesis algorithm was implemented in C as a pre-processing model to interface with a convex solver of MOSEK [4]. Further, the memory synthesis algorithm was implemented in C as a graph partitioning problem, which partitions a set of communication tasks (associated with data) into a set of sub graphs with the minimum number of cuts among the sub graphs. For a given partition with the minimum number of cuts, each sub graph is mapped to a memory block. Fig. 3 shows the synthesized number of buses, memory blocks, and the interconnection of on-chip modules with buses. There are two bridges in order to communicate modules of one bus to the modules of another bus. The synthesized bus widths are 24, 32, and 48-bit wide. Table I shows the size of each memory block and the number of cuts between memories. In the column entitled *No. of cuts* of the table, MEM1 and MEM2 have 6 cuts, MEM2 and MEM3 have 10 cuts, while MEM1 and MEM3 have no cut. i.e., the smaller the number of cuts the better the performance in terms of delay and power overhead due to the communication through a bridge. Thus, the data associated with communication tasks mapped to MEM1 has no data dependency with the data associated with the communication tasks, which are mapped to MEM3. However, the tasks that are mapped to MEM1 and MEM2, and MEM2 and MEM3 have data dependencies.

An HDTV H.264 Deblocking Filter in FPGA with RGB Video Output

Vagner S. Rosa, Altamiro A. Susin, Sergio Bampi

Informatics Institute

Federal University of Rio Grande do Sul – UFRGS

Av. Bento Gonçalves, 9500 – Porto Alegre – Brazil

vsrosa@inf.ufrgs.br, susin@eletro.ufrgs.br, bampi@inf.ufrgs.br

Abstract— This paper presents an architecture for implementing the H.264 Deblocking Filter with RGB output in FPGA, exceeding HDTV requirements when synthesized to a target FPGA. The goal of the design was to achieve the HDTV requirements, designing a deep pipelined architecture that makes a balanced use of the resources available in the target FPGA architecture. When synthesized to VirtexII-pro FPGA, the developed architecture used only 1800 logic cells and achieved 71 frames per second at 1080p HDTV resolution (1920x1080).

I. INTRODUCTION

The standard developed by the ISO/IEC MPEG-4 Advanced Video Coding (AVC) and ITU-T H.264 experts set new levels of video quality for a given bit-rate. In fact, H.264 (from this point the standard will be referred only by its ITU-T denomination) outperforms previous standards in bit-rate reduction. In H.264 an Adaptative Deblocking Filter is included in the standard to reduce blocking artifacts, very common in very high compressed video streams. In fact, most video codecs use some filtering as a pre/post-processing task. The main goal of the inclusion of this kind of filter as a part of the standard was to put it inside the feedback loop in the encoding process. As a consequence, a standardized, well tuned, and inside the encoding loop filter could be designed, achieving better objective and subjective image quality for the same bit-rate. The H.264 Deblocking Filter is located in the DPCM loop as shown in Figure 1 for the decoder (it is also called Loop Filter by this reason). Exactly the same filter is used in the encoder and the decoder. The Deblocking Filter in the H.264 standard is not only a single low pass filter, but a complex decision algorithm and filtering process with 5 different filter strengths. Its objective is to maintain the sharpness of the real images while eliminating the artifacts introduced by intra-frame and inter-frame predictions at low bit-rates, while mitigating the characteristic "blurring" effect of the filter at high bit-rates.

The complexity of the decision logic and filtering process makes the H.264 Deblocking Filter responsible for about one third [4] of the processing power needed in the decoder process.

In this paper an architecture for the H.264 deblocking filter is proposed. The architecture was developed specifically for FPGA, aiming making a balanced use of the resources (logic, registers, and memory) available in FPGA architectures. The performance goal was to meet or exceed HDTV requirements when synthesized to a XILINX Virtex II FPGA. A Xilinx Virtex II pro FPGA was used to validate the developed IP.

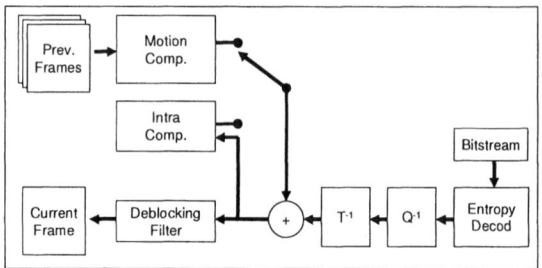

Figure 1. H.264 Decoder.

The rest of this paper is organized as follows. The section 2 describes the standardized algorithm for the H.264 deblocking filter. Section 3 presents the proposed architecture and section 4 the results obtained. Finally the section 5 presents the conclusions and future work.

II. DEBLOCKING FILTER ALGORITHM

In the H.264 standard the image is divided in small units called blocks. Each block is 4x4 pixels. The color format is YCbCr 4:2:0 (main profile), meaning the crominance (croma) components being sub-sampled to half the sample rate of the luminance (luma) in both directions. The blocks are then grouped in macroblocks which is a 4x4 block matrix for luma and 2x2 matrix for each croma component. Each block edge has to be filtered. The Deblocking Filter is applied to each decoded block of a given macroblock for luma and croma samples in raster scan order. For each block, four different edges are filtered separately, in the sequence presented in Figure 2.

For each block edge, the filter is applied to the pixel component values perpendicular to that edge. The naming

308

978-1-4244-1709-4/07/$25.00 © 2007 IEEE

conventions for the pixels around the edge are showed in figure 3. Pixel components in both the current (Q) and the previews (P) block can have values changed. Pixels already modified during a filter stage can be modified again in a subsequent filter operation (this causes some data dependencies). The filtering algorithm is adaptive, so that the pixel values, the position of a block inside the macroblock, the type of prediction employed (inter or intra), the motion vectors (inter prediction) and the quantization parameter are taken into account for the boundary strength calculation.

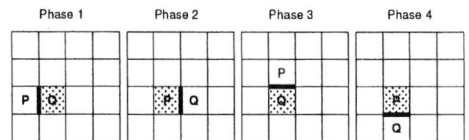

Figure 2. Edge positions for a given 4x4 block inside a 16x16 macroblock.

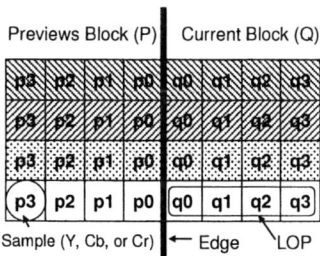

Figure 3. Filter conventions.

The boundary strength (BS) can assume five different values from 0 (no filtering) to 4 (strongest filtering) depending on the edge context and location.

The Boundary Strength for croma is the same as for the corresponding luma block, but the filters employed for luma and croma are different. The quantization parameter (QP) and the pixel values are taken into account. The parameters α and β are QP dependent and set thresholds for filtering to be applied. Saturation functions (clip and clip3) need do be used in some steps of the calculations. This makes the BS decision logic a complex set of sequential calculations. The BS calculation needs to be done for every LOP (Line Of Pixels – figure 3) pair. More details on Deblocking Filter process can be obtained in [1] and a complete flowchart in [3].

III. PROPOSED ARCHITECTURE

The proposed architecture was developed ad-hoc for a H.264 codec to be implemented in FPGA. It was developed to meet or exceed HDTV resolution requirements (1920x1080x30) in the H.264 Main Profile. The initial architectural concept was initially based on the work developed by [4], but was evolved to be a datapath block, different from the coprocessor block proposed by [4]. The primary differences from this work related to others found in the literature ([4], [5] and their references) were the FPGA focus.

The focus on FPGAs leads to some architectural decisions that leaded to a more efficient resources usage and

higher speed. In this scenario, the development of a high-depth pipeline architecture was straightforward: each Logic Element in current FPGA can behave as a look-up-table (LUT), a single bit flip-flop (register), and a carry logic at the same time. The unneeded parts of the logic elements (ex. Flip-flops in combinational logic block) are bypassed and can not be used in other blocks due to routing limitations

Some conventions on block numbering are presented as follows. The numbering starts in zero for the upper-left block of luma (Y) in a macroblock. The sequence follows first from left to right then from top to bottom. Croma (Cb and Cr) samples are numbered the same way, except Cb start from the index 16 and CR start from the index 20. Macroblocks denoting the current macroblock are colored white, Left macroblock (at the left of the current one) are colored gray, and top macroblock (above the current one) are colored black. Figure 4 illustrates the block numbering and macroblock coloring convention adopted from this point.

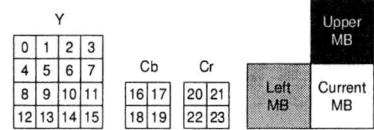

Figure 4. Luma and Croma block enumeration; Macroblock color diagram.

The proposed Deblocking Filter architecture will accept pixel data and block information from previews blocks in the flow and will produce filtered pixel data. The input data for a given luma/croma macroblock arrives in the sequence presented in Figure 5: Y first, then Cb and Cr; double Z scan for luma.

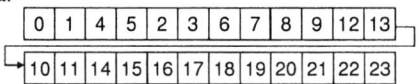

Figure 5. Block sequence arriving the deblocking filter.

In the Deblocking Filter process, the edge filter is the heart of the process. It is responsible for all filter functionality, including the thresholds and BS calculation and filtering itself. The remaining of the process is only sequence control and memory for block ordering.

The Edge Filter architecture can accept one LOP per cycle for Q and P blocks and produce the filtered Q' and P' LOP. Using this scheme, an entire block border will enter in the Edge Filter each four cycles (one block border is four LOPs tall, as illustrated in Figure 3).

The architecture developed consumes two LOPs (one for Q and other for P blocks) and their respective parameters (QP, offsets, prediction type and neighborhood information) every clock cycle, and produces two LOPs (Q' and P') every cycle. The edge filter is implemented as a 16-stage pipeline, so the filtered samples are present in the output 16 clock cycles later.

This pipelined edge filter can be then encapsulated in such way only the Q input and P' outputs are visible, as illustrated in Figure 6.

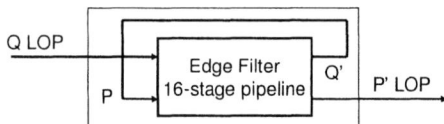

Figure 6: Edge Filter encapsulation.

Using this encapsulation makes the control logic simpler, but there is a small overhead: P data can only be fed by the Q input, so the first P data need to be fed 16 cycles before the filtering process can start. During this 16 cycles, the BS should be forced to 0, so that no filtering is applied to P data while they passes through the Q datapath. After finishing the processing of a macroblock, the Edge filter must be emptied, so a 16 cycle overhead should me tailored. Fortunately, the empting and filling stages can be overlapped, so the overhead is much lower.

Finally, the architecture of the proposed deblocking filter is presented in Figure 7. As stated in section 2, the filter has to be applied to both vertical and horizontal edges. In the proposed architecture, a single Edge Filter filters both horizontal and vertical edges. A transpose block is employed to convert vertically aligned samples in the block into horizontally aligned ones, so that horizontal edges can be filtered by the same Edge Filter.

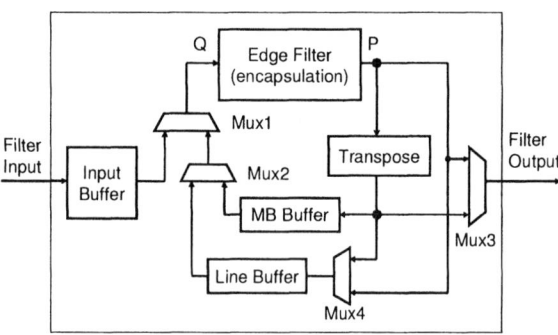

Figure 7. Proposed Deblocking Filter architecture.

The input buffer is needed only for data reordering, as input blocks arrive in a different order they are consumed in the deblocking process.

The MB and Line buffers are used to store blocks and block information data (QP, type, filter offset, state) which are not completely filtered (horizontal or vertical filtering is missing). The size of MB buffer is 32 blocks (512 samples plus 32 block information data) and the line buffer depends on the maximum frame size we want to process (7680 samples for a HDTV frame plus 480 block information data).

A. Filter operation

The filter data flow description follows. First, pixel and control data are fed to the filter toward the Input Buffer. This buffer is needed because data is read from input buffer in a different order it comes. Once a complete luma/croma macroblock is available in the input buffer, the filtering process can be started.

The Encapsulated Edge Filter entity contains a 16 stage pipeline edge filter where the input P is connected to the output Q. The mux1 and mux2 are set so that MB buffer data is fed to the input Q of the Encapsulated Edge Filter. The data read from the MB buffer is the blocks 3, 7, 11, and 15 from the left macroblock. As the data is read one LOP (Line of Pixels – a horizontal four pixels group in the block) at a time, it takes four clock cycles to read an entire block into the Encapsulated Edge Filter.

In order to read four blocks, exactly 16 cycles are needed. During this phase the filter is set to bypass (BS=0), so pixel data fed are not filtered. After, the blocks 0, 4, 8, and 12 is read from input buffer, a LOP at a time. Immediately after, the block 12 the blocks 1, 5, 9, and 13 can be fed and then the blocks 2, 6, 10, 14 and 3, 7, 11, 15. 32 cycles after the block 3 from the left MB started being read from the MB buffer, the filtered block 3 will appear at the P output of the Encapsulated Edge Filter.

The data can take 3 different destinations depending on operations pending due to neighborhood location: MB Buffer, Line Buffer or output of the filter. After filtering the vertical edge of all luma blocks the job for the horizontal edge can be done. In order to obtain the maximum throughput, the croma vertical filtering is done immediately after the luma. Then, blocks 17, 19, 21, and 23 from the left Cb and Cr macroblock are read from the MB Buffer. The Cb and Cr blocks are stacked in order to achieve the 16 LOPs needed to fill the Edge Filter pipeline. After, the croma blocks 16, 18, 20, and 22 are read from Input Buffer and then the blocks 17, 19, 21, and 23. The blocks 17 and 21 from the left Cb and Cr macroblocks, respectively are completely filtered and then can be sent to the output. The blocks 19 and 23 from the left Cb and Cr macroblocks respectively are send to the Line Buffer toward the Transposer in order to be used in the horizontal external edge filtering. The blocks from the current Cb and Cr macroblocks are stored transposed in the MB Buffer for the horizontal filtering process.

Finally, the horizontal edge filtering for luma and croma can take place (not detailed due to space limitation).

A total of 256 clock cycles are needed to process an entire 4:2:0 macroblock (24 blocks). If data is not available at the beginning of a 256 cycle operation, a bubble is inserted in the pipeline. All pipelines are emptied and the filter operation stops until there is data available in the input buffer. The stop cycle is also a 256 cycle operation.

Figure 8 illustrates the output sequence of blocks for the implemented filter architecture. Observe that for the 24 blocks output corresponding to an entire 4:2:0 luma/croma macroblock processing cycle, the output have blocks belonging to three different macroblocks.

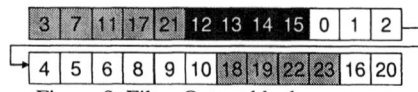

Figure 8: Filter Output block sequence.

B. RGB output

For applications that need direct RGB video output from the Deblocking Filter (eg. a low latency implementation with no B frames) an output reordering and color YCbCr to RGB color space conversion architecture was developed. Figure 9 presents the architecture developed to convert the 4x4 blocks ordered as in Figure 8 to a pixel scan YCbCr 4:2:0 and RGB output.

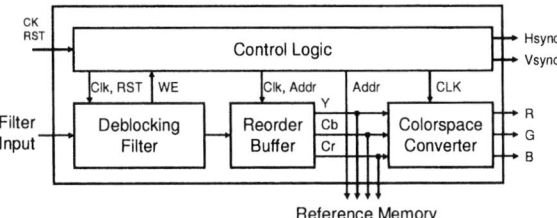

Figure 9: RGB and reference memory output.

In this architecture, the Deblocking Filter, as proposed in Figure 7, is instantiated as a building block. The Reorder Buffer is 1080x32 samples for luma and 540x16 samples for each croma component. The actual implementation uses dual port RAMs. While receiving data for reordering one line of macroblocks (16 pixel lines for luma and 8 pixel lines for each croma component), the other one can be output at output pixel rate. The control logic can throttle the clock to the deblocking filter, so the RGB output can match the correct pixel out timing, including the Hsync and Vsync periods, where no pixel output is done. The Color Space converter is a four stage pipelined implementation of the YCbCr to RGB converter, including output saturation capability to avoid arithmetic overflow and underflow in the RGB values due to the quatization error introduced in the YCbCr components during the encoder side quantization process, specially when using high QP values.

Ordered reference memory output (YCbCr) can be tapped from the reorder buffer. This simplifies memory accesses and reduces the latency when using memories with high throughput and high access latency, like DDR.

IV. IMPLEMENTATION AND RESULTS

The architecture presented in Section 3 was described in VHDL. About 4,100 lines of code were written for the Deblocking Filter, including the RGB output module presented in Figure 9. The design behavior was validated by simulation using some testbench files and data extracted from the JVT reference software using some public domain video sequences. The validated behavioral design was then synthesized, the post place and route was validated and performance results were obtained for a Xilinx Virtex2-pro FPGA.

Table 1 presents synthesis results of the Deblocking Filter including the reordered RGB output for the Xilinx XC2VP30 device.

Table 1: Deblocking Filter with RGB output.

FPGA	Xilinx XC2VP30
LUTs	4331/27392 (17%)
BRAMs	65/136 (47%)
Fmax (MHz)	135
FPS@1080p	65

The results presented in Table 1 shows the developed architecture is able to outperform the HDTV frame rate requirements. That means significant room for using a lower speed grade FPGAs and/or fitting this design in a complete design.

V. CONCLUSION

This work presented an ad-hoc architecture for H.264 Deblocking Filter process targeted to achieve HDTV requirements in FPGA. The primary contribution of this work was the deep pipeline architecture employed to better optimize the design to the resources available in a typical FPGA device. This FPGA-optimized implementation was not found in literature. The results proved it is capable to exceed the processing rate for HDTV, With RGB output, a prototyped model could be demonstrated using the RGB out available in the Digilent XUP-V2P board [7].

Future work includes the integration of this developed module to a complete encoder/decoder design and the support of High Profiles (currently only Baseline and Main profiles are supported).

REFERENCES

[1] Draft ITU-T Recommendation and Final Draft international Standard of Joint Video Specification (ITU-T Rec. H.264/ISO/IEC 14496-10 AVC), March 2003.

[2] P. List, A. Joch, J. Lainema, G. Bjotergaard, and M. Karczewicz, "Adaptative deblocking filter", IEE trans. Circuits Syst. Video Technol., Vol. 13, pp. 614-619, July 2003.

[3] A. Puri, X. Chen, A. Luthra, "Video coding using the H.264/MPEG-4 AVC compression standard", Signal Processing: Image Communication, Elsevier, n. 19, pp. 793-849, 2004.

[4] M. Sima, Y. Zhou, W. Zhang. "An Efficient Architecture for Adaptative Deblocking Filter of H.264/AVC Video Coding". Trans. On Consumer Electronics, IEEE, Vol. 50, n. 1, pp. 292-296. Feb. 2004.

[5] H-Y Lin; J-J Yang; B-D Liu; J-F Yang. "Efficient deblocking filter architecture for H.264 video coders". 2006 IEEE International Symposium on Circuits and Systems, ISCAS 2006, 21-24 May 2006.

[6] G. Khurana; A.A. Kassim; T-P. Chua; M.B.A Mi " Pipelined hardware implementation of in-loop deblocking filter in H.264/AVC" IEEE Transactions on Consumer Electronics. V.52, I. 2, May 2006. pp.536-540.

[7] Diglent Inc. XUP-V2Pro board. Available at <http://www.digilentinc.com/>

Efficient Timing Closure with a Transistor Level Design Flow

Cristiano Lazzari[1,2,3,]*, Cristiano Santos[2], Adriel Ziesemer[1], Lorena Anghel[3], Ricardo Reis[1]

[1]*PGMICRO-PPGC/UFRGS* - [2]*CEITEC*
Porto Alegre - RS, Brazil
E-mail:{clazz,amziesemerj,reis}@inf.ufrgs.br,cristiano@ceitec.org.br

[3]*TIMA Laboratory - INPG*
Grenoble - France
E-mail:lorena.anghel@imag.fr

Abstract

This paper presents a new transistor level design flow where it is possible to optimize the circuit with a wide number of logic functions and drive strengths. Different from the standard cell approach, our methodology is not limited to a previously characterized library of cells. The proposed design flow provides a virtual library with around 15,000 cells for logic synthesis and performs a transistor sizing optimization step to improve the timing of the circuit during layout generation. A transistor-level layout generator allows to explore these wide number of cells and drive strengths while optimizing the layout concerning connections and transistors. Circuits generated by our methodology were compared to the standard cell approach in which presented around 11% of delay improvement and more than 30% of power savings.

1. Introduction

The standard cell design have dominated the design of digital VLSI circuits due some virtues [1]: Standard cells hide the increasingly unpleasant details of shape-level design rules, IO pins are arranged on individual gates in geometry accessible locations, cells assemble with relative ease into row-based blocks and cells can be pre-characterized for timing and power.

However, the advent of deep-submicron technologies shifted the design paradigm from conventional logic-dominated to an interconnect-dominated design process [2]. Besides that, the demand for performance are not easily handled with the limited number of cells of a standard cell library due to the huge variance in the load capacitance. In

---
*Supported by the CAPES Brazilian Agency

order to drive the output load respecting the required timing, cells may waste a lot of power due to oversized transistors [3]. Usually, a standard cell library presents around ten versions of inverters and buffers and four or five versions of each of the other cells.

In the last years, some works have been presented in order to deal with these circuit optimization problems. Vujkovic *et al* presented in [3] a design flow in which the number of versions of each standard cell is increased. Obtained results shown an improvement of until 20% related to the EDD (energy times delay squared) measurements.

The concept of *flex cells* is presented in [4]. They proposed the identification and optimization of critical cells. The synthesis of a minimum number of cells is performed aiming at enhancing the performance of the designs. Therefore, the process must take into account both functionality and timing contexts in which each unique cell is used.

In this paper, a transistor level design flow is presented as alternative to the standard cell approach. The main idea is to generate optimized layout concerning transistor widths, polysilicon and metal routing, contact/vias reduction and diffusion gaps avoidance. Besides that, optimized layout in association with the possibility to generate a wide range of logic functions and drive strengths can mitigate the timing closure problem.

The paper is organized as the following. Section 2 presents the main features of the proposed design flow. In Section 3, the transistor level design flow is shown in details. Obtained results in comparison with the standard cell approach is presented in Section 4. Finally, conclusion remarks are given.

2. The Timing Closure Strategy

Timing closure can be a hard task due to the limited available number of logic functions and drive strengths of the standard cell libraries in traditional design flows. Libraries have up to 200 different logic functions and around four drive strengths. Inverters and buffers usually have up to ten drive strengths. These characteristics are hard limitations to the efficiency of the synthesis and sometimes can degrade the power and timing results.

In this section, an efficient timing closure strategy is presented. This strategy allows to explore a great number of possibilities concerning logic functions and drive strengths by the synthesis process.

2.1. Allowing a wide number of logic functions for synthesis

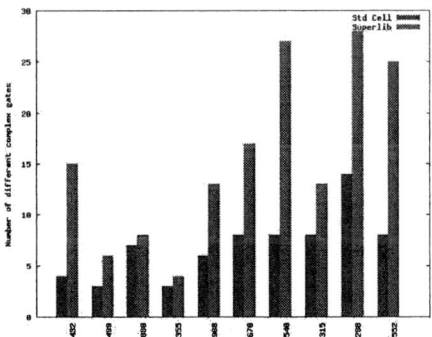

Figure 1. Number of different complex logic functions in circuits.

In a standard cell design flow, cells are generated and characterized for timing and power. The whole process takes long time and need an enormous effort from designers. As consequence, a limited number of cells and drive strengths are available for logic synthesis and technology mapping. In contrast, our design flow provides a virtual library with a big number of logic functions for technology mapping. This *superlib* is composed by around 3,500 different logic functions.

Figure 1 shows that commercial tools are able to explore the synthesis when a wide number o logic functions are available. These results were obtained with Cadence RTL Compiler [5]. Analyzing these ten circuits, we can note that circuits mapped with our *superlib* have around 52% more complex logic functions. Simple gates such as NANDs, NORs, inverters and buffers are not included in these results.

2.2. Increasing the number of drive strengths

We consider that timing closure demands the optimization of each transistor of a cell according to its output load.

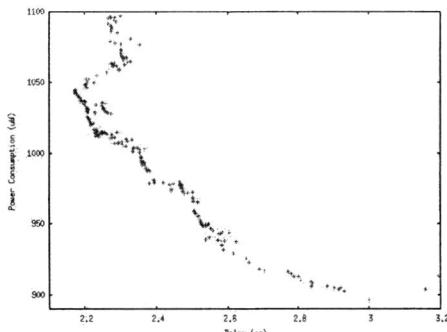

Figure 2. The power-delay points. Each point represents a possible circuit design.

As presented in [3], the capacitance per unit length varies around 35 times in a $0.18\mu m$ technology. This shown how the optimization of each cell, specially in the critical path, as function of the output load capacitance is important.

Since there is a tradeoff between performance results and CPU run time during logic synthesis according to the number of available cells of a library, the number of drive strengths of each logic function of our *superlib* were limited to four, except for inverters and buffers (8 and 32, respectively). But after a circuit be optimized and mapped to the *superlib* cells, we use a transistor sizing technique to improve its timing and power consumption. This optimization step gives the possibility of reach suited drive strengths and transistor sizes of critical cells before and after layout generation. For this, we developed a tool called **Storm** to optimize the circuits.

As result of the Storm optimization process, a set of power-delay points are generated. These points allow to choose the best power/delay tradeoff according to design specifications. Figure 2 shows power-delay points for the circuit C432. Each one of these points represents one possible circuit implementation.

2.3. Generating optimized layouts

Increasing the number of cells and drive strengths would not be possible without a transistor level layout design flow. By transistor level layout generation, we mean to generate a block or the whole circuit layout on demand. Instead of using pre-designed standard cells, we use a spice-like netlist as base to a transistor placer & router.

Figure 3 shows an example of a layout generated by our transistor level design flow. In this figure we highlighted polysilicon connections, contacts and diffusion sharing between adjacent cells.

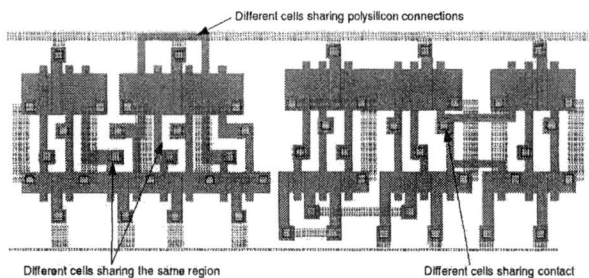

Different cells sharing polysilicon connections

Different cells sharing the same region

Different cells sharing contact

Figure 3. Example of layout generated.

3. The Transistor Level Design Flow

The proposed transistor level design flow is divided in two parts. Firstly the generation of a library is done attempting to become available a wide number of logic functions for logic synthesis. It is important to highlight that the layout of each cell is not generated at this time. Logic equations are converted to a spice-like description and simulated for timing characterization. This is done once for each technology. The second part is the design flow itself. It consists on logic synthesis/technology mapping, transistor sizing, placement, layout generation and routing. These steps are sequentially executed and at the end of the process we can return to any of them in order to improve timing or power consumption.

More details about the proposed transistor level design flow are given in the following subsections.

3.1. The *superlib* generation

The *superlib* is generated in Liberty library format and contains delay tables of different versions of each logic function. For the results presented in this paper, it was used an equation library with 3,500 different logic functions. Each logic function was implemented in four different drive strengths by sizing its transistors from 1 to 4 times the minimum possible transistor width. Inverters and buffers were implemented in 8 and 32 different versions respectively, resulting in a virtual library composed by almost 15,000 cells.

Timing characterization is done by spice simulations using Synopsys HSPICE [6]. Each cell is simulated considering five values of output load and five values of input transition time for 5x5 2D table generation. In order to reduce the number of simulation runs, we simulate only the input vectors which stimulates the biggest transistor path between supply source and the output of the cell.Thus we have only two input vectors for each cell, one for rise time and another for fall time. The whole process (more than 750,000 simulations) was automated by scripts. The generation of

the *superlib* took around 2.5 weeks considering the use of 3 Suns Fire V240.

3.2. The design flow

The design flow is shown in Figure 4. White blocks represent steps in the design flow were commercial tools are used. Steps in gray indicate tools specially developed to optimize the process.

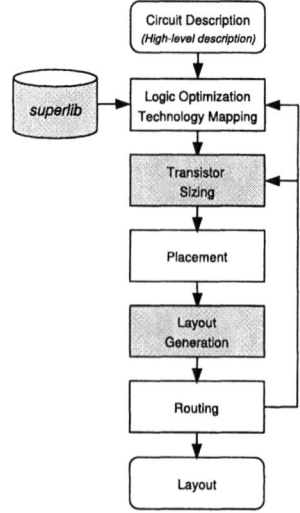

Figure 4. The proposed design flow.

Logic Synthesis/Technology Mapping

Cadence RTL Compiler [5] was used for synthesizing the high-level description. The technology mapping is done considering the cells available in the *superlib*.

Transistor Sizing

As reported in Section 2.2, the number of logic functions are widely increased in comparison to standard cell approach, but the number of drive strengths remains limited in the logic optimization step. This limitation is compensated by the possibility to optimize the circuit with a transistor sizing tool.

The Storm optimization process consists on using Nanosim e Pathmill [6] to analyze timing and power consumption of the circuit. Analyzing the reports of these tools, Storm is able to select cells belonging to the critical paths for optimization. Once a cell is selected, all transistors connecting the supply source (VDD or GND) to its output are resized.

Placement

The placement process is very similar to the standard cell placement methodology. The exception is the fact that cells area are estimated instead of pre-characterized. Cadence First Encounter [5] is used to place the cells.

Table 1. Comparison between layouts generated by the standard cell approach and the proposed transistor level design flow

Circuit	Timing (ns)			Power (uW)		
	Std Cells	Our	Gain	Std Cells	Our	Gain
C432	3.97	3.68	7.3%	4416	3726	15.6%
C499	2.36	1.89	19.9%	11881	7122	40.0%
C880	1.88	1.85	1.5%	5592	3984	28.7%
C1355	2.50	2.45	2%	12071	6965	42.2%
C1908	2.39	2.06	13.8%	9493	6007	36.7%
C3540	5.15	4.05	21.4%	21141	15235	27.9%
C6288	9.46	7.98	15.6%	211593	145660	31.1%
			11.6%			31.7%

Layout Generation

Layout generation is the most important step in our design flow. The proposed optimization process could not be implemented without a transistor-level layout generator. As shown in previous sections, transistor level layout generation allows to create any kind of logic function.

The layout generation step consists on generating layout polygons (transistor and connections) on demand according defined algorithms. We implemented simplified and high performance algorithms such as the *Eulerian* path search algorithm for place transistors and the classic maze algorithm to route. After routing, the compaction is applied to the layout in order to reduce the length of connections and reduce the area occupied by the circuit. The compaction is based on linear programming, thus polygons sizes and positions are converted to linear formulations.

Routing

The routing, just like the placement, is done by commercial tools. We use the Cadence First Encounter [5] to route the circuit. Within this tools, we are able to implement state-of-the-art placement & routing associated to an optimized layout generator.

The design flow loop

The process is enriched by using layout parasitics information. Capacitances and resistances are extracted from the layout by Cadence Diva Extractor [5]. Based on this information, a new optimization process can be done resulting in better circuits.

4. Results

Table 1 presents results of the proposed method in comparison with the standard cell approach to a commercial 0.35 μm technology. Results are very interesting because show the efficiency of our transistor level design flow. The design process was done based on high effort to meet the minimum possible delay. We note that this high effort resulted in the insertion of many buffers in the critical path.

This explains the low gain concerning the timing (around 11%) of our methodology in comparison with the standard cell approach. The power consumption gain in these circuits is between 15% and 42% due to the number of complex gates in the circuit and the optimized transistor width.

5. Conclusions

This paper presented a transistor level design flow. This design flow allows designers to mitigate the timing closure problem. Results show that this methodology is very promising. Comparisons between the transistor level methodology and the standard cell approach shown interesting results where our methodology presented around 11% of delay improvement and more than 30% power savings. These results were obtained by the effort to improve the layout quality concerning polysilicon and metal connections and by the possibility to optimized the circuit in relation to the wide number of logic functions and drive strengths presents in the library.

References

[1] P. Gopalakrishnan and R. A. Rutenbar. Direct transistor-level layout for digital blocks. In *ICCAD '01: Proceedings of the 2001 IEEE/ACM international conference on Computer-aided design*, pages 577–584, Piscataway, NJ, USA, 2001. IEEE Press.

[2] C. Cheng. Timing closure using layout based design process. available at http://www.techonline.com/community/related content/14016. April 2007.

[3] M. Vujkovic, D. Wadkins, B. Swartz, and C. Sechen. Efficient timing closure without timing driven placement and routing. In *DAC '04: Proceedings of the 41st annual conference on Design automation*, pages 268–273, New York, NY, USA, 2004. ACM Press.

[4] R. Roy, D. Bhattacharya, and V. Boppana. Transistor-level optimization of digital designs with flex cells. *IEEE Computer Society*, 38:53–61, Feb 2005.

[5] CADENCE. http://www.cadence.com. April 2007.

[6] SYNOPSYS. http://www.synopsys.com. April 2007.

Hybrid Multiplierless FIR Filter Architecture based on NEDA

J.Luis Tecpanecatl-Xihuitl, Ruth M. Aguilar-Ponce, Magdy Bayoumi
The Center for Advanced Computer Studies
University of Louisiana at Lafayette
Lafayette, LA 70504–3749
Email: luis@louisiana.edu

Abstract—**This paper presents new hybrid multiplierless finite impulse response (FIR) architecture based on New Distributed Arithmetic (NEDA). The hybrid structure is a trade off between direct form and transposed direct form that results in a reduction of the critical path and the size of the delays elements as well as the fan-out. While the multiplications involved in the hybrid structure are replaced by a butterfly adder tree. Compared with previous methods, our proposed architecture achieves an average of 20% less additions. Moreover, the design method is simple and achieves better results than previous methods.**

I. Introduction

The next generation of wireless communication systems will face important challenges such as the necessity of been always connected has resulted in a new paradigm known as Cognitive Radio. This radio must sense, adapt to the environment and learn the different configuration required. The front-ends of the new systems should be able to adapt to new parameters or reconfigure specific blocks on run-time. The front-end must be able to incorporate different communication standard. Therefore, the design of front-ends is becoming more complex. In addition, front-end must switch between different standards in very short time to avoid interruption of communication or at least the interruption should not be detectable by the user. Therefore, high-performance blocks with low area and power consumption and few parameters to configure are required for front-ends. One of these blocks is the Finite Impulse Response (FIR) filter.

The FIR filter is widely used on digital down/up converts because their linear phase, as well as filter banks utilized on OFDM systems. However, the multiplications involved make them an expensive block in terms of area, and power consumption. Additionally, the need to store a set of coefficients for each standard requires additional memory that is also power demanding.

There are several approaches to eliminate the multiplications in such filters. The multiplications are replaced by addition/subtractions and shift operations. In order to achieve this, a codification of the binary representation of the coefficients and localization of common patterns between them to reduce the number of adders/subtractions involved are applied. Heuristic optimization algorithms find the best common pattern to reduce the number of operations [1].

Our proposed approach is to employ NEw Distributed Arithmetic Architecture (NEDA) to implement the inner product us-

ing a butterfly adder tree in a hybrid structure. NEDA provides an efficient and simple structure as well as a straightforward design procedure.

The rest of the paper is organized as follows. Section II discusses multiple constant multiplication block as well as NEDA as an efficient implementation of the multiple constant multiplication. The proposed architecture is explained in section III. Section IV shows our results and comparisons with previous approaches. Finally, the conclusions are presented in section V.

II. New Distributed Arithmetic Architecture

A. Multiple Constant Multiplication Block

In a FIR filter, variable input samples are multiplied with a set of constant coefficients. This operation is known as multiple constant multiplication (MCM) block. Multiplierless filters implementations try to implement MCM using only additions, subtractions and shift operations.

Those algorithms can be classified in four categories: Digit-Based recoding Algorithm, Common subexpression elimination (CSE), graph-based, and hybrid algorithms. Digit-based recoding algorithms are a simple method of Canonical Sign Digit (CSD), even though these methods are fast they have the worst performing [1], [2], [3].

Common subexpression algorithms find common subpatterns along the rows (Horizontal CSE) or columns (Vertical CSE) in an appropriate representation of the constants. The constant are first converted to a convenient number system such as CSD. However, the main drawback is the need of additional adders to obtain the symmetry part of the coefficients when more than one column common subexpression with opposite sign in the bits exists.

Graph-based algorithms iteratively construct the graph representing the multiplier block and they are not restricted to a particular representation of the coefficients or a predefined graph topology. The main disadvantage is center on the need to consider all possible graph topologies to find the minimal adder cost for a particular integer coefficient.

Finally, the hybrid algorithms combine different algorithms for example starting from a graph topology to compute the differential coefficients and then switches to a CSE algorithm.

316

978-1-4244-1709-4/07/$25.00 © 2007 IEEE

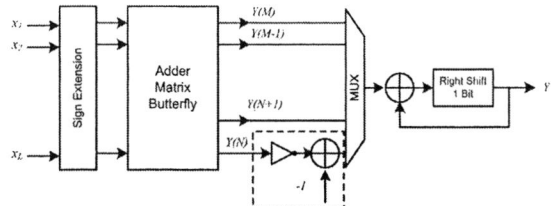

Fig. 1. NEDA Architecture to perform inner products

B. NEDA

New Distributed Arithmetic Architecture (NEDA) has been proposed as efficient implementations for Discrete Cosine Transform. NEDA reduces power consumption as well as area, but maintains a high speed without compromising the accuracy of DSP applications [4], [5]. Considering (1), where $x(t)$ is the input signal, c_i is the coefficient in a digital filter and $y(t)$ is the result from $i = 0...M$. Distributed Arithmetic considers the decomposition of the input $x(t)$ to obtain the precomputed values to be stored in a ROM memory.

$$y(t) = \sum_{i=0}^{M} c_i x(t-1) \tag{1}$$

In the other hand, on NEDA the filters coefficients are distributed allowing the input to be supplied in a parallel fashion contrary with Distributed Arithmetic where input bits are supplied in serial mode. Figure 1 shows the NEDA architecture which was proposed to perform inner products [4], [5].

The main advantage of NEDA compared with previous approaches is its simple structure. The adder butterfly matrix is a sparse matrix that represents the fixed coefficients consisting of 0's and 1's. Therefore, no codification is required. The matrix might contain redundant information. The redundancy needs to be removed by a compression scheme. Compressing the matrix leads to a reduction of the number of adders. The compression algorithm searches for similar rows. Then, only one row is consider since the result for the other rows can be obtained with the same adder array. After that, the algorithm searches for a pair of rows that shares a maximum number of 1's at the same positions. This procedure continues until is no common pattern in the rows. The detailed algorithm can be found in [5].

However, applying NEDA directly over MCM block containing all the filter coefficients achieves poor results. In some cases, these results require more adders than the previous techniques such as CSE. This problem can be overcome by employing a hybrid structure as is explained in the next section.

III. PROPOSED ARCHITECTURE

The first step on the implementation of FIR filters is the selection of the structure. The main structures are the direct form and the transposed direct form illustrated in Figure 2a, and b. Both of them have the same number of multipliers,

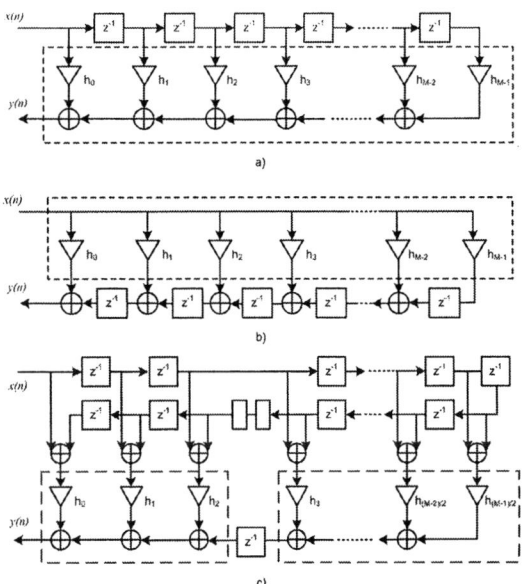

Fig. 2. a) Direct Form 2b) Transposed Form 2c) Symmetric Hybrid Form

adders and delays. However, the transposed direct form requires large registers compared with the direct form because the delays elements are introduced after the multiplication. Additionally, the transposed direct form requires a high fan-out at the input node. On the other hand, the critical path on the direct form is one multiplier and the adder tree, which is larger compared with the transposed direct form, where only two operations are involved [6], [7]. Additionally, the MCM block in both structures contains all the coefficients. Therefore, NEDA cannot be employed efficiently with these structures.

Hybrid structures had been proposed [6], [7], [8] to reduce the critical path and avoid large delay elements by combining the direct and transposed direct form. Figure 2c illustrates a hybrid structure. The filter coefficients are grouped into several sets to form blocks. These sets of coefficients are taken consecutively; (2) shows how this is performed, where $H(z)$ represents the transfer function of a digital filter. Additionally, symmetry of the FIR filter leads to a reduction of half the number of multiplications, maintaining the same number of adders. Therefore, in order to obtain an efficient implementation we consider hybrid structure and symmetry of the filter.

$$H(z) = \underbrace{h(0) + h(1)z^{-1}}_{\text{Block 1}} + \underbrace{\cdots +}_{\text{Block 2}} \underbrace{h(M-1)z^{M-1} + h(M)z^{-M}}_{\text{Block } n} \tag{2}$$

The hybrid structure reduces the critical path, reduces the high fan-out in the input node and has small size registers for the delay chain. Now, we are going to focus in the multiplication block. Based on the symmetric hybrid structure, we utilize NEDA to implement the Multiple Common Multiplier

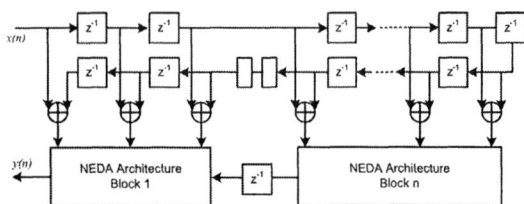

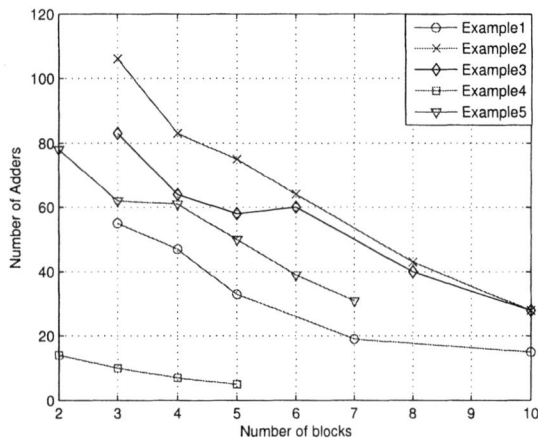

Fig. 3. Symmetric Hybrid Structure using NEDA

Fig. 4. Tendency on the reduction of the number of adders increasing the number of blocks

TABLE I
SPECIFICATIONS OF FILTERS

Filter	Passband	Stopband	#Taps	Wordlength(bits)
1	0.15	0.25	40	12
2	0.15	0.20	60	14
3	0.10	0.15	60	14
4	0.15	0.25	25	9
5	0.021	0.07	59	14

TABLE II
ADDER COST CONSIDERING ONLY THE MCM BLOCK

Example	1	2	3	4	5
Length	40	60	60	25	59
Wordlength(bits)	12	14	14	9	14
Method BHM	16	29	29	–	–
Method HARTLEY	19	35	37	–	–
Method PK	18	35	35	–	–
Method SOPOT	–	–	–	11	57
Method [3]	–	–	–	6	33
PROPOSED	**15**	**28**	**28**	**5**	**25**

number of blocks in the hybrid decomposition as well as the results reported by previous approaches [2], [3]. Our proposed architecture obtains an average reduction of 11% of the number of adders required to implement the MCM. The best reduction obtained is 25% in the example 5, while the lowest reductions are 3% and 6% in examples 1, 2, and 3.

The previous results in Table II show only the adders involved on the MCM block. However, the total number of adders involved in the implementation of the filter is the best parameter to asset the overall complexity of the structure. Table III shows the total number of adders required on each implementation obtaining an average reduction of 20%. The best reduction is 27% in the example 5, while the lowest reduction is 2% in example 1.

The data presented in Table II shows similar results compared with previous methods. However, our method presents significant savings on hardware and modularity in their components. Previous methods require adders, subtractions, and shifters to implement the MCM block, contrary on our approach, all the shifter operations are concentrated at the final stage and all preceding operations involve only adders, as can be observed from Figure 1.

NEDA represents a simple architecture to reduce the number

(MCM) Block. Figure 3 shows the proposed structure, where the boxes on dashed lines from Figure 2c are implemented by NEDA.

Filter coefficients are grouped as previously discussed in the hybrid structure. Once the coefficients have been decomposed in several blocks, the compression of the resulting matrix is performed. The resulting submatrices have a different interdependency than the original matrix formed with all the filter coefficients together. These submatrices achieved significant reduction of the number of adders required for the adders butterfly structure. As the number of subblocks increases, the number of adders decreases as show in Figure 4. Our approach utilized only adders and a shifter at the final stage to obtain the final results, compared with previous methods where adders, subtracters and shifters are utilized to implement the MCM.

IV. RESULTS

The proposed architecture was implemented using MATLAB™. The set of filters used to compare our results were obtained from [2], [3]. The specifications of those filters are presented in Table I, where the *Passband*, *Stopband* are normalized, filter length is on *#Taps* column, and *Wordlength* denotes the word size for the coefficient representation.

Table II shows the results obtained using the most efficient

TABLE III
TOTAL ADDER COST

Example	1	2	3	4	5
Length	40	60	60	25	59
Wordlength(bits)	12	14	14	9	14
Method BHM	53	88	88	–	–
Method HARTLEY	56	94	96	–	–
Method PK	55	94	94	–	–
Method KMSD	47	91	100	–	–
Method FH	51	96	106	–	–
Method FH2	53	89	95	–	–
Method SOPOT	–	–	–	36	116
Method [3]	–	–	–	30	89
PROPOSED	**46**	**69**	**69**	**23**	**65**

318   *2007 IFIP International Conference on Very Large Scale Integration (VLSI-SoC 2007)*

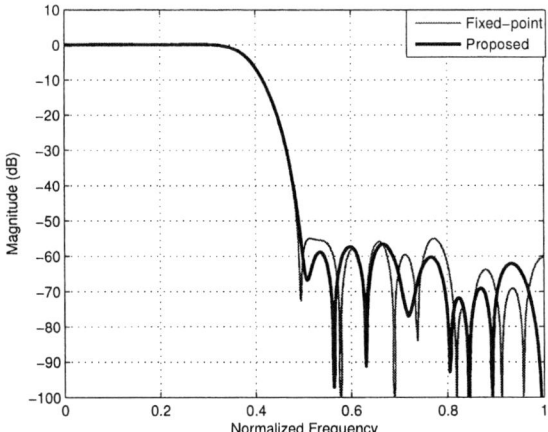

Fig. 5. Frequency Response of Example 1

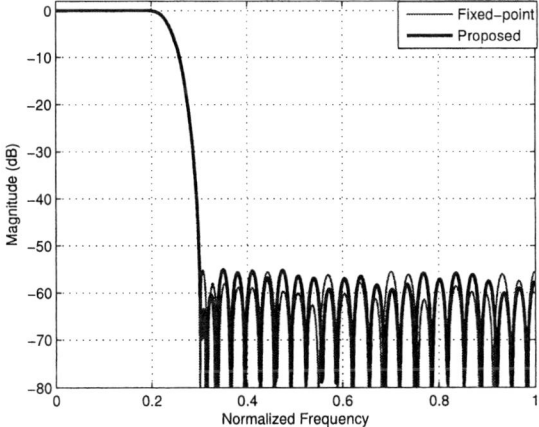

Fig. 6. Frequency Response of Example 3

of adders involved in the inner product algorithms such as FIR filters. The frequency response of the filter might be modified for the optimization procedure. Therefore, this is another important parameter to measure the performance of the design process. The frequency response of the resulting structure must be able to reach the filter specifications.

Figure 5 and 6 demonstrates that the proposed architecture does not alter the filter characteristics. The frequency response of the example 1 and 3 are show in the Figure 5 and 6, respectively. In both examples, the thin line shows the fixed-point implementation of the filter using direct form structure. The fixed-point implementation were realized using 12 and 14 bits and considering input and output quantization. The fixed-point implementation also considers adders and multiplications like a regular implementation. The bold line represents the frequency response of the proposed structure using 12 and 14

bits respectively.

V. CONCLUSION

A new hybrid multiplierless FIR structure has been proposed. A combination from the two main structures for FIR filters produces a more efficient structure. Employing NEDA architectures on hybrid structures result in a simple architecture for FIR filters. The results obtained show significant reduction without any complexity overhead. Reduction of 11% average in the number of adders on MCM block and totally of 20% average in totally adder cost. Additionally, compared with other algorithms proposed where they used additions and subtractions, this new structure offers a simple implementation using just adders in the Adder Matrix and a shifter at the end of each NEDA block.

ACKNOWLEDGMENT

The authors acknowledge the support of the U.S. Department of Energy (DoE), EETAPP program, DE97ER12220 and the Governor's Information Technology Initiative. Tecpanecatl-Xihuitl acknowledges the support from CONA-CyT, Mexico.

REFERENCES

[1] Y. Voronenko and M. Püschel, "Multiplierless multiple constant multiplication," *ACM Trans. Algorithms*, vol. 3, no. 2, p. 11, 2007.

[2] M. D. Macleod and A. G. Dempster, "Multiplierless fir filter design algorithms," *IEEE Signal Processing Letters*, vol. 12, no. 13, pp. 186–189, 2005.

[3] N. Boullis and A. Tisserand, "Some optimizations of hardware multiplication by constant matrices," *IEEE Trans. on Computers*, vol. 54, no. 10, pp. 1271–1284, 2005.

[4] W. Pan, A. Shams, and M. Bayoumi, "Neda:a new distributed arithmetic architecture and its application to one dimension discrete cosine transform," in *Proc. IEEE Workshop on Signal Processing Systems (Sips'99)*, Oct 1999, pp. 159–168.

[5] S. A.M., C. A., and M. Pan, W.and Bayoumi, "Neda: a low-power high-performance dct architecture," *IEEE Transactions on Signal Processing*, vol. 54, no. 3, pp. 955–964, 2006.

[6] K.-Y. Khoo, Z. Yu, and J. Alan N. Willson, "Design of optimal hybrid form fir filter," in *IEEE International Symposium on Circuits and Systems, (ISCAS'01)*, may 2001, pp. 621–624.

[7] O. Gustafsson, J. O. Coleman, A. G. Dempster, and M. D. Macleod, "Low-complexity hybrid form fir filters using matrix multiple constant multiplication," in *Thirty-Eighth Asilomar Conference on Signals, Systems and Computers*, nov 2004, pp. 77–80.

[8] S.-F. Lin, S.-C. Huang, C.-W. K. Feng-Sung Yang, and L.-G. Chen, "Power-efficient fir filter architecture design for wireless embedded system," *IEEE Trans. on Circuits and Systems-II:Express Briefs*, vol. 51, no. 1, pp. 21–25, 2004.

A Genetic Algorithm Based Heuristic Technique for Power Constrained Test Scheduling in Core-based SOCs

[1]Chandan Giri, [2]Soumojit Sarkar and [3]Santanu Chattopadhyay
[1,3]Dept of E & ECE, IIT Kharagpur, India; [2]Dept. of CSE, IIT Kanpur, India
Email:{[1]chandan, [3]santanu}@ece.iitkgp.ernet.in, [2]soumojitsarkar@gmail.com

Abstract

This paper presents a Genetic algorithm (GA) based solution to co-optimize test scheduling and wrapper design under power constraints for core based System-On-Chips (SOCs). Core testing solutions are generated as a set of wrapper configurations, represented as rectangles with width equal to the number of TAM (Test Access Mechanism) channels and height equal to the corresponding testing time. A locally optimal best-fit heuristic based bin packing algorithm has been used to determine placement of rectangles minimizing the overall test times, whereas, GA has been utilized to generate the sequence of rectangles to be considered for placement. Experimental result on ITC'02 benchmark SOCs shows that the proposed method provides better test time results compared to the recent works reported in the literature.

Keywords: Core-based SOCs, SOC testing, wrapper design, test scheduling, power constraints.

1. Introduction

A general problem for SOC test integration consists of the design of test access mechanism (TAM) architecture, that transports test data between SOC pin and core wrapper. Wrapper provides an interface between TAM and the core and can be operated in several modes [1]. So test scheduling is a process that determines the start and ending time of testing each core in the SOC such that the overall test application time is minimized given TAM architecture and power constraints.

Several recent works considered various aspects of the test scheduling problem. Earlier works propose methods to solve wrapper design and test scheduling as separate problems. Recently [2] and [3] proposed an integer linear programming based solution for co-optimization of wrapper design and test scheduling for SOCs. Huang et al [4] formulated the problem of SOC pin allocation to cores and test scheduling using 2-D bin-packing or rectangle packing approach. Same authors also proposed 3-D bin-packing approach [5] considering power constraints. Several other works [6,7] also consider SOC power dissipation constraints during scheduling. A heuristic approach using the sequence pair representation for test scheduling was considered in [8]. Zou et al [9] proposed test scheduling algorithm based on simulated annealing (SA) using the sequence pair representation. A B*-tree based approach has been proposed in [10] to get the test schedule. Recently in [11] SOC test scheduling with reconfigurable core wrappers is used. Ant colony optimization (ACO) based approach [12] considers the rectangle packing for test scheduling solution. A two-stage genetic algorithm (GA) based algorithm was proposed in [13] where each solution is represented by a sequence pair.

In this paper, we present a genetic algorithm (GA) based approach for the SOC test scheduling and wrapper design co-optimization problem. The primary objective of this paper is to achieve minimal test time while satisfying two constraints: 1) given number of SOC pins and 2) allowable SOC peak power consumption. Experimental results show that the proposed method obtains better test time results for the SOCs with larger number of cores than the recently proposed works.

The remainder of this paper is organized as follows. Section 2 describes the overall SOC test scheduling problem. Section 3 discusses the wrapper design optimization methods used. Section 4 discusses the test scheduling algorithm with the proposed bin-packing approach. Proposed GA formulation for selection of one rectangle from the set of rectangles of each core generated by wrapper design is presented in Section 5. We present our experimental result based on the ITC'02 benchmark [14] SOCs in Section 6. Section 7 concludes the paper.

2. Problem formulation

Let the SOC design consist of N cores, and each core C_i ($1 \leq i \leq N$) has n_i input terminals, m_i output terminals, b_i bi-directional I/Os, sin_i scan inputs and sot_i scan outputs. Also assume that maximum peak power for each core during testing is given. Power estimation for each core is a challenging job and is out of scope to discuss in this paper. Let the total width of TAMs be K and each core must be tested with P_i patterns. So the overall problem that we have to solve is as follows.

Given a set of N cores, their specific test parameters and the number of I/O pins for an SOC, maximum allowable peak power dissipation POW and peak power dissipation for each core, design the test schedule with wrapper designs for all wrapper-based cores such that overall testing time is minimized and the peak power during testing never exceeds POW.

So there exist basically two steps to solve the problem. First we generate possible optimized wrapper configurations for each core under specified TAM width. In the next step solve the test scheduling problem using the sets of optimized wrapper solutions under the maximum-allowable TAM width and power constraint.

3. Core Wrapper Design

There are two kinds of wrapper designs, balanced and unbalanced wrapper design [1]. For cores having no internal scan chains (that is, containing input and output pins only), unbalanced wrapper design is preferred since it can obtain a lower test time than the balanced one [9]. To calculate test time, T for a wrapper for different TAM widths we use the well-known formula [1] given below.

$$T = \{1 + \max(S_i, S_o)\} * P + \min(S_i, S_o) \qquad (1)$$

Where P is the number of test patterns and $S_i(S_o)$ denotes the length of longest wrapper scan chain used during scan-in(out) for a core. We consider two different approaches to the design of wrapper for cores with internal scan chains and for cores that do not have any internal scan chains. For cores with internal scan chain we use the *Design_wrapper* algorithm proposed in [3] and for cores without internal scan chain we use the scheme proposed in[9].

Using the wrapper design method for each core we can generate a set of wrapper configurations with the TAM wire usage of 1 to K, where K is maximum number of TAM channels allocated to test the SOC. Hence each wrapper configuration can be considered as a rectangle with width equal to the test time and height corresponding to the TAM channels used. From all the wrapper configurations for a core i a smaller set of wrapper configuration can be considered in the test scheduling. It is based on pareto-optimal design [3] principle.

4. Test Scheduling Problem

Suppose a SOC with N cores is to be tested using K TAM channels. Each core C_i ($1 \leq i \leq N$) is represented by a set of R_i wrapper configurations. Each wrapper configuration is represented by a pair $(W_{ij}, T(W_{ij}))$, where W_{ij} stands for the wrapper width of the j-th wrapper configuration for core C_i and $T(W_{ij})$ is the test time of the core C_i with the wrapper width W_{ij}. Also for each core peak power POW_i is given during testing. So the objective is the assignment of the core wrapper pins to the pins of SOC and obtains the test starting time and finishing time for each core such that overall test time is minimized satisfying power constraint.

This problem can be transformed into the well-known rectangle-packing problem, in which the SOC is represented by bin with width K and a set of R_i SOC wrappers for each core represented by a set of R_i rectangles with rectangle j having width W_{ij} and test time $T(W_{ij})$. We want to choose one rectangle from each set of rectangles for a core C_i and pack all the rectangles in the bin, so that height of the bin is minimized.

In this paper we treat this test problem as two different problems. For the first problem we consider the test scheduling algorithm where no power constraint is considered. For the second one we impose the power constraint on the test scheduling algorithm for the first problem and it is verified that for any time instants maximum allowable peak power POW (SOC power budget) will not exceed. To evaluate power constraint we calculate the sum of the maximum peak power for all the cores (POW_i s) at a given time instance. It is assumed that for entire test time of the core maximum peak power is same. So it is needed to calculate only the sum of the peak powers when a core is started for testing. For the sake of simplicity, we continue with the peak-power model, though the scheme can always be extended to incorporate the new power model proposed.

However, the test schedule of a SOC is feasible if no two cores are assigned to the same SOC pin at the same time instant and for each core all its wrapper pins are assigned to SOC pins for the entire time needed to test that core. Hence the following constraint is to be considered.

Constraint: At any instant of time within entire test time of the SOC the sum of the widths of all the rectangles that are tested is equal to or smaller than the total TAM width provided to test the given SOC.

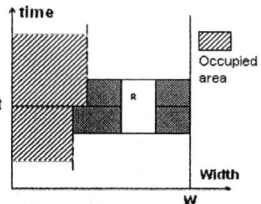

Figure 1: An optimum placement of the incoming rectangle

The following section discusses the proposed rectangle-packing algorithm.

4.1 Placement algorithm

We used a best-fit heuristic based placement or packing algorithm satisfying the above mentioned constraints (with or

without power constraint) for placing the rectangles in a bin of width W (maximum allowable TAM width) whose height is infinite (∞). The algorithm tries to place one rectangle after another each time optimizing the placement based on a cost function. The cost is rectangle specific and is calculated considering the existing profile of the already placed rectangles. The profile provides the information regarding TAM channels used in various time instants. Hence cost function depends on two parameters, TAM width utilization (U) and increase in test time (T). For a particular rectangle, the algorithm evaluates the possible placement points (time instants) on the basis of cost parameters and the point with lowest cost value is selected for placing the rectangle. For finding that optimal position we restrict our search only to the points where the profile function changes value. As shown in the Fig. 1 for the next rectangle R to be placed if we move the rectangle upwards then we are surely going to reduce the area (solid shaded region) formed by the free TAM channels. And this will continue till we reach at the time instant marked as t. Similarly for downward movement also the same time instant will provide the reduced area. So for each possible placement point (time instant) we consider two types of orientations – up and down, which denotes whether the rectangle is placed upwards (or downwards).

We use weighted sum of these two parameters to determine the final cost of a particular placement. This is

$$\text{Cost} = wt \times u(U) + (1\text{-}wt) \times t(T), (0 \leq wt \leq 1) \qquad (2)$$

Where *wt* is the weight in favour of the utilization parameter and $u(U) = U/U_{max}$ and $t(T) = T/T_{max}$. The utilization parameter is a measure of utilization of TAM lines during the entire height of the rectangle being placed. It is maximized by minimizing the number of unused TAM lines. The variable U represents the sum of unutilized TAM lines at each time instant during the testing time of the core, that is, the height of the rectangle. Suppose that the rectangle is placed between the time instants t_1 and $t_1+ h$. In this interval, there will be subintervals during which the number of unutilized lines does not change. The changes occur only at the boundaries of subintervals. For instance, may be the TAM lines used in the duration t_1 and $t_1 + \Delta$ be w_1. If W be the total TAM width used, the total number of unutilized TAM lines for this subinterval is $(W - w_1) \times \Delta$. For each such subinterval, the number of unutilized TAM lines is calculated and summed up to get the total unutilized TAM lines U. Once all such U values have been calculated, for different possible placements, U_{max} is taken to be the maximum of U values for all these placements. Thus, $u(U) = U/U_{max}$ gives a normalized value of unutilized TAM lines. The second parameter, increase in test time (T) is an obvious attempt to attach some penalty with those possible placements that result in putting the upper edge of the rectangle in consideration beyond the test time required by all the rectangles placed before it. This inclusion of such a parameter can be justified on the basis of the simple fact that it is the testing time, which we are trying to minimize. The value of *wt=0.8* has been found to work well with the ITC'02 benchmark SOCs.

For each of the possible placement points we check whether the constraint (with or without power constraint) is satisfied or not and the cost is evaluated. The point with lowest cost is selected for placement of the rectangle. This process is continued for all the cores and the last time instant provides the final test time for the selected sequence of rectangles obtained from GA.

Data Structure used:

1. L_{CORE} : A queue of representative rectangles of the cores generated by GA.
2. The current profile of the time-width plot (for previously placed rectangles) is maintained as a doubly-linked list *time_instants*, which stores the

value of the width at the particular time instant where it changes. Hence each node in this list is a tuple like <time_instant(t),width(w)> and nodes are sorted by increasing time_instant values.

3. Apart from all the above we also use a list to store the locations and orientation of the placed rectangles.

The placement procedure without power constraints is represented in Algorithm 1.

Algorithm 1:

Input: Ordered list of cores L_{CORE} in descending order according to their average area obtained from GA.

Output: Final Schedule after placing all the cores in the list L_{CORE} satisfying constraints

Step 1. Pick first core C_0 from L_{CORE} and place at 0th time instant with 'up' orientation.
 Time_instants = {<0,C_{0w}>, <C_{0h},0>} // C_{0w} and C_{0h} are the width and height of the core C_0 respectively

Step 2. While (L_{CORE} is NOT empty)
 Begin
 a). Pick core C_i from L_{CORE} in the order of the list.
 b). Find possible time instants t_k's satisfying constraints for both the orientations and calculate cost for each of these possible time instants according to equation (2).
 c). Place the core C_i at time instant t_k giving minimum cost (in case of ties, the placement with the lowest t_k-value and at such level a 'down' orientation is preferred).
 d). Update time_instants list.
 End

If power constraint is to be imposed then simple modifications in the data structure and of Algorithm 1 will do the tricks. In this case each node in the time instant list is a triplet like <time_instant(t),width(w), power_vlaue(POW_t)> and nodes are sorted by increasing time_instant values. During calculation of possible time instants [Step 2 b)] in this case not only TAM width constraints is to be met but also maximum allowable power budget POW has to be satisfied.

5. GA formulation

In this paper GA formulation is used to get sequence of rectangles generated by the wrapper design method. For each core in the SOC we have to select one rectangle from the set of rectangles (wrapper configurations) generated for that core so that total testing time will be less. The order in which we select the rectangle for each core depends on the average area of the rectangles obtained from wrapper design algorithm used. For this purpose, cores are sorted in terms of decreasing average rectangle area. Throughout the running of the algorithm, this order is maintained. Next we define the individual representation of chromosomes, mutation and crossover operators, fitness function and selection procedure.

5.1 GA formulation to get a sequence of rectangles

For this problem, chromosome structure is encoded with floating point vales <$f_1, f_2,...,f_N$> of length N depending on the number of cores present in an SOC, where f_i's are floating point values within range of 0 to 1. Each f_i is used to select one rectangle from the set of rectangles obtained for the i-th core using wrapper design process. The value f_i is multiplied by the number of rectangles for core i to identify the rectangle selected for it. For each chromosome, fitness value is calculated using the best-fit heuristic based placement algorithm discussed in Section 4. This fitness value is calculated after getting sequence of representative rectangles from each core as discussed.

To create populations for new generation 20% best fit chromosomes are directly copied and remaining 80% chromosomes are created using crossover and mutation operators. In our GA approach we used parameterized mutation and crossover operation [15]. The algorithm is run for 150-200 generations with 3000-5000 population sizes depending on the number of cores present in the SOC. Chromosome with lowest fitness value provides the solution for the concerned problem.

6. Experimental Results

All the tests were conducted on a 2.66 GHz Pentium IV machine with 512 MB memory. The proposed algorithm is implemented in C++ and we used the ITC'02 SOC benchmarks [14] to be experimented with.

Table 1 shows the comparison of our test time results with recently proposed methods. Numbers in bold represent the best result. Row "Our" indicates our method. The results reported for the proposed method are the best among five runs of the procedure. It can be noted from Table 1 that our results significantly improve upon earlier approaches. For SOCs with larger number of cores our proposed method provides better results for most of the cases of different TAM widths. The maximum CPU time required to run the algorithm for SOC p93791 containing 32 cores with 3000 population size is approximately 1812.25 sec for W=64.

Fig. 2 summarizes the results of Table 1 showing for how many cases (Taking each TAM width value as one case for each core) our results are better, equal or not better than the recently proposed schemes. It is seen from Fig. 2 that our proposed approach is better than ACO[12] for 60% of the cases, more than 50% of the cases compared to 2-Stage GA[13], more than 20% of the cases compared to B*-SA[10], and almost 50% of the cases compared to SA[9].

To get power constrained test scheduling results using ITC'02 benchmark circuits we made the assumption that the core power values given in the benchmark formats are peak power value during test. Only one benchmark (h953) among the ITC'02 benchmark set has power dissipation numbers included. Table 2 represents the results for power constraint test scheduling times for SOC h953. To compare with [6] we used the same 3 power limits for the SOC power budget, 6×10^9, 7×10^9 and 8×10^9, chosen above the maximum power consumed by one of the core 5.75×10^9. It is seen that for our proposed method provide better results than [6].

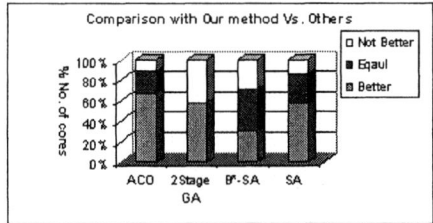

Figure 2: Comparison of results

Table 2
Test time results under power constraints for h953

Power Limit (POW)	Number of TAM wires 16-64	
	[6]	Proposed method
5753800192	122636	119357
6×10^9	122636	119357
7×10^9	119357	119357
8×10^9	119357	119357

7. Conclusion

In this paper we proposed SOC test scheduling algorithm based on 2-dimensional rectangle bin packing approach considering best-fit heuristic. A genetic algorithm scheme is also proposed for selection of representative chromosome for each core present in an SOC. The test scheduling results are obtained for ITC'02 benchmark circuits and it is seen that our proposed method provides better results than the recently proposed techniques.

ACKNOWLEDGEMENT

This work was supported in part by the Department of Science & Technology, India, under Grant SR/S3/EECE/19/2003 - SERC-ENGG, dated 05.05.2004.

Reference:

[1] Marinissen E. J., Goel S. K., and Lousberg M., *Wrapper design for Embedded core Test*, in Proc. ITC, pp:911-920, 2000.

[2] Chakrabarty K., *Test Scheduling for core-based systems using Mixed-Integer Linear Programming*, in IEEE TCAD, pp. 1163-1174, 2000.

[3] Iyengar, V., Chakrabarty, K., and Marinissen, E. J., *Test Wrapper and Test access mechanism co-optimization forsystem-on-chip*, In JETTA, Vol. 18, March 2002.

[4] Huang Y. et al, *Resource allocation and test scheduling for concurrent test of core-based SOC design*, In proc. ATS, pp:265-270, 2001.

[5] Huang, Y., Reddy, S. M., Cheng, W.-T. and Reuter, P. *Optimal core wrapper width selection and SOC test scheduling based on 3D-bin packing algorithm*, In Proc. ITC, Baltimore, pp:74-82, 2002.

[6] Xia Y., Chrzanowska-Jeske M., Wang B. and Jeske M., *Using a Distributed Bin-Packing Approach for Core-based SOC Test Scheduling with Power Constraints*, In Proc. ICCAD, pp:100-105, 2003.

[7] Pouget, J., Larsson, E. and Peng, Z., Multiple-Constraint Driven System-On-Chip Test Time Optimization, In JETTA, Vol:21, pp: 599-611, 2005.

[8] Koranne, S., and Iyengar, V. *On the use of k-tuples for SOC test schedule representation*, In Proc. ITC, pp. 539-548, 2002.

[9] Zou, D., Reddy, S. M., Pomeranz, I. and Huang, Y. *SOC Test Schdeuling Using Simulated Annealing*, In Proc. VTS, pp. 325-330, 2003.

[10] Wuu, J. –Yi. , Chen, T.-C. and Chang, Y.-W. *SOC Test Scheduling Using B*-tree Based Floor planning Technique*, In Proc. ASP- DAC, pp. 1188-1191, 2005.

[11] Larsson E., and Fujiwara H., *System-On-Chip Test Scheduling with reconfigurable core wrappers*, In IEEE TVLSI, Vol. 14, No. 3, pp:305-309, March 2006.

[12] Ahn, J.-Ho and Kang, S. *SoC Test Scheduling Algorithm Using ACO-Based Rectangle Packing*, In Proc. ICIC, pp. 655-660, August, 2006.

[13] Yu Y, Peng X Y, and Peng Y., *A Test Scheduling Algorithm Based on Two-Stage GA*, In Proc. International Symposium on Instrumentation Science and Technology, pp: 658-662, 2006.

[14] Marinissen, E. J., Iyengar, V., and Chakrabarty, K. *A set of benchmarks for modular testing of SOCs*, In *Proc. Int. Test Conf.*, pp.519–528., 2002.

[15] Melanie Mitchell, *An introduction to Genetic Algorithm*, Prentice Hall India.

Table 1
Comparison of test scheduling times in clock cycles for different ITC'02 SOC benchmarks

SOC	Solution method	Number of TAM wires						
		16	24	32	40	48	56	64
D695 (10 cores)	Our	41604	27767	20957	16913	14273	12084	10723
	ACO[12]	41737	28080	21098	17075	14310	12110	10783
	2stageGA[13]	40691	28060	20977	16894	14129	11453	10573
	B*-SA[10]	39489	26203	19773	16149	13649	11285	9885
	SA[9]	41604	28064	21161	16993	14183	12085	10723
P22810 (28 cores)	Our	428852	286352	216570	175946	147898	127071	112498
	ACO[12]	424889	289190	218035	177214	147898	130479	115791
	2stageGA[13]	438619	288565	216747	177633	148832	123857	103321
	B*-SA[10]	438619	287999	216747	178223	149592	129624	115406
	SA[9]	438619	289287	218855	175946	147944	126947	109591
P34392 (19 cores)	Our	939855	626122	544579	544579	544579	544579	544579
	ACO[12]	931588	631035	544579	544579	544579	544579	544579
	B*-SA[10]	935649	635237	544579	544579	544579	544579	544579
	SA[9]	944768	628602	544579	544579	544579	544579	544579
P93791 (32 cores)	Our	1750830	1170620	877073	704272	587117	505586	441388
	ACO[12]	1747504	1175988	891103	716112	598286	517692	452951
	B*-SA[10]	1782067	1190565	890092	707664	609580	517017	452245
	SA[9]	1757452	1169945	878493	718005	594575	509041	447974
G1023 (14 cores)	Our	30755	20498	15843	14794	14794	14794	14794
	B*-SA[10]	29765	20032	14913	14794	14794	14794	14794
F2126 (4 cores)	Our	357088	335334	335334	335334	335334	335334	335334
	B*-SA[10]	350030	335334	335334	335334	335334	335334	335334
T512505 (31 cores)	Our	10530995	10453470	5268868	5228420	5228420	5228420	5228420
	B*-SA[10]	10504020	10453470	5255380	5228420	5228420	5228420	5228420

Author Index

Abdel-Hafeez, Saleh   264
Abraham, Jacob   248
Aguilar-Ponce, Ruth M.   316
Anderson, David   13
Andre, Eric   110
Anghel, Lorena   312
Azevedo, Luciano   300
Aziz, Adnan   194

Balakrishnan, Shiv   168
Bampi, Sergio   308
Barcelos, Daniel   296
Bayoumi, Magdy   316
Bazeghi, Cyrus   60
Beck, Antonio Carlos Schneider   66
Becker, Jürgen   134
Beroulle, Vincent   206, 218
Bezerra, Eduardo   300
Bhat, Navakanta   280
Bhooshan, Rishi   292
Bonesana, Ivano   19
Brederlow, Ralf   78
Brião, Eduardo Wenzel   296
Brown, Stephen D.   172
Brusamarello, Lucas   94
Burleson, Jeff   84
Burleson, Wayne P.   258

Calazans, Ney   140
Carbonero, Jean-Louis   206
Carro, Luigi   66
Chattopadhyay, Santanu   320
Chen, Run   242
Cody, Brian   284
Cremonesi, Alessandro   122

Dalmasso, Julien   178
Datta, Basab   258
Desoli, Giuseppe   122
Dong, Yikui (Jen)   84
Donlin, Adam   134
Drechsler, Rolf   88, 304

Eddington, Chris   168
Eggersglüß, Stephan   88
Eisenstadt, William R.   264

Fanucci, Luca   236
Fey, Görschwin   88

Filho, Sérgio Johann   276
Flottes, Marie-Lise   178

Gentile, Giuseppe   236
Genz, Christian   304
Ghavami, Behnam   151
Giri, Chandan   320
Goulier, Julien   110
Greskamp, Brian   60
Große, Daniel   88
Gupta, Priti   7

Hamerski, Jean Carlo   212
Hamzaoglu, Ilker   200
Harb, Shadi M.   264
Harris, David   146
Hasler, Paul E.   13
Hessel, Fabiano   276, 300
Howard, Steve   84
Hsu, Kenneth W.   284
Hübner, Michael   134
Hwangbo, Woong   224

Jiang, Nan   146
Joannon, Yves   206
Johnson, L. G.   42

Kamala, R. V.   252
Kang, Kyungsu   224
Kastensmidt, Fernanda   78, 212
Katkoori, Srinivas   99
Keren, Osnat   25
Kim, Jaemoon   190
Kim, Jungsoo   224
Kolnik, Jan   84
Krishnan, Vyas   99
Kshirsagar, Chaitanya   280
Kyung, Chong-Min   190, 224

Lazzari, Cristiano   116, 312
Lee, Jaehwan John   157
Levin, Ilya   25
Li, Dongmei   242
Ling, Andrew C.   172
Liu, Liyuan   242
Lowrie, Scott   84
Lu, Chih-Wen   105

Author Index

MacDonald, Spencer   284
Madigan, Justin   284
Malani, Parth   230
Marcon, César   276, 300
Markan, C. M.   7
Maurine, P.   270
Mehdipour, Farhad   151
Mehdizadeh, Arash   151
Mello, Aline   140
Mesa-Martínez, Francisco J.   60
Metzler, Carolina   300
Mohd, Bassam Jamil   194
Moraes, Fernando   140
Mukre, Prakash   230

Na, Sangkown   190
Neuberger, Gustavo   78

Pacha, Christian   78
Palma, José C. S.   300
Pandey, Sujan   304
Pandini, Davide   122
Paolieri, Marco   19
Papadopoulos, L.   288
Paradis, Ken   84
Pavlidis, Vassilis F.   54
Pedram, Hossein   151
Peng, Sheng-Yu   13
Pradhan, Almitra   48

Qiu, Qinru   230

Rammohan, Srividhya   1
Rana, Vincenzo   128
Rao, Bindu P.   292
Razafindraibe, A.   270
Reckziegel, Everton   212
Reif, Carlos   300
Reis, Ricardo   78, 94, 116, 312
Renau, Jose   60
Renaudin, Marc   110
Robach, Chantal   206, 218
Robert, M.   270
Rohde, Guilherme   300
Rosa, Vagner S.   308
Rossi, Francesco   236
Rouzeyre, Bruno   178
Rovini, Massimo   236
Roy, Urmimala   280

Sahin, Esra   200
Sandionigi, Chiara   128
Santambrogio, Marco   19, 128
Santhanam, Krishna   184

Santos, Cristiano   312
Sarkar, Chandan Kumar   280
Sarkar, Soumojit   320
Schilders, Wil H. A.   31
Sciuto, Donatella   128
Sculley, Terry   248
Sen, Srimoyee   280
Serpanos, Dimitrios N.   72
Serrestou, Youssef   218
Sharma, Sameer   42
Shen, Yen-Chih   105
Sheu, Meng-Lieh   105
Shim, Heejun   224
Silva, Roberto da   94
Silveira, L. Miguel   31
Singh, Deshanand P.   172
Singh, Satnam   163
Siozios, Kostas   54
Sirakoulis, G.   288
Sotiriadis, Kostas   54
Soudris, D.   54, 288
Srinivas, M. B.   252
Stankovic, Radomir S.   25
Stevens, Kenneth S.   184
Sudhakar, M.   252
Sundaresan, Vijay   1
Susin, Altamiro A.   308
Swartzlander Jr., Earl E.   194

Tang, Hua   37
Tasdizen, Ozgur   200
Tecpanecatl-Xihuitl, J. Luis   316
Tedjini, Smail   206
Torrellas, Josep   60

Vaidya, Pranav   157
Vemuri, Ranga   1, 48
Villena, Jorge Fernandez   31

Wang, Zhihua   242
Wanger, Flávio Rech   296
Wille, Robert   88
Wirth, Gilson   78, 94
Wolf, Wayne   72
Wronski, Fabio   296

Yang, YuQing   248

Zamani, Morteza Saheb   151
Zhong, Freeman   84
Ziesemer, Adriel   116, 312
Zompakis, N.   288

CURRAN ASSOCIATES INC.
proceedings
.com

9781424417094